THE GEOLOGIC HISTORY
OF THE MOON

Astronaut David Scott at station 6, Apollo 15 landing site. Apollo 15 frame H–11514.

The Geologic History of the Moon

By DON E. WILHELMS

with sections by JOHN F. McCAULEY
and NEWELL J. TRASK

U.S. GEOLOGICAL SURVEY PROFESSIONAL PAPER 1348

A comprehensive review of lunar science and evolution from the viewpoint of historical geology, based on data from both photogeologic observations and lunar-sample analysis

UNITED STATES GOVERNMENT PRINTING OFFICE, WASHINGTON : 1987

DEPARTMENT OF THE INTERIOR
DONALD PAUL HODEL, *Secretary*

U.S. GEOLOGICAL SURVEY
Dallas L. Peck, *Director*

Library of Congress Cataloging-in-Publication Data

Wilhelms, Don Edward, 1930–
The geologic history of the Moon.

(U.S. Geological Survey Professional Paper 1348)
Bibliography: p. 283–292
Supt. of Docs. no.: I 19.16:1348
1. Lunar geology. I. McCauley, John Francis, 1932– . II. Trask, Newell Jefferson, 1930– . III. Title. IV. Series: United States. Geological Survey. Professional Paper 1348.
QB592.W5 1987   559.9'1   86-600177

CONTENTS

	Page
Preface	VII
Acknowledgments	VII
Abstract	VIII
Chapter 1. General features	1
2. The stratigraphic approach	15
3. Crater materials	25
4. Basin materials—Orientale	55
5. Mare materials	83
6. Structure	105
7. Relative ages	121
8. Pre-Nectarian system	137
9. Nectarian System	161
10. Lower Imbrian Series	193
11. Upper Imbrian Series	227
12. Eratosthenian System	247
13. Copernican System	263
14. Summary	274
References cited	283
Index	295

PLATES

[Plates are at end of book]

PLATE 1. Index to photographic coverage
2. Photographic coverage of the Moon, showing best photograph or set of photographs of each area
3. Geologic map of ringed basins
4. Maria
5. Structural features
6. Pre-Nectarian system
7. Nectarian System
8. Lower Imbrian Series
9. Upper Imbrian Series
10. Eratosthenian System
11. Copernican System
12. Geologic map of the present Moon

THE GEOLOGIC HISTORY OF THE MOON

By Don E. Wilhelms

PREFACE

The Moon held little interest for most scientists after its basic astronomic properties had been determined and before direct exploration appeared likely (Wright and others, 1963; Baldwin, 1978). Speculations about its internal structure, composition, and origin were only broadly constrained by cosmochemical data from meteorites and solar spectra, and by astronomic data about its size, shape, motions, and surficial properties (Urey, 1951, 1952; Kuiper, 1954). Most investigators who were active before the space age began in 1957 believed that significant new advances in lunar knowledge required acquisition of additional data.

One analytical technique, however, was insufficiently exploited before the 1960's. Few scientists since the geologist Gilbert (1893) had studied the lunar surface systematically from the historical point of view. Those who did immediately obtained important new insights about the Moon's postaccretion evolution (Baldwin, 1949, 1963; Kuiper, 1959). Then, the pioneering work of E.M. Shoemaker and R.J. Hackman focused the powerful methods of stratigraphy on lunar problems (Hackman and Mason, 1961; Shoemaker, 1962a, b; Shoemaker and Hackman, 1962). Stratigraphy is the study of the spatial distribution, chronologic relations, and formative processes of layered rocks. Its application to the Moon came relatively late and met resistance, but the fundamental stratigraphic approach (Albritton, 1963) was, in fact, readily transferable to the partly familiar, partly exotic deposits visible on the lunar surface (Mutch, 1970; Wilhelms, 1970b).*

Stratigraphic methods were applied systematically during the 1960's in a program of geologic mapping that aimed at reconstructing the evolution of the Moon's nearside (McCauley, 1967b; Trask, 1969, 1972; Mutch, 1970; Wilhelms, 1970b; Wilhelms and McCauley, 1971). Order was discovered among the seemingly diverse and random landforms of the lunar surface by determining the sequence in which they were emplaced. The stratigraphic sequence and the emplacement processes deduced therefrom provided a framework for exploration by the Apollo program and for the task of analyzing the returned samples.

During the 1970's, the sophisticated labor of hundreds of analysts was brought to bear on the wealth of material returned by the American Apollo and the Soviet Luna spacecraft. Our present perception of the Moon has emerged from the interplay between sampling studies and stratigraphically based photogeology. These two approaches are complementary: Photogeology contributes a historical context by viewing the whole Moon from a distant vantage point, whereas the samples contain information on rock types and absolute ages unobtainable by remote methods. Neither approach by itself, even the most elaborate program of direct surface exploration, could have yielded the current advanced state of knowledge within the relatively short time of two decades (Greeley and Carr, 1976).

This volume presents a model for the geologic evolution of the Moon that has emerged mainly from this integration of photogeologic stratigraphy and sample analysis. Other aspects of the vast field of lunar science are discussed here only insofar as they pertain to the evolution of visible surface features. Chemical data obtained by remote sensing supplement the photogeologic interpretations of some geologic units (see chap. 5), and geophysical data obtained both from lunar orbit and on the surface constrain hypotheses of the origin of many internally generated structures and deposits. Studies of the same data that treat the Moon as a whole, including speculations about the intriguing but unsolved problem of its origin, have been adequately covered in other reviews (Hartmann, 1972b, 1983; Taylor, 1975, 1982; French, 1977; Wood, 1979; Basaltic Volcanism Study Project, 1981; Cadogan, 1981; Glass, 1982).

This volume is written primarily for geoscientists and other planetologists who have examined some aspect of lunar or planetary science and who want a review of lunar science from the viewpoint of historical geology. It should also provide a useful summary for the advanced student who is conversant with common geologic terms. It may, furthermore, interest the geologist who has not studied the Moon but who wishes to see how his methodology has been applied to another planet.

The volume's organization reflects the stratigraphic practice of first identifying and interpreting geologic units, then ranking them in chronologic sequence over the whole planet. Chapter 1 presents what may be considered the raw data by briefly describing the major surface and subsurface lunar features and defining some common terms. Chapter 2 explains the general philosophy of stratigraphy by showing how it has been applied to some critical past and contemporary lunar problems. Chapters 3 through 6 are devoted to the four major classes of lunar features—craters, basins, maria, and tectonic structures—with emphasis on craters and basins. These four chapters contain descriptions and interpretations that apply to the rest of the volume; except for some new hypotheses concerning basin origin and lithospheric thickness, these chapters are based mostly on existing literature and should be of use mainly to the reader unfamiliar with lunar geology. The rest of this volume is organized around the dimension of time. Chapter 7 reviews general guidelines for ranking lunar deposits in order of relative age. Chapters 8 through 13 trace the histories of the craters, basins, and igneous materials belonging to each of the six main divisions of the lunar stratigraphic column. These six chapters constitute the core of the original contribution of this volume and contain information for all of its intended audience. The volume is summarized in chapter 14 and in 12 maps of the two lunar hemispheres (pls. 1–12) at the end of this volume.

ACKNOWLEDGMENTS

This volume is dedicated to Eugene M. Shoemaker, who, more than any other individual, was responsible for the incorporation of geology into the American space program. First, in his study of Meteor Crater in Arizona, he developed a physical theory of the impact process on which subsequent cratering investigations have been based, and established the structural and petrographic indicators of impact later used on the Earth and the Moon (Shoemaker, 1960, 1963; Chao and others, 1960). With his colleague Edward C.-T. Chao, he applied these criteria in establishing an impact origin for the previously enigmatic Ries crater (Rieskessel), Germany (Shoemaker and Chao, 1961), which has become a model for larger lunar craters. Then, in four major papers written over two years (1960–61), he (1) put lunar geologic mapping on the firm stratigraphic footing that still supports it (Shoemaker and Hackman, 1962); (2) championed the impact interpretation of large fresh lunar craters by the most convincing combination of arguments yet advanced (Shoemaker, 1962b); (3) accurately appraised the objectives and future course of lunar exploration, including the forerunner role of the small, dry, and airless Moon in comparative planetology (Shoemaker,

---

*Use of the prefix "geo-" for lunar and planetary studies has been criticized, but it is justified by: (1) applicability to all other solid bodies of the geologic principles developed for the Earth; (2) elimination of the need for new terms for every new world observed at geologically useful scales, whose number now exceeds 20; (3) the Greek etymology, which includes the meanings "land" or "ground"; and (4) two decades of usage (Shoemaker, 1962a, p. 117; Ronca, 1965; Mutch, 1970; Wilhelms, 1970b). The prefix "seleno-" is no longer used by professional lunar scientists except in some terms referring to coordinates, control points, or the global figure (selenographic, selenodetic). Although "astrogeology" was chosen as a convenient and appropriate name (Milton, 1969) for the U.S. Geological Survey's branch devoted to lunar and planetary studies, "lunar (planetary) geology" is more commonly used. "Planetology" is a broader term that includes such nongeologic sciences as atmospherics and planetary astronomy.

1962a); and (4) attacked the problem of extraterrestrial absolute ages by means of the impact-cratering rate on the Earth (Shoemaker and others, 1962a). As chairman of the Joint Working Group responsible for recommending a detailed program of scientific exploration (Hall, 1977, p. 163) and by daily contacts at the U.S. National Aeronautics and Space Administration (NASA) headquarters in 1962 and 1963, he was able to implement many of his sound and visionary concepts of exploration strategy. He has followed up his earlier interests by definitive studies of the lunar regolith, primary- and secondary-crater populations, and the impact flux from interplanetary space (Shoemaker, 1965, 1966, 1971, 1981; Shoemaker and others, 1967a, b, 1968, 1969a, b, 1970, 1979; Shoemaker and Morris, 1970). He was the geologist on the experimenter team for the Ranger program, the principal experimenter for the Surveyor television experiment, and the leader of the field geology teams for Apollos 11, 12, and 13. He has continued to guide and inspire those of us fortunate enough to have been associated with him during the development of lunar geologic investigations.

The manuscript was prepared with the help of numerous other scientists who are or were active in lunar research as members of the U.S. Geological Survey's Branch of Astrogeology. John F. McCauley contributed much of chapter 4, was codeveloper with me of many of the concepts of stratigraphy and basin formation embodied in this volume, and furnished valuable reviews of chapters 1 through 8. Odette B. James and Paul D. Spudis helped fill gaps left by my ignorance of lunar petrology and kindly supplied photographs of lunar rocks and thin sections. James submitted the first version of the section on mare-basalt samples and suggested major beneficial changes in all the sections on petrology. Spudis contributed drafts of the sections on Apollo 15 in chapter 10, read two versions of the entire manuscript, and contributed many valuable suggestions to every part. Richard J. Pike and Joseph M. Boyce contributed drafts of the sections on complex craters (chap. 3) and the D_L method (chap. 7), respectively. Baerbel K. Lucchitta and David H. Scott contributed drafts of much of chapter 6. Perceptive reviews were furnished by Maurice J. Grolier and Carroll Ann Hodges of chapters 3 and 4, respectively. Jay L. Inge prepared the shaded-relief base for the two-hemisphere maps (pls. 1–12). Donald E. Davis prepared most of the artistic illustrations. Gary D. Clow helped with the mathematics. The manuscript furthermore benefited from the first-hand experience of Survey geologists Gordon A. Swann, George E. Ulrich, and Edward W. Wolfe as leaders or members of the field geology teams for Apollos 14, 15, 16, and 17. Non-Survey experts also generously helped with sections outside my fields of research. Discussions with Gunther W. Lugmair (University of California, San Diego) and S. Ross Taylor (Australian National University, Canberra) clarified many aspects of geochronology and petrology. Carle M. Pieters (Brown University) contributed figures and a valuable review of the section on remote sensing in chapter 5.

I am most deeply indebted to Albert Estrada and David B. Snyder. Estrada provided indispensible help in preparing the photographic illustrations and gave valuable counsel about scope. Snyder read two versions of the manuscript in sequence from beginning to end and greatly improved its clarity by his careful scrutiny and insight.

This research was originally commissioned by the U.S. Geological Survey. Otherwise, the Survey's involvement in the lunar program has been entirely funded by NASA under some 30 contracts. Studies under Contracts R–66, T–1167B, T–5874A, T–66353G, W13,130, and W13,709 led most directly to this volume. The NASA personnel who monitored these contracts extended understanding support of the geologic approach to deciphering the flood of data returned so spectacularly from the Moon. Special thanks go to NASA contract monitor Robert P. Bryson for his perception and support of our approach during the most active phases of the program. Preparation of this manuscript was partly supported in 1978–80 by the Planetary Geology Program Office of NASA, Stephen H. Dwornik and Joseph M. Boyce, chiefs, under Contract W13,709. In 1977–80, both the Lunar and Planetary Programs Office and the Planetary Geology Program Office of NASA supported my research on chronology and ringed basins that is incorporated into this volume. The Geochemistry and Geophysics Office of NASA, William L. Quaide, chief, supported the Lunar Geoscience Consortium under Contract W13,130, which included my study of the maria also incorporated here.

ABSTRACT

More than two decades of study have established the major features of lunar geologic style and history. The most numerous and significant landforms belong to a size-morphology series of simple craters, complex craters, and ringed basins that were formed by impacts. Each crater and basin is the source of primary ejecta and secondary craters that, collectively, cover the entire terra. The largest impacts thinned, weakened, and redistributed feldspathic terracrustal material averaging about 75 km in thickness. Relatively small volumes of basalt, generated by partial remelting of mantle material, were erupted through the thin subbasin and subcrater crust to form the maria that cover 16 percent of the lunar surface. Tectonism has modified the various stratigraphic deposits relatively little; most structures are confined to basins and large craters.

This general geologic style, basically simple though complex in detail, has persisted longer than 4 aeons (1 aeon = 10^9 yr). Impacts began to leave a visible record about 4.2 aeons ago, after the crust and mantle had differentiated and the crust had solidified. At least 30 basins and 100 times that many craters larger than 30 km in diameter were formed before a massive impact created the Nectaris basin about 3.92 aeons ago. Impacts continued during the ensuing Nectarian Period at a lesser rate, whereas volcanism left more traces than during pre-Nectarian time. The latest basin-forming impacts created the giant and still-conspicuous Imbrium and Orientale basins during the Early Imbrian Epoch, between 3.85 and 3.80 aeons ago. The rate of crater-forming impacts continued to decline during the Imbrian Period. Beginning in the Late Imbrian Epoch, mare-basalt flows remained exposed because they were no longer obscured by many large impacts. The Eratosthenian Period (3.2–1.1 aeons ago) and the Copernican Period (1.1 aeons ago to present) were times of lesser volcanism and a still lower, probably constant impact rate. Copernican impacts created craters whose surfaces have remained brighter and topographically crisper than those of the more ancient lunar features.

1. GENERAL FEATURES

FIGURE 1.1.—Lunar nearside photographed 4 days after full Moon from the Pic du Midi Observatory, France, in August 1964. Left (west) half is illuminated at high Sun elevations, which enhance albedo (brightness) variations, and right (east) part at lower Sun elevations, which enhance relief by casting shadows. Terminator (line between dark and illuminated areas) is at long 51° E. Extensive ray systems surround craters Copernicus (upper left center) and Tycho (near bottom). Mare Nectaris, surrounded by concentric rings, is along terminator at lower right.

Note: Before 1961, lunar directions were stated with reference to their position in the sky as viewed from the Earth. Mare Orientale, the "Eastern Sea," is on the left limb of the Moon as seen in the Northern Hemisphere, that is, near the east horizon of the Earth. In 1961, the "astronautical convention" was adopted in anticipation of manned spaceflight: The direction from which the Sun rises on the Moon was henceforth called east, as it is on the Earth. The definition of north remained the same, but after 1961, more publications began orienting their figures with north at the top rather than at the bottom (as it is seen in astronomic telescopes). The 0° meridian is in the center of the nearside, and the diametrically opposite longitudal line on the farside is 180°. On most maps, longitudes increase both eastward and westward from the 0° meridian until meeting at the 180° meridian, which is also the terrestrial convention. Some lunar maps use a 360° system of longitudes, increasing eastward from the 0° central meridian. In this volume, the Orientale limb is considered the west, photographs are oriented with north at the top except as noted, nearside longitudes are less than 90°, and farside longitudes are greater than 90°.

1. GENERAL FEATURES

CONTENTS

	Page
Surface	3
Subsurface	12

SURFACE

Near full Moon, the naked eye sees a contrast between dark and light surfaces (of low and high *albedo*, respectively) that has been fancied as a "man in the Moon" or other configurations (fig. 1.1). In 1609, Galileo noted that the dark spots are smooth and the brighter areas rugged (Whitaker, 1978). These terrain types are still designated by their 17th-century names *maria* (singular, *mare*) and *terrae* (singular, *terra*; commonly known as uplands or highlands; table 1.1). The maria constitute about 16 percent, and the terrae 84 percent, of the lunar surface. Maria occupy about 30 percent of the lunar *nearside*, the hemisphere visible from the Earth; spacecraft exploration, beginning with the Soviet Luna 3 in 1959 (table 1.2; Barabashov and others, 1961; Kopal and Mikhailov, 1962, p. 1–44; Whitaker, 1963), showed that they constitute only about 2 percent of the *farside* (figs. 1.2–1.4). South of about lat 35° S., however, the proportions are reversed; the southern farside is richer in maria than is the southern nearside (fig. 1.5).

Most of the maria are approximately circular. The circular maria are bordered by annular or arcuate, commonly mountainous terra rims. The terra structures in which the circular maria lie are called *ringed basins* or, simply, *basins* (see chap. 4; table 1.1; Hartmann and Kuiper, 1962). Most large mountainous rings or arcs that do not encompass maria are also parts of basins (figs. 1.4, 1.5). Concentric rings characterize all well-exposed basins.

The maria are mostly level and smooth at coarse scales but contain local topographic relief. *Mare ridges* (*dorsa*) are long intricate welts on the mare surfaces (fig. 1.6). *Rilles* (*rimae*) are narrow troughs or grooves much longer than they are wide; they include genetically distinct sinuous, arcuate, and straight varieties. Few ridges but many arcuate and straight rilles cut the terrae as well as the maria. *Dark-mantling materials* are as dark as or darker than the maria but assume part of the topographic form of the underlying terrain (figs. 1.6, 1.7).

Craters, ranging in size from those visible with binoculars to micropits observed on returned samples, are ubiquitous on the lunar surface. At coarse scales, they are much more numerous on the terrae than on the maria. Most lunar craters have rims elevated above, and floors depressed below, the surrounding terrain. Craters smaller than about 16 to 21 km in diameter have relatively featureless interiors; larger craters are more complex, possessing central peaks, arcuate wall terraces, and other interior landforms (see chap. 3). Some crater exteriors resemble the adjacent terrain except for a short rim flank; others display coarse concentric structure near the rim crest, grooves on a lower rim flank, and numerous smaller *satellitic craters*, which are most conspicuous between about one and three crater radii from the rim (fig. 1.6). Bright *rays* radiate hundreds or thousands of kilometers from some craters (figs. 1.1, 1.6).

Most terrae appear at first glance to consist of little except large, randomly distributed, overlapping craters. However, regional differences in terra morphology emerge upon further inspection (Hackman and Mason, 1961). The central and northern nearside is characterized by ridges and grooves radial to the Imbrium basin, which contains Mare Imbrium (figs. 1.6–1.8). The terra east of the neighboring circular basin, Serenitatis, has a choppier, less regular pattern of elevations and depressions (fig. 1.7). Several regions contain particularly dense concentrations of craters that are grouped in chains or clusters. Concentric arcs of the Nectaris basin, which encompass the small Mare Nectaris, are conspicuous in the southeast quadrant (fig. 1.1). An even more conspicuous system of concentric rings, radial lineations, and satellitic craters surrounds Mare Orientale on the west limb of the Moon (fig. 1.9). *Terra plains* that are smooth and level like the maria, but lighter in color, occupy more crater floors and other depressions in and near the Imbrium- and Orientale-radial terrains (figs. 1.8, 1.9) than in any other region. These textural patterns in the Imbrium, Nectaris, and Orientale regions, as described in detail in this volume, play major roles in elucidating the geology of the Moon. They are expressions of major stratigraphic units and form the basis for interpreting less distinctive terrae.

The farside seems, at first, to be even less regularly patterned than the nearside. This difference is due mostly to the absence of maria and of the extensive circum-Imbrium radial pattern that characterize the nearside terrae. Later descriptions show that concentric and radial patterns also characterize the farside (fig. 1.4), although they are generally less extensive, less pronounced, and less well photographed than those of the nearside. The southern farside, which contains most of the farside's maria, also contains the farside's most conspicuous concentric rings and other noncrater morphologies (fig. 1.5). A major purpose of this volume is to show the basic stratigraphic order that underlies the morphologic features of the terrae.

TABLE 1.1.—*Names of common lunar landforms*

[Plural form in parentheses; the singular form is properly used adjectivally—for example, "mare basalt," not "maria basalt."]

[Latin generic terms and proper names of individual features used here are approved by the International Astronomical Union (Arthur and others, 1963; Andersson and Whitaker, 1982). Small or obscure craters have traditionally been designated by letters after the names of a nearby, more conspicuous crater (Mädler system of nomenclature); following Andersson and Whitaker, these convenient lettered names are used in this volume in preference to some new names that were invented for use on the large-scale charts made from Apollo orbital data]

Latin name	Common name	Description
Mare (maria)	Sea (not used in this volume).	Dark, smooth plains.
Lacus, palus, sinus		Small mare.
Terra (terrae)[1]	Highlands, uplands, continents.	Rugged, relatively bright (high albedo) terrain.
Mons (montes)	Mount, mountain(s)	High massif(s), generally forming arcuate ranges.
Promontorium	Promontory	Mountains partly enclosed by mare.
Rupes	Scarp	Fault in mare or high arcuate scarp in terra.
Dorsum (dorsa)	Mare ridge, wrinkle ridge.	Narrow ridge, mostly in mare.
Rima (rimae)	Rille	Narrow, elongate depression (sinuous, arcuate, or straight).
Vallis (valles)	Valley	Wide, elongate depression, commonly consisting of inconspicuous craters.
Catena (catenae)	Chain	Chain of distinct craters.
	Crater	Circular or subcircular depression, generally bounded by a raised rim.[2]
	Basin, ringed basin	Large craterlike depression containing one or more rings in addition to a rim.[3]

[1] There is no sharp distinction between an individual "terra" and the "terrae." "Terrestrial" refers to the planet Earth.

[2] Usage has established the term "ejecta," for the material thrown out of craters, as a singular noun, despite its origin as a Latin plural.

[3] The terms "mare" and "basin" are commonly confused. Genetically, these two features are only indirectly related (see chap. 2). Basins are terra structures, not all of which contain maria. Efforts to alleviate this serious semantic confusion have led to use of such terms as "mountain-bordered mare," "mare basin," "dry mare," "thalassoid" (no lava; Lipsky, 1965), and "multiringed basin." Terms for basins that contain the word "mare" are unsatisfactory, and even "basin" is misleading because these structures are characterized by huge mountainous rings as much as by the excavated depression. Nevertheless, it is probably too late to coin a new term. In this volume, the term "basin" or "ringed basin" is used for all lunar excavations at least 300 km in diameter; a more exact definition has not yet been agreed upon (see chap. 4).

FIGURE 1.2.—East limb (right edge of lunar disc as seen from the Earth) and adjacent part of farside, divided by long 90° E. Mare Crisium (C) is on nearside, and Maria Marginis (M) and Smythii (S) partly on farside. Maria also fill such craters as Lomonosov (L; 93 km, 27° N., 98° E.) and Tsiolkovskiy (T; 180 km, 20° S., 129° E.; partly in shadow; compare figs. 1.3, 1.4). Farside terrane in view, which is otherwise mostly terra, includes craters Fabry (Fa; 179 km, 43° N., 101° E.), Fleming (Fl; 130 km), Hilbert (H; 170 km), King (K; 77 km), and Pasteur (P; 235 km). A large subcircular area of light-colored plains lies between Fleming and Lomonosov. Terminator is at long 131° E.; left-hand (west) edge of photograph is at nearly the same longitude as terminator in figure 1.1. Apollo 16 frame M−3021, photographed by Apollo 16 mapping camera on Earthbound flight in April 1972.

Note: This volume includes photographs taken with three types of cameras carried in lunar orbit by Apollo spacecraft: mapping or metric (M), panoramic (P), and hand-held or bracket-mounted Hasselblads (H). All missions carried Hasselblads; Apollos 15 through 17 carried both mapping and panoramic cameras as well (pl. 2; Masursky and others, 1978).

Note: Most crater diameters and positions in this volume are from Andersson and Whitaker (1982). Basin diameters and most crater diameters used in frequency studies are from my measurements. Latitude is given before longitude throughout the volume. Coordinates differ considerably among various maps, especially on the limbs and farside. One degree of lunar latitude covers about 30 km.

FIGURE 1.3.—Part of southern farside centered on mare-filled crater Tsiolkovskiy (T). Overlaps with fig. 1.2; compare Tsiolkovskiy and Hilbert (H; 170 km, 18° S., 108° E.). Other craters: Langemak (L; 102 km; contains small mare patch), Fermi (F; 238 km), Milne (M; 262 km; fig. 1.5), Jenner (J; 72 km; fig. 1.5), Pavlov (P; 141 km), Roche (R; 146 km; superposed crater Pauli contains mare). Pl, small ringed basin Planck (325 km, 60° S., 136° E.; figs. 1.4, 1.5). Orbiter 3 frame M-121.

Note: Photographs transmitted from unmanned Lunar Orbiter spacecraft in 1966 and 1967 (table 1.2; Levin and others, 1968; Mutch, 1970) are labeled as follows: "Orbiter" followed by mission number (1-5), M (medium resolution) or H (high resolution), and frame number. Best Lunar Orbiter coverage of each area is outlined on plate 2.

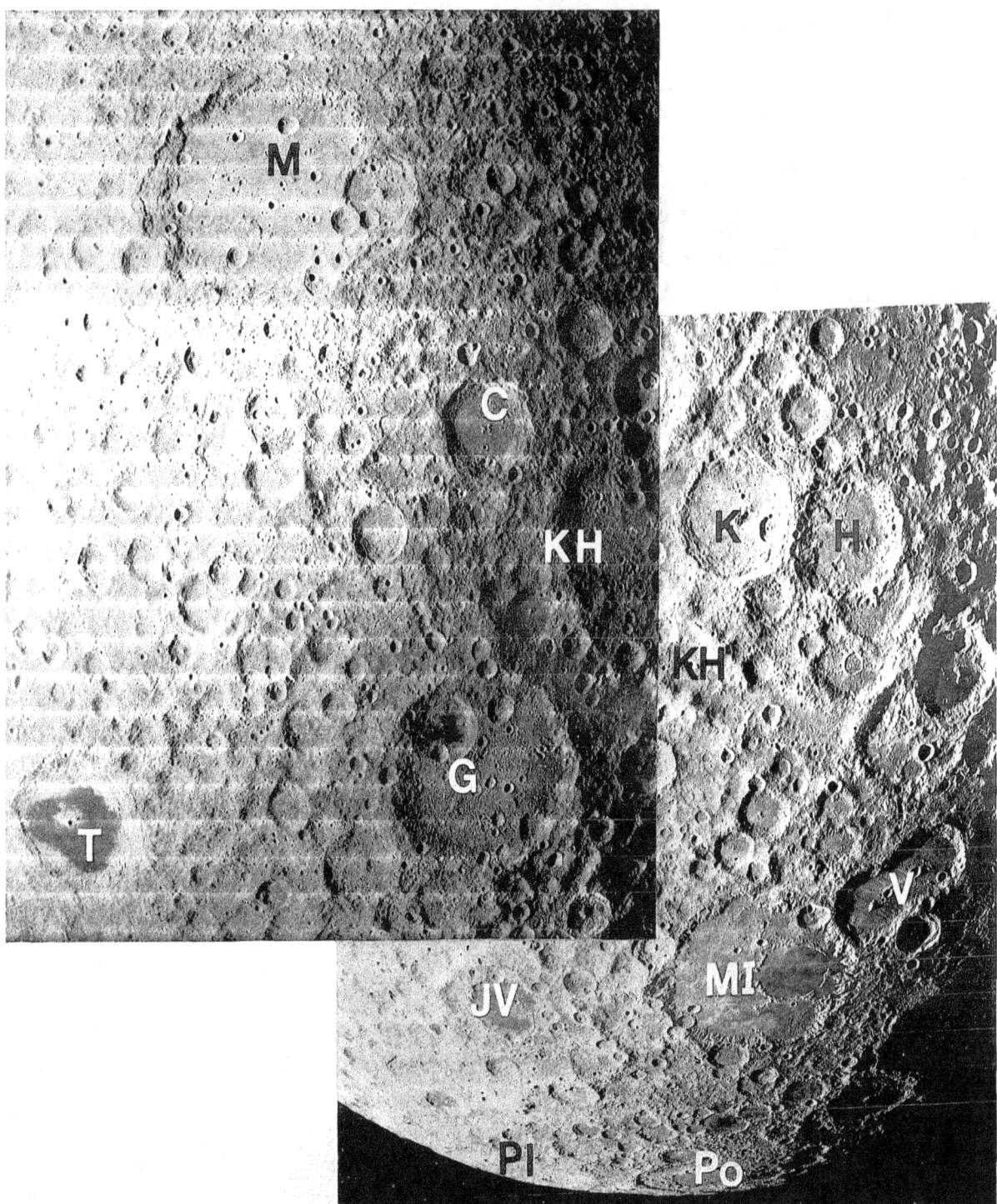

FIGURE 1.4.—Equatorial and southern farside. Most of area is heavily cratered, smoother in Gagarin (G; 272 km, 20° S., 149° E.) and north of Heaviside (H; 163 km, 11° S., 167° E.). A distinct linear trough is north of Keeler (K; 169 km). Arcuate massifs below and to left (west) of Keeler and Heaviside are parts of Keeler-Heaviside basin (KH). Maria fill Tsiolkovskiy (T), Mare Ingenii (MI, in Ingenii basin), Jules Verne (JV; 134 km), part of Poincaré basin (Po; compare fig. 1.5), and other depressions. Other craters: Chaplygin (C; 124 km) and Van de Graaff (V, double); other basins: Planck (Pl, compare fig. 1.3) and Mendeleev (M; 320 km, 6° N., 142° E.). Mosaic of Orbiter 1 frame M−115 (left) and Orbiter 2 frame M−75 (right).

FIGURE 1.5.—Southeast limb, showing concentration of maria and various ringed basins (see chap. 4). A, massifs of Australe basin, which enclose Mare Australe, consisting of many small mare matches resting in such craters as Jenner (J; compare fig. 1.3) and Lyot (L; 141 km); An, Antoniadi crater or basin (135 km); C, outer ring of Crisium basin; M, Milne (compare fig. 1.3); MS, Mare Smythii in Smythii basin (compare fig. 1.2); Pl, Planck basin (barely discernible; compare fig. 1.3); Po, Poincaré basin, containing several mare patches (compare fig. 1.4); S, Schrödinger double-ringed basin (320 km, 76° S., 134° E.). Indefinite basins (table 4.2) include Amundsen-Ganswindt (AG), Balmer-Kapteyn (BK), and Sikorsky-Rittenhouse (SR). Grooves are radial to Schrödinger and to Nectaris basin (N; basin is outside photograph). Crater Humboldt (H; 207 km, 27° S., 81° E.) contains rilles and small mare patches. Orbiter 4 frame M−9.

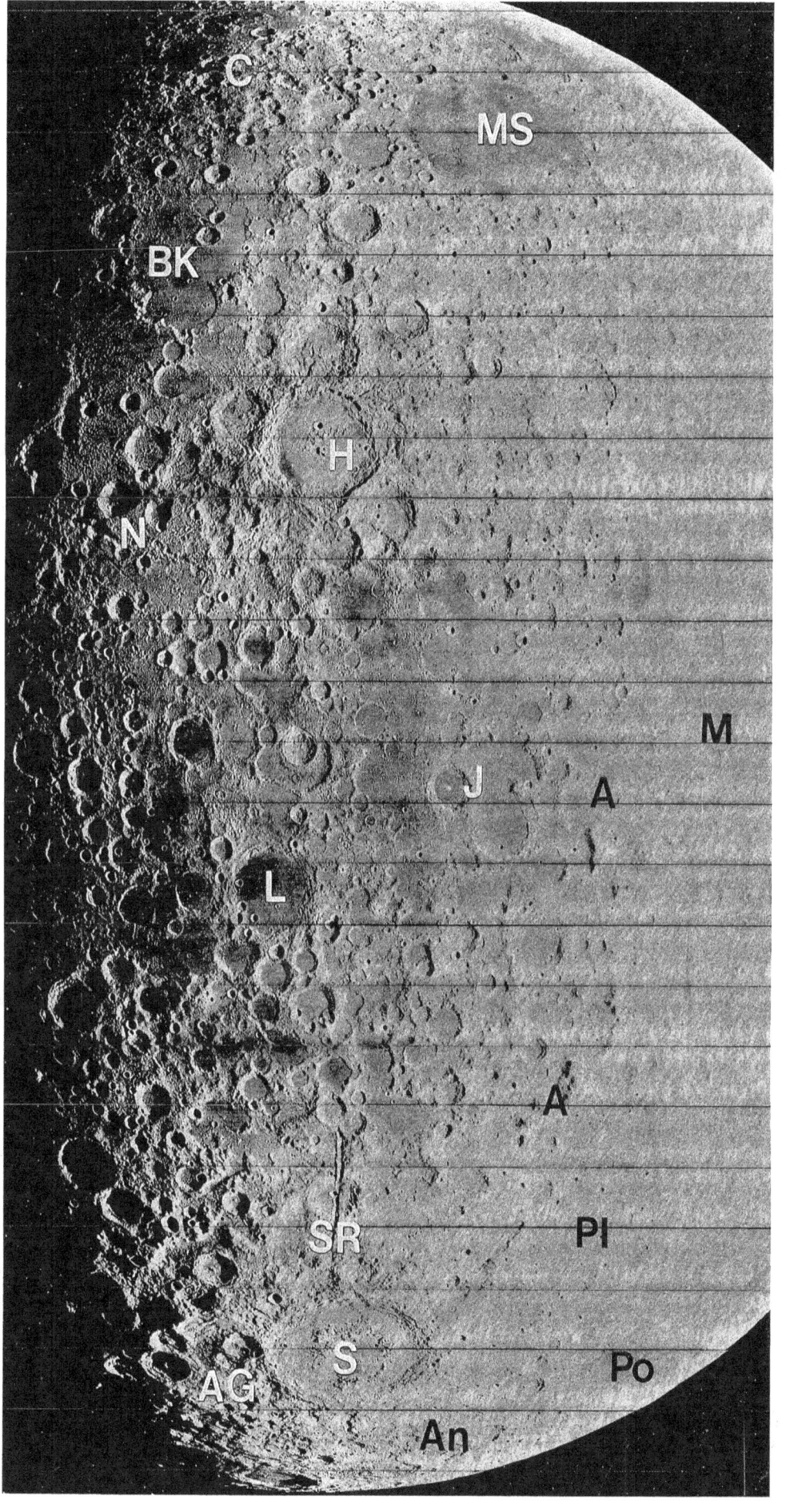

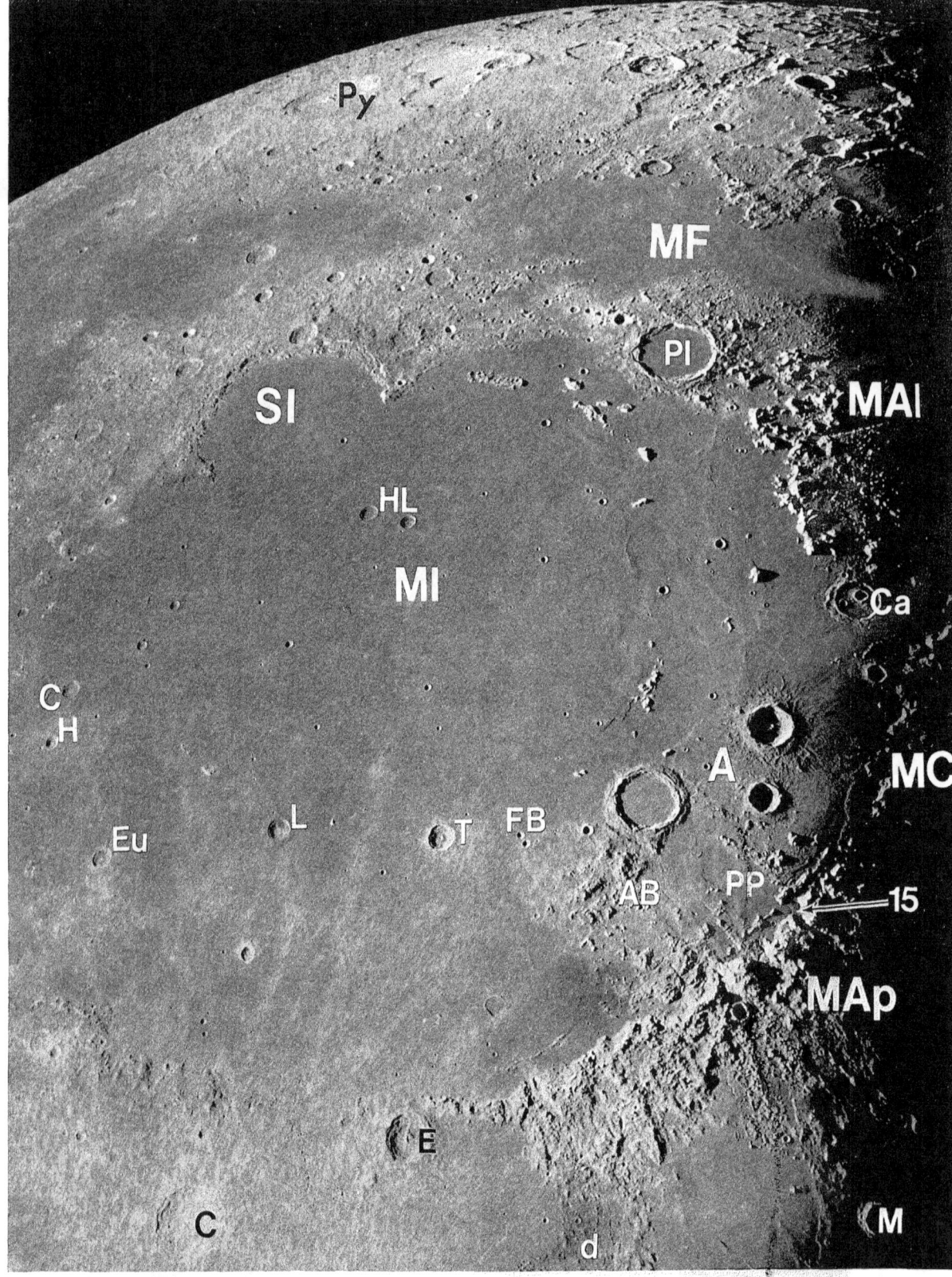

FIGURE 1.6.—Telescopic view of circular Mare Imbrium (MI) and Imbrium basin. MF, Mare Frigoris; PP, Palus Putredinis; arrow, Apollo 15 landing site. Imbrium-basin rim is composed of Montes Alpes (MAl), Montes Apenninus (MAp), Montes Caucasus (MC), and terra occupied by Iridum crater bounding mare feature Sinus Iridum (SI). Lunar stratigraphic scheme of Shoemaker and Hackman (1962) is based on relations among Imbrium basin, planar material of Apennine Bench (AB), Archimedes (left of A), mare material, Eratosthenes (E), and Copernicus (C; 95 km; compare fig. 1.1; satellitic craters are visible east of crater) (see chaps. 2, 7). "Wrinkle ridges" (dorsa) are above A; rilles are below PP. Crater pairs (see chap. 3) include Aristillus and Autolycus (right of A), Caroline Herschel and Heis (CH), Feuillée and Beer (FB), and Helicon and Leverrier (HL). Other features: Cassini (Ca), Euler (Eu), Lambert (L), Manilius (M), Plato (Pl), Pythagoras (Py), and Timocharis (T); many irregular craters north of Mare Frigoris; d, dark-mantled terra surface. Mount Wilson Observatory photograph, catalog No. 257.

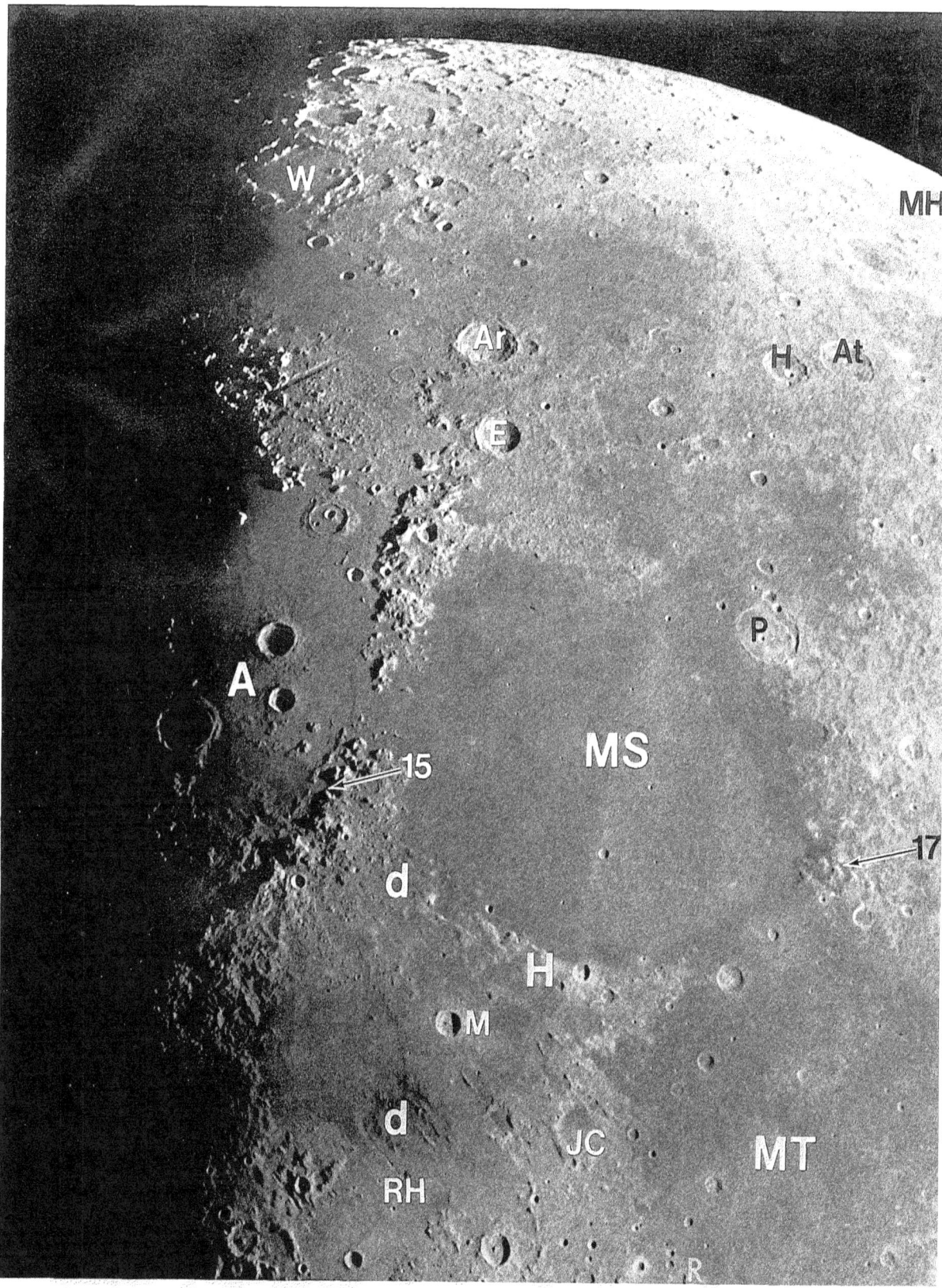

FIGURE 1.7.—Telescopic view centered on Mare Serenitatis (MS), overlapping area of figure 1.6. Lineations in Montes Haemus (H) on southern part of Serenitatis rim are radial to Montes Apenninus; nonlineate hummocky terrain adjoins Montes Caucasus and Montes Alpes. Terrain east of Mare Serenitatis consists of irregular craters and massifs. Other features: Ar, Aristoteles (87 km, 50° N., 17° E.); At, Atlas (87 km); E, Eudoxus; H, Hercules; JC, Julius Caesar; M, Manilius; MH, Mare Humboldtinaum; MT, Mare Tranquillitatis; P, Posidonius; R, Ritter (29 km, 2° N., 19° E.); RH, Rima Hyginus; densely cratered northern terrain including W. Bond (W; 158 km, 65° N., 4° E.); d, dark-mantling material. Arrows, Apollo 15 and 17 landing sites. Mount Wilson Observatory photograph, catalog No. 262.

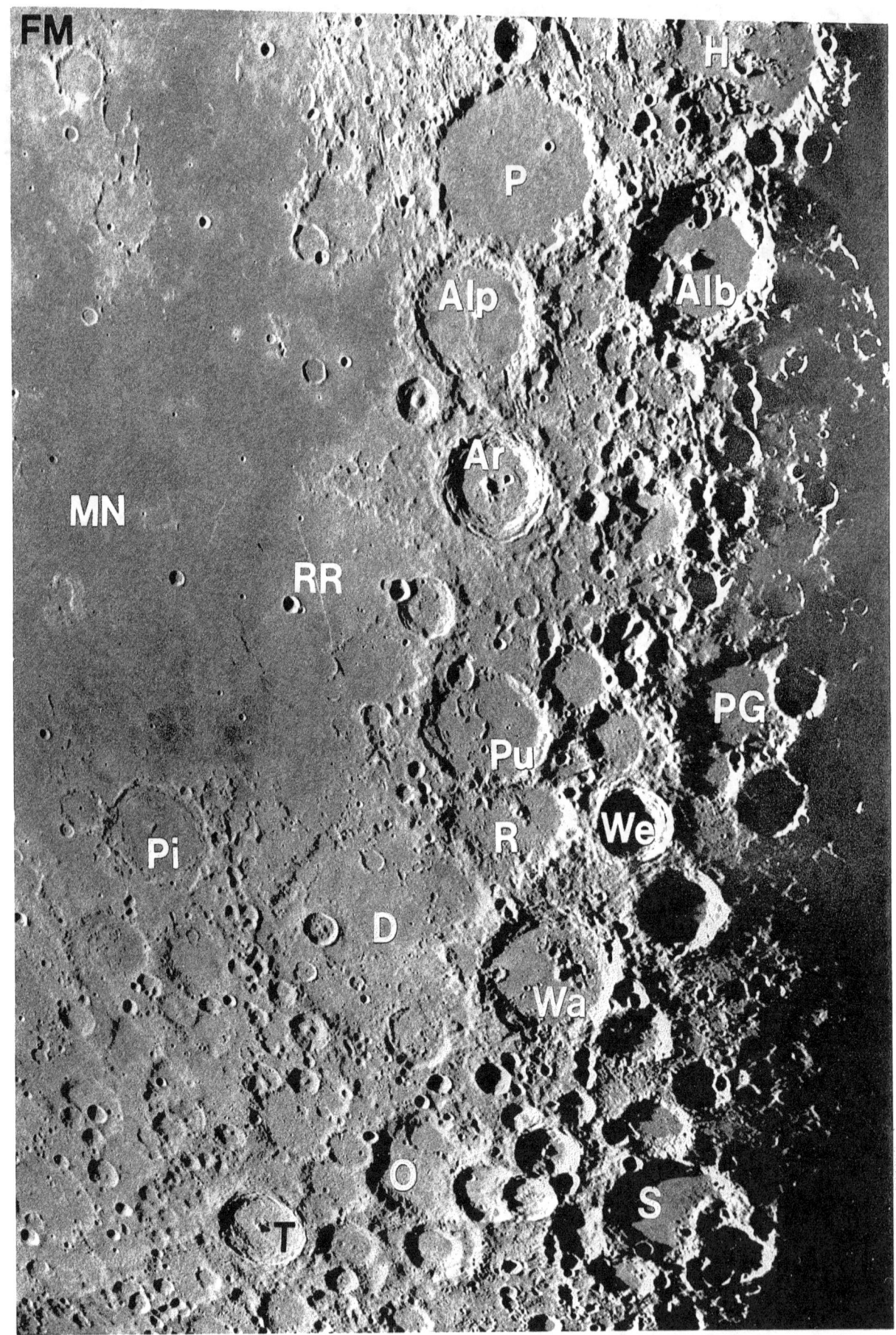

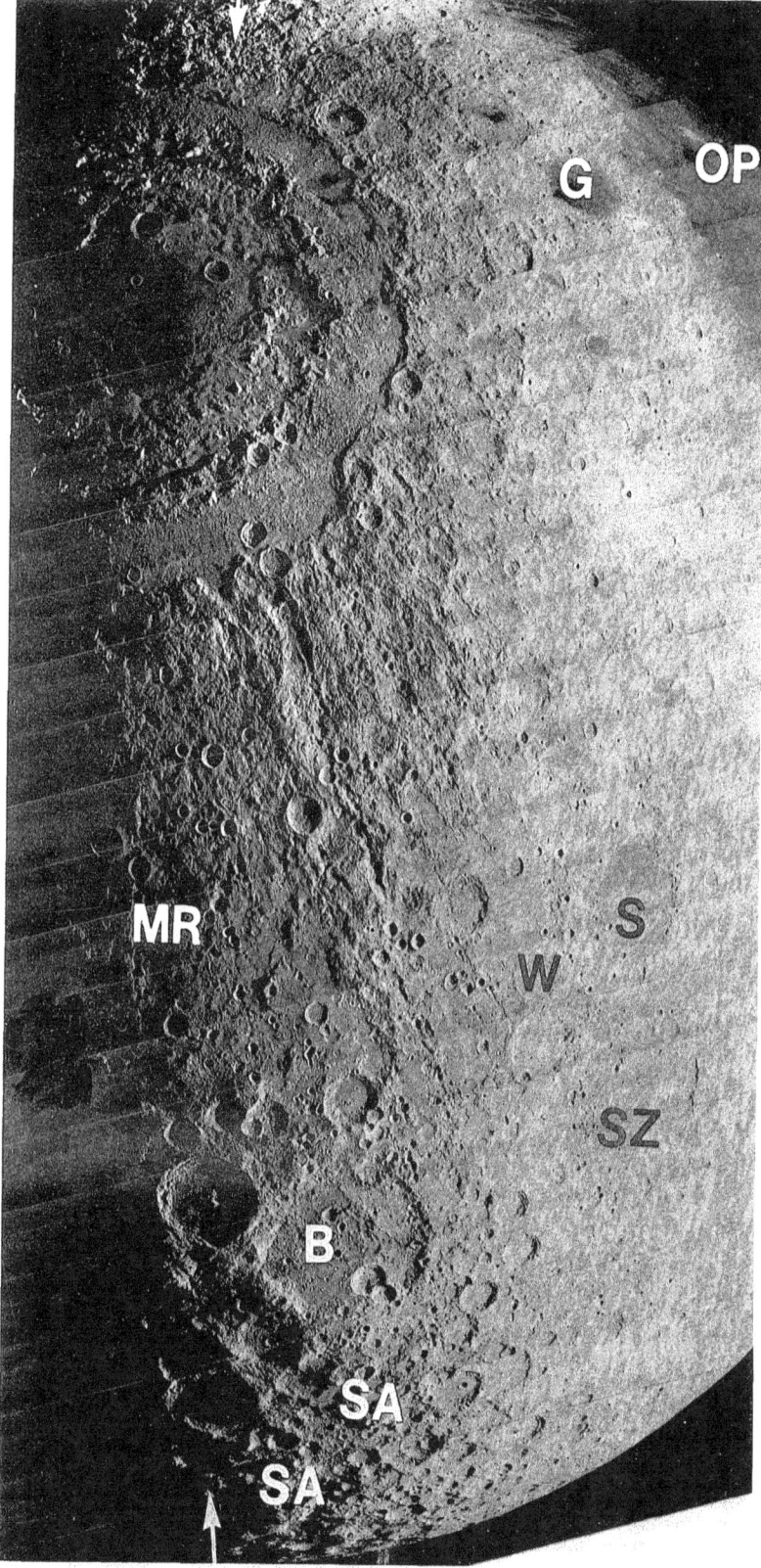

FIGURE 1.9.—Southwest limb (lower left as seen from northern hemisphere of the Earth). Arrows indicate long 90 W.; upper arrow on equator and lower on south pole. Conspicuous lineations are radial to Mare Orientale and ringed Orientale basin (centered at 20° S., 95° W.). Light-colored plains form part of terrain outside Orientale radials, for example, in Schiller-Zucchius basin (SZ), in central part of crater Schickard (S; 227 km, 44° S., 55° W.), and in and near crater Wargentin (W; 84 km). Other basins: Bailly (B; 300 km, 67° S., 69° W.), Grimaldi (G; partly mare-filled), Mendel-Rydberg (MR, barely visible), and South Pole-Aitken (SA, mountainous massifs); OP, part of Oceanus Procellarum in Procellarum basin. Footprint-shaped crater is Schiller (180 km). Orbiter 4 frame M-180.

FIGURE 1.8.—South-central nearside, including parts of "Fra Mauro peninsula" (FM) and Mare Nubium (MN), mare-filled crater Pitatus (Pi), fresh crater Tycho (T; 85 km, 43° S., 11° W.; compare fig. 1.1), moderately fresh crater Werner (We), and north-south-trending "backbone" of terra, including "chain" of large craters Ptolemaeus (P; 153 km, 9° S., 2° W.), Alphonsus (Alp), Arzachel (Ar), Purbach (Pu), Regiomontanus (R), and Walter (Wa, 140 km, 33 S., 1° E). Smooth terra plains lighter than maria fill many craters, including those in "chain" and Albategnius (Alb), Deslandres (D), Hipparchus (H), Orontius (O), Playfair G (PG), and Stöfler (S). Rupes Recta (RR, "Straight Wall") and grooves ("sculpture") in upper right quadrant are radial to Imbrium basin. Apollo 14 landing site is just beyond north edge of photograph, above letters FM. Mount Wilson Observatory photograph.

TABLE 1.2.—*Spaceflights that provided lunar data*

[Dates are beginnings of mission activities at the Moon. Ranger 7 impacted on August 1, 1964 (G.m.t.)]

Mission	Date	Type of mission	Landing location or orbital parameters
Luna 2	Sept. 1959	First impact	Flank of crater Autolycus, lat 30° N., long 0°.
Luna 3	Oct. 1959	First unmanned flyby photography	Farside and east limb; altitude, 65,000 km.
Ranger 7	July 1964	First preimpact photography	Mare Nubium (Cognitum), lat 10.6° S., long 20.7° W.
Ranger 8	Feb. 1965	Preimpact photography	Mare Tranquillitatis, lat. 2.6° N., long 24.7° E.
Ranger 9	Mar. 1965	Last preimpact photography	Floor of crater Alphonsus, lat 12.9° S., long 2.4° W.
Zond 3	July 1965	Unmanned flyby photography	Farside and west limb; altitude, 9,960 to 11,570 km.
Luna 9	Feb. 1966	First unmanned landing	Oceanus Procellarum, lat 7.1° N., long 65.4° W.
Luna 10	Apr. 1966	First unmanned orbital; gamma-ray	Perilune, 350 km.
Surveyor 1	June 1966	Unmanned landing	Mare in crater Flamsteed P, lat 2.5° S., long 43.2° W.
Lunar Orbiter 1	Aug. 1966	First unmanned orbital photography	Inclination, 12°; perilunes, 190 and 40 km.
Luna 11	Aug. 1966	Unmanned orbital photography	Perilune, 165 km.
Luna 12	Oct. 1966	do	Perilune, 100 km.
Lunar Orbiter 2	Nov. 1966	do	Inclination, 12°; perilune, 50 km.
Luna 13	Dec. 1966	Unmanned landing	Oceanus Procellarum, lat 18.9° N., long 62.1° W.
Lunar Orbiter 3	Feb. 1967	Unmanned orbital photography	Inclination, 21°; perilune, 55 km.
Surveyor 3	Apr. 1967	Unmanned landing	Oceanus Procellarum, lat 3.2° S., long 23.4° W.
Lunar Orbiter 4	May 1967	Unmanned orbital photography (global)	Inclination, 85°; perilune, 2,705 km.
Explorer 35	July 1967	Unmanned orbital magnetics	Perilune, 830 km (returned data until Feb. 1972).
Lunar Orbiter 5	Aug. 1967	Unmanned orbital photography	Inclination, 85°; perilunes, 195 and 100 km.
Surveyor 5	Sept. 1967	Unmanned landing, first chemical analysis	Mare Tranquillitatis, lat 1.4° N., long 23.1° E.
Surveyor 6	Nov. 1967	Unmanned landing	Sinus Medii, lat 0.5° N., long 1.5° W.
Surveyor 7	Jan. 1968	First unmanned landing in terra	Flank of crater Tycho, lat 40.9° S., long 11.5° W.
Zond 6	Nov. 1968	Unmanned flyby, first returned film	Altitude, ≥3,300 km.
Apollo 8	Dec. 1968	First manned orbital	Perilune, 110 km.
Apollo 10	May 1969	Manned orbital	Best perilune, 15 km.
Apollo 11	July 1969	First manned landing	Mare Tranquillitatis, lat 0.7° N., 23.4° E.
Zond 7	Aug. 1969	Unmanned flyby, returned film	Southern farside and west limb; perilune, 2,200 km(?).
Apollo 12	Nov. 1969	Manned landing	Oceanus Procellarum, lat 3.2° S., long 23.4° W.
Apollo 13	Apr. 1970	Manned flyby (aborted landing)	Southern farside and west limb.
Luna 16	Sept. 1970	First unmanned sample return	Mare Fecunditatis, lat 0.7° S., long 56.3° E.
Zond 8	Oct. 1970	Unmanned flyby, returned film	Altitude, ≥1,120 km.
Luna 17	Nov. 1970	Lunakhod 1, first unmanned rover	Sinus Iridum, lat 38.3° N., long 35.0° W.
Apollo 14	Feb. 1971	First manned landing in terra	Fra Mauro highlands, lat 3.7° S., long 17.5° W.
Apollo 15	July 1971	Manned landing, first long mission	Apennine-Hadley region, lat 26.1° N., long 3.7° E.
Luna 19	Oct. 1971	Unmanned orbital	Best perilune, 77 km(?).
Luna 20	Feb. 1972	Unmanned sample return	Crisium-basin rim, lat 3.5° N., long 56.5° E.
Apollo 16	Apr. 1972	Manned landing	Descartes highlands, lat 9.0° S., long 15.5° E.
Apollo 17	Dec. 1972	Last manned landing	Taurus-Littrow Valley, lat 20.2° N., long 30.8° E.
Luna 21	Jan. 1973	Lunakhod 2, unmanned rover	Mare in crater Le Monnier, lat 25.8° N., long 30.5° E.
Luna 24	Aug. 1976	Last unmanned sample return	Mare Crisium, lat 12.7° N., long 62.2° E.

SUBSURFACE

Spacecraft exploration has provided some information about the configuration of materials below the lunar surface. The uppermost layers of both the maria and the terrae consist of fragmental material called *regolith* (Shoemaker and others, 1967a, 1968). Regoliths are generally thinner than about 5 or 6 m on the maria and thicker than that on the terrae (Shoemaker and Morris, 1970; Cooper and others, 1974). They dominate the lunar scene at closeup scales (fig. 1.10) but are not evident in the telescopic and Lunar Orbiter photographs shown here, except as they affect albedo (figs. 1.1–1.9). The term "soil" is commonly used either as a synonym for regolith or in reference to its fine surficial material. Because this volume emphasizes regional relations, it does not discuss the regolith in detail.

The underlying *bedrock* of the maria is basaltic and has a density of 3.3 to 3.4 g/cm^3. Mare basalt typically extends hundreds of meters below the surface and locally reaches depths of about 5 km (see chap. 5). In contrast, the terra bedrock is feldspathic (plagioclase-rich) material with a lower density estimated at 2.90 to 3.05 g/cm^3 (Basaltic Volcanism Study Project, 1981, p. 671). Seismic data indicate that this terra material forms a crust 45 to 60 km thick in the west-central nearside (fig. 1.11; Toksöz and others, 1974; Koyama and Nakamura, 1979) and 75 km thick in part of the southeastern nearside (beneath the Apollo 16 landing site; Nakamura, 1981). Extrapolations of these measurements to other regions are based on elevation data, estimates of the crustal density, and models of isostatic compensation, all of which are subject to further refinement (Solomon, 1978; Thurbur and Solomon, 1978). Where measured, average elevations are higher, relative to the Moon's center of mass, on the farside than on the nearside (Kaula and others, 1974; Bills and Ferrari, 1977); the crust may be as thick as 120 km under some elevated terra on the farside (fig. 1.12; Bills and Ferrari, 1977). Most estimates of the mean crustal thickness fall within the range 74±12 km (Kaula and others, 1974; Bills and Ferrari, 1977; Kaula, 1977; Haines and Metzger, 1980; Basaltic Volcanism Study Project, 1981, p. 671; Taylor, 1982, p. 180-182). A mean thickness of 62 km corresponds to about 10 percent of the Moon's volume, of 74 km to about 12 percent, and of 86 km to about 14 percent (radius, 1,738 km).

Little is known about lunar intracrustal structure, except that seismic velocities appear to increase at about 20 to 25 km below the surface of southern Oceanus Procellarum (Toksöz and others, 1974). This discontinuity may indicate a change in physical state (open cracks above; solid, denser material below) or in chemical composition (Todd and others, 1973; Herzberg and Baker, 1980). At least the upper few kilometers of the terra crust consists of breccia, a rock type composed of angular fragments (clasts) set in a finer-grained matrix.

Beneath the terra crust and constituting all or most of the remaining lunar volume is the ultramafic lunar mantle (fig. 1.11). Its density is close to that of mare basalt and to the mean lunar density, 3.34 g/cm^3. Seismic data suggest that the mantle is fairly uniform at least to 1,100±100-km depth, although small seismic-velocity changes have been modeled (Goins and others, 1979). Seismic shear waves are attenuated below 1,100±100 km. This central part of the Moon may or may not include melted zones and (or) a chemically distinct (metallic or sulfide-rich) core (Wiskerchen and Sonett, 1977; Goins and others, 1979; Taylor, 1982).

In the past, the mechanically deformable elastic lithosphere seems to have coincided with the terra crust, which may be considered the petrologic or chemical lithosphere (see chaps. 6, 8). Today, the elastic lithosphere must include much of the mantle as well.

These basic facts or assumptions about the lunar subsurface are needed as background for later discussions of lunar tectonism and petrogenesis and for perspective on the overall constitution of the Moon. This volume, however, mostly discusses the three-dimensional form of the materials in the upper few kilometers or tens of kilometers beneath the terra and mare surfaces.

FIGURE 1.10.—Astronaut views of lunar surface.
  A. Lunar Module and astronaut at Apollo 11 landing site. Regolith consists of loose, footprint-compacted material and a few rocks. Apollo 11 frame H-5931.
  B. Wall of Rima Hadley (Hadley Rille) at Apollo 15 landing site, showing the only inplace outcrops of lunar strata visited by astronauts, overlain by thin regolith and loose boulders. Montes Apenninus, in distance, appear smooth because of cover of loose debris (compare rugged appearance of mountains in fig. 1.6). Apollo 15 frame H-12115.

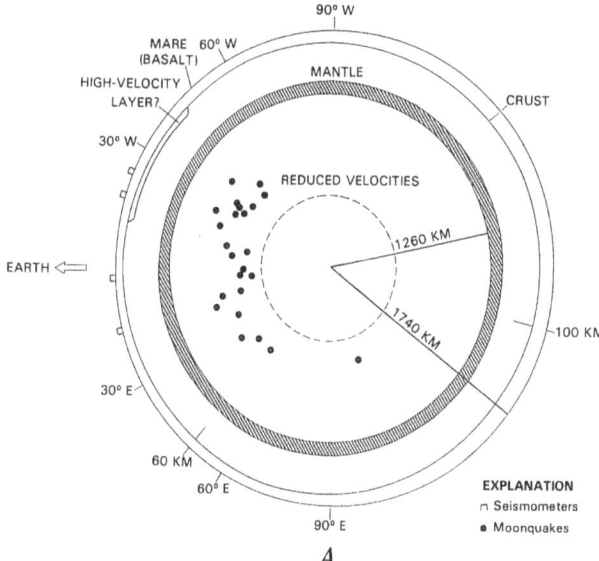
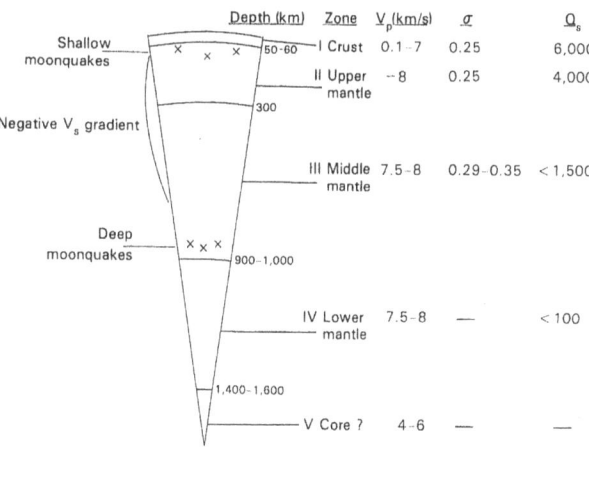

FIGURE 1.11.—Two interpretations of major features of lunar crust and mantle.
  A. Schematic equatorial slice through entire Moon. All known endogenic moonquakes are on nearside. Crust believed to be thicker on farside than on nearside. Longitudes of Apollo seismometers are shown. From Goins and others (1979, fig. 6b).
  B. Similar crustal structure but different mantle structure hypothesized by Latham and others (1978). V_p, compressional-wave velocity; V_s, shear-wave velocity; σ, Poisson's ratio; Q_s, quality factor for shear waves.

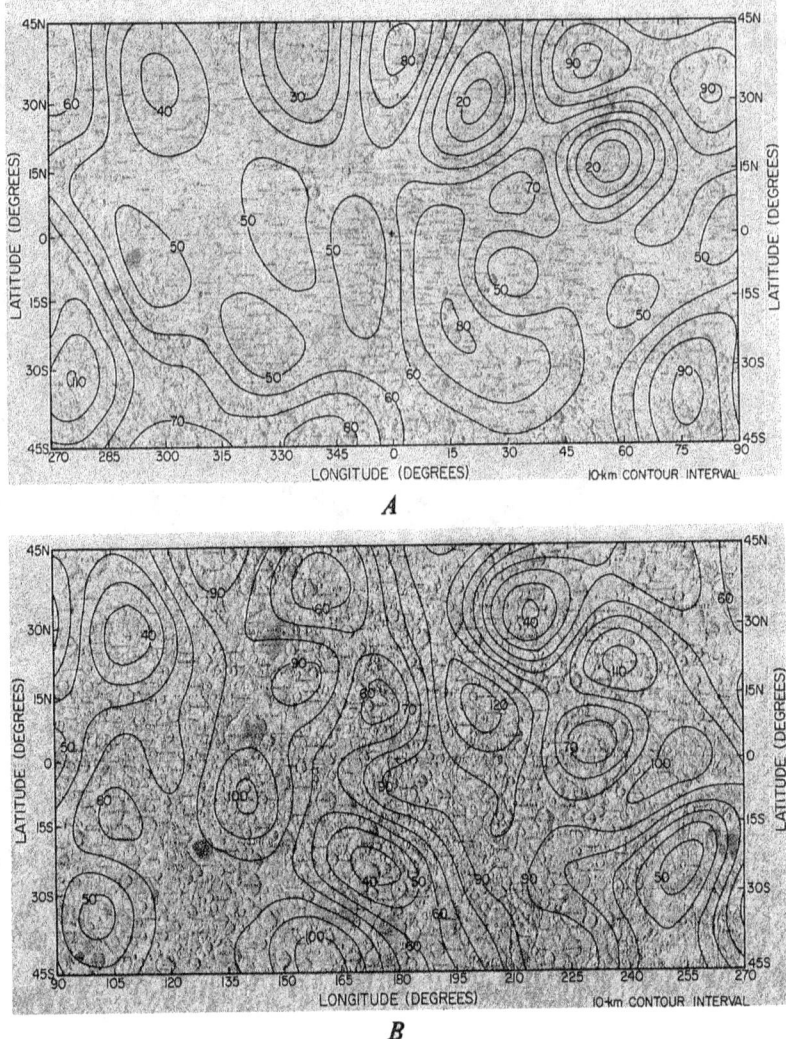

FIGURE 1.12.—Approximate crustal thicknesses (in kilometers) on nearside (A) and farside (B). From Bills and Ferrari (1977, fig. 3).

2. THE STRATIGRAPHIC APPROACH

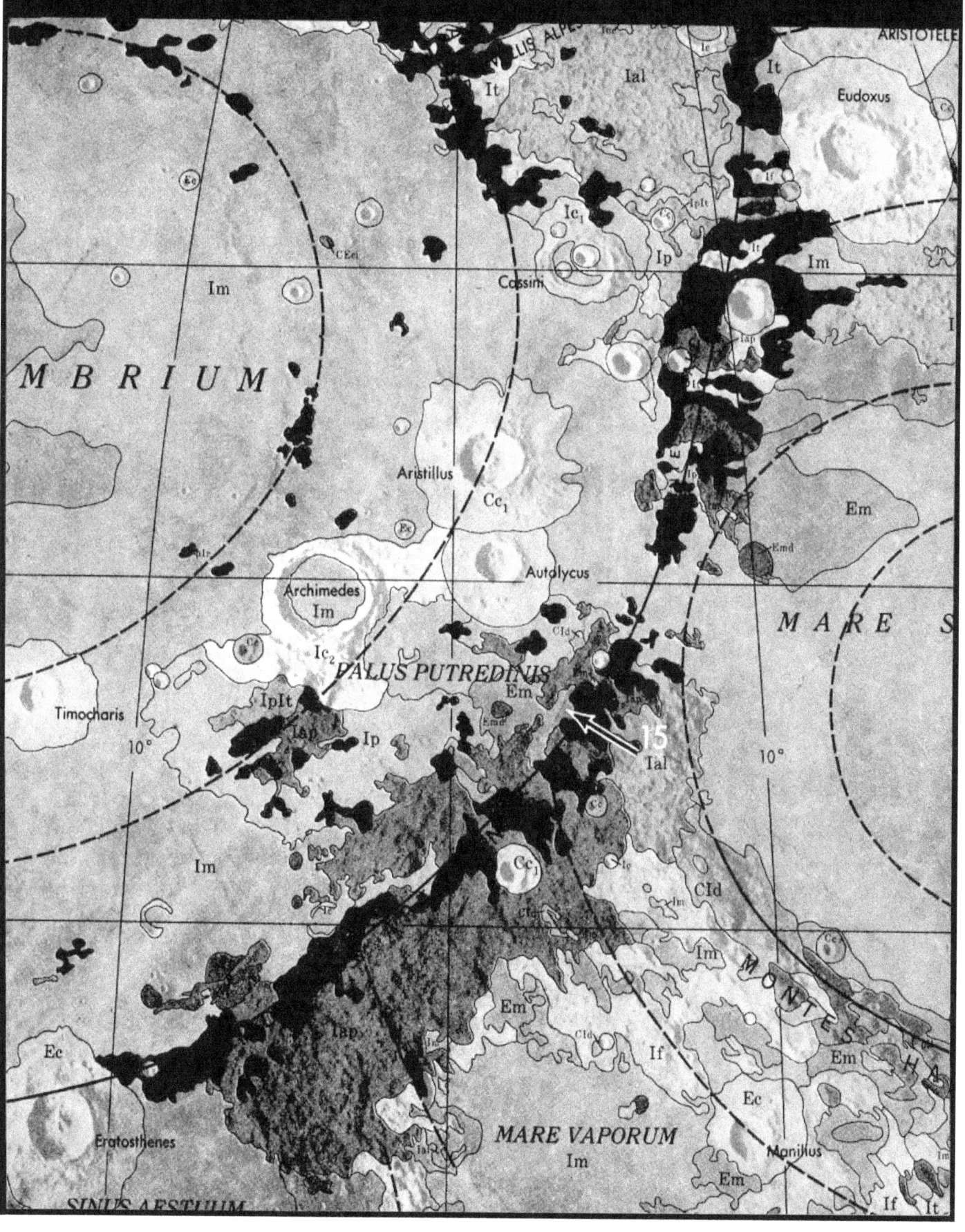

FIGURE 2.1 (OVERLEAF).—Geologic map of part of areas covered by figures 1.6 and 1.7. Important units include: Cc_1, Copernican crater materials; Cld, Copernican, Eratosthenian, or Imbrian dark-mantling materials; Ec, Eratosthenian crater materials; Em, Eratosthenian mare materials; Ic_1 and Ic_2, Imbrian crater materials; Im, Imbrian mare materials; Ip, Imbrian plains materials (Apennine Bench Formation); Ial, Alpes Formation; Iap, materials of Montes Apenninus; If, Fra Mauro Formation; IpIt, Imbrian or pre-Imbrian materials, undivided; pIr, pre-Imbrian rugged materials. Arrow, Apollo 15 landing site. From Wilhelms and McCauley (1971).

2. THE STRATIGRAPHIC APPROACH

CONTENTS

	Page
General principles	17
Case histories	17
Craters	17
Mare versus basin	19
Terra materials	19
Correlation of samples and stratigraphic units	21

GENERAL PRINCIPLES

Some earlier observers, influenced by experience with terrestrial geology, interpreted the Moon's surface in structural terms. Real or imaginary alignments of landforms were construed as faults or folds, and the present topography was explained as the product of progressive endogenic deformation of an originally simple surface (for example, Fielder, 1965). Crater and basin rims were thought to have been emplaced gradually along arcuate fissures. Chronologic sequences can be partly inferred in these structural models. A small crater inside a larger, or one crater rim that cuts across another, are signs of relative age in all but the most contrived scenarios. A sharp scarp that cuts a mare surface must be younger than the surface.

Our present understanding of lunar geology, however, has resulted from interpreting surface relations in terms of stratigraphic units (figs. 2.1, 2.2). Building upon centuries of thought that apparently began in 1669 with Nicolaus Steno (Woodford, 1965, p. 2–6), stratigraphers studying either the Earth or the Moon treat observations in terms of three-dimensional units of material. They do not consider a crater or basin as an isolated landform but as the source of a deposit. Similarly, a mare is not merely a surface with a certain color or smoothness but the upper bound of a stack of three-dimensional material layers. Each crater or mare deposit (1) is stratiform or tabular, (2) has a finite, normally varying thickness, (3) is laterally continuous over a finite area, (4) rests on other units, and (5) is bounded above either by additional units or by a free surface (fig. 2.2B; Shoemaker and Hackman, 1962; Mutch, 1970; Wilhelms, 1970b). Even topographically undistinctive terrains are composed of discrete rock units (fig. 2.3).

These geometric properties of discrete rock bodies imply that each body was formed at some instant or over some finite interval during the course of geologic time. Upon deposition, each crater and mare unit extended until it pinched out naturally or abutted against an obstacle. Interruptions of such depositional patterns as radiality of crater ejecta indicate blockage by an obstacle, superposition of a younger unit, or transection by a later structure (fig. 2.2B). Simply put, younger units overlie and thus modify older units. Age relations can be detected as far as the units extend. For the Moon, superpositional and transectional relations can generally be seen and interpreted in temporal terms much more quickly and efficiently from a photograph than on the surface. These simple observations form the basis for any understanding of lunar geologic history.

The detection and mapping of stratigraphic units and sequences do not necessarily imply that the genesis of the units is known but only that each unit or sequence was formed by a single process or related processes. A crater deposit could equally well be composed of impact ejecta or extruded volcanic material; nonstratigraphic criteria may be needed to distinguish between these origins. Matters of recognition and interpretation of units are kept separate as far as possible in photogeologic work (Mutch, 1970; Wilhelms, 1970b, 1972b). A major purpose of stratigraphic studies, however, is to help determine the origins of the units and of their constituent rocks. The debate about internal or endogenic origin versus impact or exogenic origins of lunar features is a thread running through lunar studies since their beginning.

CASE HISTORIES

Three examples illustrate the application of stratigraphic principles to problems of origin. Before direct exploration began, stratigraphic and theoretical studies had formulated pertinent questions and obtained many answers to genetic problems. Then, the major remaining problems were solved in general terms by data obtained directly from the Moon's surface by nine missions in the seven years between the flight of Apollo 11 in 1969 and the Soviet unmanned sampling mission Luna 24 in 1976 (table 1.2). The search for more specific answers, especially to the third question, the origin of terra materials, remains a field of active investigation. Later chapters of this volume further consider these "case histories" in a stratigraphic context.

Craters

The exogenic-endogenic controversy about the origin of craters probably occupied more early literature than did any other lunar topic (reviewed by Baldwin, 1949, 1963; Firsoff, 1961; Shoemaker, 1962b). The only major competitor for journal space was the Moon's surficial layer, and even that subject was commonly discussed with regard to impact-versus-volcanic arguments about crater origin (for example, Kopal and Mikhailov, 1962, p. 371–565; Salisbury and Glaser, 1964). The debate continued through the era of Ranger, early Luna, Surveyor, and Lunar Orbiter exploration (1964–68). Caldera origin was favored or entertained for several types of craters and is still favored by a few observers (Green, 1971, 1976; McCall, 1965, 1980). However, the impact origin of large fresh craters typified by Copernicus (fig. 1.6), and of most craters that share its principal features, had been settled in most minds at the beginning of the space age (Baldwin, 1965, p. 137).

Part of the solution was stratigraphic. An origin had to be consistent both with the morphology of typical crater deposits and with the spatial distribution of craters superposed on other stratigraphic units. The exterior deposits are massive, extensive, and similar in lateral morphologic gradation around craters ranging over more than five orders of magnitude in diameter. Therefore, the crater-formation process (1) released enormous energies and (2) operated similarly at all scales. These properties characterize *impacts* of cosmic projectiles, whose approach velocities range from the escape velocity of the Moon (2.4 km/s) to about 70 km/s, most typically from 16 to 20 km/s (Gault, 1974; Wetherill, 1977b). Relative to a planetary target, these velocities are *hypervelocity*, that is, greater than the speed of sound in the impacted medium (Baldwin, 1949, 1963; Shoemaker, 1962b). The hypervelocity projectiles range in size from dust particles to small asteroids. Thus, large *primary impacts* from space generate almost unlimited kinetic energies. Projectiles launched from the primary crater at lesser velocities (max 2.4 km/s) create morphologically varied *secondary craters* over great distances. The repetitive map patterns of crater deposits and the detailed morphologies of both primary and secondary craters (see chap. 3) match those of experimental impact and explosion craters much more closely than those of volcanic craters.

Stratigraphic relations, in combination with the properties expected of cosmic projectiles, also explain the spatial distribution of craters. The number of craters superposed on a given terrane is generally proportional to the age of the terrane (Gilbert, 1893; Baldwin, 1949, 1963; Öpik, 1960; Shoemaker, 1962a, b; Shoemaker and others, 1962a). For example, craters are more abundant on the older terrae than on the younger maria (figs. 1.6, 1.7). Different ages of the terranes and not different origins of the craters account for this relation. Within a given terrane, small craters are always more abundant than large craters, the sizes and frequencies are systematically related, and the most conspicuous craters are randomly scattered. This distribution is consistent with the inverse mass-frequency distribution of observable cosmic objects (Baldwin, 1949, 1963; Öpik, 1960; Shoemaker and others, 1962b; Hartmann, 1965a, b). Nonrandom distributions, which are also observed, are equally diagnostic of impact origins: Secondary craters are grouped around larger primaries. Apparently nonrandom distributions of large craters, such as the

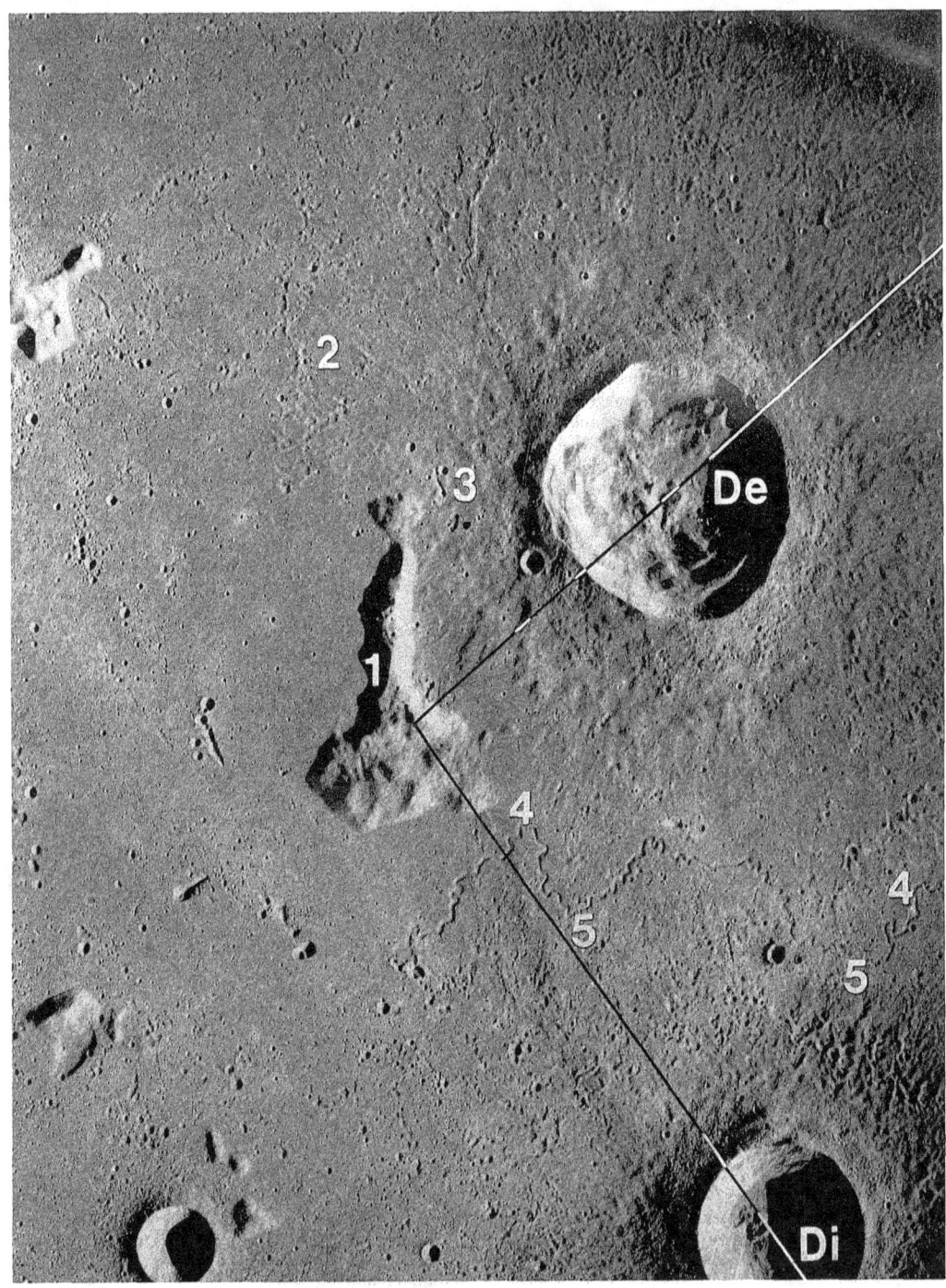

FIGURE 2.2.—Stratigraphic relations of craters Delisle (De; 25 km) and Diophantus (Di; 18 km), mare materials, and Imbrium basin. Massifs of Imbrium protrude through mare surface as islands (1). Mare unit (2) is overlain by ejecta of Delisle (3). Another mare unit containing sinuous rille (4) embays Delisle ejecta. Secondary craters of Diophantus (5) are superposed on rilled mare unit and Delisle.
A. Apollo 15 frame M–2076.

north-south-trending "chain" in the south-central nearside terra (fig. 1.8), which were ascribed to major subsurface structures in the early literature (for example, Fielder, 1965, p. 56), actually are coincidental alignments of primary craters of different ages (Baldwin, 1949, p. 158–160). When large craters of the same age are plotted together, almost all lineaments disappear (pls. 6–11). A falloff in crater density in the terra near the mare borders results from the inverse size-frequency distribution: The missing craters are mantled by ejecta of the largest members of the impact series, the ringed basins (figs. 1.6, 1.7). Thus, in any one epoch, the distributions of most large lunar craters are truly random and would require an internal generating process as random as primary impact—a prerequisite in conflict with the spatial regularities commonly cited by proponents of a volcanic or tectonic origin.

The origin of craters having apparently atypical morphologies or size-frequency distributions remained to be learned by a combination of remotely based analysis and sampling during the 1970's. Although no large craters were sampled individually, the overwhelming evidence from returned rock samples is that almost all lunar craters are of impact origin. The large degraded craters of the terrae, which apparently possess only a truncated rim flank, originally resembled fresh craters but have been eroded by impacts and deposition of later ejecta (fig. 2.3; chaps. 8, 9). Continued study has uncovered impact or modification processes that account for most of the odd landforms once ascribed to volcanism. Most elongate and irregular craters and clustered craters were formed by oblique or simultaneous primary or secondary impacts (chap. 3). Anomalous uplift of floors of impact craters in basins accounts for most remaining departures from the typical morphologies of large craters (chap. 6). Circular craters with dark halos were formed by impact excavation of dark materials from beneath lighter strata (chap. 13). A few small craters, all associated with mare or other dark deposits, may be endogenic (chap. 5). Impacts, therefore, created most lunar craters and thus are emphasized in this volume.

Mare versus basin

One of the most important products of the historical approach was the discovery that a mare and the basin that contains it are distinct features. Before the 1960's, these two features were almost universally thought to have been formed by the same process, either exogenic or endogenic; the terms "mare" and "basin" were equated. Even later, one might read that "Mare Imbrium" was created by a giant impact or that the "Imbrium basin" yielded samples of basaltic lava. There is conclusive stratigraphic evidence, however, that the mare materials are younger than the basin materials. Such craters as Archimedes that lie inside the Imbrium basin (figs. 1.6, 2.1) must be younger than that basin (Baldwin, 1949), except in the unlikely case that the basin rim grew up along internal ring fractures. The mare materials that fill Archimedes are younger still (Shoemaker and Hackman, 1962; Baldwin, 1963; Mutch, 1970; Wilhelms, 1970b). In stratigraphic terms (Shoemaker and Hackman, 1962), the sequence, from oldest to youngest, is: (1) basin material, (2) plains material of the Apennine Bench, (3) deposits of Archimedes, and (4) mare material (figs. 2.1, 2.4).

Early work also established the genetic relation of the circum-Imbrium terrane to the Imbrium basin. Gilbert (1893) and Baldwin (1949, 1978) perceived the radiality of the "Imbrium sculpture" system of grooves, and Shoemaker and Hackman (1962) added the recognition of craterlike Imbrium ejecta deposits (figs. 1.6–1.8). Mare Serenitatis fills another circular basin, which is overlain by the Imbrium sculpture or deposits (fig. 1.7). Yet Mare Serenitatis is unaffected by the Imbrium deposits, and so considerable time must have intervened after the Serenitatis basin formed and before Mare Serenitatis filled it (Baldwin, 1949, p. 210–213).

Maria and basins were, therefore, formed by different processes. Even before the Apollo missions, most workers had accepted these simple stratigraphic observations and knew that the maria are of volcanic and the basins of impact origin. Furthermore, the maria were identified as basaltic by their dark color and characteristic landforms. The basins were known to be exogenic by their similarity to craters and by the fact that only the kinetic energies of asteroidal masses impacting at cosmic velocities could supply the requisite energies of formation.

These conclusions were then confirmed by the first four Apollo landings between July 1969 and July 1971 (table 1.2). Apollo 11 returned rocks with basaltic composition and unmistakable igneous textures. Apollo 12 sampled other basalt flows half an aeon younger that could not have been generated by the same event as the Apollo 11 basalt samples.[2.1] The first nonmare mission, Apollo 14, returned entirely different rock—complex impact breccia—from deposits of the Imbrium basin (fig. 2.5A). Landing in the area covered by figure 2.1, Apollo 15 returned samples whose radiometric ages demonstrate a half-aeon age gap between the Imbrium basin and some of the mare basalt it contains (fig. 2.5B).

Terra materials

"Origin" of terra materials may mean either the chemical differentiation and igneous evolution that shortly followed the Moon's formation (see chap. 8), or the process that emplaced the visible landforms and photogeologically observed stratigraphic units. Stratigraphers, and this volume, concentrate on the timing and processes of the second, emplacement phase of the rocks' history.

Interpretations of emplacement processes and of relative ages are complementary. Recognition that the circum-Imbrium material is of impact origin enabled it to be used as a stratigraphic horizon;

[2.1] One aeon = 10^9 years, or 1 billion years in American usage. Ages in this volume have been recalculated using the radioactive-decay constants recommended in 1977 by the International Union of Geological Sciences (Steiger and Jäger, 1977). Most of them differ numerically from the ages given in the references cited in this volume.

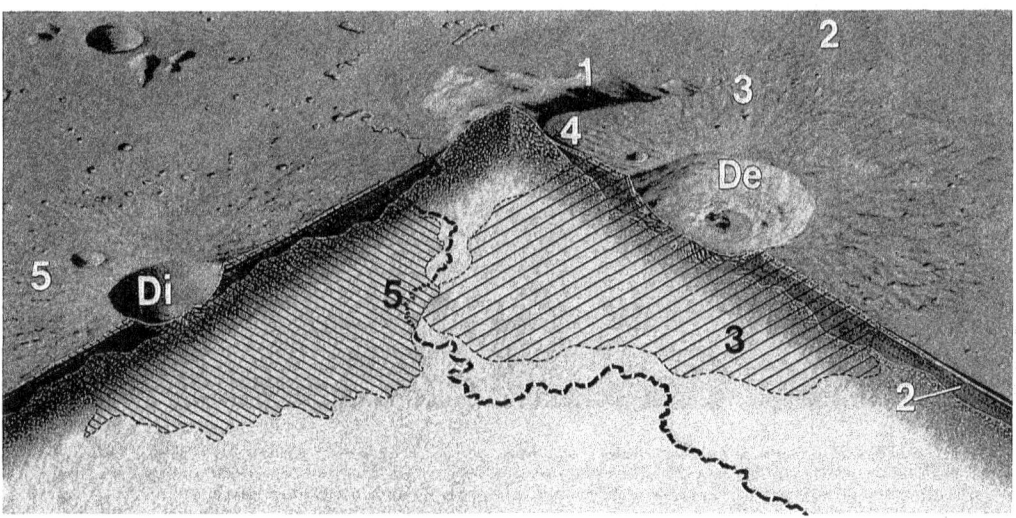

B. Cutaway view including geologic cross section drawn along bent line in A. Interpretations of age relations depend on interpretations of landforms as expressions of three-dimensional, laterally continuous units whose depositional pattern is interrupted only by blockage by older units or superposition of younger units.

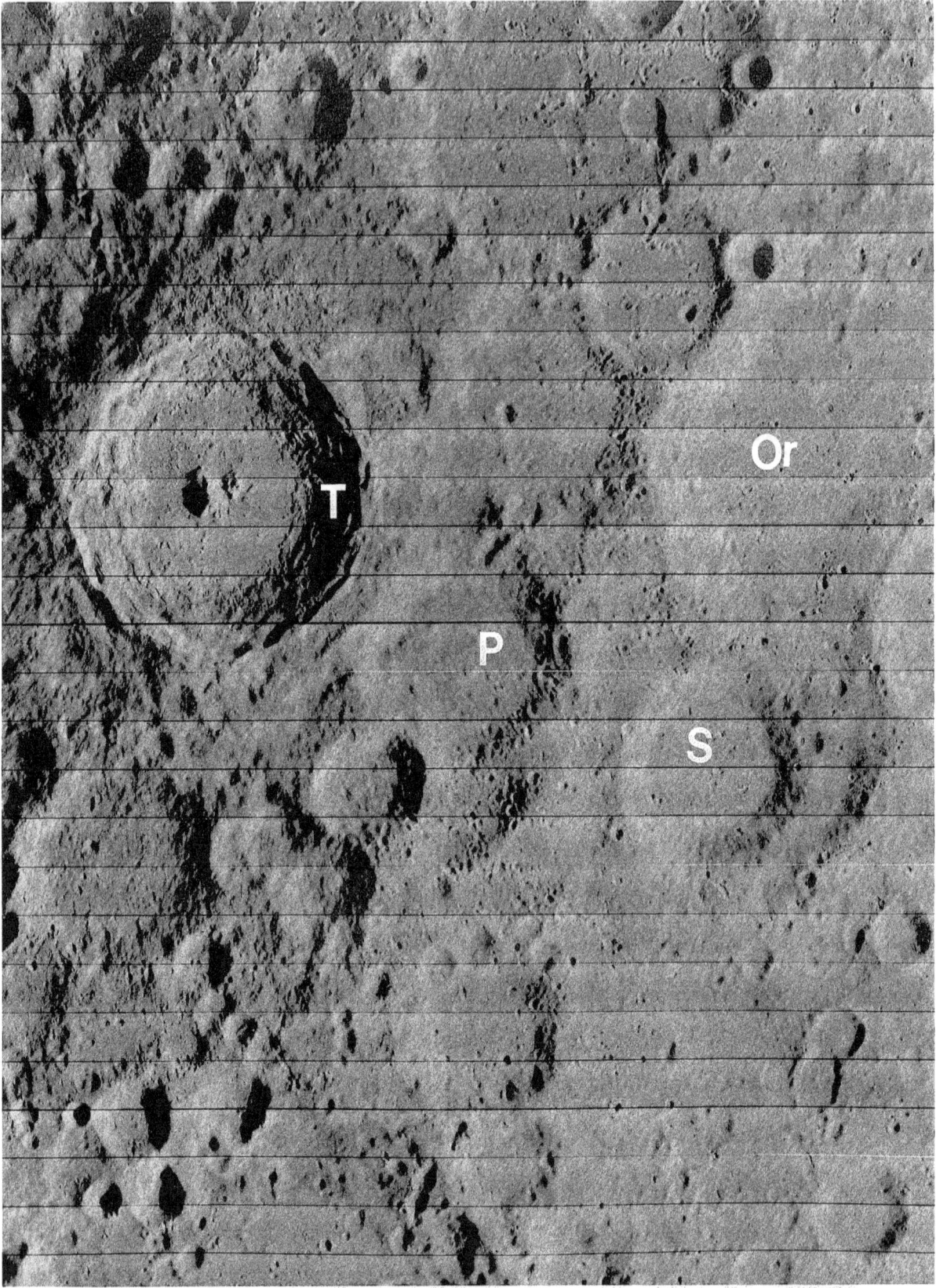

FIGURE 2.3.—Vicinity of crater Tycho (T; 85 km), showing relative degradation of lunar craters. According to the principle of uniformitarianism (Albritton, 1967; Mutch, 1970; Wilhelms and McCauley, 1971), such other craters as Orontius (Or), Pictet (P), and Saussure (S) once resembled Tycho but have lost textural details with the passage of time. Extent of Tycho deposits and secondary craters suggests that similar but now-invisible deposits surround older craters and compose intercrater terrain. Orbiter 4 frame H-119.

materials older and younger than the basin could be distinguished over much of the nearside outside the limits of the topographic basin. Moreover, the absolute ages of the spot samples collected by Apollo 14 could be extrapolated over the entire mapped extent of the unit (fig. 2.1). The samples from the Apollo 15 and 17 landing sites (figs. 2.5B,D) were also correlated with basins soon after the laboratory analyses (see chaps. 9, 10).

Some materials outside basin rims, however, were less readily interpretable. Two morphologic units sampled by Apollo 16 at the fourth major terra sampling site, the Cayley Plains and Descartes Mountains (fig. 2.5C), have proved crucial in assessing the general origin of the terrae. The history of their interpretation illustrates the interaction between photogeologic stratigraphy and its verification from actual samples of the mapped units.

The Cayley and other light-colored terra plains cover about 5 percent of the lunar terra surface and are the most distinctive terra landforms after the more craterlike ejecta of fresh basins (Wilhelms and McCauley, 1971; Howard and others, 1974). Superpositional relations and crater densities indicate that the plains-forming materials are older than the mare materials. Some of the photogeologic properties of the plains are marelike: They are smooth and level and are concentrated in depressions. In other ways, the plains-forming materials are like basin materials: They are brighter than the maria and may grade from thick in depressions to thin on adjacent more rugged terra. Accordingly, the plains-forming materials have been interpreted as both volcanic and impact deposits.

The concentration of the plains near Imbrium and their gradations with the coarse-textured basin ejecta implied lateral continuity of the two types of deposits and, thus, an impact-ejecta origin of the plains-forming material (Eggleton and Marshall, 1962; Eggleton, 1964, 1965). Similar superpositional relations of apparently isolated patches implied that all these patches belong to the same unit. The Descartes Mountains resemble coarser parts of the Imbrium ejecta blanket (Eggleton and Marshall, 1962).

Later, the same stratigraphic evidence was interpreted differently and helped support a revival of volcanic hypotheses (summarized by Wilhelms, 1970b, Wilhelms and McCauley, 1971, and Ulrich and others, 1981). The concentration of plains near basins was thought to result from marelike flooding of basin-related depressions, and the plains deposits were interpreted as younger material superposed on the blanket. Just as lateral continuity implies restricted time and mode of formation, its apparent absence may legitimately be interpreted as indicating an origin by various processes over extended times. Volcanic interpretations included (1) marelike materials brightened by longer exposure to impact cratering, or (2) materials more silicic than the basaltic maria. The plains-forming and gradational mantling materials were commonly thought to be facies of regional pyroclastic blankets, probably silicic ash-flow tuff, a highly fluid material that spreads widely and that partly conforms to the substrate (for example, Howard and Masursky, 1968; Cummings, 1972). The hummocky Descartes materials were interpreted as volcanic on the basis of their close morphologic similarities to certain terrestrial landforms (Milton, 1968a; El-Baz and Roosa, 1972; Head and Goetz, 1972; Trask and McCauley, 1972). Volcanic interpretations prevailed when Apollo landing sites were chosen (Hinners, 1972). However, a choice between mechanisms required analysis of actual samples.

One of the most significant turning points in the course of lunar geologic thinking came in April 1972, when Apollo 16 returned samples of complex terra breccia from typical patches of plains and Descartes materials (fig. 2.5C; Howard and others, 1974; Ulrich and others, 1981). As a result, volcanic interpretations were replaced by impact interpretations. Regional stratigraphic relations turned out to be better indicators of origin than did volcanic analogs. The significance of this terra sampling transcends the revised interpretations of the sampled units, because the hypothesis of impact origin could also be extended to undistinctive, previously uninterpreted terra materials peripheral to craters and basins (fig. 2.3). Like the circum-Imbrium terrane, such undistinctive terranes are coeval with the crater or basin they surround and constitute discrete stratigraphic horizons; they did not form piecemeal and do not present a hopelessly complex problem for relative dating. More and more basins have been identified since 1972, and their deposits have been increasingly recognized as similar both to those of craters and to one another (pl. 3). This series of analogous basin deposits constitutes the main stratigraphic framework of the lunar terrae.

CORRELATION OF SAMPLES AND STRATIGRAPHIC UNITS

Return of materials from the Moon has established the general origins of most lunar stratigraphic units. The maria consist of basaltic flows and pyroclastic blankets derived by internal melting (fig. 2.6). The terrae consist of complex, partly shock-melted breccia deposits that were assembled and emplaced by impacts (figs. 2.7, 2.8).

However, not all questions of origin and age have been solved by sampling. Especially serious is the fact that many returned samples have not been definitely identified with particular photogeologic units.

One problem is that all samples, except some of mare basalt from the Apollo 15 landing site (fig. 1.10B), were collected not from bedrock outcrops but from the overlying regoliths. Regoliths are composed of highly mixed materials and conceal the underlying bedrock stratigraphy and structure. Sample provenance commonly must be determined from the relative abundances of rock types at various points on the surface. This problem of correlating samples with source beds is more severe than is usually encountered on the Earth.

Second, not only the regolith but also the bedrock breccia itself is recycled from earlier deposits. A given sample thus may have acquired its chemical and textural properties during or before the impact. For example, the fact that the morphology of the photogeologically visible Cayley and Descartes units is gradational with that of the Imbrium-basin deposits does not necessarily indicate that the materials of those units are of Imbrium origin, because the morphology may have been imposed when the Imbrium impact reworked earlier materials. This situation is partly analogous to that encountered in terrestrial sedimentary conglomerate and partly unique to lunar terra breccia. The bedding and matrix fabric of a conglomerate are normally formed during sedimentation, whereas the component clasts are relicts of an earlier rock. The matrix textures of a lunar terra breccia also are generally acquired during ejection, emplacement, and subsequent cooling. The clasts, however, may have been

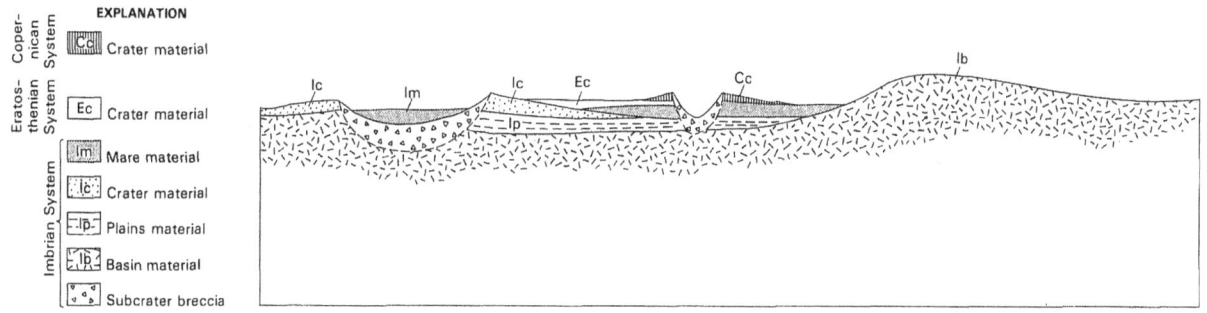

FIGURE 2.4.—Generalized geologic cross section based on area of figure 2.1, showing major lunar time-stratigraphic units.

formed either by an earlier event or in an earlier stage of the same event (see chap. 3); early-formed matrices may be broken up to become clasts in the final deposit. Intense mixing and recycling characterize the impact process.

Even a relatively uniform crystalline matrix of a terra breccia may be hard to date. It is not always clear whether impact-melted rocks (fig. 2.8) were heated sufficiently to reequilibrate the isotopes used in geochronology. Thus, the laboratory ages of the melt rocks may date an impact or an endogenic melting that preceded emplace-ment of the stratigraphic unit that contains them. Substantial petrologic, geochemical, geochronologic, and photogeologic work is required to distinguish the times of origin of the constituents of a terra breccia.

Lunar breccia poses the additional complexity that recycled pre-impact units may have been situated either in the primary target area of the new impact or outside that target area. Secondary impacts rework exterior target materials and incorporate them into new deposits, and the exterior and interior materials may have been lithologically similar before the impacts. Some investigators doubt whether the large volumes of melt rock found at two landing sites (Apollos 14, 16) are relatable to basins centered at great distances from these sites. Local origins in craters nearer the landing sites, followed by reworking during the basin impacts, are favored by many

A

C

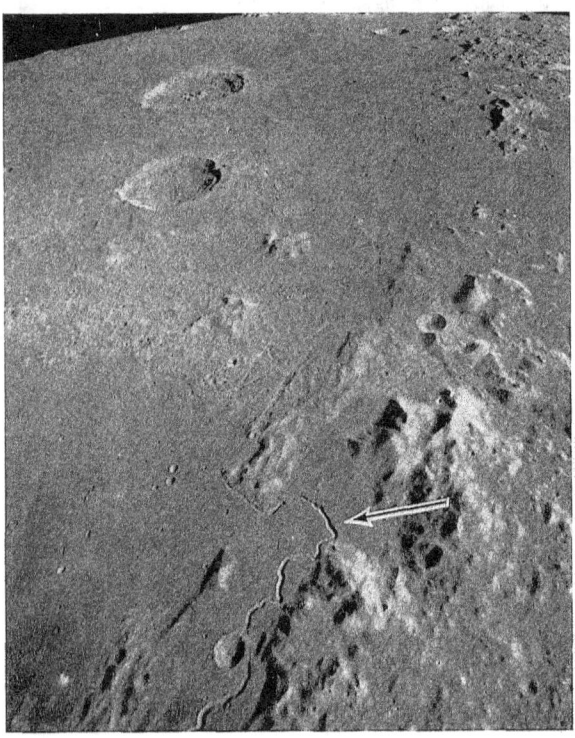

B

D

FIGURE 2.5.—Settings of four latest and most elaborate Apollo sampling missions (arrows). Each frame is an oblique view taken by an Apollo orbital mapping camera; boom of gamma-ray spectrometer protrudes into two views.

A. Region of Apollo 14 landing site on Fra Mauro peninsula. Large crater in center, to left of boom, is Fra Mauro (95 km, 6° S., 17° W.; compare fig. 1.8); Bonpland (60 km, left) and Parry (48 km, right) in foreground. View northward. Apollo 16 frame M-1419.

B. Region of Apollo 15 landing site in Palus Putredinis near Montes Apenninus. Large craters at upper left are Aristillus (55 km, 34 N., 1° E.) and Autolycus (39 km, 31° N., 1.5° E.) (compare figs. 1.6, 1.7). View northward. Apollo 15 frame M-1537.

C. Region of Apollo 16 landing site west of west rim of Nectaris basin (under boom; compare fig. 1.1). Fresh crater at bottom is Descartes A (16 km, 12° S., 15° E.), superposed on rim of crater Descartes. View westward. Apollo 16 frame M-566.

D. Region of Apollo 17 landing site in Taurus-Littrow Valley east of Mare Serenitatis. Large crater at top is Posidonius (95 km, 32° N., 30° E.; compare fig. 1.7). View northward. Apollo 17 frame M-939.

investigators, though not by me. Much of the uncertainty about the preimpact position of the constituent materials results from ignorance of the mechanics of very large impacts. Neither the size of the excavated part of basins nor the amount of melt they generate is agreed upon. These questions occupy considerable space in this volume.

In summary, the dominance of impact craters and basins on the Moon appears to be well established by analyses of the returned samples, although only a few individual units have been sampled directly. In the evolution of thought toward impact mechanisms, more and more layered rock bodies have emerged from anonymity among the Moon's seemingly chaotic features to take their place in the lunar stratigraphic column. The lithologic characterization and absolute ages of many of these units present more difficult problems.

A

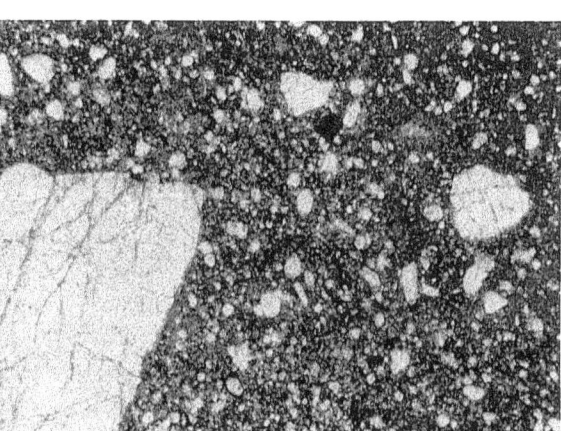

B

FIGURE 2.7.—Clast-rich "black and white" breccia (sample 15445) from station 7, Apollo 15 landing site, on flank of Montes Apenninus.
  A. "Mug shot" made when sample first arrived at LRL from the Moon.
  B. Thin section of part of sample (15445,66), showing ragged, chaotically arranged plagioclase crystals and other fragments in aphanitic matrix. Plane-polarized light; field of view, about 2 mm.

A

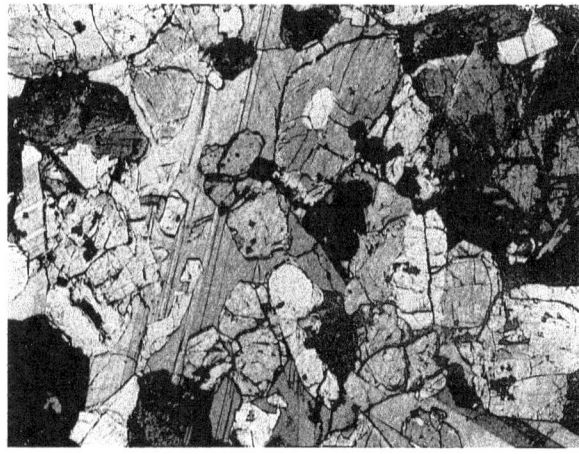

B

FIGURE 2.6.—Thin sections of typical mare basalt.
  A. Sample 12051. Laths of plagioclase (light gray), surrounded and partly enclosed by grains of pyroxene (medium gray) (subophitic texture). Plane-polarized light; field of view, 2.2 mm.
  B. Sample 15538. Laths of plagioclase (long parallel structure) partly enclose pyroxene grains poikilitically. Crossed polarizers; field of view, 2.2 mm.
  Note: The Lunar Receiving Laboratory (LRL) at the U.S. National Aeronautics and Space Administration's (NASA) Manned Spacecraft Center (MSC), Houston, Tex., assigned numbers to lunar samples upon their arrival from the Moon according to the following scheme. The numbers contain five digits, followed by a comma and additional digits if the sample has been split. The first digits represent the mission number: 10, Apollo 11; 12, Apollo 12; 14, Apollo 14; 15, Apollo 15; 6, Apollo 16; 7, Apollo 17. In some Apollo 15 and most Apollo 16 and 17 samples, the subsequent digit represents the station from which the sample was collected. For Apollo 16: 0, station 10, the region near the landed lunar module (LM) and the geophysical instruments (Apollo Lunar Surface Experiments Package [ALSEP]); 3, station 13; 7, station 11 (stations 3 and 7 had been designated in premission planning but were dropped; Muehlberger and others, 1980). Some station numbers include intrastation samples as well. The last digit of Apollo 15, 16, and 17 numbers refers to the size of the sample: 0, unsieved material; 1, 2, 3, and 4, increasingly large pieces of sieved material; 5, 6, 7, and 9, "rocks," that is, samples larger than 1 cm across. The third and fourth digits are complexly derived designations for specific samples defined in the lunar-sample catalogs prepared by the LRL; for example, the fourth digit of Apollo 16 and 17 numbers, if odd, refers to parts of large rocks and, if even, to fragments from the soil.

FIGURE 2.8.—Crystalline, igneous-appearing texture of impact-melt rock (James, 1973), sample 14310,170 from Apollo 14 landing site. Texture is partly subophitic, like that of basalt sample in figure 2.6A, and partly intersertal (fine-grained minerals in interstices of larger plagioclase crystals).

3. CRATER MATERIALS

FIGURE 3.1 (OVERLEAF).—Craters on farside area centered at 20° S., 162° E., in Keeler-Heaviside basin (massifs of ring constitute horizon). Large complex crater in center is Keeler (169 km); smaller fresh crater on east (left) wall of Keeler is Planté (38 km); degraded crater in lower left is Stratton (71 km). Small simple craters include overlapping, aligned secondary craters between Keeler and Stratton, superposed on other materials. Compare figure 1.4. Apollo 12 frame H-4961.

3. CRATER MATERIALS

CONTENTS

	Page
Introduction	27
General features	27
Typical morphology	27
Secondary-impact craters	29
Atypical craters	32
Cratering mechanics	33
Introduction	33
Shock compression	40
Cavity excavation and growth	41
Ejecta deposition	42
Secondary cratering and ground surge	42
Deformation and nonballistic ejection	43
Peak and terrace formation	43
Impact melting	44
Formation times	45
Impact breccia	45
Terrestrial analogs	45
Lunar-terra samples	45
Summary of crater-material origins	47
Introduction	47
Rim material	47
Wall material	48
Peak material	49
Floor material	50

INTRODUCTION

Stratigraphers have understandably devoted much attention to the deposits of craters, the most conspicuous lunar landforms (fig. 3.1). Much of the upper lunar crust consists of interfingering beds of crater material. Crater deposits provide more stratigraphic datum horizons for reconstructing lunar geologic history than do any other lunar materials, and large fresh craters can be relatively dated over extensive areas. Interpretations of basins and terra samples of basin deposits depend on knowledge of their smaller relatives, craters. The importance of craters and their deposits requires a detailed description here of their appearance and formative processes.

To some extent, crater materials can be mapped and relatively dated without knowledge of their origin. They obey the same basic laws of sedimentation as does terrestrial sediment, despite their radiation from randomly distributed point sources of energy (Mutch, 1970, p. 164). However, interpretation of the terrane beyond the obvious influence of a crater depends largely on knowledge of the crater's origin. On the Moon, the effects of an impact crater extend much farther than those of an endogenic crater of the same size. Craters lacking morphologies diagnostic of origin and age, therefore, have a different stratigraphic significance if they are degraded impact craters rather than separate genetic types. Thus, the question of crater origin occupies much of the next section.

Because impact is now known to be the main crater-generating process on the Moon, the rest of this chapter and this volume stresses impact craters. The section below entitled "Cratering Mechanics" discusses in detail how the distribution and interrelations of crater-material facies arose. The brecciated and melted products of impact are then described in general terms in preparation for later descriptions of the impact-generated geologic units that have been sampled. Finally, I summarize the origins of crater-material units as mapped geologically.

GENERAL FEATURES

Typical morphology

Individual primary-impact craters resemble one another more than they differ. Borrowing a term from stellar astronomy, Wilhelms and McCauley (1971) called the series of morphologically related craters the *main sequence*. The rims of these craters are nearly circular. Their floors lie below the level of the adjacent terrain. Their inner walls slope steeply, and the flank outside the raised rim crest slopes more gently. Systematic outward gradations of morphology reflect the effects of ejection and deposition of a three-dimensional, laterally continuous unit of inner-rim material and gradational outer deposits of secondary craters. Main-sequence craters are randomly scattered over a given terrain in numbers inversely proportional to crater diameter.

In general, lunar craters increase in morphologic complexity with increasing size (figs. 3.1, 3.2). The size-morphology series is not entirely gradational but undergoes a fairly abrupt discontinuity at crater diameters of about 16 to 21 km (fig. 3.3; Pike, 1974, 1980a, b, c). Smaller fresh craters have simple, smooth interior profiles, smooth and highly circular rim crests, and depth/diameter (d/D) ratios of about $\frac{1}{5}$ (table 3.1). Their floors are commonly flat or gently sloping and are evidently composed of rubbly or fine debris accumulated from the walls (fig. 3.2B). Many ejecta blankets of the younger craters of this relatively simple type display radial textures; some have subconcentric dunelike forms (fig. 3.2A).

Larger craters are more complex (figs. 3.2C–E, 3.4). In unmodified form, they have one or more of the following interior features: (1) a broad floor that is generally level but is interrupted by various hills and mounds; (2) a centrally disposed hill, peak, or peak complex; (3) single or multiple blocks or slices of material slumped from the walls; (4) continuous terraces on the wall that represent

A

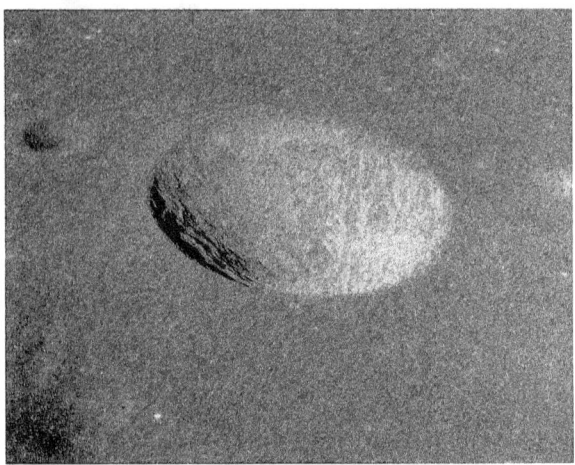

B

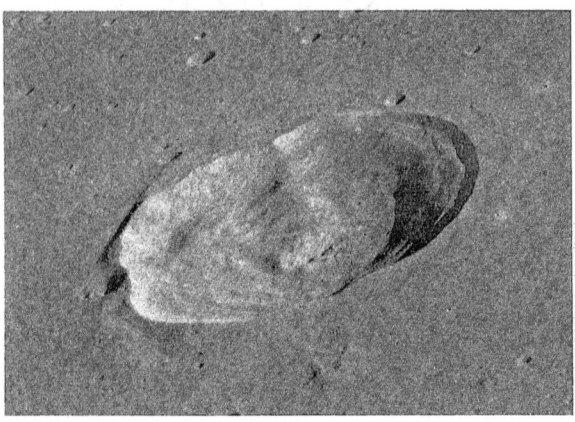

C

D

E

FIGURE 3.2.—Simple and complex lunar impact craters.
  A. Linné (2.5 km, 28° N., 12° E.). Interior profile is smooth except for minor rubble; ejecta is subconcentric and dunelike. Apollo 15 frame P-9353.
  B. Taruntius H (8.5 km, 0.5° N., 50° E.). Profile is smooth except for level floor, composed of rubble from walls; rim appears smooth, though not favorably illuminated in this photograph. Apollo 10 frame H-4253.
  C. Arago (26 km, 6° N. 21° E.). Large wall terraces, evidently formed by slumping; peak and wall merge. Apollo 10 frame H-4630.
  D. Tycho (85 km, 43° S., 11° W.). Characterized by very crisp, fresh-appearing topography much more complex than that of Arago. Floor mounds are fissured. Pools of impact melt are superposed on terraces and rim flank; radial flow texture of interior and exterior melt is also visible. Concentric inner-rim texture grades to radial outer texture. Orbiter 5 frame M-125.
  E. Hausen (167 km, 66° S., 88° W.). Floor is broader relative to diameter than in Tycho; peak is relatively smaller, though large absolutely. Possible ringlike pattern of smaller peaks is visible. Terraces and hummocky wall masses are conspicuous. Secondary craters and herringbone pattern are conspicuous north and southeast of crater. Orbiter 4 frame H-193.

TABLE 3.1.—General crater-identification criteria

[Contributions from Richard J. Pike]

Property	Primary impact	Secondary impact	Endogenic
Size	Any	Mostly <30 km in diameter; proportional to size of source	Mostly <<20 km in diameter.
Slope of size-frequency curve (cumulative)	-1.9 for postmare craters <2 km in diameter	-3.6 to -4.0	About -2 (Greeley and Gault, 1979).
Circularity[1]	Circular, <15-20 km in diameter; crenate, >20 km in diameter (0.70≤C≤0.95; median, 0.82).	Varies (0.35≤C≤0.75; median, 0.54)	Varies (0.35≤C≤0.85; median, 0.55).
Depth	Much deeper than surrounding terrain where unmodified	Shallow except far from source	Varies; floor commonly near level of surrounding terrain.
Interior profile	Simple, 16-21 km in diameter; complex, >16-21 km in diameter.	Featureless except where filled	Varies.
Rim-flank profile	Rugged near crest, then concave to 2 radii	Inconspicuous except in large craters	Varies; mostly smooth and low.
Rim-flank texture	Concentric, hummocky near crest; radial to about 1 radius.	V-shaped or linear intercrater ridges	Inconspicuous.
Ejecta distribution	Mostly symmetrical (but see fig. 3.11A)	Directed away from source	Widespread.
Mutual relations	Interference features (fig. 3.14A) or "pushthrough" (fig. 3.25) rare.	Interference features common; downrange overlap common (fig. 3.4F).	Varies.
Spatial distribution	Random on a geologic unit except for rare pairs or triplets (fig. 3.14).	Concentrated about source in clusters, chains, or loops; commonly radial to basins; on bright rays when young.	Amid dark material, on domes, or aligned on rilles.

[1]Circularity (C) is defined as the ratio of the area of the inscribed circle to that of the circumscribed circle (fitted to planimetric outline of rim crest).

wholesale circumferential failure of the rim, as opposed to the blocks or slumps or to the minor debris wasting seen in simple craters; and (5) a d/D ratio that varies with diameter, from about $\frac{1}{5}$ for small complex craters to about $\frac{1}{40}$ for the largest (fig. 3.3; Pike, 1980c). Their rims are scalloped or irregular, though still more or less radially symmetrical; some of the scallops are the source areas of the wall blocks or terraces (fig. 3.2C).

Exterior features are also better developed in large than in small craters. Rim topography adjacent to the crest and out to about half a crater radius is elevated, rugged, and commonly concentrically structured (figs. 3.2D, E). Ejecta is lower and more radially structured beyond this rugged collar. Between one and two radii from the crest, the radial pattern passes into a zone dominated by negative landforms, the secondary craters. Whereas the interior features of small and large craters differ in origin, the exterior features of all craters are similar in origin. Following Dence (1964, 1965), *simple* and *complex* are used here as technical terms for the two size-related morphologic classes.

Secondary-impact craters

Secondary-impact craters differ in most respects from their parents (table 3.1). Sizes are controlled not by the nearly unrestricted masses and kinetic energies of cosmic fragments but by the size of the primary crater and the 2.4-km/s lunar escape velocity, above which the ejected fragments would leave the Moon. Because of the lower impact velocities, rim-crest circularity is less commonly developed than in hypervelocity primaries (fig. 3.4). However, circular secondaries do form at large distances from their sources (fig. 3.5) and are difficult to distinguish from primaries if not clustered. Because the ejecta projectiles that form secondaries are larger relative to crater size than the hypervelocity cosmic projectiles, irregularities in projectile shape are more manifest in secondaries than in primaries. In addition, unbonded debris may create secondary craters (Schultz and Mendenhall, 1979). Most interior profiles of secondaries are as smooth as or smoother than those of small primaries, but shallower (Pike and Wilhelms, 1978). Ejected blocks are uncommon, and the exterior textures of individual secondaries are also smoother than those of primaries. Ejecta of grouped secondaries, however, may be texturally complex (figs. 3.4C–F, 3.6).

Spatial grouping is the main difference from primaries and is the main diagnostic characteristic of secondary-impact craters. Whereas primaries are randomly grouped, secondaries are highly concentrated. Secondaries generally occur in linear or curving chains or in patches and clusters. Only a few of the farflung projectiles may separate enough to form seemingly randomly scattered secondaries.

Secondaries may have revealed more about the cratering process than have the primary craters. Early investigators equipped with good photographs or observing the Moon visually with telescopes were impressed by the myriad small craters that are satellitic to large craters of the Copernicus type (fig. 3.4). What remains the most convincing set of arguments for the impact origin of both the satellitic and the primary craters was assembled at the beginning of the space age by Shoemaker (1962b). On the basis of an excellent telescopic photograph (fig. 3.4A), he mapped the satellitic craters of Copernicus (fig. 3.4B) and successfully accounted for their pattern by cratering and ballistic theory. He showed that the chains, loops, and clusters were probably excavated by secondary impacts of ejecta derived from certain structures in the bedrock struck by the Copernicus primary impact. His analysis was supported afterward by field study of ray loops formed from an identifiable bed in the target rock of one of the fresh Henbury meteorite craters in Australia (Milton and Michel, 1965), by laboratory experiments (Gault and others, 1968b), and by examination of high-velocity to hypervelocity missile-impact craters in natural materials (Moore, 1971, 1976).

Although primary impact of an object from space followed by secondary impact of the resulting ejecta was shown to be consistent with the lunar and terrestrial patterns of satellitic craters, other mechanisms continued to be invoked. An explosive origin by the sudden release of accumulated volcanic gases could theoretically explain the patterns except for the enormous energy required, estimated for Copernicus by Shoemaker (1962b, p. 333) at 7.5×10^{21} J. Such energies could not accumulate in a planetary crust because the weak rocks could not contain them without premature release (Taylor, 1982, p. 63). Endogenic mechanisms fail abjectly to explain the ray pattern of Tycho, part of which extends to the limbs of the Moon (fig. 3.6) and thus would require enormous internal energies or global-scale fault or fissure systems centered about a point. One advocate of such structural systems (Alter, 1963) realized that an impact must have formed Tycho but postulated that the secondaries formed endogenically along impact-opened cracks, because the rays are not exactly radial.

This offcenter relation of the rays was among the first arguments for internal origin to be finally refuted by Lunar Orbiter photography in 1967. Shoemaker (1962b, 1964; Shoemaker and Hackman, 1962, p. 290), Baldwin (1963, p. 355), the "endogenist" Firsoff (1961), and the impact-plus-endogeny proponent Alter (1963) had all noted that the nonradial rays consist of elements which are individually radial to the primary crater and which commonly originate at secondary craters (fig. 3.4A). Orbiter photographs show that the ray elements coincide with ridged ejecta of the secondary craters that was cast away from the primary crater; the secondary-crater ejecta has a herringbone or bird's-foot pattern (figs. 3.4C–F, 3.6). Septa divide many crater pairs, as they do several of the Henbury craters (Milton and Michel, 1965; Milton, 1968b). In a major advance, the ridges, herringbone pattern, septa, and even domelike features were closely

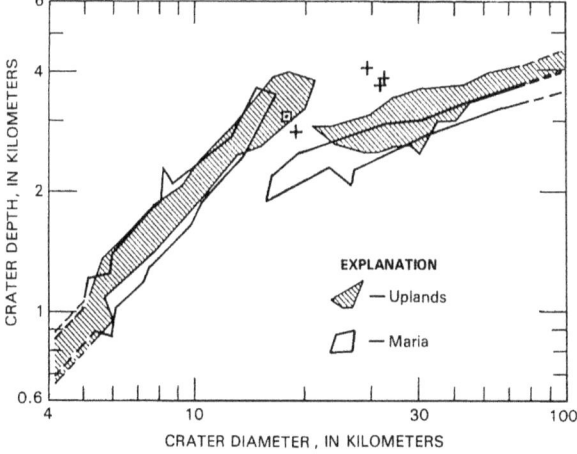

FIGURE 3.3.—Depth-diameter (d/D) ratios of simple craters (steep slopes, left) and complex craters (shallow slopes, right), including 136 craters on lunar terrae (uplands) and 203 craters on lunar maria; craters 4.2 to 95 km in diameter are plotted. Simple-to-complex transition occurs at about 21-km diameter in terrae and at 16-km diameter in maria. Complex craters show greater differences in d/D ratio in the two substrates than do simple craters. Square and crosses denote craters with transitional morphologies. From Pike (1980a, fig. 9).

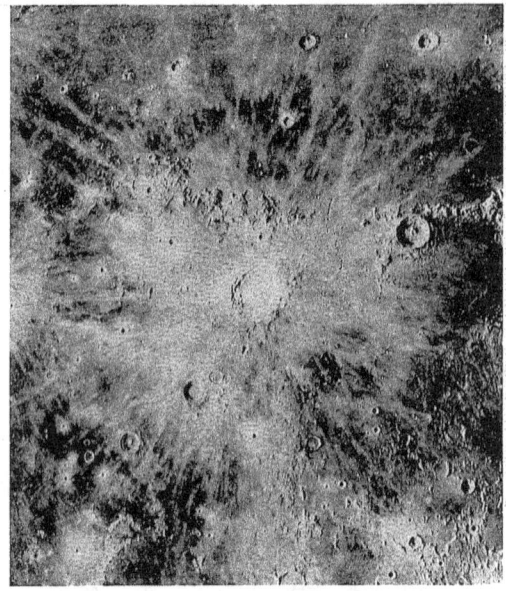

A

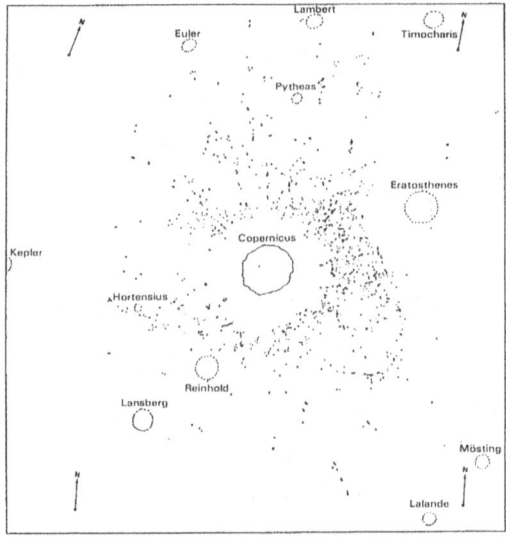

B

C

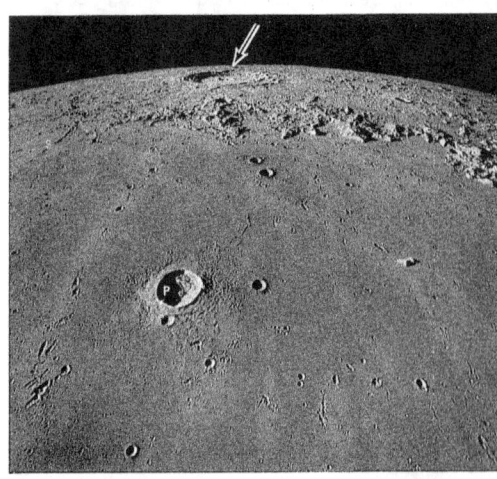

D

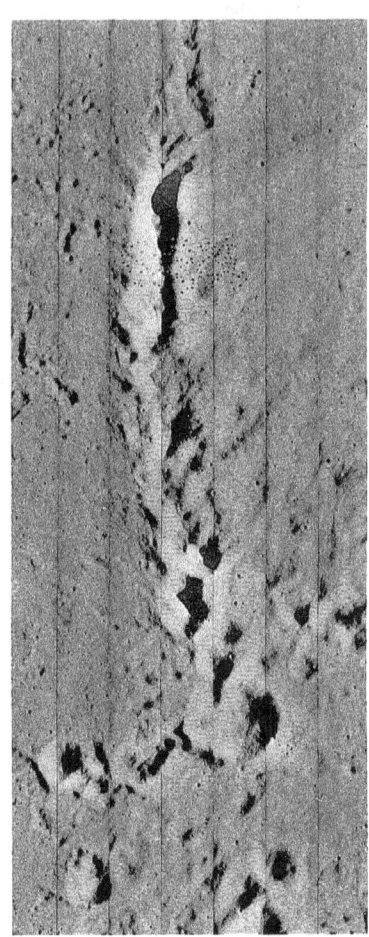

E

FIGURE 3.4.—Copernicus (93 km, 10° N., 20° W.), the classic locality for study of secondary-impact craters.
  A. Highly detailed telescopic photograph taken in 1929 by F.G. Pease with the 100-inch telescope of Mount Wilson Observatory.
  B. Map of secondary craters of Copernicus, constructed in 1960 by Shoemaker (1962b) on basis of photograph in figure 3.4A.
  C. Lunar Orbiter photograph of Copernicus and northeast exterior, showing rays and subconcentric chains of secondary craters (Rimae Stadius). Orbiter 4 frame H–121.
  D. Copernicus on horizon (arrow), Rima Stadius I (S), and other secondaries superposed on Mare Imbrium near crater Pytheas (P; 20 km). V-shaped ejecta is downrange from Copernicus at each cluster of secondaries. View southward. Apollo 17 frame M–2444.
  E. Detail of Rima Stadius I, showing V-shaped herringbone pattern. Orbiter 5 frame M–144.

imitated in laboratory experiments (fig. 3.7) by Oberbeck and Morrison (1973a, b, 1974). The intersection angles of the lunar V-shaped ridges were fully modeled by various spacings and timings of nearly simultaneous artificial impacts that caused cones of crater ejecta to interact complexly upon collision. Ironically, some of the grossly off-center satellitic chains that have the classic herringbone pattern modeled by Oberbeck and Morrison are the Stadius chains east of Copernicus (figs. 3.4C–E), which were interpreted by some of the most astute proponents of lunar impact before the Orbiter photography as volcanic (Shoemaker, 1962b, p. 302; Shoemaker and Hackman, 1962, p. 298; Baldwin, 1963, p. 378; Schmitt and others, 1967).

FIGURE 3.5.—Group of small circular craters (left center) in crater Gagarin (272 km, 20° S., 149° E.; compare fig. 1.4) are secondary to a distant crater. Largest crater superposed on Gagarin rim is Raspletin (R; 49 km), presumably a primary. Apollo 15 frame M-293.

F. Southeast sector of Copernicus. Conspicuous secondaries with herringbone pattern lie mostly beyond (to right of) one crater radius (white dots). Superposed crater in center is Copernicus H (4.3 km; see chap. 13). Orbiter 5 frame M-147.

Only the most obdurate endogenists (Green, 1971; McCall, 1980) could still believe in the volcanic origin of such craters as Copernicus and Tycho or their satellitic craters after 1967. Secondary-impact origin was also readily extrapolated to larger-scale associations, for example, the 180-km-diameter primary crater Petavius and its retinue of small craters (Hodges, 1973b). For larger craters, volcanic and tectonic interpretations continued to appear for several years. A secondary-impact origin was recognized for many, but not all, satellitic craters (max 10 km diam) of the 260-km-diameter Iridum crater (fig. 3.8; Ulrich, 1969; Scott and Eggleton, 1973). The satellitic craters of the Orientale basin (fig. 3.9) were recognized as secondary-impact craters by some geologists (Offield, 1971; Wilhelms and McCauley, 1971) but were interpreted endogenically by others well into the period of Apollo exploration (for example, Karlstrom, 1974). Internal origins were widely favored for seemingly noncompound craters larger than 5 km in diameter whose rims have such irregularities as straight segments or reentrants (fig. 3.10C; Wilhelms and McCauley, 1971).

Several systematic relations support the secondary-impact origin for all these satellitic craters, of all sizes. The size ratios and spatial relations of secondaries to primaries remain much the same around primaries ranging in diameter from 1 km (Oberbeck and others, 1974) to more than 1,000 km (ringed basins; Wilhelms, 1976; Wilhelms and others, 1978). The ratios of the largest satellitic-crater diameters to the primary-crater diameter decrease relatively little, from about 0.05 for 100-km-diameter primaries (Shoemaker, 1965, p. 121), through 0.04 for the Iridum crater (Scott and Eggleton, 1973), to about 0.02 for large basins (Wilhelms and others, 1978). Size-frequency distributions of satellitic craters also are remarkably similar over a wide size range; cumulative plots of craters satellitic to nuclear-explosion craters, large lunar primary craters (Shoemaker, 1965), and basins (Wilhelms and others, 1978) all slope between about -3.6 and -4.0. These slopes, which are much steeper than the -1.8 to -2 typical of primary craters, confirm the visual impression that secondaries in a given cluster are more nearly equal in size than are primaries in any given population. Over the entire size range, satellitic craters are concentrated at distances of one to two diameters from the primary's center (one to three radii from the rim crest; figs. 3.4, 3.8–3.10). Even the detailed map patterns of satellitic-crater fields of craters and basins are similar; chains, clusters, and loops are as characteristic of basin secondaries as they are of crater secondaries (Offield, 1971; Stuart-Alexander, 1971; Wilhelms and McCauley, 1971; Stuart-Alexander and Tabor, 1972; Scott, 1972b; Hodges, 1973a, b; Saunders and Wilhelms, 1974; Oberbeck and others, 1975; Schultz, 1976b, p. 276; Wilhelms, 1976; Ulrich and others, 1981, pl. 12). Patterns of ridges between and distal to clustered basin secondaries resemble those of crater secondaries and those formed in laboratory experiments (figs. 3.4, 3.7; Oberbeck and Morrison, 1973a, b, 1974, 1976; Oberbeck and others, 1975; Wilhelms, 1976). The only significant difference is a greater radiality of many basin-secondary groups; low impact velocities and grazing impacts create groovelike crater chains near basins (figs. 3.9A, 3.10A).

The accumulated evidence on distribution and morphology (table 3.1) leaves little doubt that secondary craters compose a large percentage of lunar craters. They probably outnumber primaries at diameters smaller than 20 km and also occur with diameters of at least 30 km (Wilhelms and others, 1978).

Atypical craters

Continued research has expanded the types of morphology ascribable to impact. Although hypervelocity impacts normally create circular craters, impacts at angles less than 10° in weak materials or about 30° for certain combinations of target material, projectile material, and velocity may generate noncircular craters (Gault and Wedekind, 1978). Elongate craters, such as Messier and Schiller (fig. 3.11), have been interpreted as volcanic or volcanotectonic. However, craters formed by artificial oblique impacts mimic their shapes (figs. 3.12, 3.13; Moore, 1976; Gault and Wedekind, 1978). Ejecta symmetry in these experiments was typically bilateral; ejecta was concentrated in lateral or downtrajectory directions, as is the Messier ejecta (fig. 3.11A). The Messier ejecta possesses such typical impact characteristics as radial ridges, secondary craters, and rays. The position of Schiller along a basin ring (fig. 3.11B) was considered as supporting the endogenic interpretation (Offield, 1971; Schultz, 1976b, p. 20). Schiller, however, consists of overlapping elliptical craters that could have been created by oblique, nearly simultaneous impact of a fragmented projectile or by a very low-angle impact of the type simulated by Gault and Wedekind (top center, fig. 3.12).

Similarities of neighboring craters have long impressed scrutinizers of the lunar surface (for example, Baldwin, 1963, p. 189; see discussion by several observers in Hess and others, 1966, p. 308–309). Examples of pairs include Sabine and Ritter, Messier and Messier A, Heis and Caroline Herschel, Helicon and Le Verrier, and Atlas and Hercules (figs. 1.6, 1.7, 3.11A, 3.14). Some of these pairs may be accidental—Atlas and Hercules probably differ in age—but too many pairs exist to be entirely coincidental. Although endogeny was commonly invoked (for example, De Hon, 1971), diagnostic features show that an impact origin is more likely. Straight septa dividing some of these pairs (fig. 3.14A) had been considered evidence for a volcanic origin until they also were found at Henbury (Milton and Michel, 1965; Milton, 1968b) and among secondary-impact craters (Oberbeck, 1971b). Many of the craters are too large to be secondaries. Pairs of large impact craters, such as East and West Clearwater, Quebec, also occur on the Earth (for example, Dence, 1964, 1965; Oberbeck, 1971b).

Two primary-impact mechanisms have been proposed to explain the lunar groupings. First, Sekiguchi (1970) showed that tidal forces may break up weak approaching bodies before they impact. Second, small bodies may orbit mutually in space. The existence of these miniature planetary systems was surmised by Baldwin (1963, p. 21, 189) and may have been substantiated astronomically (for example, Binzel and Van Flandern, 1979).

Another atypical class of craters are those called smooth-rimmed (fig. 3.15; Wilhelms and McCauley, 1971). They lack the rough rim texture of main-sequence craters and have been considered to be of nonimpact origin. Apollo astronauts called them delta-rim craters because of their equal exterior and interior slopes—another departure from the main sequence, with its shallow outer and steeper inner slopes. They were targeted for special attention during the Apollo orbital missions because they were widely hypothesized to be calderas (El-Baz and others, 1972; Evans and El-Baz, 1973). Most smooth-rimmed craters are about 20 to 40 km across and occur near

FIGURE 3.6.—Cluster of crisp-textured secondary craters of Tycho (on east rim of Ptolemaeus at 9.2° S., 1.0° E.). Fine herringbone pattern is evident above directional arrows. From Lucchitta (1977a). Apollo 16 frame P–4653.

the borders of maria, a reasonable site for volcanism. The possible significance of smooth- rimmed craters was brought home when Lunar Orbiter 4 photographed two craters inside the Orientale basin that are, therefore, of the same maximum age (fig. 3.15A; McCauley, 1968). Maunder is a typical fresh impact crater, complete with high and rough inner rim, lower and radially textured outer rim flank, deep floor, rugged central peak, wall terraces, and secondary craters. The neighboring crater Kopff is opposite in each of these properties; it has a smooth "delta" rim, elevated floor, no peak or terraces, and no obvious secondary craters.

Although the smooth-rimmed craters are atypical, they also are probably of impact origin. A multivariate analysis of 11 pairs of dimensions of terrestrial and lunar craters shows that they group with impact craters (Pike, 1980c). Kopff may have been formed by an impact in a soft substrate (Wilhelms and McCauley, 1971; Guest and Greeley, 1977, p. 115, 153), or it may be merely a premare "hybrid" crater whose rim was smoothed by volcanism and whose floor was uplifted before and after mare flooding. Other smooth rims and elevated floors are consistent with similar modifications or with a secondary-impact origin (fig. 3.15). The more typical crater Maunder is simply younger than the smooth-rimmed craters and is not affected by either volcanism or floor uplift.

In summary, we are left with impacts as the generators of most lunar craters. Chapter 5 describes the relatively uncommon endogenic craters, and chapter 6 the modification of impact craters by floor uplift. With these relatively few exceptions and a few anomalous groups or individual craters (fig. 3.16), most features of lunar craters are compatible with current impact theory.

CRATERING MECHANICS

Introduction

The basic processes that form simple impact craters are now relatively well understood after two decades of intensive research on laboratory impact craters, explosion craters, and natural terrestrial and lunar craters. This section describes an idealized sequence of events based on studies of these relatively simple craters. It adds interpretations of complex craters and suggests some of the difficulties that confront investigators trying to understand craters and ringed basins more fully. The discussion draws heavily on earlier summary works by Shoemaker (1960, 1962b, 1963), Moore and others

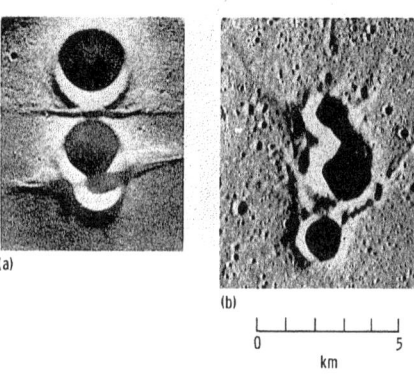

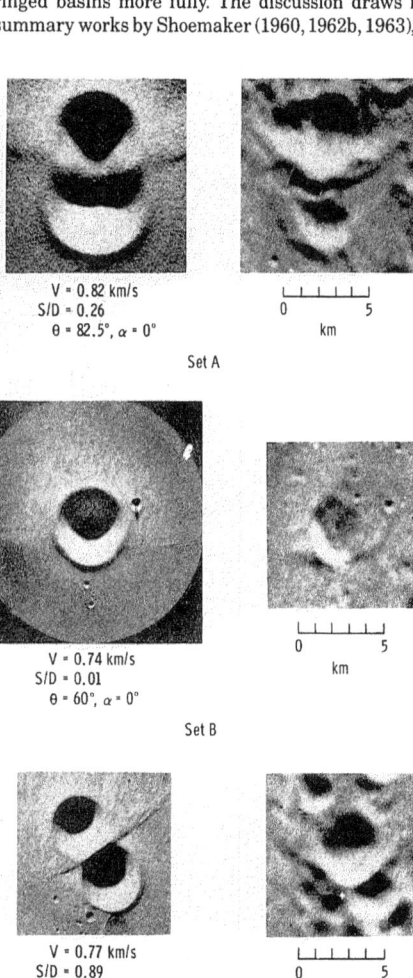

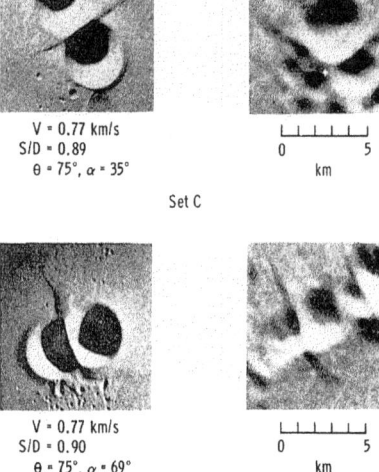

FIGURE 3.7.—Craters formed by near-simultaneous impacts in the laboratory, compared with morphologically similar secondary clusters of Copernicus (from Oberbeck and Morrison, 1973, p. 32–25). In each pair of examples, laboratory craters are on left, and lunar craters on right. V, experimental impact velocity; S/D, ratio of separation between impact points to average crater diameter; θ, impact angle of incidence measured from normal to surface; α, angle between crater axis of symmetry and flightline or radial line from Copernicus. Projectiles impacted from direction below photographs. Lunar photographs, Apollo 15 frame M-1699 (upper left) and Orbiter 4 frame H-121 (all others) (compare fig. 3.4C).

34 THE GEOLOGIC HISTORY OF THE MOON

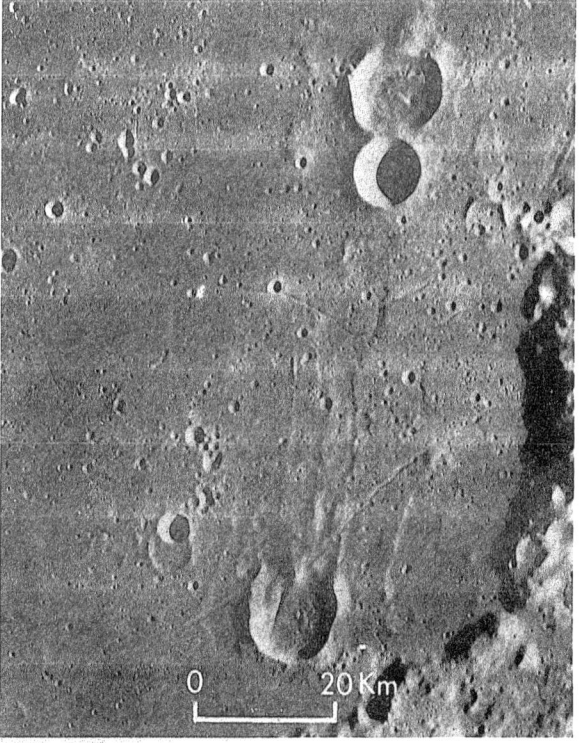

FIGURE 3.9.—Diverse Orientale-basin secondary craters north of the basin.
  A. Chain (Vallis Bohr, arrow at bottom), clustered circular craters, fissured crater floors (f), and herringbone pattern (h, near top), all formed by impact of Orientale ejecta. Head of arrow marks one basin diameter (930 km) from basin center (one radius from rim). Thick ejecta plains form scarp at p (see chap. 4). Large crater occupying most of photograph width is Einstein (170 km, 17° N., 88.5° W.). Orbiter 4 frame H–188.
  B. Domelike interference feature and radial ejecta of crater Struve L (above scale bar; 15 km, 21° N., 76° W.), 1,350 km north-northeast of Orientale center. Orbiter 4 frame H–174.

FIGURE 3.8.—Iridum crater (260 km; compare fig. 1.6) enclosing Sinus Iridum (SI). Almost entire terra in scene is blanketed by Iridum deposits or secondary craters; secondaries appear at about one radius from rim of Iridum crater. Superposed craters are Bianchini (B; 38 km), Mairan (white M; 40 km, 42° N., 43° W.), and Sharp (S; 40 km); pre-Iridum craters are La Condamine (L; 37 km, 53° N., 28° W.) and Maupertuis (black M; 46 km). Mosaic of Orbiter 4 frames H–139, H–145, and H–151 (from right to left).

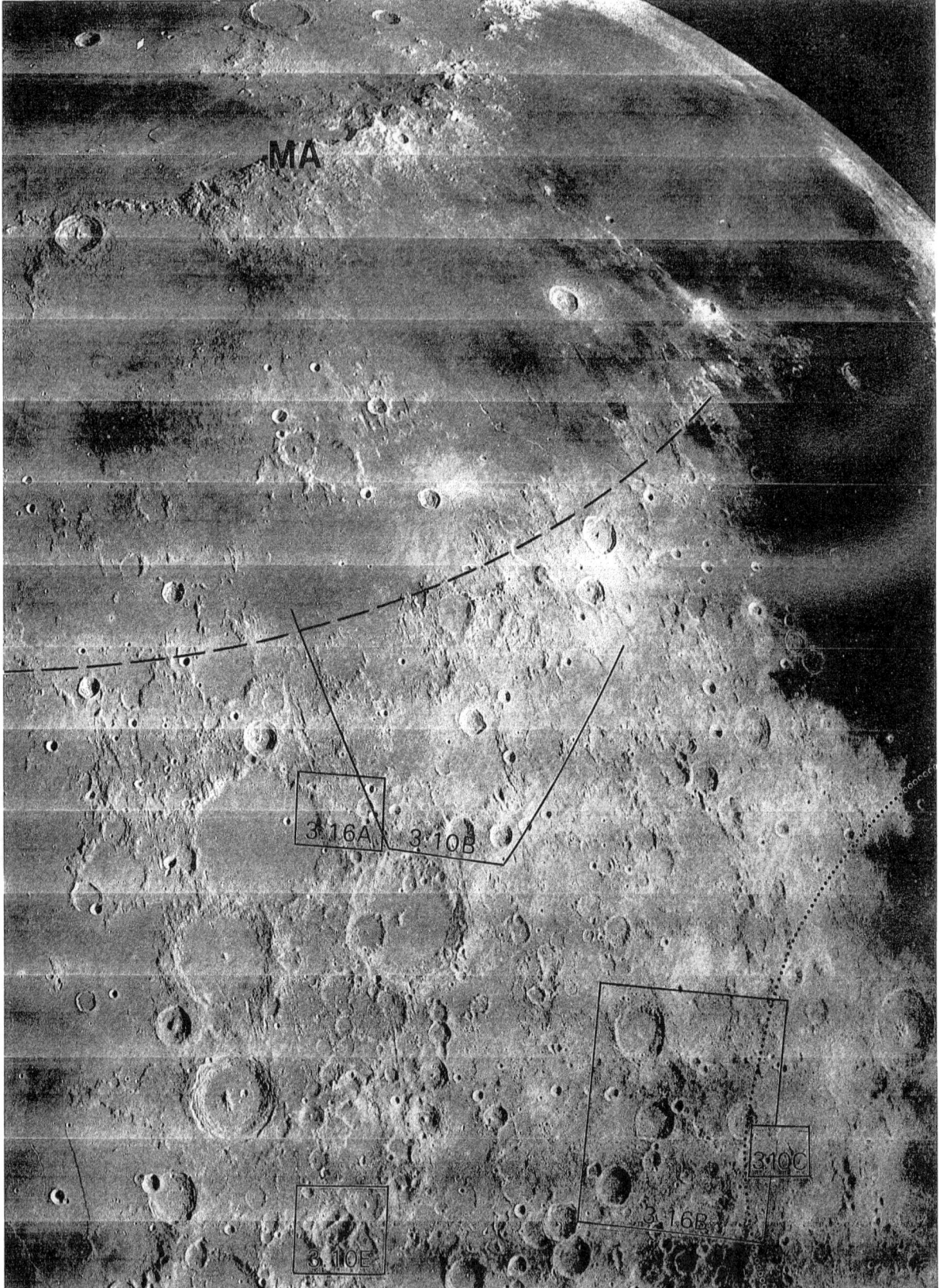

FIGURE 3.10.—Secondary craters of Imbrium and Nectaris basins.
  A. Regional view, showing distance of one basin radius (dashes, 580 km) from Imbrium rim (MA, Montes Apenninus); dotted outline, Nectaris rim. Locations of figures 3.10B, C, E and 3.16A, B are outlined; figure 3.10D lies below area of photograph. Orbiter 4 frame M–108.

B

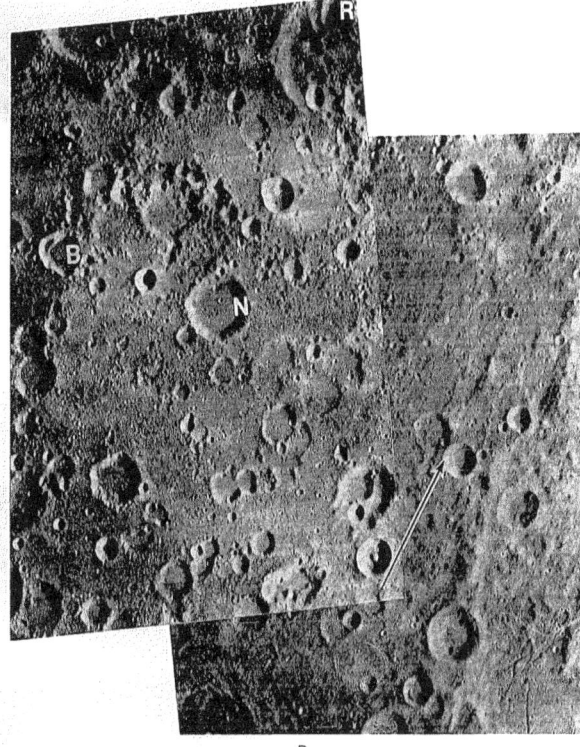

D

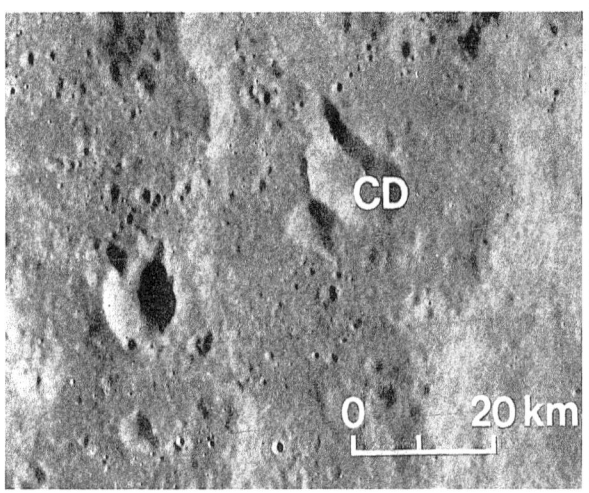

C

E

B. "Imbrium sculpture," evidently composed of coalescing elliptical craters (sc). Large crater is Hipparchus (151 km, 6° S., 5° E.). Gamma-ray spectrometer boom protrudes from right. Apollo 16 frame M–839.

C. Crater Catharina D (CD; 9 km, 19° S., 21° E.) has been considered volcanic (Wilhelms and McCauley, 1971), but is oriented and situated properly (fig. 3.10A) to be an Imbrium-basin secondary. Orbiter 4 frame H–84.

D. S-shaped group of Imbrium secondaries looping southward from rim of crater Riccius (R; 71 km, 37° S., 27° E.) to crater Nicolai (N; 42 km, 42° S., 26° E.) and back toward basin in another loop ending at crater Barocius G (B; 27 km). Arrow indicates direction to Nectaris basin; partly filled craters south of (below) Nicolai and west (left) of arrow are probably secondaries of Nectaris. Large crater Janssen (190 km; compare fig. 9.2) is partly visible in lower right. Orbiter 4 frames H–83 (right) and H–88 (left).

E. Complex morphology of Delaunay group of craters, suggesting volcanism or interference of several large basin-secondary craters (Holt, 1974). View centered at 22.5° S., 3° E. Orbiter 4 frame H–101.

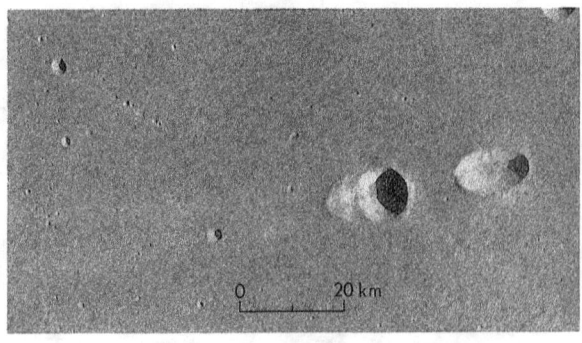

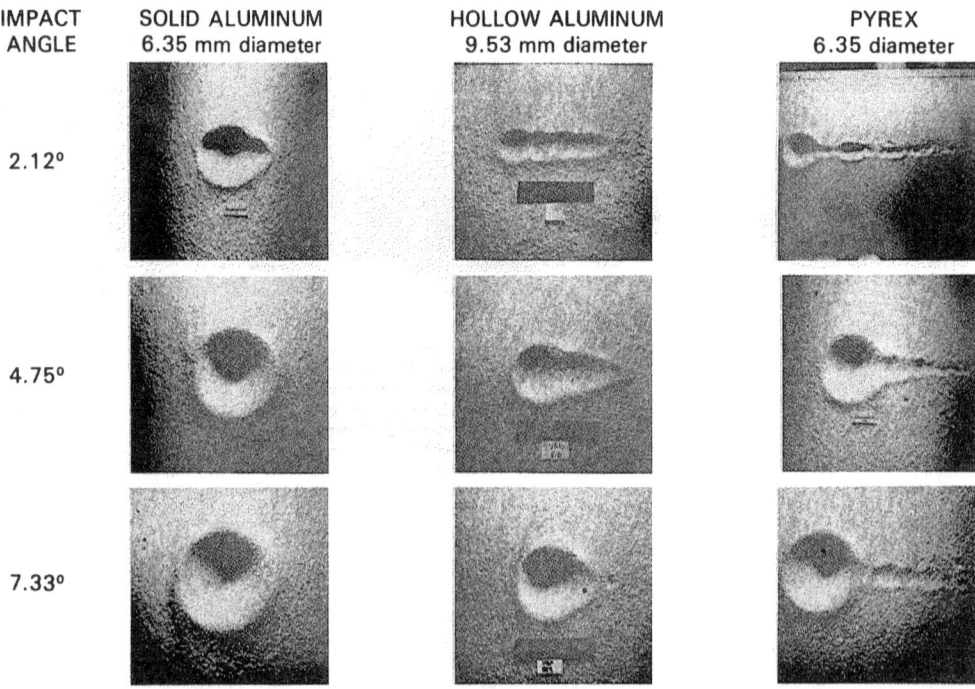

FIGURE 3.11.—Irregular primary-impact craters.
  A. Messier (right) and Messier A (2° S., 47° E.). Radial ejecta and rays are north and south of Messier, and long double rays to west (down trajectory). Apollo H-frame (number unknown).
  B. Schiller (S, footprint-shaped crater; 180-km long axis), superposed on ring of double-ringed Schiller-Zucchius basin (compare fig. 1.9); Z, crater Zucchius (64 km, 61° S., 50° W.). Orbiter 4 frame M–155.

FIGURE 3.12.—Craters formed by oblique impacts in laboratory (from Gault and Wedekind, 1978, fig. 4). Spherical projectiles are described at top; target is noncohesive quartz sand. Impact velocity, about 1.7 km/s; impact angles, above plane of target surface; trajectories, from left.

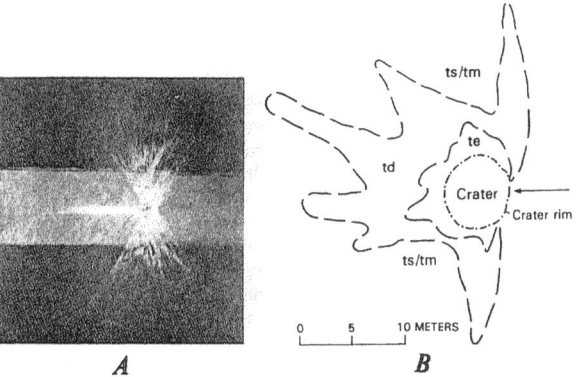

FIGURE 3.13.—Ray patterns formed by oblique impacts (from right).
   A. In laboratory (Gault and Wedekind, 1978, p. 3855). Impact angle, 5° above horizontal.
   B. By ballistic missile at White Sands Missile Range, N. Mex. (Moore, 1976, p. B8). Material units: te, thick ejecta; td, thin to discontinuous ejecta; ts/tm, scattered ejecta and target material. Arrow indicates missile trajectory; impact angle, about 45°.

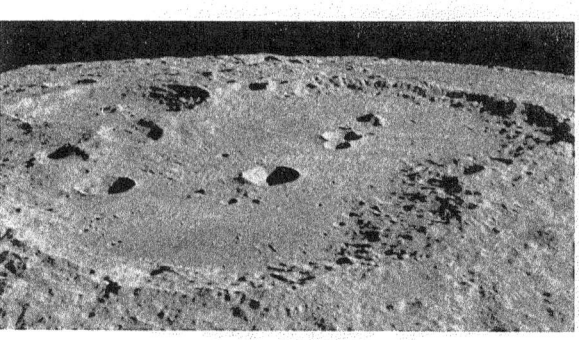

FIGURE 3.14.—Probable pairs of primary-impact craters.
   A. Bessarion B in Oceanus Procellarum (largest member of pair 12 km perpendicular to axis of pair; 17° N., 42° W.). Simultaneous impact is indicated by septum. Orbiter 4 frame H-144.
   B. Van de Graaff (230-km long axis, 27° S., 172° E.; compare fig. 1.4). Younger complex crater is superposed at upper left. View southward. Apollo 17 frame H-22959.
   C. Ritter (left, 29 km) and Sabine (right, 30 km, 1° N., 20° E.), classic lunar "calderas" at edge of Mare Tranquillitatis near parallel Rimae Hypatia. Ranger 8 frame A-34.

(1961), Baldwin (1963), Gault and others (1968b), Gault (1974), Oberbeck (1975), Roddy and others (1977), and Melosh (1980). Much of the material in the volume edited by Roddy and others (1977) was summarized by Cooper (1977).

Shock compression

An impact crater results from the meeting of an irresistible force with an immovable object (Baldwin, 1963, p. 6). A hypervelocity collision generates intense high-pressure shock waves that propagate into both the target and the projectile (Shoemaker, 1960, 1962b; Melosh, 1980). As the shock front moves downward and outward into the target, masses are set into motion with particle velocities much greater than the speed of sound in the various materials. Pressures and energies within the shock wave are commonly so great that parts of both the target and projectile are melted and vaporized around the impact zone (Gault and Heitowit, 1963). More significantly for the ultimate crater, the shock wave strongly compresses and energizes a mass of target material much greater than that of the projectile (Shoemaker, 1962b, p. 317; Gault and others, 1968b). Because peak pressures in the shock wave are orders of magnitude above the strengths of all rock materials, the materials flow hydrodynamically (Shoemaker, 1960; Gault and others, 1968b; Dence and others, 1977; Roddy, 1977). In homogeneous targets, shock waves propagate outward in approximately spherical fronts (fig. 3.17; Gault and others, 1968b). The pressures in the shock wave quickly diminish radially outward, until eventually the shock wave decays into an elastic wave (Shoemaker, 1960). The distance at which the shock wave becomes elastic is a function of original kinetic energy, projectile penetration depth, duration of contact, rock properties, and deflections due to layering and other inhomogeneities (Shoemaker, 1962b, p. 320).

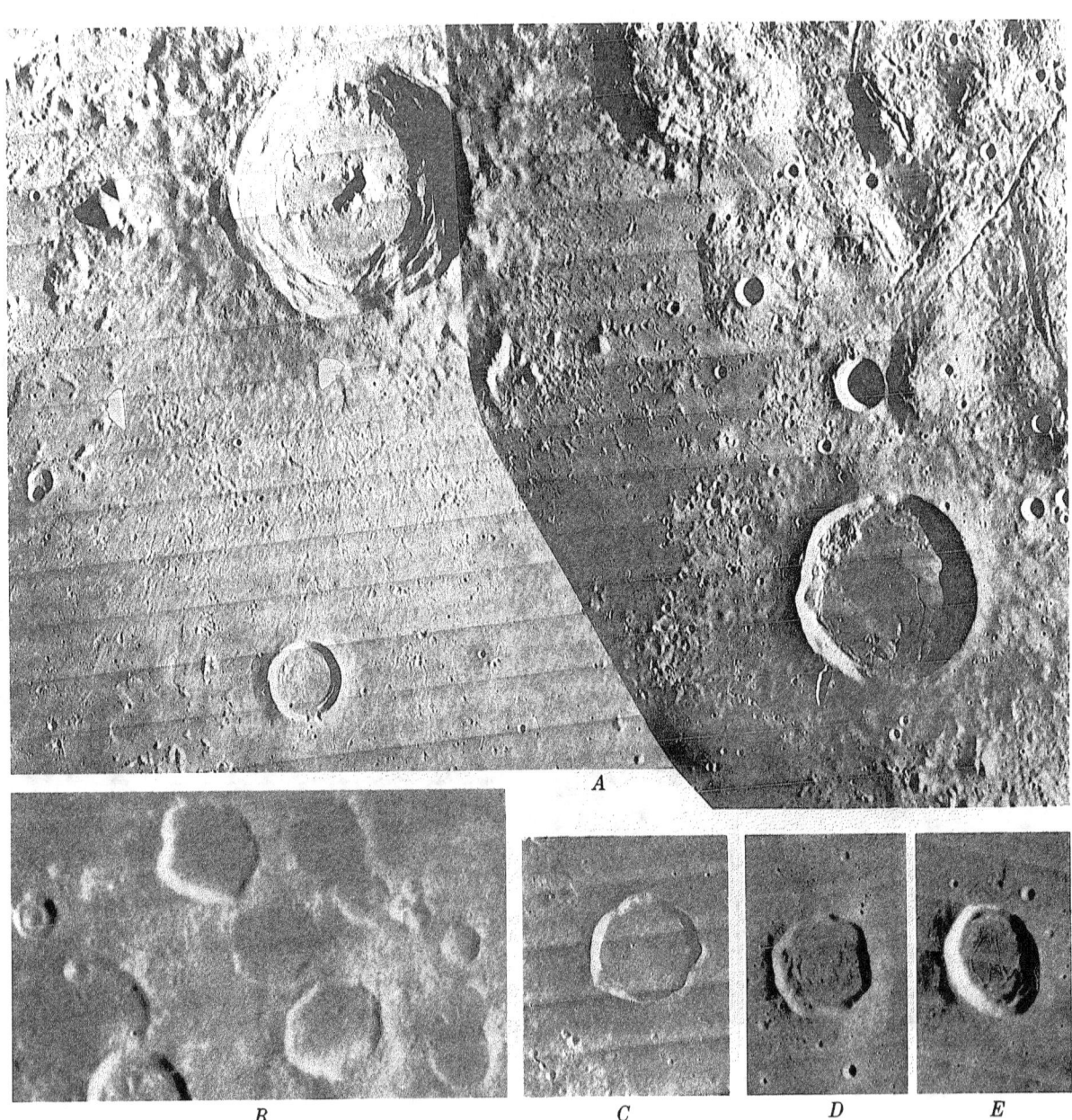

FIGURE 3.15.—Smooth-rimmed class of craters (Wilhelms and McCauley, 1971).
    A. Contrasting craters in Orientale basin: Kopff, right (42 km, 17° S., 89.5° W.), a smooth-rimmed crater; Maunder, left (55 km, 15° S., 94° W.), a typical impact crater. Orbiter 4 frames H–187 (right) and H–195 (left).
    B. Crozier-McClure group on Fecunditatis-basin rim (14° S., 51° E; each of three central craters, 21–24 km across). Clustering and smooth rims suggested caldera origin (Wilhelms and McCauley, 1971), but radial orientation and size range are also consistent with Imbrium-secondary origin. Orbiter 4 frame H–60.
    C. Gambart (25 km, 1° N., 15° W.). Dark rim material is probably volcanic, but current interpretations favor superposed pyroclastic material rather than volcanic ejecta of crater. Orbiter 4 frame H–120.
    D. Lassell (23 km, 15.5° S., 8° W.) in eastern Mare Nubium. Orbiter 4 frame H–113.
    E. Daniell (29 km, 35° N., 31° E.) in Lacus Somniorum. Volcanic origin is suggested, but not proved, by irregular rim crest, elevated floor, and mare fill. Orbiter 4 frame H–79.

Cavity excavation and growth

Very early in the sequence of events, even before the shock wave reaches the projectile's trailing edge, small amounts of both the projectile and target material may be jetted from the sides of the impact zone at velocities that may exceed the initial velocity of the projectile (fig. 3.17A; Gault and others, 1963, 1968b; Kieffer and Simonds, 1980). Most ejection, however, takes place at velocities first comparable to and then much lower than the initial projectile velocity. The agents of excavation in this main cratering-flow stage of ejection and cavity growth are rarefactions set up when the shock wave intersects the free surfaces and other discontinuities in the target and projectile. The result is a sudden decompression. Particles of material are deflected from the initially radial motions induced by the shock wave into upward and outward trajectories curving back toward the surface on the heels of the expanding shock wave (figs. 3.17B–E; Shoemaker, 1960, 1962b, p. 320; Gault and others, 1968b; Cooper, 1977, p. 37–38; Grieve and others, 1977, p. 801–804; Kreyenhagen and Schuster, 1977;

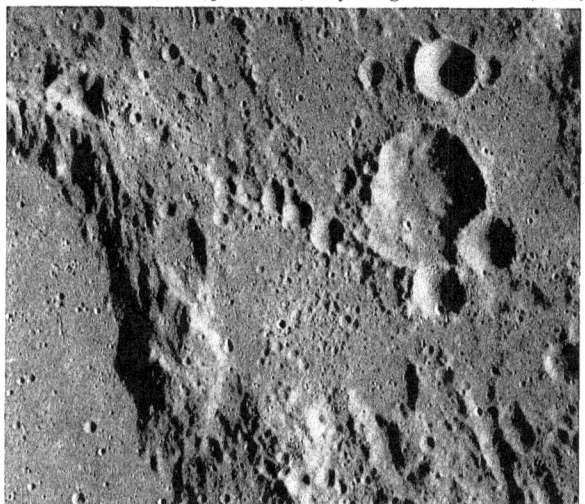

A

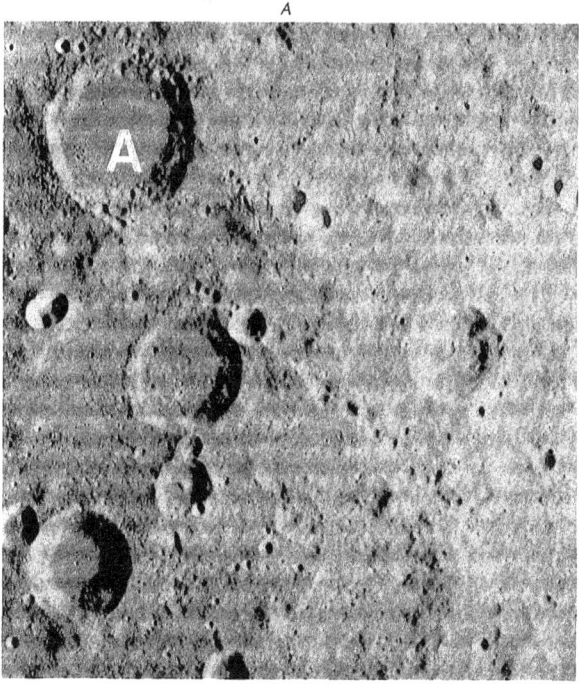

B

Roddy, 1977, p. 297; Trulio, 1977; Melosh, 1980). Thus, most impact-crater excavation is a response to the disequilibrium suddenly induced by the intensely energetic, penetrative shock wave and is not the direct result of an expansion like that induced by a true chemical explosion (Shoemaker, 1962b, p. 316). Another difference from most explosions is that the energy from a hypervelocity impact is released along the length of the path of penetration, which may be short or may be an elongate, cigar-shaped zone (Jones and Sandford, 1977, p. 1009; Gault and Wedekind, 1978). Explosions at moderately shallow depths, however, may mimic impact effects (Shoemaker, 1960, 1963; Baldwin, 1963, chap. 7; Roddy, 1968, 1977; Oberbeck, 1971a, 1977; Melosh, 1980) because their energy may be coupled into the ground similarly (Cooper and Sauer, 1977; Knowles and Brode, 1977, p. 874; Kreyenhagen and Schuster, 1977; Trulio, 1977).

Because the decompression and not the direct, intense shock compression excavates most of the cavity, ejection endures longer than the shock compression (Gault and others, 1968b). After most of the projectile material has been ejected, the cavity, lined with partly molten material, continues to grow behind the advancing shock wave by ejection of target material (figs. 3.17E, F). Though basically orderly, at least in simple craters (Shoemaker, 1960; Gault and others, 1968b), the ejection process is more subject to vagaries stemming from inhomogeneous properties and structures of the target than is the shock-compression phase. For example, inhomogeneities are the probable cause of the pattern of secondary chains and loops (Shoemaker, 1962b).

Ejection of the target material occurs approximately in the order it is enveloped by the shock wave (Stöffler and others, 1975). In general, more highly shocked materials high in the target and just outside the impact zone leave first at the highest velocities and angles (measured above the horizontal). Molten material may be shot into high ballistic trajectories. Subsequently, expanding concentric zones that include more moderately shocked materials from increasingly deep target materials are successively ejected. The ejecta forms an upward- and outward-flaring curtain of debris in the shape of an inverted lampshade (a frustum). The materials that form this curtain are sheared up along the walls of the growing cavity to the cavity lip, where they leave at angles parallel to the walls (Gault and others, 1968b; Oberbeck, 1977, p. 46; Orphal, 1977); the curtain is like an extension of the crater wall. The curtain continuously expands outward during crater growth. It apparently always remains thin, at least in small craters (fig. 3.17; Oberbeck, 1975; Oberbeck and Morrison, 1976). Typical ejection angles for the main, middle stages of experimental crater formation are 40° to 60° above horizontal but also depart from these values, depending on such properties of the target material as layering and competence (Andrews, 1977; Orphal, 1977; Wisotski, 1977).

FIGURE 3.16.—Craters of undetermined origin (see fig. 3.10A for locations).

A. Müller group on east of rim of Ptolemaeus; largest crater is Müller (22 km, 8° S., 2° E.). Large craters exhibit morphology, size, and distal overlap typical of Imbrium-secondary craters at this radial distance (800 km from Montes Apenninus); "Imbrium sculpture" above and below chain is radial to Imbrium. Small chain is also typical of secondaries in morphology and overlap but is not radial to Imbrium or any other likely source. Apollo 16 frame M–1671.

B. Abulfeda chain, parallel to Müller chain and also site of numerous craters. A, Abulfeda (65 km, 14° S., 14° E.). Orbiter 4 frame H–89.

C. Large group of craters with rectangular outlines, alternatively interpreted as originating by volcanism (Mutch and Saunders, 1972) or secondary impacts. Shape and clustering are consistent with source to lower right, but no likely source is in that direction. Group is radial to Orientale basin, but Orientale secondaries at this distance are generally sharper (compare fig. 4.6). Several other large craters are also clustered and may be basin secondaries (Schultz, 1976b, p. 276). Crater under north arrow is Asclepi (43 km, 55° S., 25° E.). Orbiter 4 frame H–100.

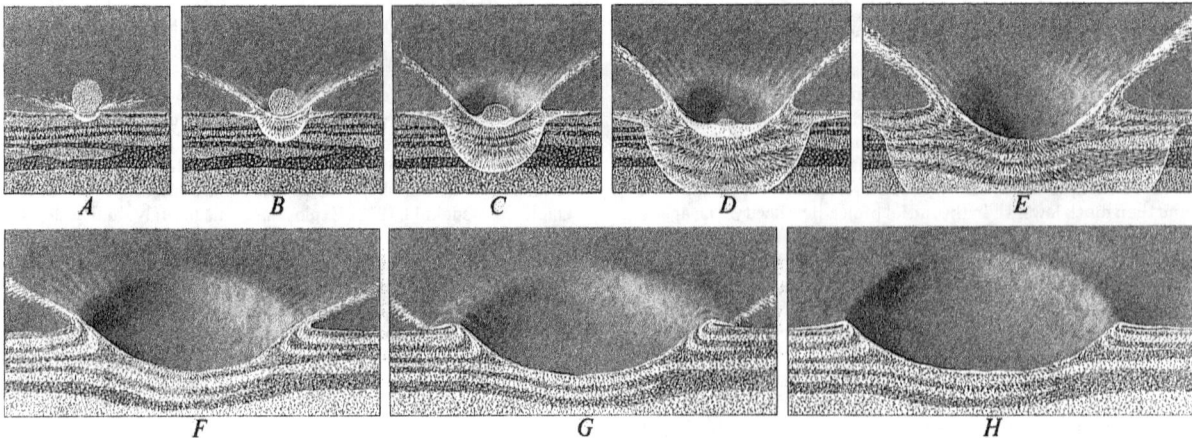

FIGURE 3.17.—Stages in formation of a simple impact crater. Drawing by Donald E. Davis, courtesy of the artist.
A. Initial contact and jetting.
B, C, D. Compressional shock wave propagates outward, and cavity grows by rarefaction behind shock wave while projectile is consumed.
E. Cavity continues growth after projectile has been consumed.
F. Maximum crater size.
G. Frustum-shaped curtain of ejected debris continues outward expansion after cavity ceases growth and overturned ejecta flap comes to rest.
H. Final crater configuration.

The cavity of a typical small simple crater ceases to expand downward when it has acquired depths of 1/5 to 1/2 of the diameter (Dence and others, 1977, p. 250–253; Knowles and Brode, 1977, p. 890–891; Orphal, 1977, p. 909). Shearing flow at the walls may then continue to broaden the crater after this maximum depth is reached (Orphal, 1977; Piekutowski, 1977; Swift, 1977). Weakly shocked and nearly undamaged ejecta derived from near the walls leaves last, during the final stages of crater excavation, under relatively low stresses and at low velocities and ejection angles (Shoemaker, 1962b, p. 335; Oberbeck, 1975; Stöffler and others, 1975; Andrews, 1977; Cooper, 1977, p. 25; Orphal, 1977). The size of ejected debris increases during cavity growth owing to decreasing shock pressures, decreasing fragmentation of the wallrock, and lower ejection velocities. Finally, ejection ceases as the tensile strength of the rock overcomes the power of the rarefaction wave to move it.

Ejecta deposition

After the cavity ceases to grow, the inverted frustum-shaped ejecta curtain continues to advance outward beyond the cavity (fig. 3.17G). The curtain continuously decreases in height while expanding in diameter, as materials at the base are deposited on the surface from the cavity rim outward. Because most ejecta is launched from a simple crater at nearly constant exit angles but at decreasing velocities, the ejecta front slopes outward at a nearly constant angle with the surface, generally 40°–50° (Oberbeck, 1975, 1977; Oberbeck and Morrison, 1976; Andrews, 1977, p. 1090–1092; Cooper, 1977, p. 38–39). As a result of this velocity distribution in the curtain, ejecta deposition occurs in approximately the reverse order of excavation. The first material to be deposited, from the base of the ejecta curtain, is the last to have been engulfed by the shock wave and to be sheared from the crater walls (figs. 3.17, 3.18). That material is lofted or barely pushed over the rim at low velocities and soon lands near the crater rim. In simple craters, the near-rim material may form a more or less coherent overturned flap in which the stratigraphic sequence of redeposited target materials is the reverse of their preimpact sequence (fig. 3.17H; Shoemaker, 1960, 1962b, 1963; Roberts, 1966; Roddy, 1968, 1976, 1977, p. 201; Stöffler and others, 1975; Moore, 1976; Oberbeck and Morrison, 1976). Some of the relief on crater rims is also due to structural upthrust and outthrust in a manner dependent on target structure and depth of energy release (Shoemaker, 1960, 1963; Roberts, 1968; Gault and others, 1968b). For a considerable time after the cavity stops growing, the curtain continues to move outward and to drop material from its lower edge into an ever-expanding but thinning ring of deposits.

Secondary cratering and ground surge

With increasing distance from the crater, the ejecta that strikes the surface forms secondary craters rather than building up a deposit. At some distance, the curtain separates into filaments of debris, whose impact creates loops and chains of secondary craters and rays. Circular secondaries are formed by the last material to impact the surface, that which was launched first at the highest velocities and in the longest, highest trajectories from sources high in the target near the impact zone.

The general picture, then, around craters as well as ringed basins is one of outward-thinning deposits of primary ejecta grading into increasingly conspicuous secondary craters and their ejecta deposits (Shoemaker and Hackman, 1962; Schmitt and others, 1967; Ulrich, 1969; Guest, 1973). The secondary impacts excavate material of the local terrain around the primary crater (Moore and others, 1974; Oberbeck and others, 1974, 1975, 1977; Morrison and Oberbeck, 1975, 1978; Oberbeck, 1975, 1977; Oberbeck and Morrison, 1976; Hörz, 1981). The amount of local material excavated increases outward. The picture is complicated around many craters and, particularly, around basins by the superposition on secondary craters of material that has flowed from points closer to the crater or basin (Shoemaker and Hackman, 1962, p. 291; Hodges, 1973b; Scott and Eggleton, 1973; Moore and others, 1974; Morrison and Oberbeck, 1975; Wilhelms, 1976, 1980). In the picture above of the regular advance of ejecta in a narrow curtain, the ejecta that forms the secondaries was still in flight when the ground-flow deposits left the crater, but the flying ejecta impacts the surface before the ground-flow deposits reach the secondary-impact zone.

The question of how the ground flow originates has been much debated because of its importance in studies of lunar-sample provenance. Chao (1974, 1977) and Chao and Minkin (1977) cited evidence

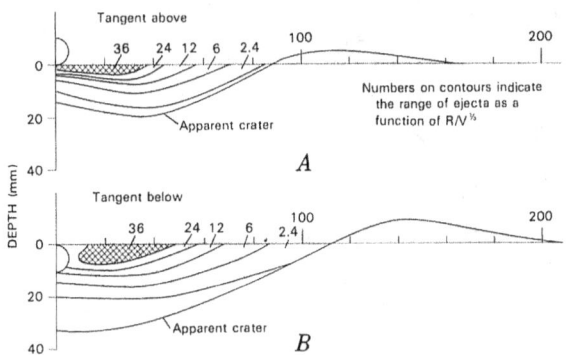

FIGURE 3.18.—Relation between sources of ejecta and range at which it is deposited (from Cooper, 1977, fig. 8). R, range; V, volume of ejecta; charge, 1.7 g of lead azide; target, dry sand. Results are similar to those of much larger experiments (Cooper, 1977, p. 25). Zones of target material nearest charge were ejected first and to greatest ranges; successively larger and more distal zones were deposited nearer rim.
A. Charge tangential above surface.
B. Charge tangential below surface, similar to energy release of impacts.

from the 25-km-diameter Ries complex crater or ringed basin, southern Germany, which suggests that the inner ejecta was pushed over the rim and moved outward along the surface by gliding and rolling. Ground flow may also have emplaced most ejecta of the 1.7-km-diameter simple crater Lonar, India (Fudali and others, 1980). In contrast, V.R. Oberbeck and his coworkers believe that all ejecta travels in ballistic trajectories and that surface flow is initiated when the ejecta impacts the surface after flight, whereupon it moves outward, together with the abundant excavated local material, under the forces of momentum and gravity. The mechanism proposed by Chao (1974, 1977) would also incorporate local material, because the surface movement occurs under confining pressure, but the proportions of local to primary material might be less than in the ballistic-impact mechanism.

To avoid prejudging the proportions of primary ejecta and local material present, use of the term "continuous deposits" is recommended over "ejecta" for the inner ground-covering material (Oberbeck and others, 1974; Oberbeck, 1975). The term "base surge" is commonly used for the outward flow (for example, Lindsay, 1976) but implies gas- or water-assisted transport, which is unlikely on the Moon. Therefore, the terms "ground surge," "debris surge," or "debris flow" are preferable (Moore and others, 1974; Morrison and Oberbeck, 1975; Oberbeck, 1975).

Deformation and nonballistic ejection

Beyond the sphere of shocked material that is launched into ballistic flight, other target material is less highly shocked but is deformed and sheared. Some of this material is pressed downward and outward along the crater floor and walls in curving paths that resemble those that precede the ballistic ejection (Dence and others, 1977; Croft, 1980). Part of this peripheral material may surge over the walls and exit the crater at very low velocities without leaving the surface.

Additional shock-damaged material is not permanently ejected. Some may be lofted above the crater and fall back inside; other material is mildly brecciated or fractured in place without significant dislocation. Both the fallback and the inplace material form a breccia lens in the bottom of the crater that grades downward and outward into a zone of fractured rock (Shoemaker, 1960, 1962b, 1963; Chao, 1977; Dence and others, 1977, p. 250–253).

The volumes of the cavity from which material was ejected and of the deformed zone may obey different scaling laws (Dence and others, 1977, p. 266–268; Croft, 1980, 1981). In general, a representative linear dimension of a crater—say, the radius or the cube root of the volume—will scale to the kinetic energy of impact (Shoemaker, 1962b; Shoemaker and others, 1962a; Baldwin, 1963; Gault and Moore, 1965; Gault and others, 1968b; Gault, 1970, 1974; Cooper, 1977, p. 16–24; Dence and others, 1977, p. 264–271; Roddy and others, 1977, p. 1133–1296; Melosh, 1980). This linear dimension is also controlled both by gravitational attraction, which inhibits lofting and flight of ejecta, and by rock strength, which influences dissipation of shock energy. Target strength becomes less important, and gravity more important, with increasing impact magnitude; gravity dominates in craters larger than about 100 m in diameter (Moore and others, 1963; Gault and Moore, 1965; Gault, 1974; Chabai, 1977; Gault and Wedekind, 1977; Gaffney, 1978). Smaller craters are created for a given impact kinetic energy when gravity dominates than when strength dominates. In large craters and basins, the excavation cavities may be "gravity craters," whereas their exterior zones may be "strength craters" (Croft, 1980, 1981).

Various terms, most of them used in more than one sense, have been applied to the excavated and nonexcavated parts of a crater. Both "transient crater" and "true crater" have been used to describe the combined zones out to the limit of the nonexcavated breccia, before modification by slumping. However, the term "transient crater" commonly denotes a stage in the cavity growth. "Apparent crater" has been used to denote the depression that is seen excluding the inplace breccia; alternatively, it has been defined as that part of the depression which lies below the precrater ground surface. Therefore, the usage of each term by a given author must be checked. In this volume, I avoid these terms because of their ambiguity and use only "excavation cavity," for the cavity from which material has been removed, either permanently or temporarily before falling back.

Peak and terrace formation

The most conspicuous indicator of internal deformation in craters is the presence of central peaks and wall terraces in complex craters (Gilbert, 1893; Baldwin, 1949, 1963; Dence, 1965, 1968; Quaide and others, 1965; Howard, 1974; Dence and others, 1977; Pike, 1980a, b, c). Diameters of complex craters are generally larger than 16 km in the maria and 21 km in the terrae (fig. 3.3; Pike, 1980a, b). Peaks increase in size and complexity in proportion to crater size up to crater diameters of about 40 or 50 km, but diminish in relative size in larger craters (figs. 3.2C–E; Murray, 1980, p. 283). Wall failure first appears as scalloping in small and rudimentary complex craters (fig. 3.2C; Head, 1976b; Settle and Head, 1979). Continuous terraces, which probably form by base failure (Grieve and others, 1977; Melosh, 1977), are observed in larger complex craters. Within the transition size range, d/D ratios also decrease markedly, and still larger craters have broad shallow floors (figs. 3.2D, E). Thus, peaks, terraces, and relatively shallow floors all seem to be somehow related. However, these features do not become evident at exactly the same diameter; for example, peaks appear before terraces (Smith and Sanchez, 1973; Head, 1976b; Cintala and others, 1977; Smith and Hartnell, 1978; Wood and Andersson, 1978; Settle and Head, 1979; Pike, 1980a, b, c).

Interpretations of the simple-to-complex transition are commonly reduced to a choice between "push" and "pull" mechanisms: Did the walls of a deep bowl-shaped cavity first collapse centripetally into terraces and push up the central peak, or did the peak originate by some type of rebound of the subcrater material that pulled the walls inward to their collapse? The material trajectories and the ultimate geometry would be similar in both cases. Some form of the "pull" model seems to be supported by appearance of peaks before terraces in the sequence of feature development. The issue largely revolves about the question of the original depth and shape of complex craters: Did they grow to d/D ratios that are *proportional* to those of simple craters and then collapse in the "modification stage" (Dence, 1968; Dence and others, 1977; Grieve and others, 1977), or did they grow *nonproportionally* and start out with shallower depths (Croft, 1978, 1980; Settle and Head, 1979; Pike, 1980b)?

A variant of the "pull" model involves a more complex excavation cavity, consisting of a deep inner part surrounded by a shallower shelf. Only the central part grows proportionally. A key feature is subhorizontal inward motion of the subcrater material during crater growth (fig. 3.19). Material beneath the central depression is severely deformed and sharply uplifted to become the central peak (Roddy, 1968, 1976, 1977, 1979; Milton and Roddy, 1972; Milton and others, 1972; Wilshire and others, 1972a; Offield and Pohn, 1979). In contrast, strata beneath the shallow floor surrounding the peaks are mildly deformed and are neither uplifted nor downdropped substantially. Support is withdrawn from the walls, which collapse most completely in large craters, where the inward motion is greatest.

Causes of the deformation and of the simple-to-complex transition remain particularly debatable (Quaide and others, 1965; Pike, 1980a, b). Some phenomenon may relate shallow depths of burst, peak formation, and shallow floors. Shallow bursts, which may form peaks in very small craters, have been documented by stratigraphic relations and the orientation of shatter cones in such terrestrial craters as Gosses Bluff, Australia (Milton and others, 1972) and Flynn Creek, Tenn. (Roddy, 1977). Low-density projectiles would release their kinetic energy near the surface because of their shallow penetration (Roddy, 1968, 1977), as would very large projectiles of any density (fig. 3.20). Large bodies penetrate to shallower depths relative to their kinetic energy than do smaller objects, and require more time to be consumed (Baldwin, 1963, p. 164–184). Thus, (1) peaks may form because of the shallow energy release; (2) more energy is directed laterally than in deeper bursts, so that craters are shallow; and (3) energy coupling is sustained over longer times, so that more of the target material is heated than in rapid energy releases (see next section below). Many properties of complex craters and ringed basins (see chap. 4) are explainable in this way.

Shallow bursts do not explain all phenomena, however, because the diameters in the simple-to-complex transition and the morphologic properties of peaks and floors differ from region to region and from planet to planet (Cintala and others, 1977; Pike, 1980a, b). Two factors are commonly proposed as modulators or even as initiators of the complex-cratering process and also of basin-ring formation: (1) differing gravitational attraction at the planets' surfaces

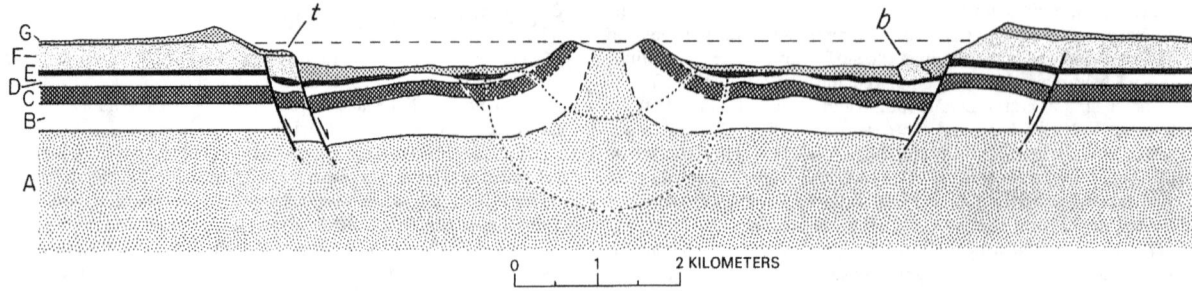

FIGURE 3.19.—Model of formation of features of complex crater, according to Pike (1980b, fig. 9). Only strata A, B, and C are exposed in central uplift, where they dip vertically or steeply. Stratum F has been stripped from crater, although it forms a slump block (b). Stratum G is composed of ejecta and intracrater breccia. Faults creating terraces (t) pinch out at shallow depths. Arcuate dotted lines indicate zones of deformation in center. Modeled after four complex craters on Earth; scale applies to average horizontal and vertical dimensions of those craters.

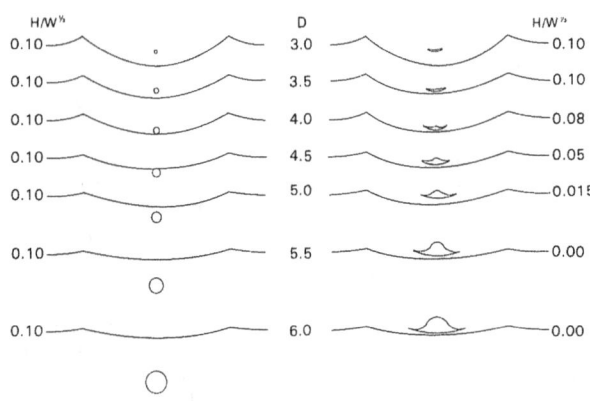

FIGURE 3.20.—Relative depth of penetration and effect on crater diameter of large and small projectiles (Baldwin, 1963, fig. 29). D, logarithm of diameter (in feet). Depth/diameter ratios are typical of lunar craters. $H/W^{1/3}$, scaled depth of burst; H, depth (in feet); W, explosive energy (in pounds of TNT equivalent). Left panel, penetration assumed beneath original ground surface to two projectile diameters for all seven sizes of craters; impact velocity is set at 10 mi/s; unlikely result of very shallow crater overlying deep burst is obtained for largest craters. Right panel, more realistic variable and shallow depths of penetration, including surface bursts for two largest sizes. Projectiles flatten during consumption by shock wave (Baldwin, 1963, p. 167).

(Hartmann, 1972a; Gault and others, 1975; Pike, 1980a, b), and (2) substrate properties, especially layering. Discontinuities between layers affect the coupling of impact energy into the target and refract and reflect the shock waves (Sabaneyev, 1962; Oberbeck and Quaide, 1967, 1968; Quaide and Oberbeck, 1968; Rinehart, 1975; Head, 1976b; Oberbeck, 1975, 1977; Piekutowski, 1977; Hodges and Wilhelms, 1978). Different stratification may explain the different simple-to-complex transition diameters between maria and terrae. Gravity or substrate properties, however, cannot be the sole factor in complex-crater formation, considering that peaks appear in all sufficiently large craters on all impacted moons and planets (Pike, 1980a, b).

In summary, shallow energy releases, strong gravitational attraction, and sharp contrasts in target stratification all seem to enhance formation of the peaks, shallow floors, and terraces characteristic of complex craters. Different combinations of these factors among planets or on a single planet, such as the Moon, yield quantitatively different results. None of these factors, however, may be the fundamental cause of complex craters. Changes in the basic physical effects of impact above some energy threshold (for example, Melosh, 1980) ultimately may be found to be more significant.

Impact melting

Estimates of the amount and stratigraphic relations of impact melt are important to the interpretations of lunar breccia presented later in this volume (chaps. 9, 10). Projectile size, impact velocity, and density of both the projectile and the target affect the partitioning of energy among mechanical excavation, deformation, and impact melting (Baldwin, 1963; Gault and Heitowit, 1963; Gault and others, 1968b, 1975; Dence, 1971; Ahrens and O'Keefe, 1972; O'Keefe and Ahrens, 1977; Kieffer and Simonds, 1980; Melosh, 1980). High-velocity impacts of small projectiles generate small amounts of very hot melt, whereas slow large impacts generate larger amounts of less thoroughly melted material (Rehfuss, 1974; O'Keefe and Ahrens, 1975, 1977, 1978; Grieve and others, 1977, p. 809; Lange and Ahrens, 1979). For the same mass and velocity—that is, the same kinetic energy—low-density projectiles generate more melt than do dense projectiles (Kieffer and Simonds, 1980). The melt traps much of the heat energy of the impact and does not contribute to the excavation process except for the melt that may have been lost by early jetting (Shoemaker, 1962b, p. 316; Gault, 1974). Therefore, in proportion to kinetic energy, large impactors melt more material and eject less fragmental material than do small impactors.

Emplacement of impact melt is one of the last processes to run its course during a cratering event but one of the first to be initiated. Melt originates in the impact zone and is pressed outward along the walls of the growing cavity (fig. 3.21). On the Moon, what appear to be flows of impact melt are superposed on crater walls and flanks, and pools of melt rest in depressions both inside and outside the rim crest (figs. 3.22, 3.23). These relations indicate that the melt is still mobile after the fragmental ejecta leaves the crater (Shoemaker and others, 1968; Guest, 1973; Moore and others, 1974; Howard, 1975; Howard and Wilshire, 1975; Schultz, 1976b, p. 228–237; Hawke and Head, 1977a). Apparently the melt sloshes over the rim along with some of the last fragmental ejecta, and locally flows back downward into the cavity after terracing ceases (Howard, 1975; Howard and Wilshire, 1975; Grieve and others, 1977; Phinney and Simonds, 1977).

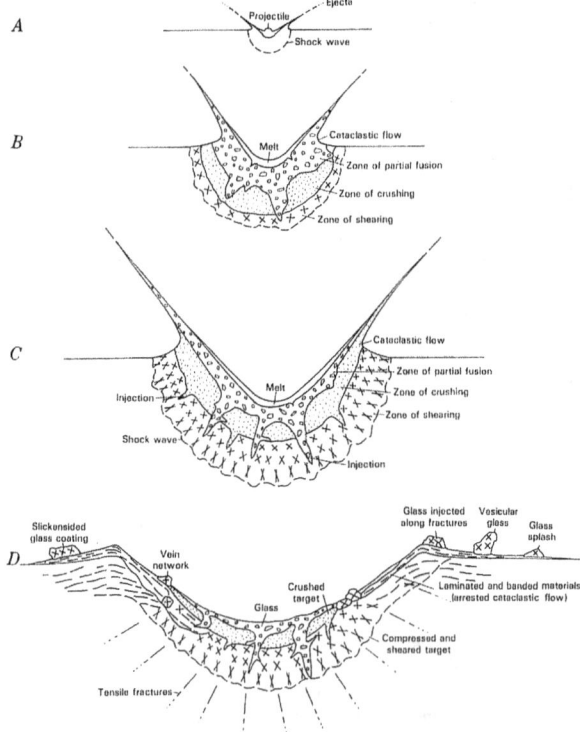

FIGURE 3.21.—Formation and mixing of shock-grade zones in a simple crater, showing four stages (Wilshire and Moore, 1974, fig. 11).

Formation times

Impact craters form in less than a minute (Melosh, 1980, p. 80). Schmidt (1981) suggested that formational time is proportional to the sixth power of the excavated volume.

In some terrestrial craters, the melt contains breccia from the peak (Simonds and others, 1976b). This observation confirms that peaks form early in the cratering sequence, as suggested above, because impact melt solidifies very quickly. In large craters on the Canadian shield, the time between impact and melt solidification was about 100 s (Onorato and others, 1976, 1978). Final cooling below temperatures able to metamorphose the melt and clasts required less than 2,000 years (Onorato and others, 1978).

IMPACT BRECCIA

Terrestrial analogs

Studies of impact-generated terrestrial materials, which began during the 1960's in conscious preparation for the Apollo landings (Chao, 1967; Engelhardt, 1967, 1971; French and Short, 1968; French, 1977, p. 155), have provided essential clues for interpretations of the returned samples. Progressive mineralogic and textural changes in the target rock have been correlated with intensity of the outward-decaying shock wave (French and Short, 1968; Dence and others, 1977; Robertson and Grieve, 1977).

Among the first impact-generated materials to attract attention were *tektites*, small, glassy, aerodynamically shaped objects found in strewn fields at several terrestrial localities. They have been thought to be lunar in origin (O'Keefe and Cameron, 1962; O'Keefe, 1963) but are much closer in composition to the target materials of terrestrial craters (King, 1976, chap. 2; Glass, 1982, chap. 6). The tektites of each strewn field probably originated during early jetting of very hot material from a terrestrial impact (fig. 3.21A; Kieffer and Simonds, 1980), were carried to various altitudes in and above the atmosphere by a fireball, and were dispersed by gases in the fireball and by stratospheric winds (Jones and Sandford, 1977).

Lunar studies have depended heavily on analogies to the Ries crater in Germany and the numerous craters or ringed basins on the Canadian shield. The diameters of the most conspicuous rings of the three largest Canadian craters, Manicouagan and West and East Clearwater, are 65, 32, and 23 km, respectively; however, the exterior ejecta has been eroded away and the excavation diameters are uncertain. The Canadian craters have furnished much data about shock metamorphism and impact melting (Dence, 1968, 1971; Grieve and others, 1974, 1977; Grieve, 1975, 1978, 1980; Floran and Dence, 1976; Simonds and others, 1976a, b, 1978a, b; Dence and others, 1977; Phinney and Simonds, 1977; Phinney and others, 1977, 1978; Robertson and Grieve, 1977; Floran and others, 1978; Grieve and Floran, 1978). The larger craters possess massive interior sheets of impact-melt rock, about 100 m thick, which texturally resembles igneous rock. The 25-km-diameter Ries crater has furnished fewer impact-melt rocks (as parts of the deposit called suevite) but abundant unshocked fragmental ejecta (mostly in the Bunte Breccia) (Engelhardt, 1967; Dennis, 1971; Chao, 1974, 1977; Gall and others, 1975; Chao and Minkin, 1977; Pohl and others, 1977; Hörz and Banholzer, 1980). Representative types of terrestrial impact materials from these and other craters have been illustrated and compared with lunar impact materials by Stöffler and others (1979).

Lunar-terra samples

Except for tektites, the whole range of shock grades is much better represented on the lunar terrae than on the eroded Earth. Terra-breccia deposits display complex shock effects, ranging from cataclasis of monolithologic rocks, through solid-state metamorphism, to intricate assemblages of cogenetic and foreign clasts within clastic and impact-melted matrices (see fig. 2.7, chaps. 8-10, and summaries by James, 1977; Phinney and others, 1977; Stöffler and others, 1979, 1980; Taylor, 1982, p. 187–201). Some individual specimens consist entirely of breccia or entirely of melt rock (fig. 2.8), and have been so designated in the literature. However, these rock types must have been parts of deposits containing both melted and unmelted materials, and so the term "breccia" is used here as a general term for lunar impact materials.[3.1]

[3.1] Only the breccia that forms deposits of the terra bedrock is discussed here. At the time of the first two Apollo landings, the only known breccia was that created in regoliths, because the terrae had not yet been visited. The regolith breccia (also called soil breccia, microbreccia, glassy breccia, vitric-matrix breccia, and numerous other names) constituted an obviously different type of material from the igneous basaltic lavas found at those sites. When the much coarser, more massive, more complex, and more extensive bedrock breccia became known, the nomenclature did not fully adapt to the differences. Some petrologic literature does not clearly distinguish between regolith breccia, formed by innumerable small impacts near the lunar surface, and bedrock breccia, formed by fewer and larger impacts (also called "megaregolith"). In this volume, "breccia" denotes bedrock breccia unless otherwise indicated, and "megaregolith" is not used because it obscures the fact that the bedrock breccia forms discrete strata.

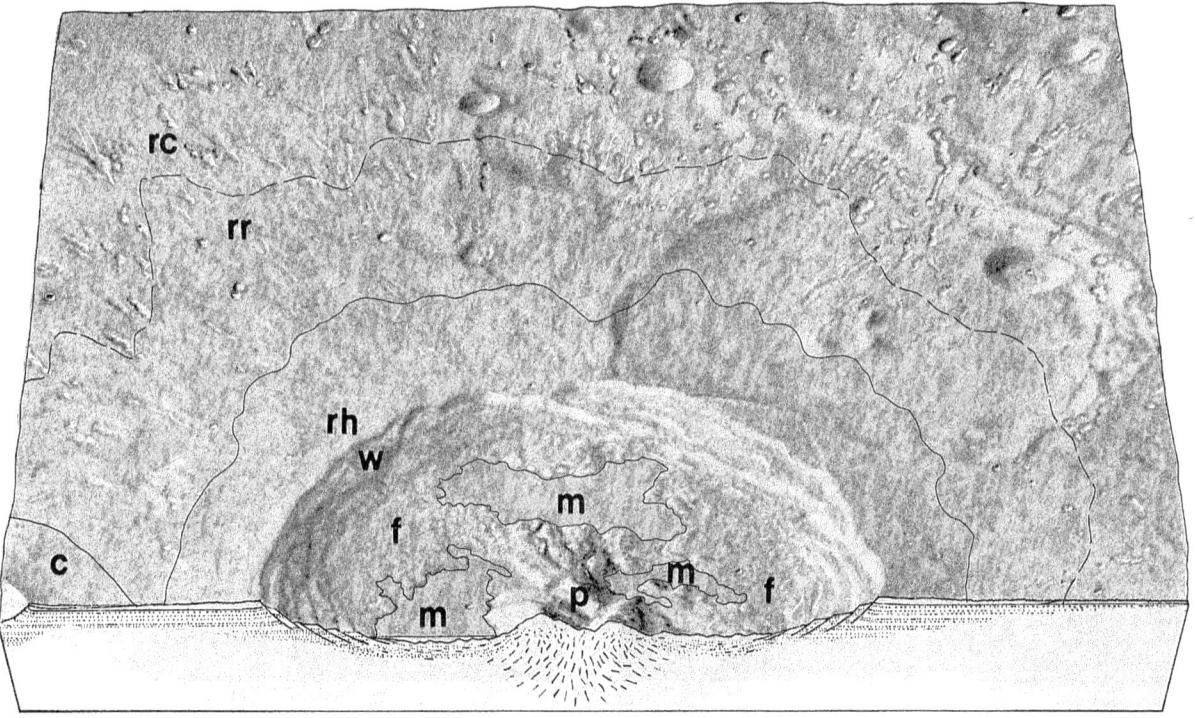

FIGURE 3.22.—Material subunits of a typical large crater, modeled after craters Theophilus and Mädler. Rim materials are three-dimensional deposits, from rim crest outward: rh, hummocky; rr, radial; rc, cratered (secondary craters). Structural units, not sharply demarcated in subsurface, are floor material (f), peak material (p), and wall material (w). Planar impact-melt rock (m) overlies other units. Material of simple crater (c) overlies unit rr. Drawing by Donald E. Davis, courtesy of the artist.

Some features of breccia deposits can be related to the outward decrease in shock intensity during the growth of a crater (fig. 3.21), whereas others illustrate the textural and compositional complexities that impact cratering induces.

Impact melts that have glassy or crystalline textures most like those of igneous basalt presumably were formed near the impact zone (for example, *ophitic* texture, in which pyroxene encloses less abundant plagioclase laths, or *subophitic* texture, in which pyroxene partly encloses a subequal volume of plagioclase laths; compare figs. 2.6, 2.8). *Poikilitic* texture in which numerous small, more nearly equant plagioclase grains and other debris are enclosed by large pyroxene crystals is considered to be a sign that many minute clasts were incorporated in the impact melt (Nabelek and others, 1978). Melt rocks with these various textures were recovered from most sampling sites, most abundantly from the Apollo 17 terra stations (see chap. 9). Melt-poor friable fragmental breccia, likely formed in the outer parts of excavation cavities, was returned from all the terra landing sites, most abundantly by Apollos 14 and 16 (see chaps. 9, 10). Some material derived from the outer parts of the excavation cavities and ejected late in the cratering sequence consists of coherent blocks (fig. 3.21D).

This zoning is muddied by the turbulent processes that characterize energetic impacts. Melted material of projectile and target that is not jetted is pressed into the growing cavity (fig. 3.21B). As the crater grows, this material, formed near the energy-release zone, is injected into the less highly shocked but fragmented rock in outer zones (figs. 3.21B–D; Dence, 1971; Wilshire and Moore, 1974; Simonds and others, 1976a, b, 1978b; Chao and Minkin, 1977; Grieve and others, 1977; James, 1977; Grieve and Floran, 1978; Stöffler and others, 1979). Thus, melt rock appears as matrices and dikes among less highly shocked fragments (fig. 3.21D). Unmelted but crushed clastic debris may be similarly injected. These processes are illustrated by *dimict breccia*, consisting of fragmental breccia and melt rock showing mutual intrusion and inclusion relations (James, 1977, p. 647; 1981; Stöffler and others, 1981).

Such mingling of shock grades can be quite thorough: Rock and mineral fragments, shocked to varying degrees, may be engulfed by impact melt (fig. 2.7). Moreover, various breccia and melt types are interbedded in the crater interiors (Stöffler and others, 1979). This bedding varies in sequence and lithology for craters of different sizes and different target materials (Stöffler and others, 1979; Stöffler, 1981), and commonly juxtaposes materials that have experienced a wide range of shock pressures. At any stage in the cratering process, the material being ejected may consist of a mixture of complete melt, fragment-laden melt, and unmelted fragments.

Shearing of material from the crater walls and transporting it outside the crater further mixes the ejecta and streaks it into bands or pods. Glass-coated striated fragments evincing shearing during ejection or transport are known (Wilshire and Moore, 1974). Some samples even appear to be accretionary bombs (Stoeser and others, 1974; Wood, 1975d; Spudis and Ryder, 1981). Mixing with the substrate material increases outward from the crater because of surface transport and increasing secondary-impact velocity of the ejecta. Lunar breccia containing clasts of melt formed in the same impact that emplaced the breccia has been referred to as "suevite"; clasts of impact-melted rock in fragmental deposits analogous to the Ries Bunte Breccia predate the deposit (Stöffler and others, 1979, 1980). However, distinguishing the two types of deposit is difficult on the Moon, where the source crater of a breccia is commonly unknown.

After the deposits come to rest, the intimate mixture of hot or superheated melt and warm or cold fragments of varying compositions in all conceivable combinations of grain size and abundance leads to further modifications of the deposit (Simonds, 1975; Simonds and others, 1976a, b). Melt and hot clasts are differentially quenched during the rapid initial cooling of the deposit. Sharp local differences in crystallinity, interface texture, and remaining content of fragments may arise. The second, much slower stage of cooling further metamorphoses the deposit and underlying materials. The terrestrial metamorphic terms "granoblastic" or "granulitic" (anhedral grains of subequal size) are applied to many lunar textures believed to have arisen from thermal metamorphism (Warner and others, 1977; Taylor, 1982, p. 198–200). Thus, the complexities arising in a single impact may be so great that they mimic the relations arising from multiple events (Simonds, 1975).

The complexities of lunar impact deposits at the scale of the outcrop or the hand specimen are not surprising in view of the complexity of each individual impact and the fact that each volume of material may be similarly reworked by many subsequent impacts. Consequently, any attempt to categorize lunar impact materials is bound to fall short of reality. The most useful in a series of attempts

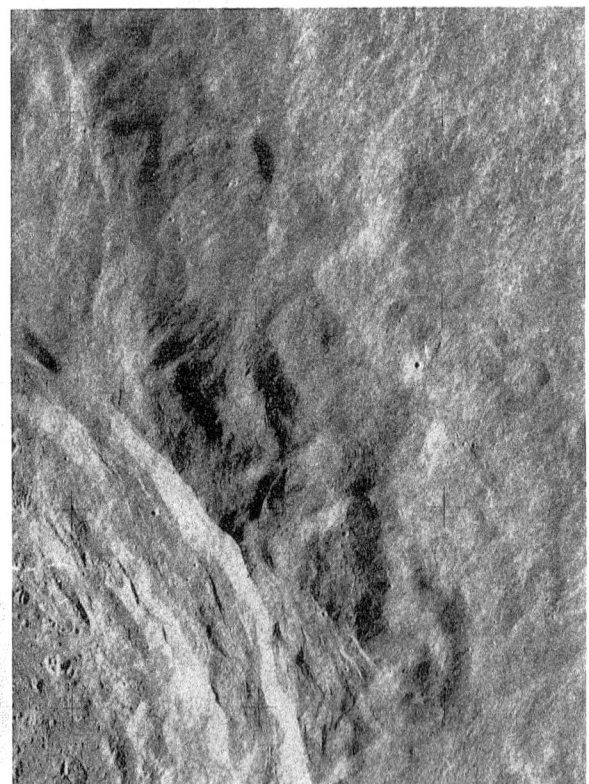

FIGURE 3.23.—Very fresh crater King on farside (77 km, 5° N., 121° E.).
A. Entire crater. "Lobster claw" central peak connects with wall and is overlain by lower wall material (enlarged in fig. 3.32). Pool of impact melt outside northwest rim is enlarged in figure 3.36. Apollo 16 frame H–19580.

B. Northeast rim, showing concentric textures of rim material downdropped along with terraces inside rim and locally lying outside rim (lower part of photograph). Gradation from concentric to radial pattern of outer rim material indicates increasing radial-flow component of outer material. Apollo 16 frame M–1579.

was probably that by Stöffler and others (1980), who compiled a long list of earlier names that should greatly aid the reader of the technical literature. Stöffler and others (1980) used textures and structures, which are signs of the process that formed the rocks, as their main basis for classification. Essentially, this classification describes matrices as either fragmental, crystallized from a melt, still glassy or partially devitrified, or metamorphosed. Clast content can be described according to need. From the geologic point of view, this classification improves upon those based on chemical composition. Many purposes are served by a still simpler breakdown into clast-rich and melt-rich (clast-poor) breccia types.

Reconstruction of the history of lunar breccia deposits severely challenges the petrologist, geochemist, and geologist. The place and process of original formation of the clasts and of the raw materials for the melts (chap. 8) must be disentangled from the process by which they arrived at the place where they were collected (chaps. 9, 10). Each clast may yield a different isotopic age, attesting to many endogenic and impact events in the sample's history, or only to incomplete migration of Ar, Rb, and Sr isotopes (chaps. 9, 10). The crater or basin in which most of them originated can only be inferred. Unsuspected complexities in the process of forming craters and, especially, basins may have shaped the materials of the lunar terrae.

SUMMARY OF CRATER-MATERIAL ORIGIN

Introduction

Around almost all impact craters, the processes described above have created similar patterns of continuous ejecta grading outward to secondary craters. The major apparent differences among craters are between the interiors of simple and complex craters. Partly understood variations in the target material, such as in structure, cohesion, and topographic relief, also affect the details of morphology; and variations in projectile velocity, density, and impact angle affect both the morphology and the proportions of melted and ejected material. Nevertheless, the basic morphologic patterns of impact craters are much more alike than those of volcanic craters.

This similarity has greatly aided the reconstruction of lunar geologic history. Stratigraphically based lunar geologic mapping has become increasingly effective as the analogies between modified and unmodified deposits have become clearer. The material subunits in and around craters were generally recognized as stratigraphic entities when they were mapped geologically in the 1960's (Mutch, 1970, p. 165–175; Wilhelms, 1970b, p. 40–42), and their mapped limits remain generally valid. The interpretations of some units, however, have changed (compare Schmitt and others, 1967, and Howard, 1975). To enable modern use of these maps, this section updates the interpretations of each class of unit on the basis of the newer research summarized above, and suggests what rock types are likely to occur in each unit and in the analogous units of basins.

Rim material

A raised rim, formed by uplift of the target and by deposition of ejecta, surrounds the crater. The geologic unit "rim material" (a member of the formation "crater material") essentially corresponds to the continuous ejecta (fig. 3.22). Smooth, clean-looking rim surfaces may have been stripped of ejecta by late-stage radial flow (Guest, 1973). On the near flanks of simple craters, ejecta materials have been deposited in inverse order of their original sequence. Original stratigraphic units are more jumbled in complex craters (Hörz, 1981). Impact melt forms ponds in depressions and coats other parts of the rim (figs. 3.23). The deposits also probably incorporate clasts and veins of impact-melt rock along with fragmental ejecta. The relative amounts of melted and unmelted ejecta depend on factors of projectile size, density, and velocity, and on such target properties as density and volatile content.

The hummocky facies of rim material (a submember) consists of late-emplaced material that was derived from near the crater wall and overlies the structurally uplifted and outthrust zone of larger craters. Blocks of relatively undeformed material are scattered on the rim (figs. 3.2A, 3.23B). The rim's generally concentric structure attests to final emplacement by outthrusting (fig. 3.23). In places, younger crater rims have breached older rims, and the hummocky materials flowed rapidly (fig. 3.24) or sluggishly (fig. 3.25) into the gap. Therefore, the deposits were emplaced along the surface. Around old craters, the hummocky rim facies, without the topographic subtleties of young craters, is the only distinguishable rim material (fig. 3.26).

Outward from the hummocky rim material, ridges in radial and herringbone patterns form a continuous-appearing wreath with irregular lobate margins (fig. 3.22). In the well-photographed complex craters illustrated here, the radial rim facies (submember) is clearly an outward gradation of the hummocky rim material (figs. 3.23, 3.24, 3.27). Surface flow after ejection is indicated by the textures curving from concentric to radial patterns (figs. 3.23B, 3.24, 3.27A). Some of the ejecta apparently was molten (fig. 3.27B). The gradation with materials at the rim crest and the content of impact melt indicate that the ejected ground-flow materials are dominated by primary ejecta. The outer part of the radial rim material, however, contains locally derived material because substrate material was incorporated either by ground flow of the ejecta under confining pressure or during ballistic emplacement of the ejecta.

FIGURE 3.24.—Lineations of northwestern ejecta of crater Tsiolkovskiy, indicating ground surge beyond concentric near-rim deposits. Smaller crater in upper left, probably filled with additional Tsiolkovskiy ejecta, is Lütke (39 km). Apollo 17 frame M–2608.

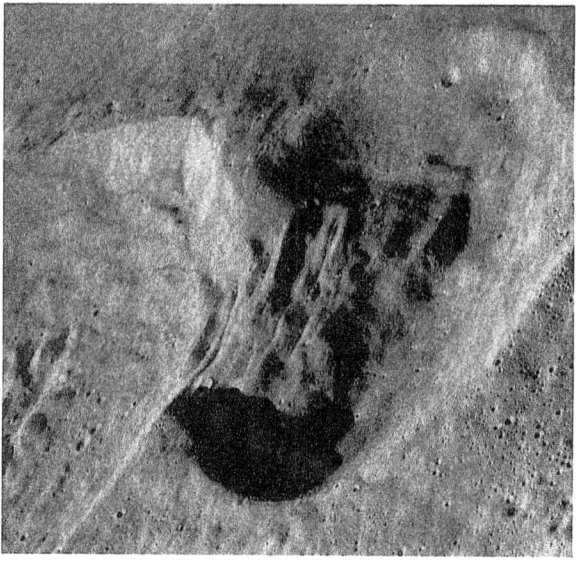

FIGURE 3.25.—Concentric ejecta of crater Shirakatsi (51 km, left), pushed through gap in rim of older or nearly contemporaneous crater Dobrovol'skiy (38 km). North of Tsiolkovskiy, centered at 13° S., 129° E. Apollo 17 frame M–2608.

At some radius, ballistically generated secondary craters appear. Much of the ejecta thrown from secondaries has a herringbone pattern that originated by intersection of cones of their ejecta (figs. 3.4C–F, 3.6, 3.7). In the ideal case, derivation of the farthest-thrown materials from near the impact zone (fig. 3.18) means that the secondaries were formed by highly shocked material from shallow depths in the primary target. Thus, whatever primary material is recovered from the secondary ejecta will contain impact melt or other high shock grades, whereas the local material in the secondary ejecta should be less shocked because of the lower kinetic energies of the secondary impacts. The proportions are not known, but locally derived material probably increases with distance from the primary crater. Some geologic maps distinguish a rim-material facies called cratered rim material (fig. 3.22) and identify the underlying unit of which it is composed (Ulrich, 1969).

The ejecta of young secondary craters commonly appears as rays or bright splashes of freshly exposed material (figs. 3.4A, C, 3.6, 3.28). Most of this ray material was probably excavated from the local terrain by secondary impacts. Many investigators of rays have suspected that fine primary ejecta also composes parts of rays (for example, Baldwin, 1963, p. 358; Schultz, 1976a, p. 204; Pieters and others, 1982). Absolute dating of Copernicus and Aristillus depends on the interpretation that the samples recovered on rays hundreds of kilometers from those craters contain primary ejecta (see chap. 13).

Wall material

Wall material is a mixed unit that includes materials of the terraces and all the slumps, debris, and impact melt that coat crater walls. Walls may expose precrater strata (fig. 3.29). Edges of the terraces may also contain precrater target rock, and tops may expose downdropped rim material. Depressions in the terrace tops, which are most common on the rears of the outward-tilted downdropped surfaces, commonly contain pools of flat-surfaced material now recognized as impact melt (figs. 3.2D, 3.23B; Howard, 1975; Howard and Wilshire, 1975; Hawke and Head, 1977a). Dribbles or cascades of additional melt or debris, commonly bounded by levees resulting from the flow, connect some terraces and the floor (fig. 3.30; Howard, 1975). These melt features, which are seen only on high-resolution photographs, merely contribute to the overall rough appearance of the walls as seen on telescopic photographs. Coarse wall hummocks are probably slumps (fig. 3.27A).

A

FIGURE 3.26.—Ejecta of moderately fresh (Eratosthenian) crater Werner (70 km, 28° S., 3° E., above), superposed on crater Aliacensis (80 km, below), whose subdued walls, raised rim wreath, and central peak were probably similar to those of Werner when first formed; Aliacensis-radial ejecta is no longer visible. Rim nearest Werner is more highly degraded than distal rim. Orbiter 4 frame H–100.

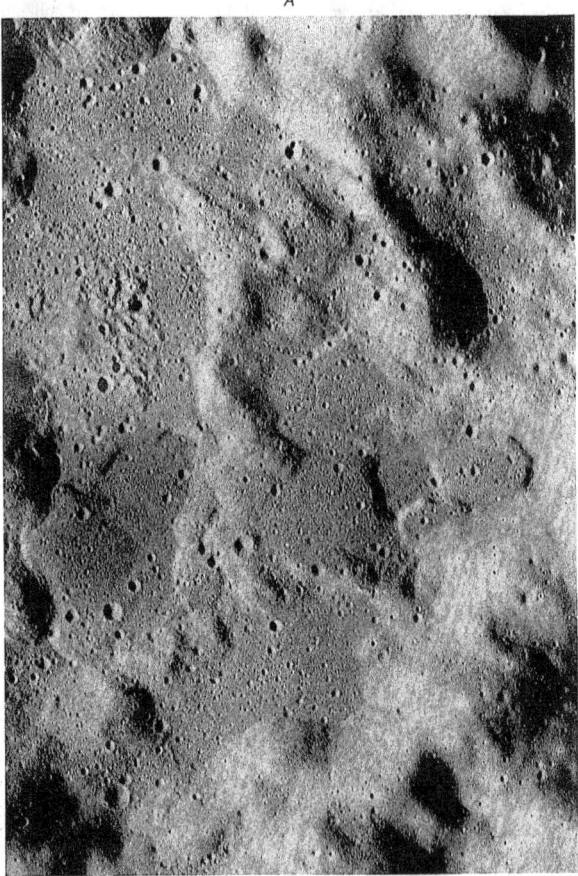

B

FIGURE 3.27.—Eastern ejecta of Tsiolkovskiy (180 km).
  A. Ground-surge emplacement is indicated by diversion from concentric to radial lineations (arrows). Impact melt is visible at upper arrow and in box at bottom. Apollo 15 frame M–757.
  B. Impact melt, partly enclosed in box in figure 3.27A, covers radial ejecta. Apollo 15 frame P–9580.

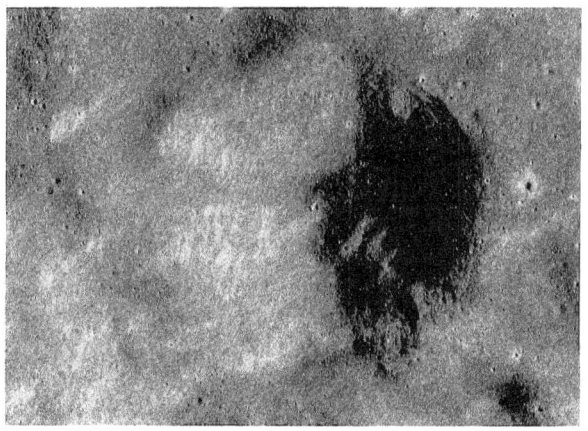

FIGURE 3.28.—Splashes of bright material dug from beneath darker surficial layers by secondary-impact craters of a crater out of scene to lower left. Crater in picture is about 5.5 km across; near west rim of Gagarin. Apollo 15 frame P-8941.

On many geologic maps, material slowly uncovered or displaced downslope on crater walls is mapped as a unit called bright slope material (Mutch, 1970, p. 166–169; Wilhelms, 1970b, p. 39). This unit was early recognized by geologic mappers and ascribed to renewal of exposure of lunar rock by downslope movement (Shoemaker and Hackman, 1962, p. 297; Shoemaker, 1965, p. 128); it was generally assigned a Copernican age regardless of the age of the underlying unit. A substantial content of fresh material has been confirmed by spectral studies (see chap. 5), and all high-resolution photographs of steep lunar slopes reveal mass-wasted debris deposits (figs. 3.31–3.33). The time of exposure of materials in the colluvium at the bases of massifs at the Apollo 15 and 17 landing sites has been dated isotopically at tens or hundreds of millions of years, in contrast to several aeons for the source rock. The colluvium and slope debris contain material foreign to the source and introduced by impacts. The oldest (darkest) slopes probably have accumulated the most exotic material (fig. 3.31).

Peak material

The geologic unit "peak material" was probably derived from greater depths than any other lunar-crater material because the material in peaks formerly lay beneath the excavated part of the crater (fig. 3.19). Peaks range considerably in morphology from low hills and single centralized pinnacles, through multiple jagged peaks clustered around a center, to dispersed smaller arrays (figs. 3.2C–E, 3.29, 3.32–3.35, 4.2). All these forms are consistent with uplift mechanisms. Peaks probably formed during crater excavation by intensive

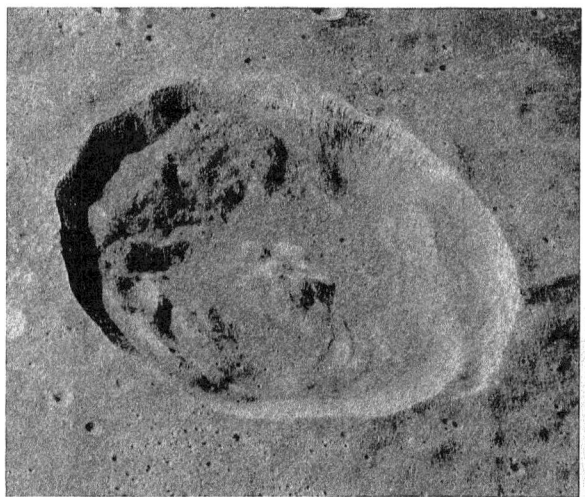

FIGURE 3.29.—Crater Euler (28 km, 23° N., 29° W.), showing wall materials weakly coalesced into terraces at bottom and uncoalesced in upper left. Precrater stratigraphic units are evident by albedo differences on upper right wall. Apollo 15 frame P-10274.

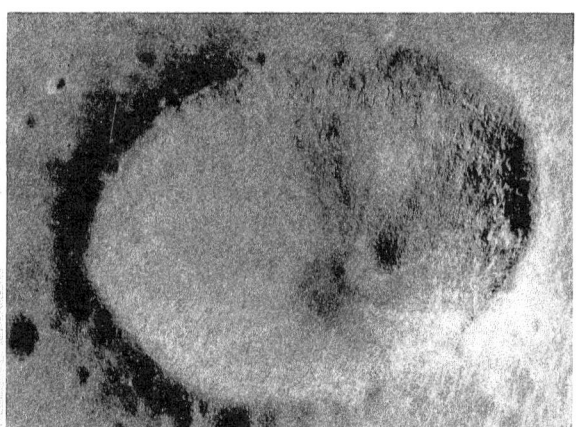

FIGURE 3.31.—Bright streaks of fresh material and darker, longer-stabilized material on wall of crater Lalande A (13 km, 6.5° S., 10° W.). Floor is composed of material already accumulated by similar downslope movement. Apollo 16 frame P-5400.

FIGURE 3.30.—Cascade of impact melt (arrow) from terrace to floor on north wall of Copernicus; w, steep part of wall just below rim. Cascade has apparently drained melt pool (right of arrow end) through groove (right of arrowhead) (Howard, 1975). Arrow is 5 km long. Orbiter 5 frame M-152.

deformation of the central part of the target. The violent and extreme uplift suggested by several lines of evidence given above seems to be supported by the observation by Murray (1980) that some peaks seem to have toppled over the crater rim (fig. 3.34). Because of the intense deformation, peaks are structurally complex. Few, if any, peaks are volcanic constructs accumulated after cessation of cratering, contrary to the interpretations presented on many geologic maps and in much other earlier lunar literature.

Floor material

Some floor materials are coeval with peak materials, some slightly younger, and some much younger. Floor materials are diverse, and photographs should be consulted during their reinterpretation. Some constitute peaklike uplifts (figs. 3.33, 3.35), which in many craters form an approximately annular pattern that foreshadows the larger rings of basins (fig. 3.2E; chap. 4). Separation of such lumps from central peaks may be more a matter of scale-dependent mapping convention than of genetic significance. Some hummocky floor materials bridge the gap between peak and wall, and may represent the toes of inward-displaced wall slivers or parts of the floor uplift (figs. 3.2E, 3.23, 3.29; Howard, 1975). Other floor materials consist of debris from the walls (figs. 3.31, 3.33).

Impact melt, formerly believed to be volcanic, partly covers many crater floors. Fissured floor materials that appeared on Lunar Orbiter 5 photographs of such young craters as Aristarchus, Tycho, and Copernicus (figs. 3.2D, 3.35) clearly were originally molten, and in the spirit of the times, many workers once considered them to be volcanic (Offield, 1971; Strom and Fielder, 1971; Pohn, 1972; Schultz, 1976b, p. 72). Some smooth floor materials (fig. 3.32) and other plains materials (figs. 3.2D, 3.36) retained volcanic interpretations longer than any other geologic-map units. The smooth "ponds" are superposed on the floor, wall, and rim, and thus were mapped not as crater materials but as special postcrater units (Milton, 1968a; Wilhelms and McCauley, 1971); they were commonly ascribed to volcanic extrusions released or localized by the impact (Strom and Fielder, 1971). However, these smooth and fissured materials have been identified as impact melt by detailed observations of the freshest and best photographed craters (figs. 3.23, 3.36; Shoemaker and others, 1968; Howard, 1975; Howard and Wilshire, 1975; Hawke and Head, 1977a; Gault and Wedekind, 1978). Impact melt veneers much of the crater interior and rim, and is ponded in favorable depressions (fig. 3.36). Fissures that continue from the floor over many of the floor uplifts indicate that melt rock coats these uplifts as well (fig. 3.35; Howard and Wilshire, 1975). The fissures and many irregular pits suggest drainage of impact melt into underlying porous breccia (fig. 3.35; Howard, 1975), not volcanism (Hartmann, 1968). Some melt flows came to rest later than the peak (fig. 3.32), in confirmation of terrestrial observations indicating that peak formation is even quicker than melt solidification. The only internally generated materials associated with craters are the mare materials that flood many craters and, possibly, small amounts of other material, such as that in the floor of the endogenic crater Hyginus (Pike, 1976).

Impact-melted floor materials probably would provide a chemically representative sample of the target rocks because target materials are extensively homogenized during impact melting (Dence, 1971; Grieve and others, 1974; Grieve, 1975; Simonds and others, 1976a, b). Conversely, the material on older crater floors may be completely unrelated to the crater that contains them or to the subcrater material (Aliacensis, fig. 3.26). Even craters that lack recognized deposits of mare basalt or ejecta may be filled by debris derived from the walls or introduced gradually as fragments by random distant impacts. Such slowly accumulated floor materials may yield random samples of large parts of the lunar surface.

In summary, geologic mapping of the materials of lunar craters has proved to be generally accurate in delineating significant units and determining mutual age relations, but the interpretations of many mapped units have changed. The absolute-age differences detected among crater-material facies by stratigraphic relations have shrunk from millions of years to minutes. The materials have moved continuously closer to an almost exclusive impact interpretation and away from volcanic or hybrid interpretations, under the prodding of sample analyses and continued photogeologic, Earth-analog, and experimental studies.

FIGURE 3.32.—Part of "lobster claw" peak of crater King (compare fig. 3.23A), showing superposition of impact melt in gap of peak. Apollo 16 frame P-4998.

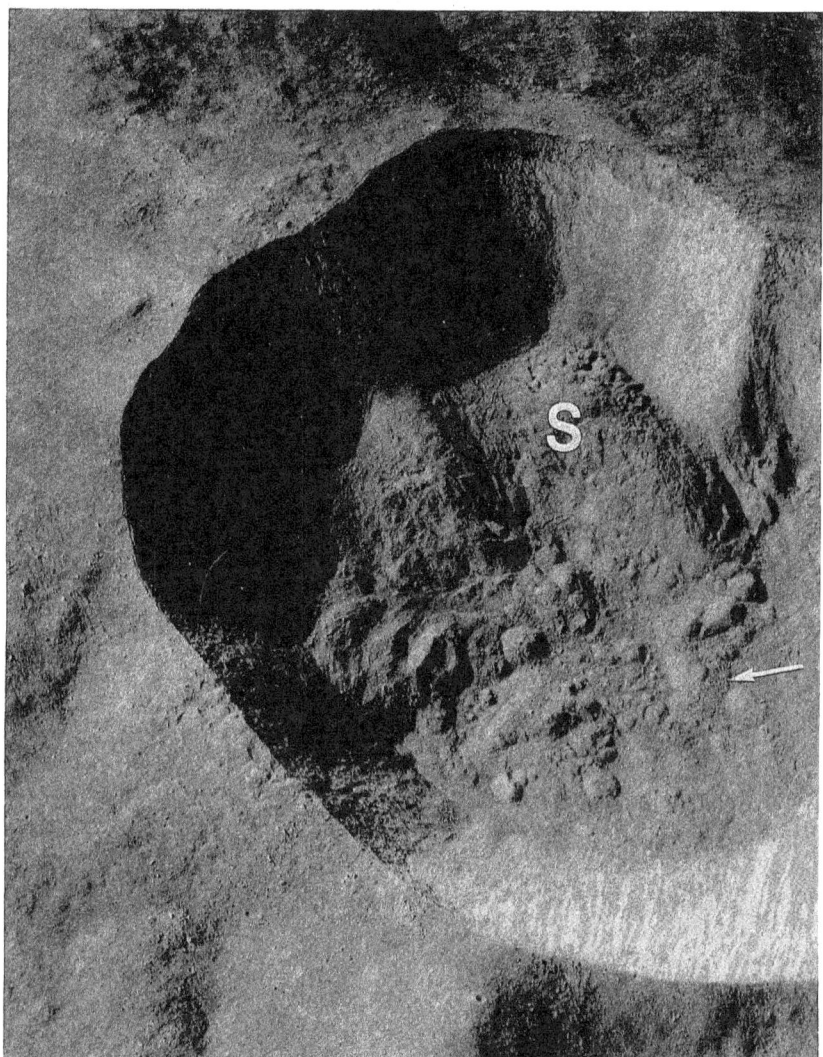

FIGURE 3.33.—Crater Proclus (28 km, 16° N., 47° E.), showing terrace-free walls, slump from upper wall (S), coarse floor mounds probably produced by floor uplift, and fissured impact melt on floor (arrow). Apollo 17 frame P-2265.

FIGURE 3.34.—Crater Anaxagoras (51 km, 73° N., 10° W.). Peak material apparently toppled over rim at arrow (Murray, 1980). Orbiter 4 frame H-128.

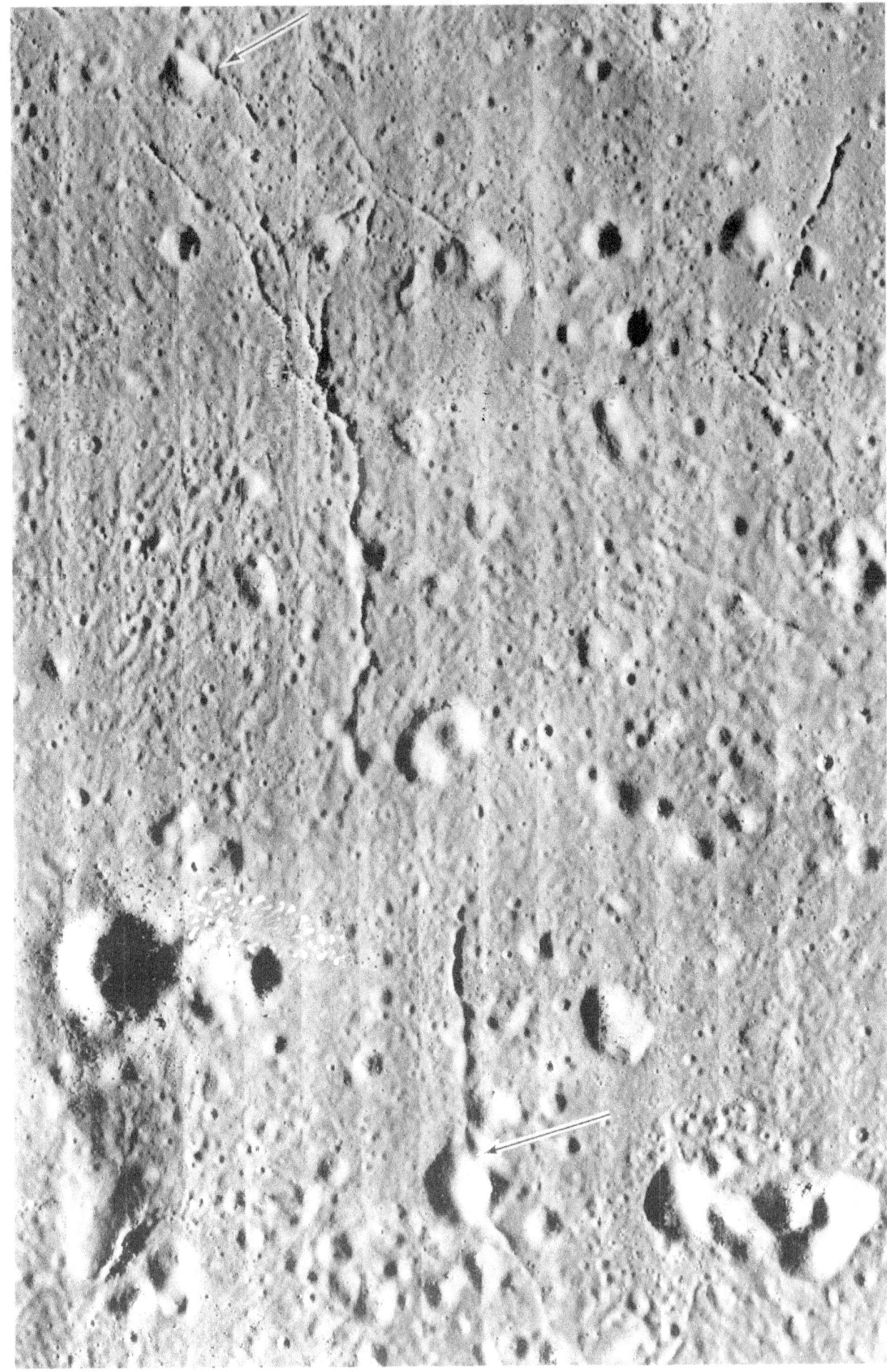

FIGURE 3.35.—Impact melt and floor mounds of Copernicus; melt is superposed on mounds at arrows. Fissures indicate shrinkage of melt and, possibly, drainage into subsurface cavities. Orbiter 5 frame H-153.

FIGURE 3-36.— Lineations and festoons indicating flow of impact melt in exterior pool of crater King (fig. 3-23A). Melt has flowed over rim material and collected in depressions (Howard and Wilshire, 1973). Apollo 16 frame P-7000.

4. BASIN MATERIALS—ORIENTALE

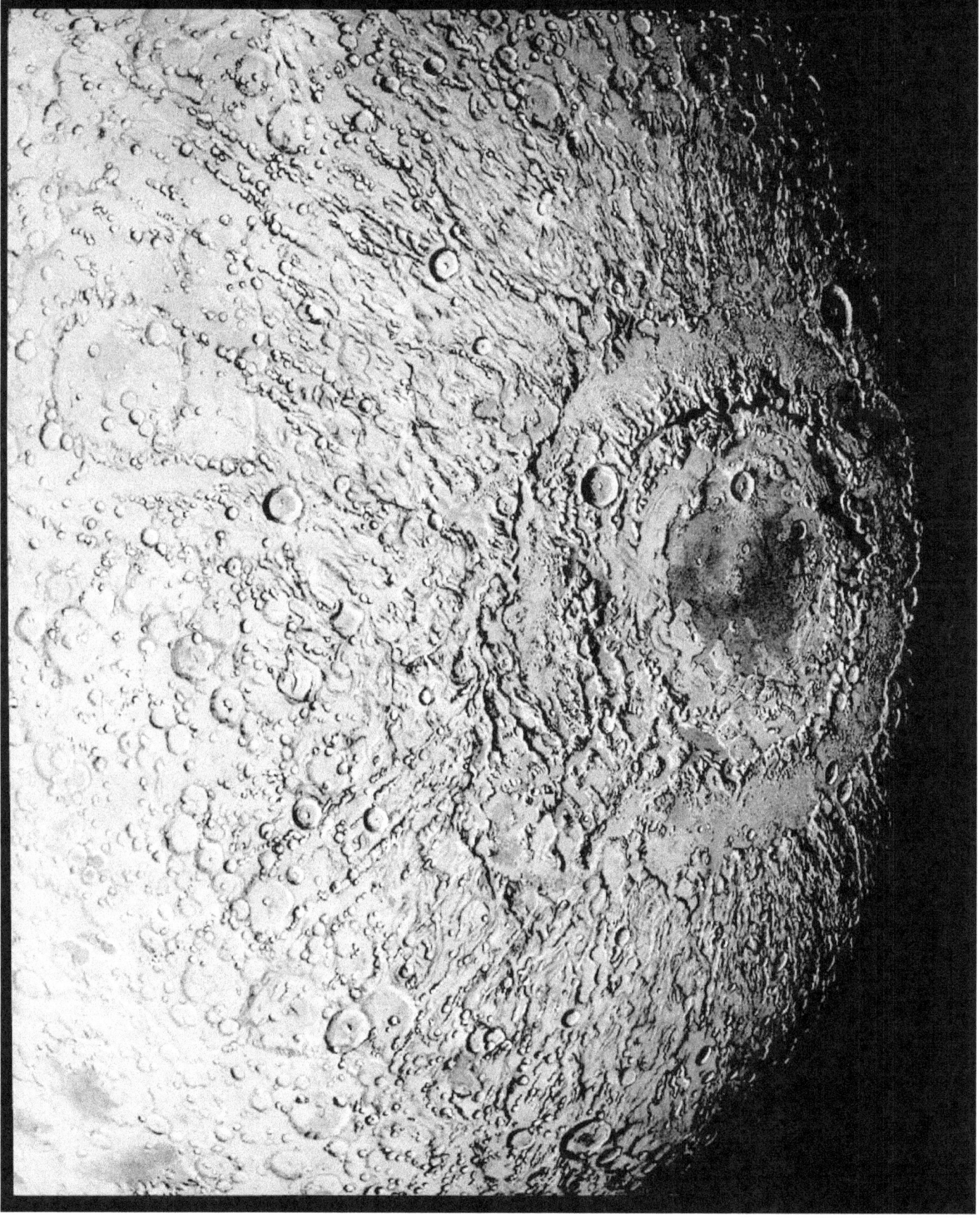

FIGURE 4.1. (OVERLEAF)—Orientale basin and its deposits. Scene centered on farside. Double-ringed basins are Hertzsprung (above) and Apollo (below, with some mare fill). Painting by Donald E. Davis, courtesy of the artist.

4. BASIN MATERIALS—ORIENTALE

CONTENTS

	Page
Introduction	57
Orientale exterior, *by John F. McCauley*	66
General features	66
Inner facies of the Hevelius Formation	66
Outer facies of the Hevelius Formation and secondary craters	67
Relation between ground-flow deposits and secondary craters	67
Plains	72
Orientale interior, *by John F. McCauley*	72
Montes Rook Formation	73
Maunder Formation	76
Topography and gravity	77
Origin of rings	77
Introduction	77
Genetic models	77
Discussion	79
Conclusions	81
Summary of basin-material origins	81

INTRODUCTION

Most simple craters, almost all complex craters, and all ringed basins were formed by impacts. Collectively, these lunar features constitute a series of impact excavations that have many common properties over the entire size range but also show many differences (figs. 4.1–4.3).

Appreciation of the importance to lunar geology of the largest members of this size-morphology series, the basins, has grown slowly. Gilbert (1893) may have been the first to perceive the Imbrium basin as the largest example of the series on the nearside. This interpretation was supported by other influential early workers who increasingly recognized the distinction between basin and mare (Baldwin, 1949, 1963; Kuiper, 1959; Hartmann and Kuiper, 1962; Mason and Hackman, 1962; Shoemaker, 1962a, b; Shoemaker and Hackman, 1962). Shoemaker and Hackman (1962) pointed out the similarity of the Imbrium-basin ejecta to crater ejecta. Several other authors enunciated the differences in style from craters, while affirming a basic similarity of origin, by stressing the linearity of the basin-radial Imbrium sculpture (Baldwin, 1949, chap. 11; 1978) and the concentricity of the ring systems (Baldwin, 1949, p. 37–46; 1963; Hartmann and Kuiper, 1962; Hartmann, 1981). A landmark study of rectified telescopic photographs by Hartmann and Kuiper (1962) documented the ring patterns and regular ring spacings of 12 telescopically visible basins, including Orientale; they also coined the term "basin" (Hartmann and Kuiper, 1962, p. 62; Hartmann and Wood, 1971, p. 3–5; Hartmann, 1981). McCauley (1964b, 1967a, b, 1968) then pointed out similarities of the circum-Orientale terrane to the Imbrium ejecta. Titley and Eggleton (1964; Titley, 1967) tentatively recognized ejecta of the Humorum basin, and Wilhelms and Trask (1965) suggested that the terrane around other basins contains basin ejecta and secondary-impact craters. Stuart-Alexander and Howard (1970) and Hartmann and Wood (1971) reviewed the photographs obtained of the whole Moon in 1966 and 1967 by the five Lunar Orbiters and described the properties, distribution, and probable origins of many additional basins and their maria.

Although these studies established the existence and the impact origin of many large basins, the morphology and extent of the exterior deposits and secondary craters were poorly known before Apollo exploration began (Mutch, 1970; Wilhelms, 1970b). Volcanic and tectonic hypotheses still competed vigorously with impact hypotheses. The first new data that would eventually lead to the current impact model were the superb photographs of the Orientale basin obtained in 1967 by Lunar Orbiter 4 (figs. 1.9, 4.4). The Orientale deposits are less obscured by later material than are those of any other large well-photographed basin. The second and vital clue came from the discovery of complex impact breccia at every sampling site in the terrae, in particular, the Apollo 16 landing site, previously thought to be underlain by volcanic materials (see chap. 2; Howard and others, 1974). The third new development was the extension of impact interpretations to the rest of the terrae, an exercise summarized in this volume. More and more basins have been recognized and established as the sources of surrounding deposits and satellitic craters (pl. 3; fig. 4.3; tables 4.1–4.3).

FIGURE 4.2 (OVERLEAF).—Crater-to-basin transition.
  A. Antoniadi (135 km, 70° S., 172° W.), with both an inner ring and a small central peak. Mare material fills depression inside inner ring and partly floods peak. Orbiter 4 frame M–8.
  B. Compton (162 km, 56° N., 105° E.), with conspicuous central peak and low but distinct inner "peak ring." Cracks indicate floor uplift. Orbiter 5 frame M–181.
  C. Petavius (177 km, 25° S., 60° E.), with massive complex central peak and vague ringlike configuration of low peaks between central peak and wall. Uplift extending to middle of wall is indicated by grabens. Slight filling by dark material is evident along walls. Orbiter 4 frame H–184.
  D. Gauss (177 km, 36° N., 79° E.), with entirely different floor structure from that of Petavius, although both craters are of same size. Orbiter 4 frame H–165.
  E. Humboldt (207 km, 27° S., 81° E.), with central peak and line of peaks. Other rugged features of original floor are higher in north than in south; cracks indicate uplift of entire floor. Light-colored plains may be later fill; dark crescents are later mare fill. Orbiter 4 frame M–12.
  F. Landau (221 km, 43° N., 119° W.), with few distinct floor features; irregularly depressed in northeast. Orbiter 5 frame M–12.
  G. Campbell (225 km, 45° N., 153° E.), with some offcenter rugged relief. An offcenter depression, as in Landau, is here filled by a small mare patch. Orbiter 5 frame M–103 (rectified).
  H. Clavius (225 km, 58° S., 14° W.), with small offcenter peaks, probably partly covered by later plains fill. Orbiter 4 frame M–124.
  I. Schwarzschild (235 km, 71° N., 120° E.), with rugged, arcuate, crescent-shaped floor topography. Orbiter 4 frame M–92.
  J. Milne (262 km, 32° S., 113° E.), with complex single to triple arcs. Milne and Schwarzschild are the most basinlike craters smaller than 300 km in diameter. Orbiter 3 frame M–121.
  K. Gagarin (272 km, 20° S., 149° E.), with indistinct ring-form structure, though larger than Milne. Orbiter 1 frame M–116.

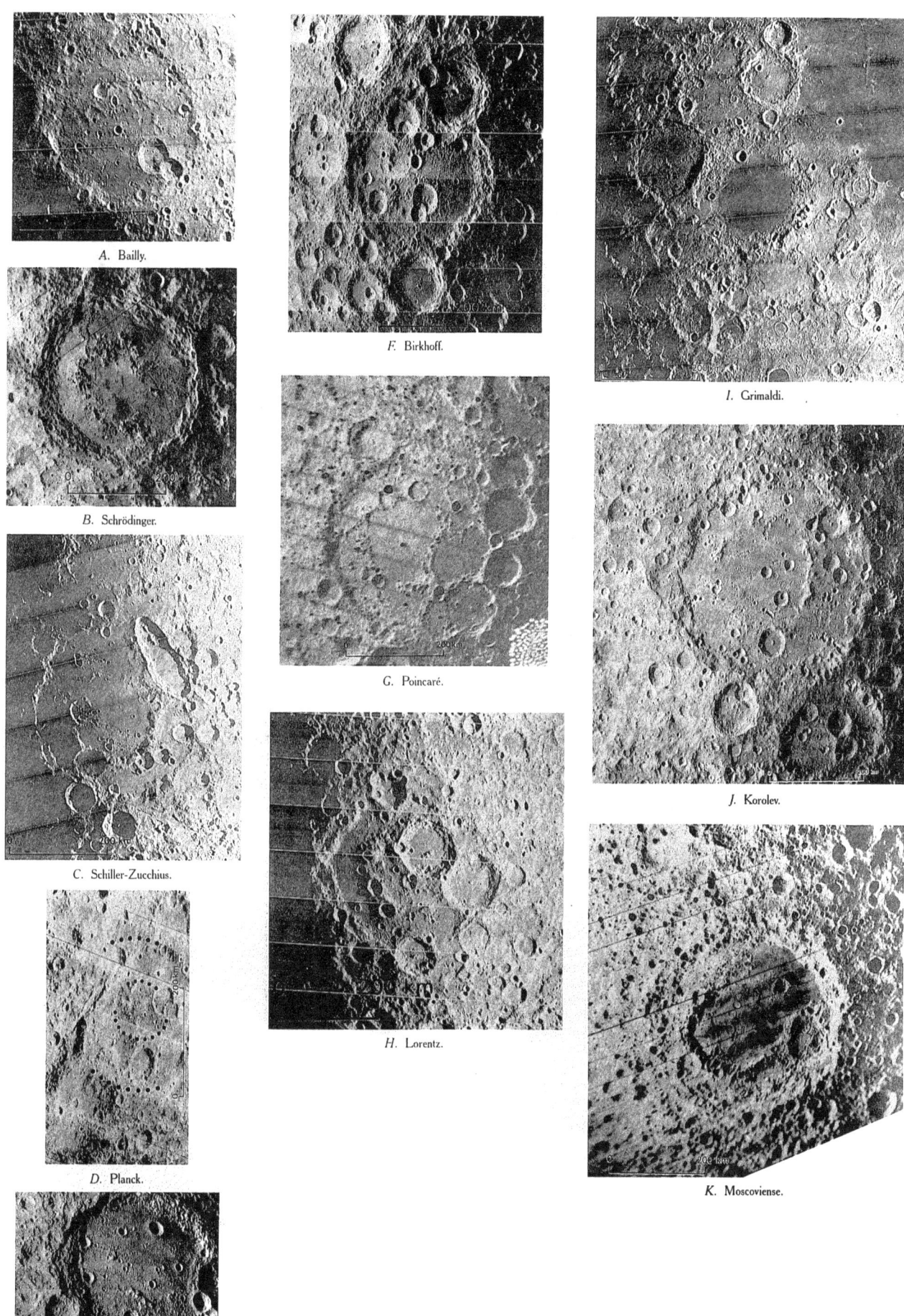

FIGURE 4.3.—Definite lunar basins at least 300 km in diameter, in order of increasing size and complexity (see table 4.1).

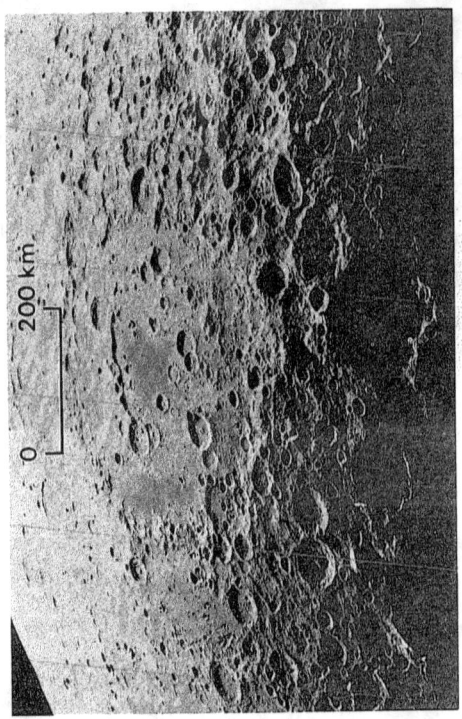

L. Apollo.

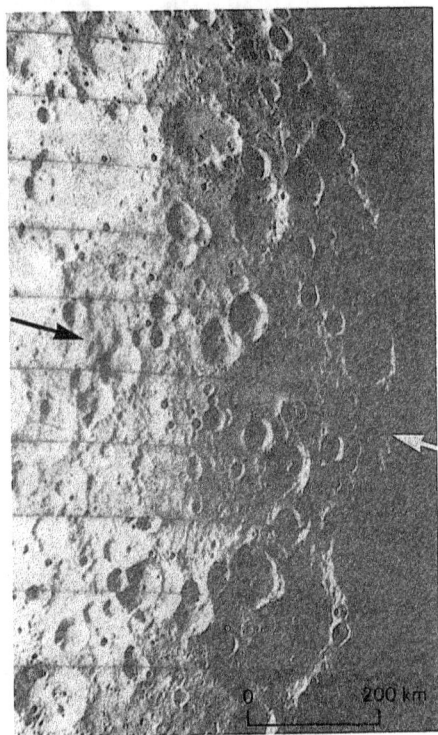

M. Coulomb-Sarton.

N. Ingenii.

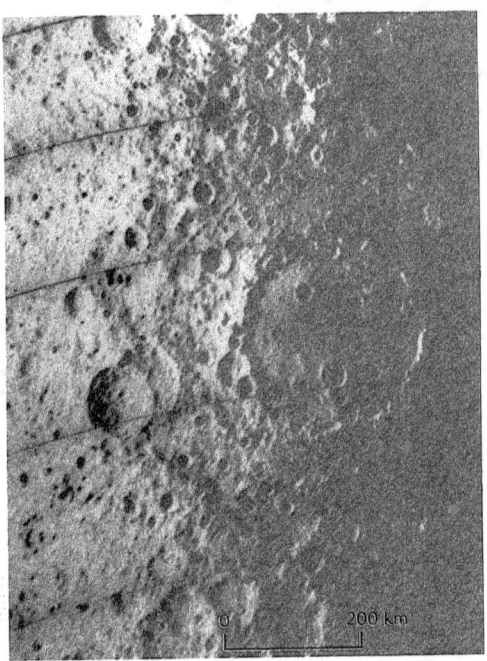

O. Hertzsprung.

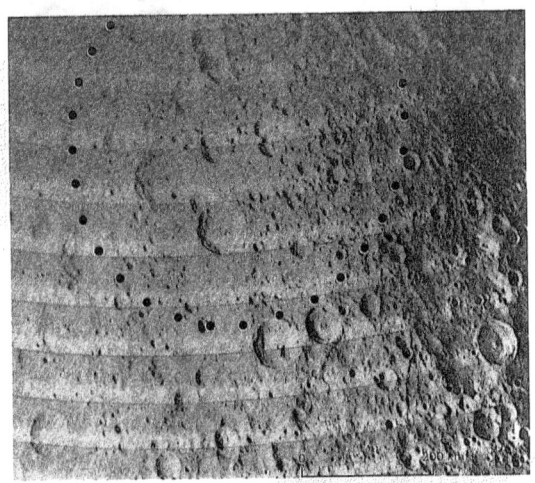

P. Freundlich-Sharonov.

Q. Humboldtianum.

R. Mendel-Rydberg.

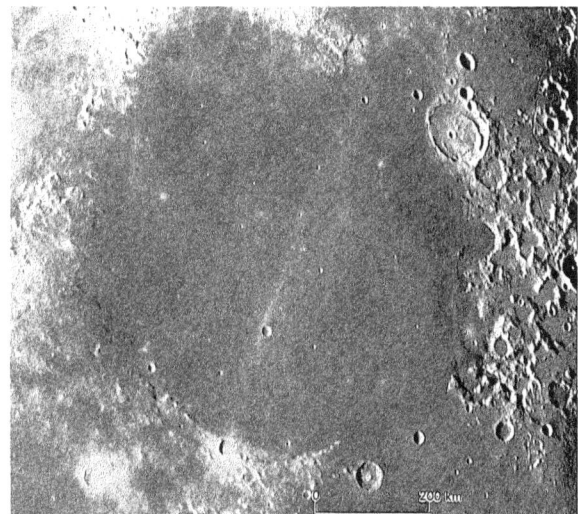

S. Serenitatis.

T. Keeler-Heaviside.

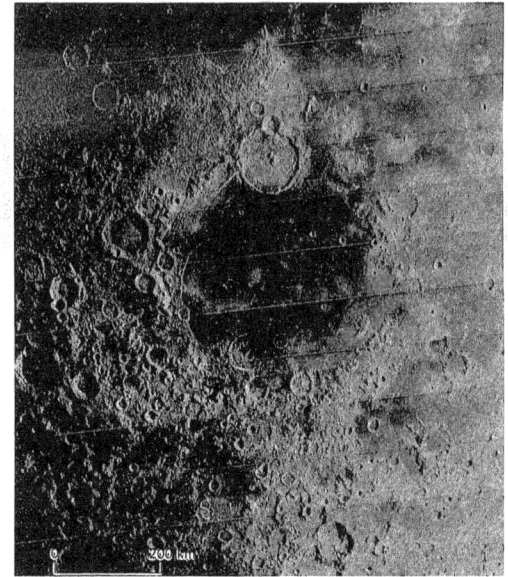

U. Humorum.

V. Smythii.

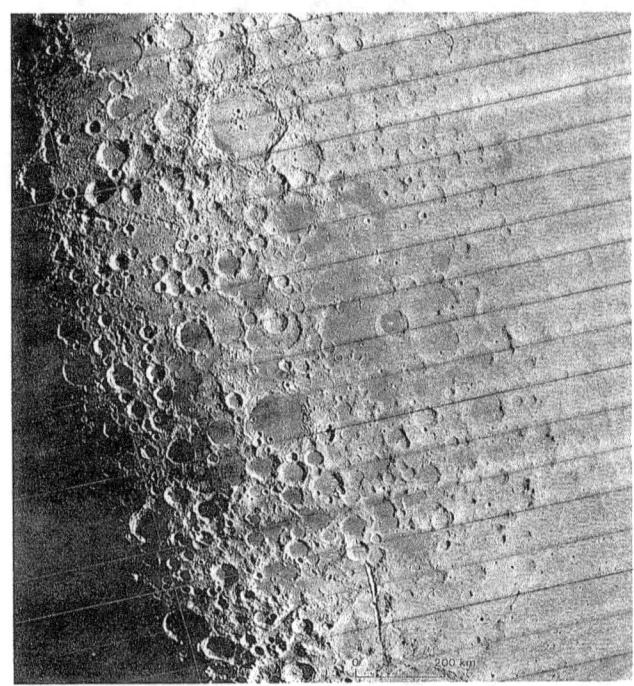

X. Australe.

W. Nectaris.

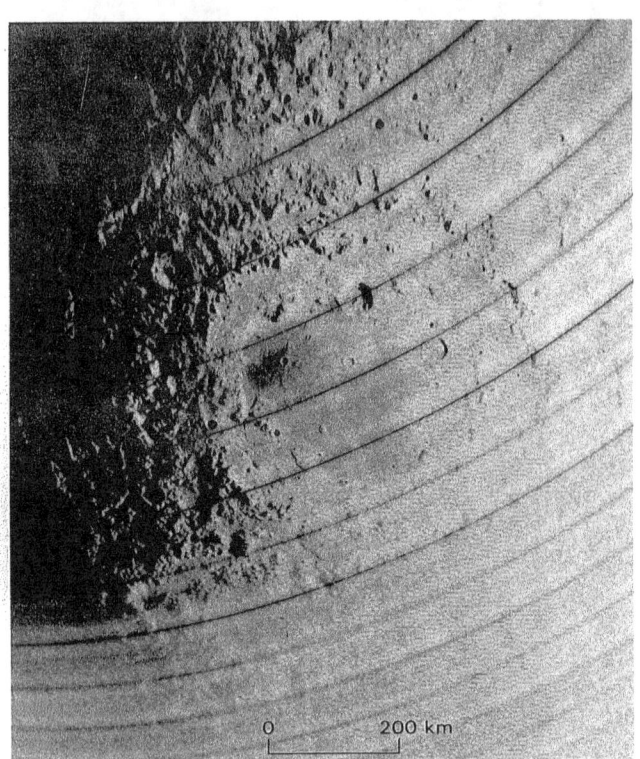

Y. Orientale.

Z. Crisium.

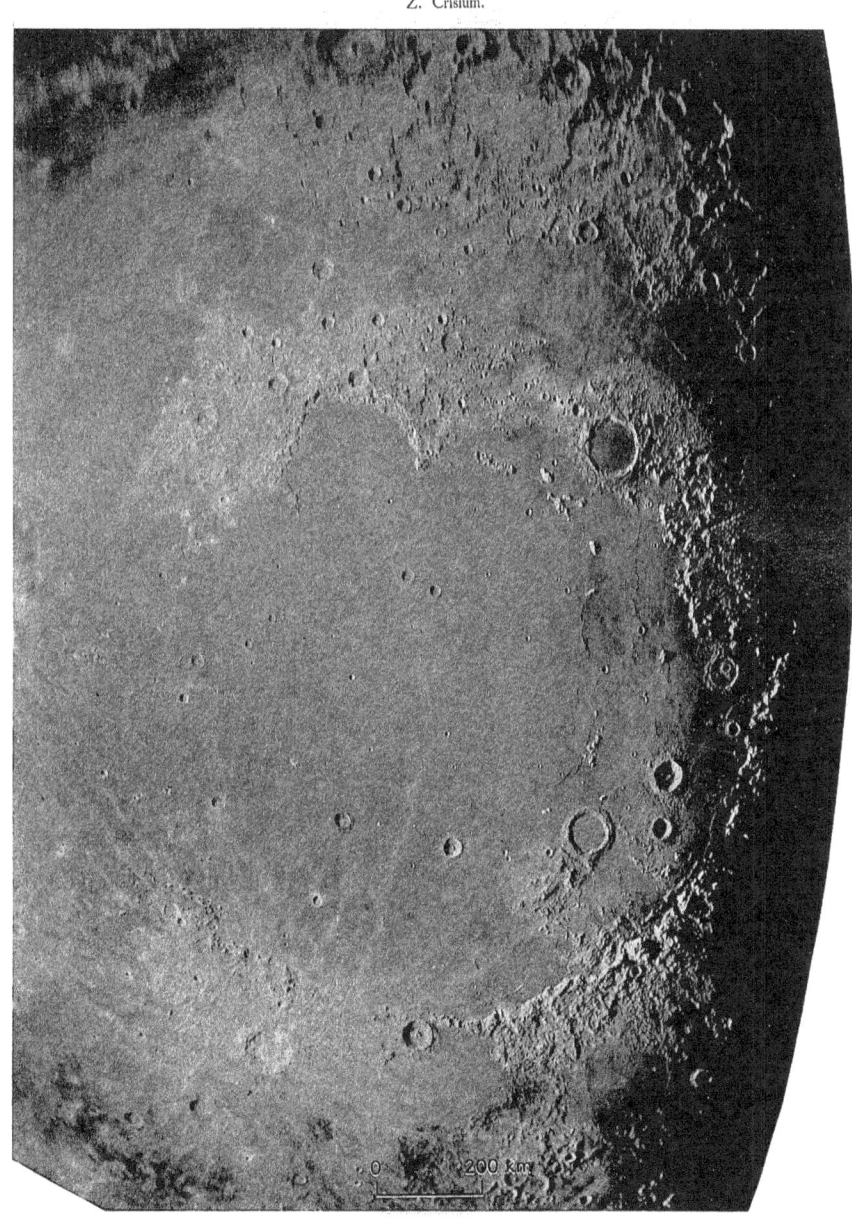

AA. Imbrium.

TABLE 4.1.—*Definite lunar basins at least 300 km in diameter*

[Basins listed in order of increasing size and complexity. Names all in capitals have deposits mapped on plate 3. Names are derived, where possible, either from the mare fill of a basin (for example, the Imbrium basin from Mare Imbrium) or from a name given by the International Astronomical Union, considering them as craters (for example, Apollo). Otherwise, they are named here from two superposed, unrelated craters:

Name used here (reference)	Earlier designation (reference)
Schiller-Zucchius (W79)	Basin near Schiller (HK62).
Coulomb-Sarton (L78)	Unnamed basin B (HW71).
Freundlich-Sharonov (S78)	---
Mendel-Rydberg (W79)	SE. limb basin (HK62).
Keeler-Heaviside (S78)	---
South Pole-Aitken (S78)	Big backside basin (common usage).

Best photographs: Photograph used in figure 4.3 is underlined; each includes one-half basin radius beyond topographic rim, where possible. HK62, telescopic photograph from Hartmann and Kuiper (1962); r, rectified; A16, Apollo 16 frame; M10, Mariner 10 frame (Mercury-Venus mission that photographed the Moon during flyby). Others are Lunar Orbiter frames.
Major rings: Diameters are listed for rings that are observed or inferred to extend more than half a basin circumference. Approximate radii are listed for partial segments that may not form complete rings. Additional small segments are plotted on plate 3. Topographic basin rim (also called main rim in descriptions) is underlined, queried where uncertain.
References: B49, Baldwin (1949); B63, Baldwin (1963); G93, Gilbert (1893); HK62, Hartmann and Kuiper (1962); HW71, Hartmann and Wood (1971); L78, Lucchitta (1978); M67, McCauley (1967a, b); S77, Scott and others (1977); S78, Stuart-Alexander (1978); SH62, Shoemaker and Hackman (1962); SH70, Stuart-Alexander and Howard (1970); T67, Titley (1967); WE77, Wilhelms and El-Baz (1977); WM71, Wilhelms and McCauley (1971); W79, Wilhelms and others (1979). Chapter numbers refer to the present volume.]

Basin	Figure	Center (lat, long)	Best photographs	Major rings Diameter (km)	Major rings Radius (km)	Description	References
Bailly	4.3A	67° S., 68° W.	4 H-179 4 M-193 4 M-179	300 150	--- ---	Degraded craterlike rim. Discontinuous inner ring, connected to rim by outward-sloping shelf. Deposits obscured by Orientale.	Chapter 9; HW71, W79.
SCHRÖDINGER	4.3B	75° S., 134° E.	5 H-21 4 M-8 4 M-94	320 155	--- ---	Fresh craterlike rim and wall. Inner ring consists of rugged crags forming nearly complete circle, joined to wall in two places by exposed shelf. Hummocky probable impact impact melt on floor. Extensive craterlike deposits, linear secondary chains.	Chapter 10; H70, HW71, W79.
Schiller-Zucchius	4.3C	56° S., 44.5° W.	4 H-160 4 M-148 4 M-136	325 165	--- ---	Outer ring craterlike though degraded; bounded inward by steep scarp, outward by gentle flank. Inner ring mountainous. No discernible deposits.	Chapter 8; HK62, W79.
Planck	4.3D	57.5° S., 135.5° E.	3 M-121 4 M-8 4 M-9 4 M-82 5 M-65	325 175	--- ---	Poorly photographed; outer ring apparently craterlike, inner ring partial. No discernible deposits.	Chapter 8; HW71, W79.
MENDELEEV	4.3E	6° N., 141° E.	1 M-115 1 M-136	330 140	--- ---	Craterlike outer rim with flat-bottomed troughs. Inner ring of discontinuous small peaks, connected to wall by outward-sloping shelf. Craterlike deposits and numerous, though subdued, secondary craters.	Chapter 9; WE77, S78.
BIRKHOFF	4.3F	59° N., 147° W.	5 M-29 5 H-29 5 M-25	330 150	--- ---	Craterlike rim. Discontinuous, mostly obscured inner ring of narrow ridges. Extensive craterlike deposits and secondaries.	Chapter 8; HW71, L78.
Poincaré	4.3G	57.5° S., 162° E.	5 M-65(r) 2 M-75 3-M-121 4-M-82	340 175	--- ---	Apparently craterlike rim. Continuous but battered lower inner ring. No deposits observed.	Chapter 8; HW71, W79.
LORENTZ	4.3H	34° N., 97° W.	4 M-189	360 185	--- ---	Craterlike rim. Low and ridgelike inner ring, continuous except where cratered. Craterlike deposits and secondary craters west of basin.	Chapter 8; HW71.
Grimaldi	4.3I	5° S., 68° W.	4 H-168 4 M-161	430 230	--- ---	Obscured by Orientale; probably craterlike rim. Discontinuous rugged inner ring. No deposits observed.	Chapter 8; HK62, WM71.
KOROLEV	4.3J	4.5° S., 157° W.	1 M-38 1 M-28	440 220	--- ---	Craterlike rim locally divided by terracing. Inner ring partly buried but detectable for almost a complete circle. Deposits well exposed to west; extensive secondaries.	Chapter 9; HW71, S78.
MOSCOVIENSE	4.3K	26° N., 147° E.	5 M-103(r) 5 M-124	--- 445 210	275 --- ---	Crude figure-8 pattern of outer and inner rings. Rim has crude echelon pattern suggestive of large-scale terracing. Inner ring complete for half of periphery; has steep inner scarp and broad, gentle outer flank. Deposits and secondaries poorly photographed but visibly reduce prebasin crater population 400 to 500 km from rim.	Chapter 9; HW71, S78.
APOLLO	4.3L	36° S., 151° W.	5 H-30 5 M-30	505 250	--- ---	Interrupted obscured main rim. Bifurcations indicate terracing. Inner ring partly continuous and bounded by steep inward scarp and gentler outer flank, partly discontinuous and rugged. Concavity or discontinuity inside inner ring indicated by mare localization. Mare patches indicate outward slope of interring shelf and possible exterior rings. Deposits and secondaries extend about 500 km southward.	Chapter 8; SH70, HW71, S78.
Coulomb-Sarton	4.3M	52° N., 123° W.	5 M-5 5 M-25(r)	530(?) 400 180	--- --- ---	Heavily obscured by Birkhoff and Lorentz deposits. Rings uncertain; 400-km-diameter ring could be the rim.	Chapter 8; HW71, L78.
Ingenii	4.3N	34° S., 163° E.	2 M-75	560(?) 325	--- ---	Outer ring questionable; consists of discontinuous but large massifs. Inner, 325-km-diameter, mare-bounding ring, more complete, though battered, may be the main rim. Deposits not directly observed but may breach south rim of Keeler-Heaviside basin.	Chapter 8; S78.
HERTZSPRUNG	4.3O	1.5° N., 128.5° W.	5 H-26 5 H-28 5 M-26 5 M-28(r) 5 M-24	570 --- 265 140	--- 205 --- ---	Craterlike rim in most sectors, irregular in northwest and indistinct in north. Half of complete and distinct innermost ring resembles crater rim; rest consists of rugged peaks. Intermediate rings conspicuous, single, and benchlike or ridgelike in most sectors; ragged, bifurcated, echelon in northwest; inconspicuous in north, where ragged ridges and hills form upland nearly at level of surrounding terrain. Deposits extensive, locally 600 km from rim; some secondary-crater lobes 900 km from rim.	Chapter 9; SH70, HW71.
FREUNDLICH-SHARONOV	4.3P	18.5° N., 175° E.	2 M-34(r) 5 M-79	600 ? ?	--- ? ?	Rim complete; seems craterlike but is poorly photographed. Ridges suggest interior ring structure, but connection into rings or terraces is uncertain. Distinct radial grooves and ridges and mantling of craters south of basin indicate deposits and secondaries extending at least 600 km from rim.	Chapter 8; S78.
HUMBOLDTIANUM	4.3Q	61° N., 84° E.	4 H-165 4 M-23 4 M-164 4 M-152 4 M-165 M10 2277	--- 600 275	600 --- ---	Figure-8 shape; each part of main rim about 600 km in diameter, equivalent in total area to a circle 700 km in diameter. Bel'kovich (200 km diam) is either an independent crater or third part of inner basin. Intermediate ring, fairly distinct in south-southwest, consists of hills in southwest, many large hills in southeast, and nearly continuous hilly area in north about at level of surrounding terrain. External 600-km-radius arcs southwest of main rim. Distinct deposits and secondary craters (<20 km diam) extend more than 600 km from southeast rim.	Chapter 9; HK62, L78.
MENDEL-RYDBERG	4.3R	50° S., 94° W.	4 H-193 4 M-193 4 M-186	630(?) 460 --- 200	--- --- 155 ---	Distinct separation of rings in south, but continuity uncertain. Faint deposits in southwest.	Chapter 9; HK62, HW71, W79.
SERENITATIS	4.3S	27° N., 19° E.	Telescopic 4 M-97	--- 740 420	450 --- ---	Ring structure indistinct in west half except in Montes Haemus. Haemus, Taurus-Littrow massifs, and other margins of mare probably form main rim (Head, 1979b). "Vitruvius front" (Head, 1974b) and other irregular elevations form long arcs with 450-km radii. Mare ridges suggest a buried 420-km-diameter ring. Basin may be double, with small northern and large southern components (Scott, 1972a; Wolfe and Reed, 1976); center as listed is that of combined components.	Chapter 9; B49.
Keeler-Heaviside	4.3T	10° S., 162° E.	2 M-75	780(?) --- ---	--- 270 170	All rings and arcs composed of large but disconnected massifs whose connection as rings and relation to excavation boundary are uncertain. Deposits not observed.	Chapter 8; S78.
HUMORUM	4.3U	24° S., 39.5° W.	4 M-143 4 H-136 4 M-136	--- 820(?) --- --- 440 325	? --- 350 280 --- ---	820-km-diameter ring defined by steep notched scarp in south and by discontinuous lower ragged, semiradial or semiconcentric ridges or massifs elsewhere. Part of most conspicuous mare-bounding ring (440 km diam) connects with 280-km-radius arcs as sloping flank, elsewhere is divided into semidistinct rings with bridges to 820-km-diameter ring. 325-km-diameter ring marked by terra shelf and mare bench. Most distinct ring separation is in south. Deposits and secondary craters distinct only in southeast.	Chapter 9; HK62, T67, WM71.
Smythii	4.3V	2° S., 87° E.	4 H-9,11 2 M-196 A16 M-3035 4 M-17	--- 840 --- 360	480 --- 330 ---	840-km-diameter ring defined by partial arcs bounded by steep inner scarp and gentle outer flank, like low crater rims. Intermediate ring or arcs indistinct. Inner, 360-km-diameter ring continuous but low. Deposits not observed.	Chapter 8; WE77.
NECTARIS	4.3W	16° S., 34° E.	4 M-71 4 M-83 4 M-84 HK62(r) 4 M-52	860 600 450 350	--- --- --- ---	Main topographic rim is well-defined scarp on west (Rupes Altai), discontinuous mare-embayed mesas on east. Intermediate ring broad and low, bifurcates in west, forms discontinuous plateaus in east. Inner ring diverse, low. Possible additional ring defined by mare border, 240 km in diameter. Deposits extend 650 to 950 km, typically 750 km, from center; secondaries extend 700 to 1,500 km.	Chapter 9; B49, HK62, WM71.
Australe	4.3X	51.5° S., 94.5° E.	4 M-9 4 M-11 4 M-118	880 --- ---	--- 275 95	Outer ring includes large massifs. Intermediate, poorly defined ring contains massifs, locally connected by rugged structure to outer ring. Inner "ring" consists of closed series of irregular elevations. No deposits observed.	Chapter 8; SH70, WE77, W79.
ORIENTALE	4.3Y 4.4 4.5	20° S., 95° W.	4 M-194(r) 5 M-19 5 M-22 See fig. 4.4	--- 930 620 480 320	650 --- --- --- ---	Most nearly complete concentric ring structure among large basins except in the west. Outermost conspicuous ring is Cordillera, bounded inwardly by steep scarp or gentle lip and topped by some ridgelike massifs. Semicircular depression is outside this ring, and a raised 650-km-radius ring may be east of basin. Extension of Cordillera west of basin merges with extensions of Rook (620 and 480 km diam) rings as rugged, irregularly structured upland. Other sectors of outer Rook ring consist of large massifs with steep inner and gentler outer slopes in a sawtooth pattern. Large section of inner Rook ring and small section of 320-km-diameter ring also have large massifs; otherwise, inner Rook ring consists of smaller massifs or a steplike lip, and 320-km-diameter ring is defined by a lip or gradual falloff. Knobby hills suggest a plateaulike central uplift. Radial structures in inner basin. Extensive ejecta and secondaries.	Chapters 4, 10; HK62, B63, M67.

TABLE 4.1.—*Definite lunar basins at least 300 km in diameter—Continued*

Basin	Figure	Center (lat, long)	Best photographs	Major rings Diameter (km)	Major rings Radius (km)	Description	References
CRISIUM	4.32, 4.9	17.5° N., 58.5° E.	4 H-191, HK62(r), 4 M-21, 4 M-60	1,060, ---, ---, 635, 500, 380	---, 450, 380, ---, ---, ---	Outer ring consists of three differently expressed segments: rimlike scarp in north, ridge bordering a trough in northwest, and irregular scarp in southeast. Elevations fall off inward from this ring, which also marks start of secondary chains. Ridges inside northwest ridge and massifs inside southeast scarp are about 380 and 450 km in radius, respectively, but are not closely concentric. Main mare-bounding rim is most conspicuous structure, divisible into 500- and 635-km-diameter rings, the inner consisting of large massifs with missing sector in east, and the outer of high but irregular massifs and other ridges and hills, separated from inner ring by deep irregular depressions. Innermost ring defined by a mare bench and small hills. Deposits and secondary craters extensive to east (1,600 km from center?), secondaries extensive to north (900 km) and south (1,100 km).	Chapter 9; HK62, SH70, WM71, WE77.
IMBRIUM	4.3AA	33° N., 18° W.	4 M-115, 4 M-139, HK62(r)	---, 1,160, 670	900, ---, ---	Center as shown and 1,160-km ring are formed by Montes Carpatus, Apenninus, and Alpes. North shore of Mare Frigoris has 900-km radius from same center and is another major rim; both parts connect with Montes Caucasus. Lineations in Caucasus radial to same center. Montes Recta and Spitzbergensis and mare ridges define 670-km-diameter ring about a more northerly center; other connections are less certain and not given here. Lineate ejecta is north of Frigoris and south and southeast of Carpatus-Apenninus (Fra Mauro Formation); knobby ejecta (Alpes Formation) is closer to rings in all sectors, imbricate on south Apenninus flank (material of Montes Apenninus). Secondary craters abundant in all exposed sectors to 1,400-2,700 km from center, typically to 2,300 km.	Chapter 10; G93, B49, B63, HK62, SH62, WM71.
South Pole-Aitken	8.8	56° S., 180°	---	2,500, ---	---, 900	Huge massifs and scarps define single rim averaging 2,200 km in diameter, or two rings about 2,500 and 1,400-2,000 km in diameter.	Chapter 8; S78, W79.

TABLE 4.2.—*Probable and possible lunar basins at least 300 km in diameter*

[Basins named by convention described in table 4.1. All basins are pre-Nectarian (see chap. 8) except possibly Sikorsky-Rittenhouse (Nectarian?). Diameter refers to outer or only observed ring. References: B63, Baldwin (1963); B69, Baldwin (1969); C74-81, Cadogan (1974, 1981); HH77, Hawke and Head (1977b); MA81, Maxwell and Andre (1981); SH70, Stuart-Alexander and Howard (1970); WE77, Wilhelms and El-Baz (1977); WM71, Wilhelms and McCauley (1971); W79, Wilhelms and others (1979); W81, Whitaker (1981)]

Basin	Map symbol (pl. 3)	Center (lat, long)	Figure	Apparent diameter (km)	Evidence and status as basin	References
Pingré-Hausen	PH	56° S., 82° W.	1.9	300	Arcuate ridges adjacent to Mendel-Rydberg; possibly part of that basin	Pingré of HK62 and W79.
Sikorsky-Rittenhouse	SR	68.5° S., 111° E.	1.5	310	Craterlike rim and inner hills. Buried by Schrödinger deposits; possibly part of South Pole-Aitken basin.	B69; "unnamed A" of WM71; W79.
Amundsen-Ganswindt	AG	81° S., 120° E.	1.5	355	Craterlike rim and trace of inner ring. Buried by Schrödinger deposits; possibly part of South Pole-Aitken basin.	B69, W79.
Werner-Airy	WA	24° S., 12° E.	8.12	500	Indistinct low ridges bordered by depressions. Doubtful basin	B63, p. 195; named here.
Balmer-Kapteyn	BK	15.5° S., 69° E.	1.5	550	Concentration of plains inside and around rugged arcuate hills. Two rings may be exposed. Probable basin.	Balmer basin of MA81; renamed here.
Flamsteed-Billy	FB	7°-8° S., 45° W.	8.10	570	Semicircular arrangement of terra islands and part of Procellarum shore. Possibly part of Procellarum basin or craters.	WM71; named here.
Marginis	Ma	20° N., 84° E.	9.3	580	Terra arcs and outline of Mare Marginis. Possibly part of Crisium basin	WE77.
Al-Khwarizmi/King	AK	1° N., 112° E.	8.11	590	Mostly elevations of crater units; one distinct ridge in northwest sector.	WE77.
Insularum	In	9° N., 18° W.	8.10	600	Islands south of Montes Carpatus, deflections of Montes Carpatus, and coarse terra structure east of Sinus Aestuum. Possibly part of Procellarum.	WM71; "south of Imbrium" of HH77; named here for Mare Insularum.
Grissom-White	GW	44° S., 161° W.	8.15	600	Elevated terra ridges and mare concentrations. Possible external structure of Apollo.	Named here.
Lomonosov-Fleming	LF	19° N., 105° E.	8.13	620	Partly complete circle of ridges; concentration of plains. Probable basin	WE77.
Mutus-Vlacq	MV	51.5° S., 21° E.	8.12	690	Distinct massifs and ridges in south; concentration of plains. Probable basin.	W79.
Nubium	Nu	21° S., 15° W.	1.8	690	Conspicuous rim east of Humorum; arcuate margin of south-central terra. Probable basin.	SH70, WM71.
Tsiolkovskiy-Stark	TS	15° S., 128° E.	1.3	700	Isolated terra ridges and elevated crater rims. Indistinct	B69, WE77.
Tranquillitatis	Tr	7° N., 40° E.	1.1, 1.7	800	Arcuate mare-terra boundary, especially in west. Probable basin	SH70, WM71.
Fecunditatis	Fe	4° S., 52° E.	11.17	990	Three possible rings: arc, 990 km in diameter, in northwest sector; main mare/terra boundary, forming circle 690 km in diameter; and circular mare ridges. Southward extension of mare mostly inside 990-km-diameter ring. Probable basin.	SH70, WM71, WE77.
Procellarum	P1, P2, P3	26° N., 15° W.	8.10	3,200	Terra ridges, mare/terra boundary, mare ridges, influence on later geology (see chaps. 5, 6, 11-13).	W81; similar to Gargantuan basin of C74-81.

Because of the pivotal role played by investigations of Orientale both before and after the Apollo 16 mission, Orientale deposits are taken as models for others and are thoroughly described in this chapter. Similarities to the Orientale deposits appear repeatedly in the rest of the terrae and reveal the presence of basin deposits in all degrees of preservation from the fresh to the barely discernible.

Although all primary-impact excavations are surrounded by basically similar ejecta and secondary craters, their interior morphologies differ sharply from the small to the large end members of the impact series (figs. 3.2, 4.2, 4.3). Morphologic complexity increases most obviously in the simple-to-complex transition at diameters of about 20 km and again at diameters of about 300 km. Like the first, this second transition has exceptions. All unburied lunar impact cavities larger than 300 km in diameter display interior rings; however, some cavities smaller than 300 km also contain rings. Antoniadi (135 km diam) and Compton (162 km diam) each have a peak centered inside an inner ring. Schwarzschild (235 km diam) has a partial ring, and Milne (262 km diam) has several. Several craters display ringlike patterns of internal peaks (figs. 3.2E, 4.2C-E). Some tectonically modified craters appear to possess a concentric structure (see chap. 6). On the other hand, several partly buried circular depressions larger than 300 km in diameter do not display the expected second ring.

These deviations muddy the definition of the term "basin." All authors agree that cavities larger than 300 km in diameter that have concentric rings instead of central peaks are basins, but opinions differ about whether this definition should be based entirely on size or on the presence of a second ring regardless of size (Wood and Head, 1976). This chapter concentrates on excavations at least 300 km in diameter, and classifies those without observed interior rings as indefinite basins (table 4.2). The problem of nomenclature for transitional features smaller than 300 km is not resolved here.

The origin of rings is among the most important unsolved problems of lunar geology. Orientale is also critical to this problem. Its four major rings are the most regularly spaced on the Moon (fig. 4.3). The 930-km-diameter Montes Cordillera ring, partly scarplike and partly topped by massifs, bounds the basin (figs. 4.1, 4.4). Two jagged massive rings, collectively called Montes Rook, have diameters of 480 and 620 km and lie inside the Cordillera. The fourth, innermost ring, 320 km in diameter, consists partly of scarps and partly of massifs. A low fifth ring, 1,300 km in diameter, outside the Cordillera has also been suggested (Hartmann and Kuiper, 1962). Each ring is 1.3 to 1.5 times larger than the one inside it, an interval that has been generalized as the square root of 2 (Hartmann and Wood, 1971). Although they are more regular than they are typical, as I will show below (compare figs. 4.4, 4.5), the Orientale rings are generally taken as the model for the spacing and formation of other rings in most of the literature, including the section below entitled "Origin of Rings."

Central to the controversy is the identity of the boundary of excavation. Some investigators, including myself, believe that the excavation cavity of basins which corresponds to the excavation cavity of simple craters is approximately the present basin itself, that is, the depression. The *topographic basin rim* that bounds most of the basin depression thus corresponds to the rim crest of simple craters. The topographic rim of Orientale is the Cordillera. In this hypothesis, the topographic rim bounds the excavation cavity and marks the inner boundary of the most voluminous ejecta. Many other investigators, however, believe that the primary, original cavity was smaller than the present basin. They identify the boundary of the excavation cavity with one of the interior rings or place it at a radial position formerly occupied by a raised ring and now by a trough. Stratigraphic questions that could be better answered by knowing the identity of the boundary include: (1) the position of the lithologic boundary between the ejecta and the interior deposits, (2) the volume of excavated material, (3) the ejecta's source depth in the terra crust or mantle, (4) the amount of melt generated by the impact, and (5) the position of the facies change between primary ejecta and ejecta of the secondary craters.

ORIENTALE EXTERIOR

By John F. McCauley

General features

The exterior ejecta deposits of Orientale form nearly concentric zones similar to the concentric facies of crater materials. In proportion to cavity size, the extent of morphologically similar facies is the same for both basins and large craters (Moore and others, 1974; Morrison and Oberbeck, 1975). A continuous deposit extends outward an average distance of 450 km, or one basin radius, from Montes Cordillera. Beyond this deposit, clustered secondary craters accompanied by their ridgelike and planar ejecta form a more varied terrane of discontinuous deposits. In some sectors, dense concentrations of secondaries extend 900 km (two basin radii) from the Cordillera; secondaries also extend beyond 1,850 km in at least one raylike string (fig. 4.6; Wilhelms and others, 1978, p. 3752), much like secondary clusters along the bright rays of young craters. The most conspicuous deposits of all types are shown on plate 3, and the continuous and discontinuous deposits are distinguished.

The continuous deposits closest to the rim and the discontinuous deposits at greater distances are collectively designated the Hevelius Formation, the first stratigraphic unit whose relation to Orientale was recognized (McCauley, 1967a, b). This definition was based on telescopic observations and was originally applied to a tract centered around lat 2° N., long 68° W., that is now considered part of the outer facies. Scott and others (1977) later expanded the definition to include the continuous, more coarsely textured blanket farther west, which had also been detected telescopically (McCauley, 1964b). The Hevelius and other Orientale materials, except the plains materials also discussed here, collectively constitute the Orientale Group (McCauley, 1977; Scott and others, 1977).

Inner facies of the Hevelius Formation

The continuous deposits of the Hevelius Formation form an enormous doughnut-shaped zone, elongate north-south and from 300 to at least 600 km wide, outside Montes Cordillera (pl. 3; figs. 1.9, 4.1, 4.4A-G; Scott and others, 1977). This zone is difficult to delineate west of Orientale because of the obliquity and poor quality of the Lunar Orbiter 5 photographs of that area (pl. 3B; fig. 4.5). However, this part of the Hevelius, called by Scott and others (1977) the inner facies of the Hevelius Formation, is generally the most distinctive unit of the Orientale Group. It displays abundant attributes of a massive ejecta deposit that acquired a distinctive textural imprint during flow along the rugged prebasin surface.

The inner, continuous Hevelius is texturally characterized by elongate ridges and irregular troughs generally radial to Orientale; figure 4.4C shows a typical locality. The ridges are curvilinear and mutually crosscutting; some ridges have nearly streamlined forms, and others appear ropy or filamentary. These landforms clearly resulted from surface flow of deposits differentially affected by obstacles. Some of the larger ridges have lobate edges that resemble the fronts of viscous flows of lava or debris (fig. 4.4D). Numerous narrow linear grooves are also visible, particularly in the lower and smoother parts of the deposits. Another characteristic are the myriads of vague elongate depressions and chains of depressions, visible only in good photographs, that are overlain by ridged and grooved materials (fig. 4.4C). The shapes, subdued appearance, and distribution pattern of these depressions indicate that they are secondary craters, partly or completely buried by slightly later flows of ejected deposits. Although these buried secondary craters contribute a negative-relief pattern, the generally larger positive forms dominate the scene. Their distribution pattern and topography argue for strong centrifugal depositional movements within the materials that make up the inner, continuous Hevelius.

The inner facies of the Hevelius Formation generally forms a continuous ground cover that deeply buries or subdues large primary craters on the pre-Orientale surface (figs. 4.1, 4.4F, G). The depth of burial of these craters generally decreases outward from the basin. Estimates based on the depth of crater burial indicate that the inner facies of the Hevelius decreases in thickness from 3 to 4 km near Montes Cordillera to no more than several hundred meters at its outer edges (Moore and others, 1974). However, these estimates assume that the buried craters had the d/D ratios of fresh craters, which compose only part of any random population of lunar craters. Theoretical models of ejecta thickness (McGetchin and others, 1973) are also very uncertain in view of the number of simplifying assumptions that underlie them (Solomon and Head, 1979, p. 1669). Calculations are hampered by uncertain cavity volumes, uncertain degree of bulking of the ejected material (Croft, 1978), and the difficulties of scaling from one gravity field to another (Gault and others, 1975). Basin deposits are visibly irregular and lobate (pl. 3), and are not equally thick all around a basin at a given radial distance. Thus, their thicknesses are poorly known but are probably less than 4 km.

The well-developed radial or semiradial trends prevail within most of the inner facies of the Hevelius except in a sector of about 50° of arc east of the basin (figs. 4.4E-G). This anomalous region has little radial texture and is part of the concentric facies of Moore and others (1974), who noted an outward gradation here and elsewhere from this concentric facies to more nearly radial material (top, fig. 4.4F). The inner Hevelius in the anomalous area (fig. 4.4F) has stubbier ridges and lobes than in more typical localities (figs. 4.4C, D, G). Large patches of smooth to rolling weakly lineate terrain are also present, some of which show numerous small ridges and troughs concentric with the basin. In addition, the depressions formed by buried secondary craters are not so large and do not contribute so significantly to the surface texture as in more typical parts of the inner Hevelius (figs. 4.4C, D, G). The long axes of many of these depressions are also concentric rather than radial to the basin. Prebasin craters are not so deeply buried as in the more strongly radial parts of the Hevelius Formation at comparable distances from the basin. All indications in this sector point to ejection of less material at lower velocities than elsewhere.

Small patches of what has been described as the fissured facies and interpreted as impact melt (Moore and others, 1974) occur within the inner facies of the Hevelius. These smooth but locally fractured deposits lie in depressions and contrast sharply with the topography of their surroundings (fig. 4.4B); they also are somewhat darker than the adjacent material. Their patterns of fissures resemble those caused by shrinkage of a coherent material. These small deposits (areas less than a few tens of kilometers on a side) are widely scattered, and are included in the inner facies of the Hevelius on the map by Scott and others (1977) and in the inner Orientale unit on plate 3. They resemble the smaller impact-melt pools on the flanks of craters (figs. 3.23, 3.27, 3.36).

In mapping the Orientale region at 1:5,000,000 scale, Scott and others (1977) distinguished a transverse facies of the Hevelius Formation, also included here in the inner Orientale zone (pl. 3). This transverse facies consists of patches of closely spaced, ropy-looking, intertwined ridges and intervening troughs arranged concentrically around Orientale. Typical dimensions of these patches are about 30 by 60 km. They are common on the far sides of crater floors (figs. 4.4F–H) 300 to 600 km from Montes Cordillera. They seem to be more abundant on the east side of the basin, but this apparent asymmetry may be due to the poor photographic coverage of the west side (fig. 4.5). These subparallel features, which have been termed "deceleration dunes" (McCauley, 1968), are good evidence for the surface flowage of Orientale ejecta. In the crater Riccioli, transverse ridges are banked up on the far side of the crater floor, beyond which a transversely textured flow lobe, 100 km long, emanates from a breech in the distal crater wall (fig. 4.4H). The influence of topography on several patches of less well defined transversely textured materials is less obvious (figs. 4.4F–H).

The inner boundary of the Hevelius Formation has been perceived both as sharply demarcated by Montes Cordillera (Baldwin, 1974a; Wilhelms and others, 1977; Hodges and Wilhelms, 1978; Murray, 1980) and, alternatively, as irregular, gradational, and not everywhere coincident with Montes Cordillera (McCauley, 1977; Scott and others, 1977). Much of the Cordillera scarp is draped by the Hevelius Formation (figs. 4.4A–F). Locally, however, sharp breaks in texture mark the inner contact of the Hevelius along the steeper parts of the Cordillera scarp, particularly where this scarp is shadowed on photographs (fig. 4.4E).

Outer facies of the Hevelius Formation and secondary craters

The outer facies of the Hevelius Formation (Scott and others, 1977), like the outer parts of all ejecta blankets, exhibits a greater variety of surface textures than the inner facies. The facies boundary is gradational but readily detectable on good photographs as the limit of the large radial lobate ridges typical of the inner facies of the Hevelius (figs. 4.4G, H). This contact is reproducibly mappable within a few tens of kilometers on the nearside but is hard to detect on the farside. Subdued, elongate, sinuous to irregular secondary-crater depressions typical of the inner facies are replaced in the outer facies by crisp secondary craters that are not overrun by ejecta. Buried prebasin primary craters are also more conspicuous in the outer than in the inner facies. Large patches of smooth plains also become abundant near the facies boundary and interrupt the radial depositional patterns. The boundary is analogous to that between the continuous inner deposits and the herringbone-patterned fields at one crater radius from the rim crest of craters.

The tract near the type area of the Hevelius Formation (McCauley, 1967a) contains faintly lineate, thin, and discontinuous material that mantles the wall of the crater Hevelius nearest to Orientale and grades to smooth plains on the crater floor. The adjacent high ground shows numerous relatively straight grooves and ridges, in contrast with the larger and more sinuous topographic forms of the continuous Hevelius. Some of these smaller and straighter grooves locally resemble mantled grabens but were probably created during emplacement of the Hevelius Formation (loc. 2, fig. 4.4H). Obvious depositional patterns are subordinate or absent, and this "lineated terra material" (Wilshire, 1973) or "grooved facies" (Moore and others, 1974) has been considered to be partly of erosional origin. However, depositional origins and erosional or secondary-impact origins are difficult to distinguish where the grooves are separated from terrain with more diagnostic features (Wilhelms, 1980).

In many places, the outer facies of the Hevelius Formation is rolling to hummocky and furrowed, particularly near the crater Crüger and eastward to Oceanus Procellarum (fig. 4.4H). This terrane generally shows only vague radial lineations and elongate depressions, as well as numerous subdued pits. Although a secondary-impact origin for this hilly and pitted terrain was mentioned (McCauley, 1973), a volcanic origin was preferred before the Apollo 16 results were obtained (Wilhelms and McCauley, 1971; McCauley, 1973; Wilshire, 1973). Continued studies have shown that the strange-textured terrane is laterally continuous with Hevelius deposits similar to those shown in figure 4.4F (Scott and others, 1977). The terrane probably consists of a pre-Orientale surface mantled by Orientale ejecta that had less radial momentum or velocity than other parts of the blanket.

Although some deposits and grooves are of primary-ejecta origin and others of doubtful origin, the outer zone of Orientale materials mapped here (pl. 3A) is clearly characterized by the products of secondary impacts. Obvious secondary craters large enough to map at 1:5,000,000 scale were considered by Scott and others (1977) as a separate facies of the Hevelius Formation, but secondary craters are not mapped separately from deposits here (pls. 3, 6–8). Secondary craters of Orientale were described by Offield (1971), Wilhelms and McCauley (1971), Moore and others (1974), and Wilhelms (1976) (see chap. 3). The Orientale secondaries overlap in chains and clusters that are scaled-up equivalents of the secondaries described around Copernicus by Shoemaker (1962b). Diameters of individual craters are as large as 28 km, but most are from 7 to 20 km (Wilhelms and others, 1978), typically about 10 km (figs. 3.9, 4.4C, G, H). Although some distant clusters have downrange herringbone patterns (fig. 3.9), these patterns are subordinate to straight ridges nearly radial to the basin (figs. 4.4G, H). Bowl-shaped to elongate gougelike features are typical of the Orientale-secondary field. Their downrange sides show wavy or partly braided deposits that partly or completely obscure the underlying topography. Many individual craters have well-developed rims and appear to be moderately fresh; others are very subdued and have been mistaken for prebasin craters (figs. 4.4G, H; Wilhelms, 1976). Orientale secondaries, including some that extend great distances from the basin (fig. 4.6), are abundant beyond a large lobe of the Hevelius Formation southeast of the basin (pl. 3; fig. 4.4G). Secondaries are especially large and abundant north of the basin (fig. 3.9) and on the poorly photographed west side (fig. 4.5), but are rare in the concentrically textured sector east of the basin (figs. 4.4E–G).

The outer facies of the Hevelius Formation, like the outer parts of all impact-crater deposits, represents a progressively thinner to more discontinuous cover that may locally extend to great distances from Orientale without leaving a perceptible imprint on the topography. The problem of identifying the outermost limit of ejecta is intrinsic to mapping all impact and explosion craters. Here, the outer contact is placed where no further linear or Orientale-radial structure can be observed on Lunar Orbiter 4 H-frames (pl. 3). Beyond this contact, only isolated secondary craters and crater clusters are recognizable as related to Orientale (figs. 4.6, 7.7, 10.5). Farflung Orientale ejecta could have reached points thousands of kilometers from the basin (Chao and others, 1975; Moore and others, 1974). It may even be concentrated in the furrowed and pitted terrain antipodal to Orientale near Mare Marginis (fig. 4.7; Schultz and Gault, 1975a; Wilhelms and El-Baz, 1977).

Relation between ground-flow deposits and secondary craters

As discussed above, numerous vague and irregular depressions, gouges, and clusters of radially lineate, buried secondaries occur within both the outer and inner facies of the Hevelius Formation (figs. 4.4C, D, F, G). Some chainlike secondary craters buried by the continuous Hevelius extend to the Cordillera ring (fig. 4.4D), as do similar chains at the double-ring Schrödinger basin (fig. 1.5). Surficial flow textures characteristic of the Hevelius Formation overlie these chains; thus, the chains were probably formed by early-arriving clots of ejecta traveling in relatively low ballistic trajectories from the basin, and were later overrun by a slower-moving ground surge (Morrison and Oberbeck, 1975; Oberbeck and Morrison, 1976). Morrison and Oberbeck (1975) suggested that most of the other radial and concentric depositional ridges of the inner facies of the Hevelius were created by interference and interaction of ejecta, as are the herringbone patterns around craters. Such an origin would imply that the ridges contain little primary basin ejecta. This process probably did form many of the ridges in the outer parts of the secondary field (figs. 3.9, 4.4G, H), but not the long sinuous ridges and grooves of the inner facies of the Hevelius, which were deflected by obstacles in a way that indicates surface flowage of thick ejecta (fig. 4.4G). Furthermore, the burial of the secondaries indicates that primary ejecta constitutes most of the inner facies; any secondary ejecta present there must have

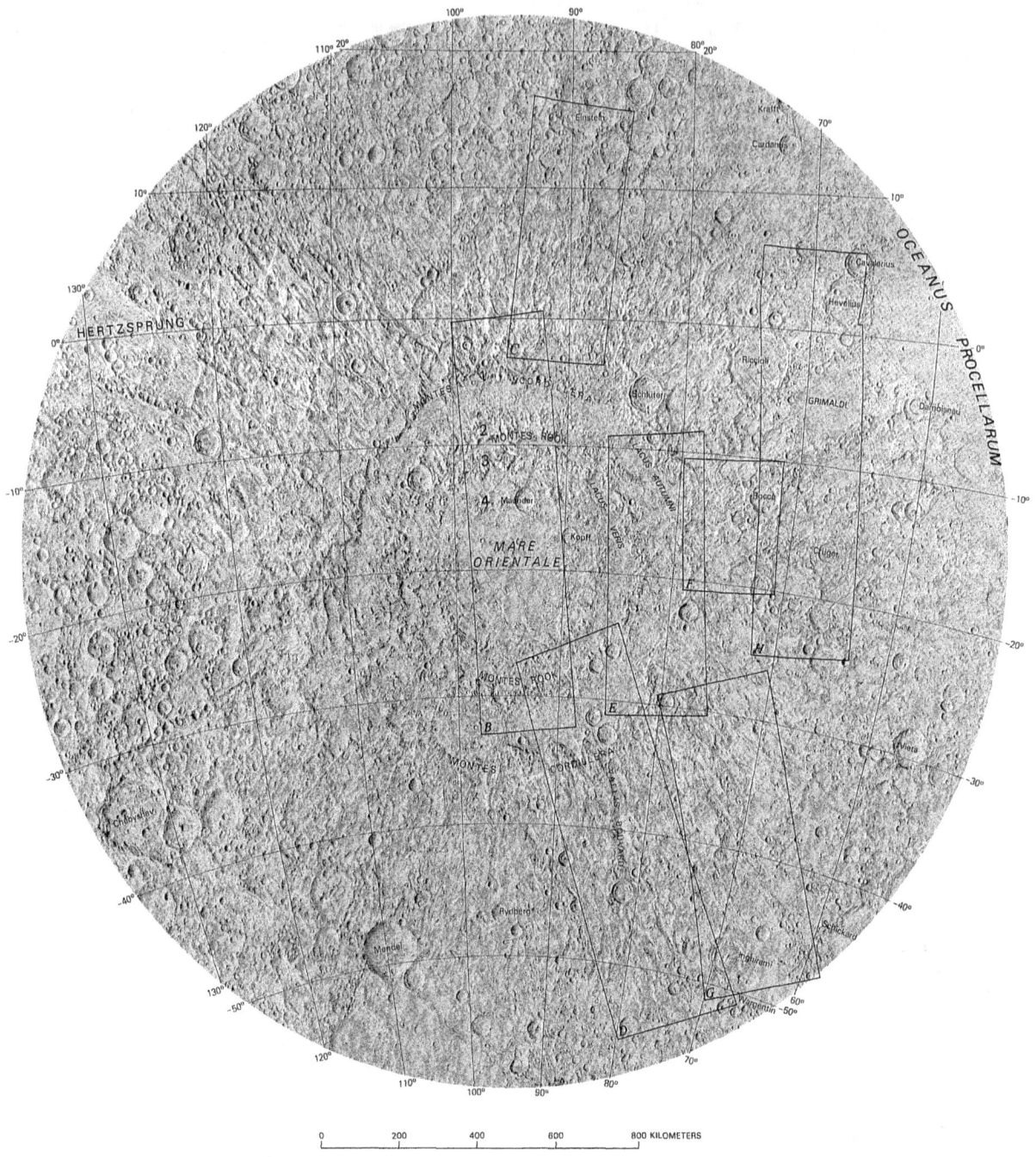

FIGURE 4.4.—Orientale basin and periphery.
  A. Basin structure. 1, Cordillera ring (930 km); 2, outer Rook ring (620 km); 3, inner Rook ring (480 km); 4, innermost ring (320 km). After U.S. Geological Survey (1978).

  B. Central Orientale basin, showing the Hevelius Formation (1), fractured impact melt (2), overlap of the Montes Rook Formation onto the Hevelius (3), Cordillera ring overlain by Hevelius (4) and Montes Rook (5) Formations, typical knobby Montes Rook Formation (6), outer Montes Rook ring massifs (7), inner Rook ring (8, 9a, 9b), massifs of 320-km-diameter ring (10), fissured domical Maunder Formation burying part of 320-km-diameter ring (11), and level Maunder Formation cut by grabens (12). Inner Rook ring massifs in south (9b) are widely separated by the Maunder Formation. Perched former higher stands of level part of the Maunder indicate subsidence (13). Orientale-basin materials are exposed from beneath shallow mare near center of basin (14). Circular structure of outer Rook ring (15) is possibly derived from prebasin crater. Crater Maunder overlies basin and mare (16). Orbiter 4 frame H-195.

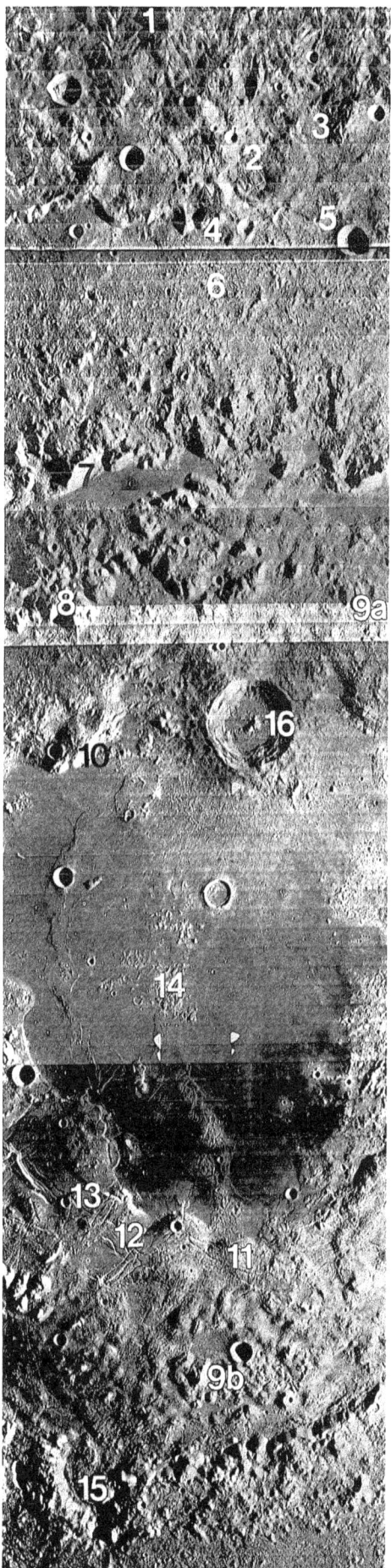
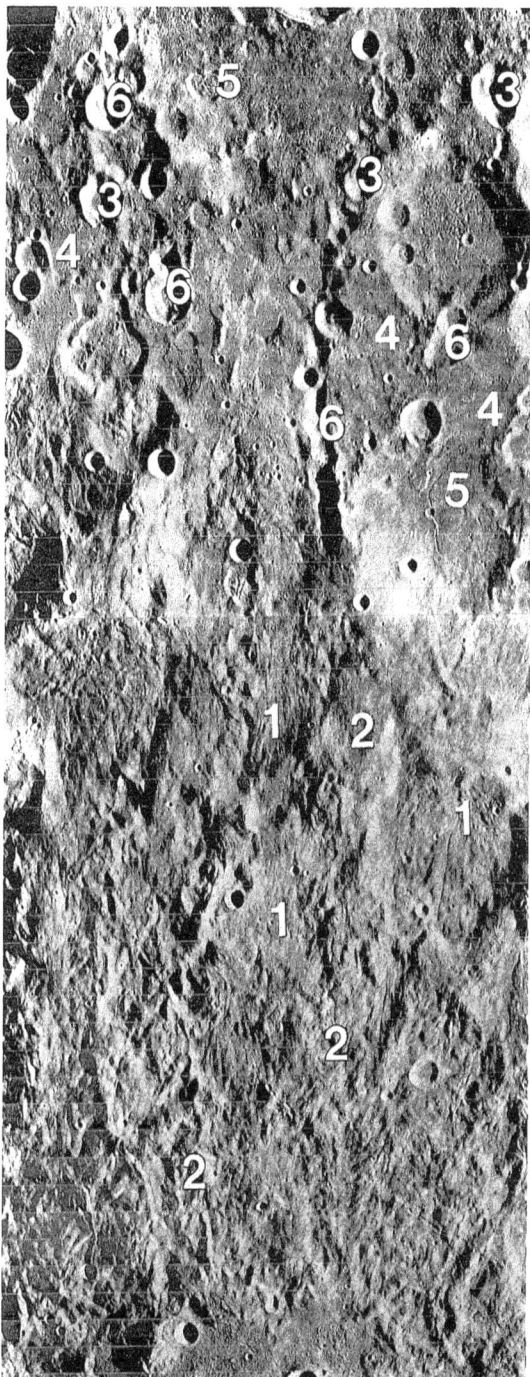

C. Typical continuous inner facies of the Hevelius Formation (bottom half of photograph). Many ridges are flow features (1); depressions are buried secondary craters (2). Large secondaries (3) emerge among smooth deposits (4) farther from basin. Fissures indicate possible melt deposits (5). Complex landforms were generated by multiple near-simultaneous secondary impacts (6). Overlaps with figure 3.9A. Orbiter 4 frame H-188.

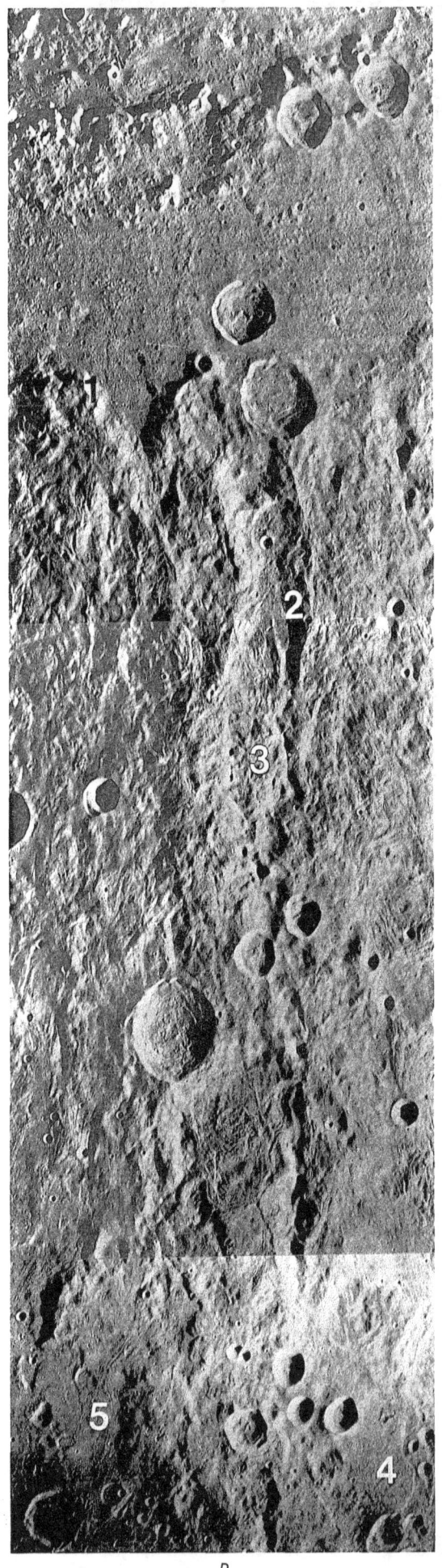

D. Cordillera ring (1) south of basin, overlain by chain of Orientale-secondary craters, Vallis Bouvard (2). Hevelius Formation mantles the secondary chain and extends outward as a thick blanket (3). Lobes of nonlineate (4) or leveed (5) material continue the Hevelius deposit outward. Orbiter 4 frame H–186.

E. Montes Rook Formation, decelerated by Cordillera ring into dunelike (1) or radially textured (2) deposits. Montes Rook or Hevelius Formation, with radial textures, overlies Cordillera scarp (3). Maunder Formation is at left (4), inside outer Rook ring (5). Fissured ejecta (6) is probable impact melt. Orbiter 4 frame H–181.

F. Radial (1) and concentric (2) textures of the Hevelius Formation directly east of Cordillera ring (3). Flow of the Hevelius was blocked by crater Rocca and piled up as dunelike mounds (4). Orbiter 4 frame H–173.

G. Transition between inner (1) and outer facies of the Hevelius Formation. Outer facies includes secondary craters (2), ridged deposits of secondaries (3), and planar deposits (4). Flow of the Hevelius was diverted by now-buried obstacles (5) and blocked by crater Inghirami (6). Some secondary craters filled by deposits look deceptively old (7). Orbiter 4 frame H–172.

H. Type area of the Hevelius Formation, centered at 1.7° N., 68.5° W. in crater Hevelius (1). Linear troughs radial to Orientale are partly filled by deposits (2). Deposits were decelerated by rim of crater Riccioli (3), but lobe of textured deposits extends northeastward of rim (4). Lobe of plains deposits inside Grimaldi (5), playout of decelerated deposits (6; same as at loc. 4, fig. 4.4F) into planar deposits (7), and pitted terrain (8) are also related to Orientale. Postbasin features: grabens (9), crater Crüger (10), and other mare-filled depressions (11). Mosaic of Orbiter 4 frames H–168 (bottom) and H–169 (top).

F

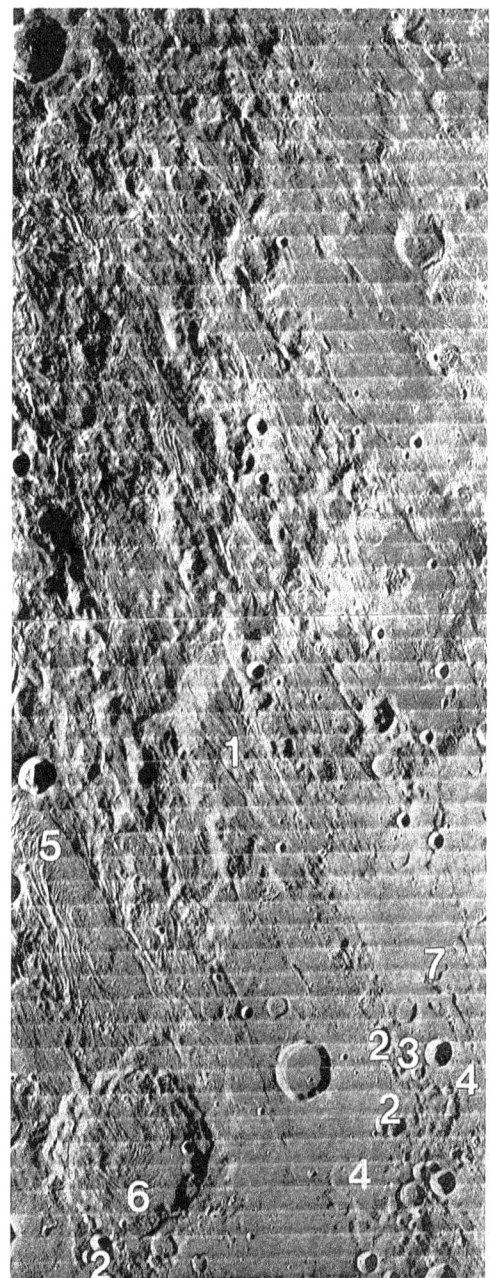

G

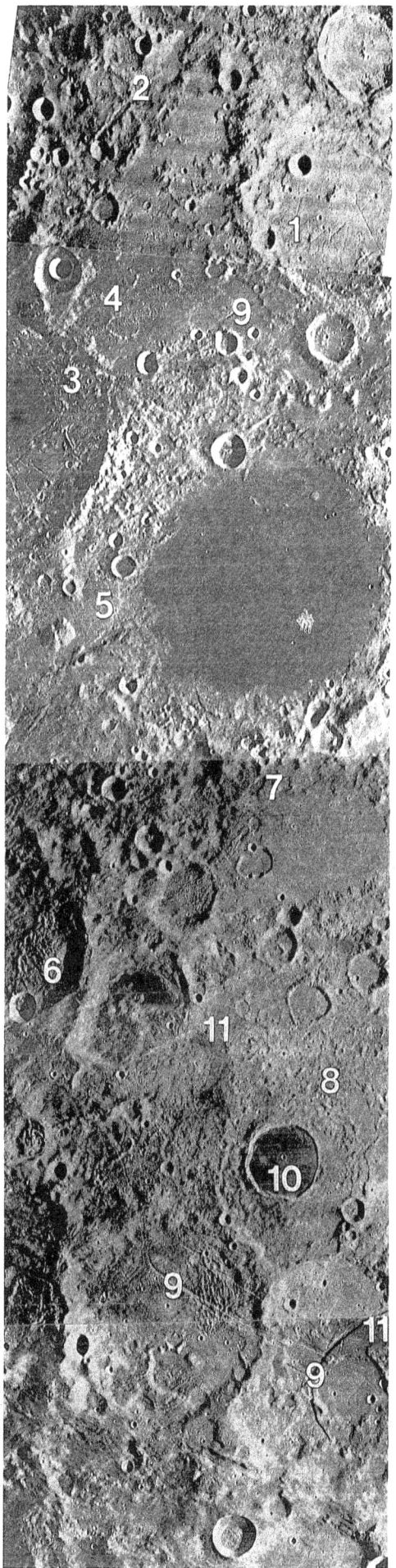

H

been overwhelmed by the primary ejecta. This question of the proportions of primary versus secondary ejecta in basin deposits is explored in discussions of the Apollo 14 and 16 samples in chapter 10.

The photographic evidence, therefore, suggests an early stage of secondary cratering followed by a ground-hugging debris surge that first inundated and then mixed with the earlier debris. The last-formed Orientale features, however, may also be secondary craters. Unusually numerous, crisp circular craters, 10 to 15 km in diameter, which at first glance appear to be primary craters younger than Orientale, are present in the west-limb region of the Moon (figs. 4.1, 4.4, 4.5). These may be very late secondary craters excavated by projectiles that were ejected from the Orientale basin at high angles and that almost reached escape velocity (as proposed by Shoemaker and others, 1968, p. 40-44, for craters on the rim of Tycho). Circular craters, presumably formed by high-velocity impacts, should be widely dispersed over the Moon not only from Orientale but also from all the older basins (figs. 4.6, 4.7).

Plains

The Hevelius Formation is gradational with the terra plains. These plains are abundant in the transition zone between the inner and outer facies of the Hevelius and occupy additional large tracts in the outer facies. These relations have been much discussed since the discovery by Apollo 16 of impact breccia in similar light-colored plains in the central lunar highlands. Volcanic origins have been discounted in all but a few areas (see chaps. 2, 10; Neukum, 1977). The principal currently competing hypotheses are an origin as primary ejecta and as ejecta from secondary craters. Relations at Orientale provide the best photogeologic clues to the plains' origin.

Some nearly planar deposits have a distinguishable relief that is an important clue to their origin. Ridges of the inner facies of the Hevelius grade into faint rises in the plains of the transition zone (figs. 4.4G, H). Flat-topped lobes with planar or faintly textured surfaces (mapped as a separate nonlineated facies by Wilhelms and others, 1979) seem to have oozed from the more rugged Hevelius (fig. 4.4D; Eggleton and Schaber, 1972); some of these smooth deposits are leveed (loc. 5, fig. 4.4D; Hodges and others, 1973; Moore and others, 1974). Fluidlike flow of the semiplanar material is indicated. A substantial content of impact-melted rock may be responsible, although fluidlike flow of completely clastic debris is not excluded. Some textured deposits are superposed on secondary craters (figs. 4.4D, F). Little secondary ejecta can be contained in textured deposits with these stratigraphic relations.

By extension, fluidlike emplacement of material segregated from the textured ejecta blanket is suggested for the completely planar deposits that have ponded in depressions (Eggleton and Schaber, 1972). This origin is even likely for such deposits as those in the crater Wargentin (fig. 1.9), whose apparently thick fill was long considered to be evidence for lunar volcanism. The light-colored deposits in Wargentin lie just beyond the zone where the continuous Hevelius grades into the plains. Although Wargentin may have been partly filled by volcanic materials, the present surface deposit is probably of Orientale ejecta (Schultz, 1976b, p. 92).

Emplacement as secondary-crater ejecta, as hypothesized for all light-colored lunar terra plains by Oberbeck and others (1974, 1975), is likely for plains that lie far from the inner facies of the Hevelius but adjacent to clusters of secondary craters (fig. 7.6). The dispute between a primary and secondary origin for terra plains, therefore, is one not of exclusive genesis but of relative contributions by the two mechanisms.

The plains illustrate that the appearances and distributions of lunar geologic units may be consistent with more than one genetic interpretation. A further example is in the south-central highlands of the nearside, where careful mapping showed that plains and mantling deposits become less conspicuous from west to east (Cummings, 1972; Mutch and Saunders, 1972). The mappers interpreted the differences as due to variations in the thickness of the deposits, which they considered to be a function of the duration of volcanism. An impact interpretation, more consistent with current understanding of the lunar terrae, is that the plains are ejected materials which thin away from their source, Orientale.

ORIENTALE INTERIOR

By John F. McCauley

All or almost all the materials of the Hevelius Formation lie outside the topographic basin rim of Orientale (pl. 3). Additional Orientale deposits occur mostly in the basin interior or on the Cordillera rim close to the basin. Two major units are named the Montes Rook Formation and the Maunder Formation (fig. 4.8); these formations complete the Orientale Group (table 4.3; McCauley, 1977; Scott and others, 1977). Morphologic and, presumably, genetic analogs occur in craters (figs. 3.2D, 3.32-3.35) but are less conspicuous than in Orientale. Their analogs in other basins are less obvious than those

FIGURE 4.5.—West and northwest sectors of Orientale basin (O) on farside (compare fig. 4.1), showing radial pattern of continuous inner facies of the Hevelius Formation (he) and secondary craters (sc). CS, Coulomb-Sarton basin; L, Lorentz basin. Orbiter 5 frame M-16.

TABLE 4.3.—*Geologic units of lunar basins*

[Fm, Formation; n/n, recognized but not named; n/r, not recognized. Most units of other basins are named informally. References: H66, Hackman (1966); H74, Head (1974c); M74, Moore and others (1974); S71, Stuart-Alexander (1971); S77, Scott and others (1977); WM71, Wilhelms and McCauley, 1971; W79, Wilhelms and others (1979)]

Location	Morphology	Orientale (references)	Imbrium (references)	Nectaris (references)	Interpretation
Inner basin	Wavy, fissured	Maunder Fm (S77); corrugated facies (H74).	n/r	n/r	Impact melt, buckled.
Do	Level, fissured	Maunder Fm (S77).	Apennine Bench Fm(?) (H66).	n/r	Impact melt, fluid, little deformed.
Shelf and inner flank.	Knobby	Montes Rook Fm, knobby facies (S77); domical facies (H74).	Alpes Fm (WM71); Fra Mauro Fm, hummocky (H66).	Hilly material (WM71)	Late low-velocity high-angle ejecta.
Rim and other rings.	Massifs	Montes Rook Fm, massif facies (S77).	Rugged material (WM71)	Rugged material (WM71)	Upthrust and covered by ejecta.
Inner flank	Concentric, wreathlike.	Concentric facies (M74)	Fra Mauro Fm, hummocky (H66); material of Montes Apenninus (WM71).	n/r	Primary ejecta pushed over rim.
Flank	Radial, coarse	Hevelius Fm, inner facies (S77); radial facies (M74).	Fra Mauro Fm (H66, WM71)	Janssen Fm (S71, WM71)	Primary ejecta, flowed on surface.
Flank depressions.	Wavy, fissured	Fissured facies (M74)	n/n	n/r	Ejected impact melt.
Flank, transition zone.	Concretic, dunelike.	Hevelius Fm, transverse facies (S77).	Hilly material (WM71)	n/r	Primary ejecta, decelerated flow.
Do	Wide valleys	n/n (Vallis Bouvard type)	n/n	Material of crater chains and clusters (WM71).	Low-angle secondary impact.
Transition zone	Lobate, smooth or leveed.	Hevelius Fm, nonlineate member (W79).	n/n	n/r	Primary ejecta, fluidlike flow.
Do	Subcircular flat-floored craters.	n/n	Irregular-crater material (WM71).	Undivided crater material (WM71).	Secondary impact, filled by surface flow.
Transition, outer zones.	Narrow grooves	Grooved facies (M74)	n/n	n/n	Eroded by secondary impact or surface flow.
Do	Subcircular bowl-shaped craters.	Satellitic-crater material (WM71, W79); Hevelius Fm, secondary-crater facies (S77).	Material of crater clusters and chains (WM71).	Material of crater chains and clusters (WM71); satellitic crater material (W79).	Secondary impact.
Do	Radial, fine, linear, or braided.	Hevelius Fm, outer facies (S77, W79).	Terra material (WM71)	Janssen Fm (S71, WM71)	Secondary-impact ejecta.
Do	Planar, smooth	Plains material or facies	Plains material	Pitted-plains material (WM71).	Primary or secondary ejecta, ponded.

of the exterior deposits because of mare flooding and other modifications. Thus, Orientale provides a unique opportunity for study of the interior materials of basins.

Montes Rook Formation

The Montes Rook Formation (or the knobby facies of that formation, Scott and others, 1977) is characterized by multitudinous smooth, equidimensional to elongate knobs, about 2 to 5 km across, set in a matrix of smooth to rolling materials (figs. 4.4*B, D, E*). In places, the formation contains weak lineations, mostly radial but locally concentric with the rings (fig. 4.4*E*); much of it has a ropy or imbricate texture. Most of the formation occurs between the Montes Cordillera and Montes Rook rings, although several large patches are exterior to the Cordillera (fig. 4.4*B*; Scott and others, 1977). Islands of lineate material, either belonging to the Hevelius Formation or intrinsic to the Montes Rook Formation, lie within the confines of the Montes Rook Formation (fig. 4.4*E*). Vague circular structures, less conspicuous than the large primary craters buried by the Hevelius Formation (fig. 4.4*B*), may be buried by the Montes Rook Formation (figs. 4.4*A, B–E*; Scott and others, 1977).

In places, the Montes Rook Formation appears to embay the coarsely lineate Hevelius (fig. 4.4*B*), and so the Montes Rook is the later unit in the basin sequence. Like the Hevelius, the Montes Rook material must have left the crater at least locally because it is so clearly draped over the Cordillera scarp (fig. 4.4*B*). Thus, it cannot be the product of inward slumping of the crater walls along the Cordillera scarp immediately or shortly after final excavation of the crater, as proposed by Head (1974c), but is a distinct type of ejecta. The Cordillera scarp is draped by ejecta, either of the Hevelius or of the Montes Rook Formation, over most of its visible extent (figs. 4.4*A–F*); thus, the scarp predates final ejecta deposition and cannot be the product of postcratering collapse as is shown on the hypothetical cross sections by Hartmann and Yale (1968), Head (1974c), and Howard and others (1974).

The coarse, knobby, weakly lineate texture of the Montes Rook Formation indicates a much smaller radial-ejection component than that of the strongly lineate Hevelius. The Montes Rook Formation probably consists of relatively coherent material ejected in relatively low-velocity, high-angle trajectories, possibly in a distinct second pulse of ejecta. Steep late-forming plumes of ejecta are observed inside the main ejecta cone in some experimental craters formed in layered target materials (Oberbeck, 1975, 1977; Andrews, 1977; Piekutowski, 1977). Blocky materials are lifted steeply from the lower, more cohesive layer and dropped onto the crater and onto the ejecta already emplaced by the earlier, larger plume. In this sense, they constitute an "overturned flap" displaying inverted stratigraphy, as on the rims of simple craters (McCauley, 1977); however, the two ejecta types do not form coherent layers on basin rims. A similar relation between two deposits has been mapped at the Ries crater, Germany (see chap. 3), which also formed in a layered target. The moderately to intensely shocked suevite, derived from the crystalline basement, was lofted and then dropped on unshocked Bunte Breccia, derived moments earlier from the overlying sedimentary beds. In both the experimental and natural craters, the earlier material exits at low angles, and the later at higher angles.

Similar phenomena but a different cause were suggested by Murray (1980), who compared the Montes Rook knobs to the coarse hummocks of probable impact melt on the floors of fresh complex craters (figs. 3.2*D*, 3.32–3.35) and suggested that they were flipped upward and slightly outward as the basin or crater floor rebounded sharply during or after the main cratering. These suggestions entail quarrying of the Montes Rook Formation from a deeper crustal layer than the Hevelius Formation, or from the lunar mantle (McCauley, 1977; Scott and others, 1977; Hodges and Wilhelms, 1978).

Scott and others (1977) mapped a massif facies of the Montes Rook Formation in addition to the knobby facies just described. The massifs of Orientale and other basins typically consist of rectilinear to equidimensional blocks, ranging from a few kilometers to as much as 100 km in length (figs. 4.4*B, D, E*). The smoothness of their surfaces (except at very high resolution) indicates slope steepness. Their backslopes are typically gentler than their inward-facing slopes (fig. 4.4*B*). Heights are estimated to be as much as 6 km above the plains at their base (Head, 1974b; Howard and others, 1974; Moore and others, 1974); this relief is greater than that between the Cordillera ring and the adjacent shelf. Scott and others (1977) considered the massif facies to consist of a stratigraphic ejecta deposit superposed on a structural uplift, as for similar massifs of Imbrium (Carr and others, 1971; Wilhelms and McCauley, 1971). This interpretation is consistent with the photogeologic appearance of fresh massifs and with the stratigraphy of crater rims. The proportions of ejecta and structurally uplifted material remain unknown.

FIGURE 4.6.—South-central terra of nearside, including secondary-impact craters of distant basins.
  A. Arrow in crater Clavius (C; 225 km, 48° S., 15°W.) points to Orientale basin, centered 2,000 km to northwest; small sharp craters around arrow are probably secondary to Orientale. Arrow in Longomontanus (L; 145 km, 49.5° S., 22° W.) points to Imbrium basin, centered 2,600 km due north; large overlapping craters around arrow are secondary to Imbrium (Saunders and Wilhelms, 1974). Similar craters on west and north rim of crater Maginus (M; 163 km, 50° S., 6° W.) also are probably secondaries of Imbrium. Right-hand arrow points to Nectaris basin, centered 1,450 km away; three craters above arrow and degraded group below end of shaft are possibly secondary to Nectaris, although lineations are not strictly radial to that basin. Mosaic of Orbiter 4 frames H-119, H-124, and H-136 (right to left).

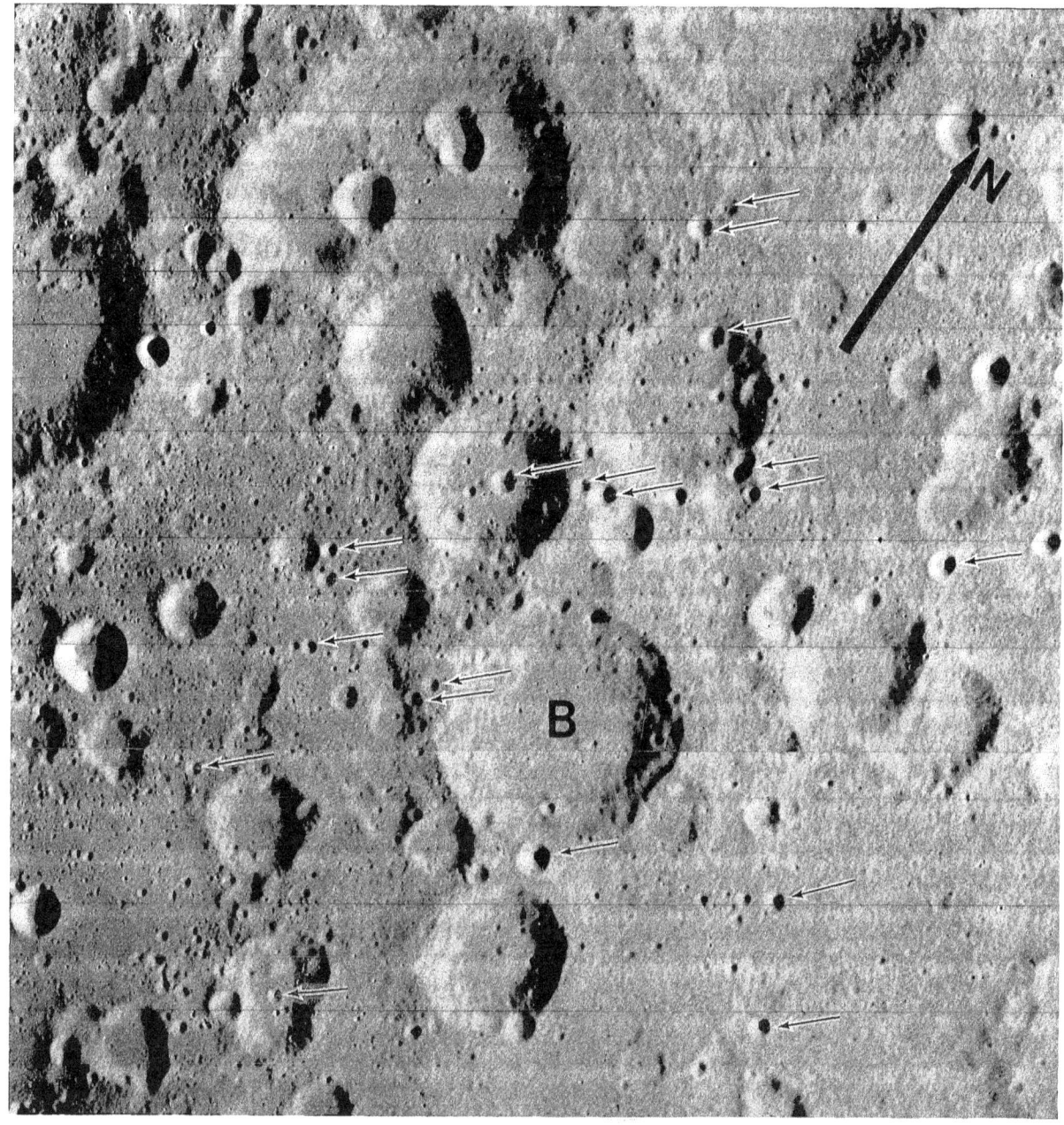

B. Area just east of figure 4.6A. Arrows point toward Orientale center and touch fresh-appearing craters that are probable Orientale secondaries. B, crater Baco (70 km, 51° S., 19° E.). Orbiter 4 frame H–100.

Maunder Formation

The Maunder Formation of the Orientale Group is named after the unrelated large impact crater Maunder on the north floor of the central Orientale basin (fig. 4.4B). This formation, like others of the Orientale Group, has been described under various names (table 4.3). The Maunder consists of smooth to rolling plains of intermediate albedo, many of which are closely fissured or fractured. As far as observed, the Maunder lies between Mare Orientale and the outer Rook ring and extends the peaks of the inner Rook ring (figs. 4.4B, E; 4.8).

The fractures are a secondary, superposed property of the unit and thus not a proper part of the definition of a rock-stratigraphic unit (see chap. 7), but they nevertheless characterize it; they led to such descriptions as "corrugated" and "crackly." These fractures are crudely concentric or radial to the basin (fig. 4.4B). Most fractures are open, downward-tapering V-shaped gashes, whereas others are larger, flat-floored complex grabens. Many fractures radiate from local centers, and others are aligned along the axes of broad gentle elongate domes, which also characterize the Maunder terrain (fig. 4.4B). The fractures are more abundant inside the inner Rook ring than close to the outer Rook ring. Other, larger radial and concentric grabens that are more pervasive probably were formed by later basin subsidence (pl. 5; chap. 6).

Texturally, the Maunder Formation is a scaled-up equivalent of the materials that cover the floors of such fresh unfilled large craters as Tycho and Copernicus (Howard, 1975; Howard and Wilshire, 1975). Though once interpreted as volcanic (Schultz, 1976b, p. 70–77), these floor materials are more convincingly interpreted as impact melts, on the basis of a comparison with terrestrial analogs and stratigraphic analysis (see chap. 3). The thickness of the Orientale melt is difficult

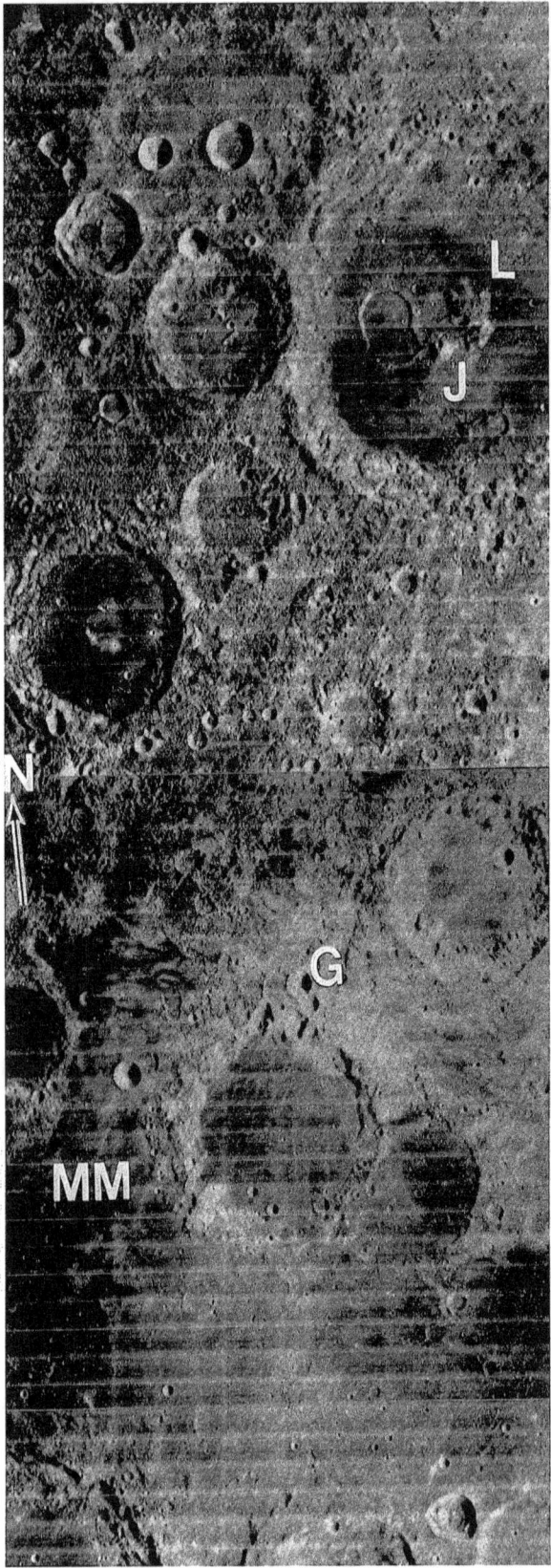

FIGURE 4.7.—Region of Orientale antipode; arrow indicates antipode (20° N., 85° E.). Most of terra is pitted in manner apparently characteristic of antipodes of young basins. MM, Mare Marginis, with bright swirls superposed; G, bright crater Goddard A (12 km), possible source of swirls; J, crater Joliot (143 km); L, mare-flooded fringe of secondary craters of Lomonosov. Mare material in quadrant below and right of Goddard A is among oldest mare units (see chap. 10). Orbiter 4 frame H–17.

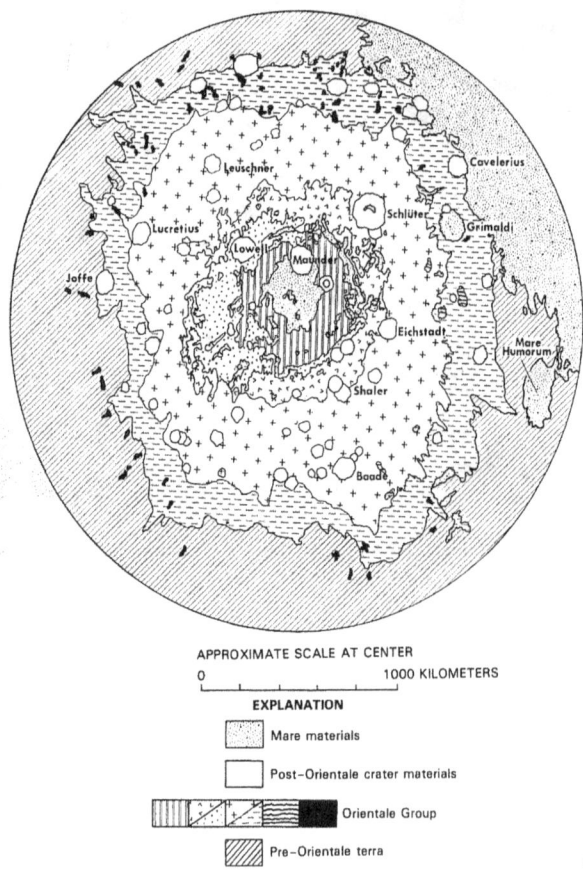

FIGURE 4.8.—Geologic map of Orientale basin, showing major units discussed in this volume. Units of the Orientale Group, from left to right, in "Explanation" are: Maunder Formation (double vertical bars), Montes Rook Formation (check marks, knobby ejecta; dots, massifs), Hevelius Formation (crosses, inner facies; dashes, outer facies), transversely structured parts of the Hevelius Formation (small areas with wavy lines), and some of the many secondary craters (black). From McCauley (1977, fig. 1; see also Scott and others, 1977).

to ascertain; Head (1974c) based an estimate of about 1 km thick on extrapolations from early work on terrestrial craters. The draping and mantling characteristics of much of the Maunder Formation suggest that its material behaved in a fluid manner. Perched shorelines locally visible (fig. 4.4B) suggest that it subsided to its present position from some higher topographic level. Sinuous rilles like those of maria (chap. 5) may mark local flowage of the melt (loc. 13, fig. 4.4B; Greeley, 1976).

The emplacement of the melt and the formation of the fracture pattern are believed to have been nearly contemporaneous with formation of the elongate ridges and domes, which are more commonly radial or concentric with the basin than equidimensional (fig. 4.4B). Mapping of excavations within experimental explosion craters shows that similar structures are compressional anticlines and synclines that formed late in the cratering sequence (Roddy, 1976; McCauley, 1977). If the domes in Orientale also formed late, the Maunder Formation may have been cracked by their uplift. The geometry of the cracks is also consistent with draping on preexisting topography (Head, 1974c), which would be their origin if the folding ended before the melt solidified.

TOPOGRAPHY AND GRAVITY

Favorably illuminated photographs (fig. 1.9) show that the elevations of the concentric troughs of the Orientale basin decrease inward and that Mare Orientale has the lowest surface. The summits of Montes Cordillera and Montes Rook appear to differ less in elevation than do the adjacent troughs. Absolute elevations are less certain because Orientale is not covered by photogrammetric-quality photographs. Preliminary measurements based on Apollo laser altimetry of the basin's north shelf and on profiles obtained from telescopic and Lunar Orbiter photographs suggest that Mare Orientale lies 2.5 to 5.5 km below, and parts of the rings and exterior terrain rise the same distance above, the 1,738-km average radius of lunar terra surfaces (relative to the center of figure) (Howard and others, 1974; Scott and others, 1977; Head and others, 1981; S.S.C. Wu, written commun., 1982). If the mare-basalt fill is less than 1 km thick (Howard and others, 1974), the top of the impact melt in the basin center lies about 3 to 6 km below the 1,738-km datum and about 6 to 11 km below the exterior.

Stairstep structure is also evident in stereoscopic photographs of the more heavily modified Crisium basin, which is a less representative basin but which lay more favorably under the Apollo 15 and 17 ground tracks (fig. 4.9). In the terra rim surrounding Mare Crisium, interring troughs decrease in elevation inward toward the mare. The mountain summits do not decrease in elevation so much (fig. 4.9); that is, the overall relief between a given summit and the adjacent trough increases inward, as it does at Orientale. Absolute elevation differences are also similar—Mare Crisium lies about 4 km below the 1,738-km datum (Sjogren and Wollenhaupt, 1976) and 3 to 6 km below the basin exterior (fig. 4.9). Relative and absolute elevations of another measured basin, Smythii, are remarkably like those of Orientale, despite these basins' substantial difference in relative age (Strain and El-Baz, 1979).

Perturbations of spacecraft orbits have revealed some important features of the Moon's gravity structure. The terrae are gravitationally bland at short wavelengths (see chap. 6; Bills and Ferrari, 1977; Ferrari, 1977). The relatively short-wavelength gravity anomalies with the largest amplitudes are positive, and are called *mascons* (for "mass concentrations"; Muller and Sjogren, 1968). The gravity anomaly of the Orientale interior is positive, but if the basalt of Mare Orientale is removed from the gravity models, a ringlike negative anomaly remains (Scott, 1974; Sjogren and Smith, 1976). The gravity structure of a given basin as a whole requires two bodies—the surficial mare basalt and a raised part of the mantle (fig. 4.10; Wise and Yates, 1970; Phillips and others, 1972; Scott, 1974; Bowin and others, 1975; Sjogren and Smith, 1976). Each basin impact removed an unknown but significant thickness of crustal material (Scott, 1974). The denser mantle then rose to compensate isostatically for the lost mass. On the basis of models for Serenitatis (fig. 4.10), the mantle may be about 20 to 25 km higher under Orientale than in nonbasin regions (Sjogren and Smith, 1976); the same value was found for the Grimaldi basin (Phillips and Dvorak, 1981). The negative anomaly, also found in many basins and fresh craters (Ferrari, 1977; Dvorak and Phillips, 1978), suggests that compensation was incomplete (Sjogren and Smith, 1976). Further alteration of the gravity structure by the later addition of mare basalt resulted in a superisostatic mascon in several basins (see chaps. 5, 6; Baldwin, 1968; Phillips and others, 1972; Scott, 1974; Solomon and Head, 1979, 1980).

Small positive anomalies displayed by Montes Cordillera (Sjogren and Smith, 1976) and Montes Apenninus of the Imbrium basin (Ferrari and others, 1978) indicate (1) a loading of the crust by material removed from the basin and (2) a lithosphere sufficiently strong to support much of this added load.

These gravity models are central to later discussions of basin-excavation depths, mare-basalt localization (chaps. 5, 11, 12), and tectonism (chap. 6). The crust-mantle discontinuity (lunar Moho) probably resembles a series of domelike swells, whose diameter and elevation are approximately proportional to diameters of the overlying basins and which are superposed where basins are superposed. The largest known basin, alternatively called Gargantuan (Cadogan, 1974, 1981) or Procellarum (pl. 3; table 4.2; Whitaker, 1981), presumably overlies the largest mantle uplift and the thinnest crust. This basin was probably responsible for thinning of the crust from the computed average of about 75 km to the observed value of 45 to 60 km under southern Oceanus Procellarum (chap. 1). This thinning had major consequences for later lunar geologic evolution (see chaps. 5, 6, 8–12).

ORIGIN OF RINGS

Introduction

The origin of basin rings has caused considerable interest and controversy because of its importance in impact mechanics and stratigraphy, and because it has proved to be such an obdurate problem (see reviews by Hartmann and Kuiper, 1962; Baldwin, 1963; Mutch, 1970; Hartmann and Wood, 1971; Howard and others, 1974; Moore and others, 1974; Brennan, 1976; Wood and Head, 1976; McCauley, 1977; Wilhelms and others, 1977; Hodges and Wilhelms, 1978; and Croft, 1981). The identity of the boundary of excavation is particularly significant. Discussions of basin-forming mechanics are generally based on knowledge of cratering mechanics derived from small craters, which can be generated experimentally and relatively easily modeled (see chap. 3). Complex craters and large terrestrial astroblemes (1) appear always to have been shallow and not to owe their complex internal structure to the collapse of a deep cavity, (2) contain peaks that seem to have formed by compression of a central zone during subhorizontal centripetal movement, (3) are terraced as a result of this centripetal movement, and (4) are surrounded by deposits and concentrated secondary craters whose extent remains approximately proportional to crater diameter (chap. 3). Some, but not all, of these properties can be extrapolated to basins.

Genetic models

The general increase in complexity with increasing size of impact cavities (figs. 4.2, 4.3, 4.11) has suggested to some investigators that the peaks and terraces of craters and the rings and shelves of basins differ in degree of development because of scale differences. Alternatively, they may have been formed by different processes. Properties of the target material, such as strength and layering, or properties of the impactor, such as impact angle, velocity, size, or shape, may be relatively more important in the larger events. The physical effects of large and small impacts may differ fundamentally in ways that are not yet clear.

A. *Large-basin models.* The main, topographic rim is the boundary of excavation (Baldwin, 1974a; Chao and others, 1975; Oberbeck, 1975; Wilhelms and others, 1977; Hodges and Wilhelms, 1978). Interior rings could have formed in two alternative ways:

1. *Nested craters.* In these models, rings are the rims of subcavities of the main excavation (Oberbeck, 1975; Wilhelms and others, 1977; Hodges and Wilhelms, 1978). Nesting is observed in small craters in the lunar regolith (Oberbeck and Quaide, 1967, 1968; Quaide and Oberbeck, 1968; Oberbeck, 1977), and probably in the Ries (Chao, 1977; Hörz and Ostertag, 1979) and other large terrestrial craters or ringed basins (Phinney and Simonds, 1977; Hodges and Wilhelms, 1978). The initiator of the dual or multiple cavities in experimental and natural terrestrial craters is a discontinuity in the target materials between an overlying weak layer and a less easily cratered substrate. Separate ejecta plumes may be ejected from the nested craters.

2. *Floor uplifts.* Various mechanisms based on cratering mechanics have been proposed. The morphologic similarity between central peaks and rings consisting of peaks (fig. 4.3B) suggests that the rings are expanded peaks (Hartmann, 1972a; Baldwin, 1974a; Head, 1974c; McCauley, 1977; Hodges and Wilhelms, 1978). Centripetal movements, akin to those hypothesized for complex craters and observed in large chemical-explosion craters (Jones, 1977; Roddy, 1976, 1977), may form ringlike structures in the innermost basin (McCauley, 1977).

A more violent, oscillatory mechanism akin to that observed in temporary craters formed in liquid has also been suggested. This "tsunami" mechanism, induced by the seismic energy of impact, was first proposed for external rings (Baldwin, 1949, 1972; Van Dorn, 1968, 1969). Baldwin (1974a) then reapplied it to rings of the cavity interior, which is more likely to be fluidized by the intense and sustained shock pressures intrinsic to impacts of large bodies (Baldwin, 1963, p. 163–169; 1974a). Dence and Grieve (1979; Grieve, 1980) recently revived the tsunami mechanism as an explanation for rings, on the basis of the high uplifts of severely weakened subcrater material that they inferred in terrestrial craters; and Murray (1980) independently proposed it on the basis of several photogeologic observations similar to those mentioned in this chapter. The oscillatory model is favored here and is described additionally below (fig. 4.12).

B. *Small-basin models.* An inner ring or the former position of an inner ring marks the boundary of basin excavation.

1. *Late formation.* Passive "megaterracing" is thought to widen the topographic basin by inward movement and rotation of a large annular tract of exterior terrain after excavation has ceased (McCauley, 1968; Mackin, 1969; Hartmann and Yale, 1968; Hartmann and Wood, 1971; Dence and Plant, 1972; Short and Forman, 1972; McGetchin and others, 1973; Gault, 1974; Head, 1974c, 1977; Howard and others, 1974; Moore and others, 1974; Head and others, 1975; Schultz and Gault, 1975a; Croft, 1981). According to this hypothesis, the rear slip surface of the terrace cuts the rim flank outside the excavation boundary and carries that boundary inward. The topographic basin rim is the site of the slip at the rear of the "megaterrace," and the migrated excavation rim becomes an inner ring.

2. *Contemporaneous formation.* A similar geometry results from centripetal movement and rotation during the excavation process

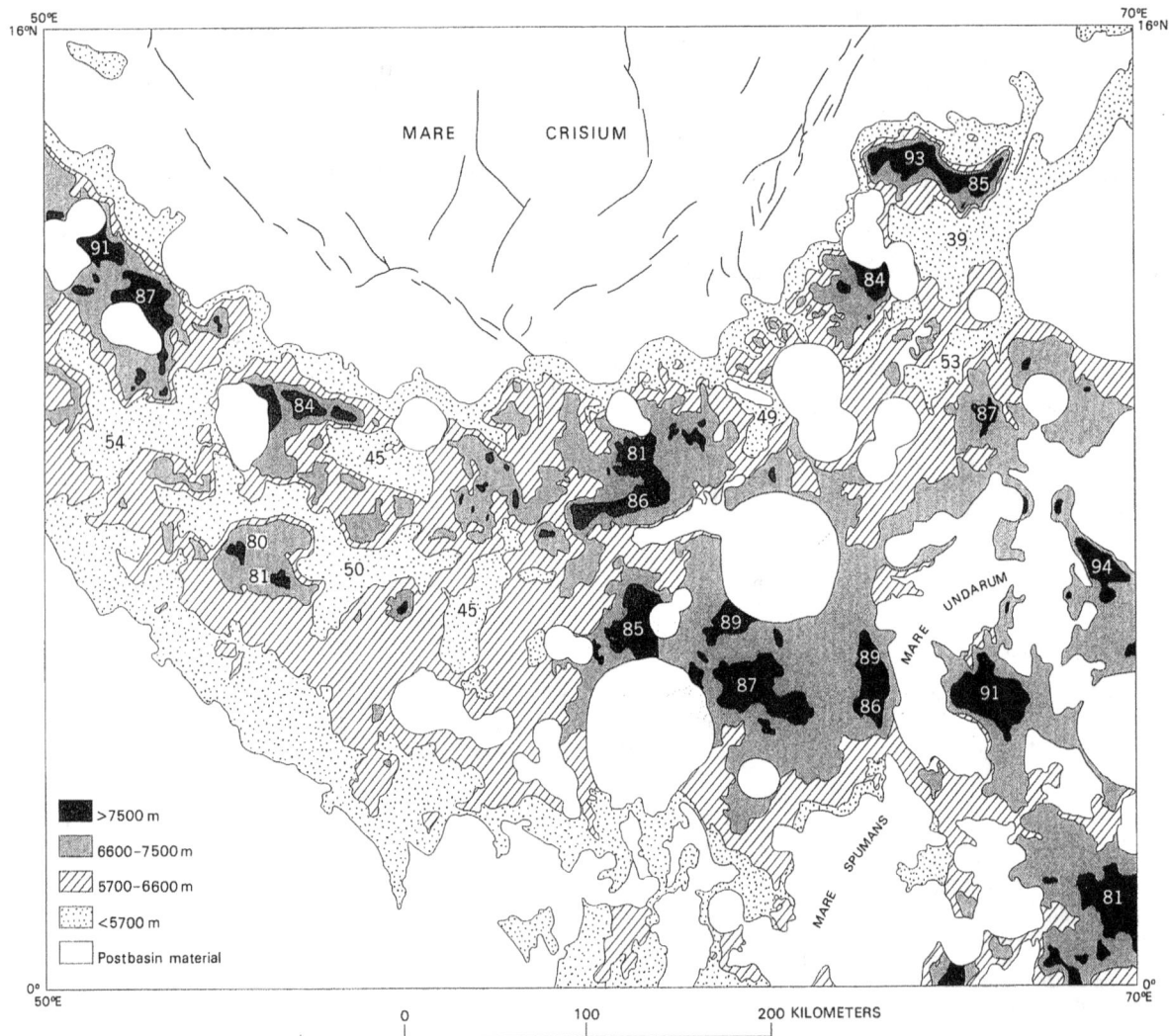

FIGURE 4.9.—Topographic map of southern Crisium basin. Arbitrary vertical datum is 1,730,000 m from Moon's center. Elevations of summits (white numbers on black) and of troughs (black numbers) are shown in hundreds of meters above this datum. Mare Crisium, 3,800 to 3,900 m; Mare Fecunditatis, 5,000 to 5,200 m; Mare Spumans, 5,500 to 5,700 m; Mare Undarum, 6,400 to 6,700 m. Postbasin materials include crater and mare deposits (Olson and Wilhelms, 1974). Topography generalized from LM 62 (1st ed., Sept. 1978), U.S. Defense Mapping Agency.

(McCauley, 1977; Scott and others, 1977). This more active process is observed in large chemical-explosion craters, where centripetal movement opens concentric exterior cracks and depresses an annular shelf between the excavation boundary and the remnant topographic rim. The rim and false rim are overlain by ejecta.

Discussion

1. The topographic rims of such double-ring basins as Schrödinger so closely resemble those of craters in morphology and in their demarcation of the inner boundary of craterlike deposits they must be equivalent in origin (Wilhelms and others, 1977; Hodges and Wilhelms, 1978). That is, the rims approximate the boundary of excavation, except as the excavation cavity was widened by visible craterlike terraces. Thus, the 160-km-diameter inner ring is a feature of the crater interior and not of the boundary of excavation. The inner ring cannot bound the Schrödinger excavation because such a cavity would be smaller than many craters, which have no second ring outside their rims.

2. Similarly, I believe that the topographic rims of larger basins, such as the Cordillera of Orientale, are also the boundaries of excavation. This more controversial conclusion is based on: (a) The abrupt demarcation of the Hevelius Formation at the Cordillera; (b) the elevated massifs along parts of the Cordillera; (c) the absence or ghostlike form[4.1] of prebasin craters inside the Cordillera, in contrast to their abundance outside the Cordillera (Baldwin, 1974a); (d) the rimlike morphology, including steep inner and gentler outer slopes and both scalloped and linear segments (Murray, 1980); (e) the qualitative similarity of both the continuous and discontinuous deposits to crater deposits in overall morphology and zoning; and (f) the proportional extent of the Orientale ejecta in comparison with crater deposits (Wilhelms and others, 1977; Murray, 1980, fig. 7). Montes Cordillera extend as a major well-defined ring around three-fourths of the basin and are not a minor, external structure. Similarly, the 570-km-diameter ring of Hertzsprung and the 600-km-diameter ring of the southern part of Humboldtianum are the boundaries of those excavations (figs. 4.3O, Q, 4.11; table 4.1). Ejecta extends an average of one basin

---

[4.1]The presence of ghostlike circular forms inside the Cordillera (Scott and others, 1977) does not obviate the interpretation that the Cordillera is the excavation boundary. If the excavations are shallow, surficial material could be stripped away without destroying the deeper structure of prebasin craters.

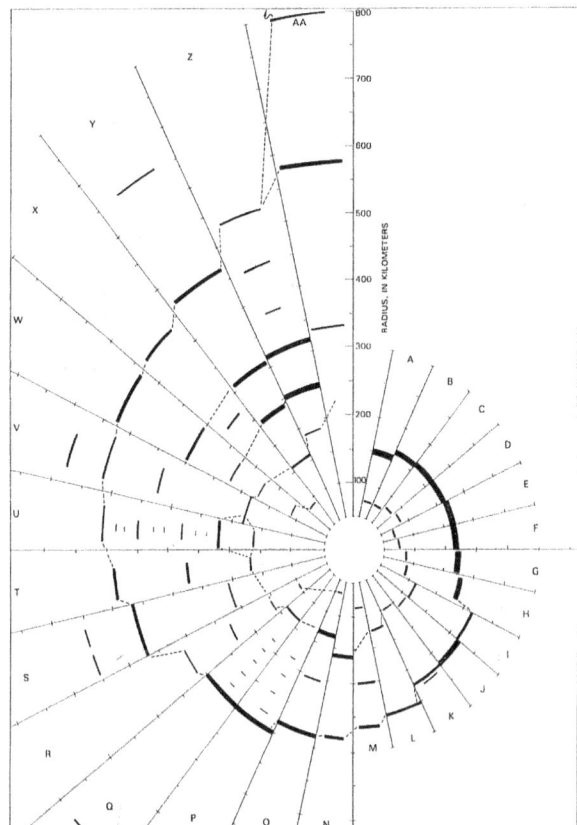

FIGURE 4.11.—Spacing of observed basin rings and ring segments, plotted by radial distance from basin center (after table 4.1, with some approximate additions from pl. 3); see table 4.1 for letter symbols. Length of arc within each sector is approximately proportional to observed circumferential extent of ring; width of line is approximately proportional to prominence. Homologous rings in different basins are connected by dots. Double-ringed structure persists through Korolev (J) or Apollo (L). Innermost and intermediate series begin in Hertzsprung (O) or, possibly, in poorly observed Coulomb-Sarton (M). These new homologous series expand in larger basins through Orientale (Y). Extensive new structure (not completely described here) appears in Crisium (Z) and Imbrium (AA).

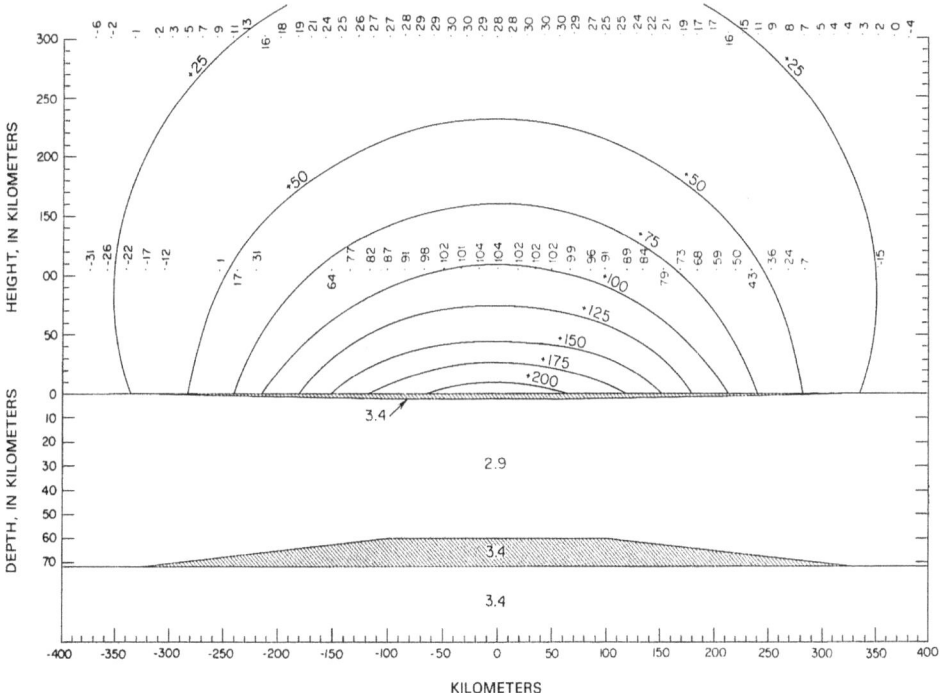

FIGURE 4.10.—Two-body gravity model of Mare Serenitatis (Bowin and others, 1975). A thin plate on surface (mare material) and a mantle uplift each have densities of 3.4 g/cm³ (shaded); terra-crustal density, 2.9 g/cm³. Values in space above mare are observed free-air anomalies from two spacecraft orbits (in milligals); curved lines are isoanomalies of computed free-air gravity.

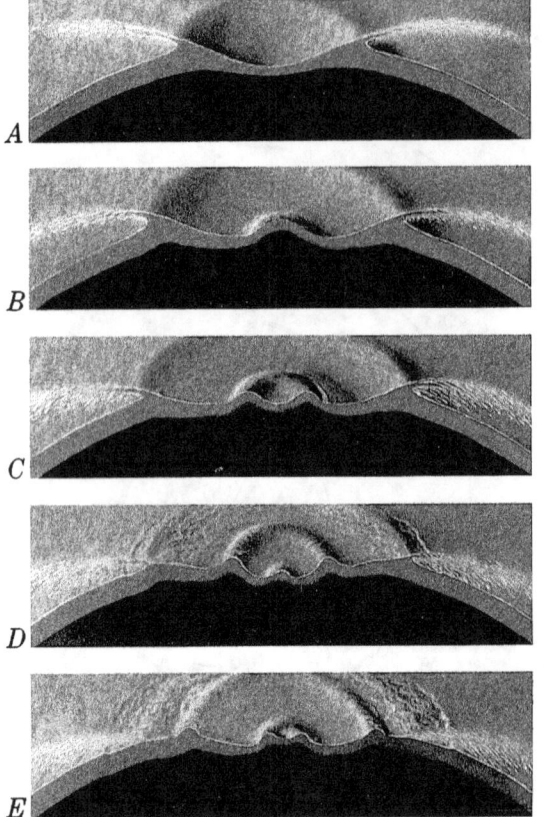
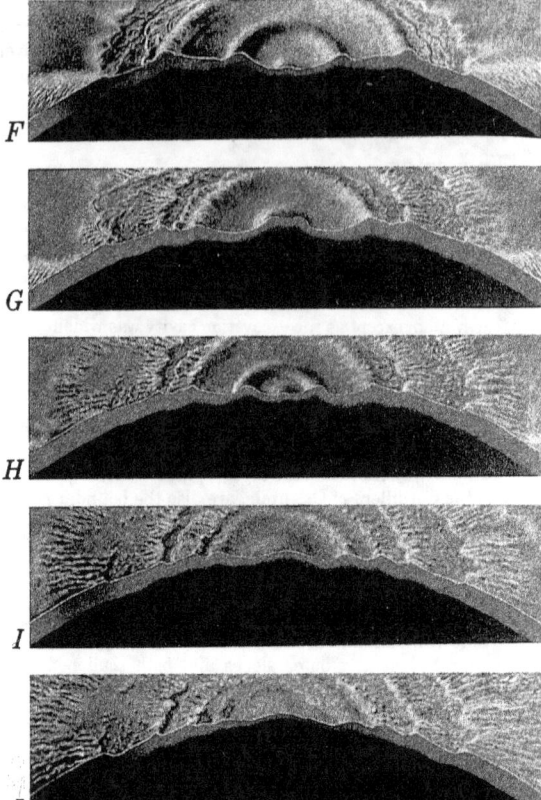

FIGURE 4.12.—Model of basin formation favored here. Drawing by Donald E. Davis, courtesy of the artist.
  A. Early stages of growth, resembling simple crater. Target rock is greatly weakened by shock wave.
  B. Cavity grows and becomes shallower relative to depth, ejection angle decreases, and central uplift begins as in complex crater.
  C. Central uplift collapses to form a temporary double-ringed basin.
  D. Center rebounds, as in a waterdrop crater.
  E. Wavelike motion enlarges inner ring shown in C and D, new innermost ring forms from central peak in D, boundary of excavation becomes lower, and ejecta blankets basin walls and exterior.
  F. Excavation boundary reaches final size, and two inner rings and a central uplift have formed.
  G. Ejecta continues to move beyond excavation; continued oscillation of interior causes surge of ejecta from intermediate ring and enlarges innermost ring.
  H. Exterior deposits are mostly in place, two rings inside excavation boundary have frozen, but innermost ring continues to grow, and center rises once again.
  I. Innermost ring has grown to final size, and central uplift relaxes into low broad mound.
  J. Final configuration.

diameter from each of these rings, which are also more rimlike than any others in the basin. Although parts of the interior rings (265 km diam in Hertzsprung and 300 km diam in Humboldtianum) are rimlike, these rings cannot bound the excavations because the similar-size rim crests of Bailly (300 km diam) and Schrödinger (320 km diam) are not surrounded by ring systems or such extensive ejecta (Wilhelms and others, 1977; Murray, 1980, p. 273).

3. Continued, passive terracing after the completion of ejecta deposition is precluded by superposition of the Montes Rook Formation on the Cordillera (McCauley, 1977). Ring formation, whatever its cause, must be an active process driven by the cratering flow.

4. Considerable evidence suggests that basins are shallow excavations (fig. 3.20; Head and others, 1975). First, they presumably were never deeper relative to their diameters than were craters. Copernicus, which is exactly $1/10$ the diameter of Orientale, now has a d/D ratio of $1/25$. According to the hypothesis presented earlier, fresh craters were not much deeper when formed than they are now (Pike, 1980a). If Orientale's original d/D ratio was $1/25$, its original, transitory depth was 37 km. This value is close to that obtained by adding the estimated present depth of 6–11 km (exterior terrain to top of impact melt) to the crustal thinning implied by the mantle uplift of 20–25 km. (Its present d/D ratio is between $1/155$ and $1/85$, a decrease that indicates substantial rebound since basin formation; furthermore, the negative gravity anomaly would be much greater than observed if no rebound had occurred.) Moreover, these ratios apply only to the central basin inside the 320-km-diameter ring; the shelves, which occupy much more area, are several kilometers shallower. That the cavity-wall slopes were gentle during excavation as well is suggested by the massive surface flows of ejecta and the linear, chainlike arrangement of secondary craters close to the rim (fig. 4.4D)—both features indicating low-angle ejection. The apparent paucity or absence of mantle material in the returned samples (chap. 8) furthermore suggests that not even the 3,200-km-diameter Procellarum basin was excavated through the 75-km-thick crust—implying an original d/D ratio of less than $1/43$ for this largest lunar basin.

5. The major interior rings (for example, Montes Rook) probably were formed by some form of active floor uplift. In morphology, they resemble central peaks more closely than they do terrace lips. Their gentle outer slopes and steep inner slopes resemble those of crater rim crests, but the jagged massifs and peaks of their most rugged parts more nearly resemble central peaks. The Rook rings resemble the inner, rugged ring of Schrödinger much more closely than they do the Schrödinger rim.

6. Even if rings and central peaks are both formed by uplifts, the uplift process probably differs. Although a transition from craters to basins is commonly stressed, the interior structure of most craters 200 to 300 km in diameter is poorly developed (fig. 4.2), and peaks actually decrease in absolute width in craters larger than about 70 km in diameter (fig. 3.2E; Murray, 1980, p. 282). Therefore, central deformation may be less intense in intermediate-size impacts than in typical complex craters and typical basins. This lull suggests different deformational processes (Hale and Head, 1979; Murray, 1980).

7. Ring uplift may occur during a wavelike deformation of the cavity floor, as follows (fig. 4.12). Highly shocked, effectively fluidized material of the cavity rebounds, probably because of general involvement of the whole floor in wavelike motion to compensate for the cavity. A smaller central zone then partly collapses when

gravity exceeds the inertial force and rock strength. If basin growth stops at this point owing to insufficient energy and all motions "freeze," the result is a double-ring basin similar to those as large as Korolev (440 km) in the size series (figs. 4.3, 4.11, 4.12C; table 4.1). In larger impacts, the collapsed part again rises in the center, and so on for as many cycles as are consistent with basin size, target strength, subsurface layering, other properties of the target, and, possibly, projectile size and velocity. Apollo, Hertzsprung, and Humboldtianum are large enough to have acquired a partial third ring (fig. 4.12E). Orientale has a fourth ring and a central uplift that signals the start of an additional cycle. Material may flow inward from outside the crater to supply the uplift, as is observed in waterdrop experiments. Sublithospheric flow of an early weak, plastic lunar asthenosphere may have enhanced wavelike ring formation (Van Dorn, 1968, 1969; Murray, 1980; McKinnon, 1981). The uplift freezes from the outside in because the inner part is the most highly fluidized. This explanation attractively accounts both for regular structure, like that in waterdrop experiments, and for irregular structure, as would be expected in heterogeneous and fractured rock materials that do not behave entirely like fluids.

8. Terrane outside the excavation is also deformed. Chapter 3 (see subsection entitled "Deformation and Nonballistic Ejection") mentions that craters may consist of two parts whose sizes are controlled by different scaling laws: an inner excavation cavity scaled to gravity, and an outer unexcavated but deformed zone scaled to rock strength (Croft, 1981). The deformed zone expands relative to the excavation cavity in proportion to impact energy, and thus may be very large for basins. External deformation, which in my model forms fewer rings than is believed by most investigators, has apparently formed arches concentric with the topographic basin rims of Humboldtianum, Orientale, and possibly other basins (pl. 3; table 4.1). Experimental analogs show how these external arches may have formed. Mounds and concentric rings were uplifted around experimental craters created by certain combinations of depth of burst and layering (Piekutowski, 1977), as well as around some large chemical-explosion craters (Baldwin, 1963, p. 122). Seismic energy is transmitted outside the cavity by elastic waves after the shock wave decays (Gault, 1974; Schultz and Gault, 1975a). The elastic waves are refracted and reflected by discontinuities between material layers (Rinehart, 1975; Cooper and Sauer, 1977) and might propagate quite far under favorable circumstances (McKinnon, 1981). Major topographic features, however, are unlikely to have formed by such exterior processes. Although several large rises and troughs in the Sinus Medii region have been interpreted as external anticlines and synclines of Imbrium (Baldwin, 1963, p. 322; Wilhelms, 1964), mapping of the Procellarum basin suggests, instead, that these and many other rings are part of that giant three-ring basin (pl. 3; Whitaker, 1981).

9. The topographic rim, which bounds the ejection, is commonly irregular in large basins. For example, Montes Caucasus east of Mare Imbrium appear to split into two distinct arcs, one continuing northwestward as Montes Alpes and the other continuing northeastward as the north shore of Mare Frigoris (fig. 4.3AA). Like the Rook ring of Orientale, the Alpes are the inner bound of knobby ejecta of Imbrium (Alpes Formation; see chap. 10). Like the Cordillera rim of Orientale, the northern Caucasus and the Frigoris shore are the inner bound of radial secondary chains and surface-flow ejecta. The Imbrium rings may have split in the middle of Montes Caucasus because the deformation and excavation were blocked by the preexisting Serenitatis massifs in the south but were freer to expand in the north (Spudis and Head, 1977) into a trough of the Procellarum basin (pl. 3). Spalling, as observed ahead of ejection in some explosion craters (Cooper, 1977, p. 40; Knowles and Brode, 1977, p. 893; Kreyenhagen and Schuster, 1977, p. 989; Maxwell, 1977, p. 1004; Piekutowski, 1977, p. 85–88), may soften up parts of the surrounding terrain for later ejection (Cintala and others, 1978, p. 3805). Ejection from the outer shelves occurred at very shallow angles, more like a surficial stripping than the conical plumes of simple craters.

Conclusions

The clear division of basin structure into rings may have been overemphasized, considering the pattern of hills, peaks, and ridges that lack evident ring structure in several sectors of Hertzsprung, Humboldtianum, Humorum, Nectaris, Serenitatis, western Orientale, Crisium, and Imbrium (fig. 4.3). Partly regular and partly random processes of floor uplift seem to be required to explain this combination of ringlike and irregular patterns. The mechanism favored here is oscillatory uplift of the cavity floor.

The conjecture about ring origin presented here has advanced some ideas as partial explanations that have previously been proposed as total explanations. Terracing has slightly widened the rim of Schrödinger and larger rims, such as the west rim of Korolev (fig. 4.3J), but decreases in magnitude with basin size, rather than increasing as predicted by megaterrace models (Murray, 1980, p. 272). In the largest basins, terracing creates only minor slump slivers, such as those northwest of Montes Apenninus (fig. 10.7; Wilhelms and others, 1977; Wilhelms, 1980). The crust-mantle interface or other layering may modulate more basic processes by enhancing ejection, ring uplift, and external deformation. Such oddities as Antoniadi and Compton (figs. 4.2A, B) must have been caused by some such local effects as layering (Hodges and Wilhelms, 1978). Oscillatory uplift may account both for regularities where enhanced and for irregularities where resisted by rock properties. Knobby ejecta of the interior and near-rim exterior may be flipped out by the oscillation. Oblique impacts probably produced asymmetric ejecta distributions (table 4.4) and may have caused internal asymmetries by distributing coupled energy unevenly, because they release their energy not at a point but along an elongate sloping zone (Gault and Wedekind, 1978). A double or triple impact excavated the Humboldtianum basin (Lucchitta, 1978), and other multiple impacts probably caused the elongation of Moscoviense and Serenitatis. External deformation did occur but created only second-order features, not the main topographic rim or other major arcs and troughs. The topographic rim is the boundary of the excavation cavity but is irregular in many large basins. In conclusion, basins are highly complex natural systems whose origin cannot be learned by straightforward extrapolation from simple cratering mechanics.

SUMMARY OF BASIN-MATERIAL ORIGINS

Orientale and similar, older basins and their deposits were formed in the following sequence of events. (1) A relatively deep, craterlike cavity developed upon impact of a large cosmic projectile; (2) this cavity expanded laterally and ejected massive amounts of material at low angles, forming gouges and thick lineate deposits; (3) the interior rebounded and ejected blocky melt-rich ejecta at higher angles and lower velocities; (4) the rebound collapsed, then oscillated further, each time to successively smaller radii, and left a ring at each hingeline; (5) the last oscillation left a central uplift (known only in Orientale). Ejecta was deposited and secondary craters were formed during events 3 through 5; then, (6) impact melt settled into place, and (7) distant secondary craters were formed.

The interpretations of the well-exposed, well-photographed Orientale deposits presented in this chapter are the principal foundation for interpreting older terra materials. Deposits of 17 basins and the 262-km-diameter transitional feature Milne (fig. 4.2J) are mapped here (pl. 3; table 4.1). The presence of rings establishes the existence of 11 basins whose deposits are not identified but which presumably are present; 17 other basins are also mapped whose existence can be disputed (table 4.2). Topographic textures indicative of primary and secondary ejecta emplacement are the first to become blurred over time. The larger secondary craters survive longer but eventually are also obliterated or can no longer be traced to their source basin. Even the mountainous rings may be so thickly buried or heavily pitted by later cratering that they become unrecognizable. Probably, therefore, fewer basins are tabulated (tables 4.1, 4.2) and illustrated here than were actually formed.

Deposits outside the topographic basin rim grade from thick to thin. When fresh, the deposits closest to the rim were coarsely textured, as is the inner facies of the Hevelius Formation at Orientale. These materials probably consist mostly of primary ejecta from the basin. At an average distance of one basin radius, secondary craters begin to dominate the scene. Secondary craters are visible around many basins whose textured ejecta deposits are invisible (see chaps.

8, 9). The inner and outer zones are here mapped separately only where their approximate transition can best be distinguished, around the Orientale, Imbrium, and Nectaris basins (pl. 3). The outer zone is dominated by secondary ejecta derived from local terrain, but the radius where the balance shifts is uncertain. The outer deposits persist for two radii on the average but thin gradually to apparent disappearance. Distinctive grooves at the basin antipodes may indicate a Moon-wide influence of large basins. Much of the barely discernible fine-scale relief of the lunar terrae must have originated by basin ejecta deposition and secondary cratering.

Deposits of both the primary and the secondary zones are lobate and asymmetric. Salients and raylike stringers of secondary craters persist outward for several basin radii in some sectors. "Bow tie" patterns suggesting oblique impacts at low angles are particularly common (pl. 3; table 4.4). Although irregularities are not so evident around old basins, they are inferrable. Therefore, the presence at a given point of the deposits of a given basin cannot be predicted, unless diagnostic textures or mantling relations are visible. Thicknesses are very uncertain.

Several types of deposits are seen only in Orientale or a few other young large basins (table 4.3). Abundant plains mark the transition from the primary to the secondary ejecta of Orientale, Imbrium, and Nectaris but are inconspicuous around smaller or older basins. The presence of degraded older plains is, nevertheless, highly probable in these positions. Planar and fissured impact melt is visible in Orientale (Maunder Formation), but in no other basin with equal certainty. At the large young basins, the presence of the knobby ejecta (Montes Rook and Alpes Formations) inside the basin, on the topographic rim, and on the earlier, thicker deposit (Hevelius Formation and the Fra Mauro Formation around Imbrium) suggests late-stage ejection during floor rebound.

Despite the incompleteness of the lunar record and the shortcomings of the data, a history of lunar basin deposits can be pieced together from superpositional relations, superposed crater densities, and degree of morphologic degradation relative to Orientale. This history constitutes most of the history of the lunar terrae. Chapters 7 through 10 show how knowledge of the mechanisms that emplaced the Orientale deposits supports the interpretations and dating of the older deposits. Breccia and impact-melted rock ascribable to impact processes (chap. 3) were recovered from basin deposits by all the terra-sampling missions (pl. 3; fig. 4.13). Apollo 14 returned melt-bearing ejecta of Imbrium corresponding to the Hevelius Formation (chap. 10), Apollo 15 sampled massifs of Imbrium (chap. 10). Apollo 17 sampled melt-rich materials from the massifs of the Serenitatis basin (chap. 9). Apollo 16 recovered materials of circum-Imbrium plains and hilly-and-furrowed materials that are analogous to Orientale deposits but whose origin is controversial (chaps. 9, 10).

TABLE 4.4.—*Basin asymmetries suggestive of impact angle*

[Azimuths measured clockwise from north. Basins grouped into four age categories (see tables 4.1, 9.3).]

Basin	Center (lat, long)	Azimuths of ejecta lobes	Azimuths of central-basin elongation	Interpreted impact angle
Orientale	20° S., 95° W.	150°–240°–330°	---	From 60°.
Schrödinger	75° S., 134° E.	150°–330°	---	From 60° or 240°.
Imbrium	38° N., 19° W.	160°	Sector 140° elevated	From 70° or 250° (320°–340°?).
Hertzsprung	2° N., 129° W.	0°–180°	---	From 90°.
Crisium	18° N., 59° E.	0°–80°(?)–155°	80°–260°	From 260°.
Humorum	24° S., 40° W.	160°	Sector 160° elevated	From 70°, 250°, or 340°.
Humboldtianum	61° N., 84° E.	0°–155°	10°–190°	From 260°.
Korolev	5° S., 157° W.	10°	---	From 100°.
Moscoviense	26° N., 147° E.	10°	30°–210°	From 100°.
Nectaris	16° S., 34° E.	145°	---	From 55° or 235°.
Apollo	36° S., 151° W.	200°–330°	---	From 85°.
Freundlich-Sharonov	19° N., 175° E.	0°–210°	---	From 105° or 285°.

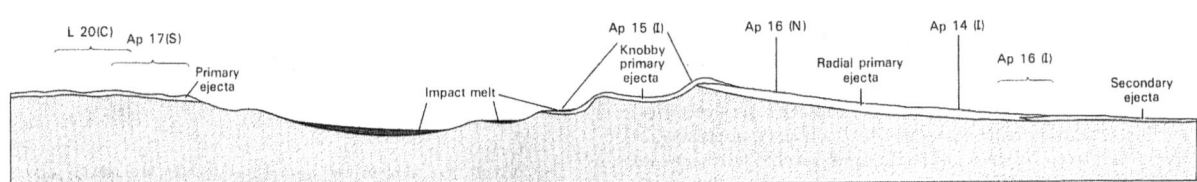

FIGURE 4.13.—Schematic geologic section of typical lunar basin (based on Orientale), showing locations of Apollo (Ap) and Luna (L) sampling sites relative to basin facies sampled (C, Crisium; I, Imbrium; N, Nectaris; S, Serenitatis). Brackets indicate uncertain positions; alternatives are given for Apollo 16, depending on the basin it sampled (see chaps. 9, 10). Right side is based on figure 4.4, left side on figure 4.5. Vertical exaggeration, ×20.

5. MARE MATERIALS

FIGURE 5.1 (OVERLEAF). —Lobate mare-basalt flows in Mare Imbrium. Apollo 15 frame M-1556.

5. MARE MATERIALS

CONTENTS

	Page
Introduction	85
Mapping properties	86
Morphology	86
Blanketing	89
Albedo	94
Color	96
Orbital chemistry	97
Other properties	98
Integration	99
Thickness	99
Returned samples	101
Introduction	101
Composition	101
Classification	101
Origin and emplacement	102

INTRODUCTION

The darker half of the Galilean dichotomy, the smooth maria, are topographically and geologically simpler than the ringed basins that contain them. The maria are the most earthlike and probably the most easily understood lunar features (fig. 5.1). They cover about 16 percent of the lunar surface, or 30 percent of the nearside and 2 percent of the farside. If younger craters were absent, maria would cover about an additional 1 percent of the Moon (dashed contacts, pl. 4). Most maria flood large basins and their peripheral depressions; they also flood craters and small basins that are superposed on larger basins. Some mare occurrences have led to the identification of basins where none were otherwise suspected. A few small maria that are not related to known basins (pl. 4; figs. 5.2, 5.3) may, similarly, indicate the presence of undiscovered basins.

Because of their resemblance to terrestrial volcanic terranes, the maria were more readily interpreted before being sampled than were the terrae. The approximately level mare surfaces were correctly taken as signs of fluid, lavalike emplacement (Gilbert, 1893; Urey, 1952; Baldwin, 1949, 1963; Kuiper, 1959; Shoemaker and Hackman, 1962). Their landforms were perceived as analogous to volcanic landforms on the Earth (Kuiper, 1959; Shoemaker and Hackman, 1962; Quaide, 1965). Although nonbasaltic composition (especially ash-flow tuff; O'Keefe and Cameron, 1962; Mackin, 1969) and such exotic origins as mobile dust (Gold, 1955) and water-laid sediment (Gilvarry, 1970) were considered, the low albedo and certain diagnostic landforms suggested basic lava to most workers (for example, Baldwin, 1963, p. 336). Basaltic composition seemed to be substantiated by Surveyor 5 and 6 analyses in 1967 (Gault and others, 1968a, p. 287), and was established by the Apollo 11 samples in July 1969. Basalt has been returned from all the mare sites visited. The rock name "basalt" can thus be used as a near-synonym for the less committal term "mare materials," even in unsampled regions.[5.1]

Stratigraphic relations of the mare-basalt flows are also readily understood in principle because they are similar to relations familiar on the Earth. The margins of the flows follow the contours of the containing basins, as would any later fluidlike deposits. Relative crater densities are consistent with the younger age of these flows.

FIGURE 5.2.—Isolated mare patch in crater Kohlschütter (arrow; 53 km, 27° N., 154° E.). Many craters of subequal size are probably secondary to Mendeleev basin (see chap. 9), outside lower left corner of scene. Boom of Apollo 16 gamma-ray spectrometer protrudes into view at right. Apollo 16 frame M-729.

[5.1] Strictly speaking, the term "maria" (singular and adjectival form, "mare") refers to the surfaces, and "mare materials" to the rock-stratigraphic units that underlie the surfaces; however, "mare" and "maria" are commonly used for the materials as well.

The main obstacle to acceptance of these obvious relations was the peculiar confusion between mare and basin (see chap. 2), which persisted until the early 1960's and is still not entirely dispelled.

By definition, mare materials (1) form generally smooth and level surfaces and (2) are dark. In some areas, mare units can be subdivided by their topographic characteristics. However, most subdivisions are based on optical, spectral, or geochemical properties commonly called remote-sensing properties. In this respect, mare units differ from impact units, because they are igneous-rock bodies whose compositional unity is more obvious than is morphology due to emplacement. The recognition of mare subunits has advanced greatly since 1972 over its rather primitive status before Apollo exploration (Head and others, 1978b). The following sections discuss the properties by which maria and their subunits are recognized and interpreted by remote observations, and the last section of this chapter briefly describes and interprets the samples of mare basalt returned from the Moon.

MAPPING PROPERTIES

Morphology

Certain areally minor, though genetically significant, morphologies interrupt the flat mare surfaces. The most significant are *lava-flow fronts*. These lobate edges resemble terrestrial lava lobes and thus indicate that at least some maria consist of volcanic lavas. The first flow fronts to be described and the largest yet known are in Mare Imbrium (fig. 5.1; Schaber, 1969). They are visible in good telescopic photographs and were shown to coincide with color boundaries, presumed to be some sort of compositional contacts (Kuiper, 1965, p. 29–32; Whitaker, 1966, p. 79–83). These observations quieted exotic hypotheses of mare origin in the minds of most workers. The Imbrium lobes range from 10 to more than 35 m in height and bound flows as much as 1,200 km long (Schaber, 1973). These great dimensions may indicate that each flow formed by rapid eruption of fluid lava from a single large vent (Schaber, 1973). The Imbrium flows are probably the only relatively recent large flows still exposed, although earlier, now-buried eruptions may have been equally large. Scarps of shorter and thinner flows are observed in Oceanus Procellarum, Mare Cognitum, Mare Vaporum, and Sinus Medii (Schaber and others, 1976).

Otherwise, lunar mare flows are either too thin or, alternatively, too floodlike (Holcomb, 1971; Greeley, 1976) to have formed lobate scarps. A pooled, massive style of some mare units is suggested by marginal terraces (fig. 5.3). Holcomb (1971) pointed out the similarity of these "bathtub rings" perched on the terrae around some maria to those left by subsidence of Hawaiian lava lakes. The terraces suggest withdrawal of ponded lava by way of observed sinuous rilles (fig. 5.4; Holcomb, 1971; K.A. Howard, in Carr and others, 1971; Howard and others, 1972; Greeley and Spudis, 1978).

Several types of positive landforms, mostly formed by extrusion of basalt, are scattered in several, and concentrated in a few, maria (pl. 4; Guest and Murray, 1976). Most *domes* are subtle circular features with low profiles, commonly topped by small smooth-rimmed craters (figs. 5.5, 5.6); they are visible in telescopic visual observations or photographs made at low Sun illuminations (Arthur, 1962). The apparent internal origin of the domes was a major factor in endogenic interpretations of the maria, and their low profiles suggested fluid volcanism characteristic of mafic magmas. Low mare domes are concentrated in Mare Tranquillitatis (Vitruvius region, fig. 5.5) and western Mare Insularum (Hortensius and Milichius regions, fig. 5.6) but occur in other maria as well. Steeper, rough-surfaced features commonly called *cones* are concentrated in the Marius Hills (fig. 5.7; McCauley, 1967a, b, 1968, 1969b; Guest, 1971) and are scattered in several maria, for example, Mare Cognitum, Mare Tranquillitatis (Wilhelms, 1972a), and the border of Mare Serenitatis (Bryan and Adams, 1973; Scott, 1973). Some cones occur in the terrae (fig. 5.8). The difference between the domes and cones presumably reflects variations in composition or in the eruptive process, and the Marius Hills were a favored site for Apollo missions in hopes that they would yield differentiated volcanic-rock types (McCauley, 1969b). The morphologies of domes or cones are mimicked by *kipukas* or *steptoes*, synonyms for protruding terra or mare deposits isolated from the rest of their geologic unit by floods of later mare materials (see fig. 6.3). The Rümker hills (fig. 5.9) apparently consist of several domes and other mare-related units partly burying Imbrium-radial terra (Guest, 1971). None of the positive lunar volcanic landforms remotely approaches a terrestrial or martian shield volcano or stratovolcano in size. Almost all the landforms listed here occur in shallow maria (Arthur, 1962, p. 322; Head, 1976a). Some 200 such landforms were recently classified by Head and Gifford (1980).

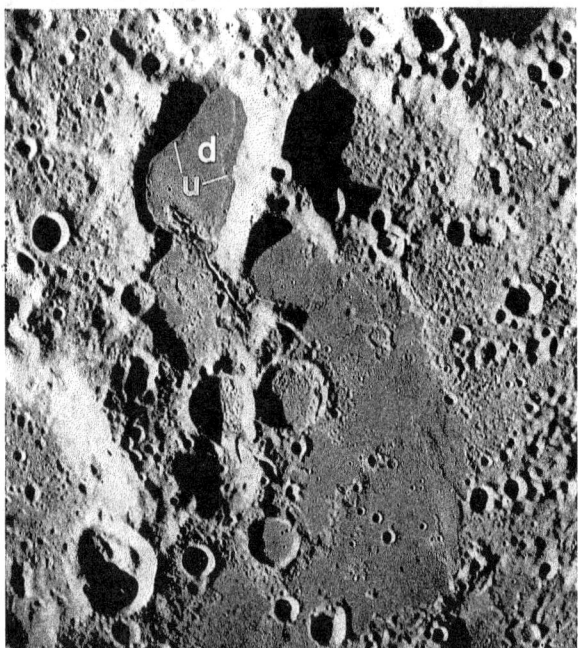

FIGURE 5.3.—Lacus Solitudinis, showing down (d) and up (u) sides of marginal terraces of mare. Scene centered at 27° S., 103° E. Apollo 15 frame M–2627.

FIGURE 5.4.—"Scablands" north of Aristarchus Plateau. Mare lava probably drained out of the region in the sinuous rille (left). Large craters in lower right are possibly secondary to Orientale basin. Scene centered at 30° N., 53.5° W. Apollo 15 frame P–349.

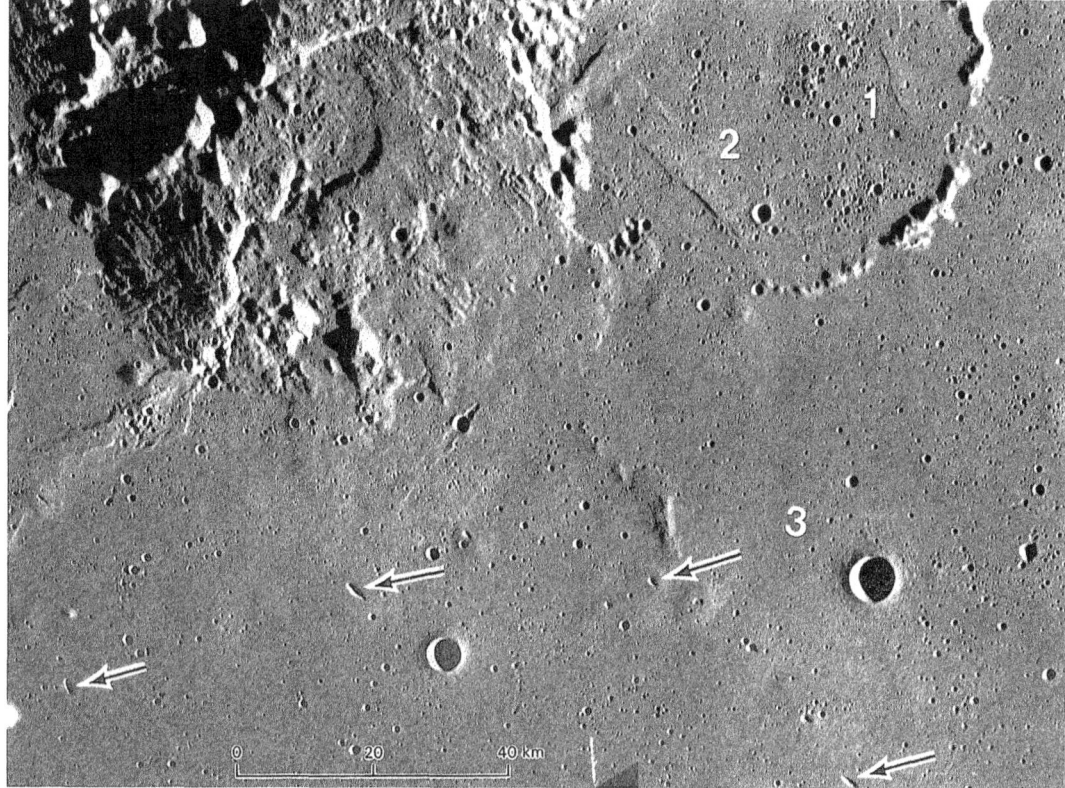

FIGURE 5.5.—Mare domes in northern Mare Tranquillitatis. Numerous elongate pits are on summits of low domes or in flat mare (arrows). Mare units are numbered in order of decreasing age. Crater below number 3 (youngest unit) is Maraldi B (7 km). Scene centered at 15° N., 35° E. Apollo 17 frame M-305.

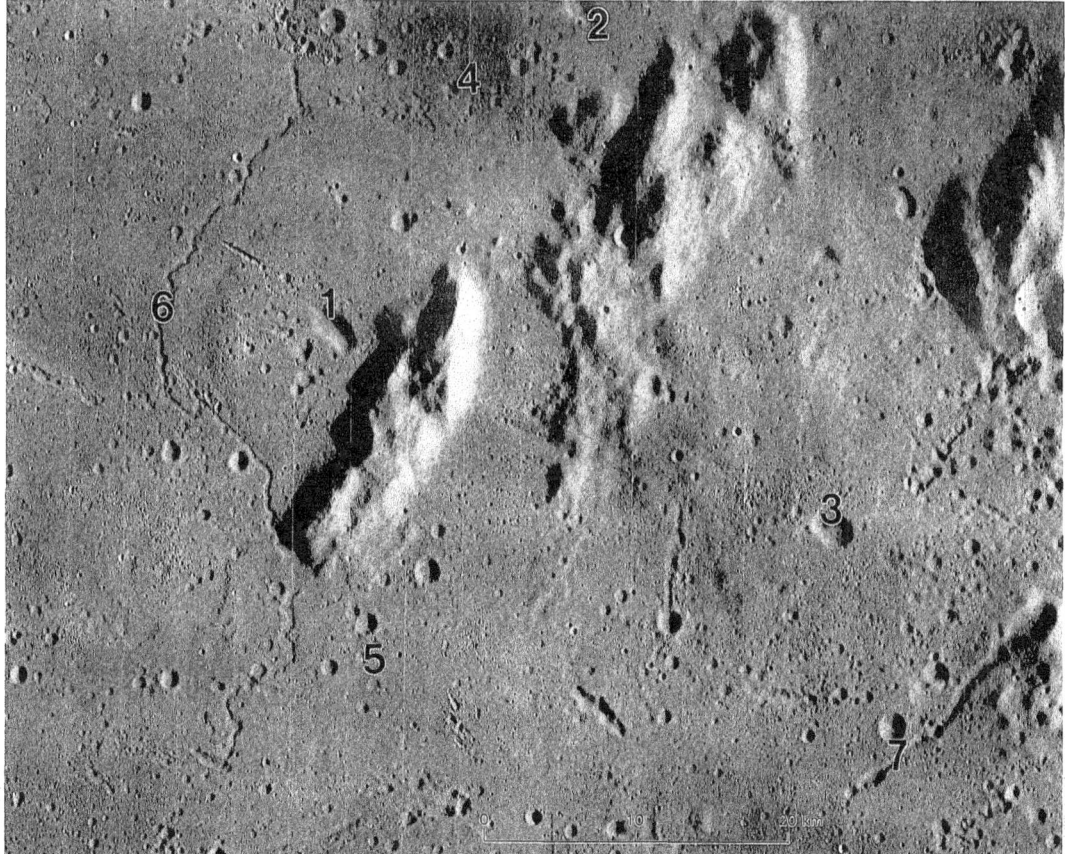

FIGURE 5.6.—Detail of mare domes and other endogenic features, including circular domes with elongate central craters (1, 2); irregular domes with circular, nevertheless probably endogenic, craters (3); concentrations of similar large, soft-appearing circular craters that could be endogenic (4, 5); sinuous rille (6); and elongate endogenic vent or collapse feature (7). Scene centered at 13° N., 30.5° W. Orbiter 5 frame M-164.

Visible source vents are rare in the maria, because the lavas probably cover their sources as do flood-basalt lavas on the Earth (Greeley, 1976). Some *endogenic craters* do appear amidst the maria, however, and are fairly numerous along their borders (figs. 5.10A–D); others lie in small patches of dark material in the terrae (figs. 5.8, 5.10E–H). Their internal origin is determined by this association with volcanic materials or by their alignment along faults. For example, random impact is unlikely to be responsible for the many craters exactly centered on what are evidently volcanic domes (figs. 5.5, 5.6). The craters or rimless pits along Rima Hyginus are similarly eliminated statistically as impact craters because they are perfectly aligned along the rille (fig. 5.10E). Although one leg of Rima Hyginus is radial to the Imbrium basin, both the rilles and the craters are too young to have been formed directly by the impact (see chap. 6; Wilhelms, 1968; Pike, 1976). Like many other endogenic craters (figs. 5.10A–D), the rille craters and the main Hyginus crater at the rille "elbow" could have formed entirely by collapse without the extrusion of any material (Pike, 1976). Single craters situated along rilles, amidst dark materials, and with elongate or irregular shapes are also probably endogenic (figs. 5.10F–H). A few craters not situated along rilles resemble secondary-impact craters, but endogeny may be assumed if the craters are highly irregular and are centered in dark deposits (figs. 5.10G, H). Although the dark ejecta of endogenic craters is commonly more extensive relative to the crater size than is the ejecta of impact craters (table 3.1), the volcanic ejecta is smoother, and the craters are all small (mostly less than 10 km diam). Despite the early popularity of a caldera origin for such large craters as the smooth-rimmed class (fig. 3.15), large calderas are not expected on the Moon; calderas form by collapse or magma withdrawal into shallow magma chambers, which probably never existed on the Moon.

Sinuous rilles are another areally minor but visually arresting type of mare feature. The Apollo 15 landing was targeted next to a large sinuous rille, Rima Hadley (fig. 5.11; chaps. 10, 11). Several other rilles were considered for landings to determine their formational

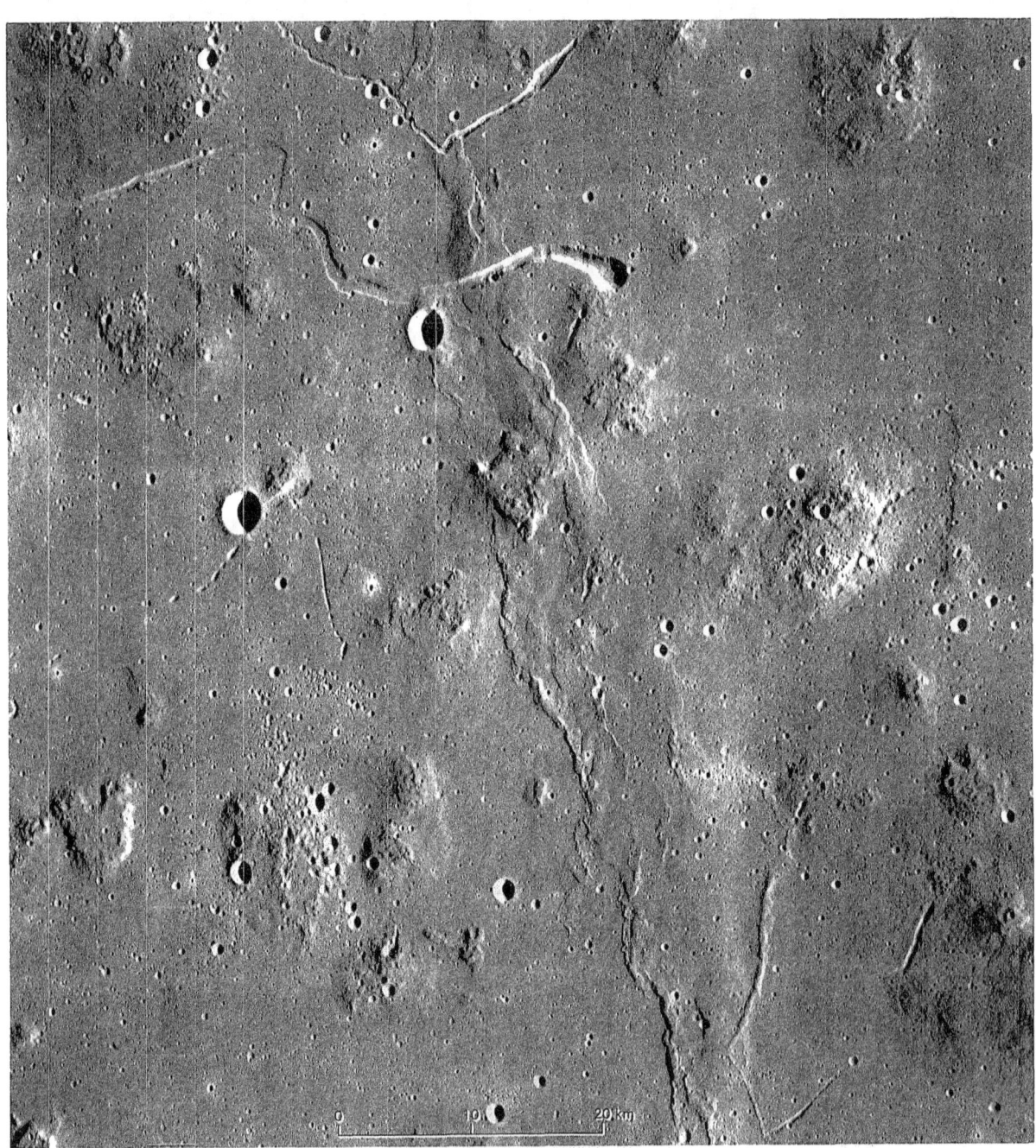

FIGURE 5.7.—Marius Hills, a complex of diverse, rough-textured domes or cones surrounded by smoother mare materials. Orbiter 5 frame M–211.

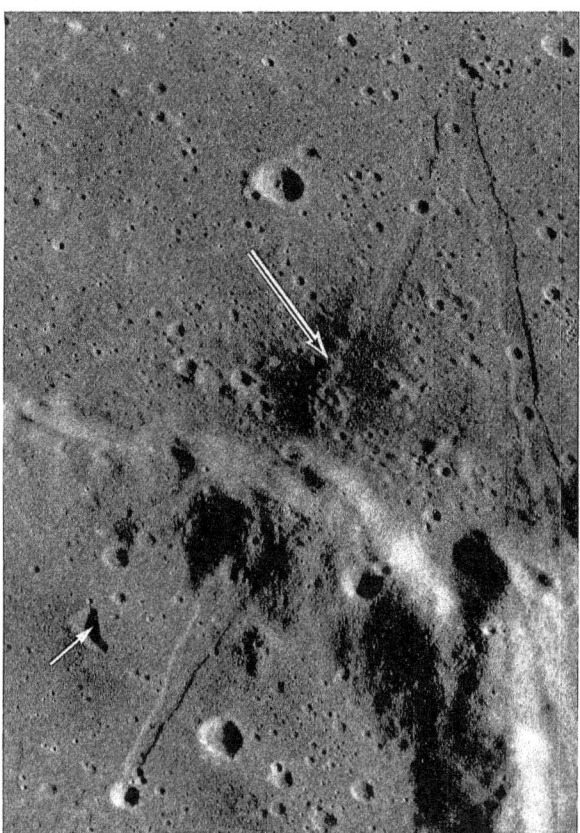

FIGURE 5.8.—Rough cones (black-and-white arrow) situated along linear rille and amidst dark-mantling material or mare lava. Smooth-profiled irregular pit (white arrow) further indicates basaltic volcanism in this terra setting. Crater rim separates craters Fra Mauro (above) and Bonpland (below). Black-and-white arrow is 5 km long; scale varies because of obliquity. Apollo 16 frame P-5425.

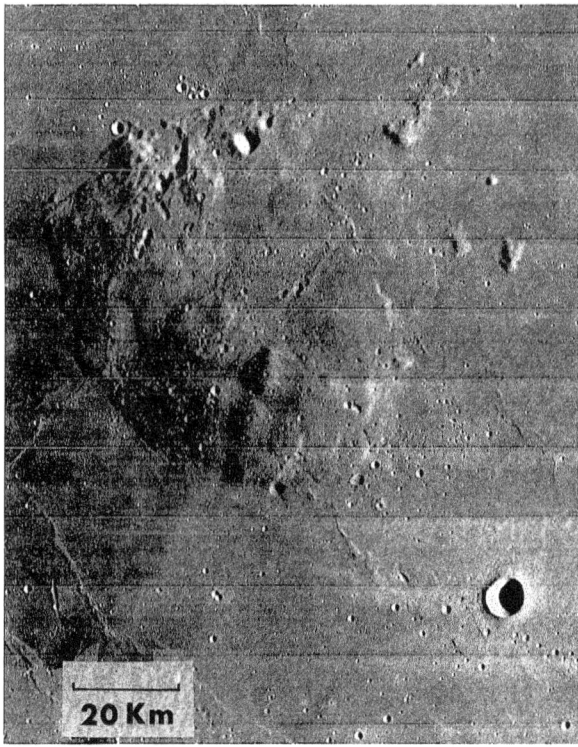

FIGURE 5.9.—Rümker hills (Mons Rümker), a complex of subdued hills in northern Oceanus Procellarum (41° N., 58° W.). Imbrium-radial lineations (upper right to lower left) suggest an origin as terra covered by thin layer of dark-mantling material.

process and to collect samples of volatile or other exotic erupted materials (fig. 5.12). Fieldwork on the Earth and at Rima Hadley has shown that these meandering grooves are probably lava channels or collapsed lava tubes, which are the sites of the last flows of molten magma in a lava unit (Kuiper and others, 1966, p. 199; Oberbeck and others, 1969; Greeley, 1971; Murray, 1971; Cruikshank and Wood, 1972; Howard and others, 1972; Guest and Greeley, 1977, p. 62–68). Very long and narrow rilles are likely to have been channels (always open); others, including Hadley, display alternately roofed and unroofed segments that indicate an origin as tubes (figs. 5.11, 5.13). The sources of some lunar rilles are in the terra or at the mare-terra contact (figs. 5.11, 5.12). Several appear to have erosionally incised the substrate, including terra material (fig. 5.12C; Carr, 1974). Sinuous rilles and other volcanic landforms of the maria were thoroughly described and illustrated by Lowman (1969), Schultz (1976b), and Masursky and others (1978).

Blanketing

The property of blanketing or mantling terra topography distinguishes another class of dark, mare-related units, the *dark-mantling materials*. They are as dark as or darker than the maria, but their observed relief is partly rugged. The largest and most numerous tracts with this seemingly anomalous combination of properties lie around the southern periphery of the Imbrium basin (pl. 4; figs. 1.6, 1.7, 5.14, 5.15). This position led to their interpretation as a dark facies of Imbrium ejecta (Shoemaker and Hackman, 1962; Hackman, 1966). Carr (1965a, 1966a, b), however, showed that the large dark areas on the southeast slope of Montes Apenninus are not composed of Imbrium ejecta. He observed that flat surfaces next to the dark ejecta are equally dark, and so the same dark material probably covers both the ejecta and the much younger mare material (fig. 5.16). Moreover, other patches of mantling material are peripheral to Mare Serenitatis (figs. 5.17, 5.18; Carr, 1965a, 1966a, b), Mare Humorum (Titley, 1967), Mare Nectaris (Elston, 1972), Mare Crisium (Olson and Wilhelms, 1974), Mare Smythii (Wilhelms and El-Baz, 1977), Oceanus Procellarum (Moore, 1965, 1967; McCauley, 1967a, b), and, in fact, all carefully scrutinized maria (pl. 4; Head, 1974a, 1976a; Wilhelms, 1980). This concentration near maria and the dark color indicate a genetic relation of the dark-mantling materials to the maria. The flat parts of maria may also contain dark-mantling materials, but these are difficult to distinguish from mare lavas.

A more specific origin for the dark-mantling materials is suggested by other physical properties. On both telescopic and spacecraft photographs, they appear fine-textured, relatively uncratered, and smoother than the underlying topography (figs. 5.15, 5.16). No blocks are visible in the 2-m resolution of the best Apollo photographs. Mantles drape differentially over mountains, low hills, and flat surfaces, and are thickest in depressions and thin or absent on sharp hills. The brightness of some hilltops surrounded by the deposits indicates nondeposition or shedding of the dark material. Shedding is the preferred explanation because some dark deposits lie at higher elevations than some bright summits. Also, the action of mass wasting is indicated by talus streaks that descend from the bright outcrops (fig. 5.16). Shedding is consistent with the propensity of unconsolidated material to slide downslope.

The morphologic properties and distribution of the dark-mantling materials suggest that they consist of volcanic fragments, *pyroclastic* material. The Marius Hills (McCauley, 1964a), the Aristarchus Plateau, and Montes Harbinger (figs. 5.7, 5.12A; Moore, 1964b) were originally interpreted as terra features covered by volcanic ash (figs. 5.7, 5.12). Carr (1965a, 1966a, b) noted that the dark deposits are situated near likely eruptive fissures in the form of rilles in the maria (fig. 5.16). Unlike lavas, few dark-mantling materials lack elongate or irregular, commonly rille-straddling craters, gashes, or dark cones that are likely source vents (figs. 5.8, 5.10F–H, 5.12, 5.16). Pyroclastic origin was favored for the dark-mantling materials in all the early photogeologic work and has been confirmed by sampling data (see chap. 11). The dark halos of many small endogenic craters, such as those in Alphonsus (fig. 5.10F), are also dark-mantling deposits. Head and Wilson (1979), however, suggested that the eruptive mechanisms differ: The extensive, regional deposits are thought to be strombolian (continuous eruptions), and the smaller halos vulcanian (intermittent eruptions).

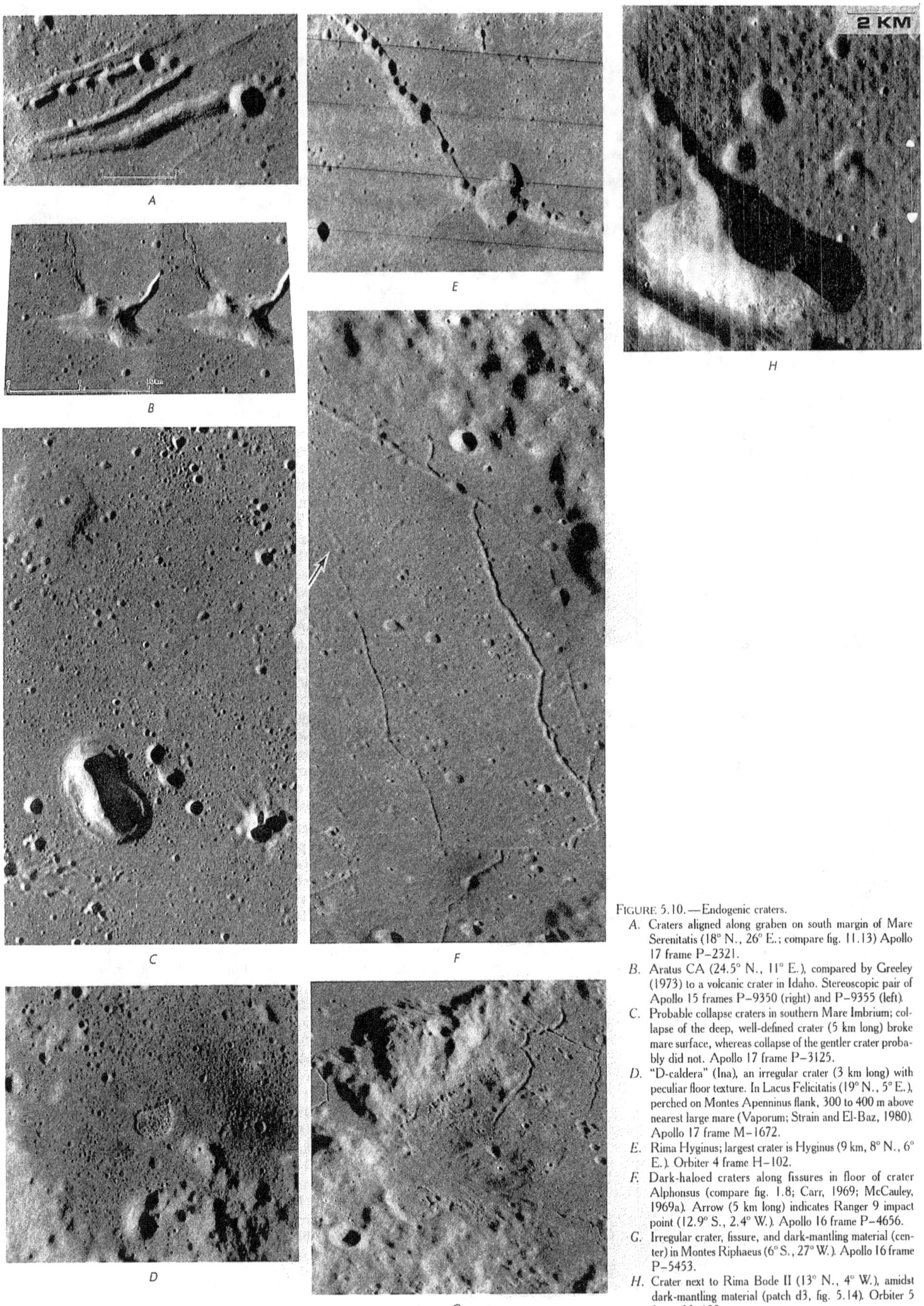

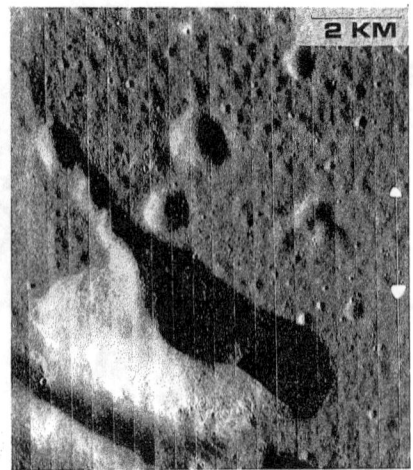

FIGURE 5.10.—Endogenic craters.

A. Craters aligned along graben on south margin of Mare Serenitatis (18° N., 26° E.; compare fig. 11.13) Apollo 17 frame P–2321.

B. Aratus CA (24.5° N., 11° E.), compared by Greeley (1973) to a volcanic crater in Idaho. Stereoscopic pair of Apollo 15 frames P–9350 (right) and P–9355 (left).

C. Probable collapse craters in southern Mare Imbrium; collapse of the deep, well-defined crater (5 km long) broke mare surface, whereas collapse of the gentler crater probably did not. Apollo 17 frame P–3125.

D. "D-caldera" (Ina), an irregular crater (3 km long) with peculiar floor texture. In Lacus Felicitatis (19° N., 5° E.), perched on Montes Apenninus flank, 300 to 400 m above nearest large mare (Vaporum; Strain and El-Baz, 1980). Apollo 17 frame M–1672.

E. Rima Hyginus; largest crater is Hyginus (9 km, 8° N., 6° E.). Orbiter 4 frame H–102.

F. Dark-haloed craters along fissures in floor of crater Alphonsus (compare fig. 1.8; Carr, 1969; McCauley, 1969a). Arrow (5 km long) indicates Ranger 9 impact point (12.9° S., 2.4° W.). Apollo 16 frame P–4656.

G. Irregular crater, fissure, and dark-mantling material (center) in Montes Riphaeus (6° S., 27° W.). Apollo 16 frame P–5453.

H. Crater next to Rima Bode II (13° N., 4° W.), amidst dark-mantling material (patch d3, fig. 5.14). Orbiter 5 frame H–122.

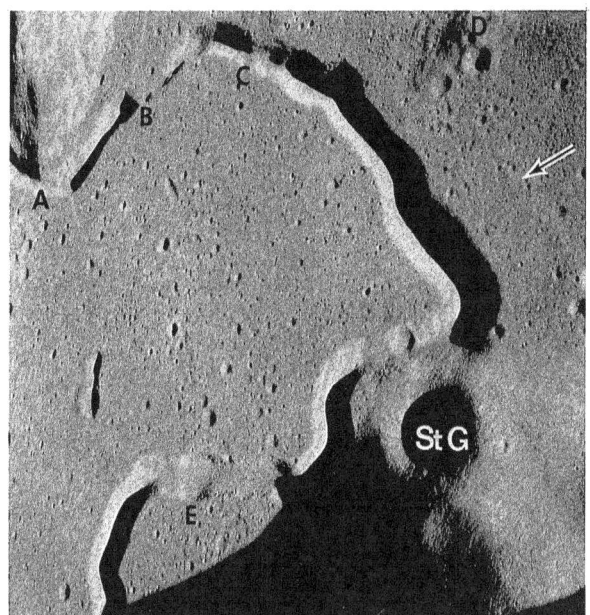

FIGURE 5.11.—Rima Hadley (Hadley rille) at Apollo 15 landing site (arrow). Rille hugs and possibly erodes terra at A, and is bridged at B and C by basalt. D, dark-mantled "North Complex," surrounded by level mare. Ledges are visible at E. StG, St. George Crater (2.25 km). Apollo 15 frame H-11720.

A

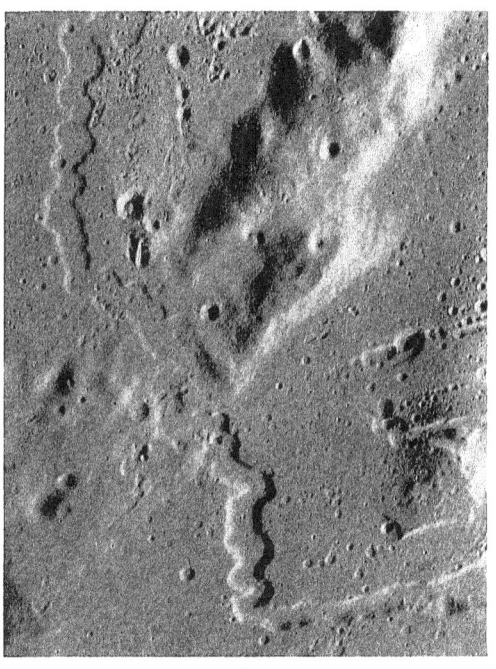

C

B

D

FIGURE 5.12.—Aristarchus Plateau-Montes Harbinger complex of sinuous rilles.
  A. Montes Harbinger and crater Prinz (P). Large fresh crater is Aristarchus (A; 40 km, 24° N., 47° W.). Arrow indicates rille in C. View southward. Apollo 15 frame M-2606.
  B. Aristarchus Plateau. Aristarchus is at left edge; H, Herodotus (35 km, 23° N., 50° W.). Arrow indicates rille in D. View southward. Apollo 15 frame M-2611.
  C. Terra ridge in Montes Harbinger (arrow in A), eroded by rille. Apollo 15 frame P-321.
  D. Small meandering rille within largest lunar sinuous rille, Vallis Schröteri (arrow in B), which probably formed in massive lava flow(s). Apollo 15 frame P-341.

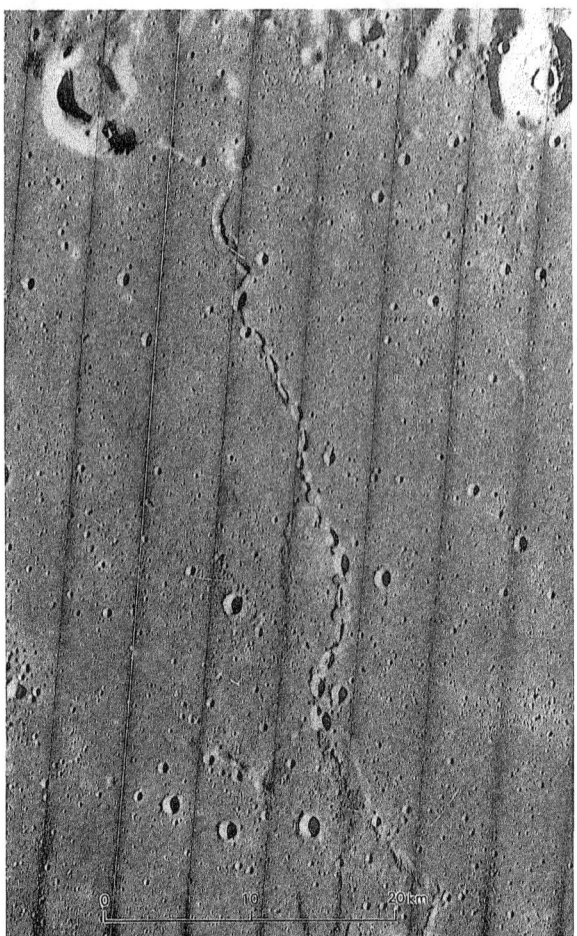

FIGURE 5.13.—Sinuous complex, consisting of ridge and elliptical and curved pits, that may have originated as a lava tube whose surficial crust collapsed intermittently along its length (Oberbeck and others, 1969). Crater in upper right corner is Gruithuisen K (6 km, 35° N., 43° W.). Orbiter 5 frame M-182.

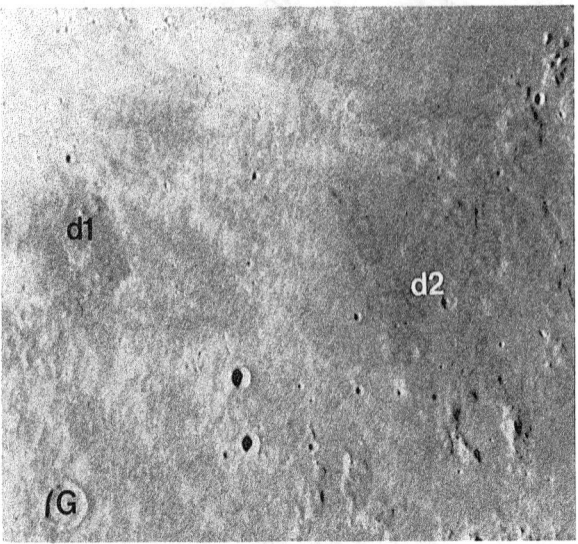

FIGURE 5.15.—Ray-crossed patch of dark-mantling material south of Copernicus (d1) and large patch south of Sinus Aestuum (d2; compare fig. 5.14). Sharp-rimmed (delta-rimmed) crater at lower left is Gambart (G; 25 km, 1° N., 15° W.; compare fig. 3.15C). Telescopic photograph.

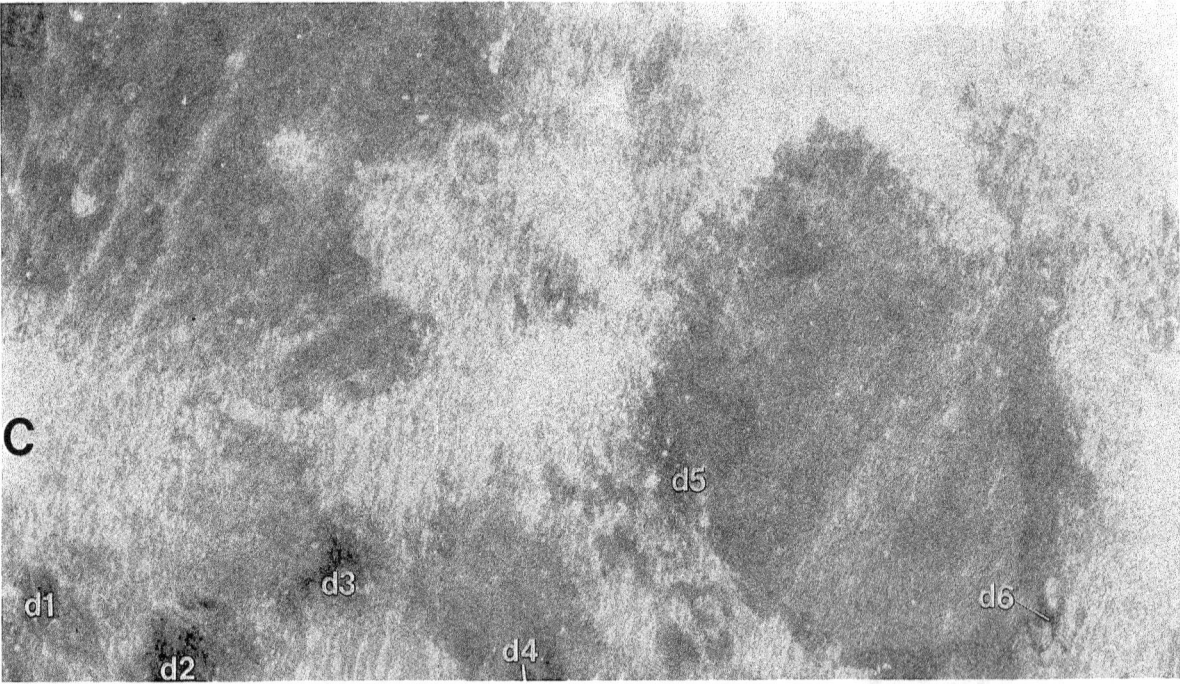

FIGURE 5.14.—Telescopic full-Moon photograph of Maria Imbrium (upper left) and Serenitatis (right). Bright crater Copernicus (C) is at left edge. Dark patches are south of Copernicus (d1), on border of Sinus Aestuum (d2, d3), on border of Mare Vaporum (d4), and along margin of Mare Serenitatis (d5, d6). Serenitatis is also bordered by other dark mare units. U.S. Naval Observatory, Flagstaff, Ariz., photograph.

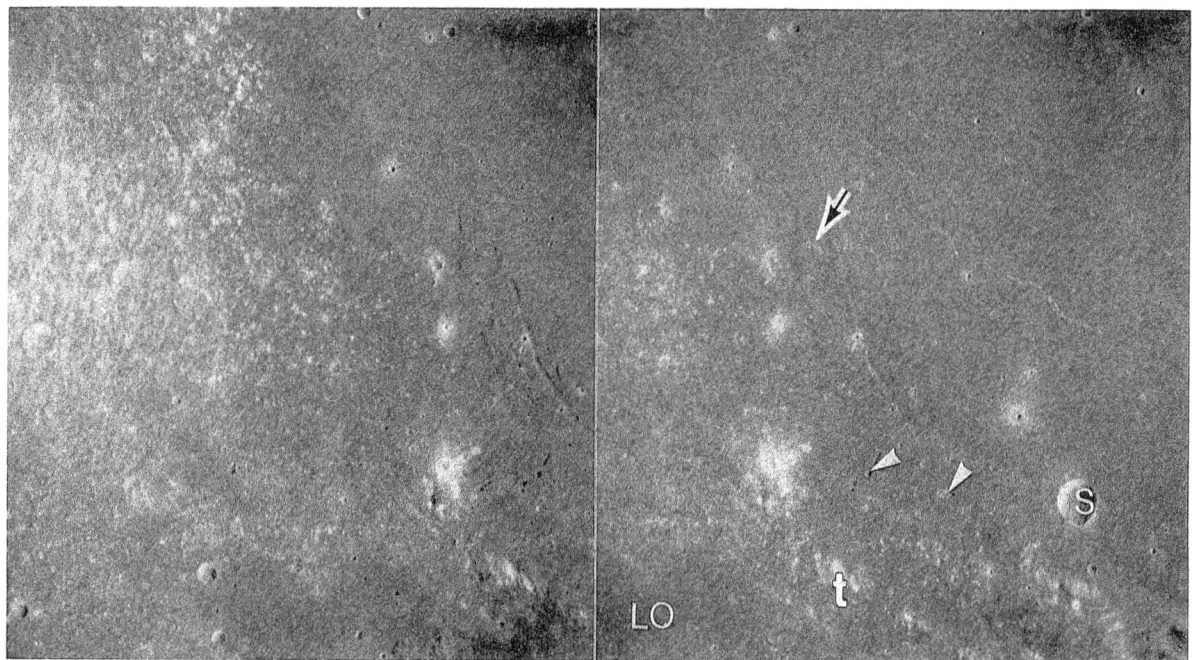

FIGURE 5.16.—Dark-mantling materials on southwest rim of Serenitatis basin (d5, fig. 5.14). Decrease in number of exposed bright terra peaks from lower left to upper right indicates increase in thickness of dark-mantling material toward the mare. Talus of dark-mantling material (t) covering brighter slope indicates shedding of superposed dark deposit. Black-and-white arrow indicates flooding of dark-mantling material by mare lava. Arrowheads denote endogenic craters. Lacus Odii (LO), which is elevated substantially (approx 1.5 km) above Mare Serenitatis, may consist of pooled dark-mantling material (Wilhelms, 1980). Arcuate rilles are Rimae Sulpicius Gallus; S, crater Sulpicius Gallus (12 km, 20° N., 12° E.). Stereoscopic pair of Apollo 17 frames M–2701 (right) and M–2703 (left).

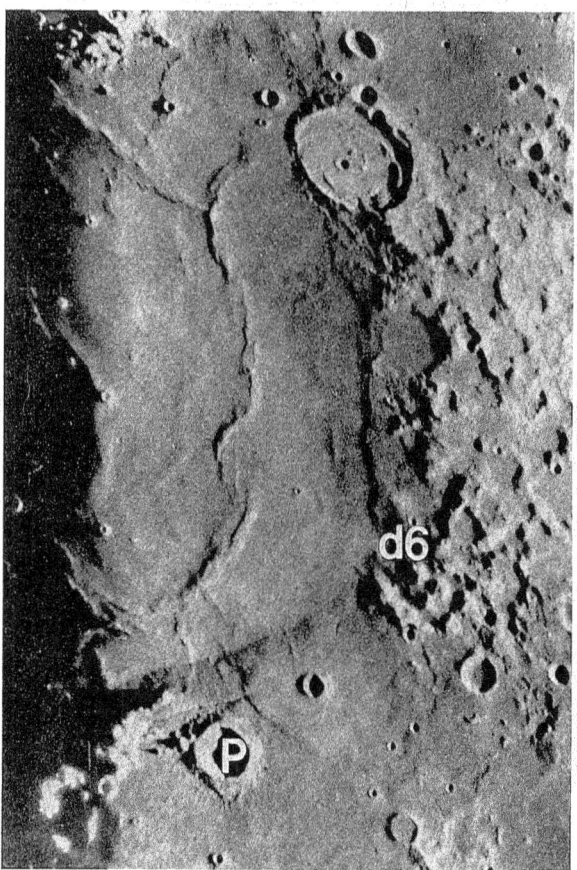

FIGURE 5.17.—Eastern Mare Serenitatis and adjacent terrain. Dark-mantling material (d6; compare fig. 5.14) is in region of Apollo 17 landing site. Central Serenitatis is depressed below darker border zone; grabens (see chap. 6) are in border zone above crater Plinius (P; 43 km, 15 N., 24 E.). Elevation of Mare Tranquillitatis (bottom) is approximately the same as that of Serenitatis border zone. Large crater Posidonius (95 km) along mare-terra contact near top of photograph has an uplifted floor (see chap. 6). Telescopic photograph.

Albedo

Because of the contrast between exposures of feldspathic terra breccia and mafic mare basalt, the maria can be delineated even by unsophisticated observations (figs. 1.1, 5.14). For example, Luna 3 was able to reveal the important fact that the farside is deficient in maria, despite poor camera resolution (Barabashov and others, 1961; Whitaker, 1963).

This observation illustrates that the mare-terra contrast shows up best when the Sun illumination is high; shadows become very rare, and true brightness differences are observed, at Sun angles higher than about 45°. The highest possible Sun angle, exactly normal to the surface, is called *zero phase*; the reflectance at zero phase is known as *normal albedo*. Zero phase is never observed from the Earth because the Moon would be eclipsed, and so quantities described by terms such as reflectance, reflectivity, brightness, or, simply, albedo are used more or less interchangeably as convenient approximations to true normal albedo. Mare albedos generally range from 0.07 to 0.10; that is, the maria reflect 7 to 10 percent of the incident visible light (Pohn and others, 1970). With minor exceptions, the total range of lunar visible albedos is about 0.07 to 0.24.

Considerable effort has been applied to quantifying albedo because of its value in distinguishing geologic units (Pohn and others, 1970). For geologic mapping of intramare units, however, high-Sun photographs with good spatial resolution are required, even if not accurately calibrated from region to region, because they show contacts between albedo units. Mare units have been extensively mapped since preparation of the geologic map of Mare Serenitatis (fig. 5.18; Carr, 1966a), where the boundaries are relatively unobscured by rays and the contrast between the dark border and the brighter interior is evident. Four Serenitatis albedo units became standards for mapping other areas (Wilhelms, 1970b, p. F30). Although the causes of these albedo differences were not known, the fact that albedo correlates to some extent with mare units was clear.

A leading hypothesis was that albedo is related to age. Several lines of evidence suggested that dark units are young and light units old. Except with respect to large craters (Dodd and others, 1963), the inward-sloping Serenitatis border mare appeared to contain fewer craters than the central mare. The dark-mantling materials of the border are superposed on some mare materials and appeared to be young. Dark-mantling or mare materials are superposed on lighter planar materials in other areas, for example, in Alphonsus, Mare

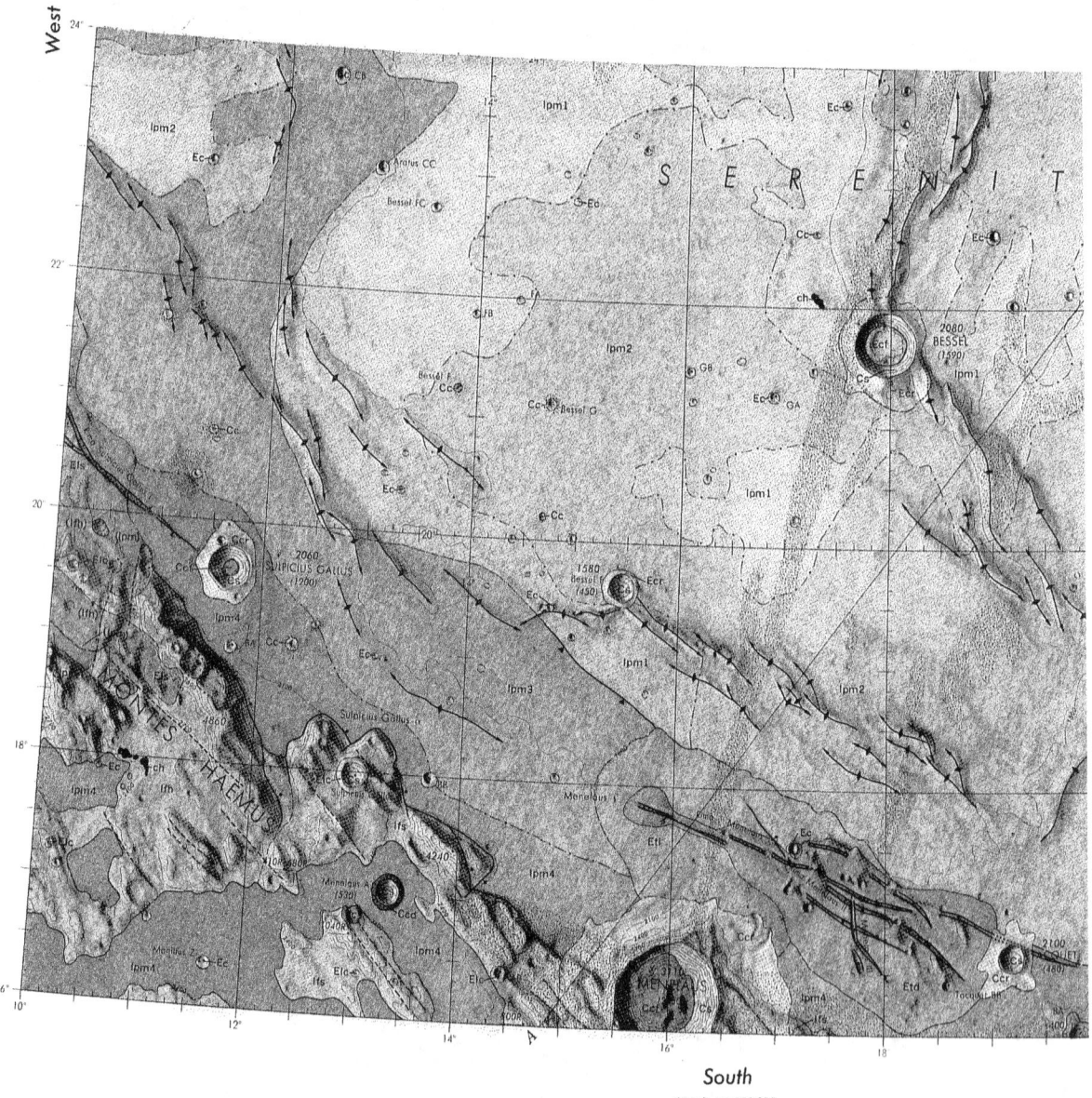

FIGURE 5.18.—Geologic map of Mare Serenitatis and margin (from Carr, 1966a). Dark mare (Ipm3, Ipm4) and dark-mantling materials (Emd, EIs) of border zone are mapped in dark tones. High-numbered units were thought to be young; they actually are young in northeast and northwest, but are old in south. Unit boundaries remain valid.

Imbrium, Palus Putredinis, Julius Caesar, and Sinus Medii (figs. 1.6, 5.10E, F, 10.35, 11.13). Moreover, crater rays are generally brighter than the maria they cross, and so cratering was assumed to be an "instant aging" process that brightened materials. Correlation of albedo units thus became a means of correlating mare ages from area to area (Wilhelms, 1970b), and albedo units were shown as age units on most lunar geologic maps.

Albedo is now known to be more a function of composition than of age, as suspected by some geologic mappers (Trask and Titley, 1966). The dark basalt of the Mare Serenitatis border that was sampled by Apollo 17 yielded some of the oldest absolute ages yet obtained on exposed mare materials (see chap. 11). Albedo is a function of exposure to the space environment and of the alteration processes that create regoliths (soils) from bedrock. It is controlled by both the amount and the composition of glasses developed from crystalline material by small impacts (Adams and McCord, 1970, 1971, 1973; Nash and Conel, 1973). These glasses are darker than their parent crystalline materials. The most significant glasses in the darkening process are those that bind other regolith fragments into the complex aggregates known as *agglutinates*. Agglutinates accumulate in the regolith as a function of exposure age. Eventually, the amount accumulated reaches a steady state in which new micrometeorite impacts destroy as much agglutinitic material as they create. The albedo is then at an equilibrium value, and the regoliths are termed *mature* (Charette and others, 1974). The observable properties of mature regoliths are relatively homogeneous over the extent of a given bedrock unit.

In mature regoliths, the composition of the agglutinates strongly affects the albedo; regoliths highest in iron and titanium are the darkest (Adams and McCord, 1970, 1973; Pieters, 1978, p. 2839–2840). Because all undisturbed mare regoliths are believed to be mature, the darkest regoliths form from the lavas that are highest in Fe and Ti. Higher contents of these elements also explain why the maria are darker than the terrae. The darkening agents may be Fe-Ti-rich glasses, Fe and Ti oxides in such opaque minerals as ilmenite that are contained in the agglutinates (Pieters, 1978), and (or) metallic Fe reduced from oxides by impacts and solar-wind sputtering (Hapke and others, 1975). Although Fe may be the greater contributor to total darkening, the differences in albedo among mare units may be due more to differences in Ti content, because Ti content varies more among mare basalt units than does Fe content (Pieters, 1978, p. 2840). In other words, all mare regoliths are dark because their glasses contain Fe and Ti, but the darkest contain the most Ti.

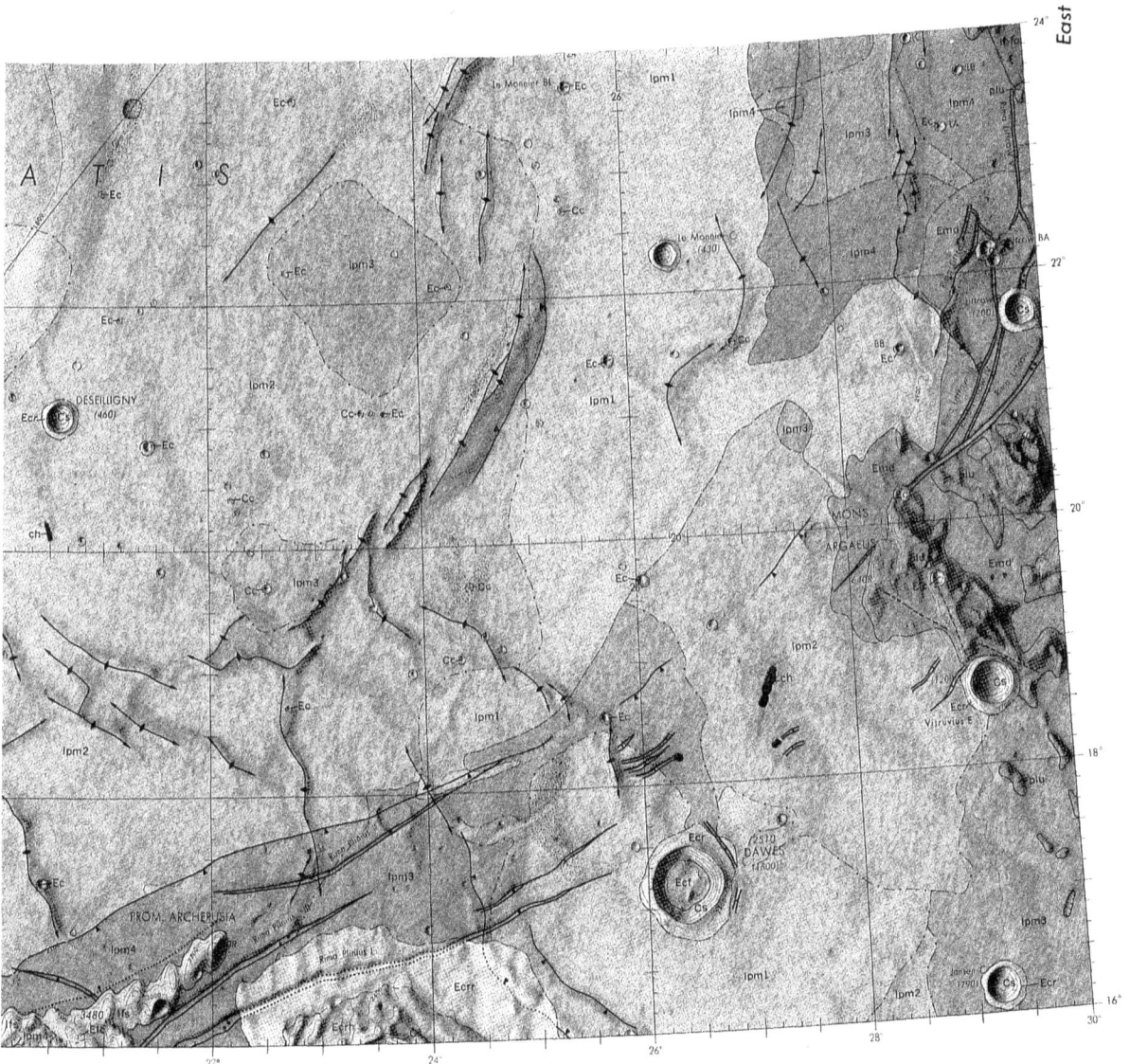

This correlation between darkness and Fe-Ti content applies only to mature regoliths. Fresh exposures of mare materials may not have been exposed long enough to accumulate impact glass in steady-state amounts. Such exposures occur in two main settings. First, new glass is shed, and nonglassy material continuously exposed, on steep slopes; the fresh material is brighter than the less disturbed material (figs. 3.28, 3.31, 5.16). In general, lunar slopes are brighter than adjacent more nearly level terrain composed of the same geologic unit (Shoemaker and Hackman, 1962). Second, relatively recent impacts exhume fresh material of mare lavas and redeposit it as bright rays. On bright, fresh lunar surfaces, more crystalline material than glass is exposed (Adams and McCord, 1970, 1971, 1973; Conel and Nash, 1970; Charette and others, 1974). Therefore, the formation of a crater ray in a given unit is not necessarily one of "instant aging" but of "instant rejuvenation." The correlation between youth and darkness demonstrated by dark-on-light superpositions is valid where the younger materials are richer in Fe and Ti, as in Alphonsus and the other examples given above, but not for all juxtapositions of dark and light.

In summary, geologic maps prepared before the mid-1970's correctly show relative ray and slope ages, and the mare units generally are correctly delineated. The mare units, however, should be interpreted as compositional, not age, units except where youth and composition correlate.

Color

Color differences are subtle on the Moon. The spectra of all areas are similar, and increase in reflectance toward longer wavelengths; that is, all lunar colors are red. Differences do exist, however, and can be detected with instrumented Earth-based telescopes and enhanced by various photographic, instrumental, and computerized image-processing techniques. Differences also appear conspicuous to the trained astronaut eye (Evans and El-Baz, 1973; Lucchitta and Schmitt, 1974). Color differences have proved to be valuable discriminators of lunar mare and dark-mantling units and, since the returned samples were analyzed, have helped extrapolate known compositions to unsampled units (table 5.1).

Reflectance spectra over wavelengths of 0.3 to 2.5 μm (ultraviolet to near-infrared) have been intensively studied during the space age for information about lunar compositions obtainable by remote means (fig. 5.19; Adams and McCord, 1970, 1973; McCord and others, 1972a, b, 1976; Adams and others, 1974; Charette and others, 1974; Pieters and others, 1974, 1975, 1980, 1982; Johnson and others, 1975, 1977; Head and others, 1978a, b; Pieters, 1977, 1978; Basaltic Volcanism Study Project, 1981, chap. 2). Areas 10 to 20 km across are measured photometrically. Absorption features due to mineral constituents of the surface material may be detected near wavelengths of 0.4, 1.0, 1.2, and 2.0 μm. These features are weak for agglutinate-rich regoliths and stronger for crystalline rocks (fig. 5.19C). To compare spectral differences, spectra are commonly divided by a standard spectrum of a relatively uniform mare surface in central Mare Serenitatis (designated MS-2; spectral class mISP, table 5.1). Small departures from this curve are then exaggerated in graphs, which are normalized to unity at 0.57 μm to eliminate albedo effects and to emphasize color differences (fig. 5.19). Most lunar spectra obtained before 1978 cover the spectral range 0.3 to 1.1 μm (fig. 5.19A, B; see references and historical background in Pieters, 1978, and Moore and others, 1980b). Spectra in the 1.0- and 2.0-μm bands have been obtained more recently (figs. 5.19C, D; McCord and others, 1981; Pieters and others, 1982).

For mapping and discriminating units, the most useful color-enhancement displays are images that show the spatial variations of simple spectral parameters (Whitaker, 1966, 1972b; Johnson and others, 1975, 1977; McCord and others, 1976). Black-and-white photographs prepared by combining a photographic negative taken at one wavelength with a positive taken at another provide much data at good spatial resolution (fig. 5.20; Whitaker, 1966, 1972b). Two major mare subdivisions, relatively red and relatively blue, can be defined. For example, Mare Tranquillitatis and the border of Mare Serenitatis are relatively blue, and central Serenitatis is relatively red. The young Imbrium flows are relatively blue, and most of the rest of Mare Imbrium is relatively red. Several intermediate colors can also be resolved.

TABLE 5.1.—*Remotely sensed properties of mare units*

[Spectral class from Pieters (1978) and Basaltic Volcanism Study Project (1981, chap. 2). Ultraviolet/visible ratio: H, high (probably more than 5 weight percent TiO_2); h, medium high; m, medium; L, low (probably less than 2 weight percent TiO_2). Albedo: D, dark (max 0.08); I, intermediate (0.03-0.09); B, bright (higher than approx 0.09); 1-μm absorption band: S, strong; G, gentle; W, weak. 2-μm absorption band: P, prominent; A, attenuated; -, undetermined.
Color (fig. 5.20): b, blue; i, intermediate; r, red.
Age (see chaps. 11-13): I1, earliest Late Imbrian; I2, middle late Imbrian; I3, latest Late Imbrian; E-I, near the Imbrian-Eratosthenian boundary; E, Eratosthenian; C?, Copernican (spectral class uncertain).
Interpretation based on Basaltic Volcanism Study Project (1981, p. 452-456)]

Symbol (pl. 4)	Spectral class	Mare	Color	Samples	Age	Interpretation
1	HDWA	Tranquillitatis, Serenitatis border.	b	Apollos 11, 17.	I2	Highest Ti.
2	HDSA	Procellarum	b	---	E, C?	High Ti, radioactivity, Fe?, olivine?
3	hDSA	Imbrium, N. Procellarum.	b	---	E	Similar to HDSA but probably lower Ti.
4	hDSP	Humorum, S. Procellarum.	b	---	E-I	Similar in Ti, Fe, and radioactivity to HDSA and hDSA, but possibly richer in pyroxene.
5	hDG-	Nubium	b	---	I3?	Medium Ti.
6	hDWA	Fecunditatis, Serenitatis border.	i	Luna 16	E, I3	Medium Ti, high Al.
7	mISP	Central Serenitatis, W. Procellarum.	i	---	I3	Low to medium Ti. Possibly richer in Fe in Serenitatis than in Procellarum.
8	mIG-	Widespread	i, r	Apollo 12	E-I	Class of generally low- to medium-Ti basalt.
9	mBG-	Nectaris, S. Fecunditatis.	i	Apollo 16?	I2	Low Ti, high Al.
10	LISP	Crisium, Roris, Frigoris.	i, r	Luna 24	I1-I3, E.	Very low Ti, high Fe.
11	LIG-	E. Imbrium, W. Serenitatis.	r	Apollo 15	I3	Low Ti, high radioactivity.
12	LBG-	Somniorum, Mortis, Imbrium border, N. Procellarum, Nubium.	r	---	I1-I3	Very low Ti, probably high Al.
13	LBSP	Frigoris, Imbrium, Nubium.	r	---	I2-E	Low Ti, high in Fe or pyroxene, high radioactivity.

Such multispectral images contain considerable compositional information. To a first approximation, the excess of ultraviolet or near-ultraviolet color over the midvisual—that is, blueness—is a measure of the content of both FeO and TiO_2 in agglutinates within mature regoliths (Pieters, 1978, p. 2835). In figure 5.20 (made from photographs filtered at 0.37 and 0.61 μm), dark areas are bluer and thus richer in these elements. The reason for the correlation (Pieters, 1978, p. 2835–2841) is somewhat complex. As discussed, Fe and Ti darken mature regoliths, and the dark materials in the regolith suppress the intensities of individual absorption bands. The more Fe and Ti in pure glass, the more ultraviolet light is absorbed, that is, the redder the spectrum. The opposite can be observed, however, for many Fe- and Ti-rich mature regoliths, because dark regoliths containing well-developed agglutinates also strongly absorb in the visible and near-infrared. The resulting greatly reduced spectral contrast causes the apparent blue color of dark mature mare regoliths (Pieters, 1978).

Low Fe and Ti contents, however, do not necessarily cause the opposite effect. Red or intermediate-color maria are lower in Ti than relatively blue maria, but their Ti content is not constrained (Pieters, 1978, p. 2842–2843). The role of Fe in the red maria is unclear. The strength of an absorption feature near 1 μm is affected both by the Fe content of the regolith glass and by the proportion and composition of the mafic minerals present. This interrelation has not yet been fully examined (Adams, 1975; Pieters, 1978, p. 2840–2841).

The dark-mantling materials have distinctive reflectance spectra (fig. 5.19D; Pieters and others, 1973; Adams and others, 1974; Charette and others, 1974). Mantles that appear black on the surface or bluish-gray to an orbiting astronaut are spectrally blue as well, and those that appear orange on the surface or reddish-brown to tan from orbit are spectrally red. These colors are due to both the composition and the crystallinity of the pyroclastic dark-mantling materials. Some mantled regions differ spectrally from any sampled materials.

The bright interiors of fresh, young craters on both maria and terrae differ spectrally from the surrounding mature regoliths because their regoliths contain less agglutinitic material and more unaltered mineral fragments (fig. 5.19C). These properties affect the general properties of the spectral continuum and allow detection of specific near-infrared absorption features (Charette and others, 1974; Pieters, 1977). Visual wavelengths and, thus, the color-difference image in figure 5.20 do not fully reveal the spectral properties of fresh craters; these properties are better revealed by spectral images in the visible and near-infrared because of the conspicuous 1.0- or 2.0-μm absorption bands of pyroxene (fig. 5.19C; McCord and others, 1976).

Orbital chemistry

Two instruments flown in lunar orbit on the Apollo 15 and 16 missions have also permitted extrapolations of sample compositions to unsampled parts of the Moon. The gamma-ray spectrometer and X-ray-fluorescence spectrometer Adler and Trombka, 1977; Basaltic Volcanism Study Project, 1981, chap. 2) are currently the only means of estimating compositions on the farside (pl. 2). Whereas the reflectance spectra help determine mineral and glass contents, the two orbital instruments determine elemental compositions independently of mineralogy.

The compositions determined are those of only the surficial layers, the uppermost tens of centimeters in the case of the gamma-ray data and the uppermost tens of micrometers for the X-ray data. Investigators are nevertheless confident that most surficial regolith materials represent underlying materials because the sharp bound-

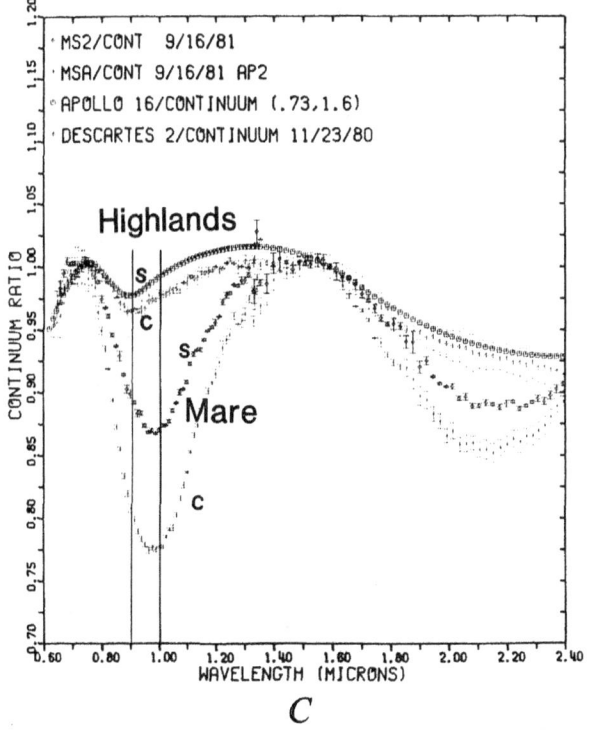

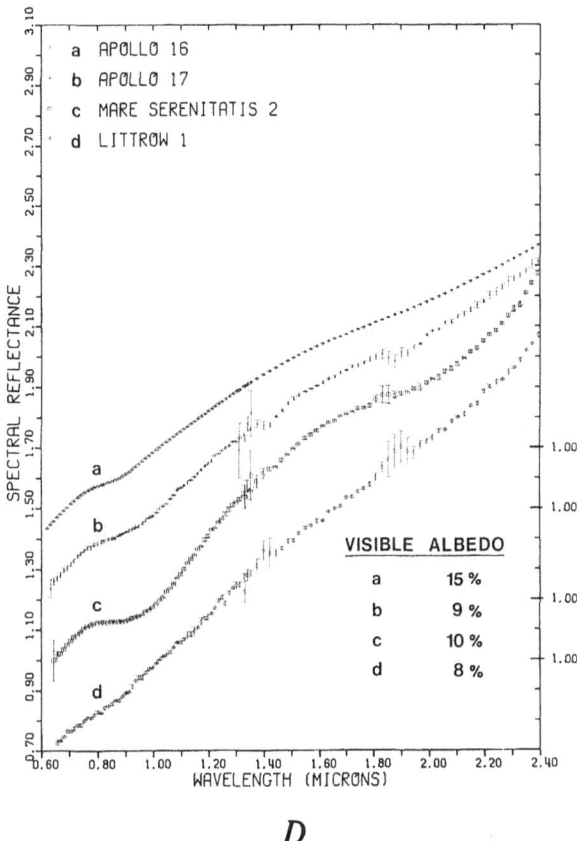

FIGURE 5.19.—Typical relative-reflectance spectra of lunar soils (surficial fine materials of regoliths) obtained by Earth-based telescopes. Courtesy of C.M. Pieters.
- A. Spectra of mare soils between 0.3 and 1.1 μm, relative to an area in central Mare Serenitatis (MS2 or MS–2) and scaled to unity at 0.57 μm. Designations to left of area names are from classification of Pieters and McCord (1976), and those to right (see table 5.1) are from Pieters (1978). After Pieters (1978, fig. 3).
- B. Additional mare spectra of same type. Slopes of plots emphasize differences of spectra from that of comparison area MS–2 (horizontal). Bandpasses of filters used in various spectral-imaging studies are shown at bottom (see Pieters, 1978, fig. 3).
- C. Near-infrared spectra of areas 5-10 km across, scaled to unity at 1.02 μm and divided by a straight-line continuum. For a given terrane (highland or mare), spectra of soils (s) and interiors of fresh craters (c) are qualitatively similar, but spectral contrast of soil spectra is reduced by alteration products in the soils. Absorption bands of highlands are centered near 0.90 μm, indicating that Ca-poor pyroxene is the dominant mafic mineral; absorption bands of maria are near 0.98 μm, indicating that Ca-rich pyroxene is the dominant mafic mineral.
- D. Spectra in the visible and near-infrared, scaled to unity at 1.02 μm: a, terra soil around Apollo 16 landing site, calibrated with returned soil sample 62231; b, Apollo 17 landing site, a Ti-rich basaltic soil; c, standard area MS2, probably a low-Ti basalt; d, pyroclastic dark-mantling material near the Apollo 17 landing site.

aries between mare and terra and among mare units, as seen on the color-difference images (fig. 5.20), have persisted since the maria formed (Kuiper, 1965, p. 13). Small fragments of nonlocal material are present in all regoliths, though not in sufficient quantity to mask the local material.

The gamma-ray spectrometer measures the natural radioactivity of K, Th, and U, as well as gamma rays emitted from other elements during nuclear reactions induced by high-energy cosmic-ray bombardment (Metzger and others, 1973, 1977; Adler and Trombka, 1977; Reedy, 1978; Basaltic Volcanism Study Project, 1981, chap. 2). The most useful results are obtained for concentrations of Th, K, Fe, Mg, and Ti (Bielefeld and others, 1976; Metzger and others, 1977; Haines and others, 1978; Metzger and Parker, 1979; Davis, 1980; Haines and Metzger, 1980). The Ti values obtained by the gamma-ray and spectral techniques agree moderately well (Basaltic Volcanism Study Project, 1981, p. 453). The gamma-ray spectrometers were limited by low spatial resolution, from 2° by 2° (3,600 km^2) to 10° by 10° (90,000 km^2) of the lunar surface; they analyzed about 22 percent of the total surface. On the basis of this areal sample, the maria are generally more radioactive than the terrae, and the western nearside maria more radioactive than the eastern (Soderblom and others, 1977).

The X-ray spectrometer detects the elements Al, Mg, and Si from the intensity of secondary X-ray fluorescence induced by solar X-rays (Adler and others, 1972; Adler and Trombka, 1977). Thus, the data are limited to illuminated parts of the lunar surface, and interpretations of the data must compensate for variations in solar activity. In all, about 10 percent of the lunar surface was analyzed. The Mg/Al ratio is most sensitive to terra-mare differences because it is substantially greater for mare materials, whereas Si content differs little between the maria and the terrae. Highly magnesian and highly aluminous mare and dark-mantling materials are also detectable by this ratio, other factors being equal (Andre and others, 1977, 1979a; Schonfeld and Bielefeld, 1978; Conca and Hubbard, 1979; Hubbard, 1979). Suitable data analysis has resolved areas as small as about 1° by 1° (900 km^2). At these resolutions, ejecta of individual craters and, therefore, the subsurface materials exhumed and redeposited in the ejecta can be analyzed (Andre and others, 1975, 1978, 1979b).

Other properties

Two other regions of the electromagnetic spectrum, accessible from the Earth or with the aid of a spacecraft, have potential value in interpretations of the maria. The first region, in the thermal infrared,

FIGURE 5.20.—Color-difference image, produced by subtracting photograph taken at 0.37 μm from one taken at 0.61 μm. Longer wavelengths (red) appear brighter and shorter wavelengths (blue) darker. Courtesy of E.A. Whitaker.

measures temperatures during a total eclipse of the Moon (Saari and others, 1966; Shorthill and Saari, 1969; Shorthill, 1973) or from orbit (Mendell and Low, 1975; Schultz and Mendell, 1978). High eclipse temperatures are valuable indicators of fresh blocky surfaces and thus of youthful craters, which may be undetectable in low-resolution photographs (see chap. 13; Winter, 1970; Thompson and others, 1974). For mare units, there seems to be a correlation between high eclipse temperatures and the youth of a unit (Moore and others, 1980b, p. B65). The knowledge gained from craters suggests that this correlation results from the coarse, blocky regoliths that thinly cover young mare units.

Lunar applications of radar have been intensively studied. Earth-based signals at 3.8 and 70 cm have yielded the most detail, including radar maps of the whole nearside disc (Zisk and others, 1974; Thompson, 1974, 1979; Moore and others, 1980b, p. B52–B61). Many factors affect radar echoes (Moore and others, 1980b, p. B53). For geologic purposes, the echoes ideally would measure depth of penetration and surface roughness at scales proportional to wavelength—larger than 70 cm for 70-cm radar and larger than 3.8 cm for 3.8-cm radar. Chemical composition, dielectric constant, electromagnetic absorption, fine-scale roughness, and regional tilt may all additionally affect the echoes. Despite the difficulty of interpretation, radar in combination with other remote-sensing techniques is a useful measure of crater youth: Depolarized radar echoes from craters decrease with decreasing concentrations of blocks visible in photographs and with decreasing infrared eclipse temperatures (Thompson and others, 1974, 1980). Radar data, infrared eclipse temperatures, and distinctive color spectra together help characterize mare units (Schaber and others, 1975; Moore and others, 1980b) and reveal the presence of block-free dark-mantling materials that might otherwise be taken for lava (Zisk and others, 1977).

During the Apollo 14, 15, and 16 missions, radar signals were relayed to and from the Moon by orbiting satellites (Moore and others, 1980b, p. B34–B41). Though technically highly successful, this "bistatic" radar experiment generated little new scientific information beyond the fact that bistatic-radar roughness correlates with the visual appearance of a terrain. Young mare units are rougher to radar than older mare units.

Integration

The examples already given show that remote-sensing techniques are most valuable when used in concert. Correlations among data sets are readily visualized when converted to a common format and displayed as false-color images showing any desired weighted combination of the data sets (Eliason and Soderblom, 1977; La Jolla Consortium, 1977; Soderblom and others, 1977). Geologic mapping requires images with good spatial resolution, including low-Sun images that enhance topographic detail, high-Sun photographs that show brightness differences (figs. 1.1, 5.14), and multispectral images (fig. 5.20). To characterize chemical and physical properties not determinable from such images, lower-resolution data from other measurements are located within each resulting map unit and then extrapolated to the entire area it covers.

Generalizations based on integration of the data, to be discussed further in chapters 11 through 13, include:
1. The western maria in the overflown strips are more highly radioactive than the eastern.
2. Ti-rich, spectrally blue lavas form at least half of Mare Tranquillitatis, the adjoining border of Serenitatis, the lobate flows of Mare Imbrium, central Mare Humorum, and much of Oceanus Procellarum. Most of these lavas are in maria that overlie the central basin and middle trough of the Procellarum basin (pl. 4). The Tranquillitatis-Serenitatis units, which include the Apollo 11 and 17 landing sites, belong to a different high-Ti spectral class (HDWA) than do the western units (HDSA, hDSA, hDSP).
3. Class hDWA occurs at several mare margins and fills most of Mare Vaporum. On the color-difference photographs, its color appears to be intermediate between red and blue and has been called "orange" (Basaltic Volcanism Study Project, 1981, p. 237). A patch at the Fecunditatis margin includes the Luna 16 landing site.
4. Red spectral classes, thought to be lowest in Ti content, are concentrated in Lacus Somniorum, Mare Frigoris, Sinus Roris, and northern Mare Imbrium; that is, they occur in diverse settings over the north half of the Procellarum basin. Apollo 15 landed within one of these classes (LIG-). Red or orange class LISP also occurs in Mare Crisium, where it probably underlies the Luna 24 landing site.
5. Red spectral class mISP forms extensive areas in central Serenitatis and the outer trough of western Procellarum. These two occurrences may differ in composition (table 5.1); they were not sampled.
6. Orange or (partly) red spectral class mIG- covers much of the area west of the central meridian and south of the strong blue-red association in Mare Imbrium. Apollo 12 landed within this belt. This class also forms much of Mare Crisium and Mare Fecunditatis. These concentrations may have little significance, however, because class mIG- is a group of low- to medium-Ti mare units that may differ compositionally (Pieters, 1978).
7. One interpretation of the orbital geochemistry is that Maria Smythii, Fecunditatis, and Crisium are all Al-rich (Conca and Hubbard, 1979; Hubbard, 1979). Orange spectral classes mBG-, characteristic of Mare Nectaris, and LBG-, common among the northern very red units, also are probably Al-rich (C.M. Pieters, oral commun., 1982).

THICKNESS

Recognition that the lunar mare surfaces are underlain by three-dimensional deposits led early to measurements of their thickness. Marshall (1961) estimated thicknesses in Oceanus Procellarum from the degree of obliteration of flooded craters, whose preburial profile was assumed to equal that of fresh craters of the same diameter (Baldwin, 1949, p. 128–138). Baldwin (1970), Eggleton and others (1974), and De Hon (1974) refined this method on the basis of better data for the profiles and rim dimensions. Mare-basalt thicknesses, including thicknesses of individual flows, also subtly affect the size-frequency distributions of craters (Eggleton and others, 1974; Neukum and others, 1975a). R.A. De Hon and his coworkers systematically extended these measurement techniques to most of the nearside maria (fig. 5.21). They found that the eastern mare materials average 200 to 400 m in thickness (De Hon and Waskom, 1976) and the western closer to 400 m (De Hon, 1979). Lenses thicker than 1,200 or 1,500 m also occur; more accurate values are hard to determine because few craters are visible in such thick mare sections (De Hon, 1979).

The accuracy of other mare-thickness determinations is uncertain. On the one hand, thicknesses will be overestimated if the craters on which they are based were highly degraded or filled before the mare flooding (Hörz, 1978), or if they have been shallowed by floor uplift (as is, in fact, observed in basins; see chap. 6). Hörz (1978) estimated that De Hon's figures are a factor of 2 too high. On the other hand, thicknesses will be underestimated if the craters are perched on older mare materials. Head (1979a) pointed out that the flooding of the present central highlands of the nearside, which must resemble the floors of some of the older basins, would require more mare material than believed by either De Hon (1979) or Hörz (1981).

A few geophysical determinations of thickness are available. The gravity data are consistent with thicknesses of 2 to 4 km in the centers of mascon maria (Sjogren and others, 1974; Solomon and Head, 1979, 1980), the largest of which are Imbrium, Serenitatis, Crisium, Humorum, Smythii, Nectaris, and Orientale (fig. 5.22; table 6.1; Solomon and Head, 1980). Large nonmascon maria include Tranquillitatis, Fecunditatis, and most of Procellarum. The 20- to 25-km-deep seismic discontinuity under Oceanus Procellarum was first interpreted as the base of the basalt section (Toksöz and others, 1974) but is now thought to represent a physical or chemical discontinuity within the terra crust (see chaps. 1, 8). Most of the mare fill of northern Oceanus Procellarum is thinner than 500 m. However, in one place in the north and in several places in the south, it reaches thicknesses greater than 1,000 m where superposed basins or large craters deepen the original Procellarum-basin floor (De Hon, 1979).

A radar sounder flown on the Apollo 17 mission directly measured depths to discontinuities in two mascon maria, Serenitatis and Crisium (Peeples and others, 1978). If certain assumptions about the dielectric constant of the overflown materials are correct, one horizon in Mare Serenitatis lies about 1 km beneath the surface, and another 1.6 km deep in the west to 2.0 km deep in the east (Peeples and others, 1978). One horizon apparently within the basalt section of Mare

Crisium lies 0.8 to 1.0 km below the surface of the shelf that separates the conspicuous ridge system and the basin rim (figs. 4.9, 5.23; Maxwell and Phillips, 1978; Peeples and others, 1978). What appears to be the same interface (Maxwell and Phillips, 1978) is 1.4 km deep inside the shelf.

Mare thicknesses have also been inferred from the geochemical experiments (Andre and others, 1979b). The craters Peirce (19 km) and Picard (23 km) penetrated the 1.4-km-thick layer in Crisium and excavated additional basalt whose Mg-rich chemistry is evident in the X-ray data (fig. 5.23B; Andre and others, 1978, 1979a, b). These craters may have excavated terra materials from beneath the mare (Head and others, 1978a). On the basis of assumed excavation depths, the total mare thickness in Crisium may be 2.4 to 3.4 km (Maxwell and Phillips, 1978).

Progressive decrease in the number of visible premare craters, the gravity data, and the radar-sounder profiles indicate that the mare materials are thickest in the mare centers (Baldwin, 1970). This relation reflects the topography of the host basin: Each concentric shelf of a basin is lower than the surrounding one, and the center is the lowest of all (chap. 4; figs. 4.9, 5.16, 5.17, 5.23). Elevations of the mare surfaces also mimic this step structure to some degree. For example, Mare Crisium lies about 2 km lower than peripheral Mare Spumans and 3 km lower than Mare Undarum (fig. 4.9). However, the mare surface drops in elevation less than the basin floor because of the inward mare thickening.

FIGURE 5.22.—Lunar gravity anomalies (from Muller and Sjogren, 1968).

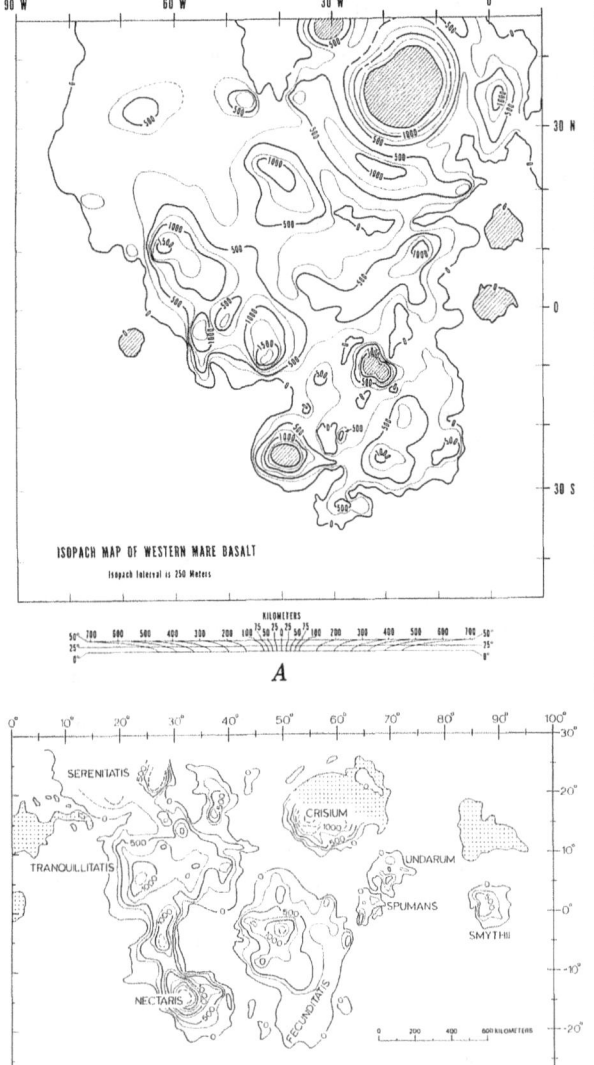

FIGURE 5.21.—Thickness of mare materials on western (A; De Hon, 1979) and eastern (B; De Hon and Waskom, 1976) nearside.

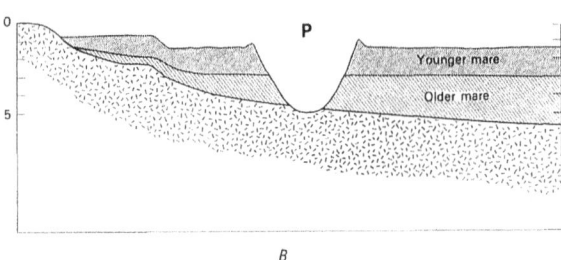

FIGURE 5.23.—Thickness of mare-basalt flows in Mare Crisium.
  A. Western Mare Crisium and part of Crisium basin, showing difference in elevation (arrows) between mare shelf and depressed center. P, crater Peirce (19 km, 18° N., 53.5° E.). Dashed lines indicate basin-concentric troughs, as interpreted by Wilhelms (1970b). Telescopic photograph.
  B. Diagrammatic cross section of Mare Crisium, based on radar-sounder profile (Peeples and others, 1978). Arrow denotes shelf indicated by arrows in A. Crater P (Peirce; also Picard) penetrates younger mare unit and excavates older mare unit (Mg-rich; Andre and others, 1978) and, possibly, terra underlying the mare section (Head and others, 1978a). Scales are diagrammatic, exaggerated vertically.

In summary, enough mare thicknesses have been measured to indicate that mare basalt is volumetrically minor. Few depths are likely to exceed 4 km; the thickest sections are in the centers of the circular maria. Thicknesses on mascon-mare shelves and nonmascon maria are no more than about 0.5 km. Likely lower and upper limits for the average mare-basalt thickness on the whole Moon are probably 0.2 and 1.0 km, respectively. Thus, the upper 75 km of the Moon would consist of 0.24 percent mare basalt if the average thickness is 1.0 km, and of only 0.047 percent if it is 0.2 km (based on coverage of 17 percent of the total lunar area of 38×10^6 km^2). Subsurface dikes and sills would add an unknown volume to these small values.

RETURNED SAMPLES

Introduction

The samples of basalt returned from the lunar maria present fewer problems than do those of breccia from the terrae, because most lavas acquire their chemical and physical properties when their magmas originate and when they are emplaced, and not from an earlier geologic unit. Most rock-size basalt samples are not shock-melted or metamorphosed. Their textures originated when the unit from which they were collected solidified. Their radiometric ages almost unambiguously date the magma's crystallization and, therefore, the unit's emplacement. No doubt remains that the sampled basalt flows originated as magmas melted within the Moon.

All nine sampling missions returned mare basalt, in greatly varying amounts. Apollos 11 and 12 landed on maria (table 1.2) and returned 16 and 36 rock-size (min 1 cm) basalt samples, respectively. The Apollo 11 material established the important fact that the maria are ancient in absolute age, though young stratigraphically. Subsequent studies showed that they were derived by partial melting from an already differentiated mantle source. Apollo 12 collected the youngest sample suite (see chap. 12) and proved that maria differ in age and composition from region to region. The remaining missions found a wide scatter in age and composition (see chap. 11), and partly clarified and partly complicated the picture of lunar-basalt petrogenesis. Apollos 15 and 17 were multipurpose missions that landed on mare surfaces and returned large amounts of basalt. Lunas 16 and 24 collected small cores of mare regolith. Even Apollos 14 and 16 and Luna 20, which landed on the terrae, obtained small bits of mare basalt. All the landing sites were near the margins of maria (pl. 4) or in the terrae (pl. 3). Therefore, only relatively thin basalt sections, and no central mascon sections, were sampled.

Composition

Lunar-mare rocks are designated "basalt" by the same general criteria used to define terrestrial basalt (Hubbard and Gast, 1971): They are dark-colored, mafic, fine-grained extrusive or shallow-intrusive rocks. Their textures are familiar from terrestrial basalt, except that they are fresher because they have not been chemically weathered (fig. 2.6). Clinopyroxene and plagioclase together compose 75 to 90 percent of most mare basalt; pyroxene is the more abundant mineral (Papike and others, 1976). Some compositional groups contain as much as 20 percent olivine and 24 percent optically opaque minerals (Fe-Ti oxides, of which ilmenite is the most abundant; Papike and others, 1976, table 3).

In elemental abundances, lunar-mare and terrestrial basalt differ substantially except in Ca content (10–11 weight percent CaO) (table 5.2; Hubbard and Gast, 1971; Taylor, 1975, 1982; Papike and others, 1976; Basaltic Volcanism Study Project, 1981, chap. 1). Lunar-mare basalt (1) contains no detectable H_2O; (2) is very low in alkalis (less than 1 weight percent), especially Na_2O; (3) is generally high in TiO_2—low lunar and average terrestrial contents are about the same; (4) is low in Al_2O_3 (8–15 weight percent) and SiO_2 (mostly 39–49 weight percent); maximum lunar and average terrestrial contents of these oxides are about the same; (5) is very high in FeO (17–22 weight percent)—higher than terrestrial basalt; and (6) is generally richer in MgO (7–18 weight percent)—the lowest lunar and average terrestrial contents are about the same. An additional important difference is in the extreme degree of reduction: Lunar-mare basalt contains essentially no Fe^{3+}; most Fe occurs as Fe^{2+}; and a minor amount of native Fe, formed by reduction at or near the surface, is always present (Basaltic Volcanism Study Project, 1981, chaps. 1, 3; Taylor, 1982, p. 292–294, 312).

Mare basalt also differs chemically from terra melt rocks of impact or volcanic origin. It is higher in FeO and MgO contents and lower in Al_2O_3 and CaO contents and Al_2O_3/CaO ratio. Its rare-earth-element patterns, as normalized to those of chondritic meteorites, generally have negative europium (Eu) anomalies, in contrast to the positive Eu anomalies of most terra materials (see chap. 8). Lunar basalt is very low in *siderophile elements*, elements that are concentrated in iron meteorites (see chap. 8). It differs further from impact melts in its lack of included debris. Lunar-mare basalt and terra materials—thus, the whole Moon—are very poor in volatiles (Wetherill, 1971; Taylor, 1975; 1982, p. 300–307).

Pyroclastic glasses, which constitute the dark-mantling materials, are partly similar and partly dissimilar to the mare lavas in bulk composition. Orange and black glasses from the Apollo 17 landing site are rich in Ti, but differ in trace elements and Mg from the high-Ti lavas in the same region (Heiken and others, 1974). Apollo 15 collected both Ti-poor green glasses and Ti-rich red glasses, neither of which are identical to the Ti-poor Apollo 15 lavas (Delano, 1979, 1980). Smaller amounts of other pyroclastic glasses are mixed in other sampled regoliths (Wood, 1975a).

Classification

Sampled mare basalt is generally classified on the basis of major-element content (table 5.2). Some authors distinguish sharply between high-Ti and low-Ti groups because the largest collections are very distinct in this element; Apollos 11 and 17 returned exceptionally Ti-rich basalt, whereas Apollos 12 and 15 returned low-Ti basalt (table 5.2; Papike and others, 1976; Taylor, 1975). The Ti gap is occupied by a few samples from the Apollo 12 landing site and small fragments from the Apollo 14 terra landing site (in breccia sample 14063, 7.3 weight percent TiO_2; Ridley, 1975). Small fragments of a very low titanium (VLT) group were found later in the Apollo 17 and Luna 24 regoliths (Papike and Vaniman, 1978). High-Al, feldspathic basalt is recognized as a distinct category, although it also was recovered in only small amounts (Ridley, 1975; Taylor, 1975, 1982; Taylor and Jakes, 1977); it characterizes collections from the eastern maria (see chap. 11) and premare basalt of the mare type (chap. 9). Small rare glass droplets and fragments found in all regoliths may represent additional magma types (Wood, 1975a; Binder and others, 1980). The continuum of spectral classes (Pieters, 1978) suggests that the gaps in Ti content and in other discriminative compositions are filled by unsampled basalt types (Papike and Vaniman, 1978). Only about a third of the observed number of spectral classes may have been sampled (Pieters, 1978).

There is now good agreement about which groups are significant, but less agreement about the classifying nomenclature. Most designations include the sampling mission. The scheme used here (table 5.2; Basaltic Volcanism Study Project, 1981, sec. 1.2.9; Taylor, 1982, p.

TABLE 5.2.—*Classification of mare-basalt samples used in this volume*

[After Basaltic Volcanism Study Project (1981, sec. 1.2.9.) and Taylor (1982, table 6.1). All values in weight percent]

Group	Distinguishing chemistry	Al_2O_3	TiO_2	K_2O	MgO
Apollo 11 high-K high Ti basalt	High K, high Ti		9–14	>0.3	7–10
Apollo 11 low-K high-Ti basalt Apollo 17 low-K high-Ti basalt	Low K, high Ti ------do------	8–10	5–9	.03–.11	
Apollo 12 ilmenite basalt	Intermediate Ti				
Apollo 12 pigeonite basalt Apollo 15 pigeonite basalt	Low Ti, high Si ------do------		1.5–5		
Apollo 12 olivine basalt Apollo 15 olivine basalt	Low Ti, high Fe, Mg ------do------				10–18
Apollo 17 very low Ti basalt Luna 24 very low Ti basalt	Very low Ti Very low Ti, high Al		<1.5	<0.04	10–11
Apollo 12 feldspathic basalt Luna 16 feldspathic basalt Apollo 16 feldspathic basalt	High Al, low Ti ------do------ High Al, low to intermediate Ti.	10–15	3–5	0.1–0.15	7–9

282–284) lists the main chemical properties that distinguish sampled groups. Most group names also include the name of a characteristic mineral observed optically (a *modal mineral*); other classifications refer to a mineral that would theoretically crystallize from a melt of the rock's composition (*norm*). For example, the terms "pigeonite (low-Ca clinopyroxene) basalt" (table 5.2), "Si-rich basalt," and "quartz-normative basalt" have all been applied to the same Apollo 12 or Apollo 15 samples. The Apollo 17 basalt suite is subdivided by trace-element contents (see chap. 11; Rhodes and others, 1976; Warner and others, 1979). Although textural terms are also used to designate some groups (chap. 11), classifications based on texture cut across those based on sampling site and chemistry (Warner, 1971). Textural variety within a basalt suite from a given site reflects differences in the crystallization and cooling histories of the lavas more than differences in composition (Lofgren and others, 1975). Some compositional differences within a given site have resulted from minor fractionation after emplacement, though fewer than in many terrestrial lavas (Taylor, 1982, p. 334).

ORIGIN AND EMPLACEMENT

The mare-basalt magmas originated by partial melting of ultramafic mantle material (mostly or entirely olivine and pyroxene) (Basaltic Volcanism Study Project, 1981, chaps. 1, 3, 4, 9). The Eu anomaly and other trace-element data demonstrate that the mantle does not consist of primitive, undifferentiated lunar material but segregated from the bulk Moon soon after it formed 4.55 aeons ago (see chap. 8; reviews by Taylor, 1975, 1982).

As discussed in chapters 12 and 13, known basalt flows were extruded until 3.5 aeons after the Moon formed. Thus, a substantial time gap separated the global differentiation that formed the mantle source from the partial melting that led to basalt extrusion. Because formation of mare basalt was a two-stage process, a source of heat in addition to that remaining from planetary accretion was apparently necessary to initiate the second-stage melting (Taylor, 1975, 1982). The amounts of radioactive K, Th, and U believed to have been present in the source regions were sufficient to generate enough heat to melt the very small observed amounts of basalt (Taylor, 1982, p. 310). These radioactive elements are most abundant in the Ti-rich basalt.

This partial melting is generally thought to have taken place between 60 and 500 km below the surface (Taylor and Jakeš, 1974; Green and others, 1975; Kesson and Lindsley, 1976; Delano, 1979, 1980; Taylor, 1982, p. 325). More precise estimates are controversial, especially concerning high-Ti basalt. Various schemes have associated basalt composition with source composition, but, again, the data are equivocal and the interpretations diverse. For example, high-Ti basalt may be derived from clinopyroxene-rich zones, VLT magmas from olivine-orthopyroxene zones, and high-Al basalt from clinopyroxene-plagioclase zones (see summary and references by Taylor, 1982, p. 330). The mantle may be crudely layered; one suggestion is that the low-Ti source lies 150 to 250 km below the surface, the high-Ti source 100 to 150 km deep, and the high-Al source 60 to 100 km deep, immediately beneath the crust (Taylor and Jakeš, 1974). Alternatively, the high-Ti source may be lower than the low-Ti source (Nyquist and others, 1977). Delano (1979, 1980) suggested that the pyroclastic glasses originated at the greatest depths, possibly 400 to 500 km below the surface, in an environment where carbon-oxygen gases may have formed (Sato, 1979).

The presence of glass and the vesicularity of many lavas are the only major indications of volatile activity on the Moon. The fact that their eruptions were driven by volatiles may explain why patches of dark-mantling materials appear where no lavas are known, as in such isolated craters as Petavius and Humboldt (fig. 4.2).

Whether or not layering exists in the mantle, the variety of sampled and unsampled compositions indicates considerable vertical and horizontal heterogeneity in the basalt-source zones. In particular, the compositional variety of basalt erupted at a given site simultaneously or within a very short time suggests that the sources were compositionally heterogeneous over short distances (see chaps. 11, 12; Green and others, 1975; Rhodes and others, 1976, 1977; Nyquist and others, 1979a; Taylor, 1982, p. 301, 320–321).

After the magmas were generated, most were modified in the original magma chambers, on the way to the surface, or on the surface (Basaltic Volcanism Study Project, 1981, p. 399–408, 498–513, 577–591). On the Moon as on the Earth (and probably all planets), early-formed crystals fractionated from the liquid magma, mantle or crustal materials were assimilated, and magmas from separate chambers mingled. Only a few magmas may have been primary in the sense that they remained unchanged after melting of the mantle source.

This volume can best contribute to the multifaceted question of mare-basalt origin by examining the geologic setting and emplacement histories of the observed mare units. Therefore, I consider why maria are concentrated in basins, especially in particular basins.

Basic hydrostatic principles suggest that extrusion of basaltic magmas is favored by (1) depressed surface elevations, (2) thin crusts and lithospheres, (3) low magma densities, and (4) deep melting sites (Solomon, 1975; Wilson and Head, 1981). Conditions 1 and 2 are most favorable in basins. A basin impact removes several kilometers or tens of kilometers of crustal material. The resulting depression is greatly reduced by an uplift of the mantle to compensate for the lost mass (chap. 4) but will partly survive if the lithosphere is strong. Addition of the ejecta also thickens the crust near the basin by a few kilometers. The depression not only collects whatever magmas are extruded but also contains most of the extrusion vents (fig. 5.24). Even if magmas melt at the same depth and ascend with equal freedom everywhere on the Moon, more would be extruded in the depressions than on the adjacent terra. A rising column of magma that would be erupted in a basin may only be intruded as dikes and sills outside the basin.

Basins may also affect extrusion in more active ways. A basin impact generates heat that may aid the melting and weaken the lithospheric barrier to magma ascent (Solomon and Head, 1980, p. 134). Subbasin structures, such as fractured or brecciated zones similar to those beneath craters, may facilitate magma movement and divert some magmas to extrusion sites high on the basin periphery (figs. 5.10D, 5.16; G, fig. 5.24).

Superposition of two or more basins or large craters enhances the likelihood of mare extrusion at a given spot. The first or the only extrusions will occur in the superposed excavations (Head, 1976a).

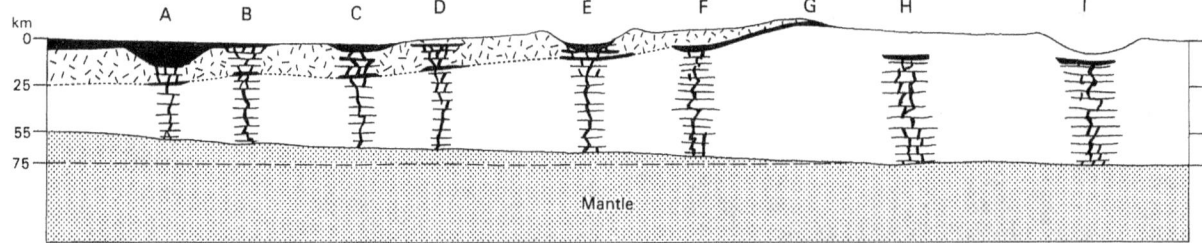

FIGURE 5.24.—Diagrammatic cross section of typical basin 1,000 km in diameter; no vertical exaggeration. Crust is about 75 km thick outside basin (see chap. 1) and is estimated at 55 km thick beneath basin center; deformed zone (above dotted line) is estimated at 25 km thick, and mantle uplift at 20 km. Magma-ascent paths, 60 km high above mantle surface, are indicated diagrammatically in nine places (A–I) as systems of dikes and sills. At A, lava was extruded onto floor of a crater superposed on the basin, and spilled beyond crater rim; at B, lava was extruded directly onto basin floor; extrusions A and B combined to flood basin center. At C, lava was extruded onto trough but did not overtop trough edges. At D, crust was too thick to allow an equal column of magma to be extruded. At E, floor of superposed crater decreased magma-ascent path by amount necessary to allow extrusion; crater interior was flooded. At F, contact of deformed zone and less severely deformed crust afforded path for lateral migration of lava, which was extruded high on basin flank at G. Crust at H and I outside basin was too thick to allow extrusion, even beneath a crater (I) equal in depth to the flooded craters inside basin.

This association is clearly illustrated on the farside, where such craters as Leibnitz, Lyot, and von Kármán and such small basins as Apollo, Planck, and Poincaré, all of which are superposed on the South Pole-Aitken and Australe basins, contain most of the farside's maria (pl. 4; fig. 5.25). In my opinion, similar superpositions of the still-larger Imbrium, Serenitatis, Tranquillitatis, Nubium, and other basins on the 3,200-km-diameter Procellarum basin cause the familiar concentrations of nearside maria.

Factors other than basin distribution have also been thought to affect mare distribution. Differential crustal thickness is commonly cited as the explanation of the 15-fold hemispheric dichotomy in mare distribution (pl. 4); the farside crust is thought to be thicker than the nearside crust (chap. 1). The Procellarum basin, however, may cause the difference. All the Moon's major maria except Crisium, Fecunditatis, Humboldtianum, Nectaris, Orientale, and Smythii lie in basins superposed on Procellarum (fig. 5.26). Of these maria, only Crisium contains a thick section of mare, and Mare Crisium is much smaller (500 km) than would be expected from the Moon's fourth largest basin (1,060 km diam; chap. 9; table 4.1).

The relative importance of magma density and melting depth is difficult to evaluate because of the many unknowns. High-Ti magmas are the densest, and high-Al magmas the least dense (Solomon, 1975). In the scheme of Taylor and Jakeš (1974), low-Ti magmas originate at greatest, and high-Al magmas at shallowest, depths. Thus, density and depth may cancel each other as factors favoring extrusion. Basin-controlled crustal thickness may play a greater role in affecting the composition of extruded lavas. High-Al magmas preferentially fill basins formed in thick crusts (Fecunditatis, Nectaris, Smythii), whereas low-Ti magmas preferentially form over thin crusts (in the Procellarum basin, especially where Serenitatis and Imbrium are superposed on that giant basin). These factors are further evaluated in chapters 11 and 12 after the historical dimension has been added.

A

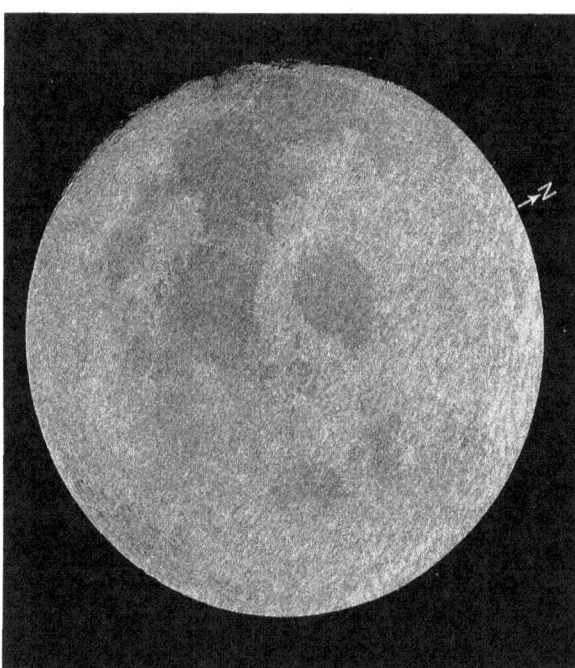

FIGURE 5.25.—Eastern nearside, east limb, and part of farside of Moon, showing spotty distribution of mare patches in sparsely flooded Australe basin (lower right edge), peripheral "lakes" concentric with Crisium basin, and other maria outside main mare concentration. Apollo 11 frame H-6665, taken during return to Earth after first lunar landing.

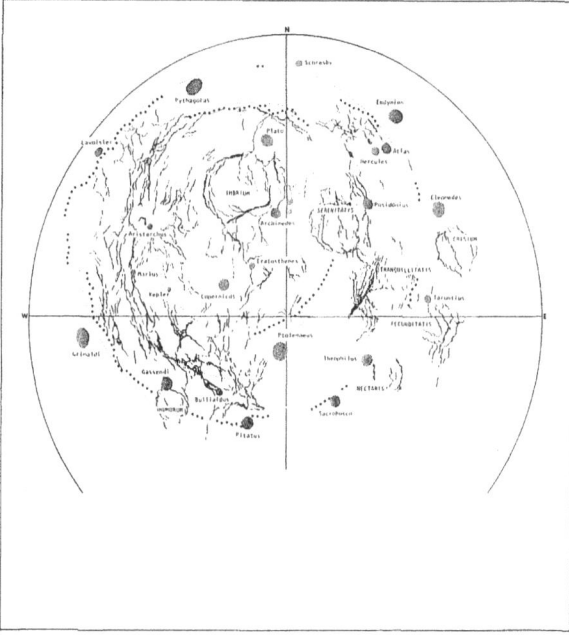

B

FIGURE 5.26.—Procellarum basin as drawn by Whitaker (1981). Courtesy of E.A. Whitaker.
A. Telescopic full-Moon photograph of nearside, showing center (P) and parts of three rings of Procellarum basin. I, center of Imbrium basin. Maria are concentrated inside largest ring of Procellarum basin. In much early work, an asymmetry in basin distribution was thought to cause the hemispheric dichotomy in mare distribution, but basins are distributed randomly (pl. 3); Procellarum may account for most of the dichotomy. Right side of photograph overlaps with left side of figure 5.25. From Whitaker (1981, fig. 2).
B. Observed scarps and terra margins concentric with basin (large dots), complete rings that best fit the topography (small dots), mare ridges (lines; see chap. 6), and major landmarks. From Whitaker (1981, fig. 1).

6. STRUCTURE

FIGURE 6.1 (OVERLEAF).—Lee-Lincoln scarp west of Apollo 17 landing site, partly covered by light-colored landslide of debris from South Massif (bottom). Scene centered at 20.25° N., 30.75° E., covers 7.5 km north-south. Apollo 17 frame P-2309.

6. STRUCTURE

CONTENTS

	Page
Introduction	107
Structures in maria and basins	107
Mare ridges	107
Arcuate rilles	111
Origin by mare subsidence	112
Crater-floor fractures	113
Straight rilles	113
General features	113
Imbrium sculpture	113
Other systems	115
Interpretation—lithospheric thickness	115

INTRODUCTION

Tectonic structures, like endogenic materials, play a smaller role in lunar geology than was once widely thought. Miscellaneous aligned landforms were interpreted as fault-controlled during much of the telescopically based mapping of the 1960's. Geologists commonly mapped faults without the objectivity that proved so valuable in mapping material units. Basin rings were generally mapped as fault-bounded, and lineament trends were commonly assigned to the "lunar grid," a hypothetical system of conjugate fractures (Fielder, 1961, 1965; Strom, 1964; Mutch, 1970, p. 247–250; Casella, 1976). High-resolution photographs, however, have substantiated the view of some astute early observers that faults are rare on the Moon (Arthur, 1962, p. 321; Baldwin, 1963, chap. 22). Except for the faults described in this chapter and for relatively minor gravitational faulting along some rings, I believe that most supposed faults are coincidental alignments of unrelated features. The lunar grid is probably an artifact of the plotting methods (Wise, 1982).

The present lunar surface, therefore, has not been extensively reshaped by endogenic forces, as has the Earth, but generally retains the pattern imposed by cratering and deposition of stratigraphic units. The dominance of lunar geology by basins and maria extends to most of the tectonic structures that do exist. Most of these structures are inside basins, and many transect both the basin and the mare (pl. 5). Some structures at mare margins that are related to dark pyroclastic materials and irregular craters have already been illustrated (figs. 5.10, 5.16). This chapter more thoroughly describes and interprets basin-related *mare ridges, arcuate rilles,* and *crater-floor fractures.*

Some *straight rilles* of fault origin also cut surfaces outside the most conspicuous mare-filled basins. These rilles have not yet been satisfactorily explained. Global deformation is commonly invoked, but this chapter presents alternative, basin-related interpretations for these structures as well. In particular, the 3,200-km-diameter Procellarum basin (Whitaker, 1981), whose importance to terra topography, lithospheric thickness, and mare localization has already been mentioned (chaps. 4, 5; pls. 3, 4; fig. 5.26), is suggested as a major control on the straight rilles as well as the ridges and arcuate faults.

STRUCTURES IN MARIA AND BASINS

Mare ridges

Deformation inside mare-filled basins is most manifest in two classes of structures—mare ridges and arcuate grabens. Mare ridges, also known as *wrinkle ridges* and now formally named *dorsa,* are linear positive features that occur in most maria (pl. 5). Most mare ridges are on the nearside because most maria are there, but even small isolated farside maria have ridges (fig. 5.2). Also, some ridges extend from mare margins into the terrae, for example, the "Lee-Lincoln" scarp at the Apollo 17 landing site (fig. 6.1) and some basin-radial ridges south of Mare Humorum (fig. 9.24B; Saunders and Wilhelms, 1974). The ridges generally occur in systems that are subconcentric and subradial to maria (pl. 5; figs. 5.17, 5.23A, 6.2). Systems in Oceanus Procellarum (fig. 6.3), which appear to be linear and parallel (Scott, 1974), are concentric with the Procellarum basin (fig. 5.26B; Whitaker, 1981).

In detail, mare ridges consist of linear segments that merge, overlap, or are arranged in echelon (figs. 6.2, 6.3); many extend as complex, branching echelon systems for hundreds of kilometers. Ridges generally have two distinct morphologic parts—a broad arch and a narrow superposed spine (Strom, 1971). Some arches are composed of overlapping domes. They range from several kilometers to as much as 10 km in width and commonly are about 100 m high; in Mare Serenitatis, some arches reach a height of 350 m (fig. 6.2; Muehlberger, 1974). Such large arches appear on all photographs, but many others appear only on photographs taken at low-Sun illuminations (less than 15°) and are not necessarily disclosed by high photographic resolutions. Many arches are asymmetric, bounded on one side by scarps or monoclinal bends that offset the mare surface by 50 to 100 m (Lucchitta, 1976). In places, the asymmetry shifts sides along the length of the ridge. Some scarps resemble flow lobes in shape. The sharp spinelike ridges commonly form shorter, narrower, tortuously braided networks along the crests and flanks of the broader arches, and may occupy 25 to 60 percent of the total arch width (figs. 6.2, 6.3; Strom, 1971). The spines are commonly 100 m high and about 200 m wide (Lucchitta, 1976). Some spines extend from the arches onto the

FIGURE 6.2.—Border zone between Mare Serenitatis and Serenitatis-basin rim, showing structures concentric with mare. View northward.
  A. West and southwest border. Rugged terra is Montes Apenninus. Grabens concentric with and truncated by mare are Rimae Sulpicius Gallus (compare fig. 5.16). Apollo 17 frame M-952.

B. East border. Two sets of grabens diverge upward from near bottom of photograph; one is closely concentric with the mare (a), and the other diverges eastward (b). L, LeMonnier (61 km, 27° N., 31° E.). Dark-mantling material at bottom (d) is truncated by lighter central mare. Arrow, Apollo 17 landing site (compare fig. 5.17). Apollo 17 frame M-940.

adjacent flat mare surfaces; others are solitary. In some places, both the arches and the narrow spines are superposed on gentle linear rises, as much as 500 m high and 25 km wide, that can be detected only on low-Sun images or by topographic data (Lucchitta, 1976).

Two main schools of interpretation of mare ridges have emerged. The volcanic school favors intrusion and extrusion of lava, controlled by global or basin-related tectonic patterns (Fielder, 1965; Quaide, 1965; Hartmann and Wood, 1971; Strom, 1971; Scott, 1974), or a variant—autointrusion of lavas into fractures (Hodges, 1973c). The tectonic school, which was founded by Baldwin (1963, p. 380–382; 1968), favors structural deformation of solid materials and is supported by the studies discussed in this chapter. Most investigators agree that many circular wrinkle ridges were formed by compaction over crater rims (fig. 6.4). Other, larger circular patterns of mare ridges are ascribed to settling of mare sections over buried basin rings; such ridges are, in fact, the only basis for locating the inner rings of Imbrium, Serenitatis, Crisium, and other deeply filled basins (table 4.1; Hartmann and Kuiper, 1962; Hartmann and Wood, 1971; Wilhelms and McCauley, 1971; Maxwell and others, 1975; Brennan, 1976).

Many exposed parts of crater rims, basin rings, and other buried terrae assume domelike landforms (figs. 6.5A, B). These landforms and the associated ridges have been interpreted as ringlike volcanic complexes. The moldinglike accumulations of material at the bases of the terra remnants resemble volcanic flows (O'Keefe and others, 1967; Strom, 1971), and their light color implies silicic composition. However, the "moldings" are common along contacts between mare surfaces and terra slopes, including those of obvious impact features (figs. 6.1, 6.5), where they consist of debris accumulated from the slopes (Milton, 1967; Offield, 1972). Therefore, these accumulations are stratigraphically younger than the ridges and the mare materials at the bases of the terra islands, whereas their source bedrock is older than the maria.

Evidence for dislocations of nonmare units (figs. 6.6, 6.7) has led to elaboration of the tectonic hypothesis. The complex ridge morphology suggests crumpling of the surface under compressional stress. Howard and Muehlberger (1973) suggested that the compression created thrust faults along intramare gliding horizons. Bryan (1973) and Maxwell and others (1975) explained the compression as a

FIGURE 6.4.—Mare ridges, formed by subsidence of mare basalt over rim of crater Lambert R (bottom; 55 km, 24° N., 20.5° W.). Unburied crater with shadowed interior above center is Lambert (30 km). Apollo 15 frame M–1010.

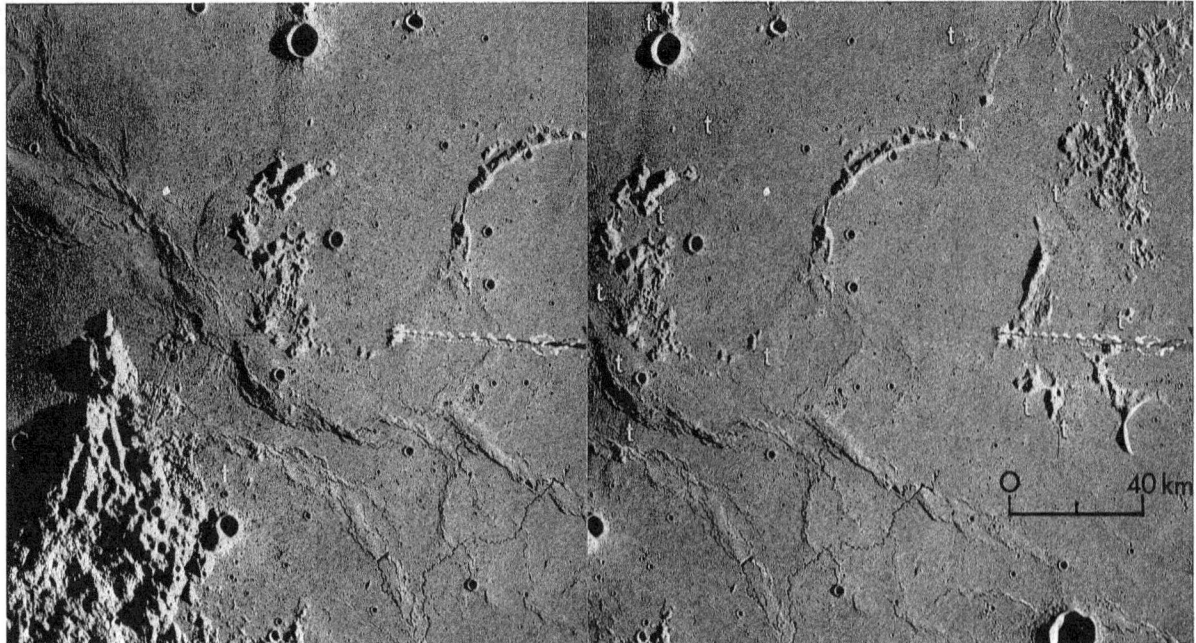

FIGURE 6.3.—Complex mare ridges and flooded terra islands in southern Oceanus Procellarum near crater Letronne (left edge of left frame; Colton and others, 1972). Some ridges and kipukas or steptoes of elevated, presumably thin maria (t) mark subsurface extensions of crater rims and other terra. Photograph covers area of lat 7.0°–13.5° S., long 32.5°–41.5° W. Object at center right of both frames is boom of Apollo 16 gamma-ray spectrometer. Stereoscopic pair of Apollo 16 frames M–2837 (right) and M–2839 (left).

buckling of the surface caused by settling of the maria into a reduced area; thick sections of mare basalt settle most, and the stresses are concentrated where the basalt thins (Maxwell and others, 1975). Lucchitta (1976, 1977b) also favored settling and suggested that the ridges lie along fault systems that include both normal and reverse faults, which are manifested by compression at the surface (fig. 6.8). In these hypotheses, accepted here, most ridges originate basically by compression resulting from vertical tectonism, although some are probably of volcanic origin (fig. 5.13).

Arcuate rilles

Arcuate rilles (*rimae*) are flat-floored, steep-walled troughs (figs. 6.9, 6.10, 6.11). Each trough consists of a floor a few kilometers wide and linear subparallel facing scarps 50 to 250 m high (Golombek, 1979). The troughs are grabens created by extension, like many similar grabens on the Earth (Baldwin, 1949, p. 197–199; 1963; Quaide, 1965; McGill, 1971; Golombek, 1979). Like the ridges, most arcuate grabens occur in parallel, crosscutting, or echelon sets; some occur

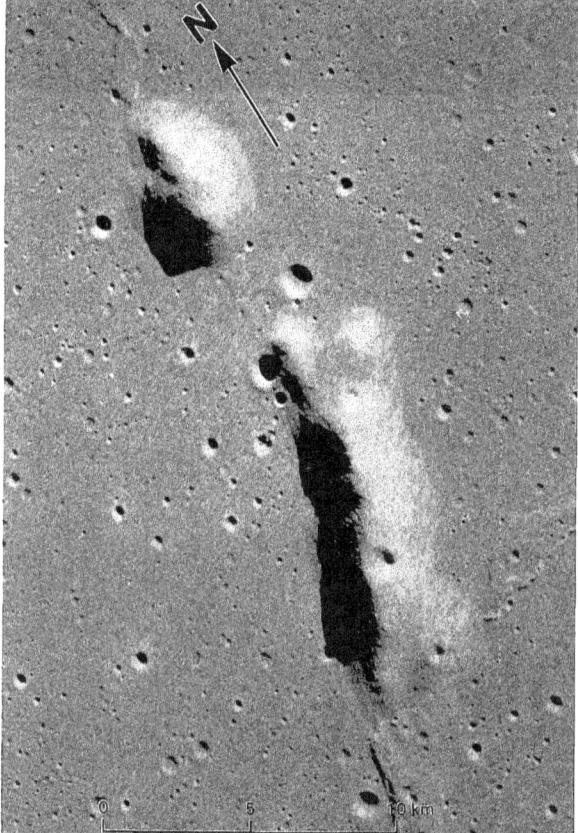

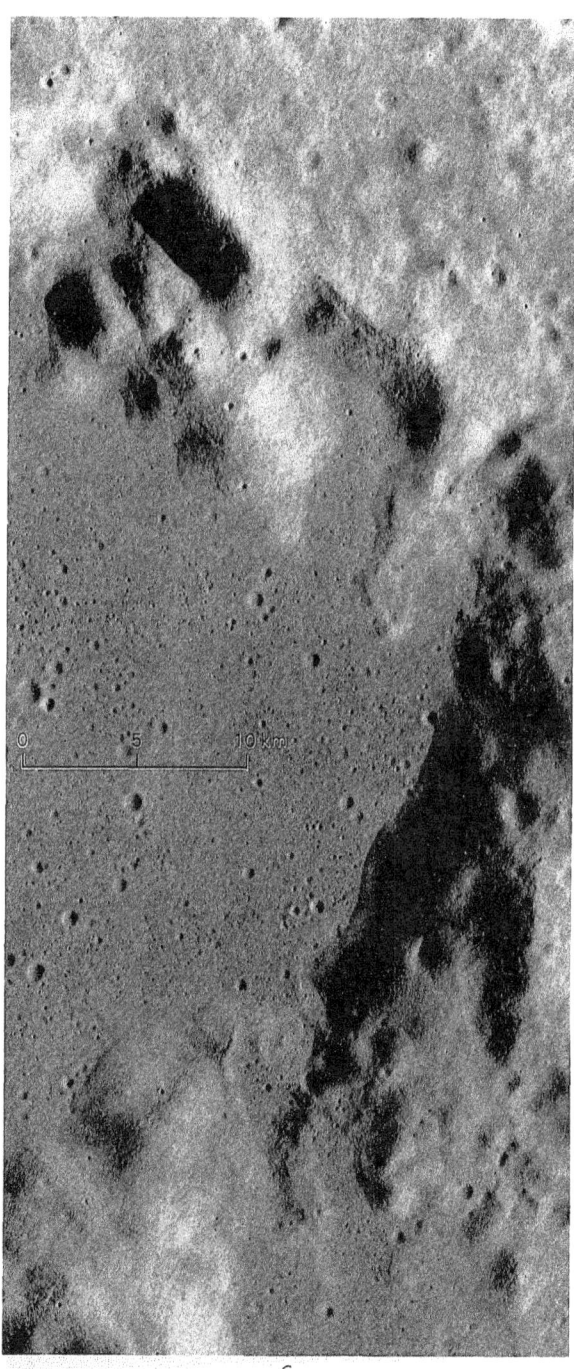

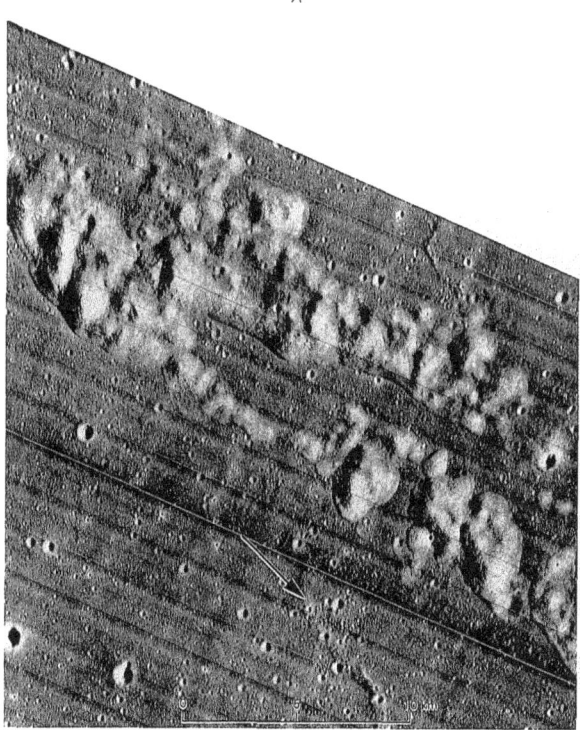

FIGURE 6.5.—Moldinglike accumulations of debris at bases of partly inundated impact features.
A. Light-colored domelike features along mare ridge are terra islands that partly deflect the ridges. Apollo 16 frame H–19244.
B. Part of crater Flamsteed P ("Flamsteed Ring") concentric with mare ridge (arrow). Orbiter 3 frame M–181.
C. Interior of crater Maraldi. Apollo 17 frame P–2302.

singly. Most grabens are concentric with circular maria and the containing basins (pl. 5; figs. 5.16, 5.17, 6.9–6.11). Moreover, they lie within the topographic-basin rim as that rim is interpreted in chapter 4 (except for some subconcentric grabens east of Serenitatis, which may be controlled by the middle Procellarum-basin rim; set b, fig. 6.2B). Grabens cut both the mare and the basin material (figs. 6.9, 6.10). With minor exceptions, they occur only in basins superposed on the Procellarum basin, particularly Imbrium, Serenitatis, Tranquillitatis, and Humorum (pl. 5), and are rare or absent in the large basins Crisium, Nectaris, and Smythii and on the farside.

Origin by mare subsidence

The recurring association of mare ridges and arcuate grabens suggests a genetic connection (Baldwin, 1963, 1968). Their geometric relation is similar in many basins: Subradial ridges innermost, arcuate rilles outermost, and concentric ridges in between. Discovery of the mascons (chap. 5; Muller and Sjogren, 1968) led to the currently prevailing interpretation, which substantiates Baldwin's (1963, p. 380–382) concept. The mare-basalt masses constitute the superisostatic loads in the mascon basins (Baldwin, 1968; Wise and Yates, 1970; Phillips and others, 1972; Bowin and others, 1975; Sjogren and Smith, 1976; Melosh, 1978; Solomon and Head, 1979, 1980). These superisostatic loads are generally smaller than they would be if no isostatic compensation had taken place since mare emplacement began (Solomon and Head, 1980, p. 136); therefore, the basalt must have subsided. This subsidence stretched the peripheral parts of the mare and the basin floor, creating the grabens, and compressed the center, creating the ridges (fig. 6.12; Baldwin, 1968; Melosh, 1978; Solomon and Head, 1979, 1980).

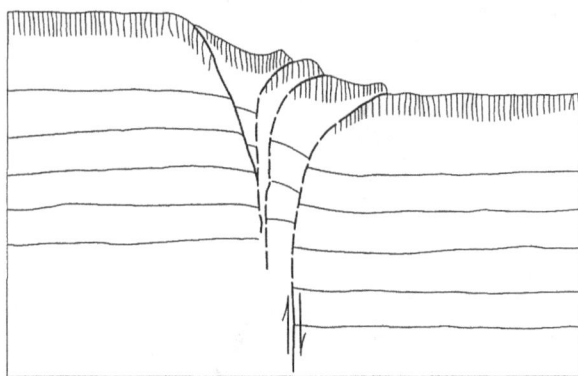

FIGURE 6.8.—Interpretation of mare ridges by Lucchitta (1976), based on models by A.R. Sanford. Thrust relations on surface originate by vertical displacements along steep faults in subsurface. Normal faulting occurs under tension in relatively raised block.

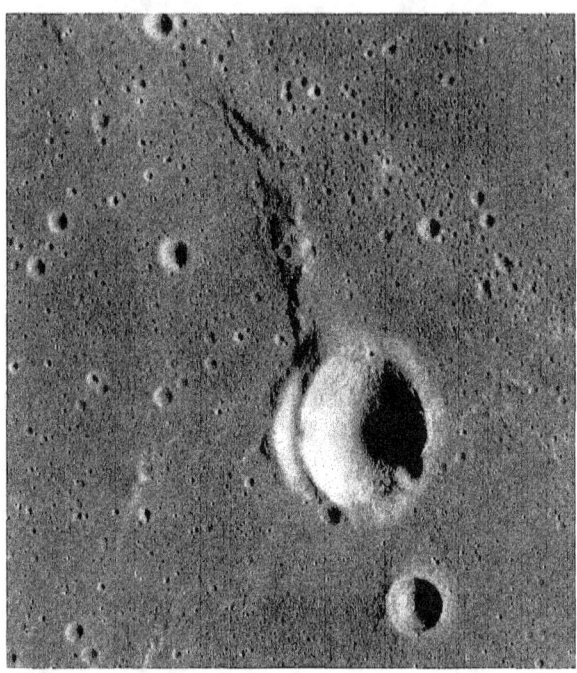

FIGURE 6.6.—Crater Bonpland D (6 km, 10° S., 18° W.) in Mare Cognitum, offset along mare ridge. Apollo 16 frame P–5429.

FIGURE 6.7.—Ridges continuous between mare and terra and appearing to thrust material from bottom over material at top. Scene centered at 7° S., 29° W., in Montes Riphaeus. View northward. Apollo 16 frame P–5452.

FIGURE 6.9.—Arcuate rilles concentric with Mare Humorum and Humorum basin, between southeastern Mare Humorum (left) and westward extension of Mare Nubium (right). Craters include Campanus (C; 48 km, 28° S., 29° W.) and more highly degraded Mercator (M; 47 km). Orbiter 4 frame H–132.

CRATER-FLOOR FRACTURES

Fractures in the floor materials of craters are among the commonest lunar tectonic structures (figs. 6.13, 6.14). They are concentrated in and near maria and basins without maria (pl. 5). Fractured-floor craters larger than about 150 km also occur in nonbasin settings, for example, Petavius and Humboldt (figs. 4.2, 9.5, 9.6). These two craters are gravitationally neutral; they have neither the negative anomalies of fresh craters nor the positive anomalies due to mare basalt (Dvorak and Phillips, 1978).

The fractured floors are higher relative to the crater rims and surrounding terrain than are the floors of typical impact craters (Pike, 1971). Many of the floors, however, occur in craters with other morphologies diagnostic of impact origin—such as central peaks, as in Gassendi and Posidonius (figs. 5.17, 6.13A), and the whole range of exterior impact phenomena including rays, as at Taruntius (fig. 6.13B). Uplift of impact-crater floors is the evident explanation. The presence of fractured floors in Petavius and Humboldt (fig. 4.2) suggests that originally negative gravity anomalies resulting from mass loss during impact excavation were erased by the uplifts. Vitello on the Mare Humorum border (fig. 6.13C) possesses a freshly fractured floor, a mantle of dark material, and an enhancement in thermal-infrared wavelengths during eclipse (Shorthill and Saari, 1969)—all features pointing to a caldera origin or, at least, to eruptions from the floor fractures (Saunders and Wilhelms, 1974). However, these features are also consistent with floor uplift (fig. 6.15), which opened fissures fresh enough to expose blocks that appear "warm" during eclipse, and with mantling by the dark material—a common phenomenon along other mare borders as well (Titley, 1967).

The uplift interpretation also suggests an explanation for a wide range of other lunar landforms (Brennan, 1975; Schultz, 1976a). The shallow tilted floors and arcuate structures of 15 craters larger than 20 km in diameter in the Smythii basin (fig. 6.13E; Wilhelms and El-Baz, 1977) suggest cylinderlike uplifts that broke free where the wall and floor meet (fig. 6.15). Especially great uplift probably formed the irregular, knobby elevated interior of Gaudibert, along the Mare Nectaris margin (fig. 6.13F; Brennan, 1975), which had been thought to be volcanic (Elston, 1972). Even small craters with such elevated interior structures as nested rims or rings of bulbous material (figs. 6.13D, G) may be modified by tectonic uplift rather than by volcanism, as is commonly believed (for example, Schultz, 1976b, p. 12–15). Twin impacts followed by uplift account for the nearly identical neighboring shallow-floored craters Sabine and Ritter along the Tranquillitatis border (fig. 3.14C), which superficially resemble terrestrial calderas more than they do impact craters (Morris and Wilhelms, 1967; De Hon, 1971).

Although many fractures and unusual landforms have been created by uplifts of impact-crater floors, a few fractured-floor craters have a less well organized fracture pattern that suggests shrinkage of a cohesive material (fig. 6.13H). Most such craters lie near but outside basins, and the fractured material may be impact melt ejected from the basins (Moore and others, 1974; Wilhelms and others, 1979), unless it is otherwise-unknown terra volcanic material (Schultz, 1976b, p. 68; Stuart-Alexander, 1978).

Igneous intrusions have been suggested as the cause of floor uplifts because the uplifts are concentrated near maria and are commonly overlain by dark mantles or small maria (Brennan, 1975; Schultz, 1976a). Intrusions seem reasonable, considering the affected craters' setting. All that is required, however, is isostatic leveling resulting from viscous relaxation of the substrate (Danes, 1965; Hall and others, 1981). The required plasticity is probably due to the weaker lithosphere beneath basins and large craters.

STRAIGHT RILLES

General features

More puzzling than the arcuate grabens and floor fractures are rilles that are here called straight by contrast with the other, curved or complexly shaped fractures. Although long stretches are straight (pl. 5; figs. 6.16–6.21), most straight rilles curve or sharply inflect as a whole. In detailed shape of profile and in their echelon patterns, straight and arcuate rilles are similar (fig. 6.16). The straight rilles are also grabens (Quaide, 1965). Some rare single fault scarps not part of a rille are also considered here.

Imbrium sculpture

Faulting initiated by the Imbrium impact was the preferred explanation during the 1960's for the "Imbrium sculpture" system of radial grooves and ridges (Kuiper, 1959; Shoemaker and Hackman, 1962; Hartmann, 1963, 1964b; Strom, 1964; Wilhelms, 1970b, p. 15; Wilhelms and McCauley, 1971). A coalescence of elliptical craters to form the grooves is evident on Apollo photographs (fig. 3.10B), however, confirming the secondary-ejecta origin proposed by Gilbert (1893) and Baldwin (1949, 1963). Nevertheless, some true faults are radial or subradial to Imbrium; those cutting the Apennine Bench, for example, probably formed in response to adjustments of Mare Imbrium and the Imbrium basin (fig. 6.10).

Vallis Alpes (the Alpine Valley, fig. 6.17) lacks the smoothly scalloped matching walls and raised lips characteristic of low-angle secondary chains. Some of its walls match but are jagged, steep, and

FIGURE 6.10.—Apennine Bench, showing grabens concentric with Imbrium basin cutting basin, bench, and mare material (lower right). Radial grabens and vent for dark-mantling material (arrow) are also present (Imbrium center is outside upper left corner of photograph). Large rugged mass is Montes Archimedes, overlain and surrounded by planar Apennine Bench Formation (see chap. 10) and by secondary craters of Archimedes, centered north of view. Mare at upper right is Palus Putredinis, west of Apollo 15 landing site. Pair of craters at left edge of bench are Feuilleé (left; 9 km) and Beer (9.5 km). Mosaic of Apollo 15 frames M–589 through M–594 (from right to left).

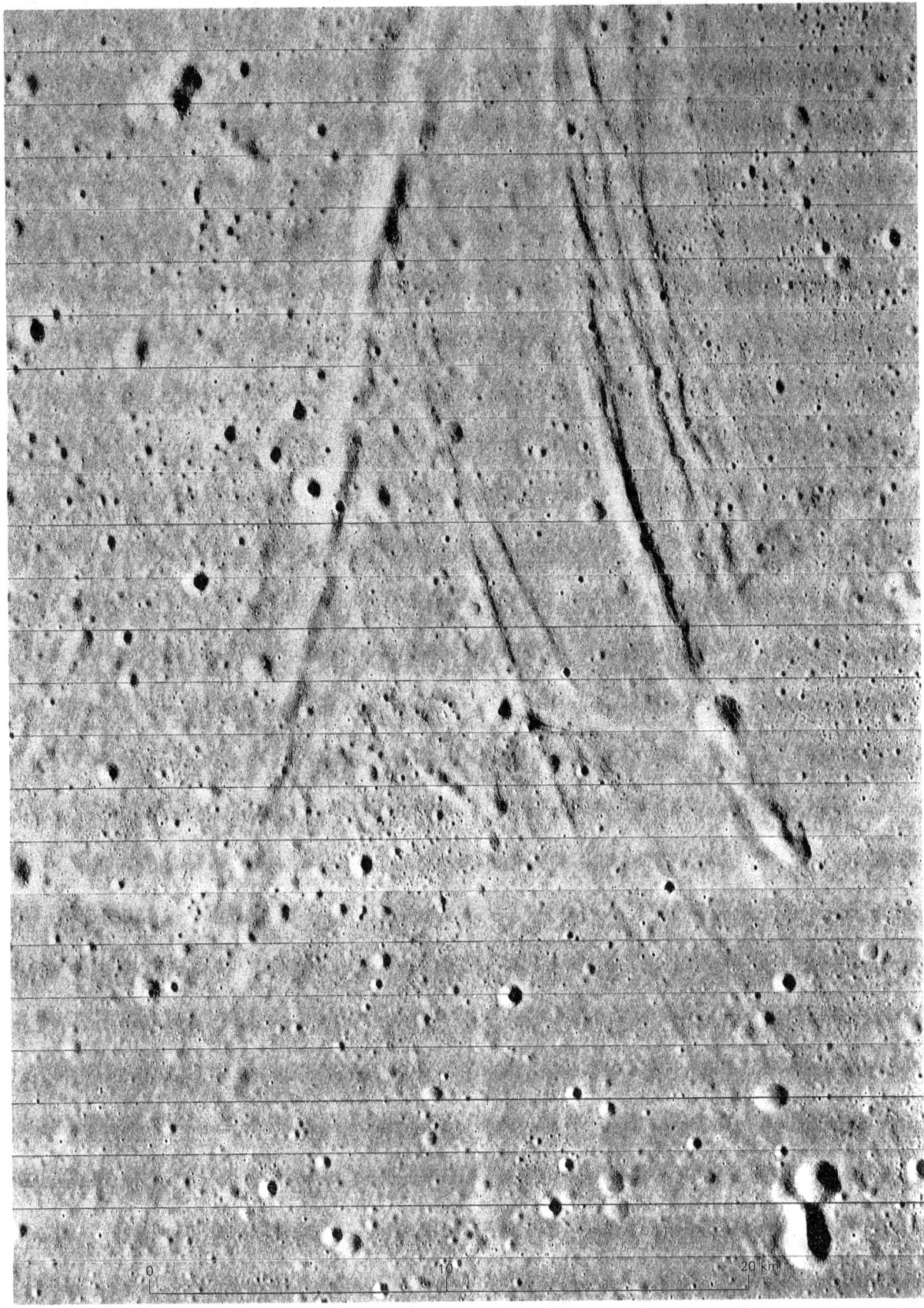

FIGURE 6.11.—Arcuate rilles in border zone of Mare Tranquillitatis (5° N., 21° E.) south of crater Arago, about 90 km east of mare edge. Intricate structure shown here is probably also common in other, less well photographed rille systems. Orbiter 2 frame M-43.

composed of linear segments. The valley is probably a graben formed shortly after the Imbrium impact, as was once proposed for all sculpture. Its radiality is consistent with an origin by isostatic doming of the subbasin mantle (see chap. 4).

Such radial or subradial faults as the Straight Wall (Rupes Recta, fig. 1.8), one leg of Rima Hyginus (figs. 5.10E, 6.16A), the Cauchy set (fig. 6.18), and several parallel grabens in southwestern Mare Fecunditatis (fig. 6.19) are more puzzling because they cut mare materials. They cannot be direct products of the Imbrium impact. If they are related to Imbrium, they may have formed by some sort of rejuvenation of radial fractures (Wilhelms, 1972a; Holt, 1974; Mason and others, 1976).

Other systems

The geometric relation of other straight rilles to basins is less apparent. Many grabens in the central and east-central equatorial zone of the nearside are concentrated along trends between 20° and 30° north of west (pl. 5). Most of these grabens belong to a set that extends 1,600 km east-southeastward from Rima Bode II (fig. 5.10H), through the eastern leg of Rima Hyginus and Rimae Ariadaeus (fig. 6.16), to the Imbrium-radial rilles in Mare Fecunditatis (figs. 6.19, 9.7). The only basin to which all these grabens are geometrically related is Procellarum, to whose center (23° N., 15° W.; Whitaker, 1981) they are radial or subradial. Their trend is paralleled in other areas by the Cauchy set of structures (also subradial to Procellarum; fig. 6.18), the short Müller crater chain (fig. 3.16A), and the 300-km-long Abulfeda crater chain (fig. 3.16B). The origin of these crater chains, which are radial to no known basin or large crater, is unclear (see chap. 3). Some grabens near Sinus Medii have a complementary trend of 20°–30° north of east (pl. 5; fig. 10.28).

At the east end of Sinus Medii is the complex Triesnecker system of grabens (pl. 5; fig. 6.20). This system as a whole is oriented north-south but includes many other trends, which mutual transection relations show to be contemporaneous (Wilhelms, 1968). East-west extension is the apparent cause.

The Moon's most complex system of (exposed) grabens occupies the southwest shore of Oceanus Procellarum from about lat 5° N. to lat 22° S. (fig. 6.21). The prevailing trends are nearly concentric with the Procellarum shoreline or radial to the general region of both the Imbrium- and Procellarum-basin centers. Other straight rilles lie along the northwest Procellarum shore, which also contains the largest lunar concentration of fractured-floor craters (pl. 5; fig. 6.14).

INTERPRETATION— LITHOSPHERIC THICKNESS

Many, if not most, of the structures described in this chapter are ascribable to thinning of the elastic lithosphere beneath basins. The subsidence of basin-filling mare basalt is aided by a thin lithosphere and inhibited by a thick lithosphere (Melosh, 1978; Solomon and Head, 1980). Solomon and Head (1980) suggested that a lithospheric thickness of about 75 km has prevented graben formation in the Crisium, Nectaris, and Smythii basins. The crust in the region of the seismic stations on the southwest nearside is 45 to 60 km thick (chap. 1). Because arcuate grabens have formed in this seismically explored region, such crustal thicknesses apparently facilitate mare subsidence and graben formation. Because the estimates of lithospheric and crustal thickness are similar in magnitude, the feldspathic crust and the elastic lithosphere were probably equivalent at the time of graben formation. The mantle constituted the more plastic asthenosphere.

Chapter 4 suggests that the Procellarum-basin impact exerted a major control over crustal (lithospheric) thickness (fig. 6.22). Consequently, arcuate grabens are restricted almost entirely to the interior of that basin (pl. 5). Even the old shallow basin Tranquillitatis and the thin basalt of Mare Tranquillitatis, which lie inside Procellarum, are cut by marginal grabens (fig. 6.11). Lithospheric thickness related to Procellarum also may be reflected in the gravity structure. Outside Procellarum, even such relatively thin maria as Nectaris, Orientale, and Smythii preserve mascons. Despite their thinness, these maria lie at low elevations (more than 3.5 km below the average lunar sphere, 1,738 km in radius; fig. 6.23; table 6.1).

Thus, the thick lithosphere apparently hindered isostatic uplift of basins, extrusion of mare basalt, sinking of mascons, and formation of grabens. Each condition was opposite inside Procellarum. Oceanus Procellarum and Mare Tranquillitatis, though areally large, have only small, local gravity highs (one part of southern Procellarum and the mare-ridge feature Lamont in Tranquillitatis; figs. 11.1, 11.9; Scott, 1974). The superisostatic loads per unit area in Imbrium and Serenitatis are smaller than would be expected from their thick mare-basalt sections (table 6.1; Solomon and Head, 1980). Apparently, the thin lithosphere inside Procellarum abetted early isostatic uplift of such basins as Tranquillitatis and permitted later isostatic sinking in response to loading even by thin basalt. The concentration of maria in Procellarum and in the superposed basins is another result of the thinner lithosphere. Although mascons are not known on the farside, the depth of the giant South Pole-Aitken basin (5–7 km; Stuart-Alexander, 1978), the paucity of farside maria except in the basins superposed on South Pole-Aitken (pl. 4), and the absence of grabens all suggest farside lithospheric thicknesses at least as great as those on the non-Procellarum nearside.

The factors that localize the straight rilles are less clear but may be further effects of the Procellarum basin. The complex system along the west Procellarum shore comprises the grabens most likely to be related to this giant basin. Widespread extension of the Procellarum margin is suggested by the trends and extent of the fracture system, which are consistent with a broad regional uplift of the mantle under the basin.

More speculatively, the many straight rilles on the central and east-central nearside may have a similar origin. Mantle uplift beneath the Procellarum basin may explain the 1,600-km-long system and such "Imbrium-radial" faults as Rupes Recta, the grabens cutting the Fra Mauro peninsula (fig. 5.8), and the Cauchy set. Continuation of mild uplift into the time of mare-basalt extrusion would account for the otherwise-puzzling transection of the maria by these faults. The Triesnecker system, whose overall orientation diverges from others in its vicinity, may have been localized by Sinus Medii, whose southeast boundary is a Procellarum ring (pl. 4; fig. 5.26). Sinus Medii may have been isostatically uplifted because of viscous relaxation of the weak lithosphere, in the manner of a fractured crater floor. Tides, relaxation of a tidal bulge (Melosh, 1977), or some other global effect would seem to be attractive alternative causes of the deformation of this near-Earth zone. To my knowledge, however, the observed distribution of lunar faults is inconsistent with any of the global mechanisms that have been thus far proposed. The Triesnecker system and the systems of long straight grabens, therefore, provide additional examples of basin-related alternatives to global origins for lunar structures.

A progressive thickening over time of the lunar lithosphere in all regions (Howard, 1970) is evidenced by increasing resistance to deformation (Solomon and Head, 1980). The arcuate and straight grabens cut terra units and old mare units (chap. 11), whereas the ridges deform both old and young mare units (Lucchitta and Watkins, 1978). Thus, the graben opening was the first deformation to cease. Mare extrusion ceased somewhat later, when the elastic lithosphere exceeded 100 km in thickness (Solomon and Head, 1979). Presumably, most previous conduits for ascent of mare magmas were shut off. Wrinkling of the central mare surfaces and crater-floor uplift were the longest enduring modifications. All these changes were presumably the result of general global cooling. Decreasing plasticity of part of the mantle added thickness to the lithosphere, which once consisted only of the terra crust. Although the mascons indicate that the superisostatic loads still exist, continued subsidence is unlikely because endogenic moonquakes do not all correlate with maria (Nakamura and others, 1979; Solomon and Head, 1979). If any asthenosphere remains today, it is too deep to affect surface deformation. The only recent deformation that has been suggested since analysis of the Apollo results is minor thrust faulting in the terrae (Binder, 1982).

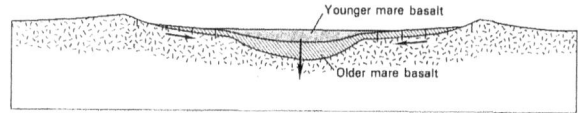

FIGURE 6.12.—Diagrammatic cross section showing stresses responsible for graben opening and ridge formation during subsidence of mare basalt in a double-ringed basin. Short vertical lines cutting older mare unit and underlying basin material denote grabens.

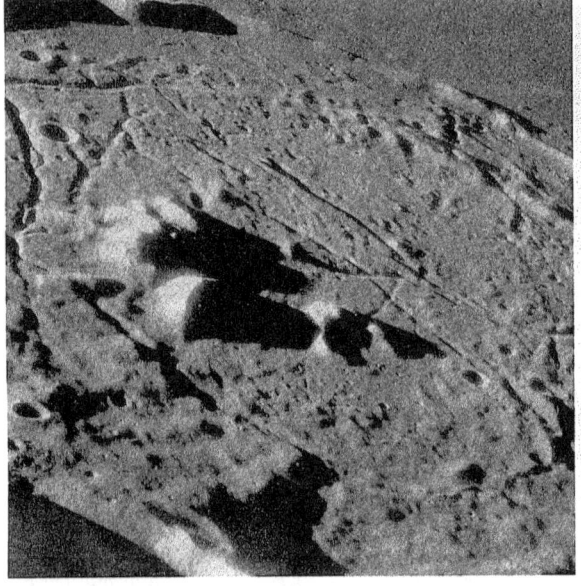

A

C

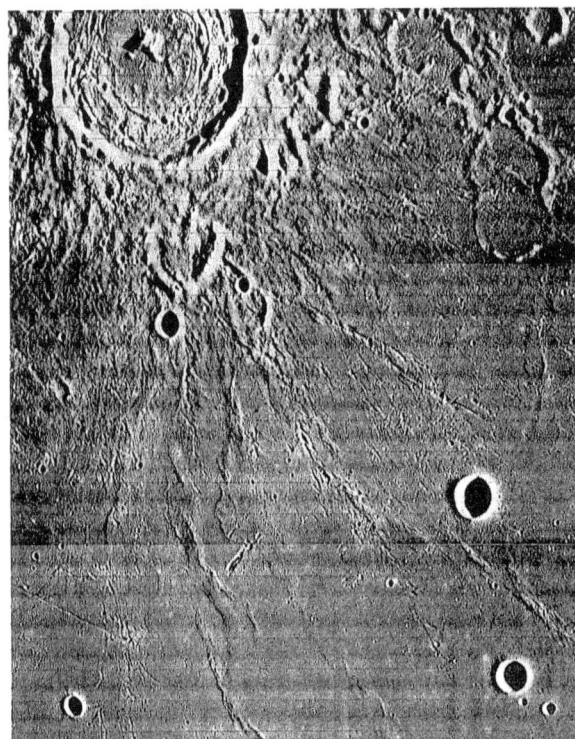

B

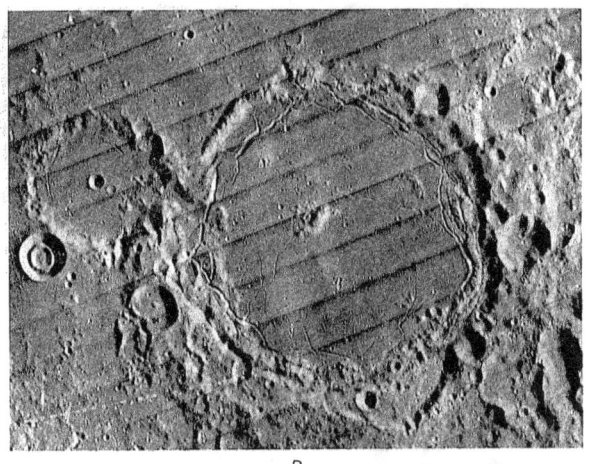

D

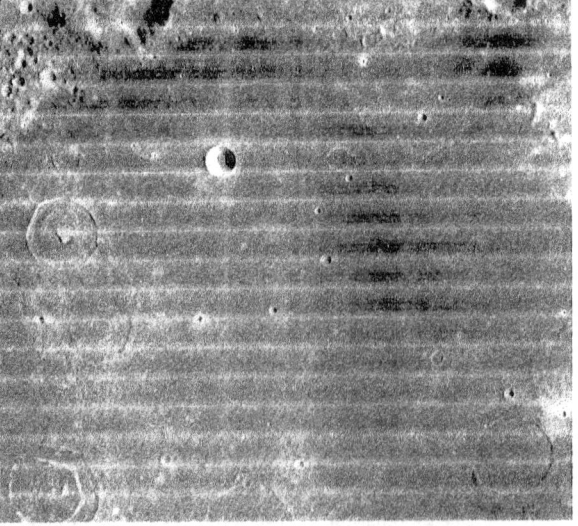

E

FIGURE 6.13.—Raised fractured floors of craters.
  A. Gassendi (110 km, 17.5° S., 40° W.), on north border of Mare Humorum. Floor is almost as high as mare, but peaks indicate impact origin. View southward. Apollo 16 frame H-19295.
  B. Taruntius (56 km, 5.5° N., 46.5° E.), on northwest border of Mare Fecunditatis, a typical fresh impact crater except for raised floor. Orbiter 1 frame M-31.
  C. Vitello (42 km, 30.5° S., 37.5° W.), on south margin of Mare Humorum, with dark-mantled rim and very fresh fractures on uplifted floor. Orbiter 4 frame H-136.
  D. Pitatus (97 km, 30° S., 13.5° W.), on south border of Mare Nubium. Concentric rilles result from uplift of crater floor and later mare fill. Small bull's-eye crater at left may have similar cause. Chain of craters at right are secondary-impact craters of Imbrium basin. Line of white dots in lower left is photoprocessing artifact. Orbiter 4 frame H-119.
  E. Group of double-ringed craters, 30 to 35 km in diameter, in Mare Smythii. Cylinderlike uplift of floor is indicated (fig. 6.15). Orbiter 4 frame H-17.

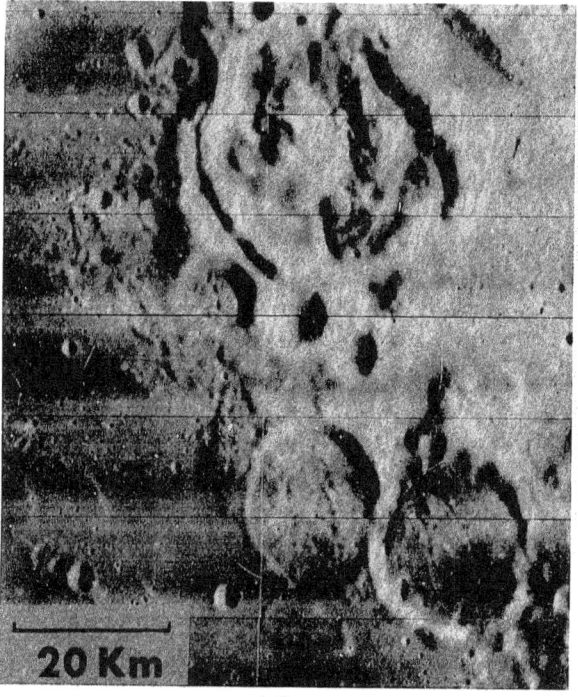

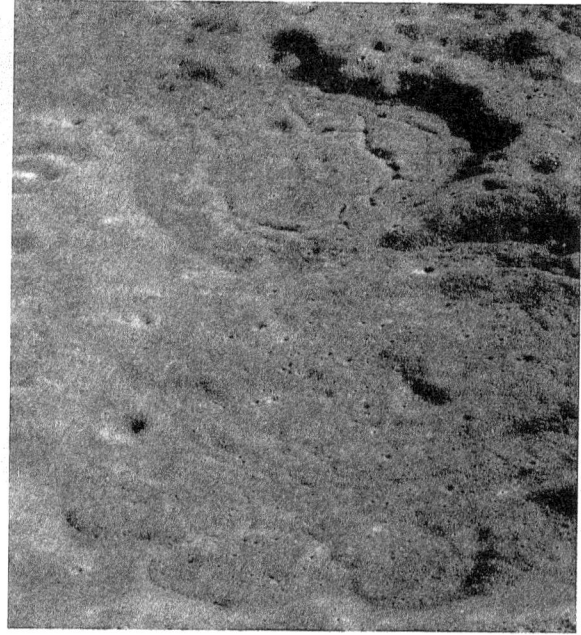

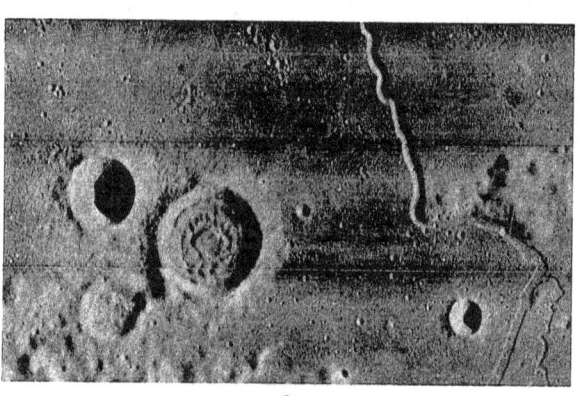

F. Gaudibert (34 km, 11° S., 38° E.) and two smaller craters on margin of Mare Nectaris. Bulbous landforms are possible evidence for viscous extrusive volcanism (Elston, 1972) but were probably created by extensive uplift (Brennan, 1975). Dark-mantling material surrounds teardrop crater along fracture in lower right crater. Orbiter 4 frame H–72.

G. Liouville DA (11 km, 46.5°, 52° W.), on border of Sinus Roris, possibly containing volcanic extrusions (Scott and Eggleton, 1973) but probably another uplift phenomenon. Sinuous rille Rima Sharp I is at right. Orbiter 4 frame H–163.

H. Tamm (38 km, 4.5° S., 146.5° E., foreground) and Van den Bos (32 km), 225 km southeast of Mendeleev-basin rim, filled by fissured, viscous-appearing material possibly emplaced as impact melt of Mendeleev. View southward. Apollo 10 frame H–4966.

TABLE 6.1.—*Physical properties of major nearside basins containing large maria*

[D, diameter (km); basin D's from table 4.1.
Average elevation: Elevation of central mare inside obvious sloping margins, in km below 1,738-km datum (S.S.C. Wu, written commun., 1983).
Mascon mass: excess mass in units of 10^{20} g (compiled by Solomon and Head, 1980, table 2).
Mascon load: superisostatic load in center of mare, in units of 10^8 dyne/cm^2 (calculated by Solomon and Head, 1980, table 2, using different mare diameters than those listed here; they took Smythii as 340 km, Grimaldi as 100 km, and Nectaris as 400 km in diameter)]

Basin	Basin D	Mare D	Ratio	Outer D of graben zone	Ratio to basin	D of 1-km isopach	Ratio to basin	Ratio to mare	Average elevation	Mascon mass	Mascon load
Imbrium (with Frigoris).	1,500	1,450	---	---	---	---	---	---	---	---	---
Imbrium (without Frigoris).	1,160	960	.83	1,160	1.00	350	.30	.35	2.5-3.5	23.0	.9
Crisium	1,060	420	.40	<420	<.40	210	.20	.50	4.0-4.5	9.4	1.3
Orientale	930	280	.30	500	.54	150(?)	.16(?)	.54	4.0	3.3	.9
Nectaris	860	300	.35	320	.37	150	.17	.50	>3.5	9.0	1.3
Smythii	840	200	.24	<200	<.24	0	0	0	4.0-4.5	5.0	1.4
Humorum	820	350	.43	750	.91	125	.15	.36	?	5.2	1.4
Tranquillitatis	775	600	.77	710	.92	150	.19	.25	1.5->2.5	0	0
Serenitatis	740	600	.81	850	1.15	300(?)	.41	.50	3.0-4.0	14.0	1.3
Fecunditatis	690	600	.87	600	.87	150	.22	.25	2.0-3.0	0(?)	0(?)
Nubium	690	600	.87	620	.90	0(?)	0(?)	0(?)	>3.0	0	0
Humboldtianum	600 (each)	120	.20	<120	<.20	0(?)	0(?)	0(?)	?	0(?)	0(?)
Grimaldi	430	150	.35	360	.84	0(?)	0(?)	0(?)	3.0-3.5	.76	1.4

FIGURE 6.14.—Numerous fractured-floor craters along west border of Oceanus Procellarum between lat 28° N. and 57° N. R, crater Röntgen (126 km, 33° N., 91° W.), superposed on Lorentz basin. Linear structure at bottom is radial to Imbrium basin. Orbiter 4 frame H-183.

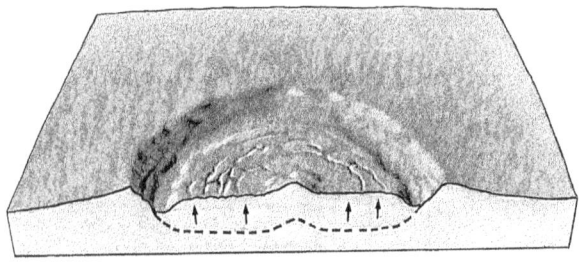

FIGURE 6.15.—Crater floor uplifted (arrows) from original position (dashes) lower than surrounding terrain.

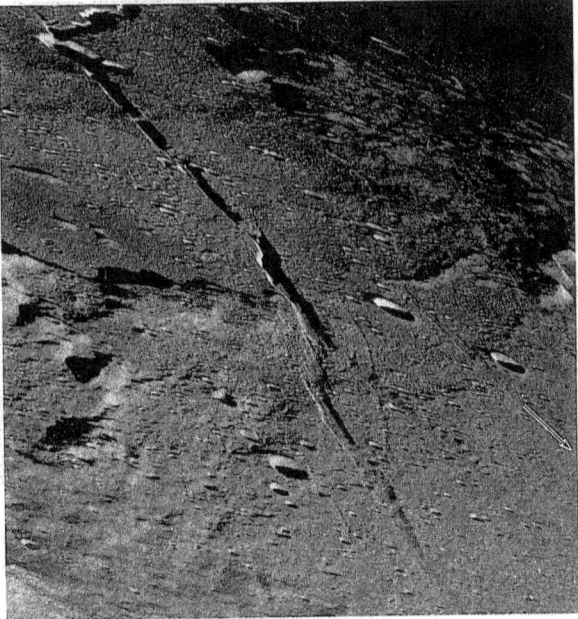

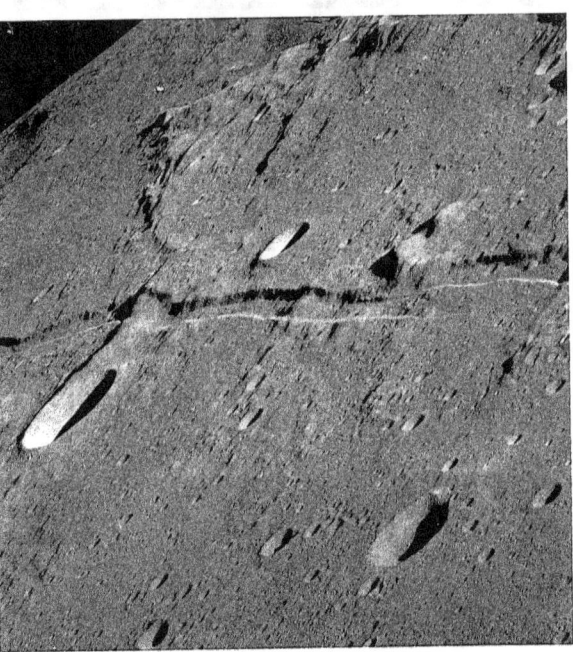

FIGURE 6.16.—Straight rilles Hyginus (A) and Ariadaeus (B). Rima Ariadaeus continues westward in B as rille indicated by arrow in A. Largest crater in B is Silberschlag (13 km, 6° N., 6.5° E.). Surface offsets and echelon patterns are obvious. Apollo 10 frames H-4648 (A) and H-4645 (B).

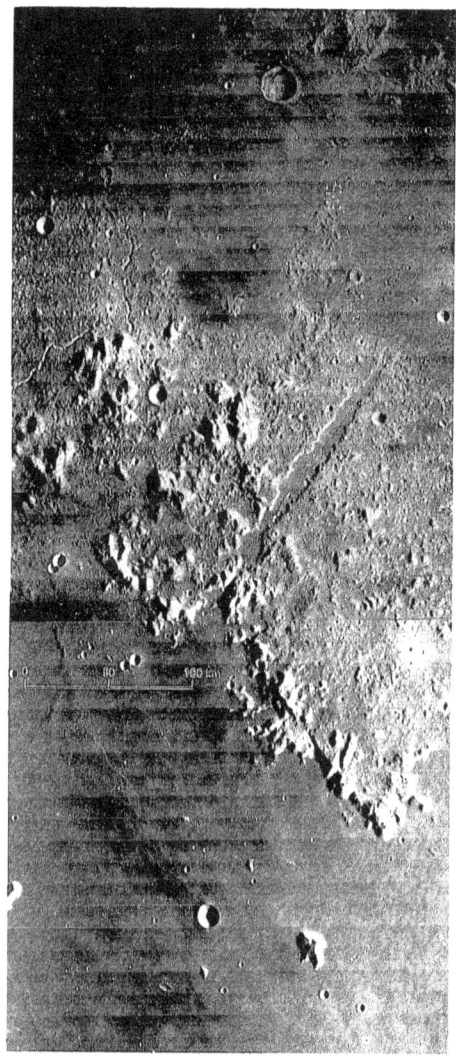

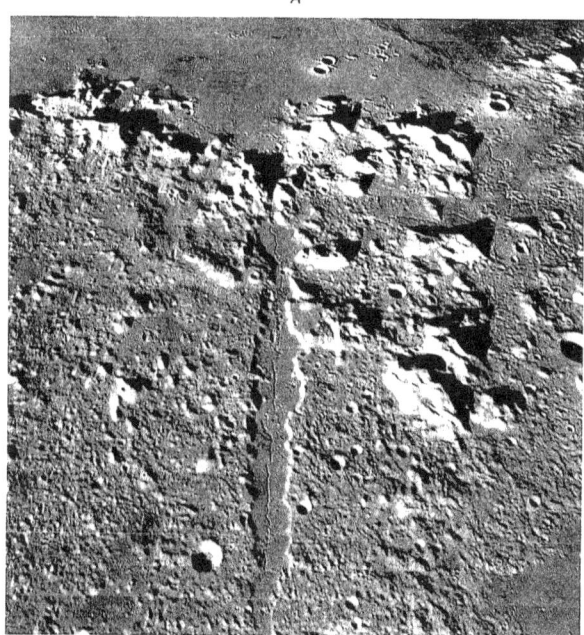

FIGURE 6.17.—Vallis Alpes (Alpine Valley), 200 km long and oriented radially to Imbrium basin. Knobby material is Alpes Formation (see chaps. 4, 10).
A. Regional setting, centered 47° N., 0°. Orbiter 4 frame H-115.
B. Detail. View southwestward. Orbiter 5 frame H-102.

FIGURE 6.18.—Mare Tranquillitatis, faulted by Cauchy rille (Rima Cauchy, right) and complex scarp (Rupes Cauchy, left), oriented radially to Imbrium basin (beyond top of photograph). Fresh crater between the two structures is Cauchy (12 km, 10° N., 39° E.). Part of crater Taruntius (compare fig. 6.13B) is visible in lower right corner. Apollo 11 frame H-6231.

FIGURE 6.19.—Grabens cutting Mare Fecunditatis, crater Goclenius (longest dimension, 72 km; 10° S., 45° E.), and mare fill of Goclenius. View southwestward. Apollo 8 frame H-2225.

FIGURE 6.20.—Triesnecker system of grabens east of crater Triesnecker (left center; 26 km, 4° N., 3.5° E.). Part of northwest branch of Rima Hyginus is at right, oriented northwest radially to Imbrium basin. Apollo 10 frame H-4816.

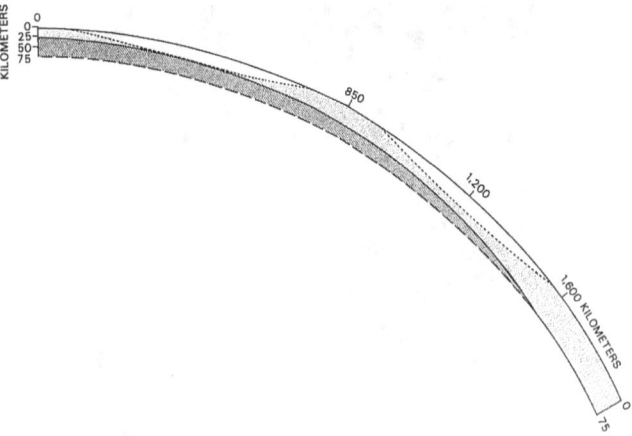

FIGURE 6.22.—Diagrammatic cross section of half of Procellarum basin, showing interpreted crustal thinning from 75 km outside basin to 25 km in basin center, caused by excavation and mantle uplift; dashed line denotes preimpact position of mantle-crust interface (lunar Moho). Ring radii (Whitaker, 1981) are shown above lunar surface; basin rim is at 1,600 km. Dotted lines denote additional excavation and crustal thinning by later basins; basin in Procellarum center penetrates crust to mantle, but same-size basin in intermediate and outer Procellarum troughs bottoms in crust. Vertical and horizontal dimensions and curvature to true scale. Surface relief and relative relief of concentric rings and shelves are barely detectable.

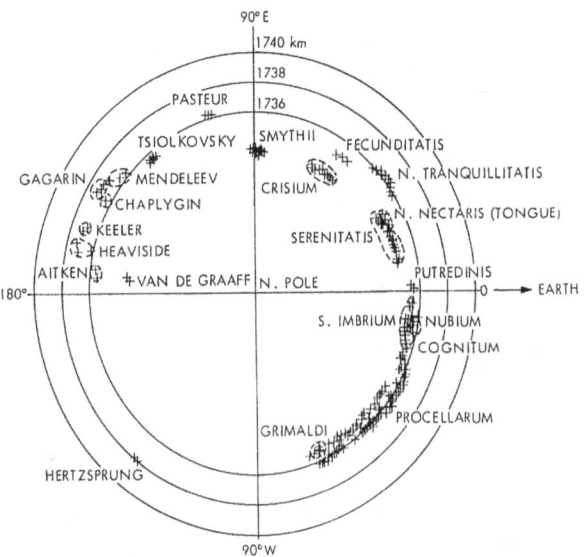

FIGURE 6.23.—Elevations of lunar maria compared with those of the terrae, most of which define a sphere 1,738 km in radius about the center of gravity (Sjogren and Wollenhaupt, 1976). Derived from laser altimetry and photogrammetry obtained during Apollo 15 and 16 missions (Kaula and others, 1973, 1974; Sjogren and Wollenhaupt, 1976). Smythii, Crisium, Serenitatis, Imbrium (S. Imbrium), and Grimaldi vary in elevation, although all have large mascons. Elevations of maria in Aitken, Tsiolkovskiy, and Van de Graaff probably reflect elevations of crater floors; Aitken is on rim of South Pole-Aitken basin, and van de Graaff well inside. Chaplygin, Gagarin, Heaviside, Hertzsprung, Keeler, Mendeleev, and Pasteur are filled by terra plains, not maria. Polar plot relative to 1,738-km-radius sphere about lunar center of gravity.

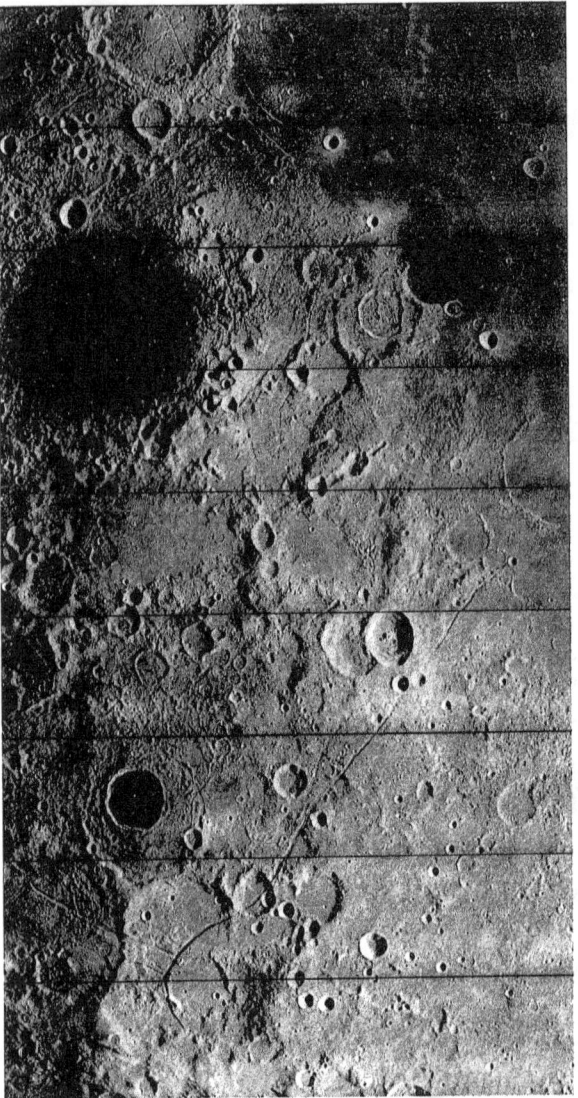

FIGURE 6.21.—Grabens along west margin of Oceanus Procellarum south of crater Hevelius (106 km, 2° N., 67° W., cut by top edge of photograph). Circular mare below Hevelius fills Grimaldi basin (compare fig. 4.4H). Orbiter 4 frame M-161.

7. RELATIVE AGES

ROCK-STRATIGRAPHIC UNITS	TIME-STRATIGRAPHIC UNIT	TIME UNIT
Crater materials: Tycho, Aristarchus, Kepler, Pytheas / Mare materials — Copernicus, Diophantus	Copernican System	Copernican Period
Mare materials — Delisle, Euler, Timocharis, Eratosthenes, Lambert	Eratosthenian System	Eratosthenian Period
Mare materials — Krieger	Upper Imbrian Series	Late Imbrian Epoch
Hevelius Formation (Orientale basin) / Volcanic materials / Crater materials / Fra Mauro Formation (Imbrium basin)	Lower Imbrian Series	Early Imbrian Epoch
Volcanic materials? / Basin and crater materials / Janssen Formation (Nectaris basin)	Nectarian System	Nectarian Period
Volcanic materials? / Basin and crater materials / Early crustal rocks	Pre-Nectarian system	Pre-Nectarian period

7. RELATIVE AGES

CONTENTS

	Page
Introduction	123
Stratigraphic nomenclature	123
Superpositions	125
Mare-crater relations	125
Crater-crater relations	127
Basin-crater relations	127
Mapping conventions	127
Crater dating	129
General principles	129
Size-frequency relations	129
Morphology of large craters	129
Morphology of small craters, by Newell J. Trask	131
D_L method	133
Summary	133

INTRODUCTION

The goals of both terrestrial and lunar stratigraphy are to integrate geologic units into a stratigraphic column applicable over the whole planet and to calibrate this column with absolute ages. The first step in reconstructing the relative stratigraphy is the identification of the material units of craters, basins, and maria (see chaps. 2–5); stratigraphic relations to neighboring units are generally recognized in the course of recognizing and defining a unit. The next step is to relate as many units as possible to areally extensive units. Units contacting one of these key stratigraphic horizons are dated directly by geometric relations, and others are dated means of the statistics and morphologies of superposed craters. In present practice, laterally extensive deposits of the Nectaris, Imbrium, and Orientale basins divide the time-stratigraphic column into four major sequences, from oldest to youngest: pre-Nectarian, Nectarian, Lower Imbrian, and Upper Imbrian through Copernican (fig. 7.1; table 7.1). The first three of these sequences, which are older than the visible mare materials, are also dominated internally by the deposits of basins. The fourth (youngest) sequence consists of mare and crater materials. This chapter explains the general methods of stratigraphic analysis that are employed in the next six chapters and on plates 6 through 11 to subdivide and calibrate the lunar stratigraphic column.

STRATIGRAPHIC NOMENCLATURE

The task of reconstructing the lunar stratigraphic column has benefited from application of the threefold code of stratigraphic nomenclature developed in North America for terrestrial geology (American Commission on Stratigraphic Nomenclature, 1970). This code is just as applicable to the Moon as to the Earth (Mutch, 1970, chap. 5; Wilhelms, 1970b).

Complete understanding of the stratigraphic nomenclature used in this volume requires some familiarity with the three major types of stratigraphic units recognized by the code (fig. 7.1). *Rock-stratigraphic* (rock) units are the observed, physical units that can be identified and mapped objectively. Ideally, they are defined by physical properties intrinsic to their emplacement process and are described additionally by other reproducibly observable properties. The basic rock-stratigraphic unit is the *formation*; formations can be combined into *groups* and divided into *members*. Most of the foregoing discussions in this volume have concerned descriptions and interpretations of rock-stratigraphic units.

Time-stratigraphic (time-rock) units include all the rock-stratigraphic units emplaced on a planet within a given time interval. The basic time-stratigraphic unit is the *system*, which can be divided into *series* (and finer subdivisions in terrestrial geology). Time-stratigraphic units are defined on the basis of specific rock-stratigraphic units. Time-stratigraphic units do not overlap, and the upper boundary of one unit is the lower boundary of the next.

The *time* intervals corresponding to the time-stratigraphic units are the third type of unit. They are defined by the corresponding time-stratigraphic units and are not physical units. *Periods* correspond to systems, and *epochs* to series.

Lunar time-stratigraphic nomenclature has changed somewhat since the concept of lunar stratigraphy was introduced by Shoemaker and Hackman (1962). The first two maps published by the U.S. Geological Survey at a scale of 1:1,000,000 (Hackman, 1962; Marshall,

TABLE 7.1.—*Time-stratigraphic nomenclature in use on lunar geologic maps of the U.S. Geological Survey*

[Diagrammatic; no absolute-age spans implied. Upper and Lower series of the Imbrian System are informal and were not recognized on all geologic maps from 1970 to 1979]

1959–63 (Shoemaker and Hackman, 1962)	1963–70 (Shoemaker, 1964; McCauley 1967b)	1970–75 (Wilhelms, 1970b)	1975–79 (Stuart-Alexander and Wilhelms, 1975)	This volume
Copernican	Copernican	Copernican	Copernican	Copernican
Eratosthenian	Eratosthenian	Eratosthenian	Eratosthenian	Eratosthenian
Procellarian	Archi- median			Upper Imbrian
	Imbrian	- -Imbrian- - -	- - -Imbrian- - -	
Imbrian	Apen- ninian			Lower Imbrian
Pre-Imbrian	Pre-Imbrian	Pre-Imbrian	Nectarian	Nectarian
			Pre-Nectarian	Pre-Nectarian

1963) used the original scheme, and the remaining 42 maps of this series used modified versions (fig. 7.2A; table 7.1). Subsequent mapping, which covers the Moon at a 1:5,000,000 scale (fig. 7.2B), has incorporated additional refinements (table 7.1).

The purpose of distinguishing rock-stratigraphic and time-stratigraphic units is illustrated by changes in concept and definition of the Imbrian System (Imbri*an* refers to the system, and Imbri*um* to the basin). Initially, Shoemaker and Hackman (1962, p. 293–294) defined the Imbrian System as equivalent to the "immense sheet" of material around Mare Imbrium now given such rock-stratigraphic names as the Fra Mauro Formation (table 4.3). They named the mare material the "Procellarian System", after the largest expanse of mare material, Oceanus Procellarum. The concept of the Imbrian Period was quickly extended to cover a longer timespan than that necessary to emplace the sheet (Shoemaker and Hackman, 1962, p. 298–299). The crater Archimedes, for example, was recognized as younger than the Imbrium basin but older than the Procellarian System (fig. 1.6). After Shoemaker and Hackman's (1962) report was prepared but before it was formally published, the Imbrian System was divided into the Apenninian Series, equated with the "regional material of the Imbrian System" (the sheet), and the younger Archimedian Series, equated with the deposits of Archimedes and other craters of similar stratigraphic position (Shoemaker and others, 1962b).

One problem with these classifications was that they equated time-stratigraphic and rock-stratigraphic units. The definition of the Procellarian System by Shoemaker and Hackman (1962) was founded on the belief, based on telescopic crater counts and consistent stratigraphic relations, that the mare materials were formed within a short time interval (Shoemaker and Hackman, 1962, p. 299; Shoemaker and others, 1962a; Baldwin, 1963, p. 309). However, continued crater counts (Dodd and others, 1963; Hartmann, 1967) and stratigraphic observations (Carr, 1966a, b; McCauley, 1967a, b; Wilhelms, 1968, 1970b) showed that they vary substantially in age. It is now clear that sections of mare basalt cross time-stratigraphic boundaries; mare and crater materials interfinger well into the Eratosthenian System (fig. 7.1; table 7.2; McCauley, 1967b, p. 436; Wilhelms, 1970b, p. F36; 1980). This extraterrestrial example vividly illustrates the rationale behind the threefold stratigraphic code, which sprang from discoveries that many terrestrial lithologic units are time-transgressive. The rapid alternation of volcanic and impact deposits could

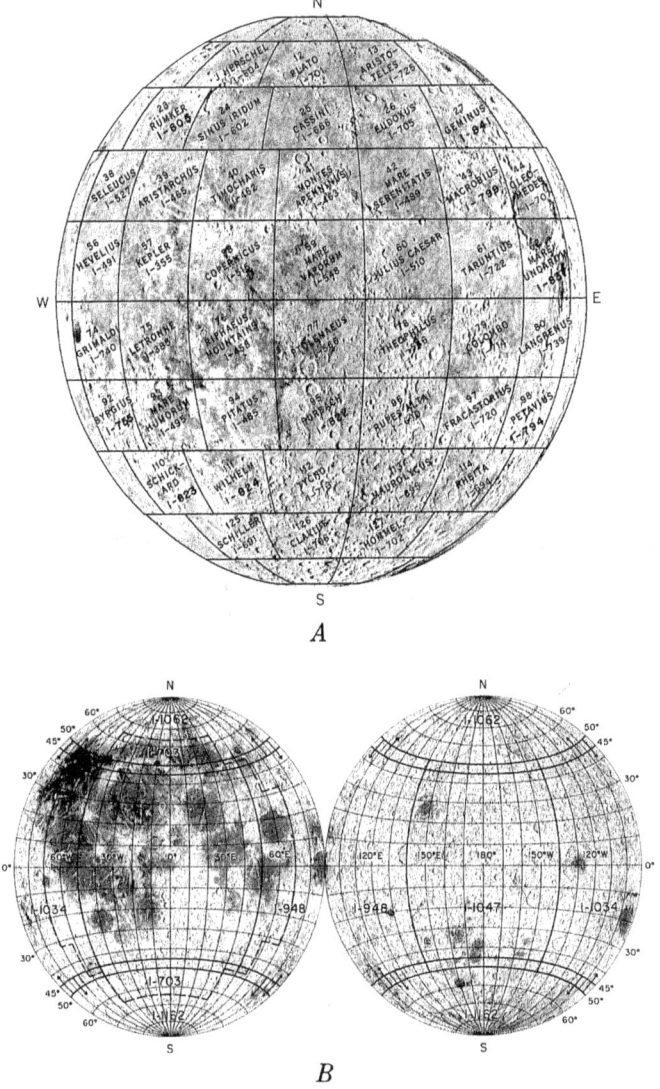

FIGURE 7.2.—Index maps of lunar geologic mapping.
  A. Nearside, showing coverage by 1:1,000,000-scale series. Number above quadrangle name refers to base chart (LAC series) by the U.S. Air Force's Aeronautical Chart and Information Center (ACIC). I-number refers to published map in U.S. Geological Survey Miscellaneous Investigations Series.
  B. Nearside and farside coverage by geologic maps at 1:5,000,000 scale: I–703, Wilhelms and McCauley (1971; same coverage as in fig. 7.2A); I–948, Wilhelms and El-Baz (1977); I–1034, Scott and others (1977); I–1047, Stuart-Alexander (1978); I–1062, Lucchitta (1978); I–1162, Wilhelms and others (1979).

TABLE 7.2.—*Selected stratigraphic units in southern Mare Imbrium and part of northern Oceanus Procellarum*

[Stated superpositional relations refer to units in actual contact. After Wilhelms (1980)]

Feature	Materials in area	Oldest superposed unit	Youngest subjacent unit
Aristarchus	Very fresh secondary craters	None	Kepler.
Kepler	do	Aristarchus	Pytheas.
Pytheas	Fresh ejecta and secondary craters	Kepler	Copernicus.
Copernicus	Large clusters of secondary craters along major rays.	Pytheas	Diophantus.
Diophantus	Fresh ejecta, small fresh secondary craters; no rays.	Copernicus	Mare unit 6.
Mare unit 6	Flow-textured lavas (D_L=140-195 m)	Diophantus	Mare unit 5.
Mare unit 5	Flow-textured lavas (D_L=170-215 m)	Mare unit 6	Delisle.
Delisle	Fresh ejecta and secondary craters; no rays.	Mare unit 5	Mare unit 4.
Euler	Fresh ejecta and secondary craters	Mare unit 5	Mare unit 4.
Mare unit 4	Flow-textured lavas (D_L=210-245 m)	Euler	Timocharis.
Timocharis	Fresh ejecta and secondary craters	Mare unit 4	Eratosthenes.
Eratosthenes	Extensive, subdued secondary craters	Timocharis	Mare unit 3.
Lambert	Slightly subdued ejecta and secondary craters.	Mare unit 4	Mare unit 2.
Mare unit 3	Smooth lavas (D_L~255 m)	Mare unit 4	Krieger.
Krieger	Asymmetric ejecta and secondary craters	Mare unit 3	Mare unit 2.
Mare unit 2	Smooth lavas (D_L=240-385 m)	Krieger	Mare unit 1.
Mare unit 1	Smooth lavas (D_L=270-385 m)	Mare unit 2	DMM.
Aristarchus plateau	Dark-mantling material (DMM)	Mare unit 1	Prinz.
Prinz	Subdued ejecta and secondary craters	DMM	Imbrium basin.
Imbrium basin	Isolated islands	Prinz	None of above.

not have been expressed by Shoemaker and Hackman's nomenclature without misuse of the time-stratigraphic concept (McCauley, 1967b; Mutch, 1970; Wilhelms, 1970b).

One possible modification would have been to redefine the Procellarian System to include crater deposits that interfinger with mare materials. The course that was taken, at a meeting of lunar stratigraphers held in Flagstaff, Ariz., in November 1963, was to drop the name "Procellarian System" and to designate mare materials by the rock-stratigraphic name "Procellarum Group" (Hackman, 1964; McCauley, 1967b; Wilhelms, 1970b). The upper boundary of the time-stratigraphic Imbrian System was to be defined by some part of the rock-stratigraphic Procellarum Group. Other rock-stratigraphic names were to be devised for Eratosthenian and Copernican mare materials when they were discovered. The contemporaneous crater materials were mapped as material units assigned to the appropriate system or series. Later, even the name "Procellarum Group" was dropped in favor of the informal rock-stratigraphic or lunar-material name "mare material," because "Procellarum Group" retained a time-stratigraphic connotation (Wilhelms, 1970b).[7.1]

The 1963 and 1970 stratigraphic schemes were used throughout most of the 1:1,000,000-scale mapping program (table 7.1). The Imbrian System included all materials from the base of the Fra Mauro Formation stratigraphically up to unspecified mare units between the premare crater Archimedes and the postmare crater Eratosthenes (figs. 1.6, 2.1; Wilhelms, 1970b, p. 23). Thus, the Imbrian System included all basin, crater, mare, and other materials emplaced from the time the Imbrium basin formed until the time its flooding was mostly complete. The amount of fill excluded from the Imbrian System was not specified in this working definition.

Chapters 10 and 11 refine the definition of the Imbrian System and define two series, Lower and Upper Imbrian, separated by the Orientale-basin materials. An earlier major change in the stratigraphic scheme divided the pre-Imbrian of Shoemaker and Hackman (1962) into the pre-Nectarian and Nectarian Systems (table 7.1; Stuart-Alexander and Wilhelms, 1975). Although the definition of the Imbrian-Eratosthenian and Eratosthenian-Copernican boundaries remains imprecise, no further serious obstacles have appeared in the lunar stratigraphic nomenclatural scheme.

[7.1]The equation of the Imbrian System or the Apenninian Series with rocks interpreted as Imbrium ejecta would have been less of a practical problem because the ejecta has not proved to be time-transgressive, as are the mare materials. However, assumptions about the time significance of rock units so commonly prove to be erroneous in terrestrial and planetary geology that the threefold code is best adhered to from the beginning of the study of any new planet (Mutch, 1970).

Several other formal rock-stratigraphic units were introduced during the mid-1960's as antidotes to the potential confusion of time and rock units and as means of clarifying local stratigraphic relations (Wilhelms, 1970b). Some of these names are still used, but present practice emphasizes Moon-wide similarities of a given class of unit by use of such informal names as "mare materials" and "crater materials." Such units may be assigned to time-stratigraphic units or subdivided by physical properties as required. Formal names are now normally given only (1) to units that are difficult to describe simply and that form important regional deposits, such as the Hevelius or Alpes Formations; or (2) to separate patches of similar-appearing deposits that are being distinguished for some reason, such as the Apennine Bench and Cayley Formations. Whitford-Stark and Head (1980), however, resuggested the practice of formally naming mare units.

SUPERPOSITIONS

Mare-crater relations

Superpositions are recognized on photographs by topographic contrasts that are not explicable by facies changes within one unit. Rugged landforms may be muted by smoother blankets, and smooth topography may be crosscut by rugged texture. Many crater deposits and mare flow units illustrate such relations. Some superpositional relations are obvious (fig. 7.3); others are established by subtle flooding of secondary craters by thin mare flows (fig. 7.4). The difference between a slightly flooded secondary crater and an unmodified secondary superposed on other lavas may be detectable only on the best photographs.

Secondary craters greatly extend the area over which the primary crater can be relatively dated. Secondaries along conspicuous rays serve best for long-distance correlations because the rays visible on favorably illuminated and processed photographs point out the source (fig. 7.4). For example, the clustered craters at the Ranger 7 impact site were tentatively identified as secondary to both Copernicus and Tycho despite their 600- and 1,000-km distances, respectively, from those primary craters (U.S. Geological Survey, 1971). Tycho, Copernicus, Kepler, and other young craters have been identified as the sources of projectiles that reached every potential Apollo landing site mapped in the equatorial belt, hundreds or even thousands of kilometers from the source craters (Carr and Titley, 1969; Carr, 1970; Grolier, 1970a, b; Cummings, 1971; Pohn, 1971; Rowan, 1971a). In principle, secondary craters can also be traced to nonrayed craters by analysis of the flight directions of the ejecta projectiles: Large secondary craters generally occur in a cluster uprange toward the primary, and herringbone patterns, radial grooves and ridges, and smooth deposits occur downrange (figs. 3.4, 3.6–3.8, 7.4). Stratigraphic relations both at the primary crater and at the outlying clusters can thus be used to date units relative to this dispersed stratigraphic datum.

FIGURE 7.3.—Crater Plato (101 km, 52° N., 9° W.), between Mare Imbrium (bottom) and Mare Frigoris (top). Mare materials fill Plato and truncate its south rim deposits (arrows), thus are younger than Plato. Fully developed rim deposits and secondary craters of Plato are preserved where superposed on terra (part of Montes Alpes); thus, Plato is younger than Montes Alpes. Secondary craters of Plato are superposed on secondary craters of Iridum crater (arrow). Orbiter 4 frame H–127.

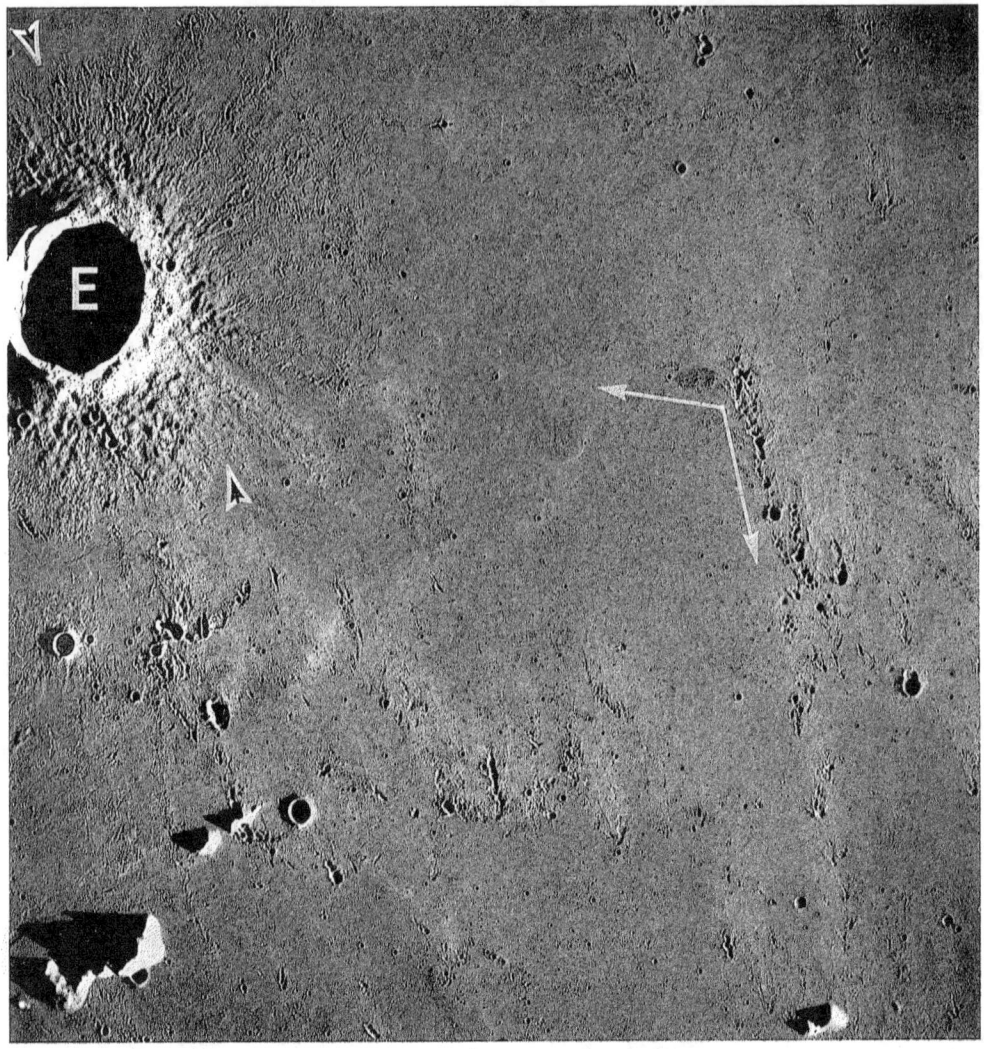

FIGURE 7.4.—Part of Mare Imbrium, including crater Euler (E; 28 km, 23° N., 29° W.) and inundated massifs of Imbrium basin (bottom, fig. 7.4A). White arrows indicate same scene in both frames; left-hand arrow points to crater Aristarchus, 650 km to west, and lower arrow to Copernicus, centered 450 km to south-southeast. Alignment of secondary-crater groups indicates these primary craters as sources of the excavating projectiles.
  A. Ejecta and secondary craters of Euler, subtly flooded by mare material (black-and-white arrowheads). Apollo 17 frame M–2291.
  B. Aristarchus secondaries superposed on Copernicus secondaries (black-and-white arrow), indicating that Aristarchus is the younger crater. Apollo 17 frame M–3093.

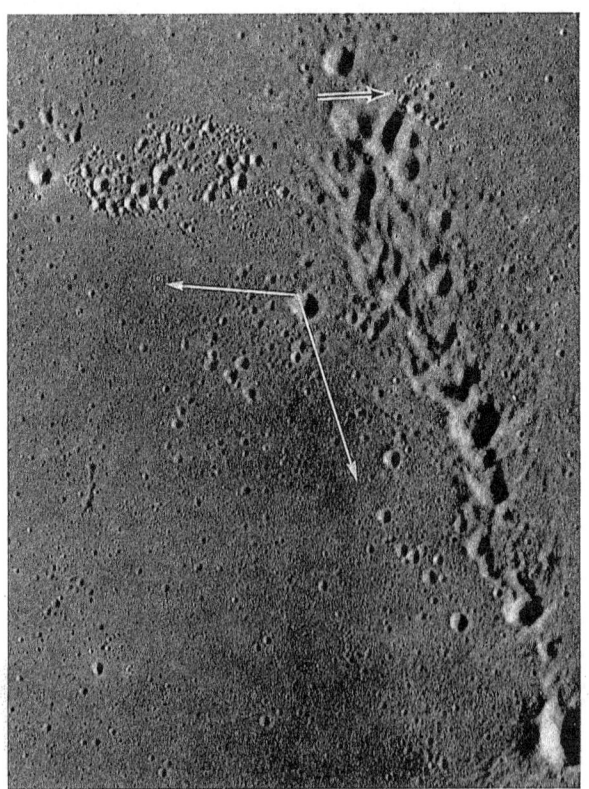

The ages of 20 mare and crater units in southern Mare Imbrium and northern Oceanus Procellarum are bracketed by the telltale marks of secondary craters and mare embayments (fig. 7.4; table 7.2; Wilhelms, 1980). This region is favored both by photographs of the necessary quality and by relatively extensive mare flows that can be used to correlate units over hundreds of kilometers (Schaber, 1973; Wilhelms, 1980). More commonly, mare materials mappable as single flow units extend only a few tens of kilometers (Schaber and others, 1976). Small units are generally dated by methods based on small superposed craters, as described below in the section entitled "Crater Dating."

Crater-crater relations

Superpositions among crater materials establish local stratigraphic sequences in all parts of the Moon. How near craters must be to allow relative dating depends on the freshness of their deposits and secondary craters. Rayed and other fresh craters can be dated relative to other craters as they can to mare units, that is, if the secondary craters happen to contact the other crater. Several pairs of rayed and nonrayed craters listed in table 7.2 are relatively dated by the chance impacts of younger secondaries upon older. Degraded craters lacking obvious secondaries can be relatively dated if one rim happens to cut out part of another, the so-called "cookie cutter" obliteration.

Figure 3.26 illustrates the more common case between these extremes, in which the radial ejecta of one crater (Werner) is superposed on an older crater lacking ejecta texture (Aliacensis). The Aliacensis rim is swamped within about one Werner radius from the Werner rim and is visibly marked to about one additional radius. If the radial textures of Werner were not visible, the degradation of the Aliacensis rim could nevertheless be detected one radius, but not two radii, from the Werner rim. These relations provide two guidelines used during preparation of the paleogeologic maps in this volume (pls. 6–11): (1) Subdual of a crater's rim within one radius of another rim indicates that the first crater is the older, and (2) ages are determinable at greater distances where textural evidence is available. Each case, however, is individual.

Basin-crater relations

Similar observations help date the extensive deposits of ringed basins relative to one another and to other units. Massif-lined rings of some basins truncate those of other basins (fig. 1.7), and the continuous deposits of younger basins swamp older basins and craters (fig. 7.5). Secondary-impact craters of basins are so common and extensive on the Moon that units more than 1,000 km from the basin rim can potentially be related by superpositions (figs. 7.6, 7.7; Wilhelms, 1976). The source of morphologic features distant from basins is even easier to determine than is the source of crater secondaries because more outer lineaments of basins are radial. Many superpositions are detectable on the basis of radial structure even in highly degraded terranes. The possibility that, around old basins, the linear or chainlike pattern of basin secondaries may be indistinguishable from the radial-flow lineations does not matter in establishing superpositional relations.

Near a basin rim, craters younger than a basin appear much fresher, of course, than those older. Farther from the rim, this difference becomes increasingly less apparent as prebasin craters are increasingly less affected by the basin deposits. Morphologic differences between buried and superposed craters caused by basin ejecta are generally detectable even around old basins out to one basin radius from the rim (the average range of the continuous ejecta) and commonly out to one diameter (the average range of abundant secondary craters). Basin deposits, however, are lobate and asymmetric, and so each case must be considered individually. Southeast of Orientale, subtle grooves and ridges radial to that basin can be used to distinguish pre- from post-Orientale units beyond 1,300 km from the Orientale rim (Wilhelms and others, 1979). Distant craters are easily datable if fortuitously struck by large projectiles (fig. 7.6). Orientale ejecta in "rays" may have excavated secondary craters as far away as the central and southeastern terra (figs. 4.6, 7.7; Wilhelms and others, 1978; Wilhelms, 1980). On the other hand, dating of some craters only a few hundred kilometers from the rim may be ambiguous. For example, the swale-and-hummock topography of the zone near the crater Crüger could be interpreted either as superposed but poorly developed Orientale-secondary craters or as well developed Orientale secondaries blanketed by Crüger ejecta (fig. 7.8). The weaker primary textures around older basins cause even more ambiguities (fig. 7.9). Some older craters may be scored or cratered in one sector but unaffected in another (figs. 7.6, 7.9). Therefore, determination of superpositional relations depends on knowledge of the whole geologic setting of the units in question.

Mapping conventions

Assignment of circumbasin materials to time-stratigraphic units older than, contemporaneous with, or younger than a given basin depends partly on interpretations of the topography, as is again well illustrated by the history of changes in mapping convention around the Imbrium basin.

During most geologic mapping at 1:1,000,000 scale, the Imbrium sculpture was thought to originate by impact-induced faulting (see chap. 6). Mappers assigned units clearly cut by the sculpture to the pre-Imbrian, and units superposed on the grooves to the Imbrian or younger systems (fig. 7.7). The grooved terrain was mapped as Imbrian where the grooves and adjacent swales were interpreted as intrinsic to the Fra Mauro Formation and thus contemporaneous with Imbrium (for example, Howard and Masursky, 1968; Wilhelms, 1968).

The newer interpretation of the rimmed outer grooves as secondary-impact chains changes the convention in detail though not in principle. Units impacted by the chains of secondaries are still mapped as pre-Imbrian, that is, either Nectarian or pre-Nectarian. A difference from earlier mapping conventions, however, results from expansion of the terrane interpreted as secondary to Imbrium. Most terrane mapped in the past as pre-Imbrian lineate material (fig. 7.7; Wilhelms and McCauley, 1971) is here mapped as Imbrium-related to emphasize the age of the surficial morphology (pls. 3, 8). Lunar Orbiter photographs have revealed similar radially lineate terrane over a vast area (see chap. 10). Lunar Orbiter and Apollo photographs, coupled with the Apollo 16 results, have also led to changed age assignments for such features as "volcanic domes," reinterpreted as parts of the Imbrium-secondary regime and thus lowermost Imbrian instead of Imbrian or younger (see chap. 10).

FIGURE 7.5.—Terrain southeast of Imbrium basin, lying just beyond lower right corner of photograph. Circular mare in top half of photograph is Mare Vaporum, about 220 km across. Linear- and knob-textured terrae are parts of Imbrium-basin ejecta. Circularity of mare suggests that it fills a Vaporum basin or crater, which was deeply buried by Imbrium ejecta before mare flooding. Several small craters (arrows) were superposed on Imbrium material before mare flooding. Graben near left-hand crater (above and parallel to arrow) is also post-Imbrium and premare (graben is part of set possibly not related to basins; see chap. 6). Rima Hyginus is in background (H). View southerly. Apollo 17 frame M–1674.

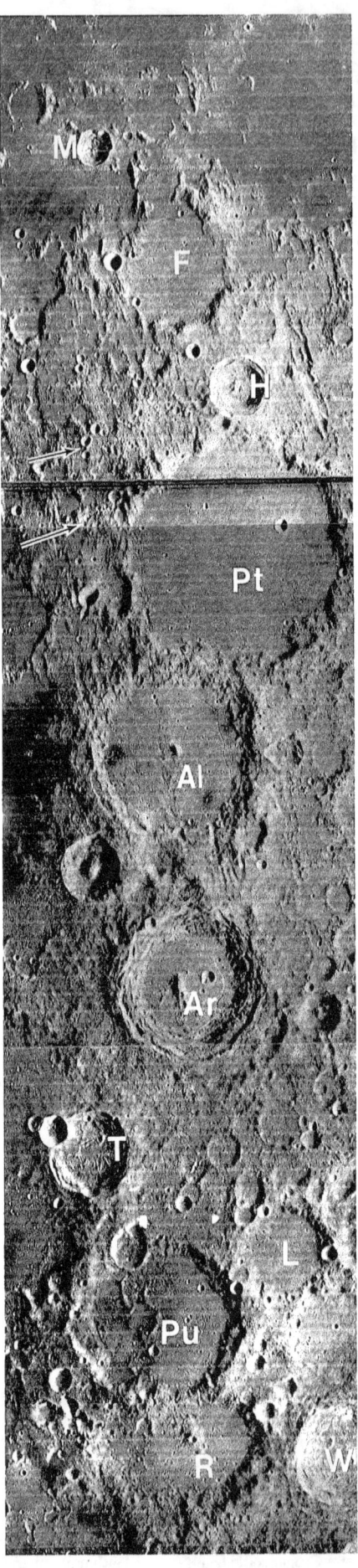

FIGURE 7.6.—Terrain southeast of Orientale basin, heavily affected by its ejecta. Large circular crater is Schickard (227 km, 44° S., 55° W.), whose rim is scored by linear ejecta of Orientale secondaries (s). Barrage of secondaries with linear or braided ejecta destroyed part of rim of crater Schiller C (SC; 49 km) but left other rim segments untouched; Schiller C is 1,475 km from Orientale center and 1,000 km from rim (Montes Cordillera). Orbiter 4 frame H–160.

In lunar geologic mapping, the age of an ejecta unit applies to the time it was emplaced, which is not necessarily the same as the isotopic ages of the contained materials.[7.2] The materials contained in Imbrium-secondary ejecta may consist partly of projectile material from the Imbrium basin, they may be identical to materials of the impacted and redistributed pre-Imbrian terrane, or they may contain intimate mixtures of the primary and redistributed "local" material. As illustrated by the Apollo 16 materials, the provenance of the constituent materials is commonly difficult to determine either photogeologically or petrographically (see chaps. 2, 9, 10). Mapping of large tracts as related to Imbrium (pl. 8) stresses the interpreted basin age and origin of the morphology over the age and provenance of the contained materials. Similar conventions are used here for all basin peripheries to show the full extent of basin-related deposits (pls. 3, 6–8, 12).

CRATER DATING

General principles

This account has shown how the Moon's extensive datum horizons, as well as less extensive strata, can potentially be ranked stratigraphically by detection of superpositional relations among their depositional features. These conceptually simple methods must be supplemented on much of the lunar surface by chronologic interpretations of small craters superposed on other units. Craters are used to correlate isolated units in the same way as fossils are used on the Earth (Mutch, 1970; McGill, 1977). Craters are thus (1) stratigraphic units in their own right and (2) time counters for other units, including other, larger crater units.

Which of these stratigraphic roles a given crater plays depends on the purpose of the investigation and on the size of the crater relative to the *steady-state* size for the cratered surface. The continuous bombardment of cosmic projectiles creates craters of all sizes, from micropits to basins. Within any given time interval, far more small than large excavations are produced. A surface becomes saturated with the smallest craters, so that new impacts destroy as many such craters as they add. Craters of these sizes are thus in a steady-state or equilibrium condition. The steady-state sizes for a surface of a given age are limited by a computable crater diameter, conventionally designated C_s (Moore, 1964a; Shoemaker, 1965; Trask, 1966; Shoemaker and others, 1967a, b; 1969; Gault, 1970; Marcus, 1970; Morris and Shoemaker, 1970; Soderblom, 1970; Soderblom and Lebofsky, 1972). Craters smaller than C_s are in all states of aging, from the very fresh to the totally flattened. Craters larger than C_s range from very fresh to only partly destroyed. C_s increases over time (fig. 7.10B; table 7.3).

The first of two dating methods described here, size-frequency counts, is based on craters larger than C_s. I describe it only briefly because it is familiar and has long been widely used (Öpik, 1960; Shoemaker and others, 1962a; Baldwin, 1963, 1964). The second set of methods is based on erosional morphology created mostly by small craters acting in the steady state. The visible craters they have degraded are assumed to have been morphologically similar when formed, within a given size range. Surfaces are assumed to have been worn down uniformly over the whole Moon in proportion to their time of exposure. The small eroding impacts were both primary and secondary; the secondaries formed in proportion to the number of primaries (Moore, 1964a; Shoemaker, 1965; Gault, 1970; Soderblom, 1970; Neukum and others, 1975a). This gradual erosion by small impacts seldom creates novel landforms but quantitatively alters the original crater morphology. Modifications that, in contrast, took place rapidly or episodically can be recognized by identifying superposed ejecta blankets, mare filling, or such anomalies as tectonically uplifted floors. In practice, the morphologically based methods are divided into those applicable to craters either larger or smaller than about 3 km in diameter.

Size-frequency relations

The use of statistical studies of craters as a dating tool is based on the assumption that primary impacts crater a surface in proportion to its time of exposure. The analyst must, therefore, (1) count only craters larger than C_s; (2) distinguish primary, secondary, and endogenic craters; and (3) eliminate buried craters from the counts (fig. 7.10; Neukum and others, 1975a, b; Young, 1977; Wilhelms and others, 1978; Basaltic Volcanism Study Project, 1981, chap. 8). For most stratigraphic purposes, craters should be counted on individual geologic units and not over whole "provinces." Where the geology is too complex or the data inadequate to attain this ideal, as is commonly the case (Hartmann, 1972c, p. 54), other methods must be applied. Thus, stratigraphic analysis should accompany any use of craters in determining ages.

Figure 7.10 summarizes the properties of size-frequency curves likely to be encountered in lunar work. The inverse relation between the masses and frequencies of impacting primary bodies produces inverse size-frequency relations of craters (fig. 7.10A). Continued bombardment by the same population of bodies moves the production curves higher on the graph, but the slopes remain the same (fig. 7.10B). The slope of the steady-state distribution also remains the same, always shallower than the production curves (fig. 7.10B). "Rollovers" of the end of a curve representing small diameters result either from inability to observe small craters or from mutual obliteration of craters in and near the steady state (fig. 7.10C). Subsequent burial removes the smaller members of a crater population and thus truncates the production curve (fig. 7.10D); this truncation is theoretically abrupt but, in fact, may resemble the effects illustrated in figure 7.10D. Resumed cratering may have the same production curve as did preburial cratering, but the offset in the curve remains visible because craters of the sizes that were buried remain relatively depleted (fig. 7.10E). Finally, craters may not all have the same production function (figs. 7.10F, G); cumulative plots of craters larger than a few kilometers across are commonly observed to slope at −1.8 (Basaltic Volcanism Study Project, 1981, ch. 8), but bends appear at smaller diameters. Plots of secondary craters differ considerably in slope from those of primaries (fig. 7.10F; Shoemaker, 1965; Wilhelms and others, 1978), and the addition of substantial numbers of secondaries to crater populations may cause the bends observed in many crater-frequency curves below diameters of about 2 km (fig. 7.10G).

Additional aspects of and problems with these methods are discussed in later chapters of this volume (see reviews by Baldwin, 1963, 1964; Hartmann, 1964a, 1972c; Chapman and Haefner, 1967; Gault, 1970; Greeley and Gault, 1970; Neukum and others, 1975a, b; Young, 1975; McGill, 1977; Wilhelms and others, 1978; Crater Analysis Techniques Working Group, 1979; Basaltic Volcanism Study Project, 1981, chap. 8).

Morphology of large craters

To some extent, the overall morphology of large craters indicates relative age and stratigraphic correlation. Baldwin (1949, p. 134; 1963, p. 191) may have been the first to show that older lunar craters are systematically shallower than younger craters of the same size. Crater classifications have been developed independently on this basis (Arthur and others, 1963; Baldwin, 1963, chap. 9; Wood and Andersson, 1978).

The morphologic system most used in the U.S. Geological Survey's lunar mapping program was based less on overall morphology than on the morphologies of parts of craters (fig. 7.11). Crater-subunit

---

[7.2] On all but the most recent geologic maps (Wilhelms and others, 1979), massif materials were exempted from the practice of assigning ages according to time of emplacement. Massifs were assigned the age of the component prebasin rock to stress the interpretation that they are structural uplifts of relatively cohesive materials not severely broken up and redistributed by the impact. For example, Imbrium massifs were mapped as pre-Imbrian massif material, and Nectarian basin massifs and rugged material as Nectarian or pre-Nectarian (Wilhelms and McCauley, 1971; Wilhelms and El-Baz, 1977). In this volume, massifs are assigned the age of the basin of which they are parts.

FIGURE 7.7.—Pre- and post-Imbrium craters, distinguishable over extensive area. Pre-Imbrian age of Flammarion (F; 75 km), Ptolemaeus (Pt; 153 km), and Alphonsus (Al; 119 km) is demonstrated by superposition of linear "Imbrium sculpture" (Gilbert, 1893); pre-Imbrian age of Purbach (Pu; 118 km), Regiomontanus (R; 124 km), and La Caille (L; 68 km) is demonstrated by superposition of more nearly circular secondary craters. Plains filling the craters also postdate sculpture and secondary craters, possibly by a very short time. Regiomontanus is 1,400 km from Imbrium rim (Montes Apenninus) and 2,100 km from center. Craters Herschel (H; 41 km), Arzachel (Ar; 97 km), Thebit (T; 57 km), and Werner (W; 70 km) lack Imbrium effects and thus are post-Imbrian (Imbrian and Eratosthenian). Arrows indicate possible secondary craters of Orientale basin, centered 2,600 km in direction of arrow shafts (south of west). Mösting (M; 25 km, 0.5° S., 6° W.) and the craters superposed on rim of Thebit (Thebit A and Thebit L) are Copernican. Orbiter 4 frame H-108.

TABLE 7.3.—*Stratigraphic criteria for lunar time-stratigraphic units*

[C_s, limiting diameter of the steady state; all values approximate, estimated from the data of Boyce and Johnson (1977) and Moore and others (1980b, table 1), from rollovers in the curves in figure 7.16, and from the approximate formula $D_L=1.7C_s$ (Moore and others, 1980b, p. 88).
D_L, diameter of largest crater theoretically eroded to 1° interior slopes (see text); n/a, not applicable because method is invalid for the stated system or series (see chap. 10).
Subunits: Approximate diameters of craters datable by evaluation of their material-subunit morphologies.
Crater frequencies (cumulative); see chapters 8 through 13 for the basis of the stated values; n/a, not applicable because craters of this size are not useful for age determinations on the system or series.
Stratigraphic superpositions: X, stratigraphic relations determinable; radii, distance from the rim crest within which the relations are generally determinable; n/a, not applicable because no units are known.
Stated values of C_s, D_L, and crater frequencies are the ranges for the entire system or series; they overlap for adjacent systems or series because of uncertainties, and depend partly on the substrate (in parentheses)]

System or series	C_s (m)	D_L (m)	Subunits (crater diameter, in km)	Crater frequency (number per km²)		Stratigraphic superpositions		
				≥ 1 km	≥ 20 km	Crater-mare	Crater-crater	Crater-basin
Copernican System	?	<165 (mare), ≤200 (crater)	>3	<7.5×10⁻⁴ (mare), <1.0×10⁻³ (crater)	n/a	X	>2 radii in rays	n/a
Eratosthenian System	<100 (mare)	145–250 (mare)	>3	7.5×10⁻⁴ to ~2.5×10⁻³ (mare)	n/a	X	Mostly ~2 radii	n/a
Upper Imbrian Series	80–300 (mare)	230–550 (mare)	>3	~2.5×10⁻³ (mare) to ~2.2×10⁻²	2.8×10⁻⁵	X	~2 radii for large craters	n/a
Lower Imbrian Series	320–860 (basin)	n/a	>5	~2.2–4.8×10⁻² (basin)	1.8–3.3×10⁻⁵	n/a	None observed	X
Nectarian System	800–4,000? (basin)	n/a	>20	n/a	2.3–8.8×10⁻⁵	n/a	1 radius	X
Pre-Nectarian system	>4,000? (basin)	n/a	>20–30	n/a	>7.0×10⁻⁵	n/a	1 radius	X

morphology was extensively applied as an age criterion before basin deposits and their secondary-crater fields were completely mapped, and is still used in areas lacking good stratigraphic-datum planes. The method, developed by Pohn and Offield (1970), was the initial means for extending the Imbrium-region stratigraphy to the rest of the Moon (Offield and Pohn, 1970). Correlations are best made by comparing the morphologies of isolated craters with those of craters previously dated relative to one of the regional datum planes. The methods work best for primary-impact craters (for which they were intended) and less securely for secondaries, which vary greatly in morphology around a given source and do not initially resemble primary craters, except where situated far from the source.

Pohn and Offield (1970) first divided craters into three size classes based on the planimetric shapes of the rims of fresh craters of each class: More than 45 km diameter, round with distinct rim crenulations; 20 to 45 km, polygonal; 8 to 20 km, round (fig. 7.11; Mutch,

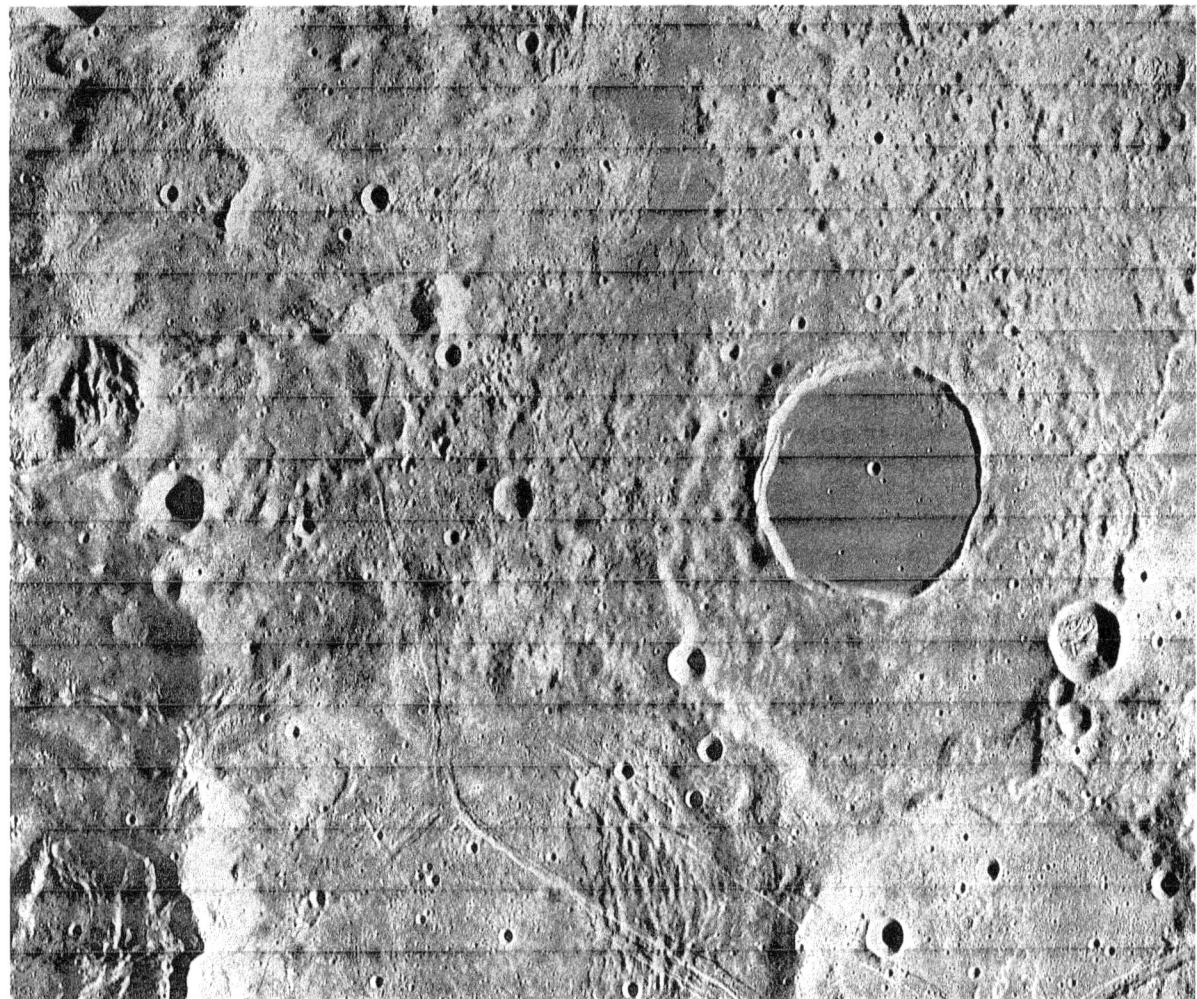

FIGURE 7.8.—Large mare-filled crater Crüger (46 km, 17 S., 67 W.), 370 km from Montes Cordillera. Dunelike topography to southwest in large crater Darwin is decelerated surface-flow ejecta of Orientale (see chap. 4), but stratigraphic relation of other hummocky ejecta to Crüger is unclear. Orbiter 4 frame H-168.

1970, p. 272). The first two classes are complex; the third, simple. Because of these differences in the initial morphologies, a slightly different morphologic continuum is definable for each size class. Craters from 1 to 8 km in diameter can also be ranked by morphologic criteria (Offield and Pohn, 1970, p. C164). Craters in each class are ranked in order of age by an arbitrary decimal scale, 0.0 (oldest) to 7.0 (youngest). The seven criteria for estimating age illustrated in figure 7.11 still appear to be valid, except that interior radial channels do not occur even in many old craters.

This classification was widely used and generally successful. The sequence of eight ringed basins and the relative ages of several mare provinces derived by Offield and Pohn (1970) agree with those given in this volume. Plains units were found to be only slightly older than old mare units—a correct result that was misinterpreted on some geologic maps to mean that a terra type of volcanic plains is coeval with the mare type (for example, Rowan, 1971b). Offield and Pohn (1970) believed the light plains to be volcanic, but their age ranks showed the plains to be consistent with Orientale and Imbrium provenances. They also stated that "Numerous craters not obviously related to the Orientale and Imbrium basins are determined to be of the same age as the basins and are tentatively identified as basin secondaries" (Offield and Pohn, 1970, p. C168)—an important result subsequently extended to craters that "look" older and younger than the basins (Wilhelms, 1976; Oberbeck and others, 1977; Wilhelms and others, 1978).

This volume also uses morphology as an age criterion where necessary. The next six chapters give the morphologic characteristics and a list of typical craters for each age. No attempt is made, however, to establish type craters or formulas for recognizing crater ages. Such an attempt would mislead more than help, in view of the many variables of photographic resolution, illumination, and, above all, geologic setting, that affect apparent crater morphology. Instead, the type craters are described and illustrated in their geologic context throughout the volume.

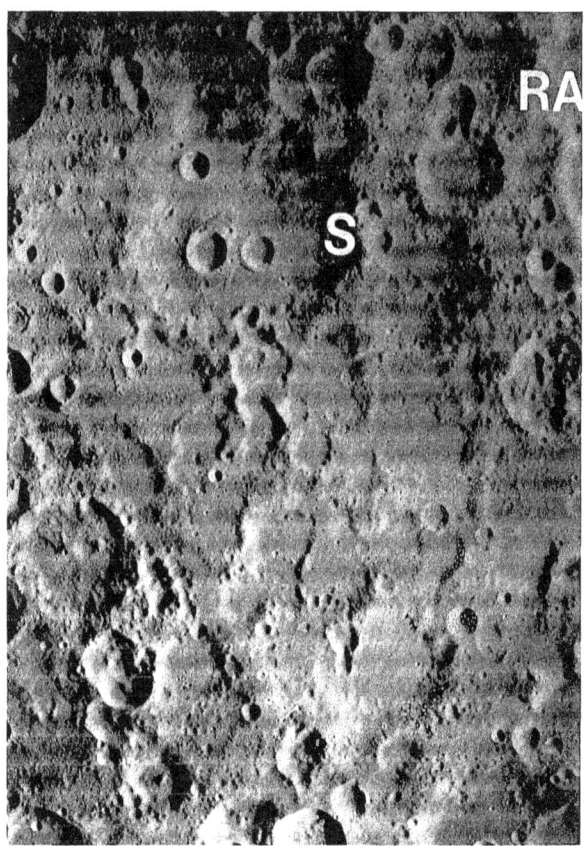

FIGURE 7.9.—Rim of pre-Nectarian crater Sacrobosco (S; 98 km, 24° S., 17° E.), lineate radially to Nectaris basin; basin rim (Rupes Altai, RA) is only 100 km northeast. Unlabeled west rim of Sacrobosco is less conspicuously affected by Nectaris material. Most smaller craters are probably secondaries of Imbrium and Nectaris basins. Orbiter 4 frame H-89.

Morphology of small craters

By Newell J. Trask (modified from Trask, 1971)

To fill the need for stratigraphic correlations of units of interest to early Apollo landings, Trask (1969, 1971) devised a system based on the morphology of small craters (less than 3 km diam). These ubiquitous features were and remain necessary tools for dating mare units because of the uneven distribution of stratigraphically discrete and extensive mare flows or crater deposits at closeup scales. The craters on the mare surfaces range in morphology from sharp to indistinct. The sharpest craters have well-defined rim crests and bright, blocky ejecta blankets. Less sharp craters have rounded rim crests, little in the way of an ejecta blanket, and few surrounding blocks. The most indistinct craters are very shallow depressions and lack a well-defined rim crest. Figure 7.12 shows examples of this continuum of crater types for various crater diameters. Where craters of the same approximate size are superposed, the sharper are superposed on the less distinct, indicating that the sharper craters are the younger. These observations suggest that most new craters are sharp and distinct and become progressively more subdued and indistinct over time; in other words, craters in the lower frames of figure 7.12 once looked like those in the upper frames. The sharpest craters (top, fig. 7.12) have all the characteristics of impact craters, including rays and ray loops. Thus, nearly all craters in the 0.05- to 3-km diameter range were probably formed by impacts. Older surfaces should, therefore, have more craters than younger surfaces and should also have more older-appearing craters than younger surfaces. Such a relation between the number of craters on the surface and the number of old-appearing, subdued craters is, in fact, observed.

Observation of these relations on the entire suite of Lunar Orbiter photographs led Trask (1969, 1971) to the following working hypotheses. (1) When first formed, craters appear sharp and fresh. (2) All craters undergo modification to less sharp and more subdued forms by erosion of the rim crest and infilling of the floors. (3) Smaller craters disappear sooner than larger ones. (4) There exists for any surface a critical crater diameter (C_{s1}, C_{s2}, C_{s3}, fig. 7.10B), which increases over time for a given surface; craters smaller than C_s that are as old as the surface have had time to be completely destroyed, whereas those larger have not.

These hypotheses, in conjunction with the assumption that the rate of crater degradation is approximately the same everywhere on the maria, formed the basis for assigning geologic ages to craters in geologic mapping of potential Apollo landing sites (Trask, 1969, 1971). By concentrating on the oldest craters, surfaces could also be dated. Figures 7.12 and 7.13 illustrate this system of assigning ages. The horizontal lines in both figures represent equal points in time, or isochrons. Accordingly, a gentle depression 50 m in diameter, for example, is of the same age as a 1-km-diameter crater that is only moderately subdued; and a 50-m-diameter moderately subdued crater is of the same age (Copernican) as a 1-km-diameter crater with well developed rays. The degree of crater degradation implied by this assumed equivalence seems reasonable in the light of theoretical studies on the extent and mechanism of crater subdual by the impact of small particles (Soderblom, 1970). The number, spacing, and names of the time intervals on the vertical axes of figures 7.12 and 7.13 were chosen to be consistent with the stratigraphic scheme used in regional mapping. In that scheme, the base of the Imbrian System is defined by the Fra Mauro Formation, upon which gentle depressions as much as 1 km in diameter are superposed; therefore, 1-km-diameter gentle depressions are placed at the base of the Imbrian System in the scheme of figures 7.12 and 7.13. Similarly, the dividing line between the Eratosthenian and Copernican Systems generally is placed between craters with bright halos and craters without bright halos in the diameter range 5 to 10 km in both small- and large-scale mapping, although individual cases may depart from this rule. The six numbered subdivisions of the Copernican System were chosen solely for convenience in the detailed site mapping and are not used at regional scales. The approach and assumptions are similar to those applied in the system for determining relative ages of craters larger than 8 km in diameter that was described above.

Trask (1969, 1971) examined high-resolution Lunar Orbiter 1, 2, 3, and 5 photographs of parts of the maria to determine whether systematic differences exist in the relative ages of the oldest superposed craters as determined by this scheme. The isochrons labeled "Eratosthenian mare" and "Imbrian mare" in figure 7.13 refer to two mare surfaces with different crater populations and different apparent relative ages. Craters 700 m in diameter on the Imbrian mare range in morphology from indistinct gentle depressions, barely mappable as craters, to fresh excavations; craters of the same size on the Eratosthenian mare range from strongly subdued but still quite distinct to fresh. Thus, craters of the same size may be older on the Imbrian than on the Eratosthenian mare, as expected (fig. 7.14). The Apollo 12 landing site (fig. 7.14B) falls near the Eratosthenian-Imbrian boundary in this system. Still younger mare surfaces not visited have gentle, indistinct depressions no smaller than 300 m in diameter (fig. 7.14A). The most highly degraded craters 10 m in diameter on Copernican 6 surfaces would be bright and blocky. In summary, a search for the largest craters barely visible on a surface can be used to date surfaces.

A difficulty with this system and all others based on morphology is the likelihood that craters of different origins may have different initial morphologies (Trask, 1969). Secondary-impact craters near their sources will begin their existence with the subdued rim crests and shallow depths characteristic of older primary craters of their size. Attempts were made in mapping of the Apollo landing sites to distinguish secondary craters by their clustering, alignment, and elongation; this system was applied to the secondaries cautiously and with allowance for the probable initial differences.

Another difficulty is that the physical characteristics of the materials in which craters form influence their sizes and morphologies. Craters in such low-cohesion materials as regolith, pyroclastic glass, and fragmental terra breccia will appear more subdued even when first formed than will craters in basaltic bedrock (see chaps. 12, 13;

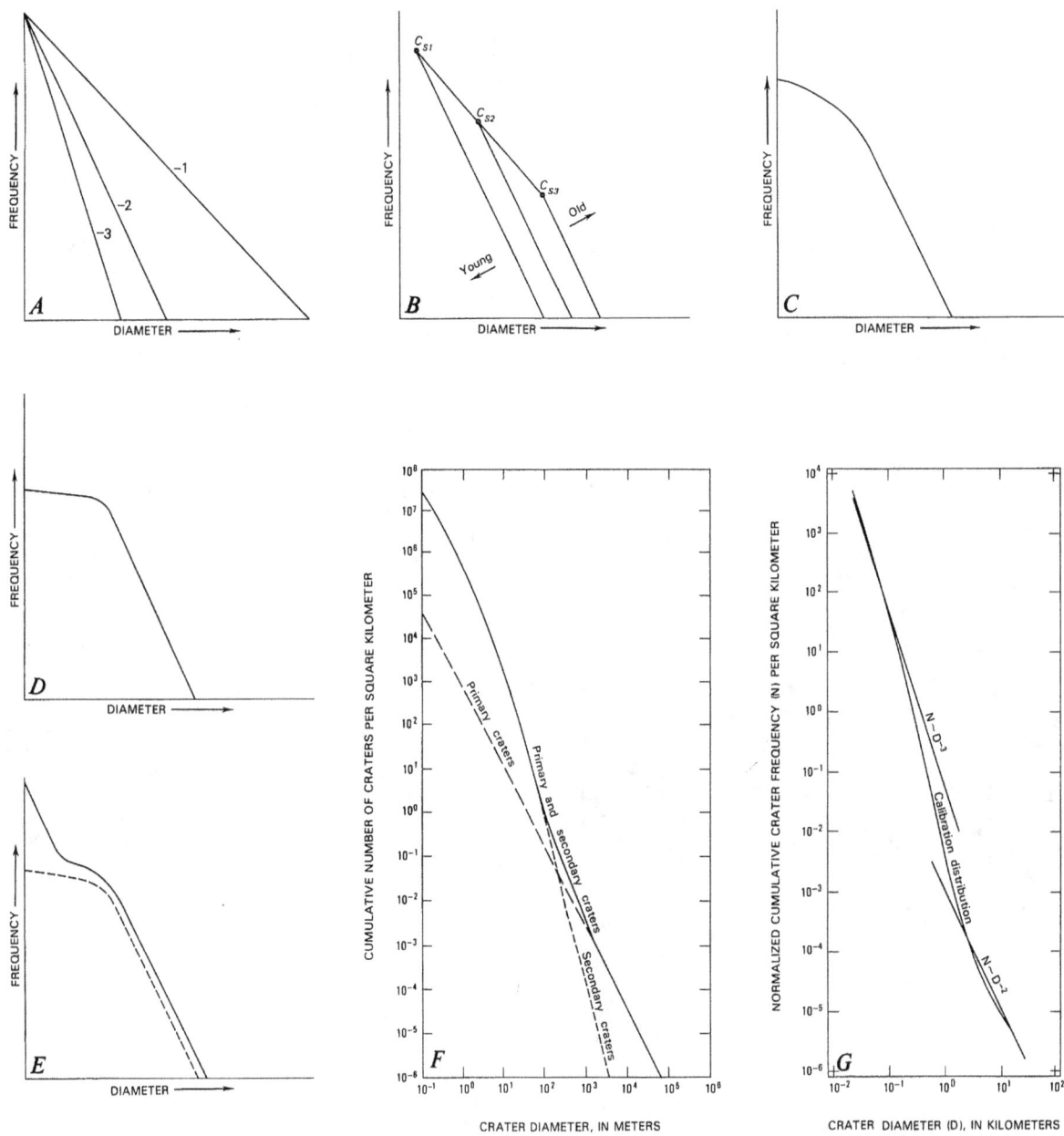

FIGURE 7.10.—Idealized cumulative crater-frequency curves representing different geologic and observational conditions.
  A. Crater-production functions with indicated slopes.
  B. Distributions of three different ages but same production function. Limiting diameters of steady state for all three ages (C_{s1}, C_{s2}, C_{s3}) define a curve sloping more gently than production functions; if production slope is -3, slope of steady-state distribution is -2 (Trask, 1966).
  C. Gradual rollover for frequencies of small craters, caused either by (1) inability to observe small craters with available photographic resolution or by (2) removal of small craters by mutual interactions of primary- and secondary-crater populations.
  D. Production function (slope, -2), truncated by burial of smallest and shallowest craters. Rollover is gradual because diameter cutoff is not precise under natural conditions.
  E. Frequency of D (dashed), altered by renewed cratering.
  F. Idealized plots of secondary craters (steeper slope, about -3.6) and primary craters (slope, about -2). After Shoemaker (1965).
  G. "Calibration distribution" of Neukum and others (1975a), showing different slopes in three diameter ranges. Steep slope in middle regime probably results from mixtures of secondaries and primaries, as in F.

Lucchitta and Sanchez, 1975). Moreover, craters formed in low-cohesion material degrade more quickly than those in cohesive bedrock. These effects are complex in craters formed in interlayered rock types, such as regolith and basalt (Young and others, 1974; Young, 1975, 1977; Schultz and others, 1977; Chapman and others, 1979). Other postformational modifications, such as downslope movement dependent on angle of slope, will also cause variations in morphologies and frequencies among contemporaneous craters of the same size (Trask, 1966, 1969). Despite the need to consider so many variables, the general validity of this dating system in any one area is supported by the superpositions of craters assigned young ages on craters assigned older ages, and by the densities of small superposed craters.

D_L method

A quantitative refinement of the above model has been extensively applied by the U.S. Geological Survey in dating geologic units too small to be dated reliably by crater counts (Soderblom and Boyce, 1972; Boyce and Dial, 1973, 1975; Schaber, 1973; Boyce and others, 1974, 1975; Boyce, 1976; Moore and others, 1980b, p. 3–18). Craters 200 to 2,000 m in diameter are generally examined. The method is based on a theoretical model of crater erosion operating in the steady state (Soderblom, 1970; Soderblom and Lebofsky, 1972). As in the method of Trask described above, the largest crater almost destroyed on a surface is sought. However, because a search will not necessarily disclose that crater, the crater's size is calculated. On a given surface, craters somewhat larger than C_s will be less nearly obliterated than the craters of diameter C_s. The hypothetical larger craters selected are those whose walls slope at 1°; their diameters are D_L (in meters). As a rule of thumb, $D_L = 1.7 C_s$.

Two related techniques are used to determine D_L values. The technique originally devised (Soderblom and Lebofsky, 1972) first determined the largest crater whose slope is eroded to the Sun-illumination angle of the photograph being examined. In practice, this crater (diameter, D_s) is bracketed by finding the largest unshadowed crater and the smallest diameters at which all craters are clearly shadowed. D_s is then converted mathematically to D_L (D_L is thus a theoretical diameter). In the second technique (Boyce and Dial, 1975), which can be applied to photographs of lower quality than the first, the crater (diameter, D_m) that is shadowed to the midpoint of its floor is bracketed. The upper bound on D_m is the diameter for which all craters of that size and larger are shadowed to the midpoint; its lower bound is the largest diameter of those craters that are shadowed halfway or slightly less. D_m is also converted mathematically to D_L.

D_L values of geologic units are quoted extensively in this volume. The method appears to have yielded correct results for mare units (Wilhelms, 1980) but less secure results for light-plains units (see chap. 10). Its successful application depends on several conditions: (1) Photographic quality sufficient to distinguish true shadows and true diameters; (2) Sun-illumination angles between 8° and 20° above horizontal (Soderblom and Lebofsky technique), or between 8° and 30° (Boyce and Dial technique); (3) a sample area of at least 100 km² for young surfaces (Sun angle, 10°) and 1,800 km² for terra plains (Sun angle, 20°; Boyce and others, 1975; Moore and others, 1980b, fig. 7); (4) the same initial crater shapes; (5) crater-wall slopes of from 8° to 25°, the interval over which the underlying theoretical model, assuming erosion by small impacts, applies; (6) craters exposed, not blanketed by later deposits; (7) craters having a slope of -3 on cumulative size-frequency plots when formed; (8) craters formed on level surfaces (detectable on the photographs); and (9) inclusion of craters superposed on only one geologic unit. Condition 7, which may not apply to primary craters larger than 2 km in diameter (fig. 7.10G), probably explains why D_L and size-frequency values are inconsistent for light plains and other units characterized by large superposed craters (chap. 10; fig. 7.15; Neukum and others, 1975a). Condition 9 may be the one most frequently violated, because measurements have commonly been made at the center of rectangles laid along a grid rather than within the confines of units mapped geologically. Thus, small units may have been missed.

Most of the pitfalls of the D_L method, or any other attempt to determine the ages of lunar and planetary materials, can be avoided by commonsense and examination of a surface from the viewpoint of the sequence of events through which it was shaped.

SUMMARY

Lunar stratigraphic units are relatively dated by a combination of superpositional criteria and methods based on superposed craters. Pairs of adjacent crater, basin, or mare units can normally be dated by straightforward observations of superpositions and transections based on the simple concept that the younger units modify the older. In favorable cases, secondary-impact craters extend the areas in which these observations can be applied.

Craters play a dual role as individual stratigraphic units treated by these relations and as time counters for other units' ages. Crater dating rests on the assumption, implied by the law of superposition, that the craters overlying a given stratigraphic unit have size-frequency distributions and morphologies typical of stratigraphic intervals younger than the unit. Old units support more craters than do younger units. Topographically sharp craters are superposed on all units, but large subdued craters only on old units (unless the subdual results from burial by still younger deposits).

The properties of the inverse size-frequency relation result in a hierarchy of dating methods (table 7.3). Direct observation of craters smaller than C_s is not useful in dating because fewer craters are observed than were formed. However, observation of the cumulative effect of these craters on morphologies is useful in dating. Craters slightly larger than C_s are dated by the method of Trask described above, or by the D_L method. The determined ages bracket the ages of small planar units on which the craters formed. Still larger craters (more than 3 km diam) are dated by assessments of the degradation of their subunits. Ages of whatever units underlie or overlie these craters can then be bracketed. Sufficiently extensive units can be dated not only by individual craters but also by statistical evaluation of the superposed craters larger than C_s. In general, because C_s increases over time, increasingly larger craters must be depended on for dating of increasingly older units.

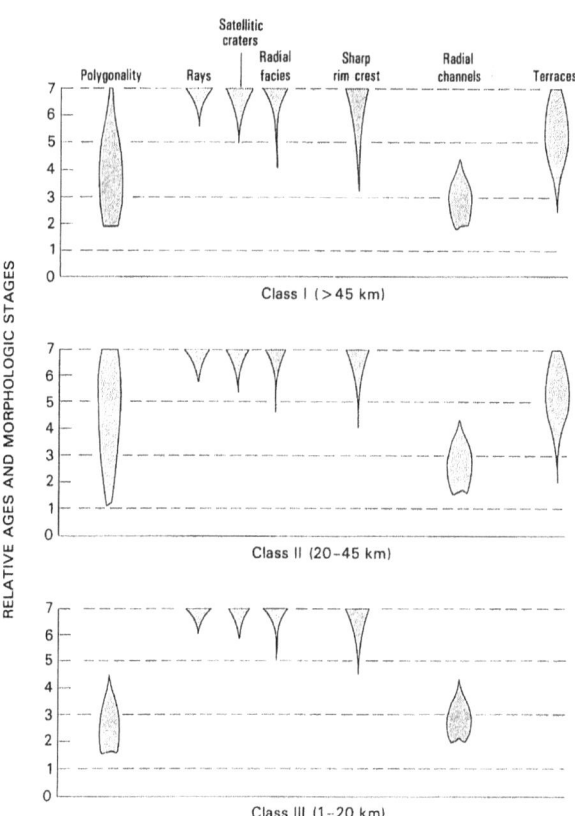

FIGURE 7.11.—Changes in diagnostic crater features with age (Pohn and Offield, 1970). Numbers 0 through 7 denote seven morphologic stages corresponding to relative ages, from oldest to youngest. Width of symbol indicates relative degree of development of each feature at each stage.

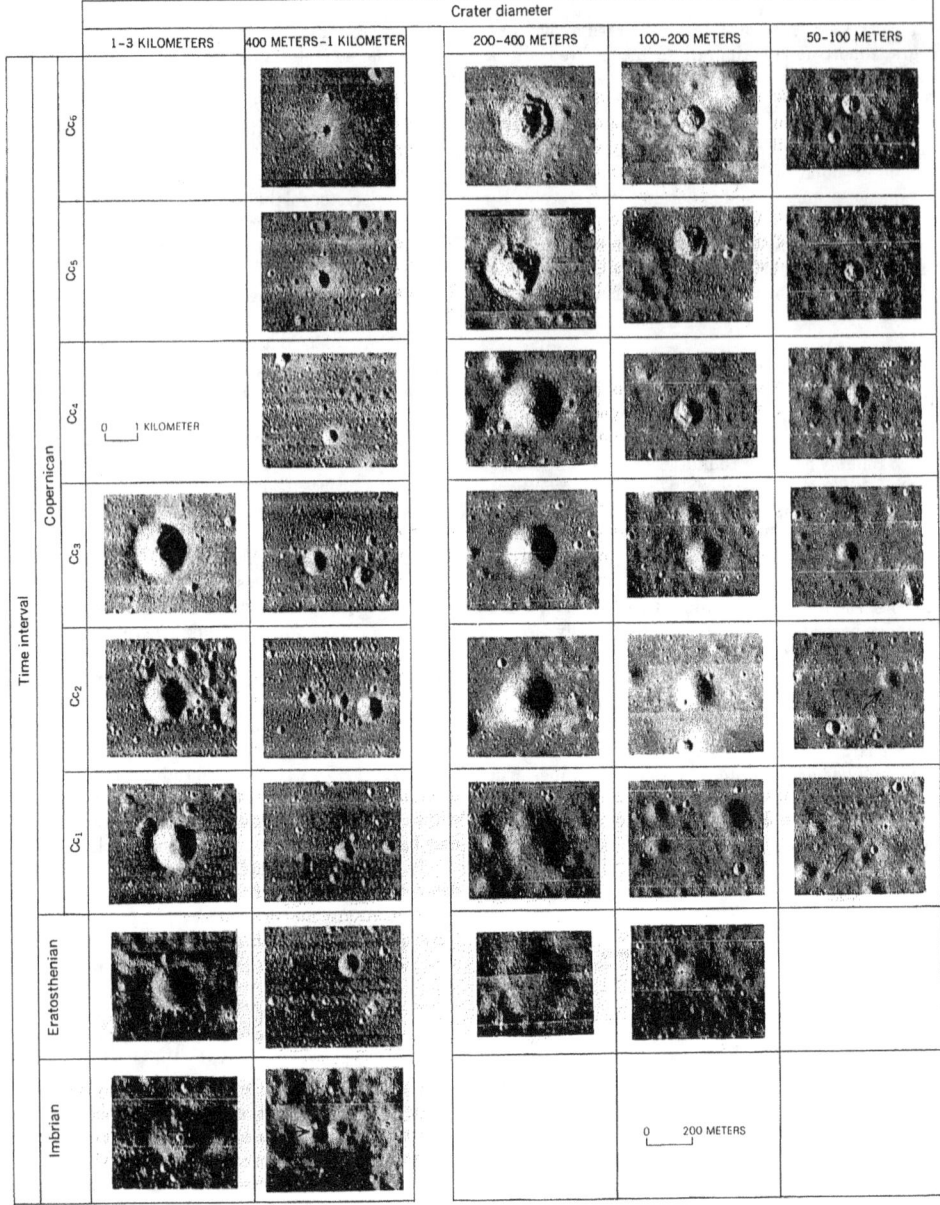

FIGURE 7.12.—Craters smaller than 3 km in diameter, arrayed by age and size. Sun-illumination angles are mostly 19.5° above horizontal; a few are from 18.2° to 23°. Craters of each size decrease in sharpness with age. From Trask (1971, fig. 2).

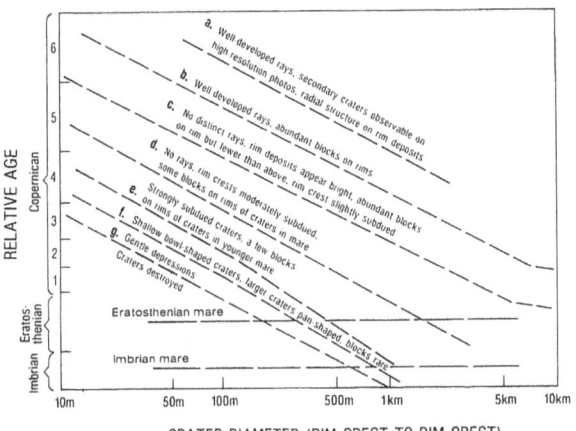

FIGURE 7.13.—Assumed relations between sizes, properties, and relative ages of craters from 0.01 to 10 km in diameter. Categories are intergradational; horizontal lines are isochrons indicating crater populations on Eratosthenian and Imbrian mare materials. From Trask (1971, fig. 3).

Thus, there is a substantial practical difference in dating the six divisions of the lunar stratigraphic column considered here (fig. 7.16; table 7.3):
1. Copernican units are dated by the frequencies and morphologies of craters so small as to be visible only on the best photographs.
2. Eratosthenian crater units photographed at high resolution can also be dated by size-frequency counts. Eratosthenian mare units and some impact-melt pools can be dated by the D_L method.
3. Late Imbrian mare units are readily accessible to the D_L method.
4. Only a few crater units of Late or Early Imbrian age are large enough to support enough craters larger than C_s for statistically valid size-frequency counts. These craters are dated mostly by the degradational morphology of their own subunits.
5. The Early Imbrian Orientale and Imbrium basin materials are extensive enough, young enough, and well enough photographed to be dated by craters over a large diameter range (fig. 7.16).
6. The steady-state size increases for older units (fig. 7.16; table 7.3), and only craters larger than 20 km in diameter are thought here to be valid as counters for the whole Nectarian and pre-Nectarian age range. Morphologies of individual Nectarian and pre-Nectarian craters larger than about 20 km in diameter supplement the size-frequency counts.

The next six chapters show in detail how the lunar stratigraphic column is constructed by these methods.

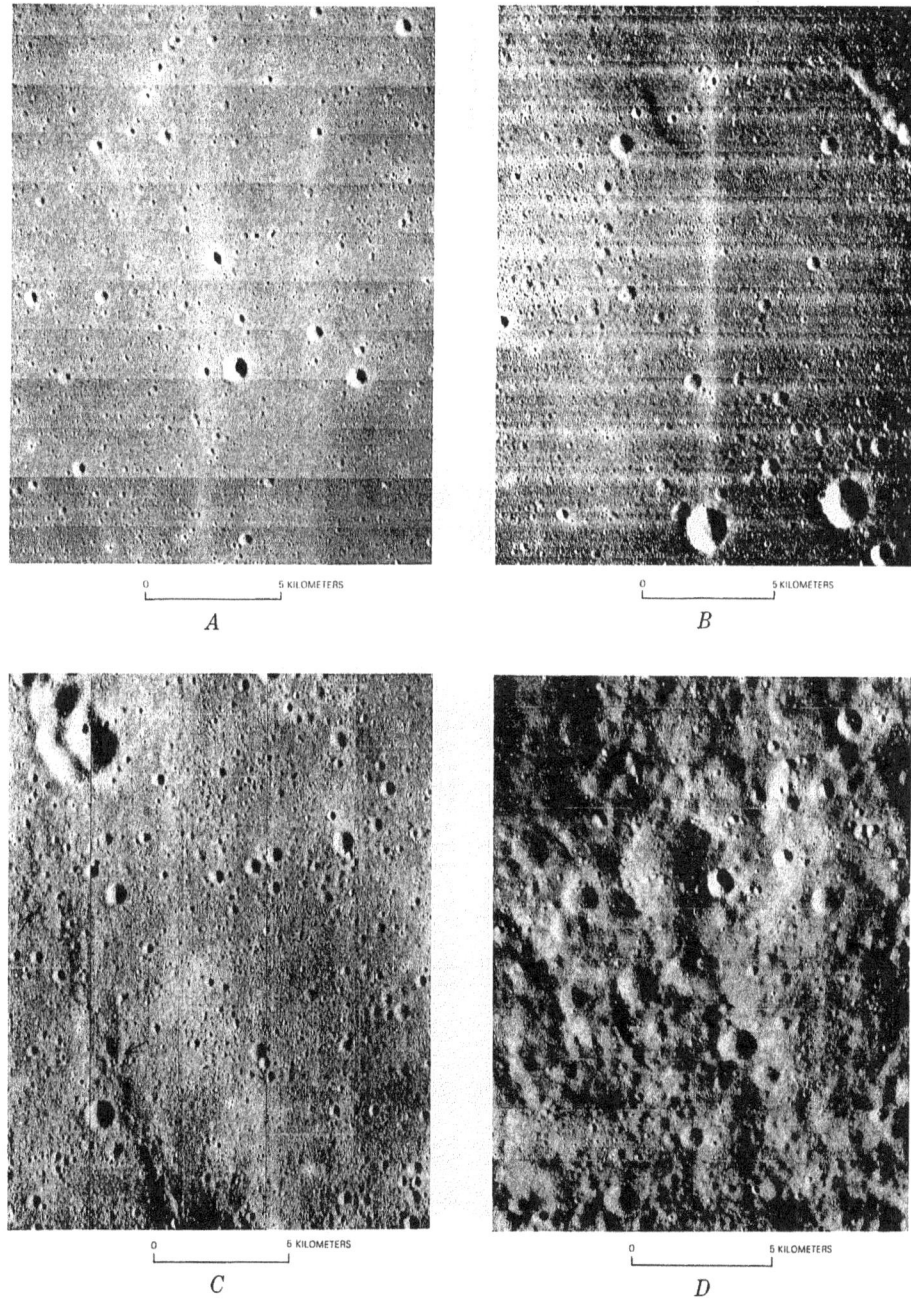

FIGURE 7.14.—Surfaces of four ages, showing differences in crater populations. Sun angles are all from 21.3° to 22.8° above horizontal. From Trask (1971, fig. 4).
    A. Young Eratosthenian mare in southern Oceanus Procellarum, characterized by small sharp craters and much level intercrater terrain.
    B. Older Eratosthenian mare near Apollo 12 landing site; larger and more subdued craters are present.
    C. Imbrian mare near Apollo 11 landing site, characterized by high density of subdued craters and little noncratered terrain. Arrow indicates low swell discussed by Trask (1971).
    D. Fra Mauro Formation near type area. Saturation of surface by large subdued craters suggests that many more craters formed than are now visible.

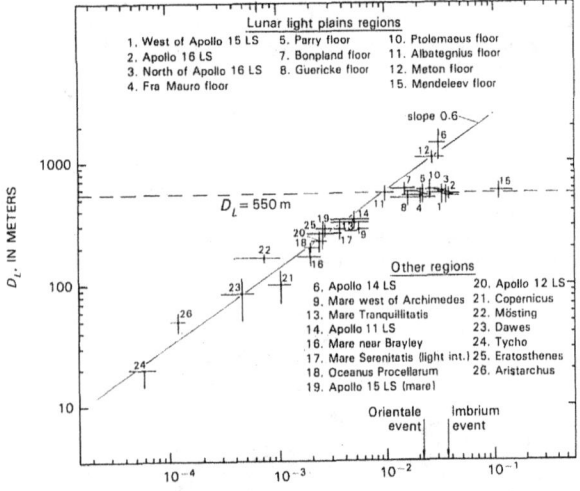

FIGURE 7.15.—Comparison of D_L values and cumulative crater frequencies, normalized to crater diameters of 1 km (Neukum, 1977). For many plains whose D_L value is 550 m, D_L values and crater frequencies do not correlate. "Event" refers to basin materials; Imbrium "event" counts were made on Montes Apenninus (Neukum and others, 1975a). LS, landing site.

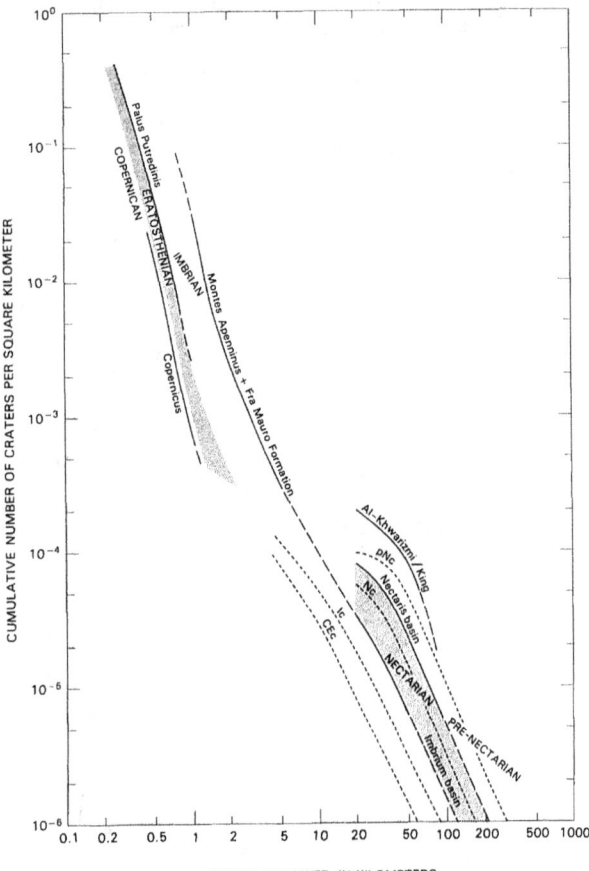

FIGURE 7.16.—Cumulative size-frequency distributions of craters in each time-stratigraphic system. Curves are solid within the diameter range where frequencies are most accurately determined, and long-dashed where extrapolated. Short-dashed curves represent average frequencies of primary-impact craters of only the age indicated: pNc, pre-Nectarian; Nc, Nectarian; Ic, Imbrian; CEc, Copernican and Eratosthenian, undivided (after Wilhelms and others, 1978). Other curves include all primary craters accumulated on the units. Where photographic data permit, successively smaller craters can be counted on successively younger units (many CEc and Ic craters smaller than 4.5 km in diameter are visible, but average frequencies were not determined). Numbers of smallest plotted Nc, pNc, Nectaris, and Al-Khwarizmi/King craters are diminished relative to larger craters of same age (compare fig. 7.10C). Rollover in curve indicating this deficiency is sharper and extends to larger diameters in pNc than Nc (rollover is less pronounced for Al-Khwarizmi/King than for pNc because former curve is plot of all accumulated craters, including CEc and Ic, which have relatively linear distributions). Al-Khwarizmi/King curve is steep at largest plotted diameters because of small statistical sample (see supplementary table). Imbrium-basin material is only lunar unit for which statistically valid frequencies in entire diameter range are available. Counts on Copernicus from Neukum and König (1976); counts on Montes Apenninus and on Fra Mauro Formation (combined here) from Neukum and others (1975a, b); Imbrium- and Nectaris-basin counts of craters at least 20 km in diameter from present study. Data and system boundaries further discussed in chapters 8 through 13.

8. PRE-NECTARIAN SYSTEM

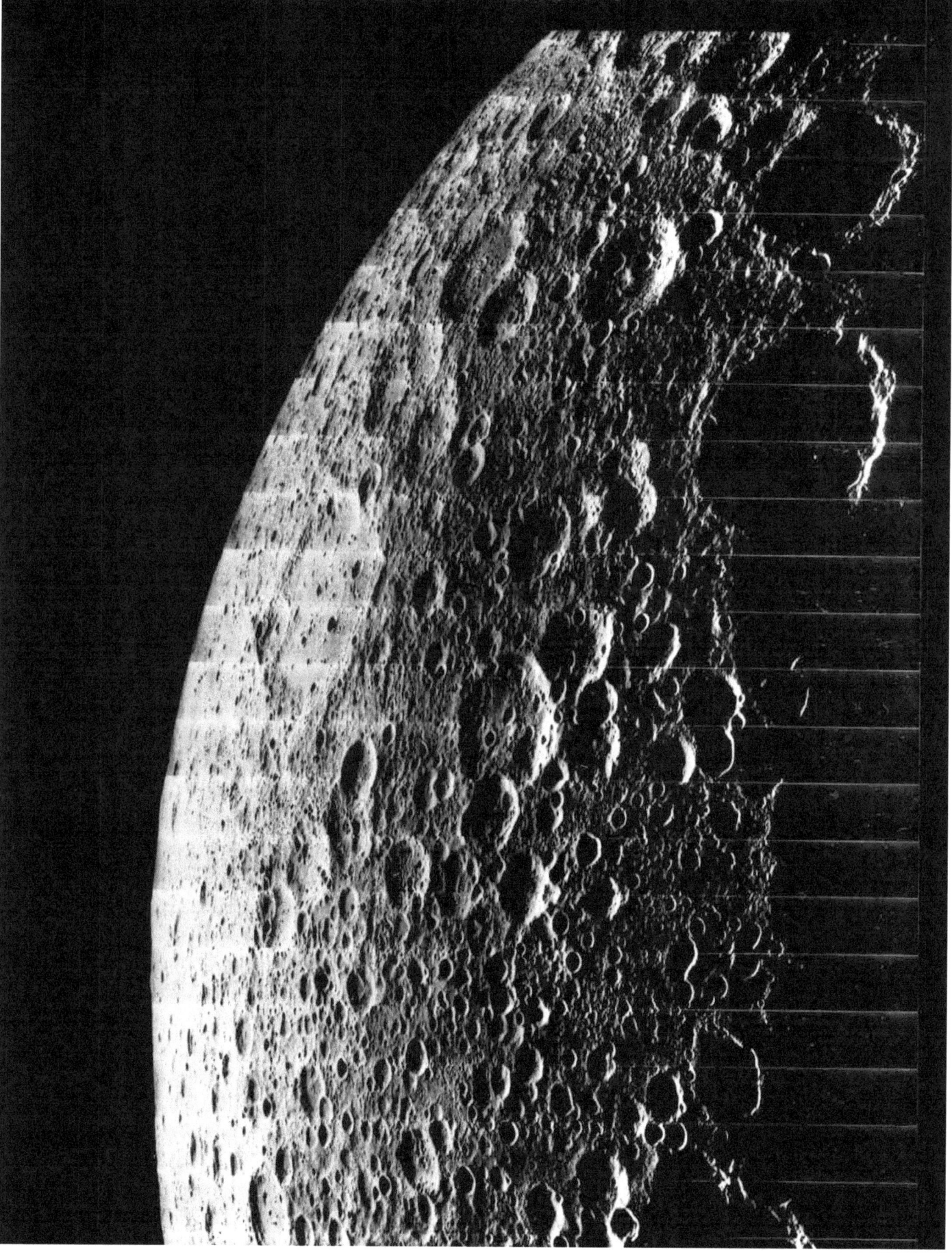

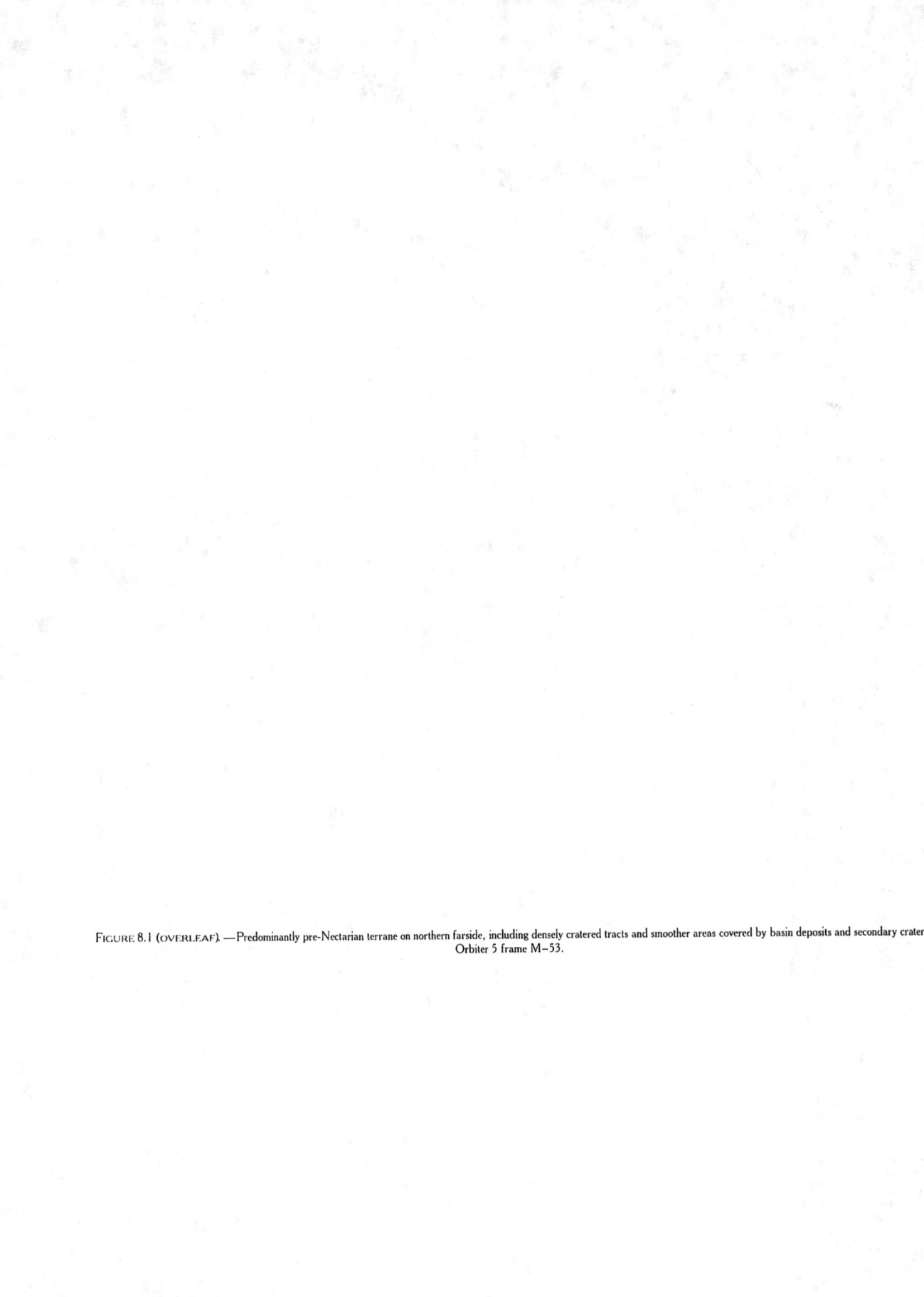

FIGURE 8.1 (OVERLEAF). —Predominantly pre-Nectarian terrane on northern farside, including densely cratered tracts and smoother areas covered by basin deposits and secondary craters. Orbiter 5 frame M-53.

8. PRE-NECTARIAN SYSTEM

CONTENTS

	Page
Introduction	139
Crustal petrology and geochemistry	139
Composition	139
Igneous or impact origin?	140
Early crustal petrogenesis	142
Magma ocean	142
Multiple plutons	142
Crustal zonation	143
Basin and crater materials	143
Introduction	143
Recognition criteria	145
Basins	145
Craters	147
Volcanic rocks	156
Chronology	156

INTRODUCTION

The lunar pre-Nectarian system[8.1] provides our clearest look at the early Solar System. The barrage of cosmic bodies that must have affected all moons and planets is richly recorded by pre-Nectarian basins and large craters, which are packed more densely than those of any other age (pl. 6; fig. 8.1). Their deposits cover a triangular area of the southern nearside and large parts of the farside. Smaller craters presumably formed in proportional abundance but were degraded by other craters (see chap. 7). Fine distinctions among facies of basin and crater deposits, which are observed in rayed craters and the Orientale basin, have similarly been blurred but can commonly be detected by comparisons with the younger analogs. Any maria that may have formed must have been even more completely devastated by the rain of impacts. Faults and folds dating from the pre-Nectarian are unknown, although they may have formed then as they did later. Therefore, efforts to understand early lunar and Solar-System history depend largely on interpreting the generally subdued and modified deposits of pre-Nectarian basins and craters.

Another look into the distant past is provided by the samples returned by the Apollo and Luna spacecraft (pl. 3). When fully evaluated, data from the sample analyses should show much about the composition of Solar-System material, the bulk composition of the Moon, and the processes by which a body of the Moon's size and composition differentiates. A major problem confronting analysts is distinguishing the effects of these early chemical and petrologic processes from those of the later impacts which formed the deposits that contain the samples. By definition, all materials formed since the origin of the Moon until the formation of the Nectaris basin are pre-Nectarian (Stuart-Alexander and Wilhelms, 1975). Accordingly, both the stratigraphic units that are still visible and the plutonic rocks of the early terra crust that were originally created by igneous processes are pre-Nectarian.

It is important to stress this distinction between the internal origin and impact emplacement of lunar terra rocks. The various groups of scientists that have investigated these two phases in the rocks' history have not always communicated effectively. Most petrologists and geochemists have concentrated on the endogenic phases, and their interest in the origin and chemistry of the Moon and of its early crust is reflected in the most commonly used lunar petrologic nomenclature. For example, names derived from terrestrial igneous petrology have been applied, inappropriately, to rocks melted and emplaced by impacts (Smith, 1974, p. 233; Irving, 1975, p. 364; Prinz and Keil, 1977). Stratigraphers are more interested in the times and processes of assembly and deposition of the breccia beds. These events in the rocks' histories are emphasized here and in the breccia-classification scheme discussed above in chapter 3 (Stöffler and others, 1979, 1980). The early crust is not described comprehensively here (see Taylor, 1982) but is treated mainly from the standpoint of its role as raw material for later stratigraphic units.

Unfortunately for those interested in the early Moon, no rocks of the early crust are preserved in outcrop, and only small fragments believed to be relatively unaltered samples of that crust have been recovered from the breccia deposits formed and emplaced by later impacts. Still more seriously for attempts to decipher the early Moon, no deposits identifiable with a specific pre-Nectarian stratigraphic unit were sampled. All the terra landings took place on Nectarian and Imbrian deposits because operational factors prevented Apollo and Luna from landing on the farside and southern nearside tracts where pre-Nectarian materials are exposed (pl. 6). Therefore, the photogeologically decipherable record and the sample record for the pre-Nectarian are connected much more tenuously than for later time-stratigraphic systems. Correlation of pre-Nectarian absolute and relative time scales and determination of the rates of impact and volcanic activity are difficult problems confronting stratigraphic and laboratory analysis.

CRUSTAL PETROLOGY AND GEOCHEMISTRY

Composition

The lunar terrae are composed of a crust of material that differentiated from the bulk Moon early in lunar history. The crustal composition was determined during early pre-Nectarian time by processes that are basically igneous, and was later mixed and otherwise modified by impacts. This mixing aids attempts to determine the average crustal composition, although it obscures whatever geochemical provinces may once have existed. Data from unmanned Surveyor analyses (Gault and others, 1968a; Turkevich and others, 1969; Phinney and others, 1970; Turkevich, 1971), returned-sample analyses, and orbital geochemical sensing (see chap. 5) currently combine to indicate an average terra composition rich in Al (25–29 weight percent Al_2O_3) and Ca (approx 14 weight percent CaO) (Taylor, 1975, 1982; Basaltic Volcanism Study Project, 1981, p. 666–678).

[8.1] To indicate that the pre-Nectarian system is not a formal stratigraphic name, the term is not capitalized (see introduction to the section in this chapter entitled "Basin and Crater Materials").

This chemistry is reflected in the low density of the crust (2.90–3.05 g/cm^3; Haines and Metzger, 1980; Basaltic Volcanism Study Project, 1981, p. 671) and in the fact that highly calcic plagioclase (mostly greater than 90 percent anorthite; Smith and Steele, 1975) is the most abundant sampled mineral. The next most abundant minerals in terra samples are low-Ca pyroxenes (mostly orthopyroxene, with subordinate pigeonite), olivine, and augite (high-Ca clinopyroxene). Aluminous or chrome spinel, ilmenite, silica minerals, troilite, and Fe metal are common accessories, and many other trace minerals are present (Smith, 1974; Frondel, 1975; Smith and Steele, 1975; Taylor, 1975, table 4.2; Cadogan, 1981, chap. 3). Lunar rocks vary far less in mode (optically visible minerals) and in mineral chemistry than do terrestrial rocks.

A series of rocks commonly called the ANT suite composes most of the returned terra material, where "ANT" refers to the plutonic-rock types *a*northosite (min 90 percent plagioclase), *n*orite (plagioclase and orthopyroxene), and *t*roctolite (plagioclase and olivine; table 8.1; Prinz and Keil, 1977; Stöffler and others, 1980). The plagioclase-rich composition of the terrae was first recognized in studies of small fragments in the Apollo 11 mare regolith (Wood, 1970; Wood and others, 1970), and the ANT suite was defined on the basis of small fragments in the Luna 16 and 20 regolith samples (Keil and others, 1972; Prinz and others, 1973; Taylor, G.J., and others, 1973; Prinz and Keil, 1977). Because the importance of impact mixing and melting was not fully appreciated during these early analyses, compositional types were commonly delineated incorrectly on the basis of mixed, brecciated, fine-grained, or even glassy rocks as if they were plutonic igneous rocks containing large optically identifiable crystals. The name "anorthosite" has been especially misused for materials that are not anorthosites in either the compositional or textural sense.

In ANT terminology, the "average" terra rock is an anorthositic norite or a noritic anorthosite (approx 70 percent normative plagioclase; Taylor, 1975, 1982). Unfortunately for the nonspecialist, this composition is also referred to as "highland basalt." This term was coined during study of regolith glasses (Reid and others, 1972a) and is one of several terms containing the word "basalt" that, especially in literature of the early 1970's, refer not to crystalline extrusive rocks but to magmas whose existence was predicted on the basis of glassy fragments (Apollo Soil Survey, 1971, 1974; Reid and others, 1972a, b; Prinz and others, 1973). Early workers commonly stated that "highland basalt" represents the most important primary magma of the terrae.

ANT rocks occur at all terra-sampling sites, though greatly varying in amount from the small fragments returned by Apollos 11, 12, and 14 and Lunas 16 and 20 to many kilograms from the Apollo 15, 16, and 17 landing sites. Earthlike granite and granodiorite are absent, but rare fragments of "granite" and "quartz monzodiorite" have been found (Ryder and others, 1975a; Ryder, 1976); these materials consist dominantly of quartz and potassium feldspar and contain varying amounts of relatively Fe-rich pyroxene and olivine and of relatively sodic plagioclase.

The most abundant type of highly differentiated lunar terra material is known, confusingly, by many partial synonyms, of which KREEP and KREEP-rich rock are the most nearly appropriate (table 8.1). The major-element composition of this material is not well defined, and the term refers primarily to an assemblage of minor and trace elements present in characteristic proportions (Meyer, 1977; Warren and Wasson, 1979c). The catchy acronym "KREEP" (coined by Hubbard and others, 1971, p. 343) refers to potassium (K), rare-earth elements (REE), and phosphorus (P); other diagnostic trace elements are Ba, Rb, Th, U, and Zr. These distinctive elements share the properties of large ionic radius (they are large-ion lithophiles [LIL]) and exclusion from the major mineral phases ("incompatible"). Although the name "KREEP basalt" commonly appears (Hubbard and Gast, 1971; Hubbard and others, 1971; Meyer and others, 1971), very few KREEP-rich samples have igneous textures. The term "KREEP" is properly used in a chemical, not a lithologic, sense for these diversely textured rocks. Major normative minerals of KREEP-rich materials are similar to those of the ANT rocks, except that plagioclase is more sodic (mostly less than An$_{88}$) and orthopyroxene and pigeonite are richer in Fe. Thus, both the minor- and major-mineral phases indicate advanced differentiation according to the Bowen (1928) reaction series. The incompatible elements are the longest to survive in residual liquids of magmas from which crystals have separated, and are the first to appear in new partial melts.

TABLE 8.1.—*Major-oxide compositions of major types of lunar terra material*

[All values in weight percent. Others, mostly TiO$_2$, Na$_2$O, K$_2$O, and Cr$_2$O$_3$ (Taylor, 1982, table 5.5); still other trace elements characterize KREEP (Warren and Wasson, 1979c). After Taylor (1975, p. 233-253; 1982, p. 201-233, tables 5.5, 8.2; S.R. Taylor, oral commun., 1982); data on high-K KREEP from Reid and others (1972b).
ANT: Compositions generally range from anorthosite (A, left) to norite (N, right) or, for MgO, to troctolite. Gabbro is synonymous with norite, except that high-Ca pyroxene characterizes gabbro and low-Ca pyroxene characterizes norite (Taylor, G.J., and others, 1973). In the terrae, low-Ca pyroxene is more abundant than high-Ca pyroxene.
KREEP: Compositions range from low-K (left) to high-K (right). KREEP-rich materials are also called KREEP basalt, Fra Mauro basalt (Reid and others, 1972b), nonmare basalt (Hubbard and Gast, 1971), high-Al basalt, alkali high-Al basalt (Prinz and others, 1973), norite (Wood, 1972a), KREEP norite, and many other names and acronyms (Taylor, 1975, p. 215, 272, table 5.8). Low-, medium-, and high-K KREEP-rich materials are commonly referred to as low-, medium-, and high-K Fra Mauro basalt (LKFM, MKFM, and HKFM, respectively).]

Material	ANT (A-N)	KREEP (low-high K)	Average terra
SiO$_2$	44-50	47-53	45.0
Al$_2$O$_3$	36-15	23-16	24.6
CaO	19-10	12-10	15.8
FeO	1-11	10	6.6
MgO	1-15	11-6	6.8
Others	≤.7	≤3.5	≤1.2

A common classification recognizes three categories of KREEP-rich materials that are similar in major-element bulk composition but differ in the proportions of the distinctive minor elements, exemplified by potassium: high K (more than 1 weight percent K$_2$O), medium K (approx 0.5 weight percent K$_2$O), and low K (approx 0.1 weight percent K$_2$O)(Taylor, 1975, p. 234–237). Also, the Fe/Mg ratio and SiO$_2$ content increase as the K content increases. Low-K KREEP is the most abundant category and, in fact, may be the second most abundant component of the terra crust after the ANT suite (approx 20 percent; Taylor, 1975, p. 234, 252). Low-K KREEP occurs in glasses or fragment-laden impact melts in some abundance at all landing sites except the mare-dominated Apollo 11 site (James, 1980). It was first suggested as a magma type on the basis of glassy fragments (Reid and others, 1972b). As a legacy of the large Apollo 14 collection, these three categories of material are commonly called high-K, medium-K, or low-K Fra Mauro basalt (HKFM, MKFM, LKFM) (Apollo Soil Survey, 1971; Reid and others, 1972b). However, this volume avoids the term "basalt" for nonvolcanic rocks and restricts the name "Fra Mauro" to the rock-stratigraphic unit the Fra Mauro Formation.

Igneous or impact origin?

A petrologically and stratigraphically important question is how many terra rocks have compositions that result from igneous processes and how many have compositions that were assembled by impacts. Until about the time of the Third Lunar Science Conference in 1972, most investigators seemed to favor an origin of certain Apollo 14 samples with KREEP-rich compositions and textures indicating crystallization from melts as endogenic volcanic basalt ("Fra Mauro basalt"; fig. 2.8). The current consensus, however, is that almost all terra melt rocks with basaltlike textures originated as impact melts (Dence and Plant, 1972; Green and others, 1972; Taylor and others, 1972; James, 1973; Irving, 1975, 1977; McKay and others, 1978, 1979). The main clues are (1) trace elements indicating contamination by meteorites and (2) small-scale textural and compositional heterogeneity, especially among clasts.

Although impact-created rocks dominate the terrae, some igneous rocks from the early pre-Nectarian crust were recovered (fig. 8.2; Warren and Wasson, 1977, 1979b, 1980a, b; Norman and Ryder, 1979). A diligent search has been conducted for samples of these "pristine" rocks, which have also been referred to as meteorite-free, endogenic,

endogenous, indigenous, primitive, and plutonic (see references in Warren and Wasson, 1980a, p. 81–82). The following are the main criteria for pristinity that currently are most widely accepted.
1. A low content of *siderophile elements*, many of which are characteristic of iron meteorites and chondrites (for example, Hertogen and others, 1977; Morgan and others, 1977). The highly siderophilic elements Au, Ir, Os, Pd, and Re probably accompanied metallic Fe during early differentiation of Solar-System material and thus were concentrated in planetary cores and small Fe-rich bodies. A few investigators believe that the silicates of the Moon are intrinsically richer in these elements than is generally believed and that pristine rocks low in siderophiles are impact-melt rocks from which siderophiles have been removed by extraction of metal (Delano and Ringwood, 1978; Wänke and others, 1978). Most investigators, however, believe that high concentrations of these elements indicate meteoritic contamination, although their absence does not necessarily prove pristinity (Irving, 1975, 1977; Warren and Wasson, 1977; Anders, 1978; Norman and Ryder, 1979, p. 533–536; Taylor, 1982, p. 222).
2. Igneous, especially *cumulus*, textures. Impact melts rarely cool slowly enough to develop cumulus textures, which are conspicuous in layered intrusions on the Earth, where such factors as gravity, liquid density, and fluid currents have produced quasi-sedimentary features (Jackson, 1971; Jackson and others, 1975; Raedeke and McCallum, 1980). Thus, lunar cumulus textures indicate pristinity (fig. 8.2D). Such textures furthermore suggest that the rock's composition arose by igneous processes. In contrast, metamorphic textures commonly form in deposits of impact melt and breccia whose compositions arose by mechanical mixing of preexisting rocks. Volcanic textures may form either in true volcanic rocks or in impact melts, and thus are ambiguous when not supplemented by other clues, such as siderophile content.
3. Grain size larger than 3 mm, indicative of slow cooling in a plutonic environment. In many pristine samples, however, the original igneous textures and the coarse grain size have been obscured by subsequent shock-induced cataclasis.

It seems clear today that the bulk compositions of most terra rocks are determined by impact-induced mixing. Plots of bulk major-element compositions of lunar terra rocks reveal that individual endogenic cumulates and other pristine samples vary much more in composition than do obviously impact-generated polymict breccias and regoliths containing siderophiles (Warren and Wasson, 1977; 1979b, p. 601; 1980b; Norman and Ryder, 1979, p. 554; Taylor, 1982, p. 202). The "anorthositic norite" or "highland basalt" average terra composition, granulitic impactite (Warner and others, 1977; Bickel and Warner, 1978), very high alumina "basalt" (Dowty and others, 1973, 1974b; Hubbard and others, 1974), and low-K KREEP are among the compositional types that fall in the middle of such plots away from known pristine rocks. Low-K KREEP is found mostly (Ryder and others, 1977) or entirely in siderophile-bearing fragment-laden melts

A

B

C

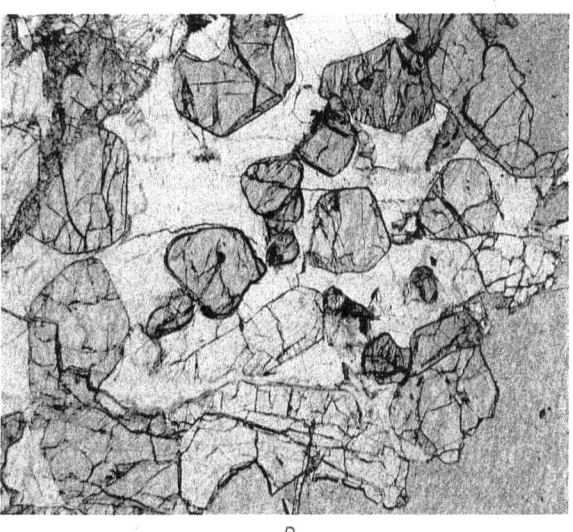

D

FIGURE 8.2.—"Pristine" terra samples.
    A. Cataclastic anorthosite (sample 60025), as photographed in Lunar Receiving Laboratory after arrival from Apollo 16 landing site. Little texture is evident. Scale cube is 1 cm on a side in this and all other "mug shots."
    B. Thin section of sample 60025. Most optically observable crystals (modes) are of plagioclase. Texture is seriate. Crossed polarizers; field of view, about 2 mm.
    C. Coarse-grained granulitic troctolite (sample 76535) from North Massif, Apollo 17 landing site. Major modes are plagioclase (58 percent) and olivine (37 percent), with accessory orthopyroxene (4 percent) (Dymek and others, 1975).
    D. Thin section of pristine pink spinel troctolite clast (sample 67435,14). Original igneous cumulate texture is largely preserved (Prinz and others, 1973). Euhedral to subhedral olivine (gray) is enclosed poikilitically by plagioclase (white). Plane-polarized light; field of view, about 2 mm.

(Meyer, 1977; Hess and others, 1977; McKay and Weill, 1977; Reid and others, 1977; Shih, 1977; Warren and Wasson, 1979b). This extensive impact mixing is to be expected from the geologic style of the Moon because each new impact mixes and melts rocks, most of which already are complex breccia deposits with compositions far removed from those of their endogenic origins. New hybrid compositions are formed by each event, and continued impacting creates a hybrid average of the original pristine compositions. Mixing models attempt to reconstruct the observed compositions of regoliths and bedrock-breccia deposits from a set of end members believed to represent original magma types (for example, Hubbard and Gast, 1971; Schonfeld and Meyer, 1972; see recent critiques by Ryder, 1979, and Taylor, 1982, p. 201–205).

In summary, the composition of the lunar terra crust was derived by mixing pristine ANT compositions with smaller amounts of highly differentiated KREEP-rich material. Although a few intact fragments of the early crustal rocks were recovered, the textures and structures of almost all returned terra samples were created by impacts. Many of the samples assembled by impacts retain the early endogenic compositions, whereas others have been compositionally as well as texturally altered during the impact reworking. Problems still persist in identifying the sources of specific samples and determining the relative amounts of pristine and impact-mixed materials. However, enough clues about the early crust have been assembled from this obscure record to reconstruct the following partial plutonic history of the early crust.

EARLY CRUSTAL PETROGENESIS

Magma ocean

Analysis of the small fragments from the Apollo 11 regolith led to what is still the leading model for the origin of the early crust (Smith and others, 1970; Wood and others, 1970). The terra crust and the mantle source of mare-basalt magmas are widely thought to have separated from a very large volume of magma popularly called a *magma ocean* (Walker and others, 1975; Wood, 1975b). The first hint of an early widespread differentiation of primitive materials came from REE abundances. Europium is overabundant in "average" terra rocks and underabundant in mare basalt relative to chondritic REE abundances (for example, Taylor, G.J., and others, 1973; Smith, 1974; Taylor, 1975, p. 154–159). This relation is consistent with an efficient mechanical separation of the ultramafic mare-source material, as cumulates, from a magma from which large amounts of plagioclase had already crystallized and extracted Eu relative to the other REE's. Plagioclase has large lattice sites that accept the large divalent (reduced) ion that only Eu among the REE's possesses.

"Oceanic" depths of 200 to 400 km are commonly favored, although lesser or greater depths are not excluded (half the lunar volume lies above, and half below, 360-km depth) (Walker and others, 1975; Hubbard and Minear, 1975; Herbert and others, 1977; Taylor, 1978; Longhi, 1980; Binder, 1982). Because of its lesser density, the plagioclase-rich terra-crustal material presumably floated in the magma ocean, whereas ultramafic cumulus crystals sank to form the upper mantle that became the mare source (for example, Smith and others, 1970; Wood and others, 1970; Taylor and Jakeš, 1974; Taylor, 1975, 1978; Wood, 1975b, c; Herbert and others, 1977; Ryder and Wood, 1977; Longhi and Boudreau, 1979; Warren and Wasson, 1979a, 1980a). The divide between the sunken and floated materials is generally assumed to be the seismic discontinuity at the base of the crust. If the ocean was 360 km deep and the crust is 75 km thick, the crustal material constituted about a quarter of the original oceanic volume.

The origin of KREEP in general and its role in "magma oceanography" in particular have generated a very large literature because of its persistent appearance in lunar samples and its exotic highly differentiated chemistry (see reviews by Taylor, 1975, 1982; Irving, 1977; Meyer, 1977; Warren and Wasson, 1979c; James, 1980; Vaniman and Papike, 1980). Its origin has been ascribed to partial melting of ANT material or, as commonly in more recent literature, to a late stage in fractionation of the magma ocean. KREEP has been considered to be derived from a minor liquid accumulated in dispersed niches throughout the magma ocean or the crust, or, alternatively, concentrated in certain regions. The question of its origin is difficult to resolve because because most or all analyzed KREEP is a chemical constituent incorporated into other rocks, rather than a distinct rock type itself.

Multiple plutons

Study of pristine rocks, many of which are thought to be relics of the primordial crustal differentiation, has shown that the magma-ocean model outlined above is too simple. Many of the pristine crustal rocks, especially those from the Apollo 17 landing site, are cumulate norite and troctolite that apparently formed by settling in their parent magmas (Norman and Ryder, 1979, p. 538). Data from the orbital geochemical instruments, as interpreted by mixing models, indicate that the topmost part of the "average" crust can contain no more than about 60 percent anorthosite (Haskin and Korotev, 1981) and that compositions range laterally from highly feldspathic to highly mafic (Spudis and Hawke, 1981). Moreover, the orbital instruments covered only a small percentage of the Moon and are subject to calibration problems (see chap. 5). Therefore, mafic materials may be more abundant than once thought.

Compositional subtleties of the mafic and anorthositic ANT rocks suggest that more than one magma system may have built the terra crust. For example, if the $Mg/(Mg+Fe)$ ratios in mafic minerals (olivine, pyroxene) in the pristine rocks are plotted against the $Ca/(Ca+Na+K)$ ratios in plagioclase, two distinct trends are definable (fig. 8.3; Steele and Smith, 1973; Taylor, G.J., and others, 1973; Warner and others, 1976; Warren and Wasson, 1977, 1979a, b, 1980a; Wood, 1977; Binder, 1980; James, 1980; Longhi, 1980; Raedeke and McCallum, 1980; Taylor, 1982, p. 248–251). One trend consists of highly feldspathic rocks (anorthosite and troctolitic anorthosite) that contain relatively calcic, uniform plagioclase and relatively Fe-rich, compositionally varying mafic minerals; these rocks are termed *ferroan anorthosite* because of the Fe-rich mafic minerals (Dowty and others, 1974a). The other trend consists of more mafic rocks (troctolite, spinel-bearing troctolite, norite, gabbronorite, and a few ultramafic rocks) that have highly varying plagioclase and relatively magnesian, compositionally varying mafic minerals; these rocks are referred to as Mg-rich plutonic rocks or the *Mg suite*. In an individual rock of the Mg suite, Mg-rich mafic minerals are associated with Ca-rich plagioclase, and Fe-rich mafic minerals with more sodic plagioclase. These associations are what is to be expected in typical cotectic igneous differentiation (Bowen, 1928). The anorthosite was at first thought to be more anomalous (McCallum and others, 1975)

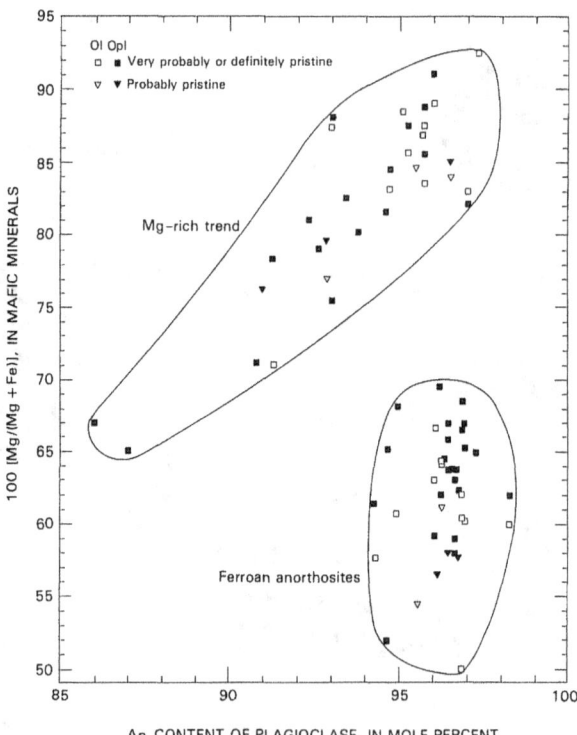

FIGURE 8.3.—Ferroan-anorthosite and Mg-rich trends in ANT rocks, separated by a distinct hiatus. Examples shown are for olivine (ol) and orthopyroxene (opx) in pristine rocks (Warren and Wasson, 1980a).

because Fe-rich mafic minerals are associated with Ca-rich plagioclase (Wood, 1975b; 1977, p. 48). However, the same type of mineralogic variation has been discovered in anorthosite of the Stillwater layered igneous complex (Raedeke and McCallum, 1980). No simple magma system is likely to have generated both trends simultaneously (Wood, 1975b; Longhi and Boudreau, 1979; Warren and Wasson, 1979a, b, 1980a, b; Raedeke and McCallum, 1980).

To accommodate the intricacies and the global scale that both seem required, several authors have proposed "separate pluton" models, in which layered Mg-suite plutons (Jackson and others, 1975; Raedeke and McCallum, 1980) intruded from below into an already-floated crust of ferroan anorthosite (Binder, 1980; James, 1980; Warren and Wasson, 1980a). Alternatively, the early pre-Nectarian crust may never have been an "ocean" but only a system of magma "pods" or "ponds" (Wetherill, 1975b, 1976; Simonds, 1979). Complex petrologic schemes dating from the accretionary period (Simonds, 1979; Smith, 1982) or operating in an early convecting magma ocean (Longhi, 1980) have also been proposed. Impacts must have bombarded the primitive Moon catastrophically; they would have pierced any solid crust that formed on an "ocean" and have stirred and mixed the liquid, influencing the early geochemistry and petrogenesis (Wetherill, 1975b; Wood, 1975b; Hartmann, 1980; Minear, 1980; Simonds, 1979; Warren and Wasson, 1980a; Taylor, 1982, p. 248–251). Although the magma-ocean concept is still favored by most petrologists and geochemists, it is becoming increasingly modified and complex.

Crustal zonation

Possible layering or, more likely, crude vertical zonation of the crustal material has been studied by a combination of seismic, petrologic, and impact investigations. Seismic data indicate a discontinuity in the crust at about 20- or 25-km depth (Toksöz, 1974; Toksöz and others, 1974; Goins and others, 1979) that is commonly ascribed to an interface between low-density, relatively porous crustal rock and denser sealed material of the same composition (Todd and others, 1973; Bills and Ferrari, 1977; Haines and Metzger, 1980). Alternatively, or additionally, this seismic feature could represent a compositional discontinuity between more feldspathic material above and more mafic material below (Herzberg and Baker, 1980; Warren and Wasson, 1980a). A distinct KREEP-rich layer has also been suggested, either between layers of anorthosite and more mafic material (Ryder and Wood, 1977) or at the base of the crust (Warren and Wasson, 1979c).

Because basins probe the subsurface, the surface distribution of ejecta compositions should help in reconstructing any layering or zonation that exists. The large size of basins might suggest excavation as deep as the mantle. Mineral assemblages in some spinel-bearing Mg-suite rocks have been cited as requiring pressures available only in the deep crust or mantle of the Moon (Herzberg, 1978; Herzberg and Baker, 1980). However, alternative phase-equilibria schemes allow an origin as shallow as 25 km, well within the crust (Warren and Wasson, 1979a; Herzberg and Baker, 1980). Although most terrestrial dunite is probably of mantle origin, at least one Mg-suite lunar dunite probably originated in the crust, judging from its association with, and mineralogic similarity to, troctolite sample 76535, whose cooling and reequilibration depths were probably in the range 10–30 km (Gooley and others, 1974; Dymek and others, 1975). Orbital geochemistry suggests crustal composition even for Imbrium-basin ejecta, which I believe was derived from the cavity of the Moon's largest known basin, Procellarum (fig. 8.4). Therefore, little, if any, mantle material was brought to the surface in impact ejecta, and shallow basin excavations restricted to the crust are implied.

Crustal thickness and surface composition seem to be related; the more aluminous the material, the thicker the crust (Schonfeld, 1977; Haines and Metzger, 1980). That is, the Pratt (unequal density) model of isostasy seems to be more nearly applicable to the terrae than is the Airy model (Solomon, 1978). The Airy (equal density) model was assumed in arriving at the commonly stated conclusion that the farside crust is thicker than the nearside crust (Kaula and others, 1974). Most investigators have nevertheless concluded that the farside crust is thicker (see chap. 1), even under Pratt or mixed isostatic models (Haines and Metzger, 1980).

This difference in crustal composition and thickness, however, may not be hemispheric in extent. The orbital geochemistry (Haines and Metzger, 1980) suggests that the crust is both thinner and more magnesian beneath the two largest lunar basins, Procellarum and South Pole-Aitken, than in other regions where data are available. Seismic data (see chap. 1) suggest that the crust is 50 or 60 km thick under southern Oceanus Procellarum and about 75 km thick under the highly feldspathic Apollo 16 landing site, which is near the Procellarum rim (pl. 3). Crustal thicknesses are similar on the southern nearside and western (long 90°–180° E.) farside (fig. 1.12; Bills and Ferrari, 1977). The X-ray spectrometer detected higher Al/Mg ratios on the farside and on parts of the eastern and southeastern nearside than in the Imbrium-Procellarum region (Adler and others, 1972, 1973; Hubbard and others, 1978). No data known to me contradict the conclusions that the thick aluminous crust is restricted to the rims and exteriors of the two giant basins and that the thinner, more magnesian crust is concentrated in basin interiors. Additionally, the gamma-ray spectrometer detected the high radioactivity characteristic of KREEP in the Imbrium-central Procellarum region (Metzger and others, 1973, 1977), where I would expect the thinnest part of the lunar crust.

The sample data support the conclusion that the giant basins control the crustal thickness and, thus, the compositional zone that was accessible to later impacts (fig. 8.4). The greatest depths were reached by the Imbrium basin, whose ejecta (Fra Mauro Formation) is correspondingly the lunar deposit richest in KREEP (see chap. 10). Mg-suite rocks and low-K KREEP, probably derived from the same or shallower depths, apparently compose most of the Imbrium (Apollo 15; chap. 10) and Serenitatis (Apollo 17; chap. 9) rims. The Apollo 16 and Luna 20 landing sites are the richest in feldspathic ANT materials (chap. 9), as befits their derivation from the thick crust of the outer Procellarum shelf or exterior terrain.

This picture, though internally consistent, is subject to revision or rejection on the basis of better data. Lateral zonation of primordial origin also is possible. Only four major sample suites from four landing sites have been collected from the terrae, and each suite added complicating detail to the lunar petrogenetic picture. On the basis of the complexities generally uncovered when geologic systems are examined closely, fine-scale local intrusions only crudely organized into petrologic provinces might be expected. The pattern that seems to reflect global layers exposed by large basins might be fortuitous.

BASIN AND CRATER MATERIALS

Introduction

The severely deformed, chemically and texturally mixed, siderophile-rich materials found in the Nectarian and Imbrian deposits attest to a high rate of impacts during pre-Nectarian time as well. More specific information on the characteristics of individual pre-Nectarian units is sparse because none were sampled directly and because the chemical and petrologic criteria for recognizing such units are ambiguous. Attempts have been made to identify individual impacting projectiles with specific basins by means of distinctive siderophile signatures (Hertogen and others, 1977; Morgan and others, 1977), but the success of these attempts is limited by knowledge of the prebasin stratigraphy. The study of pre-Nectarian impact deposits thus requires stratigraphic analysis.

Originally, pre-Nectarian and Nectarian were combined as pre-Imbrian (Shoemaker and Hackman, 1962). This designation sufficed for most old units mapped at 1:1,000,000 scale because the mapping was confined to the central nearside, which is dominated by the Imbrium basin (pl. 3). If they were subdivided at all, pre-Imbrian craters were numbered according to the morphologic criteria established by Pohn and Offield (1970).

The desirability of stratigraphically dividing the pre-Imbrian emerged during mapping of the nearside regions farthest from Imbrium. While mapping south of Mare Nectaris, Stuart-Alexander (1971) recognized a lineate surface as the expression of Nectaris-basin deposits and secondary-impact craters similar to those of the Orientale and Imbrium basins (fig. 8.5); she named the deposit the "Janssen Formation" (Stuart-Alexander, 1971). At first, the Janssen was used to subdivide pre-Imbrian crater materials into numbered pre- and post-Nectaris categories descendant from Pohn and Offield's

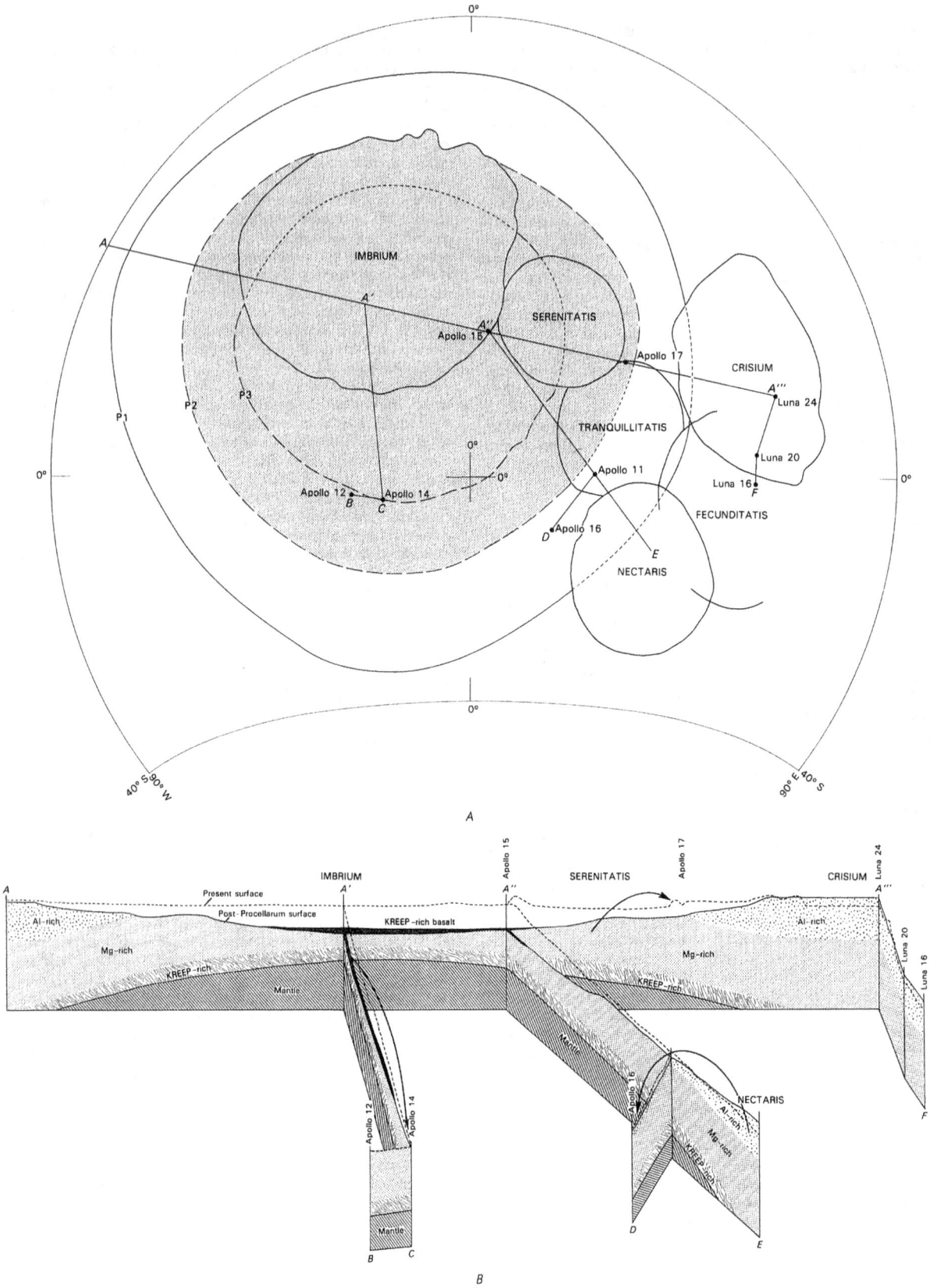

FIGURE 8.4.—Hypothetical crustal layers intersected by impact basins.
  A. Index map based on plate 3, showing basins from which samples were obtained. P1, P2, P3, rings of Procellarum basin (dotted where inside younger basins; inner rings dashed). Shading, Mg-rich terrane after formation of Procellarum and before formation of other basins.
  B. Diagrammatic geologic fence diagram, suggesting crustal and mantle configuration after Procellarum impact, subsequent uplift, and partial fill by hypothetical KREEP-rich volcanic basalt (black). Three gradational crustal layers are intersected by Procellarum-basin surface: upper Al-rich material is exposed outside basin and on outermost shelf (between P1 and P2); Mg-rich material is exposed on intermediate shelf (between P2 and P3) and in central basin; and lower-crustal KREEP-rich material is near surface (or may be exposed) near center of basin. Nectaris and Crisium basins were formed mostly in Al-rich terrane (Apollo 16 and Luna 20 samples, respectively); Serenitatis basin was formed in Mg-rich terrane (Apollo 17 samples); Imbrium basin was formed in Mg-rich terrane (Apollo 15 samples) and also ejected large amounts of KREEP-rich material and (or) KREEP-rich basalt (Apollo 14 samples). Arrows indicate paths of ejecta. Curvature of surface trace and of the Moon is ignored; vertical exaggeration is arbitrary.

numeric designations (Stuart-Alexander, 1971; Wilhelms and McCauley, 1971; Scott, 1972b). In 1975, the pre-Imbrian was formally divided into the pre-Nectarian and Nectarian (Stuart-Alexander and Wilhelms, 1975). This division is shown on all subsequently published geologic maps, where appropriate, and is used here. The utility of this subdivision was proved as the limb, farside, and polar regions were mapped further and large numbers of pre-Nectarian and Nectarian units were identified (Scott and others, 1977; Wilhelms and El-Baz, 1977; Lucchitta, 1978; Stuart-Alexander, 1978; Wilhelms and others, 1979). The term "pre-Imbrian," however, is still used where more convenient, as in places on the nearside where the finer distinction cannot easily be made. Because neither "pre-Nectarian" nor "pre-Imbrian" has a formally defined base, such terms as "system" and "period" used in conjunction with them are informal and not capitalized.

Recognition criteria

A start in establishing a pre-Nectarian age is to examine terrain covered by the deposits and secondary craters of the Nectaris basin (fig. 8.5). Deposits overlain by these materials are, by definition, pre-Nectarian. Pre-Nectarian basins in direct contact with Nectaris-basin materials include Australe, Fecunditatis, Mutus-Vlacq, Tranquillitatis, and Werner-Airy (pl. 6; table 8.2). Craters lightly nicked by Nectaris secondaries but not deeply buried by deposits are type examples of pre-Nectarian craters (figs. 8.5C, D).

Units not in contact with Nectaris-basin materials can potentially be dated by comparing the size-frequency distributions of superposed primary craters with those of craters superposed on Nectaris (fig. 8.6; table 8.2). Size-frequency curves lying entirely above that for Nectaris demonstrate a pre-Nectarian age. One curve for a basin here considered to be pre-Nectarian (Apollo) crosses below the Nectaris curve, but only at the statistically weak end representing the frequencies of large craters.

The differences among the many curves demonstrate the fact that the lunar terrae do not have some single characteristic size-frequency distribution but are composed of basin blankets, each of which has reset the visible crater population of the surfaces it covers. The slopes of the cumulative crater-frequency curves decrease and become more highly curved from the youngest to the oldest surfaces (fig. 8.6). Some of the crater-frequency curves approach a slope of -1 for craters as large as 60 or 70 km in diameter, a slope substantially lower than that of -2 or -1.8 which is commonly believed to represent the production frequency for younger craters (for example, Basaltic Volcanism Study Project, 1981, chap. 8). Although the -1 slope has not been modeled in studies of crater saturation (Woronow, 1977, 1978), it nevertheless probably results from mutual obliteration of pre-Nectarian and Nectarian craters.

Several factors in addition to this destruction of craters limit the counts to craters at least 20 km in diameter. Meaningful counts are restricted to (1) superposed craters and (2) primary craters. A superposed crater useful in dating the basin, and a subjacent crater whose inclusion in the counts would be erroneous, may differ only slightly in appearance. Numbers of secondary craters generally do not indicate time of exposure because such craters are generated in bursts when their parent primaries form. The superposed-buried and primary-secondary distinctions can be made only in well exposed, well photographed terrains. Some counts of pre-Imbrian craters are inevitably restricted to small areas because of burial by later deposits; the buried craters are visible but are useless for dating (fig. 8.7). Photographic quality on the farside is inadequate for interpreting craters smaller than 20 km in diameter on many basins, especially at latitudes poleward of about 40°.

The results of these natural and artificial shortcomings are indistinct separations of the curves for old basins and statistically poor curves for the smallest and the youngest (least cratered) units. No individual craters are dated here by crater densities except the 265-km-diameter crater-basin transitional feature Milne (figs. 4.2J, 8.11). For most pre-Nectarian craters, the small superposed craters that are required for statistically significant counts have been erased by mutual obliteration and degradation on slopes. Several counts of superposed craters are listed in table 8.2 as "poor," meaning that they contain few total craters or lack small craters because of burial. These counts are not plotted in figure 8.6 except for the marginally useful one for the Coulomb-Sarton basin.

Pre-Nectarian surfaces may be heavily or only moderately cratered, depending on age and happenstance position relative to younger secondary clusters. Although densities of superposed secondary craters are almost useless for dating, clustered secondaries are very useful for establishing relative age if they can be traced to their source basins (fig. 8.7A). Once a basin is established as pre-Nectarian, superposition of its secondary clusters on another basin or crater unit establishes the pre-Nectarian age of that unit as well. Where available, superpositions prevail over crater counts as age criteria. Secondary clusters can also be used to date small units, such as those of individual pre-Imbrian craters that are inaccessible to size-frequency studies. Numerous stratigraphic relations between pre-Nectarian and younger basins and between pre-Nectarian basins have been established by such superpositional relations (table 8.2).

Basins

This volume divides 30 pre-Nectarian basins into 9 age groups (table 8.2). Each group is headed by one basin whose relative age seems to be well established by crater densities or superpositional relations. Additional basins are tentatively placed in the groups on weaker grounds.

The first group consists of the giant South Pole-Aitken and Procellarum basins. South Pole-Aitken is characterized by saturation of superposed impact structures as large as basin size (fig. 8.8; table 8.2). The history of its discovery illustrates the need for mapping large areas. When Hartmann and Kuiper (1962) realized that basin rings are prevalent on the Moon, they predicted the presence of a large farside basin on the basis of telescopic photographs showing huge mountains near the Moon's south pole. This prediction went unnoticed when Soviet Zond photographic altimetry and Apollo laser altimetry revealed a large depression as much as 5 to 7 km below the average farside surface (see review by Stuart-Alexander, 1978). Some massifs had been photographed by Apollo 8, but their connection with a basin was not immediately perceived (fig. 8.9; Wilhelms and others, 1969). Wilhelms and El-Baz (1977) mapped other massifs without knowing of their connection with the basin. Several observers then independently discovered that mountains over a huge area of the southern farside and southernmost nearside form a giant ring (pls. 3, 6; fig. 8.8; Howard and others, 1974; Schultz, 1976b, p. 306; Stuart-Alexander, 1978; E.A. Whitaker, written commun., 1976). Now, there is little doubt about the existence of this "Big Backside Basin," although the number and exact diameters of the rings are uncertain (Wilhelms and others, 1979).

Regional studies also suggest the presence of the Procellarum (Whitaker, 1981) or "Gargantuan" (Cadogan, 1974, 1981) basin. Some of the profound effects of this 3,200-km-diameter basin (pl. 3; fig. 8.10; Whitaker, 1981) on the later geologic evolution of the Moon have already been described. (1) Its excavation thinned the crust considerably, possibly by 50 km, and the mantle rose in isostatic compensation (figs. 6.22, 8.4). (2) As a result, the terra surface was lower, and the elastic lithosphere thinner or weaker, during mare extrusion and structural deformation in the Procellarum region than elsewhere on the Moon. (3) Their relation to Procellarum accounts for numerous concentric ridges and troughs of the terrae, which, like the mountains now known to be parts of South Pole-Aitken, would be anomalous if not part of a basin (fig. 8.10). (4) Redistribution of crustal materials by the Procellarum impact may be an important factor in the distribution of anorthositic, Mg-suite, and KREEP-rich terra materials (fig. 8.4). (5) More speculatively, Procellarum ejecta and, possibly, ejecta of the South Pole-Aitken basin may account for the suggested thickening of the farside crust (Cadogan, 1974; proposed previously in somewhat different form by Wood, 1973). (6) The Procellarum basin is the reason for the existence and circular shape of Oceanus Procellarum (pl. 4; fig. 5.26), which otherwise would have had to form by a special "continental" type of mare-basalt volcanism otherwise unknown on the Moon. (7) Other maria, contained in basins superposed on Procellarum, are more numerous, larger, and deeper than maria in equally large or larger basins outside Procellarum (pl. 4). (8) Isostatic compensation has gone further inside Procellarum than on the thicker lithosphere outside. (9) More uplifted crater floors lie in and near Procellarum than elsewhere (pl. 5). (10) Arcuate rilles, opened by mare subsidence, are abundant inside but rare outside the basin (pl. 5); straight rilles may also be controlled by this largest lunar feature (see chap. 6). Additional effects of Procellarum on mare volcanism are described in chapters 11 through 13.

Age group 2 (table 8.2) rather arbitrarily consists of eight basins whose existence is uncertain; their collection into a single group is based on the assumption that if these basins were younger (and if they exist), they would be more obvious. The −1 slope of the cumulative size-frequency distribution of the numerous craters counted on Al-Khwarizmi/King (figs. 8.6, 8.11) characterizes that ancient basin. If this basin does not exist, the −1 slope characterizes other ancient lunar terrane, possibly South Pole-Aitken deposits, which probably extend into this region. Two of the age-group 2 basins, Flamsteed-Billy and Insularum, were identified on the basis of circular patterns of terra islands in the western maria (pl. 3; fig. 8.10; Wilhelms and McCauley, 1971). This criterion for basin identification arose from the discovery that seemingly isolated elevations generally are parts of basin rings.

The basins of age groups 3 and 4 are also so heavily degraded that they have lost most traces of ejecta, and superposed and prebasin craters are difficult to distinguish in their peripheries. Nevertheless, they are still likely to be true basins. Saturation by large craters appears to affect the size-frequency plots significantly; the slopes of some curves for craters believed to be superposed approach −1. Because of this saturation or deep burial by later basin or mare deposits, ranking of basins by age within one of these older groups is not considered feasible, even where crater frequencies appear to differ substantially (for example, Lomonosov-Fleming versus Mutus-Vlacq).

Identification of Mutus-Vlacq and Lomonosov-Fleming as basins is another result of mapping with the impact model in mind. Extensive tracts of "pitted plains" of pre-Imbrian age are concentrated southwest of the Nectaris basin (pl. 7; Wilhelms and McCauley, 1971; Scott, 1972b). These plains are centered within a partial circle of massifs and ridges (fig. 8.12; Wilhelms and others, 1979). This concentration of plains suggests the presence of a basin, Mutus-Vlacq, regardless of the plains' origin (see chap. 9). Some of the Mutus-Vlacq massifs had been identified in earlier mapping but were considered to be "volcanic domes" (Wilhelms and McCauley, 1971; Mutch and Saunders, 1972) before the Apollo 16 results downgraded the significance of terra volcanism. Similarly, Lomonosov-Fleming was identified by mapping mounds, ridges, and scarps that form a circle 620 km across and that encompass extensive light-colored plains (fig. 8.13; Wilhelms and El-Baz, 1977). Although no interior ring is observed in either of these basins, the term "ringed basin" is appropriate because the presence of another ring or rings beneath the plains fill is predictable by analogy with better exposed basins.

Although the indistinct Nubium, Fecunditatis, and Tranquillitatis basins cannot be accurately dated by superposed craters, they are subjectively ranked with the other age-group 3 basins on the supposition that they would be equally conspicuous if not flooded by mare materials. Fecunditatis probably has at least two exposed and one buried ring and thus has the most elaborate visible ring structure of the poorly observed basins listed in table 4.2. The 680-km-diameter ring, overlain by the young fractured-floor crater Taruntius (Wilhelms, 1972b), is the most conspicuous, but another ring, about 980 km in diameter, is probably the main topographic rim.

Age group 4 consists of three well-established basins. The diameters, numbers of rings, and age relations of Keeler-Heaviside and Ingenii, however, are somewhat uncertain because several nearby massifs may belong to the South Pole-Aitken basin (figs. 1.4, 8.7).

Although the two basins assigned to group 5 are poorly photographed or obscured by younger deposits, their deposits apparently have diminished the crater densities within a basin radius or diameter of the rim, as expected around basins. Smythii contains a distinct mare (fig. 8.11) and a mascon (table 6.1) but has only indistinct rings. Coulomb-Sarton (figs. 4.3M, 8.14) was discovered shortly after photography of the farside by the Lunar Orbiter spacecraft and was designated "unnamed basin B" (Hartmann and Wood, 1971). The rings are vague, and the identity of the topographic rim uncertain (table 4.1). Wood and Head (1976) withdrew the basin from the list of lunar basins. Massifs and abundant interior plains do exist, however, and the basin is here considered to be definitely established (Lucchitta, 1978).

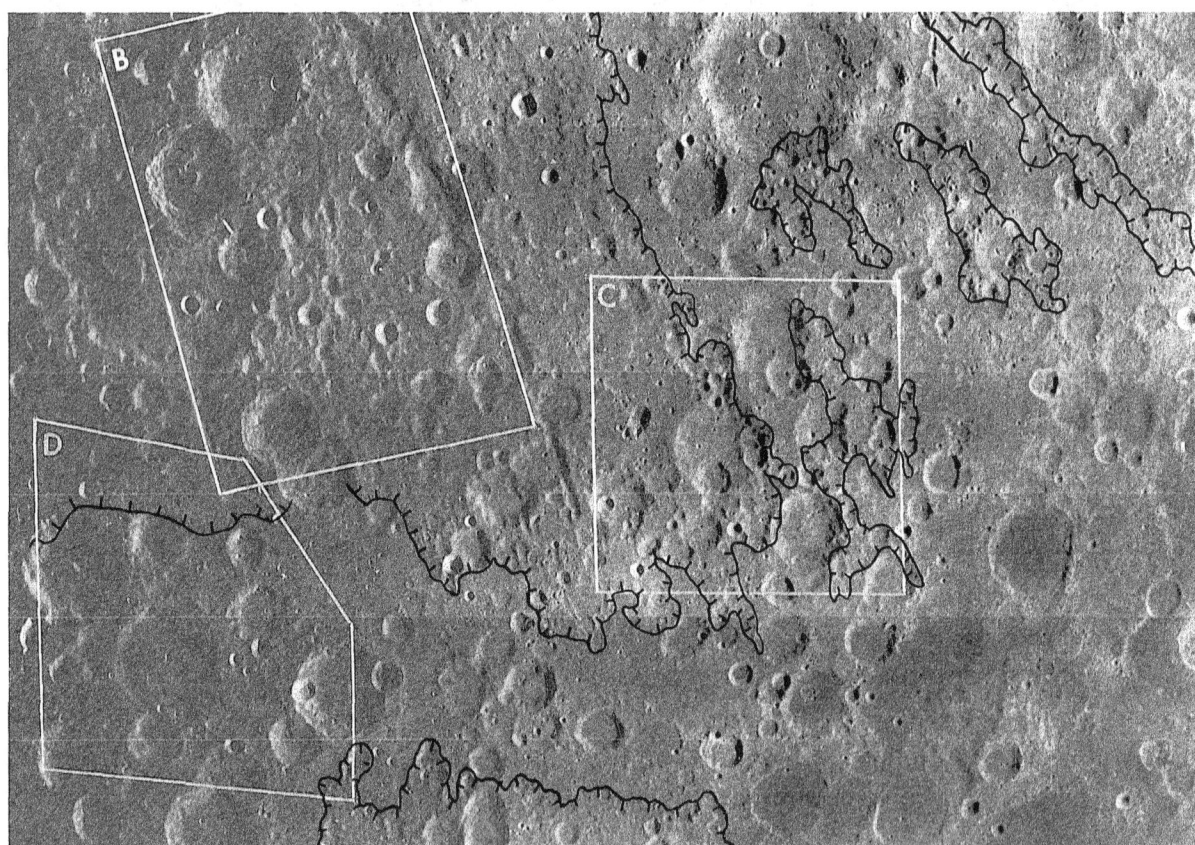

FIGURE 8.5.—Boundary between pre-Nectarian units and Nectarian System southeast of Nectaris basin (Stuart-Alexander, 1971; Stuart-Alexander and Wilhelms, 1975; Wilhelms and El-Baz, 1977; Wilhelms and others, 1979).

A. Regional view showing concentrations of textured deposits and secondary craters of Nectaris basin (hachured black outlines) and areas of B through D (white outlines). Orbiter 4 frame M–52.

The well-dated basins of the four youngest age groups (table 8.2) have observable exterior deposits and secondary craters where well exposed and well photographed. The craterlike rims of Birkhoff and Lorentz are surrounded by ejecta and secondary craters to distances of at least one basin diameter (fig. 8.14). Secondary craters of Apollo are visible in the moderately well photographed zone south of the basin (fig. 8.15; Wilhelms and others, 1979), although they are covered by younger materials in the north and northeast. Apollo, Birkhoff, and the large crater or small basin Milne were considered to be Nectarian because of their evident ejecta (Wilhelms and El-Baz, 1977; Lucchitta, 1978; Wilhelms and others, 1979); the crater counts, however, indicate a pre-Nectarian age (as proposed for Apollo by Stuart-Alexander, 1978). Grimaldi ejecta appears to have smoothed its periphery, although it is too heavily obscured by Orientale deposits to demonstrate textured ejecta (figs. 4.3*I*, 4.4*H*; McCauley, 1973). The Planck and Schiller-Zucchius basins similarly lack textured ejecta but have smoothed the surrounding terrain and reduced the crater densities.

Freundlich-Sharonov is an especially instructive basin. The basin rings and interior are poorly photographed (fig. 4.3*P*), and their pattern emerged only upon mapping (Stuart-Alexander, 1978). Stuart-Alexander (1978) noted that superposed craters are relatively sparse in the surrounding terrain. A good photograph of the south basin periphery reveals relatively uncratered terrain lineate radially to the basin (fig. 8.7*A*). Analogy with Orientale indicates that the lineations are not tectonic or volcanic but were formed by the deposition of Freundlich-Sharonov ejecta. The basin would probably not have been discovered or interpreted as a typical impact basin without knowledge of other basins.

The progress of basin discovery suggests that additional pre-Nectarian basins will be found. Moreover, many basins probably formed before the South Pole-Aitken and Procellarum basins but are now completely obliterated. Basin deposits probably covered all parts of the Moon many times over in the manner displayed by the visible basins.

Craters

In the absence of reliable crater counts, morphologic comparisons are used to date stratigraphically isolated craters. These

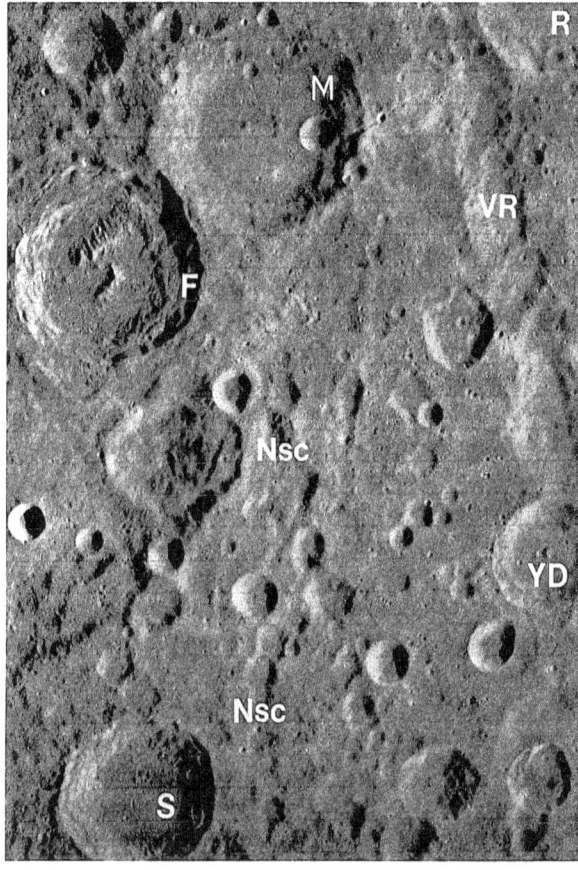

B

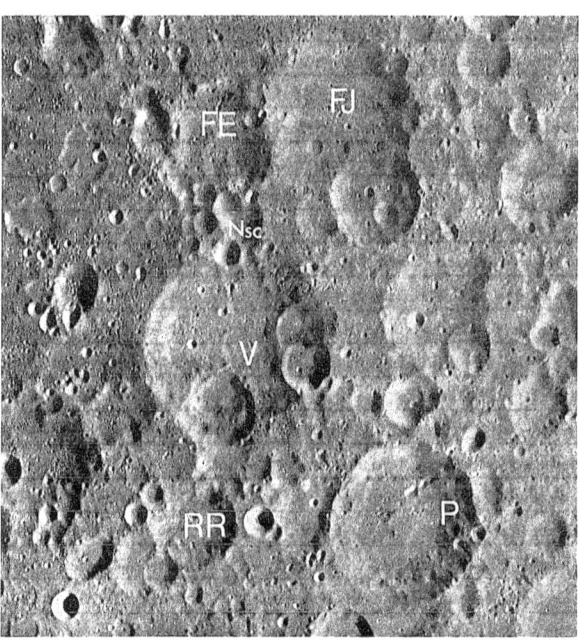

C

B. Poorly exposed pre-Nectarian deposits. Part of Vallis Rheita (VR), a secondary chain of Nectaris basin, overlain by Nectarian craters Rheita (R; 70 km, 37° S., 42° E.) and Young D (YD; 46 km). Other Nectaris-basin secondary chains (Nsc) are overlain by Nectarian primary crater Steinheil (S; 67 km). Crater Metius (M; 88 km) also is probably Nectarian because it interrupts pattern of Nectaris-basin deposits. Fabricius (F; 78 km) is much younger, probably Eratosthenian. Orbiter 4 frame H–71.

C. Moderately well exposed pre-Nectarian craters, including Vega (V; 76 km, 45° S., 63° E.), Fraunhofer E (FE; 42 km), Fraunhofer J (FJ; 63 km), Reimarus R (RR; 35 km), and Peirescius (P; 62 km), overlain by circular and subcircular Nectaris-basin secondary craters (Nsc). Vega and Peirescius are good type examples of pre-Nectarian crater morphology because large rim segments are preserved. Orbiter 4 frame H–52.

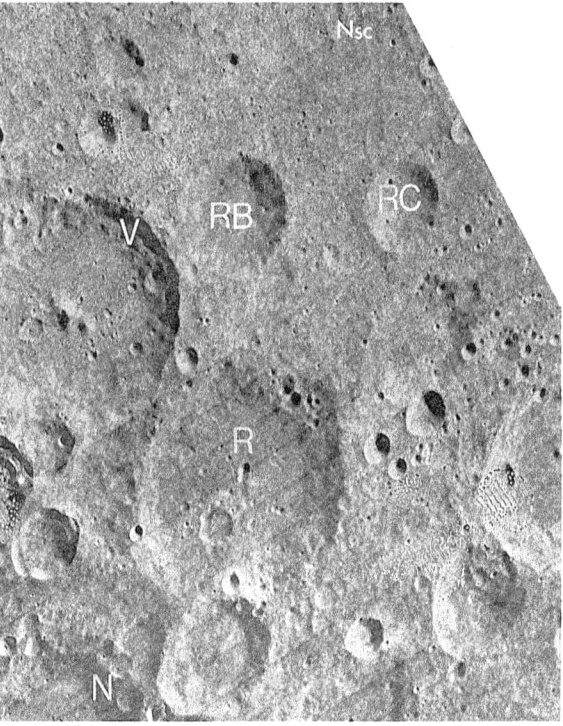

D. Well-exposed pre-Nectarian terrane, lightly pitted by Nectaris-basin secondary craters, affording additional typical pre-Nectarian morphologies. Subdued pits are superposed on pre-Nectarian craters Nearch (N; 76 km, 58.5° S., 39° E), Rosenberger (R; 96 km), Rosenberger C (RC; 47 km), Rosenberger B (RB; 33 km), and probably Vlacq (V; 89 km). Well-defined part of Nectaris-basin secondary field is in upper right (Nsc). Orbiter 4 frame H–83.

TABLE 8.2.—*Pre-Nectarian basins*

[Age groups in order of increasing age, each group arbitrarily headed by the largest basin of the group. Diameters from table 4.1. Parenthetical basins, existence not definitely established; parenthetical areas, buried; parenthetical densities, poor sample. Superposed craters are at least 20 km in diameter unless otherwise noted. Underlying and overlying basins determined from directly observed superpositional relations; n.d., no relations detected]

Age group	Basin	Diameter (km)	Superposed craters			Underlying basin		Overlying basin		Remarks
			Number	Area (10^6 km^2)	Density (craters/10^6 km^2)	Name	Figure	Name	Figure	
9	Apollo	505	57	0.480	119	(Grissom-White)	8.15	Orientale, Hertzsprung, Korolev.	9.21	---
						South Pole-Aitken	9.21			
	Grimaldi	430	15	(0.154)	(97)	Procellarum	1.9	Orientale	1.9	---
8	Freundlich-Sharonov	600	81	.629	129	Keeler-Heaviside	8.7	Moscoviense	8.7	---
								Mendeleev	---	
								Korolev	9.21	
7	Birkhoff	330	50 (≥21 km)	.401	127	Lorentz, Coulomb-Sarton.	8.14	Imbrium	8.14	---
								Hertzsprung	8.14, 9.25	
	Planck	325	9	.082	(110)	South Pole-Aitken	1.3	Schrödinger	1.5	Small craters buried.
						Poincaré	1.5			
	Schiller-Zucchius	325	16	(0.143)	(112)	n.d		Orientale	1.9, 7.6	Do.
								Humorum	9.24	
	(Amundsen-Ganswindt)	355	11	(0.102)	(108)	South Pole-Aitken(?)	1.5	Schrödinger	1.5	Deeply buried.
6	Lorentz	360	33	.208	159	Coulomb-Sarton	8.14	Orientale, Imbrium	10.2	---
								Birkhoff	8.14	
5	Smythii	840	74	.445	166	n.d		Crisium	1.5	---
	Coulomb-Sarton	530	43	(0.296)	(145)	n.d		Orientale, Imbrium, Hertzsprung, Birkhoff, Lorentz.	8.14	Small craters buried.
4	Keeler-Heaviside	780	69	.371	186	Ingenii(?)	8.7	Mendeleev	1.4	---
						South Pole-Aitken	8.7	Freundlich-Sharonov	8.7	
	Poincare	340	32	(0.168, partly)	(190)	South Pole-Aitken	1.5, 8.7	Schrödinger, Planck	1.5	Small craters buried.
	Ingenii	650	37 (≥22 km)	.228	162	South Pole-Aitken	8.7	Keeler-Heaviside(?)	8.7	Small craters obscured.
3	Lomonosov-Fleming	620	63 (≥23 km)	.356	177	n.d		Imbrium, Crisium, Humboldtianum.	8.13	Do.
	Nubium	690	---	---	---	n.d		Imbrium	1.8, 8.10	Deeply buried.
								Humorum	8.10	
	(Mutus-Vlacq)	700	80	.336	225	n.d		Imbrium, Nectaris	8.12	---
	Tranquillitatis	800	---	---	---	n.d		Imbrium	11.1	Deeply buried.
								Serenitatis	1.7, 5.17	
								Crisium	---	
								Nectaris	9.2, 11.1	
								Fecunditatis	---	
	Australe	880	129	(0.608, partly)	(>212)	South Pole-Aitken(?)	1.5	Schrödinger	1.5	Complex burial pattern.
								Nectaris	1.5, 9.5, 9.6	
	Fecunditatis	990	---	---	---	Tranquillitatis	---	Crisium	9.5	Deeply buried.
								Nectaris	9.2	
2	(Al-Khwarizmi/King)	590	63	.320	197	n.d		Smythii	8.11	---
	(Pingré-Hausen)	300	---	---	---	n.d		Orientale, Mendel-Rydberg.	1.9	---
	(Werner-Airy)	500	---	---	---	n.d		Imbrium, Nectaris	9.2	---
	(Balmer-Kapteyn)	550	---	---	---	n.d		Crisium, Nectaris	1.5	---
	(Flamsteed-Billy)	570	---	---	---	n.d		Orientale, Imbrium	8.10	---
	(Marginis)	580	---	---	---	n.d		Crisium, Humboldtianum.	9.4	---
	(Insularum)	600	---	---	---	Procellarum	8.10	Imbrium	8.10	---
	(Grissom-White)	600	---	---	---	South Pole-Aitken	8.15	Hertzsprung, Apollo	8.15	---
	(Tsiolkovskiy-Stark)	700	---	---	---	n.d		Milne	1.3	---
								Mendeleev	1.4	
1	South Pole-Aitken	2,500	---	---	---	n.d		Orientale, Mendel-Rydberg.	1.9	---
								Schrödinger, Planck, Poincare, Australe, (Sikorsky-Rittenhouse), (Amundsen-Ganswindt).	1.5	
								Hertzsprung, Korolev, Apollo.	9.21	
								Nectaris	9.6	
								Keeler-Heaviside, Ingenii.	8.7	
								Grissom-White)	8.15	
	(Procellarum)	3,200	---	---	---	n.d		Most nearside basins.	8.10, 10.2	---

craters are compared with typical pre-Nectarian craters in the circum-Nectaris zone. Very few craters smaller than 20 km in diameter have been recognized as unequivocally pre-Nectarian because of severe degradation and the ambiguous significance of morphology (Wilhelms and others, 1978). In general, pre-Nectarian craters larger than 20 km in diameter that are degraded by the accumulated impact flux and not by catastrophic deposition of later deposits have lost fine-scale textures but still retain the overall shape of fresh craters (figs. 8.5B–D, 8.7C, 8.11, 8.14B, C; table 8.3). A short, sloping rim flank is still visible, but not the fine radial texture of the outer flanks that is displayed by younger craters; pre-Nectarian radial textures are visible only around basins, where they correspond to chains of secondary craters or, possibly, to large ridges of primary ejecta (fig. 8.7). The general morphology, but little sharp detail, of the crater rim-wall terraces is visible; the terraces are commonly amalgamated, and their edges cut by subradial notches. Floors are shallower, and rims more rounded, than in younger craters. Despite the degradation, many complete and well-defined rim segments are preserved. Because of this preservation of the overall morphology of the main impact sequence, even in older craters, good practice is to compare morphologies not only of the whole crater but also of small areas of the same size on each crater.

Like crater densities, morphologies of craters can be used as dating tools only if their geologic setting is also considered. Later deposits may degrade craters differentially (fig. 8.5). Moreover, typical "pre-Nectarian" morphologies may result from burial of a Nectarian crater by Nectarian or Imbrian basin deposits. For example, the subdued craters shown in figure 9.21 that are buried by deposits of the Nectarian Hertzsprung basin could be either Nectarian or pre-Nectarian. Much sharper, unburied craters could be pre-Nectarian (figs. 8.7C, 8.14B, C).

TABLE 8.3.—Representative pre-Nectarian craters

[Cross rules divide table into craters smaller than 60 km, 60 to 119 km, and at least 120 km in diameter; interiors only are mapped in plate 6. F-S, Freundlich-Sharonov basin; N?, possibly Nectarian]

Crater	Diameter (km)	Center (lat)	(long)	Figure	Remarks
Daedalus U	30	4° S.	175° E.	8.7	Older than F-S.
Coriolis L	32	2° S.	173° E.	8.7	Younger than F-S.
Rosenberger B	33	52° S.	46° E.	8.5	---
Reimarus R	35	48° S.	64° E.	8.5	Pitted by Nectaris basin.
Daedalus R	41	8° S.	75° E.	8.7	Older than F-S.
Fraunhofer E	42	43° S.	62° E.	8.5	Pitted by Nectaris basin.
Krusenstern	47	26° S.	6° E.	9.27	Mantled by Nectaris basin.
Rosenberger C	47	52° S.	42° E.	8.5	---
Blanchinus	61	25° S.	31° E.	9.27	Mantled by Nectaris basin.
Peirescius	62	47° S.	68° E.	8.5	Typical.
Apianus	63	27° S.	8° E.	9.27	---
Fraunhofer J	63	42° S.	64° E.	8.5	Pitted by Nectaris basin.
Barbier	67	24° S.	158° E.	8.7	Relatively fresh.
Riccius	71	37° S.	27° E.	3.10D, 11.5	N?
Licetus	75	47° S.	7° E.	10.31	---
Nearch	76	59° S.	39° E.	8.5	Typical.
Vega	76	45° S.	63° E.	8.5	Do.
Birkhoff X	77	62° N.	150° W.	8.14	---
Mutus	78	64° S.	30° E.	8.12	---
Esnault-Pelterie	79	48° N.	141° W.	8.14	N?
Cyrano	81	21° S.	158° E.	8.7	Relatively fresh.
Rabbi Levi	81	35° S.	24° E.	11.5	---
Zagut	84	32° S.	22° E.	11.5	---
Vlacq	89	53° S.	39° E.	8.5	---
Barrow	93	71° N.	8° E.	10.15	N?
Playfair G	94	24° S.	7° E.	1.8, 9.27	---
Fra Mauro	95	6° S.	17° W.	2.5A, 10.19	N?; covered by the Fra Mauro and Cayley Formations.
Rosenberger	96	55° S.	43° E.	8.5	---
Schlesinger	97	47° N.	139° W.	8.14	---
Sacrobosco	98	24° S.	17° E.	7.9	Typical.
Purbach	118	26° S.	2° W.	1.8, 7.7	Do.
Goldschmidt	120	73° N.	3° W.	10.15	N?
Orontius	122	40° S.	4° W.	1.8, 2.3	Typical.
Messala	124	39° N.	60° E.	9.3	---
Regiomontanus	124	28° S.	1° W.	1.8, 7.7	---
Furnerius	125	36° S.	60° E.	9.5	Typical.
Stöfler	126	41° S.	6° E.	1.8, 10.31	---
Szilard	127	34° N.	106° E.	8.13	---
Paschen	133	14° S.	140° W.	9.21	---
Stebbins	135	65° N.	143° W.	8.14	N?
Fowler	136	43° N.	145° W.	8.14	---
Curie	139	23° S.	92° E.	8.11, 10.46	---
Hirayama	139	6° S.	94° E.	8.11	---
Lyot	141	50° S.	84° E.	1.5, 10.46	---
Joliot	143	26° N.	93° E.	4.7, 9.4, 9.28	---
Hipparchus	151	6° S.	5° E.	1.8, 3.10, 10.27	---
Ptolemaeus	153	9° S.	2° W.	1.8, 7.7, 10.28	---
W. Bond	158	65° N.	4° E.	1.7, 10.6, 10.15	---
Richardson	161	31° N.	100° E.	8.13	---
Heaviside	163	11° S.	167° E.	1.4, 8.7	---
Maginus	163	50° S.	6° W.	4.6A	Typical.
Einstein	170	17° N.	89° W.	3.9A	---
Tsander	171	6° N.	149° W.	9.21	---
Fabry	179	43° N.	101° E.	1.2	---
Von Kármán	179	44° S.	176° E.	8.15	---
Janssen	190	45° S.	42° E.	3.10D, 9.1, 9.2	Covered by the Janssen Formation.
Galois	207	14° S.	152° W.	9.21	---
Landau	221	43° N.	119° W.	4.2F, 8.14	---
Campbell	225	45° N.	153° E.	4.2G, 12.9	---
Von Kármán M	225	47° S.	176° E.	8.15	---
Schickard	227	44° S.	55° W.	1.9, 7.6	---
Deslandres	234	33° S.	5° W.	1.8	Typical.
Pasteur	235	11° S.	105° E.	1.2, 8.11	Do.
Leibnitz	236	38° S.	179° E.	8.15	---
Fermi	238	20° S.	123° E.	1.3	---
Milne	262	31° S.	113° E.	1.3, 4.2J, 8.11	Ring arcs.
Gagarin	272	20° S.	149° E.	1.4, 3.5, 4.2K	Typical.

Despite the problems, the results of dating basins by superpositional relations, crater counts, and assessments of superposed crater morphology are generally consistent. No known craters with pre-Nectarian morphology are clearly superposed on Nectarian basins. This general internal consistency is the outcome of a series of iterations in which the morphologic criteria for pre-Nectarian age were repeatedly reassessed. Anomalous rebound of craters inside basins and anomalous degradation by deposits of younger craters were detected by geologic mapping. Thus, some assignments of crater age in this volume differ from those shown on earlier geologic maps, including those by Wilhelms and El-Baz (1977) and Wilhelms and others (1979). Pre-Nectarian craters are subdivided on the paleogeologic map (pl. 6) as exposed (solid black), buried (solid rim-crest outline), or possibly Nectarian (dashed rim-crest outline).

A total of 1,208 pre-Nectarian craters larger than 30 km in diameter, 456 of which are unburied by pre-Nectarian or younger deposits, are mapped here (pl. 6). They vary greatly in areal density because the units that underlie them vary in age and because the thickness of overlying deposits affects the number of buried craters

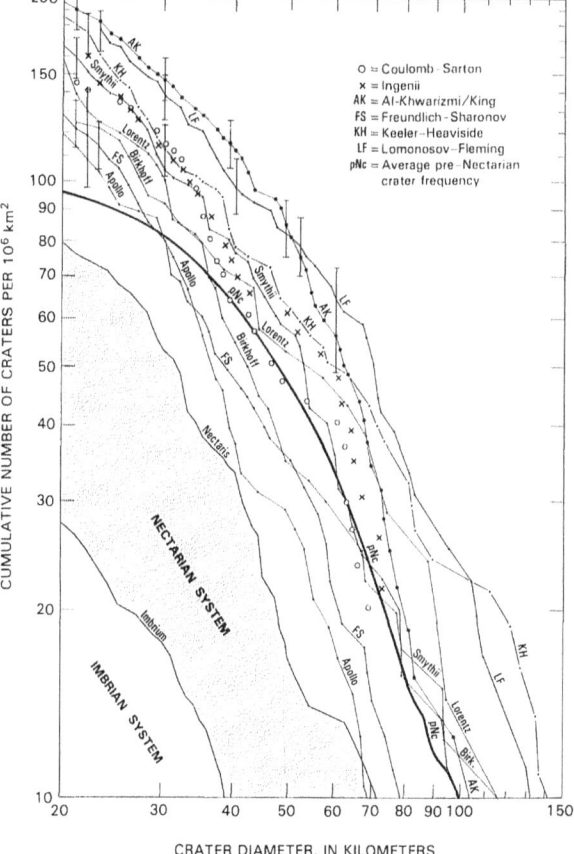

FIGURE 8.6.—Cumulative size-frequency distributions of craters at least 20 km in diameter superposed on pre-Nectarian basins. Materials of Nectarian and Imbrian Systems are indicated. Curve pNc (from Wilhelms and others, 1978) includes only pre-Nectarian craters, whereas each basin curve includes all postbasin primary craters of sizes indicated; curve pNc is average for best exposed pre-Nectarian terranes. Error bars represent square root of plotted point, added to and subtracted from value of point (compare table 8.2). Data tabulated in supplementary table.

FIGURE 8.7.—Stratigraphic relations between pre-Nectarian units on farside.
  A. Radial lineations (arrows) of Freundlich-Sharonov basin (FS), superposed on Keeler-Heaviside basin (named for unrelated craters Keeler [K; 169 km, 10° S., 162° E., probably Lower Imbrian] and Heaviside [H; 163 km, pre-Nectarian]; compare figs. 4.3P, S); areas of B and C are outlined. Orbiter 2 frames M–75 (lower left) and M–34 (upper right); overlaps figure 1.4.

that can still be seen. A density of 77 unburied pre-Nectarian craters per million square kilometers (422 craters min 30 km diam in 5.45×10^6 km^2) was determined for terrain of the oldest basins (age groups 1–5, table 8.2) and other heavily cratered terrain. The younger pre-Nectarian basins (age groups 6–9) display a much lower average density of 10 such craters (34 in 3.24×10^6 km^2). Extrapolation of the larger value to the whole Moon (area, 38×10^6 km^2) suggests that a minimum of 2,940 craters larger than 30 km in diameter formed during pre-Nectarian time. Assuming that their deposits were as extensive relative to crater size as are those of young craters, the deposits covered about 110 percent of the Moon's surface area. Based on the Nectarian frequency (see chap. 9), which is less affected by saturation, the actual number of pre-Nectarian craters of this size that formed while the 28 pre-Nectarian basins of groups 2 through 9 were forming was closer to 3,400. A still greater, unknown number formed earlier.

B. Pre-Nectarian terrane pitted by probable secondary craters of Freundlich-Sharonov basin (sc) and other, smaller secondaries. Craters older than Freundlich-Sharonov include Daedalus R (DR; 41 km) and Daedalus U (DU; 30 km); Coriolis L (CL; 32 km) may be younger. A much younger crater chain secondary to a distant impact crater or basin, probably Orientale (centered 2,800 km to east), is also superposed on Coriolis L and Daedalus U. Crater Daedalus S (DS; 20 km) is also young, probably Imbrian. Orbiter 2 frame H–33.

C. Craters Cyrano (81 km) and Barbier (67 km), similar to pre-Nectarian craters in figure 8.5 and of a morphologic type absent in B, are beyond range of Freundlich-Sharonov deposits; thus, Cyrano and Barbier are probably pre-Nectarian and possibly older than Freundlich-Sharonov basin, despite moderately sharp topography. Orbiter 2 frame H–75.

FIGURE 8.8.—Pre-Nectarian South Pole-Aitken basin, centered around lat 56° S., long 180°. Airbrush drawing by Donald E. Davis, courtesy of the artist.

FIGURE 8.9.—South Pole-Aitken massifs photographed by Apollo 8. Arrow indicates crater Mohorovičić A (20 km, 16° S., 163° W.) atop one arcuate massif; other massifs along lat 22° S. occupy horizon. At time of Apollo 8 mission, relation of the large mountains to a basin had not yet been recognized. View southeastward. Apollo 8 frame H–2319.

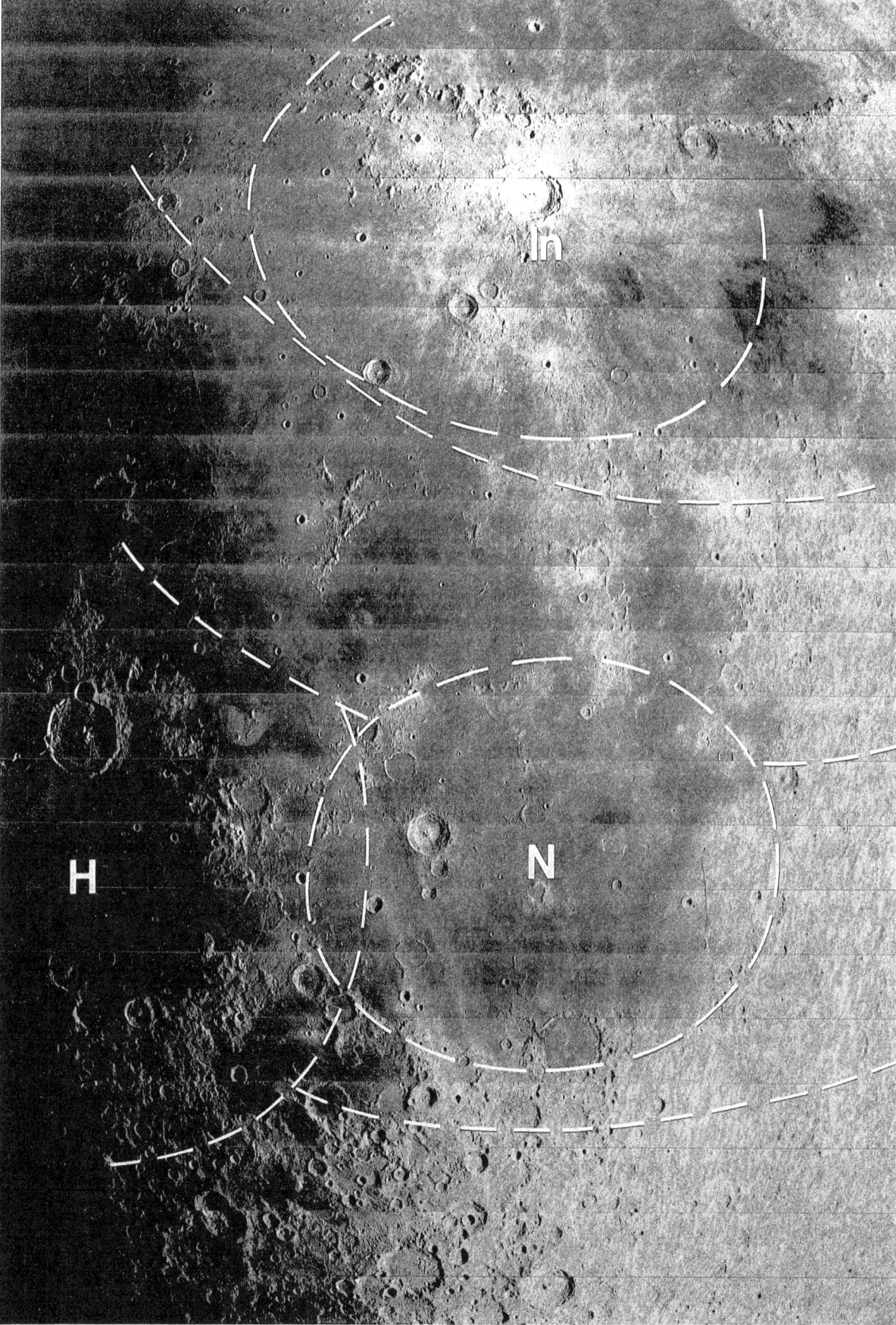

FIGURE 8.10.—Part of southern Oceanus Procellarum region, including Nectarian Humorum basin (H), pre-Nectarian Insularum (In) and Nubium (N) basins, and three rings of pre-Nectarian Procellarum basin (Whitaker, 1981); central Procellarum ring is expressed here mostly by mare ridges (compare fig. 5.26). Shallowness of mare is indicated by many islands belonging to basins and craters that are superposed on Procellarum basin (pl. 3). Orbiter 4 frame M-125.

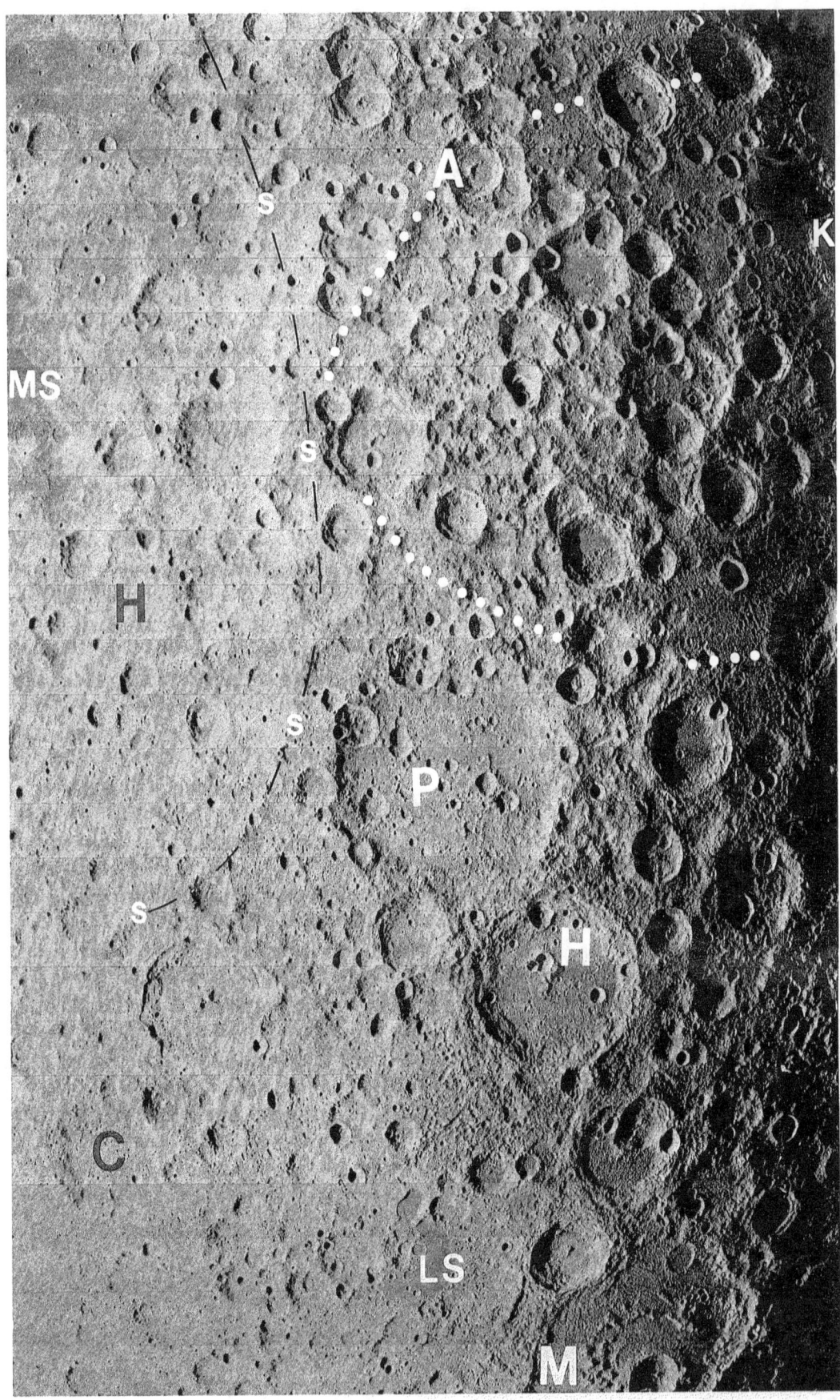

FIGURE 8.11.—Large section of western farside, centered on craters Pasteur (P; 235 km, 12° S., 105° E., Pre-Nectarian) and Hilbert (H; 170 km, 18° S., 108° E., Nectarian), both filled by cratered plains. White dots, indefinite pre-Nectarian basin Al-Khwarizmi/King (Wilhelms and El-Baz, 1977), named for craters Al-Khwarizmi (A; 65 km, 7° N., 107° E., Nectarian) and King (K; 77 km, 5° N., 121° E., Copernican); most conspicuous part of rim is southwest of letter A. MS, Mare Smythii; s and black dashes, rim of pre-Nectarian Smythii basin. Lacus Solitudinis (LS) is not related to known basins. Pre-Nectarian crater or small basin Milne (M; 265 km, 31° S., 113° E.) has basinlike ring structure but is omitted from table 4.1 because of its size. Orbiter 2 frame M–196.

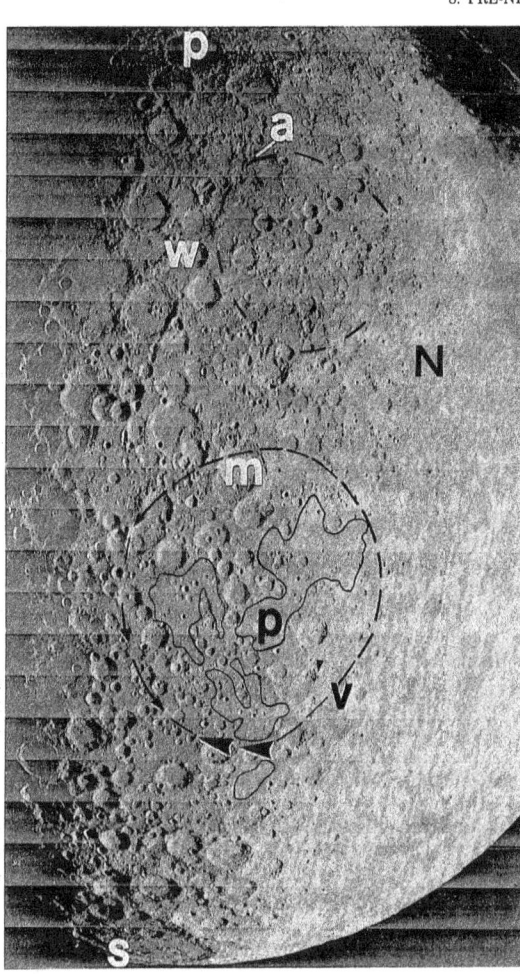

FIGURE 8.12.—Pre-Nectarian Mutus-Vlacq and Werner-Airy basins.
  A. South-central nearside and area around south pole including Imbrian Schrödinger basin on farside (s; 320 km, 76° S., 134° E.). Black dashes, rims of Mutus-Vlacq (below) and Werner-Airy (indistinct, above). Arrowheads, prominent massifs of Mutus-Vlacq; those indicated by larger arrowheads are also shown in B. Black p, Nectarian plains concentrated in Mutus-Vlacq basin. N, Nectaris-basin rim (Rupes Altai). Pre-Nectarian craters include Ptolemaeus (white p; 153 km, 9° S., 2° W.) and Vlacq, (v; 89 km, 53.5° S., 39° E.); Nectarian craters include Maurolycus (m; 114 km, 42° S., 14° E.) and Airy (a; 37 km, 18° S., 5.5° E., possibly Imbrium-basin secondary crater); Werner (w; 70 km, 28° S., 3.5° E.) is Eratosthenian. Orbiter 4 frame M-95.
  B. Detail of Mutus-Vlacq massifs. Large arrows indicate same massifs as large arrowheads in A. Small arrow indicates hill possibly belonging to an inner ring. M, Mutus (78 km, 63.5° S., 30° E., pre-Nectarian). Orbiter 4 frame H-82.

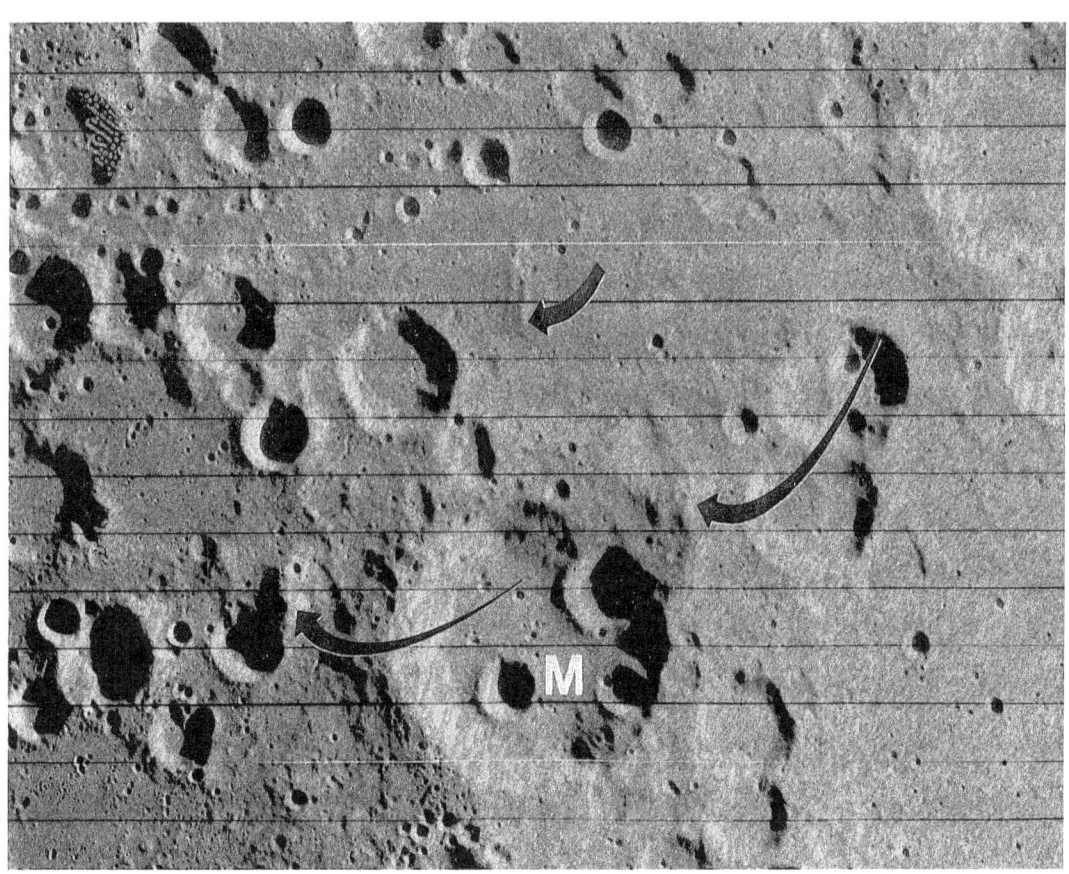

VOLCANIC ROCKS

The discovery of ancient volcanic rocks on the Moon would be of great interest from the standpoints of both petrogenesis and stratigraphy. The unforewarned geologist might think that such rocks have been found in abundance because the terms "highland basalt," "KREEP basalt," and numerous other "basalts" are so common in the literature. However, no pristine samples of terra composition are known to be both pre-Nectarian in age and volcanic in origin; volcanic KREEP basalt of Early Imbrian age is the oldest currently recognized terra-type volcanic material (see chap. 10). Fragments of mare basalt have been recovered from breccia deposits emplaced in the Nectarian and Imbrian Periods, and a few have been radiometrically dated (table 9.5; chaps. 9, 10). Whether they are pre-Nectarian depends on the absolute age of the Nectaris basin (see chap. 9); none are of definite pre-Nectarian age.

Moreover, no plains or other terrane of pre-Nectarian age sufficiently resemble maria to be definitely established as volcanic. Plains in such basins as Mutus-Vlacq (fig. 8.12) may consist of pre-Nectarian basalt underlying Nectarian ejecta, but this is uncertain (see chap. 9). All pre-Nectarian interbasin terrane is ascribable to basin deposits and is here mapped accordingly (pl. 6).

Despite the absence of firm evidence for pre-Nectarian volcanism, mare and KREEP basalt may have been extruded during the period. The extensive early plutonism suggests accompanying volcanism, and some of the recovered mare-basalt clasts are, at least, very close to pre-Nectarian in age. Another line of indirect evidence for volcanism is the concentration of KREEP revealed by gamma-ray spectrometers in the Imbrium-Procellarum region. The Procellarum basin may have unroofed deep-seated KREEP-rich material that was then excavated by further impacts; alternatively, however, KREEP basalt may actually have been extruded volcanically into the basin (fig. 8.4B; Wood, 1972b; Cadogan, 1974, 1981). These pre-Nectarian extrusions could have been recycled by later impacts and mare extrusions, so that the KREEP chemistry appears to be concentrated today in deposits of Archimedes, Aristarchus, Copernicus, and other craters in the region. Continued sampling might well disclose pre-Nectarian volcanic rocks here and elsewhere on the Moon.

CHRONOLOGY

The earliest pre-Nectarian event was the origin of the Moon. Before the space age, the Moon's age was inferred from U-Th-Pb ages of the Earth and meteorites to be 4.55 ± 0.07 aeons, assuming that all these bodies formed contemporaneously (Patterson, 1956). Absolute dating of lunar rocks based on radiometric ages and assumed initial isotopic ratios (model ages) is consistent with 4.55 aeons ago as the approximate time of lunar origin (Tatsumoto and others, 1973, 1977; Tera and Wasserburg, 1974; Nyquist, 1977; Tilton and Chen, 1979).

Some recovered clasts of ANT rock yield very old isotopic ages (table 8.4). Ferroan anorthosite is difficult to date because it contains very little Rb, Nd, and Sm; its low initial $^{87}Sr/^{86}Sr$ ratios are consistent with, but do not require, very primitive ages (Papanastassiou and Wasserburg, 1975, p. 1483; Tatsumoto and others, 1977). More precise ages appearing to be crystallization ages have been obtained for pristine Mg-suite rocks (table 8.4). They range as old as 4.51 ± 0.07 aeons for inclusions in olivine of troctolite sample 76535 from the Apollo 17 landing site (Papanastassiou and Wasserburg, 1976). The analysts were uneasy about the similarity of this age to that of the Moon, but the age may, indeed, have been set by igneous melting in the Moon shortly after its accretion (Carlson and Lugmair, 1981b). A later event in the same rock's history at about 4.26 aeons ago is recorded by Sm-Nd mineral isochrons and $^{40}Ar-^{39}Ar$ plateaus. If the 4.51-aeon age is an artifact, this later age may date the crystallization of the troctolite, or it may date the end of a long cooling episode during which isotopic reequilibration was possible (Gooley and others, 1974; Lugmair and others, 1976; Carlson and Lugmair, 1981b).

Interpretation of such internal isochrons as dates of igneous melting, along with evidence for very early geochemical fractionations (Rb-Sr, Pb-U, Sm-Nd), leads to the conclusion that the differentiation which formed the lunar crust began during or immediately after accretion (for example, Papanastassiou and Wasserburg, 1971a, 1975, 1976; Tera and Wasserburg, 1974; Lugmair and others, 1975; Nunes and others, 1975; Herbert and others, 1977; Nyquist, 1977; Tatsumoto and others, 1977; Lugmair and Carlson, 1978; Carlson and Lugmair, 1981a, b). How long this differentiation lasted and the extent of igneous activity after crustal formation but before extrusion of the visible mare basalt are unanswered questions. The geologically important date of when the crust solidifed enough to support basin-and-crater topography probably is between 4.4 and 4.2 aeons ago; the

FIGURE 8.13.—Pre-Nectarian Lomonosov-Fleming basin (dashes outline rim), named for unrelated craters Fleming (F; 130 km, 15° N., 109.5° E., Nectarian) and mare-filled Lomonosov (L; 93 km, 27.5° N., 98° E., Lower Imbrian). Pre-Nectarian craters include Richardson (R; 161 km) and Szilard (S; 127 km). Nectarian and Imbrian plains occupy much of basin interior, especially between arrows and east of Fleming. Groovelike crater chains (arrows; above, Catena Artamonov; below, Catena Dziewulski) are radial to and, probably, secondary to the Imbrium basin, centered 3,000 km away on opposite side of Moon. Apollo 12 frame H–8296.

TABLE 8.4.—*Radiometric ages of samples of pristine pre-Nectarian lunar rocks*

[Sample 15455, Apollo 15 (Ryder and Bower, 1976); numbers beginning with 6, Apollo 16 (Ulrich and others, 1981); numbers beginning with 7, Apollo 17 (Wolfe and others, 1981). Numbers after commas represent splits of sample. Mineralogy: cpx, clinopyroxene; ol, olivine; opx, orthopyroxene; pl, plagioclase. Ages calculated using decay constants recommended by the International Union of Geological Sciences (IUGS) (Steiger and Jäger, 1977).
References: B75, Bogard and others (1975); CL81a, Carlson and Lugmair (1981a); CL81b, Carlson and Lugmair (1981b); C75, Compston and others (1975); HS75, Husain and Schaefer (1975); HW75, Huneke and Wasserburg (1975); J77a, Jessberger and others (1977a); L75, Leich and others (1975); L76, Lugmair and others (1976); N76, Nakamura and others (1976); N79b, Nyquist and others (1979b); N81a, Nyquist and others (1981a); PW75, Papanastassiou and Wasserburg (1975); PW76, Papanastassiou and Wasserburg (1976); S74, Stettler and others (1974)]

Sample	Source rock and lithology (minerals in volume percent)	Age (aeons)	Method	Reference	Remarks
15455	Norite clast (60-65 pl, 35-40 opx) from black-and-white breccia.	4.48±0.17	Rb-Sr	N79b	Isotopic systems disturbed (CL81b).
67435	Tiny plagioclase clasts from 354-g polymict breccia.	4.35±0.05	Ar-Ar	J77a	---
67667	7.9-g rake sample of lherzolite (58 ol, 21 pl, 15 opx, 5 cpx).	4.18±0.07	Sm-Nd	CL81a	Crystallization or impact-excavation age.
72255	Plagioclase from 2-cm "Civet Cat" norite clast from boulder 1, station 2.	3.93±0.03 4.08±0.05	Ar-Ar Rb-Sr	L75 C75	Possibly as old as 4.36±0.13 aeons (C75).
72417	Dunite (93 ol); part of clast 72415-72418 from boulder 3, station 2.	4.45±0.10	Rb-Sr	PW75	Imprecise whole-rock isochrons.
73255, 27,45	Granulated 0.9-g norite clast (53 pl, 40 opx) from 394-g polymict breccia 73255.	4.23±0.05	Sm-Nd	CL81b	Crystallization age; not disturbed.
76535	156-g rake sample of troctolite (37-60 ol, 35-58 pl, 4-5 opx).	4.16±0.04 4.26±0.02 4.26±0.06 4.27±0.08 4.51±0.07	Ar-Ar Ar-Ar Sm-Nd Ar-Ar Rb-Sr	HW75 HS75 L76 B75 PW76	Crystallization or impact-excavation age. Do. Do. Do. Crystallization age or artifact.
77215, 37	Norite clast from station 7 boulder.	3.98±0.04 4.33±0.04 4.37±0.07	Ar-Ar Rb-Sr Sm-Nd	S74 N76 N76	Crystallization age at least 4.33 aeons (CL81b).
78236	Shocked norite clast (opx+pl) from station 8 boulder.	4.34±0.05 4.38±0.02 4.39 4.43±0.04	Sm-Nd Rb-Sr Ar-Ar Sm-Nd	CL81b N81a N81a N81a	Crystallization age is probably close to 4.34 aeons (CL81b).

older date is more common in the literature (for example, Tera and Wasserburg, 1974). Taylor (1982, p. 246), however, suggested that crystallization proceeded rapidly until 4.3 aeons ago and that intense impact bombardment kept the crust thoroughly mixed and chemically requilibrated until about 4.2 aeons ago. The picture of igneous petrogenesis that is emerging furthermore suggests that the differentiation was complex. Local zones may have solidified early, especially near the surface, and local hotspots probably persisted until late. Early-formed crust was probably intruded by later plutons, dikes, and sills. This complex, late-ending history (Nunes and others, 1975; Nakamura and Tatsumoto, 1977, p. 2312; Carlson and Lugmair, 1981b) seems more likely than the simpler scenarios that appear in much of the literature.

Some pre-Nectarian clasts record probable postcrystallization impacts. The 4.26-aeon event in the history of troctolite sample 76535 may have been an impact that ended the cooling (Papanastassiou and Wasserburg, 1976; Carlson and Lugmair, 1981b). Dunite sample 72415–72418 was shocked at least once, probably during the pre-Nectarian, before incorporation into the Apollo 17 melt-rich breccia deposits (Dymek and others, 1975). The more abundant Apollo 17 norite samples were also shocked before final emplacement (Nakamura and Tatsumoto, 1977; see summary by James, 1980). The shock events may have been the formation of one of the pre-Nectarian basins in the Serenitatis target area, such as Tranquillitatis (pl. 6). Similarly complex, extended histories have been detected in other primitive lunar rocks (table 8.4). The later events indicate that the isochrons have been disturbed, and so the pristine ages are not accurate except as minima.

The impact processing manifest in samples from younger deposits generally supports the high early impact rate indicated by the abundance of observed pre-Nectarian basins and craters. For example, widespread thermal metamorphism at shallow depths, probably caused by accumulation of hot ejecta blankets, is suggested as the origin of the granulitic textures of many impact-mixed rocks (Warner and others, 1977; Bickel and Warner, 1978; James, 1980). However, more specific information about pre-Nectarian chronology applicable to the stratigraphic record has not been forthcoming from the rocks. Attempts to calibrate the impact flux by means of the amounts of siderophile elements or impact-melt rock are hindered by uncertainties about projectile-to-ejecta ratio or energy partitioning between impact melt and excavation (for example, Lange and Ahrens, 1979).

Another potential method of establishing the emplacement ages of pre-Nectarian units is the classic one of calibrating crater frequencies with absolute ages. However, the relative ages of ancient surfaces are difficult to determine precisely by crater frequencies because they are saturated to large crater diameters (Baldwin, 1969; Hartmann, 1975); all the frequency curves will have similar intercepts at 20-km diameter and will differ only by subtle differences in the steady-state diameters. In figure 8.16, an envelope of curves is fitted to (1) the approximately known relative ages and well known absolute ages of the Imbrian and Eratosthenian mare-basalt samples (see chaps. 11, 12), (2) the well known relative and absolute ages (3.85 aeons) of the Imbrium basin (chap. 10), (3) the well known relative age and the estimated absolute age (3.92 aeons) of the Nectaris basin (chap. 9), and (4) the highest crater frequency found in this study, that within the limits of the possible Al-Khwarizmi/King basin. The boxes for the Imbrium, Nectaris, and Al-Khwarizmi/King surfaces are bounded below by the actual observed cumulative frequencies of all craters larger than 20 km in diameter and above by projections to 20-km diameter from the parts of the curves representing craters large enough to have escaped steady-state obliteration (fig. 8.6). Proper comparison of the boxes requires comparing the same edge of each. This extrapolation suggests that the Al-Khwarizmi/King crater population is between 3.97 and 4.03 aeons old and that, accordingly, the basin itself, part of pre-Nectarian group 2, is equally old or somewhat older.

A steep exponential decline like that plotted in figure 8.16 is only one of many possible models of the early impact rate (Hartmann, 1965a, 1966, 1972c; Baldwin, 1971; Shoemaker, 1972, 1981; Wetherill, 1975a, 1977a, b, 1981; Basaltic Volcanism Study Project, 1981, chap. 8). Possibly, the pre-Nectarian impact rate was closer to the Nectarian rate (19,000 craters and 157 basins per aeon; see chap. 9) than is generally assumed. In this case and if, additionally, the pre-Nectarian crater size-frequency distribution was originally the same as the Nectarian (small craters not yet destroyed), 3,400 craters from 30 to 300 km in diameter, as well as the 28 basins of pre-Nectarian groups 2 through 9, formed from 4.1 to 3.92 aeons ago. A similar figure of 4.06 aeons for the age of the oldest group-2 basins is derived by assuming a constant rate of surface blanketing by basin deposits (Nectarian basin deposits one basin diameter in radius theoretically cover 41 percent of the Moon, and group 2–9 pre-Nectarian deposits cover 84 percent). Thus, the models of steeply declining and constant impact rates bracket the age of the oldest group-2 basins and accompanying craters between about 4.1 and 4.0 aeons. The younger age is more likely if the early impact rate declined as steeply as shown in figure 8.16, and may even be too old if the impacts declined still more steeply. On the other hand, 4.1 aeons may be too young if the Nectarian Period lasted longer than 0.07 aeon. I tentatively adopt 4.1 aeons as the age of the oldest group-2 basins because of this intermediate position among the likely estimates.

The two giant basins of group 1, Procellarum and South Pole-Aitken, probably formed before all the other visible basins and all the visible craters. Because these basins are probably saturated by impact structures of basin size, their ages are very difficult to estimate without more calibration points or better models of the impact rate than are available. The two models of the impact rate discussed here imply different scenarios in the early era when these basins formed. If the impacts declined steeply all through pre-Nectarian time, far more basins formed than are observed. This scenario is, in fact, consistent with the extreme impact processing observed in the lunar crustal rocks and the relative paucity of "pristine" samples. If extrapolated to the early pre-Nectarian, however, such steep curves also imply that an unrealistically great mass of material impacted the Moon since its accretion (R.B. Baldwin, oral commun., 1983). This difficulty is avoided if basins formed throughout pre-Nectarian and Nectarian time at an approximately constant rate. About 14 basins in addition to the two giant basins would then have formed from 4.2 to 4.1 aeons ago—still enough for considerable impact processing of the crust.

In summary, the many uncertainties preclude accurate estimates of any pre-Nectarian event after the origin of the Moon 4.55 aeons ago. The many great impacts that must have preceded 4.2 aeons may have struck either a solid or a mushy crust; the geologic record does not specify which, but frequencies of craters and basins are, at least, consistent with the 4.2-aeon date of crustal solidification given earlier. Procellarum, South Pole-Aitken, and at least 14 now-obliterated basins formed between crustal solidification and the oldest of the 28 pre-Nectarian basins of groups 2 through 9. The group-2 basins began about forming 4.1 aeons ago if the Nectarian Period lasted 0.07 aeon (from 3.92 to 3.85 aeons ago) and if the impact rate was approximately constant during pre-Nectarian and Nectarian time. These conservative estimates of 4.2, 4.1, and 3.92 aeons for the major milestones of the pre-Nectarian period are consistent with available data but may be considerably altered by better data on the age of the Nectaris basin or the early impact rate. At least, pre-Nectarian basins and craters clearly formed in such abundance that they repeatedly destroyed earlier features and reworked earlier crustal materials.

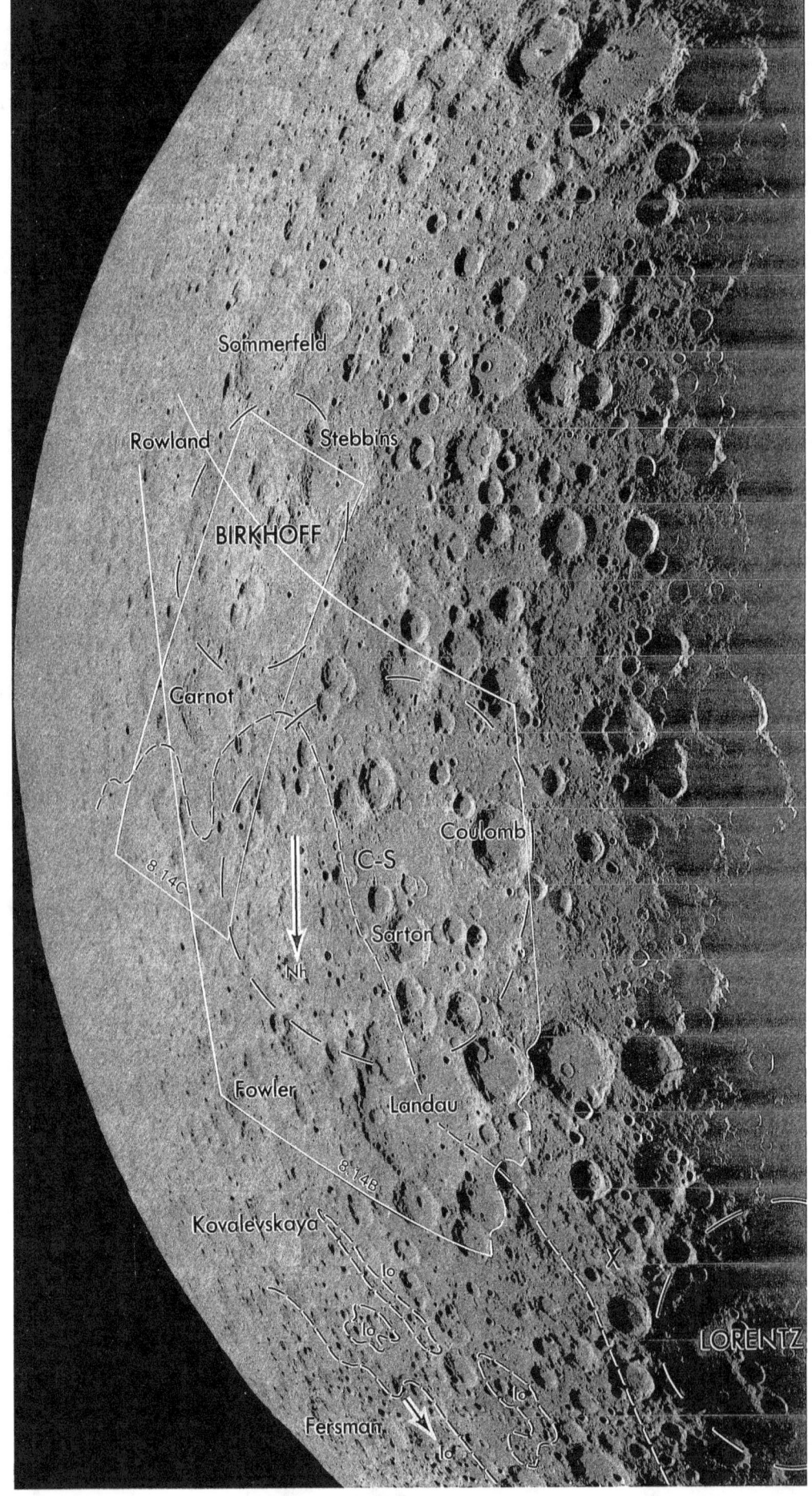

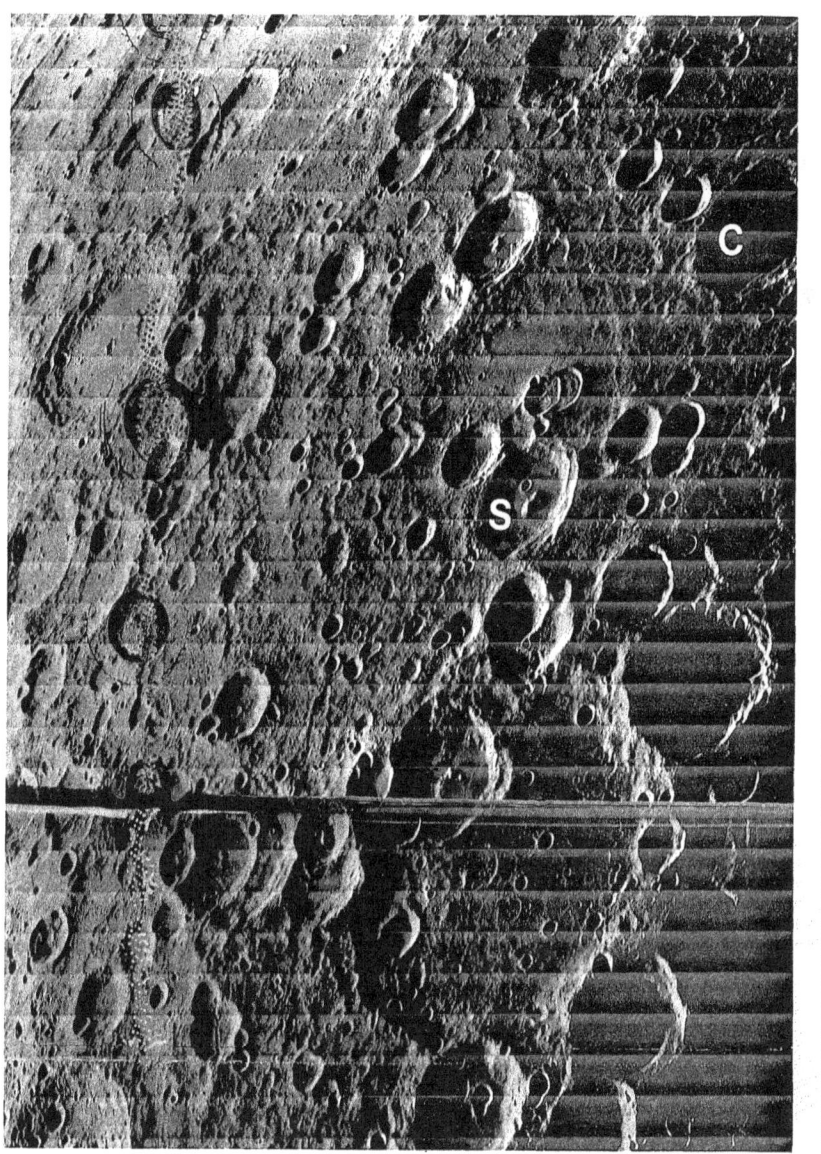

B. Detail of Coulomb-Sarton basin and superposed craters, including Coulomb (C) and Sarton (S). Orbiter 5 frame H–24.

C. Craters superposed on Birkhoff include Stebbins (S; 135 km, 64.°5 N., 142.5° W.) and Birkhoff X (BX; 77 km, 62° N., 149.5° W.), probably pre-Nectarian; and Carnot (C; 52.5° N., 144° W.), fresher appearing than the others, probably Nectarian. Numerous pits on floor of Birkhoff are probably secondary craters of Stebbins, Carnot, and other large nearby craters. Birkhoff deposits or secondary craters overlie pre-Nectarian craters Fowler (F), Schlesinger (Sch; 97 km, 47.5° N., 138.5° W.), and probably Esnault-Pelterie (E-P; 79 km, 47.5° N., 141.5° W.); sharper rim of E-P suggests, alternatively, degradation only by Carnot deposits and superposition on Birkhoff deposits. Secondary craters of Orientale and (or) Hertzsprung basins are conspicuous in Fowler (compare A). Orbiter 5 frame H–29.

FIGURE 8.14.—Stratigraphic relations of pre-Nectarian basins Birkhoff (59° N., 145.5° W.), Coulomb-Sarton (530? km, 52° N., 123° W.), and Lorentz (360 km, 34.5° N., 97° W.) on northeastern farside.
  A. Widely spaced dashes outline basin rims. Interior of Coulomb-Sarton basin (C-S) is pitted by small subequal subdued craters probably secondary to Birkhoff and Lorentz, and is overlain by radial chains secondary to Nectarian Hertzsprung basin (Nh), centered in direction of arrow 1,350 km south of crater Fowler (136 km, 43° N., 145° W., pre-Nectarian); dashed line encloses additional secondaries and other outer deposits of Hertzsprung. Craters superposed on C-S basin include Landau (221 km, 43° N, 119° W., pre-Nectarian), Coulomb (89 km, 54.5° N., 114.5° W., Nectarian), and Sarton (69 km, 49.5° N., 121° W., Nectarian). Overlap of old craters by Lorentz ejecta is visible at X. Sharp small craters in lower left near crater Fersman (143 km, 18° N., 125° W., pre-Nectarian) are secondary to the Imbrian Orientale basin (Io; centered in direction of arrow) and to Hertzsprung. Kovalevskaya (111 km) is probably Upper Imbrian. Orbiter 5 frame M–8.

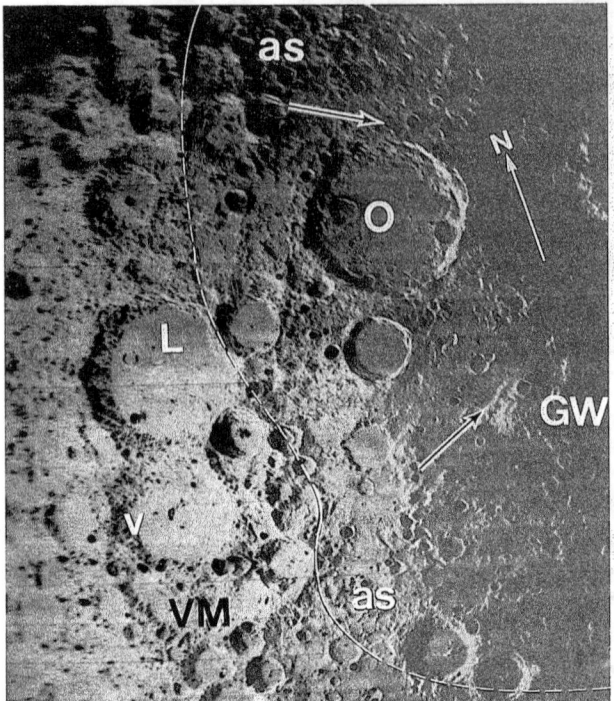

FIGURE 8.15.—Region west and southwest of Apollo basin, including Apollo-secondary craters (right of dashed line; most evident examples at letters as); black-and-white arrows point toward basin center. Center of indefinite basin Grissom-White (GW) is also indicated, but basin could be part of Apollo's peripheral structure. Pre-Nectarian craters include Leibnitz (L; 236 km), Von Kármán (V; 179 km, 44° S., 176° E.) and Von Kármán M (VM; 225 km). Oppenheimer (O; 206 km, 35° S., 166° W.) is believed to be Nectarian, though poorly photographed. Orbiter 5 frame M-43, rectified at University of Arizona's Lunar and Planetary Laboratory, Tucson, Ariz.

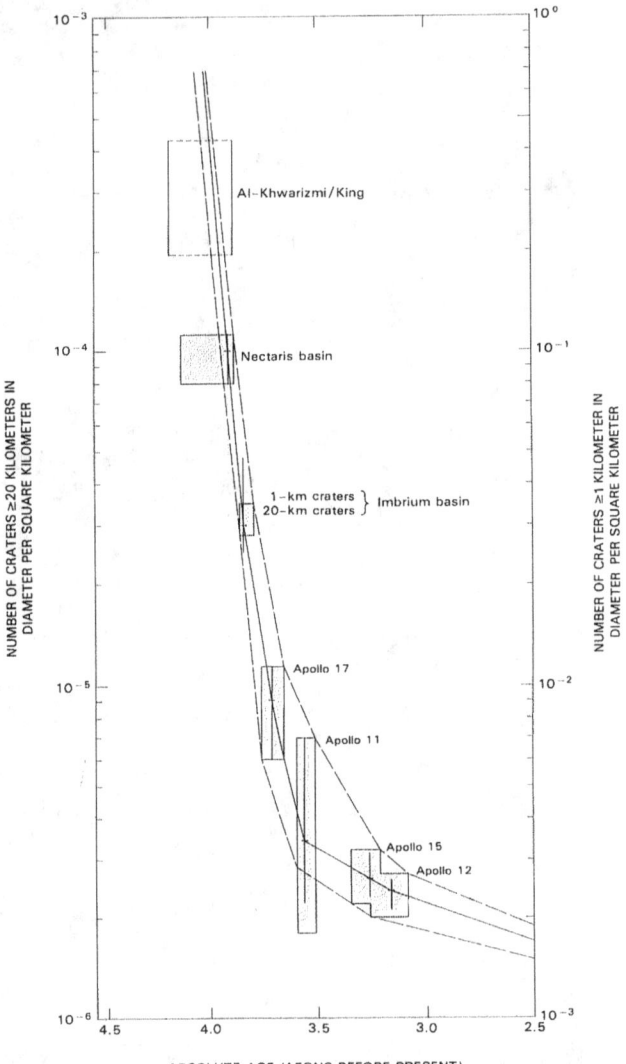

FIGURE 8.16.—Cratering rate during pre-Nectarian, Nectarian, and part of Imbrian time. Boxes represent observed crater densities (height) and absolute ages (width) given in later chapters; crosses represent my estimates for most likely values. Craters larger than 20 km in diameter (left vertical axis) were counted on Imbrium-basin and older units, and small craters (right vertical axis) on Imbrium-basin and younger units; Imbrium basin (fig. 10.4; table 7.3) provides calibration between cumulative frequencies of craters at 20- and 1-km diameters. Straight segments of curves drawn through corners of boxes enclose likely envelope of cratering frequencies as a function of time (dashed); heavy solid line is drawn through centers of crosses. Crater densities, but not absolute ages, are available for Al-Khwarizmi/King.

9. NECTARIAN SYSTEM

FIGURE 9.1 (OVERLEAF).—Deposits and secondary craters of Nectaris basin southeast of basin rim (upper left). View centered on Vallis Rheita (VR), a Nectaris-basin secondary-crater chain, and crater Rheita, a superposed Nectarian primary-impact crater (R; 70 km, 37° S., 47° E.). Pre-Nectarian terrane is at lower left; largest crater is Janssen (J; 190 km, 45° S., 42° E.). Compare figures 8.15 and 9.2. Orbiter 4 frame M–52.

9. NECTARIAN SYSTEM

CONTENTS

	Page
Introduction	163
Nectaris basin	163
General features	163
Nectaris component in Apollo 16 samples	164
Absolute age	168
Crisium basin	170
General features	170
Luna 20 samples	171
Serenitatis basin	171
General features	171
Relative age	173
Petrology of Apollo 17 samples	173
Emplacement process	174
Absolute age	177
Other basin and crater materials	178
Basins	178
Craters	180
Volcanic rocks	190
Chronology	190

INTRODUCTION

The Nectarian System includes all materials of the Nectaris basin and all younger lunar materials emplaced before the deposition of Imbrium-basin materials (Stuart-Alexander and Wilhelms, 1975). Thus, the Nectarian Period comprises the timespan between the Nectaris and Imbrium impacts. This system contains fewer rock-stratigraphic units than does the pre-Nectarian but is nevertheless widespread on the Moon (pl. 7). Nectarian basin materials cover a large area on the nearside and farside centered on the east limb (figs. 9.1–9.6). Other concentrations are on the farside around the Orientale deposits. Nectarian craters, formed at a declining but still frequent rate (fig. 8.16), are scattered randomly wherever younger deposits are absent or thin. Nectarian plains are more evident than are pre-Nectarian; photogeologic and sample data to be discussed here suggest that some Nectarian plains may be partly volcanic in origin.

Samples were collected from as many as three Nectarian basins. The Apollo 17 mission collected large amounts of material of the Serenitatis basin, which is probably Nectarian. Small fragments returned by Luna 20 from a spot on the Crisium-basin flank probably were emplaced in the ejecta of that Nectarian basin. Apollo 16, arguably, returned materials of the Nectaris basin. As always, the term "Nectarian" in this context refers to the photogeologically observed stratigraphy (see chaps. 2, 7, 8). Most of the materials contained in Nectarian units originated endogenically during the pre-Nectarian, and, additionally, had been extensively reworked by pre-Nectarian impacts before emplacement during the Nectarian Period. In general, the samples can be identified with specific Nectarian stratigraphic units better than with the pre-Nectarian units in which they once resided. The possible Nectaris-basin provenance of the important and controversial Apollo 16 sample suite is discussed in this chapter, and alternative interpretations are weighed in chapter 10.

NECTARIS BASIN

General features

Deposits of the Nectaris basin clearly separate pre-Nectarian and younger Nectarian deposits over a wide area (pls. 3, 7; figs. 8.5, 9.1, 9.2, 9.5, 9.6). At 860 km in diameter, the Nectaris basin is only 70 km smaller than the Orientale basin (table 4.1), and probably resembled Orientale before modification by later basin and crater deposits. The west and south sectors have a moderately well defined ring structure spaced like the Orientale rings (fig. 9.2A). In fact, Nectaris was chosen as the type basin by Hartmann and Kuiper (1962).

The landscape south and southeast of the Nectaris basin is a degraded equivalent of the Hevelius Formation and the secondary craters south and southeast of Orientale (figs. 4.4, 9.1; Stuart-Alexander, 1971; Wilhelms and McCauley, 1971; Stuart-Alexander and Tabor, 1972; Stuart-Alexander and Wilhelms, 1975; Wilhelms and El-Baz, 1977). Thick mantling of pre-Nectarian terrane is evident from the paucity of large craters in comparison with the region farther south (figs. 8.5, 9.1). Like the circum-Orientale terrane, the Janssen Formation (fig. 9.2; Stuart-Alexander, 1971) and much other circum-Nectaris terrane are strongly lineate radially to the basin center. Vallis Rheita (figs. 9.1, 9.2) and similar troughs (fig. 9.5; Hodges, 1973b), consist of subtle overlapping or tangential elliptical craters and evidently were formed by secondary impact, as was Vallis Bouvard on the south Orientale-basin flank (fig. 4.4D). Subcircular satellitic craters of Nectaris (figs. 9.1, 9.2B) also resemble the secondary craters of Orientale in shape, cluster geometry, and distance from the source basin, although they appear less crisp. At Nectaris, the primary-secondary transitional zone that is reproducibly mappable east of Orientale and is also detectable at Imbrium (chap. 4) has been blurred by "aging" and secondary cratering from Imbrium (fig. 3.10A); nevertheless, an attempt has been made here to identify the contact

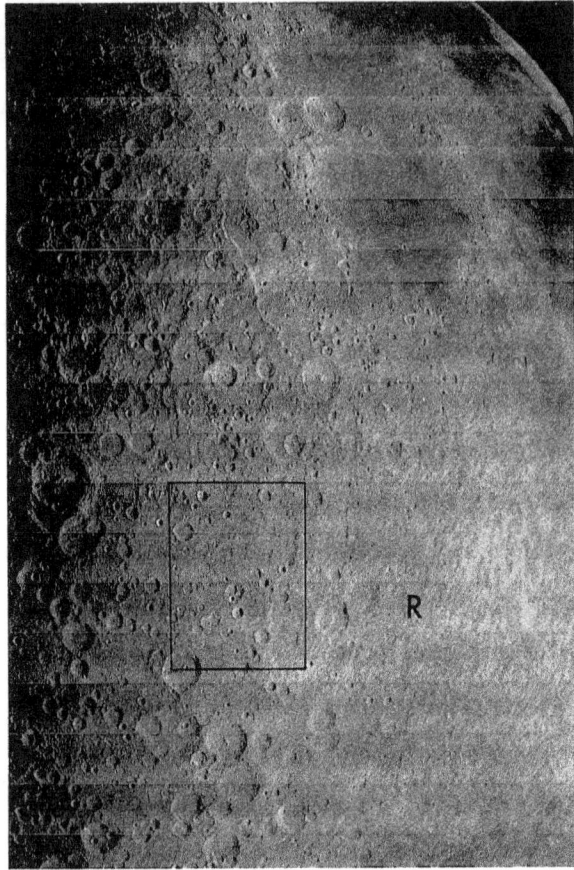

between the two Nectaris facies (pls. 3, 7).

The north and east sectors of Nectaris have an overall arcuate map pattern, but individual rings are inconspicuous (fig. 9.7). These sectors resemble western Orientale in this respect (fig. 4.5), although the Nectaris uplifts are more flat-topped than jagged. The northern and eastern Nectaris ejecta is largely buried by younger deposits (pl. 7). Any secondary chains that were formed in these sectors have been buried by younger deposits of the maria and of the post-Nectaris Crisium, Serenitatis, and Imbrium basins (pls. 3, 7; figs. 9.3–9.6).

More light-colored plains are concentrated inside the Nectaris basin than in other settings except the Orientale- and Imbrium-basin peripheries (fig. 10.34). This setting suggests an analogy with the Maunder Formation of Orientale, that basin's interior impact melt. Hummocky material on the northern Nectaris terra is similar in topography and relative position to the Montes Rook Formation and may be the Nectaris equivalent (fig. 9.7; Elston, 1972; Wilhelms, 1972c). However, this hummocky material lies on the distal sides of Imbrium-secondary craters and thus may consist of their decelerated ejecta (Wilhelms, 1980, p. 45).

Light-colored plains are also widespread south of the basin (pl. 7). Wilhelms and McCauley (1971) and Scott (1972b) interpreted the large patches of cratered or "pitted" plains southwest of the basin (fig. 9.2B) as volcanic in origin by analogy with the younger, Imbrian plains also thought to be volcanic before the Apollo 16 analyses. The subequal "pits" and larger, also subequal, flat-floored depressions were thought to be volcanic because of their grouping. The superposition of the pitted plains and smoother Imbrian plains on the larger craters was interpreted as analogous to the flooding of craters by mare materials. An alternative explanation, however, seems to be required by the Apollo 16 results. The plains grade with the lineate Janssen Formation, as do circum-Orientale or circum-Imbrium plains with the lineate ejecta of those basins. Therefore, the stratigraphic relations here can be interpreted by the impact model of lunar-terra formation. The large grouped craters are probably Nectaris secondaries, and most of the smaller craters or "pits" are probably Imbrium secondaries. The plains were likely emplaced by one of the ejecta-deposition mechanisms whereby a surge of debris arrives at a site after the secondaries form. This concentration of plains may result from catchment of the Nectaris ejecta by the Mutus-Vlacq basin (fig. 8.12; Wilhelms and others, 1979). In addition, a marelike volcanic deposit, extruded into the Mutus-Vlacq basin during pre-Nectarian time, may underlie the plains and contribute to their level topography (see section below entitled "Volcanic Rocks").

The terrain west of Nectaris differs considerably from that in the other sectors. Rupes Altai (the Altai scarp) constitutes a continuous, well-defined topographic rim (fig. 9.2A). Despite its proximity to this rim, the western region contains little evidence of Nectaris ejecta. Thick, strongly lineate ejecta is visible at the same distance from the southwest and southeast rim sectors. Part of the reason for the difference is that the western Nectaris deposits were modified by Imbrium projectiles when the strong Imbrium-radial pattern of craters and grooves was formed (figs. 3.10, 9.2A). Additionally, Nectaris ejecta may have been deposited less thickly here than in the south.

Nectaris component in Apollo 16 samples

The thickness of the western Nectaris ejecta is important in the interpretation of samples collected by the Apollo 16 mission. The astronauts landed 60 km west of the Kant Plateau, a broad part of the Nectaris-basin rim crest about 100 km wide (pl. 3; figs. 2.5C, 9.8). At Orientale, thick ejecta is visible in the most nearly analogous position, 160 km east of the Cordillera scarp, even though this point is in

FIGURE 9.2.—Materials of Nectaris basin.
  A. Regional view showing Mare Nectaris and Nectaris basin (upper right quadrant) and area of B (outlined). Letter R is below crater Rheita and right of Vallis Rheita. Orbiter 4 frame M–83.
  B. Detail of Janssen Formation (Nj), plains deposits (Np), Nectaris-basin secondary craters (Nnsc), and Imbrium-basin secondary craters (Iisc). Nj, type area of the Janssen Formation (Stuart-Alexander, 1971), which partly buries pre-Nectarian crater Janssen; arrow is radial to center of Nectaris basin. Craters Nicolai (42 km, 42° S., 26° E.) and Pitiscus (82 km, 50° S., 30° E.) are Nectarian. Nj and Np thickly to thinly bury Nnsc; Iisc overlies Nectarian units (Scott, 1972b). Orbiter 4 frame H–83.

the relatively "excluded" zone of ejecta (pl. 3; fig. 4.4F). Thus, Nectaris ejecta probably either underlies or composes the surface materials at the Apollo 16 landing site. However, the search for Nectaris materials has been inconclusive.

Two photogeologic units were sampled by Apollo 16. The first is the Cayley Formation, which forms the Cayley plains (figs. 9.8, 9.9). This formation was emplaced during the Imbrian Period (chap. 10); however, it may consist of Nectaris materials reworked during its emplacement. The Cayley was sampled in the central and southern parts of the region (all stations except 11 and 13, fig. 9.9). Station 10, near the landed lunar module (LM) and the geophysical instruments (Apollo Lunar Surface Experiments Package [ALSEP]), is probably the most intensely examined part of the Moon (Muehlberger and others, 1980; Schaber, 1981). Nevertheless, no outcrops or large boulders of the Cayley were found. A regolith 10 to 15 m thick (Cooper and others, 1974; Freeman, 1981) obscures the internal stratigraphy of the Cayley. The Cayley may contain coherent, lithologically uniform strata or lenses (Hodges and Muehlberger, 1981), or may consist of loosely aggregated poorly sorted clastic debris. Some of the mechanisms described in chapter 10 could have emplaced a Cayley Formation consisting of poorly sorted Nectaris debris during the Imbrian Period.

Regolith samples show that impact-melted materials are common in the Cayley Formation (James and Hörz, 1981). Their sources and ages, however, are major problems. A tight compositional cluster (approx 29–31 weight percent Al_2O_3) and a looser cluster (17–21 weight percent Al_2O_3) suggest to Ryder (1981) that two groups of melt rocks came from two distinct melt sheets. The first group is richer in Al than the average terra (25–26 weight percent Al_2O_3; Taylor, 1982) and is mostly subophitic to intergranular in texture, like much volcanic basalt but resulting from thorough impact melting (lithologic type SM, compositional type V, table 9.1). The low-Al samples are the richest in KREEP, are laden with fragmental material, and are typically poikilitic in texture (lithologic type PM, compositional type K). The poikilitic texture results from the relatively mafic composition and from the presence of many clasts, which form nuclei for crystallization of the melt (Lofgren, 1977; Nabelek and others, 1978). In addition, many melt rocks have Al_2O_3 contents intermediate between the two clusters (lithologic type IM, compositional type I); their compositions are commonly described inappropriately as "very high alumina (VHA) basalt" (Hubbard and others, 1973, 1974).

The largest samples of the Cayley Formation were excavated by South Ray Crater, 680 m wide and 130 m deep, which is 2 million years old (table 13.2). Materials probably ejected from South Ray were sampled at stations 4 and 8 (Reed, 1981; Sanchez, 1981; Ulrich and Reed, 1981; Ulrich and others, 1981b). Among other materials is a distinctive dimict (dilithologic) breccia that apparently characterizes the Cayley Formation (James and Hörz, 1981). These "black and white" rocks consist of cataclastic anorthosite and fine-grained impact-melt rock of "VHA" composition. The dimict breccia seems to be analogous to terrestrial dike-breccia deposits formed in the outer parts of impact craters (fig. 3.21). During cratering, melt intruded into cataclastic feldspathic material in the walls of a growing crater or basin and then was quenched, fragmented, and intruded by other clastic feldspathic material (James, 1977, 1981). Fragments of similar anorthosite and melt rock that also abound in the regolith probably originated in dimict breccia.

The melt rocks of the Cayley Formation may have been emplaced while hot as beds or pods (Hodges and Muehlberger, 1981), or they may have been emplaced cold in isolated boulders that South Ray and other craters happened to penetrate. If individual impacts generate uniform melt sheets, one melt type, at most, is likely to have been emplaced hot when the Cayley was emplaced (Ryder, 1981). Thus, reworking of an already complex deposit is suggested. Nonetheless, large single impacts can generate great lithologic and compositional complexities (see chap. 3). It is still unclear how many melt sheets are represented by these melt rocks.

The other unit sampled by Apollo 16 is material of the Descartes Mountains (fig. 9.9; Milton and Hodges, 1972). For convenience, this unit is here called the Descartes Formation, although it has not been defined formally. The Descartes is divided into two gradational facies. The northern or Smoky Mountain facies is grooved radially to the Imbrium basin. Grooves in the southern or Stone Mountain facies curve away from this Imbrium-radial direction into a more nearly transverse trend approximately where the formation contacts the Kant Plateau (figs. 2.5C, 9.9). This embankment of the Descartes Formation against the Nectaris-basin rim has been considered as evidence for a partial Nectaris origin of the topography (Head, 1972; Wilhelms, 1972c). Thus, both the Imbrium and Nectaris basins may have contributed to the morphology of the Descartes Formation; the same is true for the materials of which it is composed.

Large sample collections were made at stations 11 and 13 on Smoky Mountain from ejecta of the 1-km-wide, 50-million-year-old (table 13.2) North Ray Crater (fig. 9.10; Stöffler and others, 1981; Ulrich and others, 1981b). Station 11 is on the crater rim, and station 13 is about one crater diameter from the rim (fig. 9.9). North Ray penetrated a 50-m-high mass-wasted ridge that is part of Smoky Mountain (300–400 m high overall). The 230-m depth of North Ray probably ensures that the Smoky Mountain material was excavated (Ulrich, 1973; Ulrich and others, 1981b). The station 11 and 13 samples are mostly of fragmental feldspathic breccia whose fragmental melt-poor matrix is composed of material similar to that of the larger clasts (fig. 9.10). About half the clasts are of cataclastic anorthosite (Minkin and others, 1977; James, 1981), a quarter of granulitic breccia, and a quarter of feldspathic fragment-laden impact melt (lithologic types CA, G, and FM, respectively, table 9.1). The granulitic breccia is thermally metamorphosed (Warner and others, 1977; James, 1980, 1981). The relation of the FM-type melts to the breccia deposit in which they are found is not known. They may have been cogenetic with the fragmental breccia (as in suevite at the Ries; see chap. 3; Stöffler and others, 1979) or may predate the deposition of the breccia. Determining which is the case here would be an important step in interpretation of the Descartes Formation, especially considering the rarity of impact melts in the breccia matrices; such melts are generally the most valuable for dating impact deposits.

Small amounts of the Stone Mountain facies may have been sampled at stations 4 and 5, where feldspathic material like that at stations 11 and 13 was recovered. Contamination by Cayley material

---

Note: Figures 9.3 through 9.6 constitute a series of overlapping oblique photographs taken by Lunar Orbiter 4 of areas covered poorly from lower altitudes earlier in the mission (pl. 2); Sun illumination is opposite from that in other nearside photographs.

FIGURE 9.3. (p. 166, left)—Part of northeast quadrant of nearside, including Mare Crisium and Crisium basin (bottom; centered at 17.5° N., 58.5° E.). Major Nectarian craters are Cleomedes (Cl; 126 km) and Endymion (E; 125 km; pl. 7; table 9.4). Arrows indicate secondary-crater chains radial to Crisium (C) and Imbrium (I) basins; terra south of dashed line is covered by deposits of Crisium basin, except where covered by younger deposits of Cleomedes, Geminus (G; 86 km, Eratosthenian), and Tarentius (T; 56 km, Copernican). A, Atlas (87 km, Lower Imbrian); H, Hercules (69 km, Eratosthenian); M, Messala (124 km, pre-Nectarian, partly covered by Crisium deposits). Orbiter 4 frame H–191.

FIGURE 9.4. (p. 166, right)—Part of northeast quadrant near east limb, including two distinct rings of Humboldtianum basin (300- and 600-km diameter, centered at 61° N., 84° E.) between Mare Humboldtianum (H) and crater Endymion (E; compare fig. 9.3). Arrows indicate Humboldtianun-radial or -subradial directions; arrow I is also radial to Imbrium (fig. 9.3), and arrow below crater Zeno (Z) is also radial to Crisium. Zeno is superposed on Humboldtianum basin and overlain by Crisium-basin secondary-crater cluster (compare fig. 4.3Q). Deposits and clustered craters between Gauss (G; 177 km, a Nectarian crater with fractured floor) and Joliot (J; 143 km, pre-Nectarian) are probably outer materials of Humboldtianum basin; deposits in lower third of photograph may be related to Crisium (C). Basin deposits underlie Nectarian craters Gauss, Berosus (B; 74 km), and Neper (N; 137 km, 9 N., 85 E.), Lower Imbrian crater Hahn (Ha; 84 km), and mare basalt (M; Mare Marginis) (compare fig. 4.7). Orbiter 4 frame H–165.

FIGURE 9.5. (p. 167, left)—Terrain south of Mare Crisium (C) and east of Nectaris basin. Black-and-white arrows are radial to basin centers and denote large secondary-crater chains of Crisium basin (c) and Nectaris basin (N; lower chain is part of Vallis Rheita); g, graben. Outer deposits of Crisium (Nco) and Nectaris (Nno) include secondary craters and terra-mantling deposits. Luna 20 landing site (L20) is on Crisium-basin rim (compare fig. 11.16). Nj, Janssen Formation, partly covers pre-Nectarian crater Furnerius (F; 125 km, 36° S., 60° E.). Np, plains of Nectarian age; INp, deposit in pre-Nectarian Balmer-Kapteyn basin, probably consisting of Nectarian plains overlain by Imbrian plains. y, deposits younger than Nectarian. Primary and secondary deposits of floor-fractured Lower Imbrian crater Petavius (P; 177 km) are superposed on plains, Crisium-secondary chain (c; Vallis Palitzsch), and crater Adams (A; 66 km); Adams and Pontecoulant (Po; 91 km, 59° S., 66° E.), both Nectarian, overlie basin deposits. L, Langrenus (132 km), an Eratosthenian crater superposed on Crisium deposits and Mare Fecunditatis. Orbiter 4 frame H–184.

FIGURE 9.6. (p. 167, right)—Eastward continuation of geology shown in figure 9.5. Southward extent of Crisium-basin deposits (Nco) is obscured by large, floor-fractured, Upper Imbrian crater Humboldt (H; 207 km, 27° S., 81° E., unrelated to Humboldtianum basin); clustered craters between H and arrow N (indicating Nectaris center) may be Crisium-basin secondaries. C, direction to center of Crisium; Nno, outer deposits of Nectaris basin. Po, Pontecoulant (91 km, 59° S., 66° E.); m, part of Mare Australe. Orbiter 4 frame H–178.

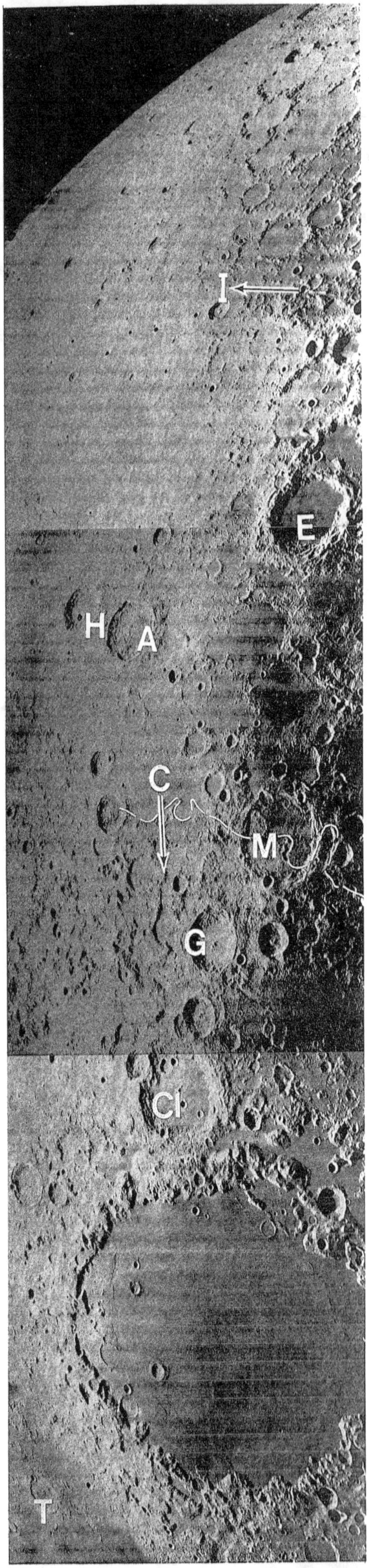
I
E
H
A
C
M
G
Cl
T

I
E
H
Z
G
B
Ha
J
M
C
N

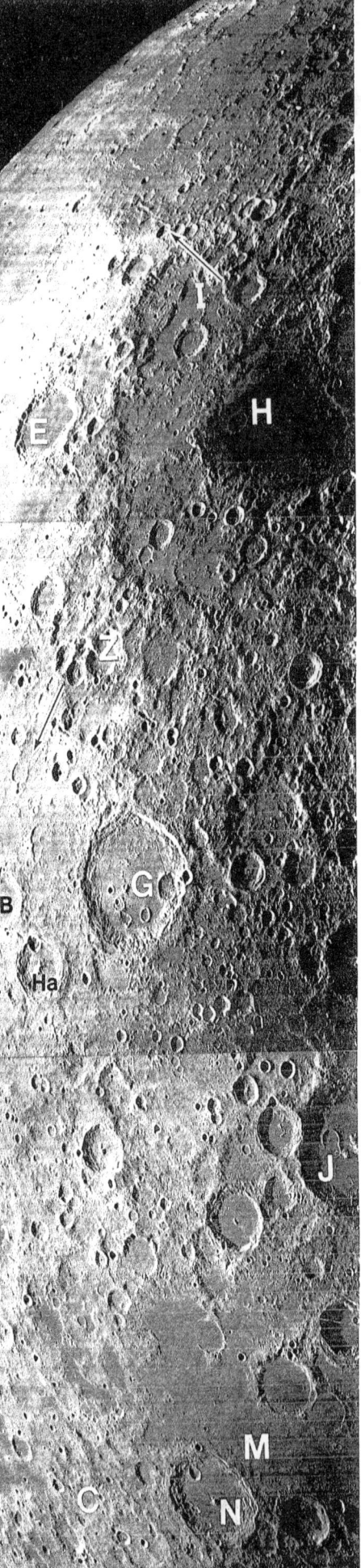

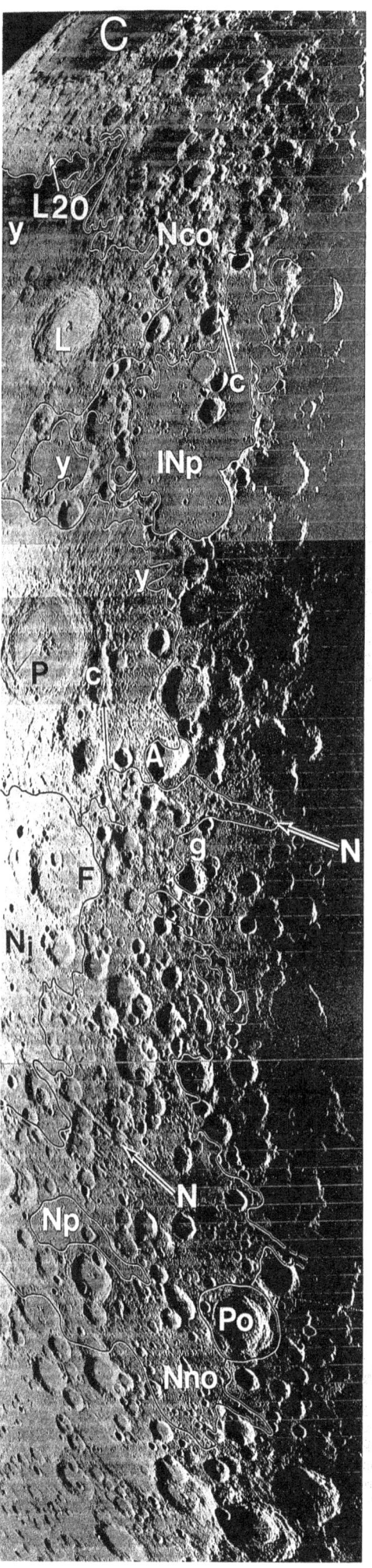

(or a similarity of Cayley and Descartes material), however, is likely at those stations, considering that many large samples consist of dimict breccia with a 2-million-year exposure age (Sanchez, 1981).

In summary, the Cayley and Descartes Formations appear to differ in several respects in addition to morphology (James, 1981; James and Hörz, 1981; Stöffler and others, 1981). (1) The North Ray Crater samples (probable Descartes) are more aluminous than the South Ray Crater samples (Cayley). (2) The Cayley appears to contain more impact-melt rock, whereas the Descartes is characterized by friable fragmental feldspathic breccia. (3) More of the Cayley impact-melt rocks are of low-K KREEP and "VHA" composition. (4) The Cayley apparently has a higher magnetism as measured on the surface, a possible indication of buried melt-rich breccia (Strangway and others, 1973). Finally, (5), the Cayley is characterized by sharper craters, more blocks ejected onto the surface from craters, and a thinner regolith (Ulrich, 1973), properties consistent with the Cayley being a more cohesive unit than the Descartes. A possible explanation for these lithologic and topographic differences is that the Descartes Formation consists mostly of Nectaris ejecta and the Cayley of melt-rich Imbrium ejecta.

This complex region is still not well understood. Neither the Cayley nor the Descartes Formation can be said to be well characterized lithologically or stratigraphically. Apart from a boulder composed of lithified regolith (sample 62195), the only boulders sampled are those from North and South Ray Craters. The observed differences may not be qualitative but may reflect relative proportions of the same constituents (O.B. James and M.M. Lindstrom, oral commun., 1983). Analyses of the returned samples have stressed their highly feldspathic composition, and only now are beginning to address problems of the origin and emplacement age of the breccia deposits.

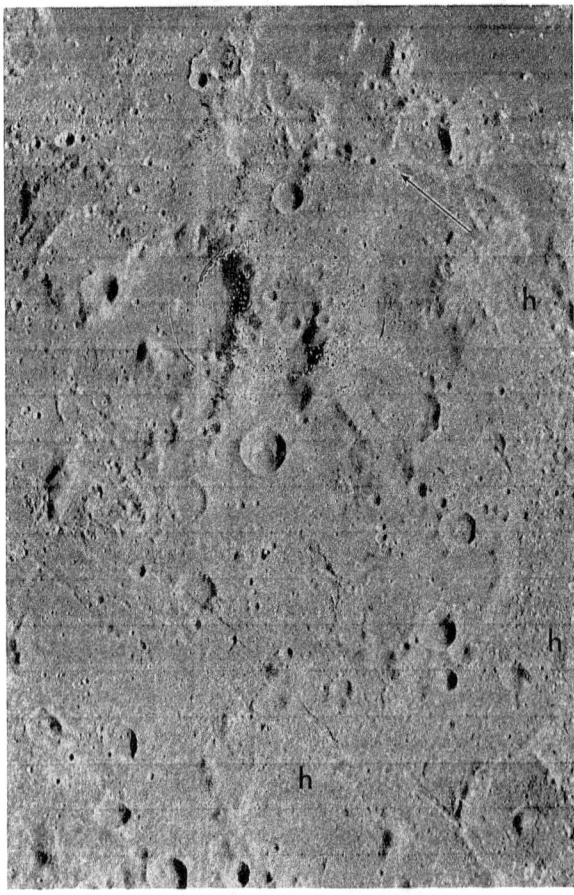

FIGURE 9.7.—North rim of Nectaris basin. Hilly terrain (h) is possible Nectaris equivalent of Montes Rook and Alpes Formations. Topographic freshness, however, suggests emplacement by Imbrium-basin secondary craters (arrow is radial to Imbrium). Imbrium-radial grabens are in lower third of photograph (see chap. 6). View centered at 3° S., 37.5° E. Orbiter 4 frame H–72.

TABLE 9.1.—^{40}Ar-^{39}Ar ages of Apollo 16 samples

[After James (1981).
Sample numbers (Ulrich and others, 1981): 60-, station 10, central part of area near Lunar Module and Apollo Lunar Science Experiment Package (ALSEP); 61-, station 1, central area on rim of 300-m-diameter Flag Crater; 615-, chips from rake sample 61500; 63-, station 13, on ejecta of North Ray Crater, about 800 m from rim; 635-, 2- to 4-mm fines from rake sample 63500; 64-, station 4, on Stone Mountain; 65-, station 5, on Stone Mountain; 66-, station 6, near base of Stone Mountain; 67-, station 11, on southeast rim of North Ray Crater; 676-, 2- to 4-mm fines from rake sample 67600; 677-, 2- to 4-mm fines from rake sample 67700; 68-, station 8, on ejecta of South Ray Crater, 3.3 km northeast of rim; 685-, fragment from rake sample 68500; identical to DM (see below). Digits and letters after commas refer to splits of sample.
Lithologies: CA, cataclastic (mechanically disrupted) anorthosite, found as clasts in feldspathic fragmental breccia, as one of the major lithologies in dimict breccia, and as monolithologic fragments in the regolith; some recrystallized. DC, anorthosite in dimict breccia. DM, impact-melt rock in dimict breccia; texturally diverse; mostly fine grained, compositionally uniform melt-rock component of the breccia. FM, feldspathic fragment-laden impact-melt rock; diverse texture, nonpoikilitic. G, granulitic breccia (granulitic impactite of Warner and others, 1977); probably recrystallized or partially melted after heating of earlier breccia (Warner and others, 1977; James, 1980, 1981); compositionally heterogeneous; granulitic (granoblastic) texture consists of small equant crystals lacking well-developed crystalline forms. IM, intergranular impact-melt rock; fine texture: intergranular texture consists of pyroxene filling voids between plagioclase crystals. PM, poikilitic impact-melt rock: poikilitic texture consists of large single grains (commonly of pyroxene) enclosing many small crystals (commonly of plagioclase). SM, subophitic impact-melt rock; fine texture, grading to intergranular; glassy residuum also present: subophitic texture consists of pyroxene filling voids and partly surrounding plagioclase crystals.
Al_2O_3 contents from Ryder and Norman (1980) and James (1981).
Surface unit is unit in which sample was emplaced at surface and from which it was collected (Ulrich and others, 1981): Ca, Cinco a crater: local, in regolith developed from underlying unit; NR, rim or other ejecta of North Ray Crater; SR, ejecta of South Ray Crater.
Source unit is photogeologic unit in which sample was emplaced near the landing site: C, Cayley Formation; D, Descartes Formation.
Ages recalculated from original values, using International Union of Geological Sciences (IUGS) decay constants (Steiger and Jäger, 1977), except for those determined by Schaeffer and Husain (1973, 1974), Schaeffer and others (1976), and Schaeffer and Schaeffer (1977); reevaluation of the monitor used by these analysts nearly compensates for the changes in decay constants (O. A. Schaeffer, in James, 1981, p. 210).
References: H73, Huneke and others (1973); J74, Jessberger and others (1974); J77a, Jessberger and others (1977a); K73b, Kirsten and others (1973b); M78, Maurer and others (1978); P75, Phinney and others (1975); S73, Stettler and others (1973); S76, Schaeffer and others (1976); SH73, Schaeffer and Hussain (1973); SH74, Schaeffer and Husain (1974); SS77, Schaeffer and Schaeffer (1977); T73, Turner and others (1973); TC75, Turner and Cadogan (1975)]

Sample	Lithology	Al_2O_3 (wt pct)	Surface unit	Source unit	Age (aeons)	References
60015,22	CA	35±	SR	C	3.50±0.05	SH74, P75.
68415,49	SM	25-30	SR or local	C	3.74±0.04	S73.
68415,10	SM	27	SR or local	C	3.80±0.04	H73.
68415,50	SM	25-30	SR or local	C	3.80±0.06	K73b.
67603,1-ME	PM	27	NR	D	3.83±0.05	M78.
65015,61	PM?	20±	?	?	3.86±0.04	K73b.
63503,17-LN	IM	23	NR	D	3.87±0.04	M78.
65055,12B	SM	?	SR?	C?	3.89±0.02	J77a.
67703,14-NF	G	29	NR	D	3.89±0.04	M78.
67703,14-NG	G	28	NR	D	3.89±0.04	M78.
67016,100	?	30±	NR	D	3.89±0.05	TC75.
67603,1-MQ	G	28	NR	D	3.90±0.06	M78.
65055,12A	SM	?	SR?	C?	3.90±0.02	J77a.
64536,3	DC	?	SR or Ca	C	3.91±0.01	J77a.
60315,6,1	PM	17±	SR or local	C	3.91±0.02	S76.
67703,14-NI	PM	25	NR	D	3.91±0.04	M78.
67703,14-ND	SM	24	NR	D	3.91±0.04	M78.
64536,7	DC	?	SR or Ca	C	3.92±0.02	J77a.
65015	PM	20±	?	?	3.92	J74.
63503,17-LE	G	33	NR	D	3.92±0.04	M78.
63503,17-LC	G	25	NR	D	3.93±0.04	M78.
68416,34	SM	25-30	SR	C	3.94±0.05	K73b.
63503,17-LK	IM	26	NR	D	3.94±0.05	M78.
68503,16,33	IM	23	Local	C	3.95±0.06	SH73.
63503,17-LI	G	31	NR	D	3.96±0.04	M78.
67603,1-MD	FM	32	NR	D	3.96±0.05	M78.
60315,19	PM	<20	SR or local	C	3.97±0.03	K73b.
65785,13	SM	20-25	Local	D	3.97±0.02	SS77.
67075	CA?	>30	NR	D	3.98±0.05	T73.
68503,16,1	PM	?	Local	C	4.00±0.05	SH73.
66043,2,5	PM	?	Local	D?	4.01±0.05	SH73.
61503,1,11	PM	?	?	?	4.02±0.02	SS77.
61503,1,18	IM	23	?	?	4.02±0.01	S76.
68415,10	Plagioclase	25-30	SR?	C	4.03±0.04	H73.
63503,17-LH	FM	33	NR	D	4.09±0.07	M78.
67603,1-MM	FM	31	NR	D	4.10±0.05	M78.
67703,14-NL	FM	32	NR	D	4.12±0.05	M78.
63503,17-LL	FM	33	NR	D	4.13±0.04	M78.
63503,17-LF	FM	33	NR	D	4.13±0.1	M78.
67703,14-NN	FM	32	NR	D	4.14±0.04	M78.
60025,86	DC?	35±	SR	C	4.17-4.21	SH74.

Absolute age

Absolute ages determined by the ^{40}Ar-^{39}Ar method on materials obtained from the Cayley and Descartes Formations do not clearly indicate a difference in provenance consistent with the suggested lithologic differences. Pre-Imbrian ages (older than 3.85 aeons) are more common than Imbrium-basin ages. The following interpretation suggests an age for the Nectaris basin (James, 1981).

The dated material that is most critical with respect to a Nectaris-basin age was derived from the feldspathic fragmental breccia deposits penetrated by North Ray Crater (table 9.1). Coarse fines (2–4 mm) of melt type FM collected by rake from the North Ray regolith yielded a cluster of six dates between 4.09 and 4.14 aeons and one younger date of 3.96 aeons (Maurer and others, 1978). Because the young fragment (67603,1–MD) is compositionally and texturally identical to the apparently older clasts, 3.96 aeons is a maximum age for the FM-type materials (James, 1981). The older ages were probably inherited from an earlier event and were not reset when the feldspathic fragmental breccia deposits were assembled. Their close clustering suggests an actual event at about 4.1 aeons. Wetherill (1981) suggested that the older event was the Nectaris impact, whereas Maurer and others (1978) and James (1981) proposed that it was a pre-Nectarian crater impact(s).

The other chronologically significant type of coarse fines derived from the feldspathic fragmental breccia deposits is granulitic breccia. The ages of six fragments range from 3.89 to 3.96 aeons, a range that also includes the age of the youngest FM-type melt (Maurer and others, 1978). This range may reflect analytical inaccuracies and may record a single metamorphic episode generated by a major source of near-surface heat, probably a hot basin-ejecta blanket (Warner and others, 1977). The geologic setting of the samples suggests the Nectaris ejecta as the source of this heat. The average of the age cluster, 3.92 aeons, or the youngest age, 3.89 aeons, which is also the age of a dark clast and a plagioclase separate from sample 67016 (Turner and Cadogan, 1975), may date the Nectaris impact (Jessberger and others, 1974; Turner, 1977; Maurer and others, 1978; James, 1981).

The wide range of isotopic ages presents obvious problems for determining the emplacement ages of the Cayley and Descartes Formations. An age occupies almost every 10-million-year interval from 4.14 to 3.86 aeons ago, and four ages are younger (table 9.1). The inevitably long list of uncertainties includes: (1) analytical errors; (2) incomplete reequilibration of argon isotopes by impacts; (3) addition or removal of argon by later events; (4) varying degrees of equilibration of clasts and matrices in the same sample, which are commonly analyzed together; (5) failure to collect thoroughly reset materials that do date an event; (6) uncertainty whether the recorded events are the crystallization of plutonic rocks, metamorphism (granulitic breccia), the main emplacing impact, or minor later impacts; and (7) uncertainty whether the photogeologic unit underlying the sampling stations was actually sampled (especially questionable for the

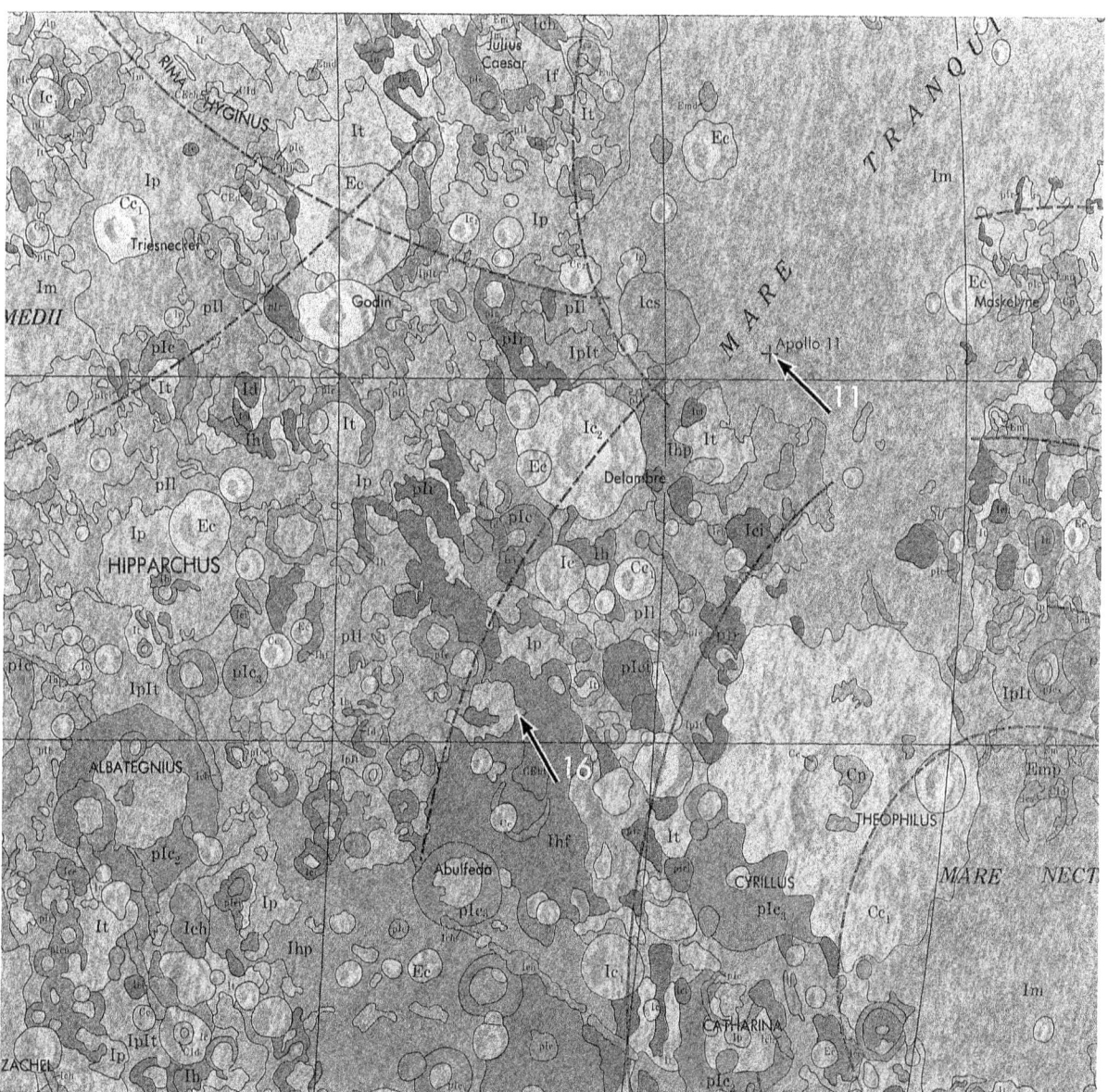

FIGURE 9.8.—Geologic map of part of nearside (Wilhelms and McCauley, 1971), including Apollo 11 (see chap. 11) and Apollo 16 landing sites. Im, mare basalt of Imbrian age (Apollo 11); Ip, plains material of Imbrian age (Cayley Formation; Apollo 16); Ihf, hilly and furrowed material of Imbrian age (Descartes Formation; Apollo 16). Arcs indicate basin rings as interpreted at time of map's preparation.

Descartes Formation). The problems are illustrated by sample 60315 from the Cayley, which yielded ages of 3.91 to 3.97 aeons (table 9.1). Therefore, the dating may not be accurate; the ages listed in table 9.1 may date no real event at all.

These ages, however, are at least consistent with formation of the Nectaris basin at about 3.92 ± 0.03 aeons, and this date is tentatively accepted in this volume (fig. 8.16). The basin is unlikely to be younger than that age range because no samples of granulitic breccia or type-FM melt rock recovered from the Descartes Formation have yielded younger ages (James, 1981). It is unlikely to be older than that age range because some type-FM melts were reset during that time interval, because metamorphism of granulitic breccia requires a major heat source not available after the Nectaris impact, and because the impact-rate curve (fig. 8.16) would have to bend to fit both the available Imbrian ages and an older age for Nectaris (see chap. 8). The ages older than 3.92 ± 0.03 aeons (table 9.1) date pre-Nectarian events, if they date any discrete events. The few younger ages may have been introduced or partly reset during the Imbrium impact by the mechanisms described in chapter 10.

CRISIUM BASIN
General features

Crisium has already been referred to in connection with basin topography, ring origin, mare-basalt thickness, and structural modification because quantitative data were obtained by overflights of

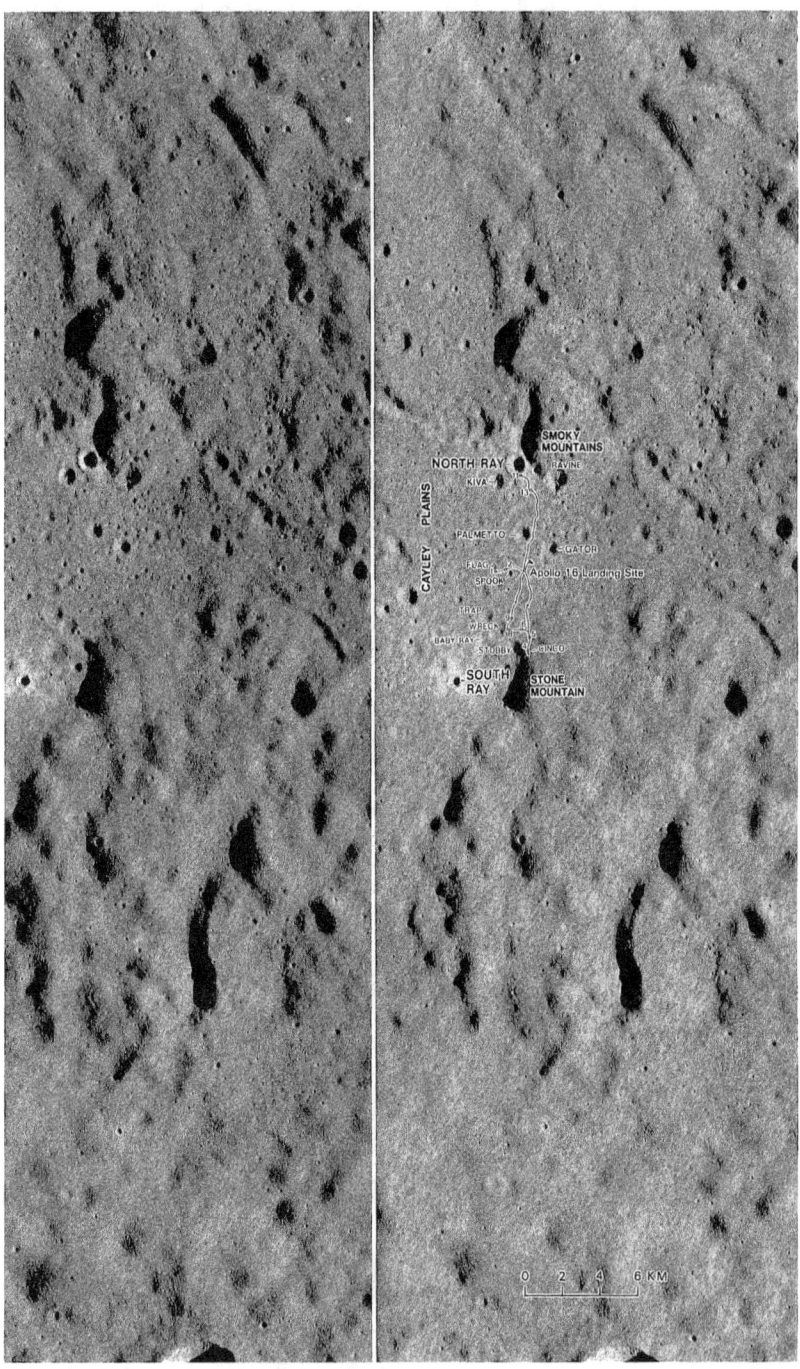

FIGURE 9.9.—Region of Apollo landing 16 site. Facies of the Descartes Formation in the Smoky Mountains is furrowed radially to Imbrium basin; facies in Stone Mountain is furrowed transversely to Imbrium and radially to Nectaris basin. Lines, astronaut traverses; numbered circles, sampling stations. Stereoscopic pair of Apollo 16 frames P–4558 (right) and P–4563 (left).

Apollo 15 and 17 instruments (pl. 2; figs. 4.9, 5.23). As in many large basins, the identity of its topographic rim is unclear. The massive mare-bounding rim, about 635 km in diameter, would seem to be the logical choice as the topographic rim and, therefore, the boundary of excavation. It probably does bound a deep cavity. Incomplete but locally conspicuous arcs, however, form a near-circle, 1,060 km in diameter, whose massifs are as high as or higher than those of the 635-km-diameter ring (figs. 4.3Z, 4.9, 9.11). These arcs are locally rimlike (raised crest, steep inner slope, gentler outer flank) and mark the inner boundary of conspicuous secondary chains (figs. 9.3, 9.5, 9.6). Thus, the outer zone may have been partly excavated. A large basin excavation is consistent with the many topographic complexities within the 1,060-km periphery, and that diameter is favored here as the excavation diameter of Crisium.

Another fact favoring a large excavation is the probable great extent of the ejecta and secondary craters. Crisium-secondary craters extend as far as 900 km northward and 1,100 km southward of the basin center (pl. 7; figs. 9.3–9.6). One group of secondaries seems to be superposed on the crater Zeno, which, in turn, is superposed on the Humboldtianum basin (fig. 9.4). The extent of the continuous ejecta is less clear. Hummocky terrain occupies some sectors close to the basin, as does the somewhat similar, though topographically sharper, Montes Rook Formation of Orientale (fig. 9.11; Scott and Pohn, 1972; Scott and others, 1972; Head, 1974b; Wilhelms, 1973, 1980). Crisium ejecta probably covers terrain in the poorly photographed zone east of Crisium, which is smoother than the adjacent more heavily cratered pre-Nectarian terrane (fig. 9.4). An extent of Crisium ejecta as great as 1,600 km east of the Crisium center is suggested by a concentration of radially arranged craters 10 to 20 km in diameter.

The observed pattern of Crisium ejecta is three-pronged, rather like that of the crater Messier (fig. 3.11A). A further similarity with Messier is the elliptical outline of Crisium, elongate approximately east-west. The east sector of the mare-bounding ring is almost missing (fig. 4.3Z). An oblique impact from the west, possibly by a fragmented body, is suggested as the cause of these asymmetries (table 4.4).

Luna 20 samples

The Luna 20 mission returned a small core of regolith material from a part of the Crisium flank that is outside the mare-bounding rings and that is here considered to lie inside a buried part of the main rim (pls. 3, 7; fig. 9.5). Most of the fragments are of aluminous ANT rock not unlike the Apollo 16 suite; some are of low-K KREEP composition (Prinz and others, 1973; Taylor, G.J., and others, 1973). They have been dated at 3.84 ± 0.04 aeons (Podosek and others, 1973), an age consistent either with a late Nectarian relative age of Crisium or with an Imbrian age of superposed Imbrium-basin or Imbrian-crater material. In view of the small size of the sample and the difficulties encountered in dating the larger Apollo 16 and 17 sample suites, no firm conclusions should be based on this age despite its apparent analytical precision.

SERENITATIS BASIN

General features

Although Mare Serenitatis is one of the Moon's most conspicuous circular maria (fig. 1.7), the Serenitatis basin is indistinct and was not even listed as a basin by Hartmann and Kuiper (1962). Its west sector has been nearly destroyed by the Imbrium basin, and its north rim is covered by hummocky Imbrium ejecta (Alpes Formation; figs. 1.7, 10.12). The map pattern of the north rim and the presence of two mascons suggest that Serenitatis consists of a smaller northern basin and a larger southern basin (Baldwin, 1963, p. 320; Scott, 1972a, 1974; Reed and Wolfe, 1975; Wolfe and Reed, 1976; Head, 1979c). Although two independent impacts have been proposed, the considerable evidence for multiple simultaneous crater impacts (see chap. 3) and basin impacts (Humboldtianum, Moscoviense, possibly Crisium) suggests such an origin here as well. Montes Haemus, the south topographic rim of Serenitatis or "South Serenitatis," is conspicuous, though covered by lineate and smooth Imbrium ejecta (figs. 1.7, 9.12).

The east sector is of most interest because part of it was sampled by Apollo 17. The terra east of Mare Serenitatis consists of angular massifs that are partly concentric and partly radial to the mare (fig. 9.13). A general ring form is visible, but no single ring stands out. The largest massifs surround the Taurus-Littrow valley, in which Apollo 17 landed. The vague ring containing these massifs resembles the outer Rook ring of Orientale (Head, 1974b, 1979c; Reed and Wolfe, 1975; Solomon and Head, 1979). Although parts of an indistinct ring east of the Littrow massifs and ending near the crater Vitruvius (fig. 9.3) are higher than the massifs (Head, 1979c), this "Vitruvius front" is not sharply delineated like Montes Cordillera of Orientale. Therefore, eastern Serenitatis resembles the irregularly structured western Orientale more closely than it does the regularly ringed eastern Orientale (fig. 4.1). As is common for large basins, the east boundary of the excavation cavity of Serenitatis is thus poorly defined.

One possibility is that the Vitruvius front is the excavation boundary; small amounts of material may have been ejected from the zone between the Littrow and Vitruvius rings, as proposed above for Crisium and in chapter 4 for other large basins. The most straightforward interpretation, however, is that the excavation boundary passes along the east edge of Mare Serenitatis through the largest massifs of the Littrow ring (figs. 5.17, 9.12, 9.13). I tentatively favor this interpretation despite the similarity of the Littrow ring to the outer Rook ring, which I believe lies inside the basin excavation. A diameter of about 740 km—190 km smaller than Orientale—fits the largest Littrow and Haemus massifs. This identification of the rim is strengthened by Whitaker's (1981) interpretation of the Vitruvius ring as a part of the middle ring of the Procellarum basin (pl. 3; fig. 9.11)—an interpretation that also explains why the terra east of Serenitatis is so irregular and why no relation of rings to basins is obvious.

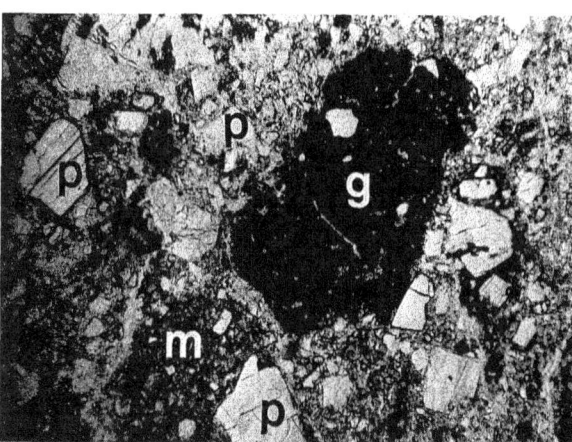

FIGURE 9.10.—Feldspathic fragmental breccia (sample 67015) collected from rim of North Ray Crater.
  A. Whole sample. Melt-rich clasts are dark; light material includes feldspathic matrix and clasts. Sides of cube (letter N), 1 cm. Courtesy of Lunar and Planetary Institute, Houston, Tex.
  B. Thin section of sample 67015,88. Seriate-textured matrix consists of cataclastic plagioclase and minor pyroxene. Clasts include glassy-matrix breccia (g), clast-laden, crystalline impact-melt breccia (m), and monomineralic plagioclase (p). Plane-polarized light; field of view, 1.5 mm. Courtesy of P.D. Spudis.

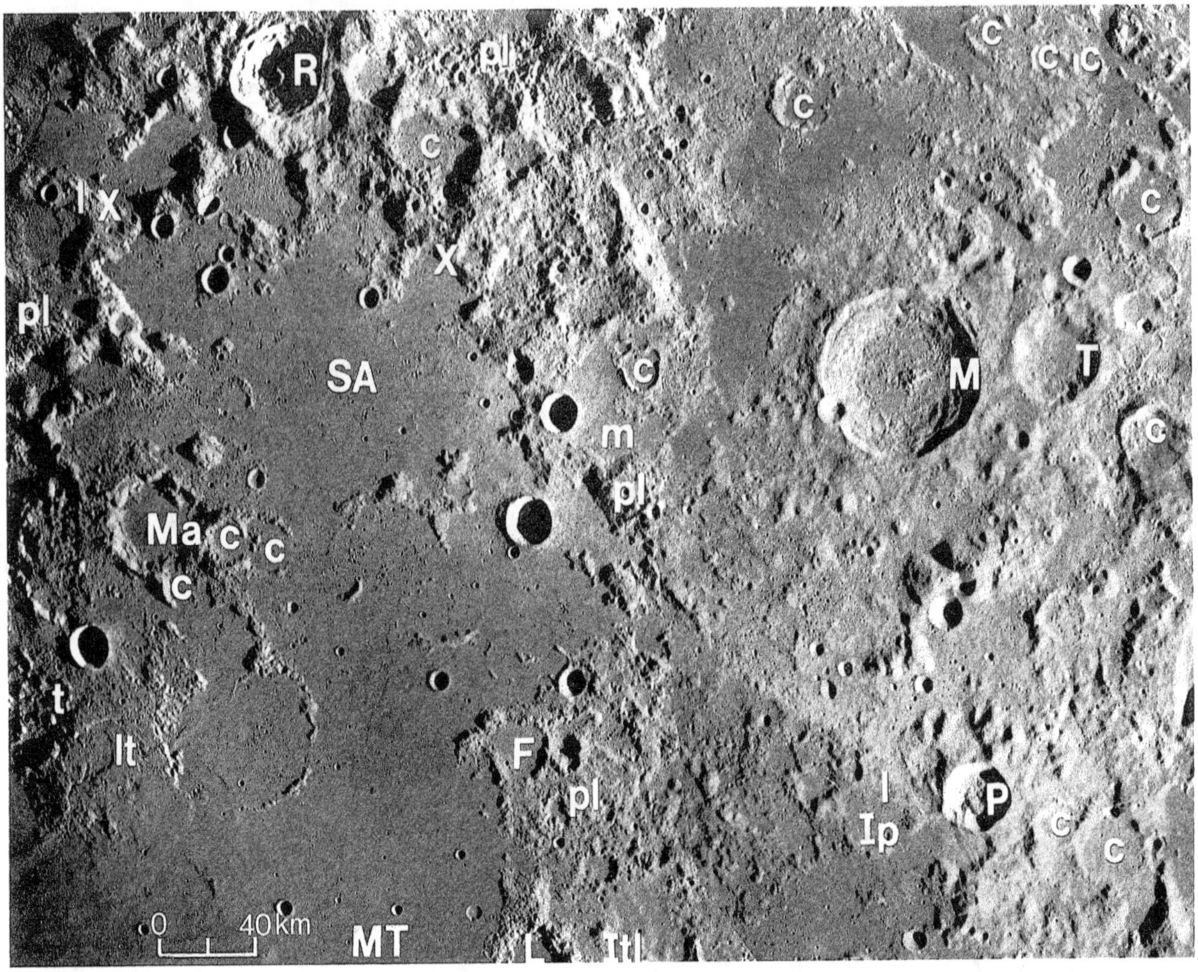

R
pl
c
c
c
c
c
c
I
X
X
pl
SA
c
M
T
m
pl
c
Ma
c
c
c
t
It
F
pl
I
Ip
P
c
c
0
40 km
MT
L
Ttl

A

M
Pe
P
Pi

B

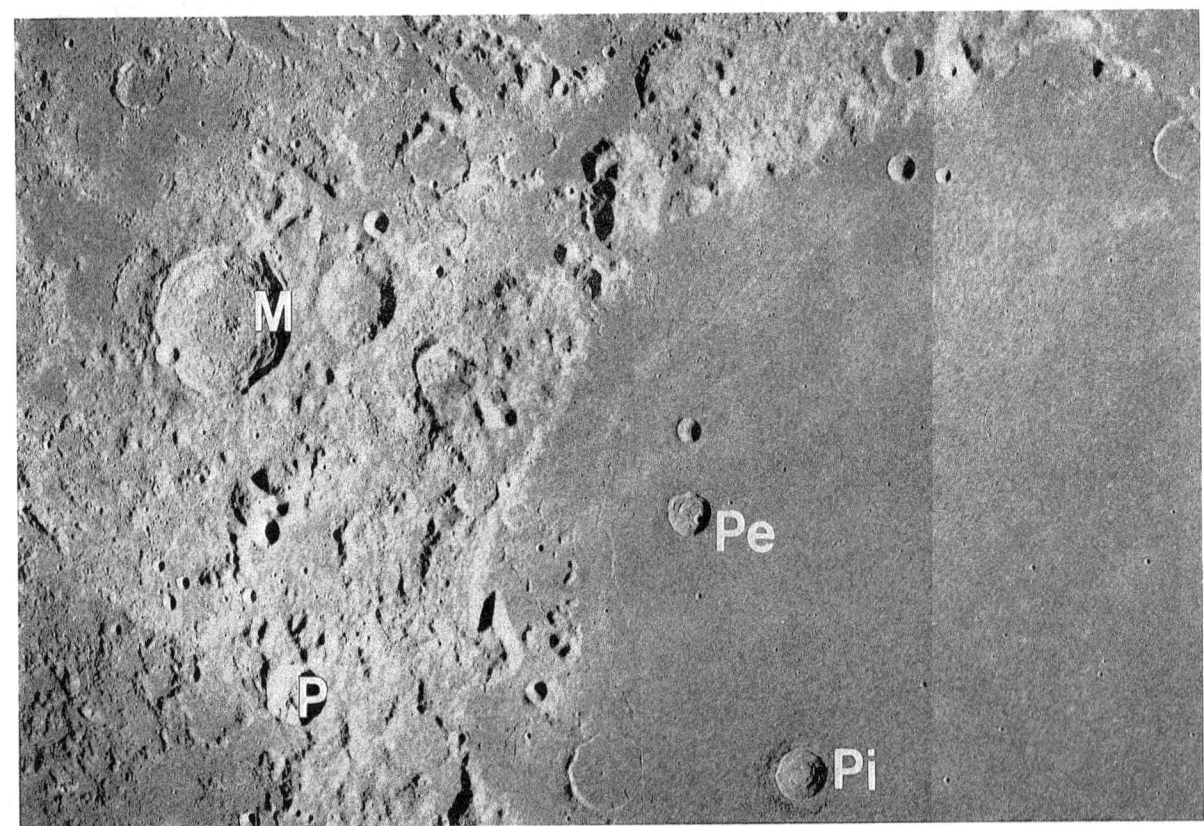

Relative age

Serenitatis was long considered to be an old basin because of its highly degraded appearance (Stuart-Alexander and Howard, 1970; Head, 1974b). Numerous degraded craters in the terra east of the mare were thought to be superposed on the basin (figs. 1.7, 5.17, 9.11, 9.14). The interpretation of these craters and the position of the rim are critical to the important question of the relative age of the Serenitatis basin. The clustering and nearly subequal size of some of the craters suggest an origin as secondary craters of the Imbrium basin (for example, Littrow; figs. 9.11, 9.13; Wilhelms, 1980). Almost all the other craters resemble the degraded prebasin craters visible close to the rims of the Orientale and Nectaris basins (figs. 4.4, 7.9). Therefore, these craters are older than Serenitatis, provided that the Serenitatis rim lies west of the craters near the edge of the mare. Only three primary craters larger than 20 km in diameter visible east of Serenitatis are clearly younger than Serenitatis and older than Imbrium (Kirchhoff, Maraldi, Römer A; figs. 9.11, 9.14). Le Monnier (61 km, 27° N., 31° E.; fig. 6.2B) also seems likely to lie in this intermediate stratigraphic position, but it is only a "ghost" crater whose age is uncertain.

Several additional lines of evidence are consistent with a relatively young age of the Serenitatis basin. (1) The near-destruction of the west sector by the Imbrium basin, a clear indicator of relative age, has no significance for absolute age; the two basins could be almost contemporaneous. (2) Serenitatis secondaries may be superposed on Crisium ejecta and the Crisium rim (figs. 9.14, 9.15; Wilhelms, 1976). (3) No Crisium ejecta is texturally evident near the Taurus-Littrow massifs (figs. 9.11, 9.13); if Serenitatis is not the younger of the two basins, the Crisium pattern of ejecta must have been highly asymmetric (which, in fact, is possible, as shown by the analogy with Messier). (4) The substantial thickness of the basalt in Mare Serenitatis indicates little filling or isostatic rebound since formation of the Serenitatis basin. (5) Apollo 17 astronaut R.E. Evans described the massifs of Serenitatis as much fresher and more boulder-strewn than those of Crisium (Evans and El-Baz, 1973, p. 16); although morphology has not proved to be a very precise age indicator for basin ages, at least this evidence helps counter reliance on Crisium's apparently fresh overall morphology as an indicator of its youth.

In summary, the roughness of the terra east of Serenitatis is due to Procellarum and Serenitatis massifs, Imbrium secondaries, Imbrium deposits (fig. 9.11), Serenitatis secondaries, and pre-Serenitatis craters. None of these features establishes an old age for the basin. A relatively young Nectarian age for Serenitatis is favored here.

Petrology of Apollo 17 samples

The Apollo 17 landing site (figs. 9.16, 9.17) was chosen mainly to provide samples of terra outside strong influence of the Imbrium basin and in a known geologic context (Scott and others, 1972; Hinners, 1973). Petrologists, geochemists, and geochronologists held out the hope that primitive terra would be sampled. Geologists wanted information on basins other than Imbrium, on basin-formation processes in general, and, as always, on dates of specific events. The mare and dark-mantling materials were also to be sampled (see chaps. 5, 11), but Serenitatis-basin materials were the main objective.

FIGURE 9.11.—Mare Crisium and circum-Crisium terrain. Primary-impact craters common to both scenes include Macrobius (M; 64 km, 21° N., 46° E., Lower Imbrian) and Proclus (P; 28 km, late Copernican; compare fig. 3.33). Trough concentric with Crisium is west (left) of Macrobius. Hummocky material, possibly ejected from Crisium or Serenitatis basin, is between Proclus and Mare Tranquillitatis (MT).

A. Nectarian primary craters include Maraldi (Ma; 40 km, 19° N., 35° E.) and Tisserand (T; 37 km); Römer (R; 40 km) is Copernican. Craters labeled "c" are believed to be secondary-impact craters of Serenitatis or Imbrium basin because of compound shape, alignment with basin radials, varying, diverse morphology of ridges, grooves, and mounds, and concentration in size range 10-20 km. Deposits of Imbrian age (Ip, Itl) and fine-scale morphology indicated by other symbols (I, It, pI, t), possibly related to Imbrium basin, are superposed on rugged features of Crisium and Serenitatis basins (Wilhelms, 1980). SA, Sinus Amoris; MT, Mare Tranquillitatis; m, other mare patches. Rugged terra along north-south midline of photograph is probably part of Procellarum basin (pl. 3). From Wilhelms (1980, fig. 19). Apollo 17 frames M-195 (right) and M-302 (left).

B. Crisium-basin massifs, continuous in west and discontinuous (but poorly illuminated) in east. Eratosthenian craters Peirce (Pe; 19 km, 18° N., 53.5° E.) and Picard (Pi; 23 km, 15° N., 55° E.) penetrate mare (see chap. 5). Mosaic of Apollo 17 frames M-274, M-278, M-281, M-286, M-289, M-293, and M-294 (from right to left).

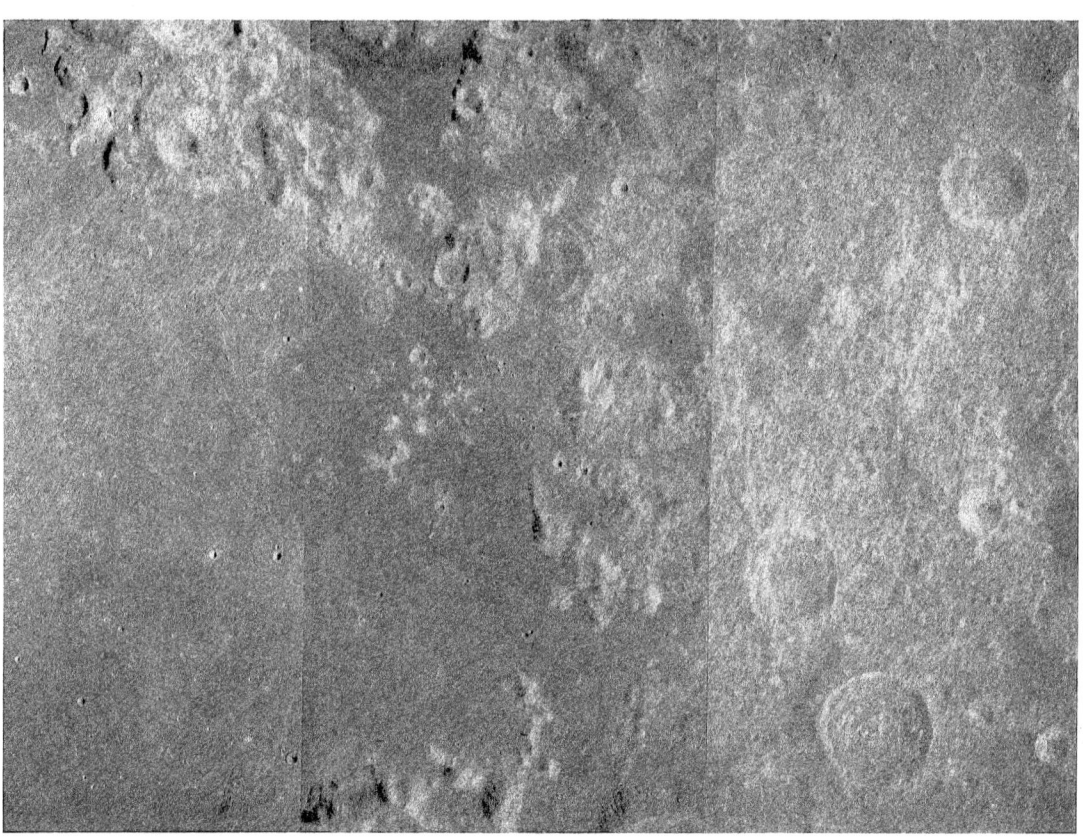

B

Most of the sampled material is fragment-laden impact melt with a wide variety of textures but a narrow range of chemical compositions. Matrices are rich in low-K KREEP (Ryder and Wood, 1977; Winzer and others, 1977). Their prevalent textures are the types generally called poikilitic, ophitic, or subophitic, which indicate igneous-like crystallization from hot melts (fig. 9.18; Simonds and others, 1974). Other textures are aphanitic, crystallized from rapidly cooling melts (fig. 9.19; James and others, 1978). Most lithic clasts are of granulitic breccia, but clasts of Mg-suite cumulates are unusually abundant (for example, norite sample 77215 from a 2-m boulder, 156-g troctolite sample 76535, and dunite sample 72415–72418; table 8.4). Many of the Mg-suite cumulates probably came from a single layered pluton (Dymek and others, 1975; Jackson and others, 1975; James, 1980). Multiple samples were collected from five melt-rock boulders, three of which were sampled at station 2 on a colluvial apron at the base of South Massif, and one each at stations 6 and 7 on basal aprons of North Massif (Dymek and others, 1976). All but one of the boulders left telltale tracks indicating their source ledges, which crop out two-thirds of the way up the massifs (fig. 9.20). Smaller samples of fragment-laden melt rock were taken from station 3, on a light-colored landslide derived from South Massif (figs. 6.1, 9.13, 9.16, 9.19). Some material of an additional terrain, the "sculptured hills," may have been collected at station 8. The sculptured hills have a hummocky surface resembling the Montes Rook Formation at Orientale or the Alpes Formation of Imbrium. The amounts collected are too small, however, to assure that they fairly represent that unit (Wolfe and others, 1981).

Emplacement process

The fragment-laden melt rocks have generally been interpreted as part of the Serenitatis-basin ejecta (Wood, 1975d; Winzer and others, 1977). This interpretation is consistent with the similarity of the massifs to those of Montes Rook, many of which are surrounded by the probably melt-rich Maunder and Montes Rook Formations (see chap. 4). The ledges presumably compose either the massifs proper or a thick superposed deposit. Superposition of Serenitatis ejecta is consistent with photogeologic observations here and at Orientale and Imbrium (figs. 4.4E, 10.7). However, the abundance of Imbrium-radial material in the vicinity is consistent with superposition of Imbrium ejecta on the Serenitatis massifs (fig. 9.13; Wilhelms, 1980).

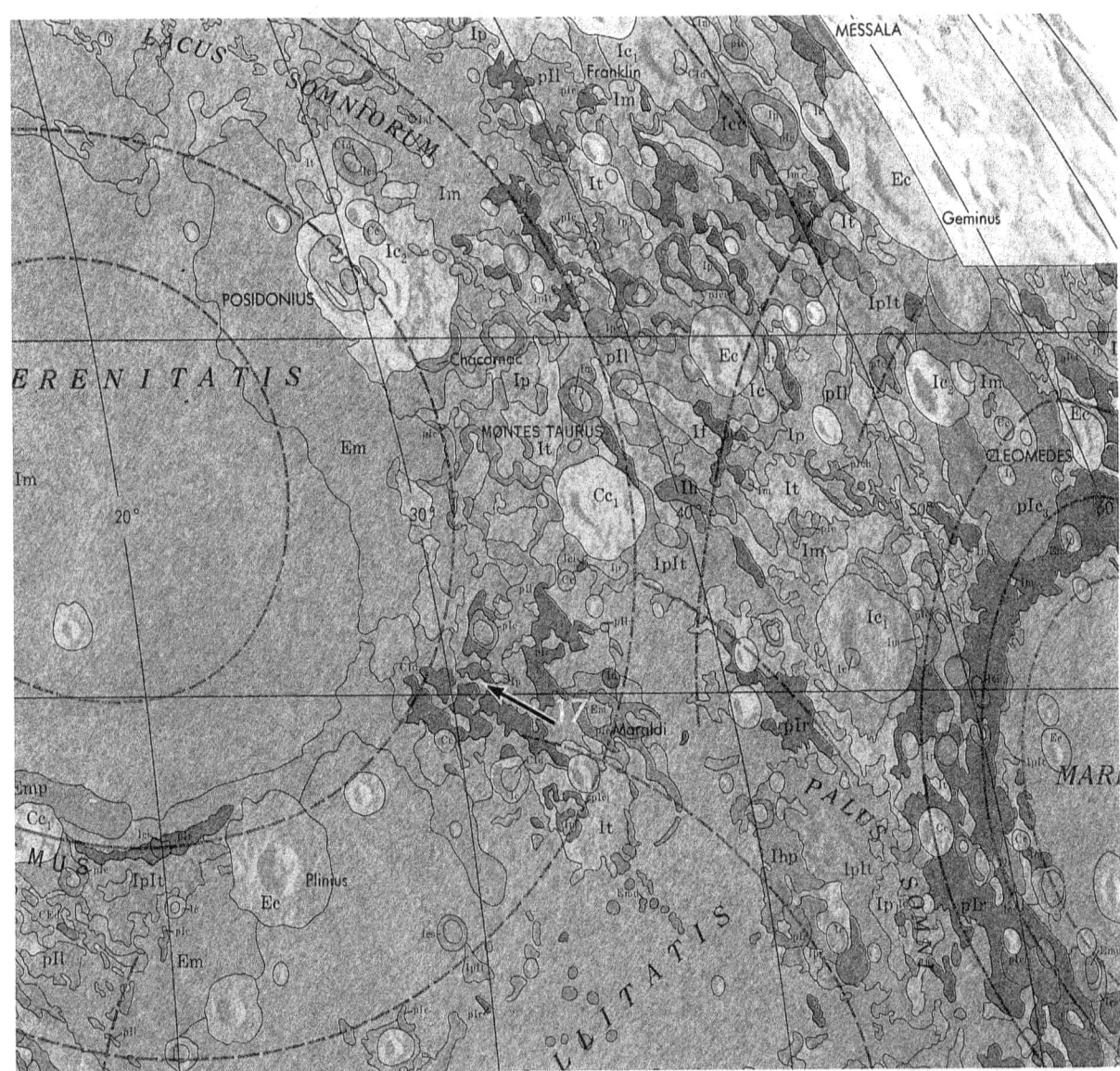

FIGURE 9.12.—Part of geologic map of nearside (Wilhelms and McCauley, 1971), including Apollo 17 landing site (arrow). Site is on unit Cld, dark-mantling material of Imbrian to Copernican age, shown by mission results to be Imbrian (see chap. 11). Adjacent massifs (dark color, pIr, pre-Imbrian rugged material) are parts of Serenitatis-basin ring, interpreted here as main basin ring (table 4.1), which continues as Montes Haemus near left edge of map. Crisium-basin massifs (same color and symbol) are at right edge of map (compare fig. 9.11). Possible rings of Tranquillitatis basin are also shown. Ages of mare units (Em, Im) have been partly revised since publication of map (see chap. 11). Lower left corner of figure adjoins upper right corner of figure 9.8.

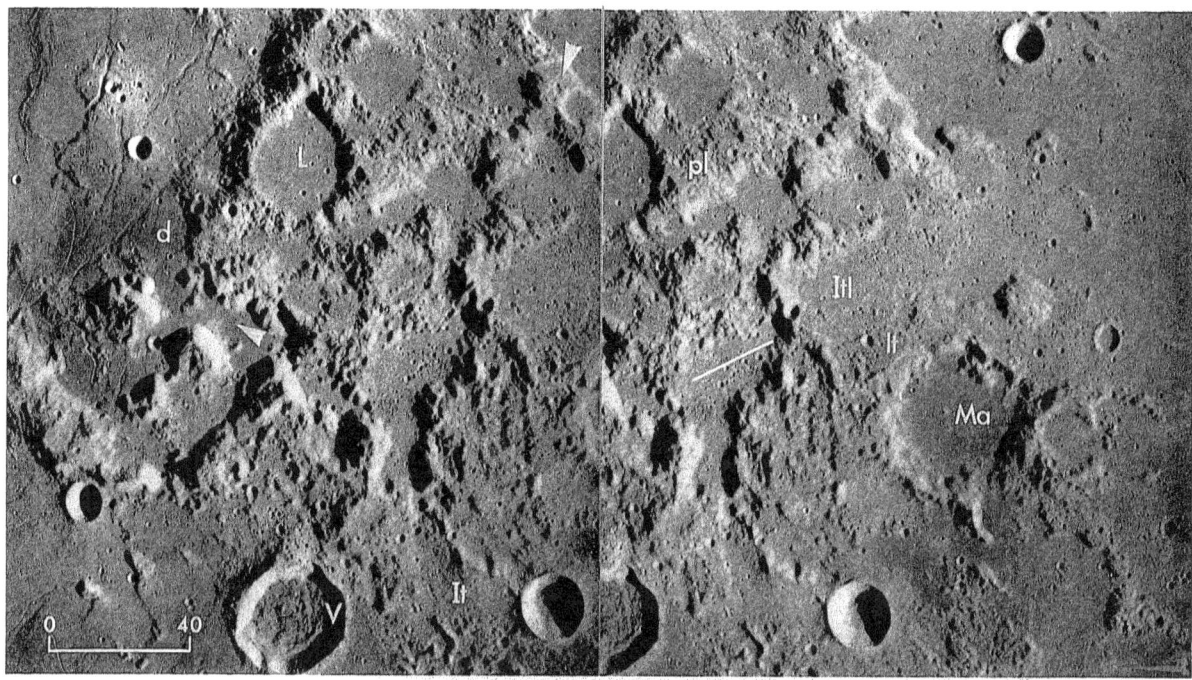

FIGURE 9.13.—Apollo 17 landing site (lower arrowhead; compare figs. 2.5D, 6.2B). South Massif is below and left of arrowhead; North Massif is above arrowhead. d, dark-mantling material. Units radial to, and believed related to, Imbrium basin (It, Itl, lt, pl; fig. 9.11A; Wilhelms, 1980) are superposed on much of rugged terrain but not on massifs. Lineations parallel to white line are not obviously related to any basin. L, Littrow, possible Imbrium-basin secondary crater. Ma, Nectarian crater Maraldi (compare fig. 9.11A); V, Upper Imbrian crater Vitruvius. Ragged line of massifs extending between Vitruvius and upper arrowhead is "Vitruvius front" (Head, 1974b). From Wilhelms (1980, fig. 18); overlaps with figure 9.11A. Stereoscopic pair of Apollo 17 frames M–444 (right) and M–446 (left).

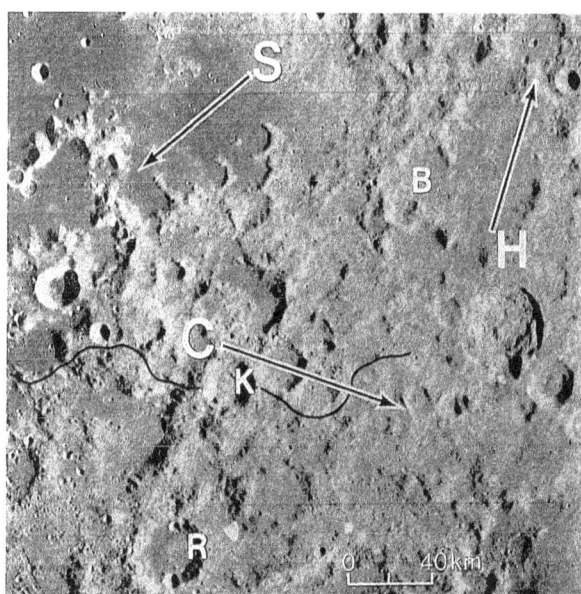

FIGURE 9.14.—Terrain northeast of Serenitatis (arrow S), northwest of Crisium (arrow C), and south-southwest of Humboldtianum (arrow H). Craters above (north of) black line are probable Serenitatis-basin secondaries, as indicated by radiality of clusters and associated linear deposits. Lineate terrain south of black line is not closely radial to any basin but is subradial to Imbrium and Crisium. Humboldtianum-basin secondary origin of irregular chain or compound craters around arrow H (including Berzelius, B) is suggested by approximate radiality. Kirchhoff (K; 25 km, 30° N., 39° E.) and Römer A (R; 35 km, 28° N., 37° E.) are probably the only post-Serenitatis, pre-Imbrian primaries in view. From Wilhelms (1976, fig. 8). Orbiter 4 frame H–74.

FIGURE 9.15.—Southwest rim of Crisium basin. Large crater groups aligned along arrow are probably secondaries of a basin; they are 1,000 km and 2,000 km from centers of Serenitatis and Imbrium basins, respectively. Degraded morphology favors a Serenitatis source; superposition then shows that Crisium is the older basin. P, Proclus (28 km; compare fig. 9.11). Apollo 11 frame H–6230.

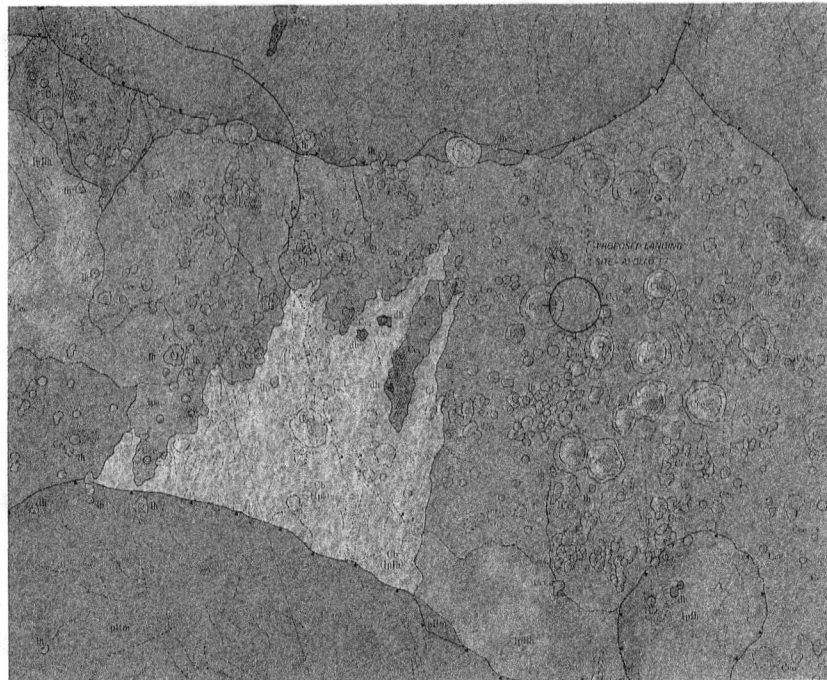

FIGURE 9.16.—Premission geologic map of Apollo 17 landing site by B.K. Lucchitta (in Scott and others, 1972). Landing took place within proposed circle. Sampled units, in order of decreasing age: pItm, pre-Imbrian terra-massif material; IpIh, Imbrian or pre-Imbrian hilly material ("sculptured hills"); Ips, Imbrian smooth plains material (subfloor basalt; see chap. 11); Ec, Eratosthenian crater material; Cce, craters now believed secondary to Tycho (chap. 13); Cb, bright Copernican material (light mantle or landslide); dh, dark-halo craters (now known to be Copernican impact craters; see chap. 13).

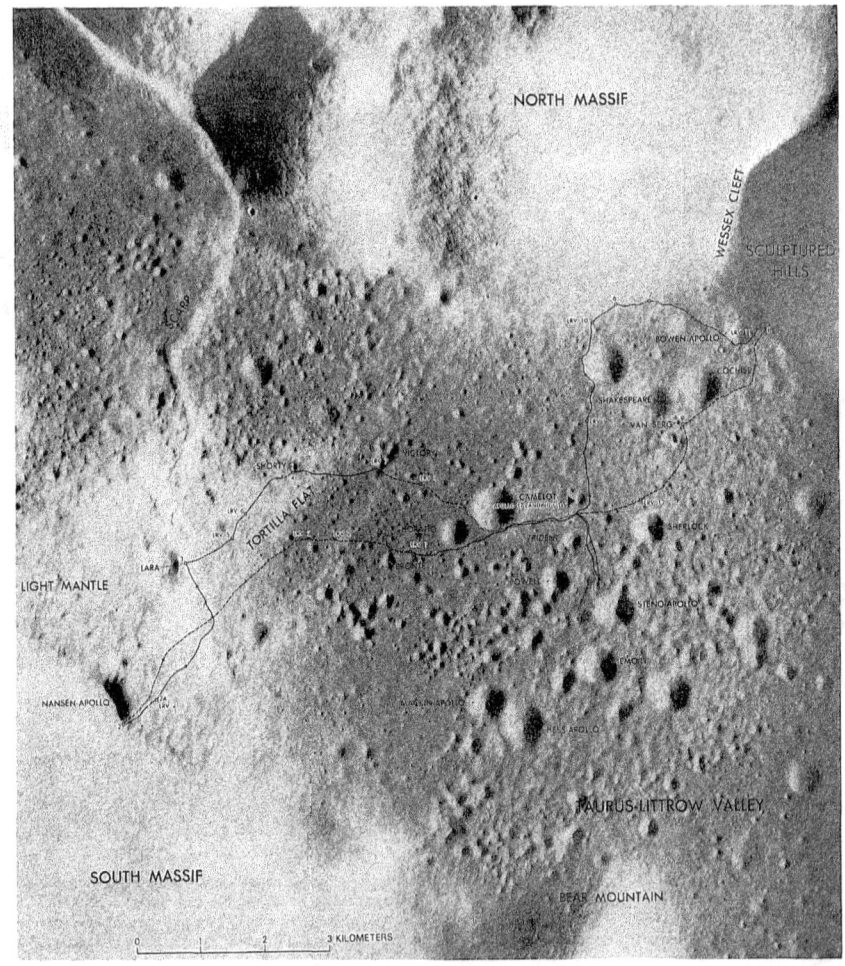

FIGURE 9.17.—Astronaut traverses and locations of sampling stations at Apollo 17 landing site.

The general chemical uniformity of the sample suite has led most of the authors cited above to conclude that the Apollo 17 samples were emplaced in a single impact, whereas others cite differences in chemistry, clast content, or apparent assembly process as evidence for multiple impacts (Ryder and others, 1975b; Chao and others, 1976; Minkin and others, 1978; Spudis and Ryder, 1981). The argument that impact melting and mixing of facies homogenize target materials of terrestrial craters (Grieve and others, 1974, 1977; Simonds and others, 1976a, b; Phinney and Simonds, 1977; Grieve and Floran, 1978) is commonly used to defend the multiple origins of lunar sample suites evaluated as "different." Extensive mixing does occur, but not total homogenization. Only the zone near the projectile is totally melted; in impacts of large, relatively slow projectiles, much additional material is incompletely melted and partly preserves the inhomogeneities of the target (see chap. 3). Furthermore, segregation of constituents according to density or volatility might create diversity during impact ejection and deposition. Therefore, the observed diversity is consistent with a Serenitatis origin for all rock-size samples collected by Apollo 17.

Absolute age

A better understanding of the geology and the availability of suitable materials have led to a clearer picture of the Apollo 17 absolute ages than for the Apollo 16 samples. The Apollo 17 ages also scatter considerably (table 9.2; Jessberger and others, 1977b, p. 2578; 1978; James and others, 1978; Wolfe and others, 1981) because of incomplete outgassing and isotopic reequilibration, as is usual in terra breccia deposits. In fact, dozens of sophisticated Ar-Ar analyses by stepwise heating of the same breccia sample (73215) yielded ages spread over 300 million years (Jessberger and others, 1977b). The laser-melting technique of determining the Ar isotopes resolved apparent age differences between parts of the same sample only 0.1 mm apart (Müller and others, 1977). However, the same careful

FIGURE 9.18.—Vesicular poikilitic impact melt (sample 76215) from station 6 boulder, North Massif, Apollo 17 landing site. Probably part of Serenitatis melt-rich ejecta.
A. Hand specimen. Sides of cube, 1 cm. NASA photograph.
B. Thin section of sample 76215,55. Poikilitic texture is well developed in pyroxene oikocryst (bottom center); incipient subophitic texture is around vesicles (black). Crossed polarizers; field of view, about 2 mm. Courtesy of P.D. Spudis.

FIGURE 9.19.—Slab cut through center of aphanitic, clast-laden impact melt (sample 73255) from station 3, Apollo 17 landing site (James and others, 1978). Vesicular rind and nonvesicular interior suggest origin as melt bomb.

TABLE 9.2.—^{40}Ar-^{39}Ar ages of Apollo 17 terra samples

Stations (Wolfe and others, 1981): 2, at contact between light mantle (landslide) and north base of South Massif; 3, on light mantle near lower slope of Lee-Lincoln scarp; 6, on south slope of North Massif; 7, near base of North Massif, 0.5 km east of station 6.
Samples (Wolfe and others, 1981): All are from boulders or are impact-melt samples larger than 1 cm. 722-, station 2, boulder 1; 724-, station 2, boulder 3; 732-, station 3, rocks collected from rim of 10-m-diameter crater; 760-, station 6, subboulder 5; 762-, station 6, subboulder 1; 763-, station 6, subboulder 2; 77-, station 7; numbers 770-, 771-, and 772- each refer to a separate sample from a 3-m boulder.
Lithologies: Samples 72255, 73215, and 73255 are light gray, others are "blue-gray" or "green-gray"; the various colors are commonly thought to derive from different source beds.
Ages: Calculated using International Union of Geological Sciences (IUGS) decay constants (Steiger and Jäger, 1977).
References: CT76, Cadogan and Turner (1976); E79, Eichhorn and others (1979); H78, Huneke (1978); L75, Leich and others (1975); M77, Müller and others (1977); J77b, Jessberger and others (1977b); J78, Jessberger and others (1978); P75, Phinney and others (1975); S79, Staudacher and others (1979); S78, Stettler and others (1978); TC75, Turner and Cadogan (1975)

Station	Sample	Lithology	Age (aeons)	References
2	72255	Bulk fragment-laden melt rock	3.95±0.03	L75.
	72435	do	3.86±0.04	H78.
3	73215	Felsite clast	3.86±0.04	J77b, J78.
		Melt-derived groundmass	3.91±0.03	M77, J77b.
		Bulk fragment-laden melt rock	3.92±0.04	J77b.
		Clasts of fragment-laden melt rock	3.96-4.17±0.04	J77b.
	73235	Bulk fragment-laden melt rock	3.90 +0.04/-0.08	TC75.
		do	3.92±0.04	P75.
	73255	do	3.86±0.04	E79.
		do	3.88±0.03	J78.
		do	3.89-3.92±0.02	S79.
		Felsite clast	3.89±0.03	S79.
	73275	Bulk fragment-laden melt rock	3.90±0.05	TC75.
6	76015	Poikilitic melt rock	3.87±0.04	CT76.
		Mineral concentrates	3.86±0.04 to 3.90±0.06	CT76.
	76215	Poikilitic melt rock	3.88±0.04	CT76.
	76235	Clasts	3.87-3.89±0.06	CT76.
	76255	do	3.96±0.04	CT76.
	76275	Melt rock	3.96±0.04	CT76.
	76295	do	3.89-3.90±0.04	CT76.
	76315	Clasts	3.91-3.92±0.04	CT76.
		Melt rock	3.92±0.04	CT76.
7	77075	Melt-rock dike in large norite clast	3.91±0.04	S78.
	77115	Bulk fragment-laden melt rock	3.84±0.03	S78.
	77135	Poikilitic melt rock	3.83±0.04	S78.
	77215	Plagioclase from norite clast	3.92±0.03	S78.

application of sophisticated Ar-Ar techniques that disclosed the problems also probably disclosed the youngest materials that have been completely reequilibrated—the aphanitic groundmass of a felsite clast from station 3 (sample 73255; Eichhorn and others, 1979; Staudacher and others, 1979). The age of this material is 3.87±0.04 aeons.

The applicability of these precise measurements to Serenitatis has been questioned because of the uncertain provenance of the best-studied samples, which were collected at station 3 from float on the landslide and not from a known source bed, as were most of the boulders. This provenance might suggest superposition on the massifs of a post-Serenitatis deposit, possibly from Imbrium. Sample 73255 resembles a bomb that could have arrived as an isolated ballistic projectile from Imbrium or elsewhere (Spudis and Ryder, 1981). Spudis and Ryder (1981) cited the homogenization model as showing that the differences between the station 3 samples and the boulder samples are too great for both groups to have been assembled by the same impact.

The consortium responsible for studying the station 3 materials, however, considered a genetic relation of those materials to the boulder samples to be established by similar compositions (James, 1976; James and others, 1978). Moreover, one boulder sample from station 2, 72435, is coeval with the youngest age from sample 73255, 3.86±0.04 aeons (Huneke, 1978). Other boulder samples yield older ages, but so do parts of the aphanatic samples, as noted. Therefore, an age of 3.86 or 3.87 aeons is accepted here as that of the Serenitatis basin. This age is consistent with tentative evidence for the relative youth of Serenitatis within the Nectarian System.

OTHER BASIN AND CRATER MATERIALS

Basins

Twelve basins, including Nectaris and Serenitatis, are considered to be of probable Nectarian age (pl. 7; table 9.3). Direct superpositional relations establish many age relations between pre-Nectarian, Nectarian, and Imbrian basins and among the Nectarian basins. An especially clear sequence is Apollo (pre-Nectarian)-Korolev (early Nectarian)-Hertzsprung (late Nectarian)-Orientale (Imbrian) (figs. 9.21A, C). Two, possibly three, age groups of Nectarian basins are recognized (table 9.3). Within a single group, the statistical error ranges of the counts of superposed craters overlap (fig. 9.22).

The older group consists of the Nectaris, Mendel-Rydberg, Moscoviense, and Korolev basins. The Nectaris deposits, which constitute the base of the Nectarian System, are very extensive and the superposed craters correspondingly numerous. The density of 79 craters at least 20 km in diameter per million square kilometers (fig. 9.22) can thus be taken as expressing the base of the system in size-frequency terms.

The deep burial by Orientale deposits of the Mendel-Rydberg basin has reduced the crater population on that basin (figs. 4.1, 4.3R); because the curve lies close to that for Nectaris, the Nectarian age of

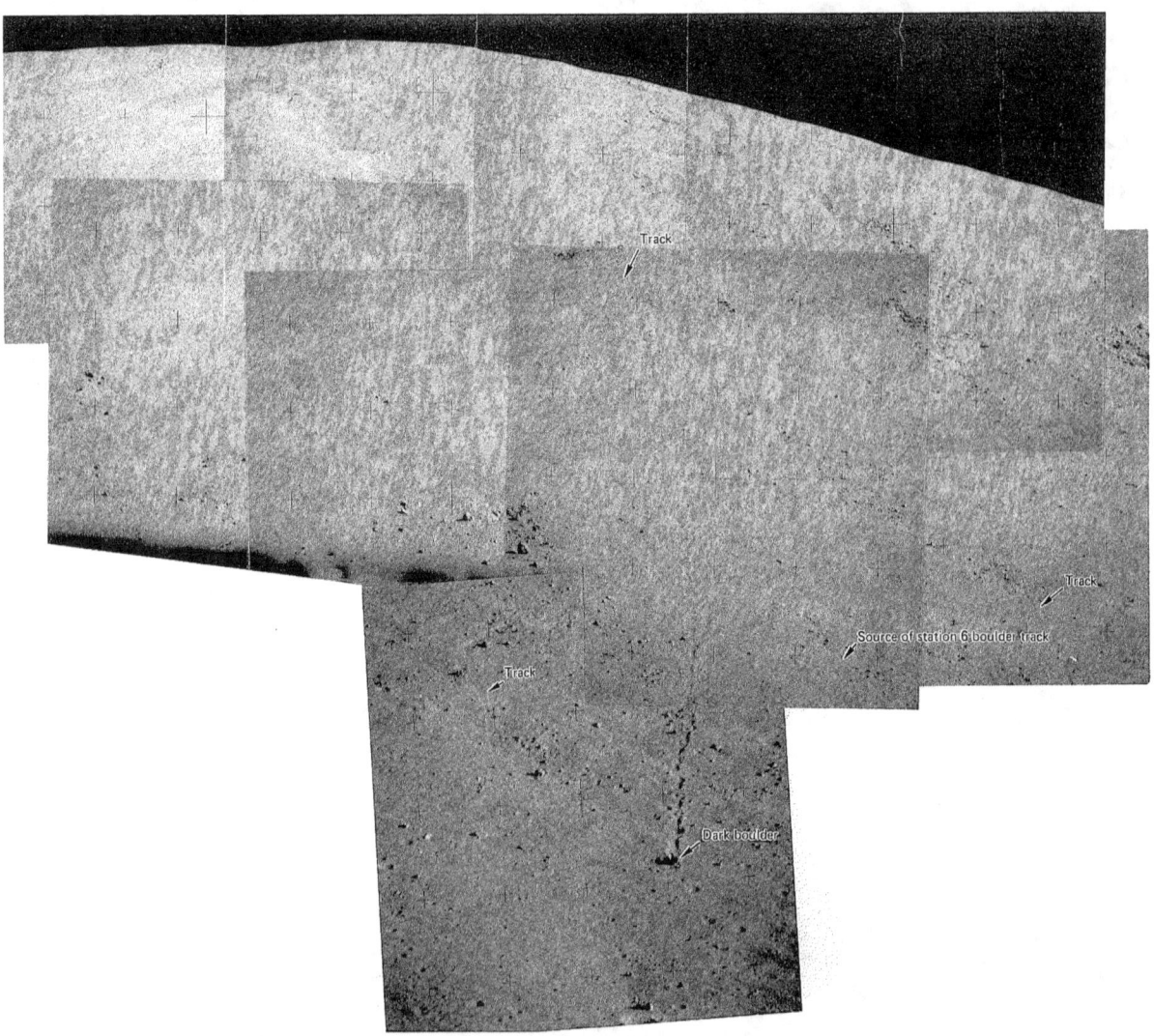

FIGURE 9.20.—Boulder tracks on North Massif (from Wolfe and others, 1981, fig. 145).

9. NECTARIAN SYSTEM

TABLE 9.3.—*Nectarian basins*

[Basins are ranked in order of increasing age; ranking is by stated density of craters or by criteria given in text; ranking of groups is more certain than ranking within a group. Diameters are from table 4.1. Parenthetical basins, existence not definitely established; parenthetical areas, buried; parenthetical densities, poor sample. Superposed craters are at least 20 km in diameter. Underlying and overlying basins determined from directly observed superpositional relations; n.d., no relations detected]

Age group	Basin	Diameter (km)	Superposed craters			Underlying basin		Superposed basin	
			Number	Area (10^6 km^2)	Density (craters/ 10^6 km^2)	Name	Figure	Name	Figure
2?	Bailly	300	3	(0.096)	(31)	n.d		Orientale	1.9.
2?	(Sikorsky-Rittenhouse)	310	2	(0.075)	(27)	South Pole-Aitken	1.5	Schrödinger	1.5.
2	Hertzsprung	570	51	.883	58	Korolev	9.21	Orientale	8.14, 9.21, 9.25.
						Birkhoff	8.14, 9.25		
						Apollo	8.15, 9.21		
						Lorentz	8.14		
						Coulomb-Sarton	8.14, 9.25		
						(Grissom-White)	8.15		
						South Pole-Aitken	9.21		
2?	Serenitatis	740	9	.108	(83)	Crisium	9.11, 9.14, 9.15.	Imbrium	1.7, 9.11.
						Humboldtianum	9.14		
2	Crisium	1,060	45	.843	53	Humboldtianum	9.4	Imbrium	9.11.
						Smythii	1.5	Serenitatis	9.11, 9.14, 9.15.
						Lomonosov-Fleming	---		
						Fecunditatis	9.5		
						Tranquillitatis	9.11		
						Marginis	9.4		
2	Humorum	820	24	.428	56	Nubium	8.10	Orientale, Imbrium	9.23.
2	Humboldtianum	avg 700	32	.515	62	Moscoviense	9.23	Imbrium	9.4, 10.6.
								Serenitatis	9.14.
								Crisium	9.4.
2?	Mendeleev	330	36	.569	63	Moscoviense, Freundlich-Sharonov, Keeler-Heaviside.	1.4	n.d	
1	Korolev	440	88	1.113	79	Freundlich-Sharonov, Apollo, South Pole-Aitken.	9.21	Orientale, Hertzsprung.	9.21.
1	Moscoviense	445	53	.609	87	Freundlich-Sharonov.	1.4	Mendeleev	1.4.
								Humboldtianum	9.23.
1?	Mendel-Rydberg	630	18	(0.247)	(73)	(Pingré-Hausen), South Pole-Aitken.	1.9	Orientale	1.9.
1	Nectaris	860	102	1.286	79	Mutus-Vlacq	8.12, 9.2	Imbrium	3.10, 9.2.
						Fecunditatis	9.2		
						Tranquillitatis	9.2, 11.1		
						Australe	1.5, 9.6		
						Werner-Airy	8.12		
						South Pole-Aitken	---		

Mendel-Rydberg is doubtful. A small patch of Mendel-Rydberg ejecta apparently is exposed southwest of the basin, near the crater Fizeau (pl. 7; Wilhelms and others, 1979).

Moscoviense and Korolev ejecta or secondary craters blanket pre-Nectarian craters over distances averaging a basin diameter and extending, in salients, more than two basin diameters (pls. 3, 7; figs. 9.21C, 9.23). Large tracts can thus be dated relative to these basins. The southwestern ejecta of Korolev is distinctly lineate (fig. 9.21D). Radial alignment of some fresh secondary chains inside Korolev suggests an Orientale origin, and so the adjacent plains may also be of Orientale origin (fig. 9.21B). The crater densities on Korolev and Nectaris are similar, but they overlap only where the Nectaris curve is kinked. Although more craters per unit area are superposed on Moscoviense than on Nectaris, most of the surplus craters are small. The curves as a whole thus indicate that both the Korolev and Moscoviense basins are younger than Nectaris.

The definite basins of the younger group are Humboldtianum, Humorum, Crisium, and Hertzsprung. Bailly and Serenitatis are probably members of this group, although they cannot be dated reliably by superposed craters. Humboldtianum ejecta covers much of the northern part of the east lunar limb (fig. 9.4). Conspicuous radial lineations and secondary craters extend 600 km south of the topographic rim, which is a clearly defined ring 300 km in radius in this sector. An exterior arc 600 km in radius is covered by the ejecta (Wilhelms and El-Baz, 1977). Humboldtianum is a double or triple basin whose parts form a figure-8 (pl. 3; fig. 9.4; Lucchitta, 1978). A ring 150 km in radius is conspicuous in both parts of this figure-8; a poorly defined ring between the topographic rim and the 150-km-radius ring is visible in the southern part. The northern part of the figure-8 consists of irregular uplifts that do not clearly define rings (Lucchitta, 1978). Crater counts (fig. 9.22) are consistent with the age relation demonstrated by superposition of Crisium secondaries on Humboldtianum (fig. 9.4).

The Humorum basin is also irregular in most sectors (figs. 4.3U, 9.24). The most complete ring, 440 km in diameter, bounds Mare Humorum (table 4.1). A conspicuous and rimlike scarp almost twice as large (410 km radius) and resembling Montes Cordillera lies south of this mare-bounding ring (fig. 9.24B). This scarp is the topographic rim in its sector and so is here considered to be the boundary of excavation, despite lesser expression in other sectors. Although an 820-km diameter is, furthermore, consistent with the considerable extent of conspicuous deposits and secondary craters southwest of the basin (fig. 9.24B), this large size cannot be confirmed because Humorum deposits are buried in other sectors. Sufficient numbers of superposed craters are exposed to date Humorum as part of the younger group but not to establish its exact place in the group. Humorum and Crisium appear to be nearly contemporaneous.

The fresh-appearing Hertzsprung helps clarify the geology of the other basins (fig. 9.25). The secondary craters extend as far as 1,300 km north of the 570-km-diameter ring. This relation and the near-completeness and rimlike morphology indicate that the 570-km-diameter ring bounds the excavation (see chap. 4). The interior rings probably formed by floor deformation by the oscillatory-uplift mechanism (stage E, fig. 4.12). Hertzsprung may be the youngest Nectarian basin. Although the ends of the cumulative size-frequency curves representing small craters lie above those for Crisium and Humorum, the large-crater end lies below (fig. 9.22). The large-diameter end is considered to be more reliable because some small craters were probably missed in the counts of poorly photographed parts of Crisium and Humboldtianum.

The craterlike Mendeleev basin is surrounded by lineate ejecta and secondary craters in the well photographed southern peripheral zone (fig. 9.26A). Because of this morphologic freshness, it is mapped as late Nectarian on plate 7. Fairly distinct crater counts suggest, however, that Mendeleev may be intermediate in age between the two main groups.

Several factors limit the further resolution of Nectarian age differences by crater counts. The small Bailly basin, the questionable Sikorsky-Rittenhouse basin, and the extensively buried Serenitatis basin present insufficient exposed surface to be dated by superposed craters. Even for the well-exposed large basins, crater counts are statistically no more definitive than for pre-Nectarian basins because relatively few craters are available for dating the Nectarian surfaces. The statistics of some of these counts could be improved by adding craters smaller than the 20-km-diameter cutoff used to prepare figure 9.22, but the necessary primary-secondary and superposed-buried distinctions cannot be made for other, poorly photographed basins.

The inability of these criteria to distinguish between some basins may reflect natural groupings. Wetherill (1977a) suggested that large impacts might cluster in time because related projectiles derived by breakup of a single large body from the outer Solar System are consumed by collisions with the Moon and terrestrial planets in relatively brief spurts (millions or tens of millions of years). The substantial gap in crater densities between the youngest large Nectarian basin (probably Hertzsprung; fig. 9.22) and the oldest Imbrian basin (Imbrium) is strikingly suggestive of such clustering of large impacts.

Craters

Many craters can be identified as Nectarian (table 9.4), and some of these craters can be dated within the system relative to the Nectarian basins (pl. 7). Craters superposed on Nectaris-basin materials but nicked by Imbrium-secondary craters provide type examples of Nectarian crater morphology (figs. 8.5, 9.1, 9.2, 9.27). Nectarian craters larger than 20 km in diameter have generally deeper floors and rougher primary topography of rim and walls (as opposed to superposed topography) than do pre-Nectarian craters. Central peaks also are more commonly exposed in Nectarian craters, although they may be submerged if subjected to deposits from a nearby crater or basin (figs. 9.21E, 9.27). Some of the largest Nectarian craters have radial ejecta textures and secondary craters (fig. 8.11). Apparently, typical pre-Nectarian crater morphologies are nowhere superposed on Nectarian surfaces, except possibly for some marginal craters that appear to be superposed on outer parts of the Korolev and Moscoviense basins, here considered to be early Nectarian (fig. 9.21D), and for a few disputed nearside craters that could be pre-Nectarian (pl. 7). Although identifiable Nectarian craters smaller than 20 km in diameter are more numerous than identifiable pre-Nectarian craters of this size (Wilhelms and others, 1978), their morphologies vary too greatly to be useful for dating isolated craters. Their profiles range from cone-shaped to flat-floored, and their numbers appear to dwindle greatly at diameters of about 5 km because of mutual obliteration near the steady state (table 7.3).

TABLE 9.4.—*Representative Nectarian craters*

[Cross rules divide diameter ranges mapped differently in plate 7: smaller than 30 km, unmapped; 30 to 119 km, interiors mapped; 120 km and larger, exterior deposits mapped where exposed. L, long axis of elongate crater; LI?, possibly Lower Imbrian; pN?, possibly pre-Nectarian]

Crater	Diameter (km)	Center (lat)	Center (long)	Figure	Remarks
Kirchhoff	25	30° N.	39° E.	9.14	---
Römer A	35	28° N.	37° E.	9.14	---
Tisserand	37	21° N.	48° E.	9.11	---
Maraldi	40	19° N.	35° E.	6.5C, 9.11, 9.13	---
Crookes D	41	10° S.	163° W.	9.21	---
Young D	46	44° S.	52° E.	8.5	Typical.
Mercator	47	29° S.	26° W.	6.9	---
Playfair	48	24° S.	8° E.	9.27	---
Schiller C	49	55° S.	49° W.	7.6	---
Kohlschütter	53	14° N.	154° E.	5.2	---
Timiryazev	53	6° S.	147° W.	9.21	Post-Hertzsprung, LI?
Epigenes	55	68° N.	5° W.	10.15	---
Korolev M	58	9° S.	157° W.	9.21	---
Congreve	58	0°	167° W.	9.21	Relatively old.
Congreve U	59	0.5° N.	171° W.	9.21	Relatively young.
Mechnikov	60	11° S.	149° W.	9.21	Pre-Hertzsprung.
Lebidinskiy	62	8° N.	164° W.	9.21	Relatively young.
Sechenov	62	7° S.	143° W.	9.21	Pre-Hertzsprung.
Abulfeda	65	14° S.	14° E.	3.16B	Typical.
Al-Khwarizmi	65	7° N.	106° E.	8.11	Do.
Adams	66	32° S.	68° E.	9.5	---
Steinheil	67	49° S.	47° E.	8.5	Typical.
Rheita	70	37° S.	47° E.	8.5, 9.1	Do.
Berosus	74	34° N.	70° E.	9.4	---
Aliacensis	80	31° S.	5° E.	3.26, 9.27	Typical.
Zhukovskiy	81	8° N.	167° W.	9.21	Relatively old.
Pitiscus	82	50° S.	31° E.	9.2	Typical.
Mersenius	84	22° S.	49° W.	9.24A	Do.
Metius	88	40° S.	43° E.	8.5	Do.
Rocca	90	13° S.	73° W.	4.4F	LI?, pN?
Inghirami	91	48° S.	69° W.	4.4G, 10.42	---
Pontecoulant	91	59° S.	66° E.	9.5, 9.6	Typical.
Kibal'chich	93	3° N.	147° W.	9.21	Post-Hertzsprung, LI?
Pitatus	97	30° S.	14° W.	1.8, 6.130, 10.30	Atypical.
Hevelius	106	2° N.	67° W.	4.4H, 6.21	Faulted.
Carnot	117	52° N.	144° W.	8.14	pN?
Alphonsus	119	15° S.	2° W.	1.8, 5.10F, 7.7, 10.27	---
Chaplygin	124	6° S.	150° E.	1.4	Typical.
Endymion	125	54° N.	57° E.	9.3, 9.4	Mare fill.
Cleomedes	126	28° N.	56° E.	9.3	Floor fractured.
Fleming	130	15° N.	110° E.	1.2, 8.13, 9.28	Typical.
Albategnius	136	11° S.	4° E.	1.8, 10.27-10.29, 10.36	---
Neper	137	9° N.	85° E.	9.4	Typical.
Longomontanus	145	50° S.	22° W.	4.6A	---
Hilbert	170	18° S.	108° E.	1.2, 1.3, 8.11	---
Gauss	177	36° N.	79° E.	1.2, 4.2D, 9.4	Floor fractured.
Clavius	225	58° S.	14° W.	4.2H, 4.6A	Typical.
Van de Graaff	234(L)	27° S.	172° E.	1.4, 3.14B, 10.32	Double, pN?
Schwarzschild	235	71° N.	120° E.	4.21, 10.6	Ring arcs.

Plate 7 shows a total of 560 unburied Nectarian craters larger than 30 km in diameter within a total area, covered by both pre-Nectarian and Nectarian deposits, of 19.53×10^6 km^2. As expected from stratigraphic relations, the crater density is higher on the pre-Nectarian terrane than on the Nectarian basin deposits (35 versus 22 craters/10^6 km^2, respectively). The density on the pre-Nectarian terrane is the maximum Nectarian density; extrapolating it to the whole Moon (area, 38×10^6 km^2) indicates that a total of 1,330 craters at least 30 km in diameter were formed during the Nectarian Period. If the number of Nectarian basins is 12, 111 such craters were formed between each basin impact; if there are 10 basins (excluding Mendel-Rydberg and Sikorsky-Rittenhouse), 133 such craters per basin were formed.

FIGURE 9.21.—Stratigraphic relations of lower Nectarian Korolev basin.
  A. Korolev (K), massifs of ancient pre-Nectarian South Pole-Aitken basin (black arrows indicate massifs in fig. 8.9), northern interior of upper pre-Nectarian Apollo basin (A; letter athwart Nectarian crater Barringer, 69 km, 28° S., 150° W.), and deposits of upper Nectarian Hertzsprung basin (H; centered in direction of white arrows). Massifs near antipode of Serenitatis basin (S) are grooved, as near Orientale and Imbrium antipodes (see chap. 10); however, grooves may be secondary craters of Korolev, Hertzsprung, and Apollo. Areas of B through E are outlined; black-and-white arrow indicates chain shown in B and is subradial to Orientale basin. Orbiter 1 frame M-28.

B. Detail inside Korolev basin, showing probable Orientale-basin secondary chain and fine Orientale-radial lineations. Large crater is Korolev M (58 km, 9° S., 157° W., Nectarian), filled by Imbrian plains deposits. Orbiter 1 frame H-38.

C. Enlargement of part of A, showing stratigraphic relations of Korolev and Hertzsprung basins and adjacent craters. Degradation of pre-Nectarian craters such as Galois (G; 207 km), Paschen (P; 133 km), and Tsander (Ts; 171 km) indicates burial by both Korolev and Hertzsprung deposits. Such Lower Nectarian craters as Mechnikov (M; 60 km) and Sechenov (S; 62 km) are too sharp to be older than nearby Korolev but are clearly overlain by Hertzsprung-basin secondary craters (trending northeast-southwest). Such fresh-appearing upper Nectarian craters as Kibal'chich (K; 93 km) and Timiryazev (Ti; 53 km) are superposed on Hertzsprung deposits. Sharp Orientale-basin secondaries (Iosc) cross all other units.

D. Korolev basin and part of periphery. Line separates inner Korolev ejecta (Nki; clearly radial southwest of basin) from outer deposits containing many secondary craters (Nko). Fresh-appearing Nectarian craters superposed on Korolev include Congreve U (CU; 59 km, 0.5° N., 171° W.) and Lebedinskiy (L; 62 km); sharper Engelhardt (E; 43 km) may be Lower Imbrian. Congreve (C; 58 km) and Zhukovskiy (Zh; 81 km, 8° N., 167° W.) are degraded Nectarian craters superposed on Korolev deposits but have morphologies and crater densities resembling those of pre-Nectarian craters. Sharply textured Doppler (D; 100 km, 13° S., 160° W.) is probably Lower Imbrian. Most plains inside Korolev are Nectarian (Np), but some lightly cratered patches are Imbrian (Ip), possibly of Orientale origin. Np and Ip may overlie basalt flows. Orbiter 1 frame M–38.

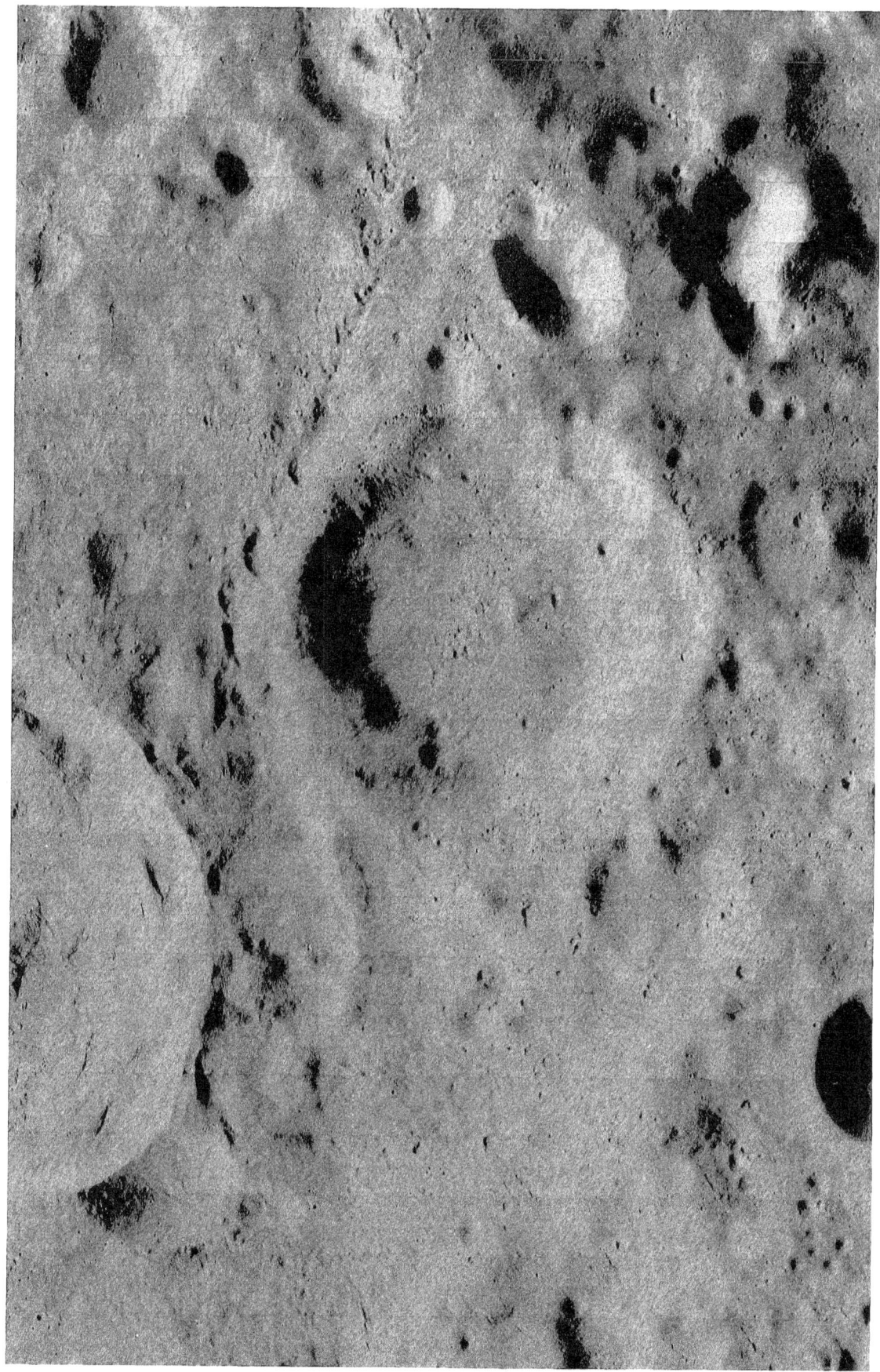

E. Crookes D (center; 41 km), overlain by deposits of Crookes (49 km, 10° S., 165° W., truncated by left edge). Superposition of Crookes D on Korolev demonstrates a Nectarian age; however, subdual because of partial burial, especially of rim sector nearest Crookes, would otherwise suggest a pre-Nectarian age. Freshness of Crookes indicates a Copernican age. Orbiter 1 frame H–30.

FIGURE 9.23.—Secondary-crater chains of upper Nectarian Humboldtianum basin (H, Mare Humboldtianum, outlined) truncate chains subradial to early Nectarian Moscoviense basin (M; arrows just inside shadowed basin rim). Because deposits thin away from source basins, visible large pre-Nectarian craters are less numerous near basins than in belt midway between them. C, Compton (162 km, 56° N., 105° E., Lower Imbrian). Orbiter 5 frame M–158.

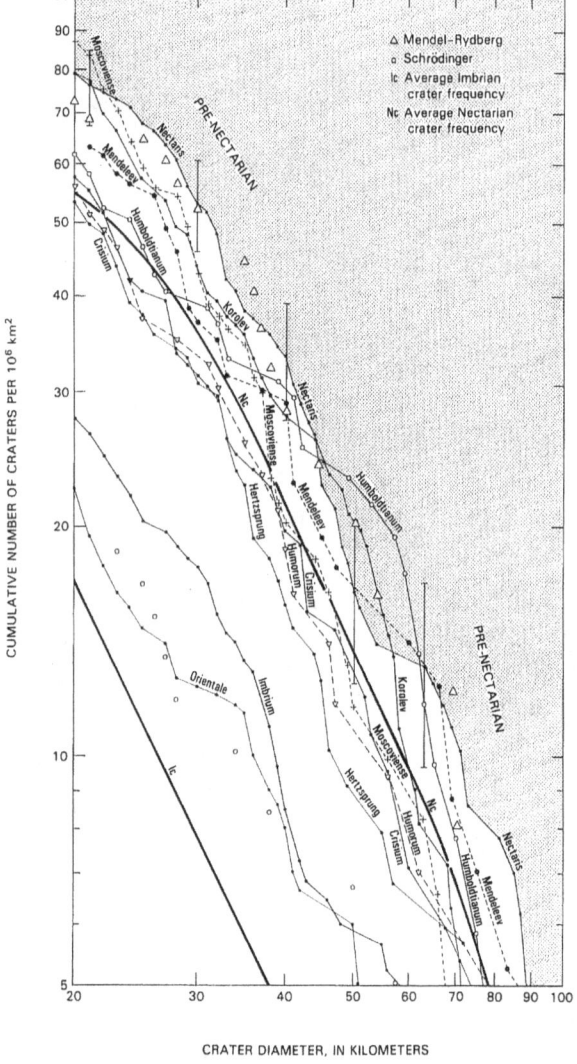

FIGURE 9.22.—Cumulative frequency distributions of craters at least 20 km in diameter superposed on Nectarian and Imbrian basins. Part of pre-Nectarian field (fig. 8.6) is in upper right. Curves for Nectarian and Imbrian craters (Nc and Ic) are from Wilhelms and others (1978). Error bars (constructed as in fig. 8.6) are for Nectaris basin; others are larger and omitted for clarity.

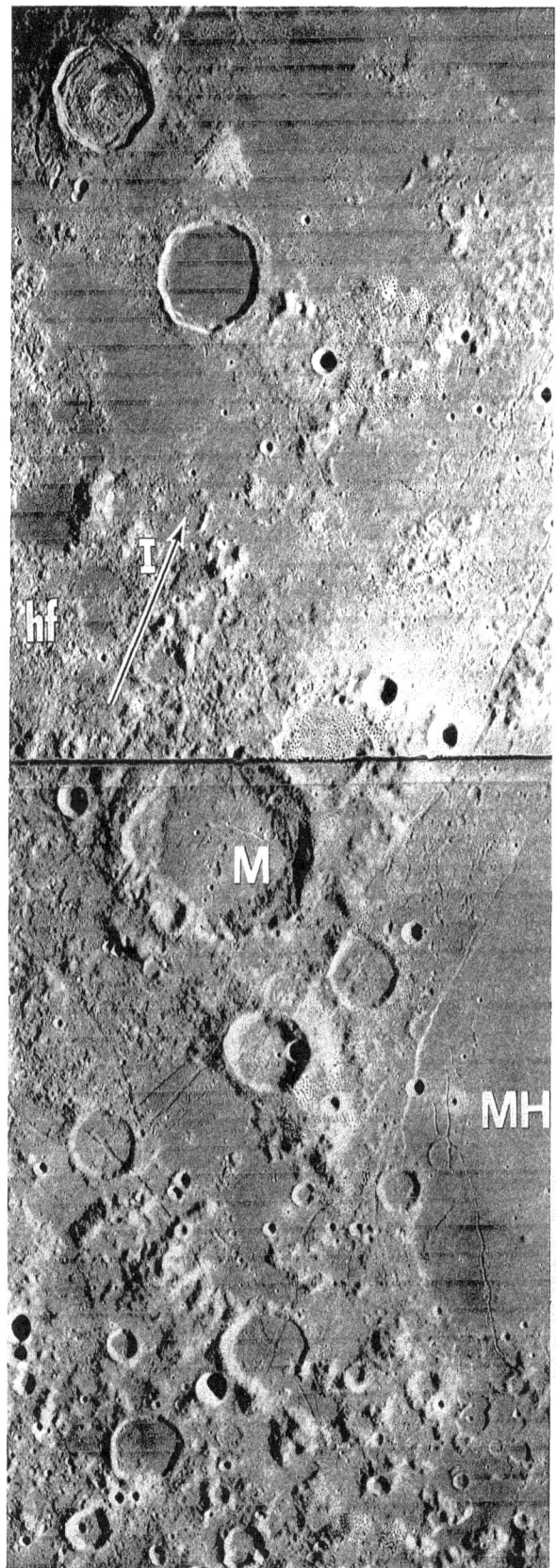

B. Terrain south-southeast of Humorum, centered in direction of arrow. Arrow marks probable secondary-impact craters of Humorum, including a subradial valley resembling Vallis Bouvard (compare fig. 4.4D). Major scarps of Humorum are at 1 and 2. H, Hainzel A (53 km, 40° S., 34° W., Eratosthenian); V, Vitello (42 km; see chap. 6). Orbiter 4 frame H–136.

FIGURE 9.24.—Upper Nectarian Humorum basin. MH, Mare Humorum.
  A. Hilly and furrowed terrain (hf) is probably a composite of Humorum basin ejecta, Imbrium-basin secondary pits and lineations, and, possibly, thin Orientale deposits; arrow (I) is radial to Imbrium. Nectarian crater Mersenius (M; 84 km, 22° S., 49° W.) is superposed on Humorum basin but is marked by Imbrium-radial lineations. Humorum-basin massifs are conspicuous south of (below) Mersenius. Orbiter 4 frame H–149.

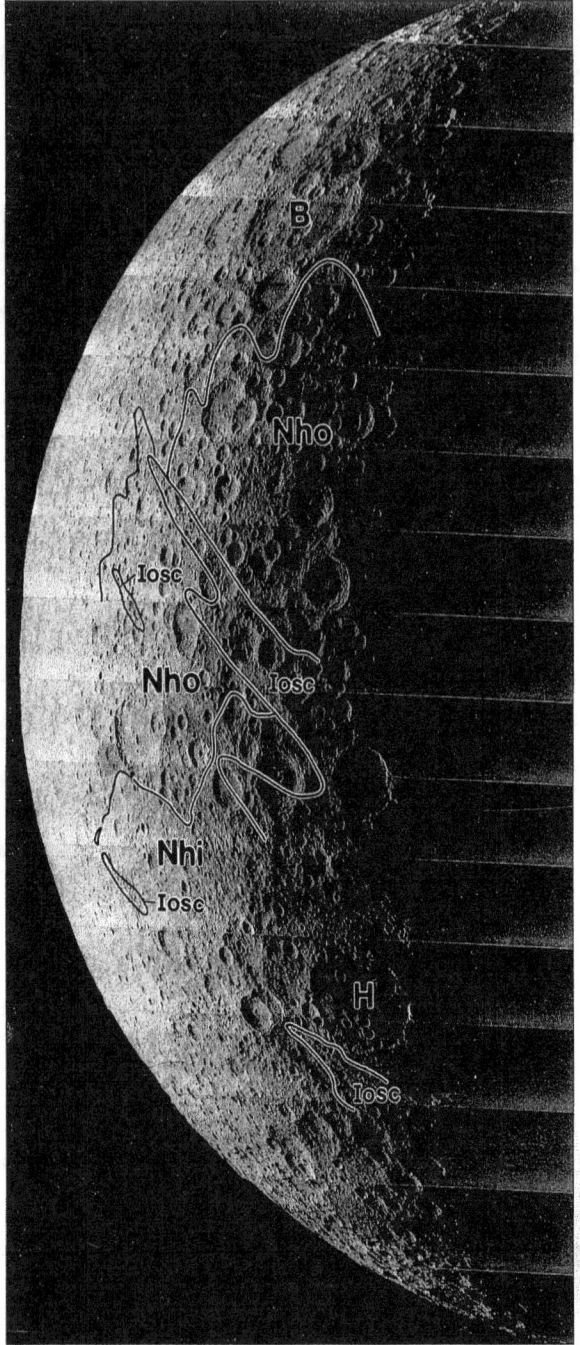

FIGURE 9.25.—Upper Nectarian Hertzsprung basin (H). Inner (Nhi) and outer (Nho) deposits are superposed on pre-Nectarian basin Birkhoff (B) and overlain by Orientale-basin secondaries (Iosc) (compare figs. 4.5, 8.14). Orbiter 5 frame M-28.

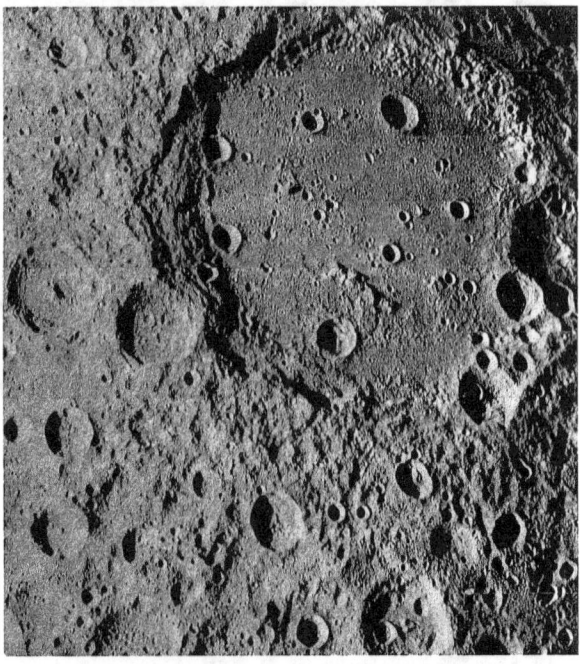

FIGURE 9.26.—Craterlike Nectarian basin Mendeleev, filled with Nectarian plains.
 A. Part of radial ejecta and secondary field south of basin (between rim and lower edge of picture). Prebasin, mostly pre-Nectarian craters diminish in number and expression near rim. Orbiter 1 frame M-136.
 B. Interior plains. Crater chain (Catena Mendeleev) is radial to, and undoubtedly secondary to, crater Tsiolkovskiy, 750 km south in direction of chain. Apollo 16 frame M-875.

FIGURE 9.27.—South-central nearside (centered at 27° S., 5° E.). Apianus (Ap; 63 km) is lightly scored by lineations radial to Nectaris basin (Rupes Altai is 400 km in direction of arrow) and thus is probably pre-Nectarian. Playfair G (PG; 94 km), Krusenstern (K; 47 km), and Blanchinus (B; 61 km) are heavily mantled by probable Nectaris deposits and thus are pre-Nectarian; their rims are also nicked by Imbrium-basin secondaries (sc). Intercrater terrain probably consists mostly of Nectaris-basin deposits. Crater Aliacensis (Al; 80 km) would appear moderately fresh if it were not degraded by deposits of Werner (W; 70 km, Eratosthenian; compare fig. 3.26); Imbrium-basin secondary craters (sc) superposed on rim constrain age of Aliacensis to Nectarian. Playfair (P; 48 km) is also moderately fresh and probably Nectarian. Orbiter 4 frame H–101.

VOLCANIC ROCKS

Volcanism began on the Moon before the emplacement of the visible maria. This conclusion was reached even before any samples were examined, based on the reasoning that generation of basaltic magmas cannot have started coincidentally with the Orientale impact. Only post-Orientale maria can have remained exposed because all earlier maria were obscured by impacts (Hartmann and Wood, 1971; Shoemaker, 1972; Hartmann, 1973; Ryder and Taylor, 1976; Basaltic Volcanism Study Project, 1981, chap. 8). The high pre-Orientale impact rate must have kept lunar surfaces so thoroughly churned that extensive basalt flows did not long remain exposed, and Orientale and each earlier large basin impact must have covered most of the Moon with debris. Nor did the impacts trigger the volcanism; Orientale would be filled more deeply if it had triggered its own volcanic fill.

This well-founded inference is confirmed by physical evidence. Mare basalt has been found as fragments in bedrock breccia deposits (table 9.5; Ryder and Taylor, 1976; Ryder and Spudis, 1980). The Apollo 14 collection contains the most numerous of these samples, possibly because it contains material ejected from the already-flooded Procellarum and Insularum basins. This collection includes four large and several small samples of high-Al basalt, identified by cotectic compositions and the absence of siderophiles as endogenic, and by other chemical characteristics as related to mare basalt (Ridley, 1975; Taylor, 1975, p. 136; 1982, p. 300). Dated samples of high-Al basalt include the 251-g sample 14053 and 45-g sample 14072, found as individual rocks, and clasts from breccia sample 14321 (table 9.5). Additional high-Al basalt was found by Apollo 14 as small clasts in breccia samples (for example, fragments in sample 14063 with intermediate Ti contents; Ridley, 1975) and as fragments in the regolith (Ryder and Spudis, 1980). Apollo 17 fragment-laden impact-melt rocks also contain scattered small fragments of high-Al mare basalt (James and McGee, 1980). One sampled boulder contains clasts of basalt that may be intermediate between KREEP and mare basalt (sample 72275; Ryder and Spudis, 1980). The fact that relatively few samples of volcanic basalt have been discovered in terra breccia deposits is not surprising in view of the small volume of the crust composed of mare basalt today (see chap. 5). A major impact on any part of today's Moon would generate breccia much richer in terra material than in mare basalt.

The pre-Imbrian basalt fragments have been dated at from 3.85 to 3.99 aeons (combined range of ages obtained by Ar-Ar and Rb-Sr methods, table 9.5; Ryder and Spudis, 1980). If the Nectaris basin formed 3.92 aeons ago, these ages span the boundary between the Nectarian and pre-Nectarian; most are Nectarian.

The extent of Nectarian and pre-Nectarian volcanism is unclear. If the heat source was entirely radiogenic (Taylor, 1982), the ancient basalt units were less extensive than the visible maria, because radioactive U, Th, and K are not abundant in the Moon. If, however, the Moon retained considerable primordial heat during its first aeon (Hubbard and Gast, 1971; Wood, 1972b; Toksöz and Johnston, 1974; Hubbard and Minear, 1975; Ryder and Taylor, 1976; Solomon and Head, 1979, 1980), only a small rise in temperature might have been sufficient to melt considerable magma.

Photogeologic evidence is consistent with considerable premare volcanism. Scattered patches of Nectarian plains have been identified in many parts of the terrae on the basis of crater densities (pl. 7; figs. 9.26B, 9.28, 9.29; Neukum and others, 1975b; Wilhelms and El-Baz, 1977; Stuart-Alexander, 1978; Lucchitta, 1978; Wilhelms and others, 1979). Some Nectarian plains, including the concentration in the Mutus-Vlacq basin already discussed (figs. 8.12, 9.2B), surround Nectarian basins and are stratigraphically related to the lineate deposits in the same way the circum-Orientale plains are related to the Hevelius Formation. That is, they are younger than the ejecta and secondary craters, but by an unknown amount. If they were part of the ejecta-deposition process, they are momentarily younger; if they are volcanic, they may be substantially younger.

Many of the plains may, in fact, have originated by a combination of volcanism and ejecta deposition. Eggleton and Marshall (1962) ascribed the planar light-colored surfaces south of the type area of the Fra Mauro Formation (figs. 10.18, 10.19) and elsewhere in the central lunar highlands to covering of premare surfaces by Imbrium ejecta. Wilhelms and El-Baz (1977) noted that the Nectarian and Imbrian plains in the Lomonosov-Fleming basin (fig. 9.28) are too smooth and extensive to be easily explained as basin ejecta and suggested that they were volcanically emplaced. Schultz and Spudis (1979) showed that the Lomonosov-Fleming plains are spotted by many dark-haloed craters that probably excavated dark material from beneath a lighter-colored cover. Therefore, the plains may consist of basaltic materials overlain by a thin cover of Crisium, Imbrium, or Orientale ejecta. Orientale ejecta may have covered basaltic plains in the Schiller-Zucchius basin near Orientale (Hartmann, 1966; Schultz and Spudis, 1979), which have diffuse patches darker than most terra plains (fig. 7.6; Offield, 1971). Numerous other plains, mostly inside basins of pre-Nectarian (Mutus-Vlacq) and Nectarian (Hertzsprung, Korolev) age, may also consist partly of premare basalt (pl. 7). Original emplacement in the style of the currently visible mare-basalt flows is indicated.

Geochemical sensing suggests that the dark halos contain both mare and KREEP basalt (Schultz and Spudis, 1979; Hawke and Spudis, 1980). Dark-haloed craters superposed on light-colored plains of Nectarian and Imbrian age (Wilhelms and El-Baz, 1977) within the possible Balmer-Kapteyn basin (table 4.2) display the telltale high radioactivity of KREEP (Haines and others, 1978), an observation suggesting impact exhumation of Nectarian or pre-Nectarian KREEP basalt (Hawke and Spudis, 1980). Mare basalt was probably exhumed from beneath the plains north of Taruntius and in many other localities (Schultz and Spudis, 1979; Hawke and Spudis, 1980). Therefore, basalt with a range of compositions may have been extruded far back into lunar history. The extent of this volcanism, however, remains uncertain.

CHRONOLOGY

If the age of 3.92 aeons favored here for the Nectaris basin is correct, then the Nectarian Period began 3.92 aeons ago and ended 3.85 aeons ago when the Imbrium basin was formed (see chap. 10).

One of the most frequently cited hypotheses in the lunar literature is that a "cataclysm" formed most or all basins within an interval of only 0.1 aeon (Tera and others, 1974). This cataclysm is generally conceived as a resumption of heavy bombardment after a lull following the heavy accretionary bombardment. The main basis for this hypothesis is that radiometric ages of terra rocks are concentrated around 3.8 to 3.9 aeons. Serenitatis looks like a relatively old basin; yet it has an absolute age that falls within this interval. The following data, presented here and in chapter 8, are consistent with the contrary view that the cratering rate declined relatively steadily (Baldwin, 1974b; Hartmann, 1975). (1) Numerous ages do predate 4.0 aeons (table 8.4; Baldwin, 1974b). (2) A concentration of late dates would be the natural consequence of a steadily declining cratering rate because of repeated impact reworking (Hartmann, 1973, 1975).

TABLE 9.5.—*Absolute ages of some pre-Imbrian volcanic clasts*

[After Ryder and Spudis (1980).
Stations C2, C', and C1 are on rim flank of Cone Crater at the Apollo 14 landing site; 2(1), station 2, boulder 1, Apollo 17 landing site.
Ages are calculated using International Union of Geological Sciences (IUGS) radioactive decay constants (Steiger and Jäger, 1977), except the age determined by Husain and others (see headnote, table 9.1). Ar-Ar ages are $^{40}Ar-^{39}Ar$ plateau ages.
References: C72, Compston and others (1972); C75, Compston and others (1975); H72, Husain and others (1972); M73, Mark and others (1973); M75, Mark and others (1975); PW71, Papanastassiou and Wasserburg (1971b); S73, Stettler and others (1973); T71, Turner and others (1971; see also Turner, 1977); Y72, York and others (1972)]

Sample	Station	Lithology	Age	Method	Reference
14053	C2	251-g high-Al mare-type basalt.	3.88±0.04	Ar-Ar	S73
			3.88±0.04	Rb-Sr	PW71
			3.89±0.05	Ar-Ar	T71
			3.92±0.08	Ar-Ar	H72
14072	C'	45-g high-Al mare-type basalt.	3.91±0.09	Rb-Sr	C72
			3.91±0.14	Rb-Sr	M75
			3.93±0.12	Rb-Sr	M73
			3.98±0.05	Ar-Ar	Y72
Clasts in 14321	C1	High-Al mare-type basalt clasts (sample 14321 is diverse 9-kg breccia).	3.85±0.05	Ar-Ar	Y72
			3.87±0.05	Rb-Sr	PW71
			3.88±0.05	Ar-Ar	Y72
			3.93±0.05	Ar-Ar	Y72
			3.93±0.12	Rb-Sr	M73
			3.96±0.06	Rb-Sr	M73
			3.96±0.08	Rb-Sr	C72
Clasts in 72275	2(1)	Basalt intermediate between mare and KREEP types.	3.87±0.04	Rb-Sr	PW71
			3.93±0.04	Rb-Sr	C75

(3) No pre-Nectarian basins were sampled (fig. 9.30). (4) A smooth cratering curve (fig. 8.16) is fixed by a 3.92-aeon age for Nectaris, a 3.85-aeon age for Imbrium, and the ages of mare-basalt samples; this curve is consistent with a relatively young Nectarian relative age (table 9.3) and a 3.87-aeon absolute age for the Serenitatis basin.

"Minicataclysms" forming some of the basin age groups are possible, but no cataclysm in the sense generally meant is recorded in lunar rocks. Instead, the record of early lunar impacts reveals a continuous barrage of large projectiles that only during the Nectarian Period began to approach less than cataclysmic proportions.

FIGURE 9.28.—Lomonosov-Fleming basin (compare fig. 8.13). F, Nectarian crater Fleming (130 km, 15° N., 110° E.). Dark halo (arrowhead) indicates basalt buried beneath light-colored Nectarian and Imbrian plains of probable basin origin. Mare fills crater Lomonosov (L; 93 km, 27° N., 98° E.), and Lomonosov ejecta is apparently covered by older mare in crater Joliot (under letter J); thus, Lomonosov is Lower Imbrian (compare fig. 4.7). Mare at left edge is Mare Marginis. GB, Giordano Bruno (22 km, 36° N., 103° E.), one of the Moon's youngest craters of its size. Apollo 16 frame M-3004.

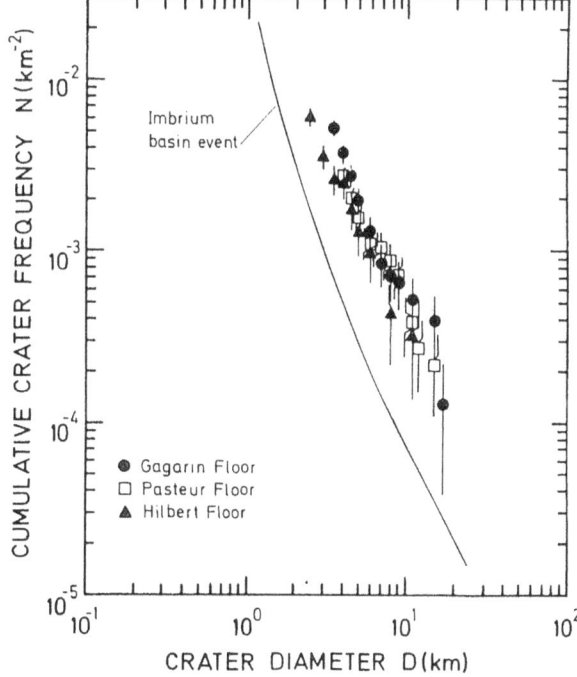

FIGURE 9.29.—Crater size-frequency curves by Neukum (1977), showing ages of plains in craters Gagarin, Pasteur, and Hilbert relative to Imbrium-basin deposits ("event"). Older, probably Nectarian frequencies are a factor of at least 7 higher than the Imbrium-basin frequency. Containing craters are pre-Nectarian (Gagarin and Pasteur) and Nectarian (Hilbert) (compare figs. 3.5, 8.11).

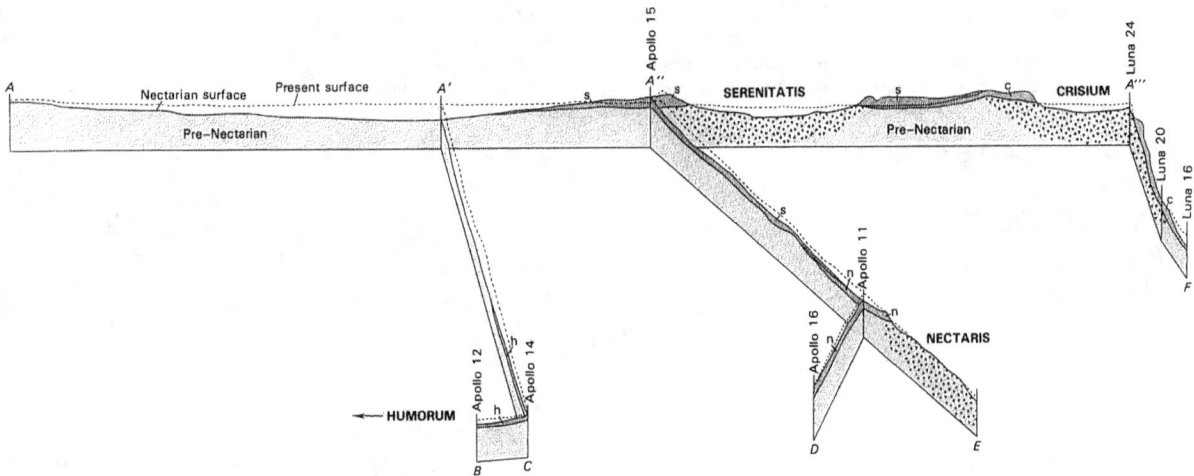

FIGURE 9.30.—Geologic cross sections based on figure 8.4A, showing deposits of the Nectarian basins Nectaris (n), Humorum (h), Crisium (c), and Serenitatis (s). Triangles, subbasin breccia (overlain by impact melt, not shown). Materials of the later Imbrium basin are derived partly from Serenitatis deposits and partly from pre-Nectarian terrane. Depression created by Procellarum basin (see fig. 8.4B) is still evident, though reduced by superposition of other pre-Nectarian deposits. Dotted lines, present surface.

10. LOWER IMBRIAN SERIES

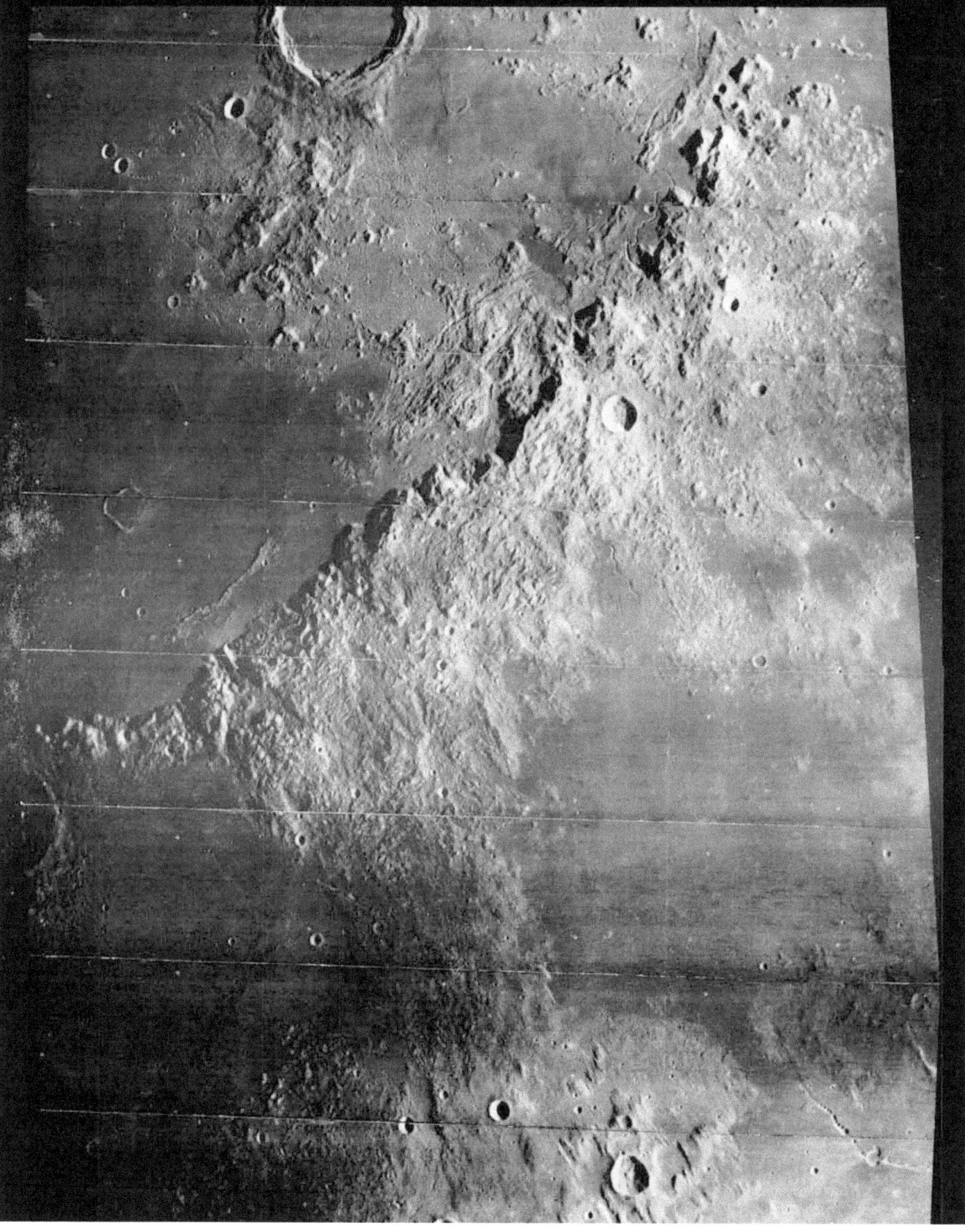

FIGURE 10.1 (overleaf).—Units of the Imbrian System where first recognized (compare figs. 1.6, 2.1). View centers on massive, rugged Montes Apenninus, part of the Imbrium-basin rim. Imbrium-basin ejecta grades from coarsely hummocky forms on Apennine flank to smoother drumlinlike or whaleback forms near crater Ukert (near bottom; 23 km, 8° N., 1.5° E., probably Lower Imbrian). Archimedes (at top; 83 km, 30° N., 4° W.), which is superposed on rugged Montes Archimedes and plains of the Apennine Bench (Hackman, 1966), is flooded by mare basalt. Massive slumping from Montes Apenninus is indicated by parallel structure and outlines of ridges northwest of Apennine front. Type area of top of Imbrian System is in bay of Mare Imbrium between the Apennine Bench and crater Eratosthenes (truncated by left edge of photograph). Orbiter 4 frame M–102.

10. LOWER IMBRIAN SERIES

CONTENTS

	Page
Introduction	195
Definition	196
Interior and near-rim Imbrium-basin deposits	196
Introduction	196
Apennine Bench Formation—Apollo 15 samples	197
Montes Apenninus massifs—Apollo 15 samples	198
Material of Montes Apenninus	202
Alpes Formation	203
Fissured plains	203
Fra Mauro Formation	204
Regional relations	204
Setting of the Apollo 14 landing site	204
Sample provenance	204
Petrology	205
Emplacement process	211
Absolute age	212
Conclusions	212
Imbrium-basin secondary craters	213
Light-plains materials	215
Cayley and Descartes Formations	216
Introduction	216
Imbrium-basin origin	217
"Local" origin	219
Conclusions	220
Other basin and crater materials	221
Recognition criteria	221
Frequency	222
Other terra materials	222
Mare materials	223
Chronology	224

INTRODUCTION

The extensively exposed and relatively well sampled Lower Imbrian Series is the key to the ages and origins of most materials of the lunar terrae. Deposits of the Lower Imbrian Orientale and Imbrium basins constitute the Moon's most extensive laterally continuous stratigraphic horizons (pl. 8). The Imbrium basin dominates the surface visible from Earth and was the first basin to be treated stratigraphically (see chaps. 2, 4; fig. 10.1). Deposits of the Orientale basin dominate the terrae on both sides of long 90 W. and serve as models for other basin materials, as previous discussions have made clear. Thus, the Lower Imbrian Series dominates a large part of the lunar surface, despite the small number of events that emplaced its rock units (pl. 8). Both groups of basin deposits make excellent stratigraphic markers for dividing the lunar stratigraphic column (table 7.1) because of their great extent and clear expression.

Because of their early discovery and prominence, the Imbrium-basin deposits attracted more sampling missions than did any other class of terra units. The curtailed Apollo 13 and the successful Apollo 14 flights were sent to a point on the Fra Mauro Formation, the most distinctive Imbrium-basin unit. Exploration of the Imbrium rim at Montes Apenninus was the main objective of the Apollo 15 mission. The Cayley and Descartes Formations, though considered to be volcanic when selected as targets for the Apollo 16 landing (see chap. 2), are Lower Imbrian photogeologic units emplaced by impact and may also contain Imbrium-basin material.

In terms of total volume of returned material, however, rocks emplaced during the Early Imbrian Epoch are less well represented than Nectarian and pre-Nectarian materials. Apollos 14 and 15 together returned much less terra material than did Apollo 17. Arguably, the abundant Apollo 16 collection may include more material processed in the Nectarian than in the Imbrian Period (see chap. 9). Effects of igneous crustal processes active in pre-Nectarian time have profoundly affected all sample collections (see chap. 8).

The problem of determining which compositions and textures of the Apollo 14, 15, and 16 samples were created during the Early Imbrian events that emplaced them and which were created earlier occupies considerable space in this chapter. The first half of the chapter is devoted to the Imbrium basin and to the samples returned from its deposits by the Apollo 14 and 15 missions. I develop the viewpoint that the Apollo 14 materials owe their properties to their emplacement in the Fra Mauro Formation, as originally proposed but not currently so widely believed. A similar analysis for the more controversial Cayley and Descartes Formations sampled by Apollo 16 is less conclusive. Resolution of the remaining questions about the Apollo 14 through 16 rocks would significantly aid in dating large areas on the absolute and relative time scales and in correlating compositions with sources in the crust. Nonsampled Lower Imbrian units, including the Orientale basin that already has been thoroughly described, are given less attention here.

DEFINITION

Deposits of the Imbrium basin and the radial system of grooves called "Imbrian(um) sculpture" have long served as the main reference for dividing materials into Imbrian and pre-Imbrian on the nearside (see chap. 7; Gilbert, 1893; Shoemaker and Hackman, 1962). In August 1967, photography of the Orientale basin at the end of the Lunar Orbiter 4 mission added a major stratigraphic horizon to the Imbrian System as it had previously been defined. Superpositional relations (fig. 10.2), morphologic freshness (figs. 4.4, 10.3), and size-frequency distributions (fig. 10.4) all indicate that Orientale is younger than Imbrium. Orientale is older than the nearby mare surface, which would have been swamped by the basin ejecta if it predated Orientale (figs. 10.3, 10.5). The Orientale deposits constitute excellent stratigraphic markers because of their vigorous topographic expression and clear relations with other units over a wide area (pls. 3, 8). They probably would have been used as a stratigraphic datum to define a major time-stratigraphic boundary, had they been in a better position for study during the telescopic phase of the geologic-mapping program. This key horizon has, in fact, been used on some geologic maps to divide informal lower and upper series of the Imbrian System (for example, McCauley, 1973).

The two series of the Imbrian System are here defined formally. The *Lower Imbrian Series* includes all materials from the bottom of the Fra Mauro Formation to the top of the Hevelius Formation (fig. 7.1). The type area of the base of the Lower Imbrian Series is the type area of the Fra Mauro Formation, lat 0°–2° S., long 16°–17.5° W. (figs. 10.1, 10.19; Wilhelms, 1970b). The type area of the top of the series is the type area of the Hevelius Formation, near lat 2° N., long 68° W. (fig. 4.4*H*; McCauley, 1967a). A reference area for the top of the series is bounded by lat 2° S.–2° N., long 75°–80° W. (fig. 10.3). The *Upper Imbrian Series* includes all materials above the top of the Hevelius Formation—that is, younger than the Orientale basin—to the top of the Imbrian System as defined in chapter 11. The *Early Imbrian Epoch* thus began at the moment of the Imbrium impact and ended when the Hevelius Formation came to rest. The *Late Imbrian Epoch* began when the Hevelius came to rest and ended when the mare materials that define the top of the Imbrian System solidified.

INTERIOR AND NEAR-RIM IMBRIUM-BASIN DEPOSITS

Introduction

Imbrium is the Moon's largest well-preserved basin (table 4.1) but is not entirely understood. Its basaltic fill, Mare Imbrium, obscures an area larger than the entire Orientale topographic basin (table 6.1). Interior deposits and large parts of interior rings are visible only at the Apennine Bench (figs. 1.6, 2.1, 4.3*AA*, 10.6). Concentric mare ridges help somewhat in locating buried rings (Hartmann and Kuiper, 1962; Brennan, 1976). The southwest, west, and northwest sectors of the topographic rim are also buried by basalt flows,

FIGURE 10.2.—Superpositions of secondary craters of the Orientale basin on those of the Imbrium basin.

    A. Regional view of northwest nearside limb, including north pole (black-and-white arrow). Orientale basin is at bottom. Black arrow parallels Imbrium sculpture (northeast-southwest) and marks superposition indicated in *B*. White L, pre-Nectarian Lorentz basin (360 km, 34° N., 97° W.); P, Eratosthenian crater Pythagoras (130 km, 64° N., 63° W.); black L, Copernican crater Lichtenberg (20 km, 32° N., 68° W.), embayed by Copernican mare basalt (see chap. 13). Numerous craters with uplifted floors are near terra-mare boundary (see chap. 6). Area of figure 10.3 is outlined. Orbiter 4 frame M-189.

    B. Detailed view of Orientale-secondary craters, showing herringbone-patterned ejecta interrupting Imbrium-radial chains (arrow at bottom). Orbiter 4 frame H-189.

10. LOWER IMBRIAN SERIES

pyroclastic materials, and deposits of such large craters as Iridum (figs. 3.8, 5.12). The connections between rings in the north and northeast are partly buried (pl. 8). Therefore, the identity not only of the boundary of excavation but even of the topographic rim is uncertain in the half of the basin periphery from the southwest to the northeast sectors. In the east and south, Montes Apenninus and Carpatus (figs. 10.6, 10.7) constitute the topographic rim and, in my opinion, the excavation boundary (see chap. 4). These ranges are equivalent to Montes Cordillera in position and morphology (Wilhelms and McCauley, 1971; M'Gonigle and Schleicher, 1972; Brennan, 1976) and in origin (Baldwin, 1974a; Wilhelms and others, 1977; Hodges and Wilhelms, 1978).

The best exposures of Imbrium-basin materials are on Montes Apenninus, in large tracts southeast of that range, and north of Mare Frigoris. This section describes the Imbrium materials from the inside out and interprets them by comparison with their Orientale equivalents.

Apennine Bench Formation—Apollo 15 samples

Hackman (1964, 1966) named the level or wavy materials between Montes Archimedes and Montes Apenninus the "Apennine Bench Formation" (figs. 2.1, 6.10, 10.1, 10.7). Page (1970) extended this name to other, similar plains inside the Imbrium-basin rim that underlie deposits of Imbrian craters and the mare (figs. 2.4, 10.8). Overlap of

FIGURE 10.3.—Orientale-secondary craters and plains deposits, partly flooded by mare basalt (arrows); old mare unit is at lower arrow (see fig. 10.2A for location). Reference area for typical exposure of inner facies of the Hevelius Formation (2° N.–2° S., 75–80° W.) is enclosed by corner markers. Secondary craters of Krafft (K; 51 km, 17° N., 73° W.) and Cardanus (C; 50 km) are superposed on Orientale deposits and old mare, but are flooded on west by younger mare; similar ages and sizes of these two craters suggest twin impacts (see chap. 3). Ejecta of Copernican crater Olbers A (OA; 43 km, 8° N., 78° W.) is superposed on Orientale deposits and partly covered by small patch of very young mare. Olbers (O; 75 km, probably Nectarian) and Riccioli (R; 146 km, probably pre-Nectarian) are buried by Orientale deposits. Orbiter 4 frame H-174.

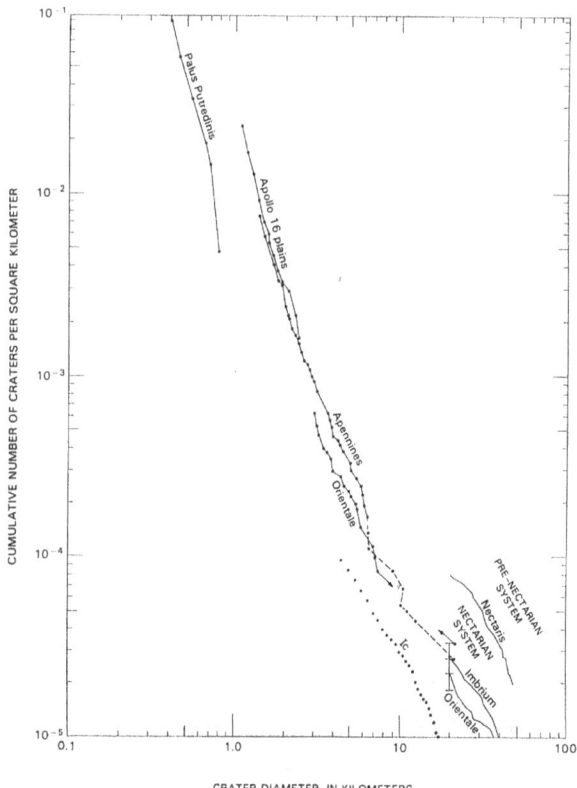

FIGURE 10.4.—Cumulative size-frequency distributions of craters superposed on major Imbrian units. Counts of craters larger than 20 km in diameter on deposits of the Imbrium and Orientale basins are from this study; Nectarian and pre-Nectarian fields are added for comparison (figs. 8.6, 9.22). Other curves are plotted from data of Neukum and others (1975a); data points for the Apennines are from two counts by Neukum and others (1975a), which have a common point at 3 km. Curve for Imbrian craters (Ic) is from Wilhelms and others (1978).

the Apennine Bench Formation on the rugged Imbrium-basin topography shows that the plains are at least slightly younger than the basin (figs. 10.7, 10.8).

As for other terra-plains deposits, interpretations of origin have shifted back and forth between volcanic and impact. E.M. Shoemaker and R.J. Hackman (oral commun., 1963) proposed that the Apennine Bench Formation consists of Imbrium-basin ejecta or impact melt. The impact-melt interpretation is supported by position inside the basin and by morphologies indicating subsidence of a fluidlike material (fig. 10.7; Wilhelms, 1980). The Apennine Bench Formation and the Maunder Formation of Orientale are similar in these respects (table 4.3). Volcanic interpretations, which are also consistent with the fluidlike appearance, were favored in the era when lunar light-colored plains in general were commonly thought to be volcanic (Hackman, 1966; Wilhelms, 1970b; Carr and others, 1971; Wilhelms and McCauley, 1971).

A volcanic origin is supported by discovery among the Apollo 15 samples of fragments of possible terra-type volcanic rock. The Apennine Bench Formation probably underlies the mare basalt in the region of the landing site (figs. 10.7, 10.8; Carr and others, 1971), although the formation is not exposed within range of the astronaut traverses (fig. 10.9). The distribution of KREEP-rich fragments in the regolith (Basu and McKay, 1979) is consistent with derivation from beneath the mare (Spudis, 1978b). Their Th-rich medium-K KREEP chemistry matches the composition detected by the orbiting gamma-ray spectrometer over Apennine Bench exposures west of the landing site (Spudis, 1978a; Hawke and Head, 1978b; Metzger and others, 1979; Clark and Hawke, 1981). Because the fragments have igneous textures and no meteoritic siderophile elements, many investigators have concluded that they are true endogenic igneous rocks (fig. 10.10; Irving, 1975, 1977; Dowty and others, 1976; Ryder, 1976; Meyer, 1977; Basaltic Volcanism Study Project, 1981, p. 274–278). A further argument for a volcanic origin is that the samples were probably derived from a source more radioactive than any now present beneath the landing site (as shown by the Apollo 15 heat-flow experiment; Langseth and others, 1976); an Imbrium melt sheet would incorporate material similar to the present subbasin material (Spudis, 1978a). An episode of KREEP volcanism in the Early Imbrian Epoch is indicated, possibly triggered by the Imbrium impact (Spudis, 1978a).

The volcanic interpretation is still not universally accepted. Taylor (1982, p. 214–216) pointed out that the absence of siderophiles and fragments does not prove an igneous origin and suggested that Montes Apenninus are as rich in KREEP as is the Apennine Bench. Some interpretations of the orbital data, however, favor a higher Th content for the bench (Metzger and others, 1979).

Three geochronologic methods have yielded ages for the medium-K KREEP-rich samples 15382 and 15386 averaging 3.85 aeons (table 10.1). This average age probably dates the emplacement of the KREEP-rich basalt and constrains the age of the Imbrium basin to 3.85 aeons or older, regardless of whether the KREEP-rich liquids were melted by impact or endogenic heat.

Montes Apenninus massifs—Apollo 15 samples

Steep rugged massifs that are among the Moon's highest mountains form the crest of Montes Apenninus; they probably were uplifted structurally and overlain by ejecta (figs. 10.1, 10.7, 10.8, 10.11; Carr and others, 1971; Wilhelms, 1980). Large slump masses lie as far as 60 km northwest of embayments in the mountains, which were evidently the sources of the slumps (figs. 6.10, 10.7, 10.8; Wilhelms and others, 1977; Wilhelms, 1980). Parts of the Apennine Bench Formation overlie some of these slumps (Spudis, 1978a; Wilhelms, 1980). Most investigators currently believe that Montes Apenninus are an exterior scarp of an Imbrium basin smaller than its topographic rim. Others, including me (see section in chap. 4 entitled "Origin of Rings"), regard an origin as the excavation boundary as much more likely because the Apennines are a massive elevated range of mountains, the largest on the lunar nearside, and because they mark a sharp discontinuity in the textures of basin deposits (figs. 10.1, 10.6, 10.7).

The identity of the boundary of excavation in other sectors is less clear. The Apennines and Montes Carpatus, another obvious part of the topographic-basin rim, define a circle 1,200 km in diameter with Montes Alpes and an oval 1,500 km long with the north shore of Mare Frigoris (pl. 3; figs. 4.3AA, 10.6). The Apenninus-Carpatus ring apparently divides in Montes Caucasus (figs. 10.6, 10.12). The southern Caucasus continue as Montes Alpes, and the northern Caucasus as the Frigoris shore (Wilhelms and McCauley, 1971). In the interpretation given in chapter 4, the large oval bounds the Imbrium cavity. The hypothesis that the outer parts of large basins are differentially excavated may explain the divergence; the Frigoris sector was excavated to a larger diameter than was the Carpatus sector, which was blocked by the Serenitatis rim. Lineations in the northern Caucasus (figs. 10.6, 10.12) point to an impact point at the center of the Carpatus-Alpes circle at lat 33° N., long 18° W. (table 4.1).

The origin of Montes Archimedes, the rugged elevation south of the crater Archimedes, is ambiguous (fig. 10.1). It lies at the same distance from the center as does Montes Alpes (pl. 3; Wilhelms and McCauley, 1971) but also resembles the slumps derived from the Apennine scarp. Thus, it may be either part of an Imbrium-basin ring or a slump.

One of the major objectives of the Apollo 15 mission in July 1971 was to collect samples of Montes Apenninus. According to impact models based on simple craters (see chap. 3), this basin-rim material might have come from deep within the Moon. Primitive materials or, at least, pre-Imbrian materials might have been uplifted and exposed

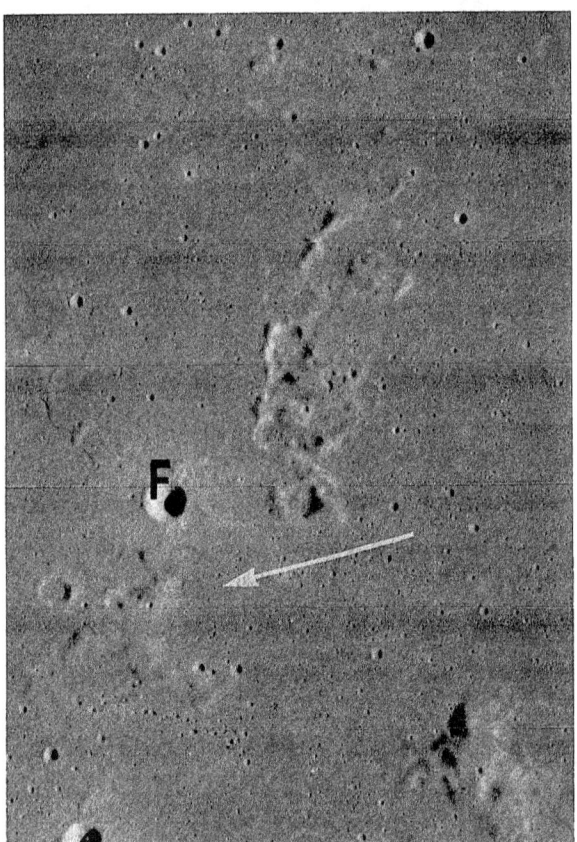

FIGURE 10.5.—Partly buried crater group (above arrow) in Oceanus Procellarum (2° S., 40.5° W.), identified as secondary to Orientale basin by ejecta orientation subradial to Orientale (centered 1,700 km to west in direction of arrow) and subequal size (about 5 km; compare crater F, Flamsteed FB, 5 km). Orbiter 4 frame H–143.

FIGURE 10.6.—Imbrium basin, bounded by Montes Alpes (Al), Caucasus (white C), and Apenninus (Ap). Imbrium-radial crater chains (black-and-white arrows) north of Mare Frigoris (F) are superposed on deposits of Nectarian Humboldtianum basin (H; 61° N., 84° E.; Lucchitta, 1978) and on many craters, including pre-Nectarian W. Bond (W; 158 km) and Barrow (B; 93 km; compare fig. 10.15). Humboldtianum deposits cover probable Nectarian craters Nansen (N; 122 km, 81° N., 95° E.) and Schwarzschild (S; 235 km, 71° N., 120° E.), and are overlain by Nectarian crater Endymion (E; 125 km, 54° N., 57° E.), Lower Imbrian crater Compton (black C; 162 km, 56° N., 105° E.; compare fig. 9.4), and Copernican crater Hayn (87 km, north of H). Fewer small prebasin craters are visible near both basins than along upper left edge of photograph. White arrow, north pole. Orbiter 4 frame M–116.

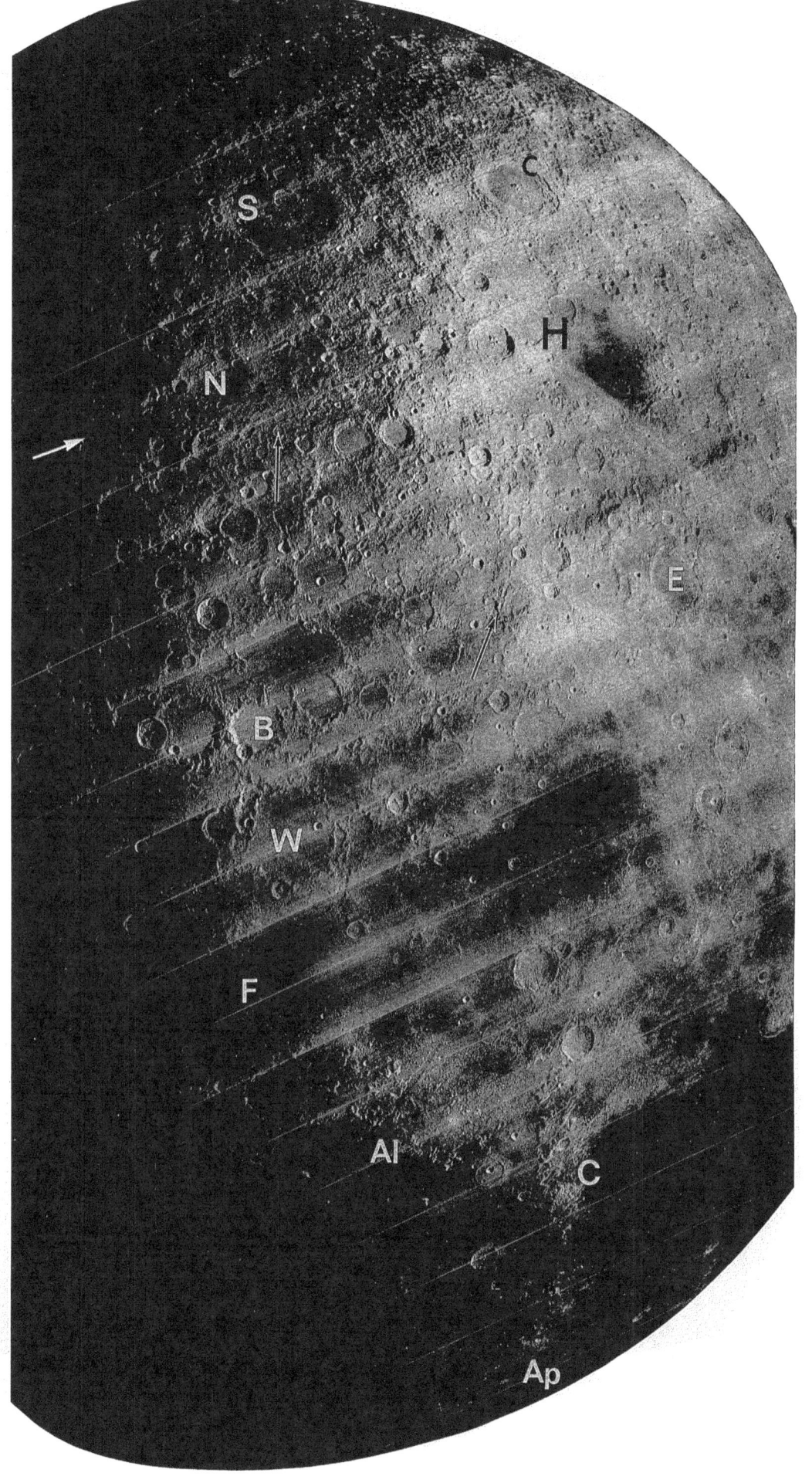

TABLE 10.1.—*Absolute ages of Imbrium-basin materials*

Stations: Apollo 14 (see fig. 10.21): Traverse EVA 1 was within 150 m of the Lunar Module (LM); stations C1, C', and C2 are on the rim flank of Cone Crater along traverse EVA 2; F, 400 m east of LM on EVA 2; G, 250 m east of LM on EVA 2. Apollo 15 (see fig. 10.9): Station 7 is at Spur Crater along the front of Montes Apenninus.

Rock types: CMR, crystalline melt rock of uncertain origin (volcanic basalt or clast-poor or -rich impact-melt rock); CPM, clast-poor impact-melt rock; FMB, Fra Mauro breccia (clast-rich impact-melt rock); xln, crystalline. Lithologic descriptions in much of the geochronologic literature are inadequate for categorizing samples further; for example, the term "basalt" is commonly used for impact-melt rocks (see chap. 8).

Ages calculated using International Union of Geological Sciences (IUGS) radioactive-decay constants (Steiger and Jäger, 1977), except those determined by Husain and others (1972) (see headnote, table 9.1). Most Ar-Ar ages are ^{40}Ar-^{39}Ar plateau ages. Age determinations with error ranges of ±0.10 aeon or greater and those obviously outside the likely range (for example, those of Upper Imbrian mare-basalt grains) are omitted.

References: AD74, Alexander and Davis (1974); AK74, Alexander and Kahl (1972); C72, Compston and others (1972); CL79, Carlson and Lugmair (1979); H72, Husain and others (1972); K72, Kirsten and others (1972); M72, Murthy and others (1972); M74, Mark and others (1974); M79, McKay and others (1979); N74, Nyquist and others (1974); N75, Nyquist and others (1975); PW71, Papanastassiou and Wasserburg (1971b); PW76, Papanastassiou and Wasserburg (1976); S73, Stettler and others (1973); T71, Turner and others (1971); T72, Turner and others (1972); T73, Turner and others (1973; summarized in Turner, 1977); WP71, Wasserburg and Papanastassiou (1971); Y72, York and others (1972)]

Sample	Station	Rock type	Age	Method	Reference
		Apollo 14 young group			
14001,7,3	EVA 1	2-4 mm CMR	3.81±0.03	Rb-Sr	PW71
			3.84±0.04	Ar-Ar	T71
14066	F	510-g FMB			
,21,1.02		Whole rock	3.84±0.03	Ar-Ar	AD74
,21,1.01		Xln matrix	3.86±0.03	Ar-Ar	AD74
,21,2.04,2		White xln clast[1]	3.88±0.03	Ar-Ar	AD74
,21,2.01		Breccia clast[1]	3.90±0.02	Ar-Ar	AD74
14073	G	10-g CPM (plagioclase separates)	3.80±0.04	Rb-Sr	PW71
			3.82±0.05	Ar-Ar	T72
14150	G	4-10-mm fines			
,7,3		CPM	3.82±0.02	Rb-Sr	M79
,7,2		CPM	3.83±0.01	Rb-Sr	M79
14161	EVA 1	2-4-mm fines			
,34,2		CMR	3.80±0.08	Ar-Ar	K72
,34,6		CPM	3.84±0.04	Ar-Ar	K72
,34,5		CMR[1]	3.87±0.08	Ar-Ar	K72
,34,4		CMR[1]	3.90±0.06	Ar-Ar	K72
14167	EVA 1	2-4-mm fines			
,6,1		CMR	3.79±0.03	Ar-Ar	H72
,6,7		CMR	3.82±0.03	Ar-Ar	H72
,6,3		CMR	3.83±0.05	Ar-Ar	H72
,8,1		CMR[1]	3.88±0.04	Ar-Ar	T71
,9		CMR[1]	3.91-3.95	Ar-Ar	Y72
14257,12,1	EVA 1	2-4-mm fine	3.87±0.06	Ar-Ar	H72
14270,1-7	EVA 1	26-g FMB	3.83±0.05	Ar-Ar	AK74
14276	EVA 1	13-g CPM	3.80±0.04	Rb-Sr	WP71
14303,13,R5,1	EVA 1	Clast from 898-g FMB	3.85±0.04	Ar-Ar	K72
14310	G	3.4-kg CPM	3.78±0.03	Ar-Ar	H72
			3.79±0.04	Rb-Sr	PW71
			3.81±0.05	Ar-Ar	T72
			3.82±0.06	Ar-Ar	S73
			3.83±0.04	Ar-Ar	T71
			3.85±0.04	Rb-Sr	C72
			3.85±0.05	Ar-Ar	Y72
			3.85±0.06	Rb-Sr	M72
			3.86±0.03	Rb-Sr	M74
		Apollo 14 old group (pre-Imbrian)			
14053	C2	High-Al basalt	3.88-3.92	See table 9.5.	
14072	C'	do	3.91-3.98	Do	
14321	C1	do	3.85-3.96	Do	
		Apollo 15 group			
15382	7	KREEP-basalt fragment	3.82±0.02	Rb-Sr	PW76
			3.84±0.05	Ar-Ar	S73
			3.85±0.05	Ar-Ar	T73
15386	7	do	3.85±0.08	Sm-Nd	CL79
			3.86±0.04	Rb-Sr	N75
15434,73	7	do.?	3.83±0.04	Rb-Sr	N74
15455	7	Impact melt of breccia matrix.	3.86±0.04	Ar-Ar	AK74

[1]Probably older than the Fra Mauro Formation.

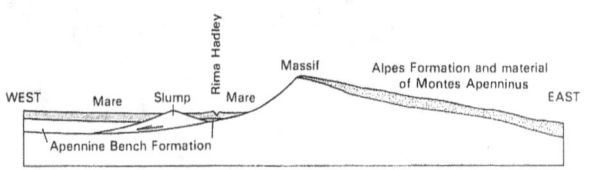

FIGURE 10.8.—Diagrammatic geologic cross section of region of Apollo 15 landing site, based on figure 10.7. Indicated history: (1) uplift of massifs and ejection of the Alpes Formation and "material of Montes Apenninus," (2) slumping of large blocks from Apennine front, (3) emplacement of the Apennine Bench Formation (either impact melt or volcanic basalt), and (4) emplacement of mare basalt.

FIGURE 10.7.—Region of Apollo 15 landing site (arrow). Mare basalt of Palus Putredinis (PP) is overlain by secondary craters and rays of Autolycus, centered north of area. Sinuous rille is Rima Hadley. Iab, Apennine Bench Formation; s, slumps from Apennine front; C, Conon (22 km; compare fig. 10.15). Stereoscopic pair of Apollo 15 frames M-413 (right) and M-415 (left).

in the Imbrium-basin massifs. Materials of the Serenitatis basin are likely to be included because that basin's west rim is cut off by the Apennines and thus lay in the target area of the Imbrium impact (pl. 3; fig. 9.30; Wilhelms and McCauley, 1971; Spudis and Head, 1977).

Fewer samples than expected were obtained from the massifs. Most that were collected are small and of uncertain geologic context because outcrops are covered by thick colluvium (fig. 10.11). The largest samples are two "black and white rocks" (15445, 15455), totaling 1.17 kg in mass and representing a 1-m boulder sampled at station 7 (figs. 10.9, 10.13; Swann and others, 1972). Dark, aphanitic impact-melt rock of low-K KREEP composition encloses light-colored clasts of mafic ANT rock (Reid and others, 1977; Ryder and Bower, 1977). The ANT rock was crushed or fragmented in one or several episodes (James, 1977). Some of the clasts are composed of ancient, apparently "pristine" noritic and troctolitic rock (table 8.4; Nyquist and others, 1979b). The impact-melt rock is similar in composition to impact-melt rock from the Taurus-Littrow massifs of the Serenitatis basin, though richer in Mg (Taylor, S.R., and others, 1973). This similarity may support the photogeologic observation that Serenitatis material was part of the Imbrium-basin impact target. Alternatively, it may indicate that low-K KREEP is a common crustal constituent (Ryder and Wood, 1977).

The lithologic similarity to the Serenitatis melt-rich deposit has led to suggestions that the black-and-white rocks are part of a melt-rich deposit of the Imbrium basin (Ryder and Bower, 1977; Ryder and Wood, 1977). The ^{40}Ar-^{39}Ar gas-retention age of 3.86 ± 0.04 aeons (table 10.1; Alexander and Kahl, 1974) probably dates the assembly of the breccia deposit and, therefore, the Imbrium impact.

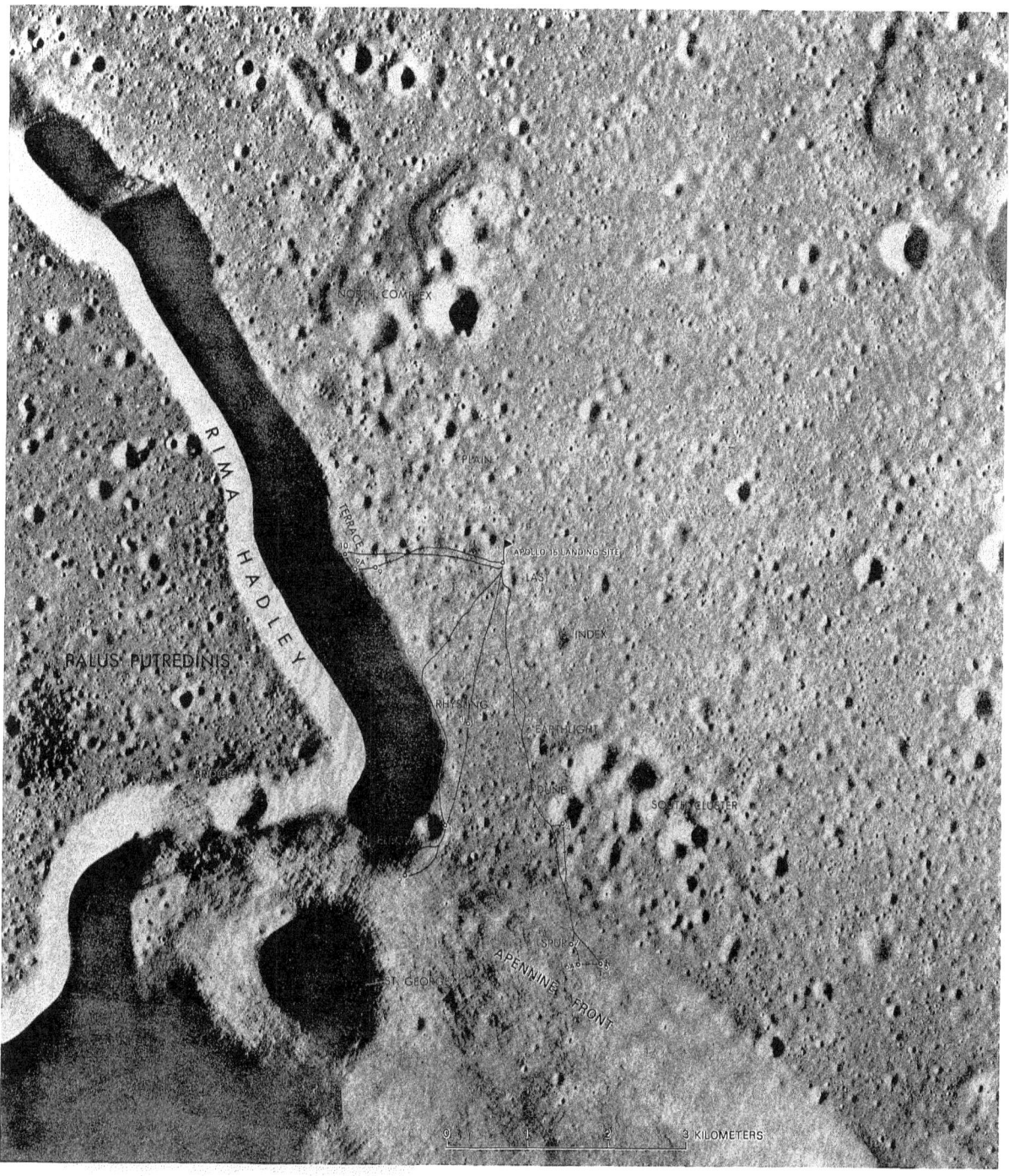

FIGURE 10.9.—Area of Apollo 15 astronaut traverses. Numbered circles are sampling stations. Prepared by U.S. Defense Mapping Agency; base from Apollo 15 frames P–9370 and P–9377.

Material of Montes Apenninus

Four textural types of Imbrium-basin ejecta lie on the southeast Apennine flank (fig. 10.14; table 4.3). One type has been given the informal name "material of Montes Apenninus" (Wilhelms and McCauley, 1971), not to be confused with the massif material. This unit is characterized by closely spaced imbricate slivers concentric with the Apennine crest (figs. 10.1, 10.7, 10.14). Its position near the Apennine crest and the coherent morphology of the blocks are consistent either with structural deformation along the excavation rim or with low-velocity lofting or pushing of late ejecta over the rim (Wilhelms, 1980). The unit resembles the concentric ejecta facies of the Orientale basin in morphology, relative position, and areal extent (figs. 4.4E, F; Wilhelms, 1980).

FIGURE 10.10.—Thin section of crystalline, probably volcanic KREEP-rich sample 15386,8, showing basaltic intersertal to subophitic texture (Dowty and others, 1976). Fragment lacks clastic inclusions and meteoritic siderophile elements (Irving, 1975, 1977; Dowty and others, 1976). Crossed polarizers; field of view, about 2 mm.

FIGURE 10.11.—Part of Montes Apenninus, showing paucity of outcrops and blocks on slope, and colluvial apron at base (a). Apollo 15 frame P-9804.

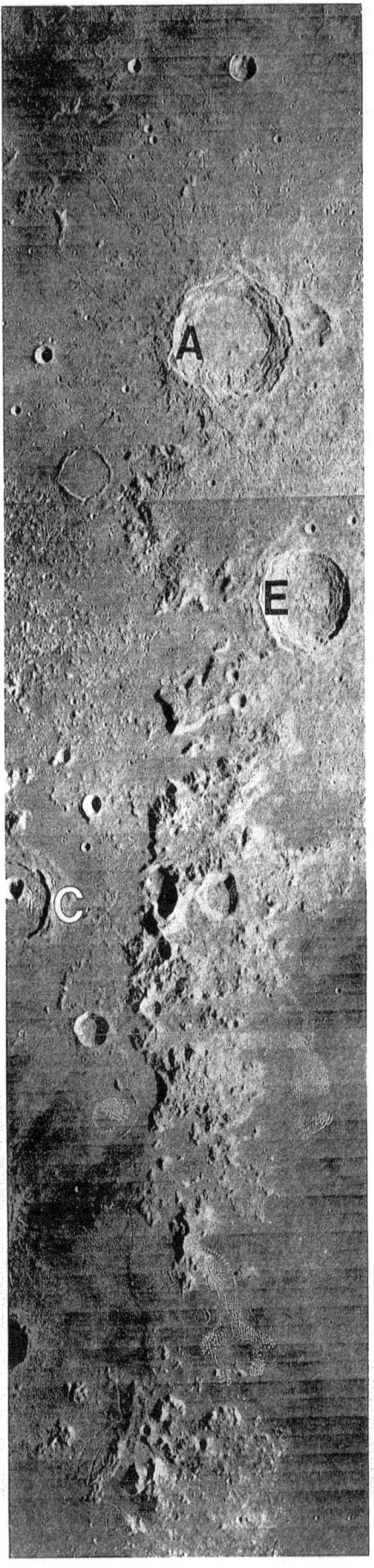

FIGURE 10.12.—Montes Caucasus, showing lineations oblique to trend of mountains and radial to center of Imbrium basin. Hummocky Alpes Formation is above crater Cassini (C; 57 km, 40° N., 5° E., probably Lower Imbrian) and west of Eudoxus (E; 67, 44° N., 16° E., probably Copernican), and in part of Montes Apenninus at bottom. A, Aristoteles (87 km, 50° N., 17° E., probably Eratosthenian). Orbiter 4 frame H-103.

Alpes Formation

Another, more extensive hummocky Imbrium-basin deposit is the Alpes Formation (Page, 1970). In contrast with the material of Montes Apenninus, the Alpes Formation consists of randomly oriented knobs and larger tracts of smoother material (figs. 10.1, 10.12, 10.14–10.16). This morphology closely resembles that of the Montes Rook Formation of Orientale (figs. 4.4B, E). The two units are also similar in spatial relation to massifs that are considered to be parts of major rings (Montes Alpes and the outer Montes Rook ring), which lie inside the topographic rims of their respective basins. The Alpes Formation, however, extends even farther from the Imbrium rim—as far as 600 km in the east—than does the Montes Rook Formation from Montes Cordillera (pl. 8; fig. 4.8). Both units were probably emplaced as low-velocity, late-stage ejecta from the inner parts of the basins. The Alpes Formation is the more extensive because the Imbrium basin is larger than Orientale.

Fissured plains

A third type of material on the Apennine flank forms level or wavy pools resting in depressions in the other units (fig. 10.14; Wilhelms, 1980). Some of these pools, as well as smooth veneers on rougher material, are cut by fissures of the types that suggest shrinkage of a formerly molten material (see chap. 3). This Apennine material is thus likely to be impact-melt rock like that occupying comparable positions on the flanks of craters and the Orientale basin (fig. 4.4B).

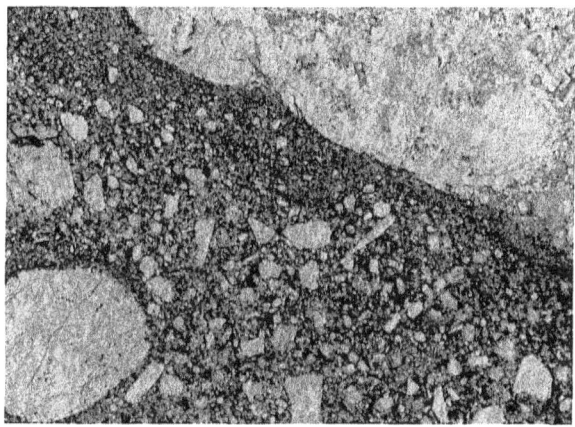

FIGURE 10.13.—"Black-and-white" breccia sample 15455 from rim of Spur Crater, station 7, Apollo 15 landing site. Black component is impact melt, probably formed by Imbrium impact (Ryder and Bower, 1977). White component is pre-Nectarian anorthositic norite (table 8.4).
A. Hand specimen after arrival from the Moon.

B. Thin section of 15455,29, showing chaotically distributed ANT clasts in melt matrix. Plane-polarized light; field of view, about 1 mm.

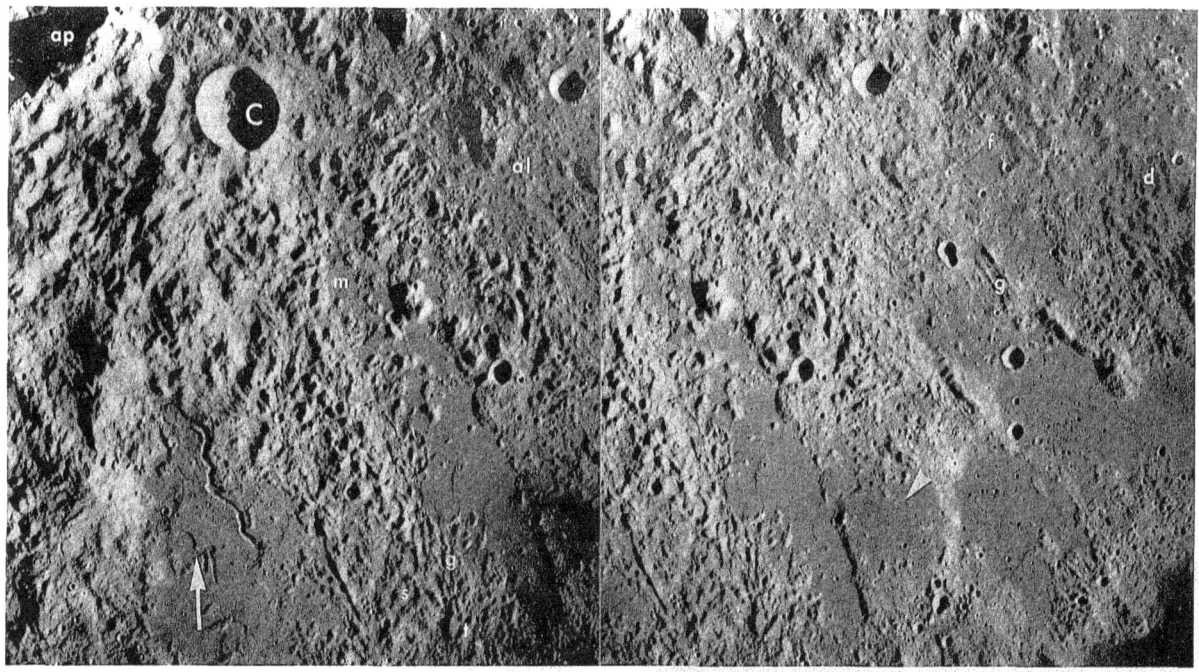

FIGURE 10.14.—Types of Imbrium ejecta on flank of Montes Apenninus. ap, northwest-facing Apennine scarp. Concentric "material of Montes Apenninus" is closest to scarp; farther out are knobby Alpes Formation (al), radially grooved terrain (g), smooth terrain (s), transverse fissures (f), and level ponds of probable impact melt (m). Arrows, irregular craters in mare, including "D caldera" (arrowhead in right frame; chap. 5; compare fig. 5.10D); d, dark-mantling material. C, fresh crater Conon (22 km; compare fig. 10.7). Stereoscopic pair of Apollo 17 frames M-1821 (right) and M-1823 (left). From Wilhelms (1980, fig. 13).

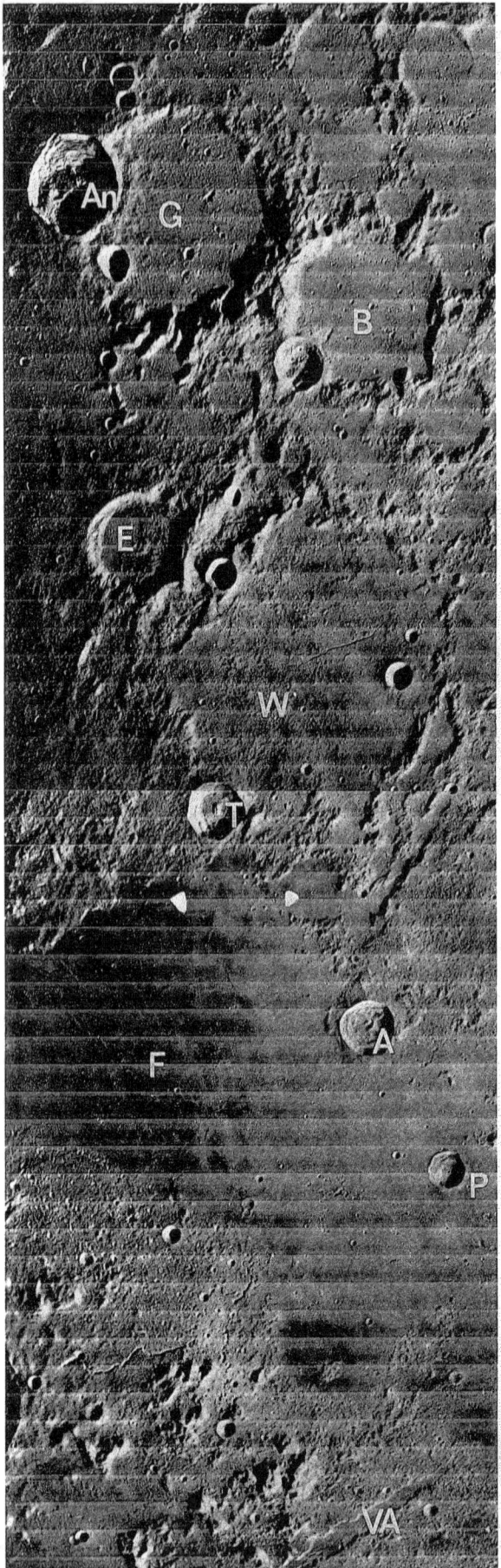

FRA MAURO FORMATION

Regional relations

A fourth unit on the Apennine flank, better developed farther out, is the Fra Mauro Formation (figs. 10.15–10.20; Eggleton, 1964, 1965; Wilhelms, 1970b). The Fra Mauro Formation extends 600 to 800 km from the rim of Imbrium in all exposed sectors (pls. 3, 8), and grades outward from coarsely hummocky to smooth and wavy (figs. 10.1, 10.14–10.16). The Fra Mauro Formation obliterates almost all prebasin craters to distances of 350 to 600 km from the Apennine crest. Thus, the Imbrium-basin deposits resemble exterior deposits of craters, whose continuous radial facies also extend beyond a coarse-textured zone out to 1.0 to 1.5 radii from the crater rim (chap. 3; Moore and others, 1974; Oberbeck and others, 1974). This similarity to craters was the chief basis for concluding that the Fra Mauro Formation is primary ejecta from the Imbrium basin (Shoemaker and Hackman, 1962).

The Fra Mauro Formation also resembles the Hevelius Formation of Orientale (fig. 4.4), except that the Fra Mauro includes more equidimensional hummocks and fewer distinct ridges. The continuous facies of the Hevelius extends 300 to 600 km from the rim of Montes Cordillera, 465 km in radius, which is the probable boundary of excavation of Orientale (see chap. 4). These similarities in morphology and extent relative to the diameters of the sources indicate that the Fra Mauro, Hevelius, and crater deposits form a related class of impact deposits. The inner and much of the outer facies of the Hevelius Formation include thick deposits that appear to have flowed along the surface (fig. 4.4), and the evident thickness and blockage by obstacles indicate that the Fra Mauro is also a major ground-flow deposit (Morris and Wilhelms, 1967; Wilhelms, 1980). Grooves and ridges indicative of the flow are also present (figs. 10.15–10.17, 10.19, 10.20). Small hills that have been interpreted as volcanic (fig. 10.17; Wilhelms, 1968; Wilhelms and McCauley, 1971) may, instead, be "deceleration dunes" formed like those of the transverse facies of the Hevelius Formation (figs. 4.4G, H).

Setting of the Apollo 14 landing site

The Apollo 14 landing site (figs. 10.18–10.21; lat 3.7° S., long 17.5° W.) was selected to obtain samples of primary ejecta from the Imbrium basin (Eggleton and Offield, 1970). Ridges and grooves radial to Imbrium and typical of the Fra Mauro Formation characterize the area around the landing site and most of the northern part of the north-south-elongate terra on which it is located (figs. 10.19, 10.20). The site is 550 km south of Montes Carpatus, which constitutes the south topographic rim of Imbrium and the boundary of the excavation cavity (fig. 10.18). Taking the Imbrium impact point as lat 33° N., long 18° W., the excavation cavity has a radius of about 540 km in this sector. Secondary craters are buried near the landing site (fig. 10.20; Eggleton and Offield, 1970) and are invisible farther north. Despite appearances that suggest dominance by primary ejecta at one-radius distances, the currently most hotly debated issue about the Apollo 14 rocks is the proportion of primary ejecta to nonbasin material in the sample collection.

Sample provenance

The samples returned by Apollo 14 were collected on two traverses outside the Lunar Module (LM) in February 1971 (fig. 10.21; Swann and others, 1971, 1977; Sutton and others, 1972). The first

FIGURE 10.15.—Several facies of Imbrium ejecta and an extensive stratigraphic section of older and younger units. Knobby, nonlineated Alpes Formation (bottom) is transected by Vallis Alpes (VA). All pre-Imbrian craters north of Mare Frigoris (F; compare fig. 10.6) are transected by Imbrium sculpture and filled by Lower Imbrian plains deposits. W. Bond (W; 158 km, 65° N., 4° E.) is also filled by hummocky and braided Fra Mauro Formation. W. Bond, Barrow (B; 93 km), and Goldschmidt (G; 120 km) are probably pre-Nectarian. Epigenes (E; 55 km) has a higher wall, despite its smaller size, and so it is probably Nectarian (not covered earlier by Humboldtianum deposits). Protagoras (P; 22 km) and Timaeus (T; 33 km) are not affected by Imbrium basin but are peripherally flooded by Imbrian mare basalt, and so they are Imbrian; Protagoras is degraded, possibly Lower Imbrian; Timaeus is fresh, Upper Imbrian. Archytas (A; 32 km) is fresh and superposed on mare, and so is probably Eratosthenian. Anaxagoras (An; 51 km, compare fig. 3.34) is very fresh and bright-rayed (see W. Bond floor), and so it is Copernican. Orbiter 4 frame H-116.

traverse (extravehicular activity [EVA]), west of the landing site, was relatively short but resulted in the collection of many fragments from the regolith and some large rocks. Most of the hand specimens of coherent rocks were collected on the second EVA, whose main goal was the ejecta of Cone Crater, 1.1 km northwest of the landing site (figs. 10.21–10.24). Cone is a Copernican crater, 25 million years old (table 13.2), 370 m across, and at least 75 m deep (Swann and others, 1977, pls. 1, 2). Interpretation of the rocks' ultimate provenance depends on whether Cone Crater sampled the Fra Mauro Formation, the underlying bedrock, or only the regolith built on the Fra Mauro Formation. Estimates of regolith thickness at the landing site range from 5–12 m, predicted from observations of small craters (Eggleton and Offield, 1970), through 8.5 m, inferred from seismic-refraction measurements (Cooper and others, 1974), to about 35 m, estimated theoretically on the basis of superposed craters (Moore and others, 1980b, p. 13–14). Because Cone Crater lies on a ridge crest, the regolith that it penetrated is unlikely to be thicker than even the largest of these values. Thus, Cone Crater probably penetrated the regolith and excavated the bedrock unit, which is identified as a 19- to 76-m-thick layer with a seismic velocity of 299 m/s (fig. 10.23; Kovach and Watkins, 1972; Chao, 1973).

Interpretations of what bedrock material the Cone impact excavated from beneath the regolith depend partly on interpretations of the ridge. The ridge is 50 to 100 m high (Sutton and others, 1972; Swann and others, 1977, pls. 1, 2). Hawke and Head (1977b) proposed that it is underlain by a large pre-Imbrian crater, in which case Cone might have excavated some of the crater material. Eggleton and Offield (1970), however, interpreted the ridge as one of the group that is characteristic of the Fra Mauro Formation and estimated a Fra Mauro thickness of 100 to 200 m on the basis of the ridge relief. In this interpretation, Cone probably did not penetrate beneath the Fra Mauro Formation. Even if the ridge is underlain by a pre-Imbrian crater, a substantial thickness of the Fra Mauro must overlie it (fig. 10.23; Chao, 1973). Because the bulk of a crater's ejecta comes from the upper part of the target material (Stöffler and others, 1974), the hand specimens of coherent rock from Cone's ejecta are likely to consist of Fra Mauro bedrock (Swann and others, 1977). Many of the smaller fragments could have been derived from regolith that overlay the Fra Mauro before the Cone impact; these fragments would also consist mostly of Fra Mauro material, recycled by impacts. However, they probably would also include bits of pre-Imbrian material dug from beneath the Fra Mauro Formation by craters older and larger than Cone (Head and Hawke, 1975; Swann and others, 1977, p. 28).

Samples important for interpreting and dating the Fra Mauro Formation were also collected, on both EVA's, from relatively smooth terrain near the LM and at stations A, B, E, F, and G (fig. 10.21). Before the mission, this smooth unit was interpreted alternatively as a facies of the Fra Mauro Formation or as a superposed volcanic deposit (Eggleton and Offield, 1970). It is clearly not volcanic and appears to be sufficiently similar in composition and lithology to the Cone Crater material as to be thought part of the Fra Mauro Formation (Swann and others, 1977). However, the relations among the parts of the landing-site region are somewhat unclear because of limited data gathered in the field.

Petrology

The coherent rocks from the Cone Crater ejecta exhibit flow banding, large-scale stratification, and strong thermal effects, features that suggested to early analysts an origin in a vigorously

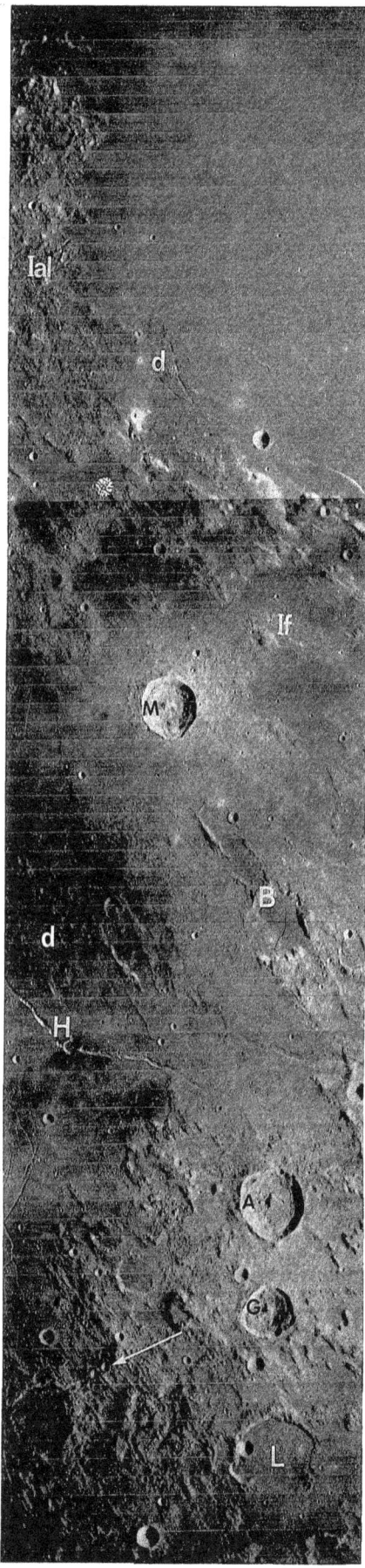

FIGURE 10.16.—Outward gradation of Imbrium deposits. H, Rima Hyginus (compare figs. 5.10E, 6.16A). Knobby ejecta (Alpes Formation, Ial) grades to smoother Fra Mauro Formation (If) in and near Montes Haemus. Surface flow of Imbrium ejecta apparently was deflected by Imbrium-secondary chain Boscovich (B; Boscovich Formation was interpreted as volcanic by Morris and Wilhelms, 1967). Terrain south of Hyginus, consisting of pre-Imbrian craters (for example, Lade [L; 56 km]; compare fig. 10.17), is scored by Imbrium lineations and is mapped either as pre-Imbrian lineated terrain (Wilhelms and McCauley, 1971) or as part of Imbrium zone (this volume). Arrow, possible Orientale-basin secondary chains; Orientale is centered 3,100 km to west in direction of arrow. d, dark-mantling deposits near Sulpicius Gallus (compare fig. 1.7) and north of Rima Hyginus. Fresh craters include Manilius (M; 39 km, 14.5° N., 9° E., probably latest Imbrian), Agrippa (A; 44 km, younger than mare and probably Eratosthenian), and Godin (G; 35 km, younger, more sparsely cratered, brighter, and with a higher reading in thermal infrared than Agrippa; probably Copernican). Overlaps with figures 10.1, 10.35, and 10.38. Orbiter 4 frame H-97.

FIGURE 10.17.—Lineated terrain at top (north) is Fra Mauro Formation, probably superposed on Imbrium-secondary chains. Crater Lade (bottom; 56 km, 1° S., 10° E.) is filled with hummocky deposit, probably consisting of Fra Mauro material that flowed over rim and decelerated. Apollo 11 frame H-4552.

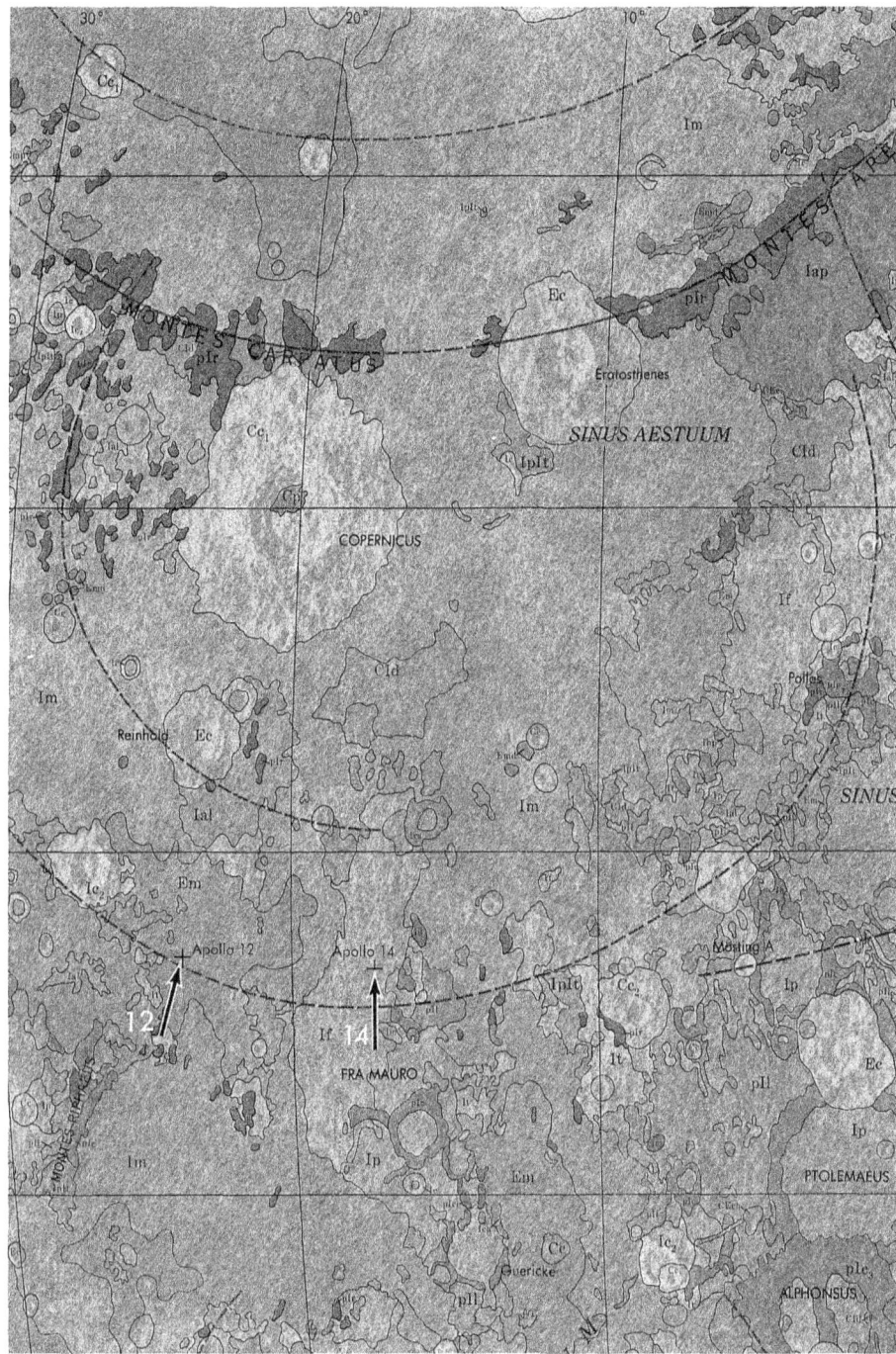

FIGURE 10.18.—Part of geologic map of lunar nearside (Wilhelms and McCauley, 1971), including Apollo 12 and 14 landing sites (arrows). If, Fra Mauro Formation; Em, Eratosthenian mare material (Apollo 12; see chap. 12). Montes Carpatus, 550 km north of Apollo 14 landing site, constitutes south rim of Imbrium basin. Ring passing near landing sites is part of Insularum basin in one interpretation, but may be ring of Procellarum basin (see chap. 8; pls. 3–7).

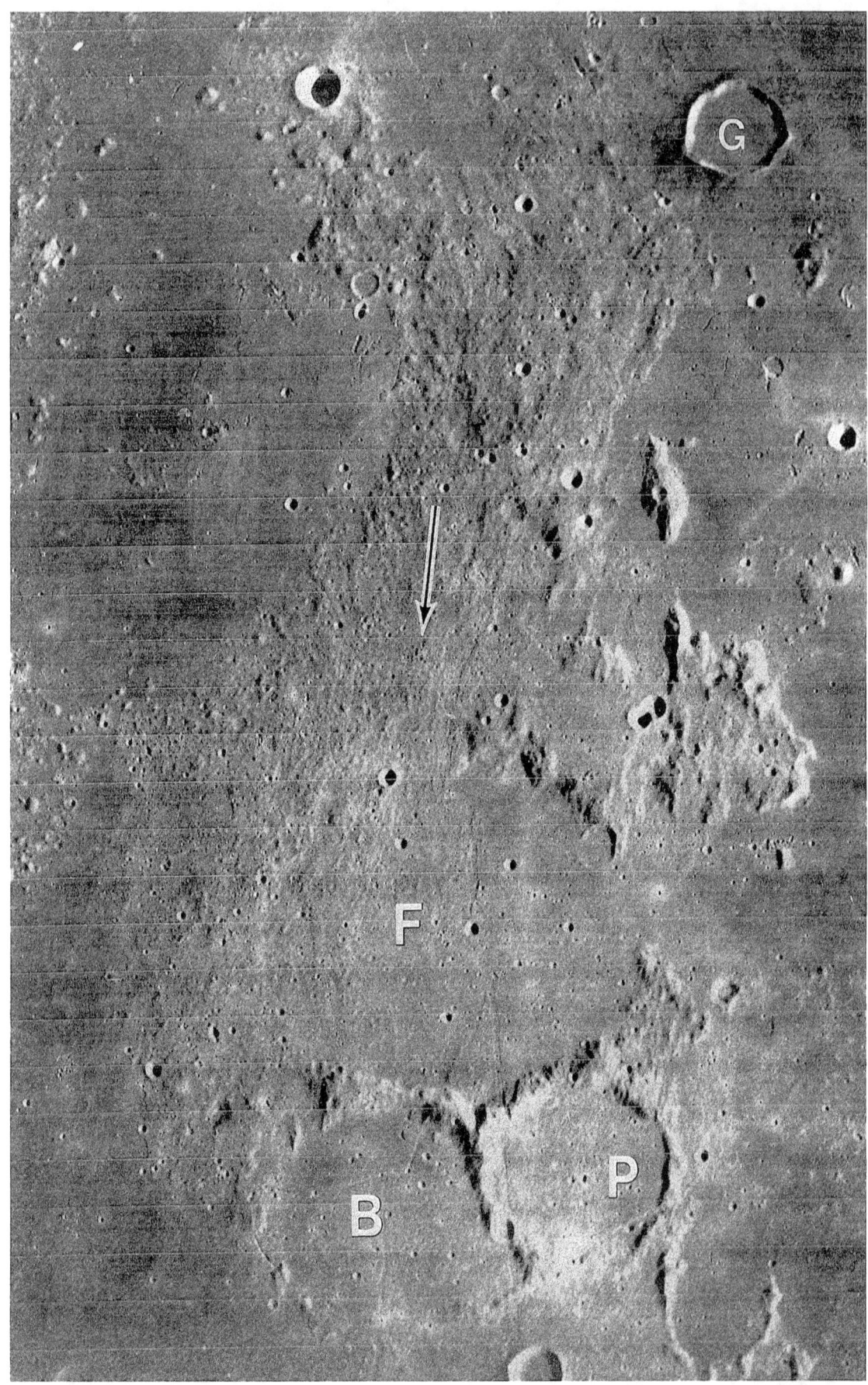

FIGURE 10.19.—Type area of the Fra Mauro Formation. Arrow, Apollo 14 landing site. Fra Mauro Formation, Imbrium sculpture, and plains deposits overlie pre-Nectarian or Nectarian craters Fra Mauro (F; 95 km), Bonpland (B; 60 km), and Parry (P; 48 km); Fra Mauro Formation is overlain by Imbrian crater Gambart (G; 25 km, 1° N., 15° W.). Orbiter 4 frame H–113.

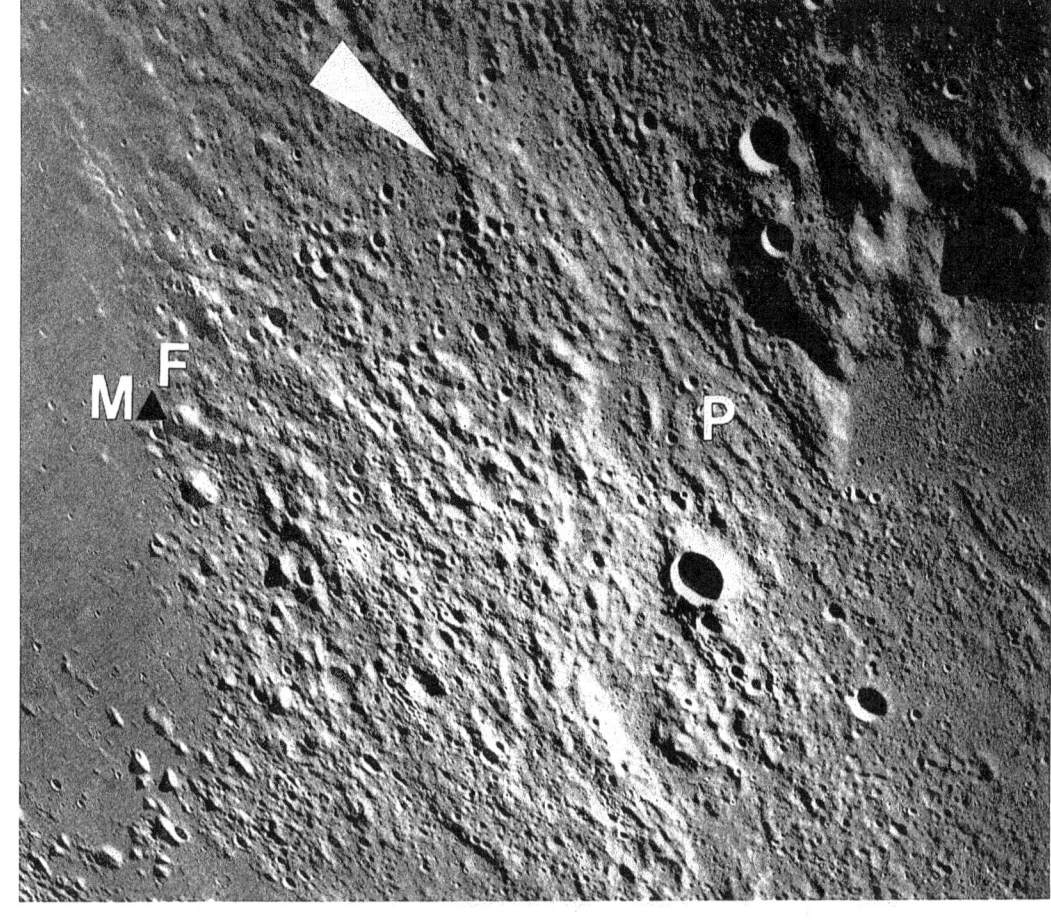

FIGURE 10.20.—Vicinity of Apollo 14 landing site, including Fra Mauro Formation (F) in sharp contact with mare material (M) at triangle. The Fra Mauro Formation is strongly lineated radially to Imbrium basin (compare fig. 10.19). Cone Crater is barely visible at tip of arrow. P, buried pre-Imbrian crater. Apollo 12 frame H-7597.

FIGURE 10.21.—Apollo 14 astronaut traverses. Pennant, landing point; ALSEP, geophysical instruments; lettered and numbered circles, sampling stations. Numerical scale refers to original map. Airbrush drawing by U.S. Geological Survey, published by U.S. Defense Mapping Agency for NASA.

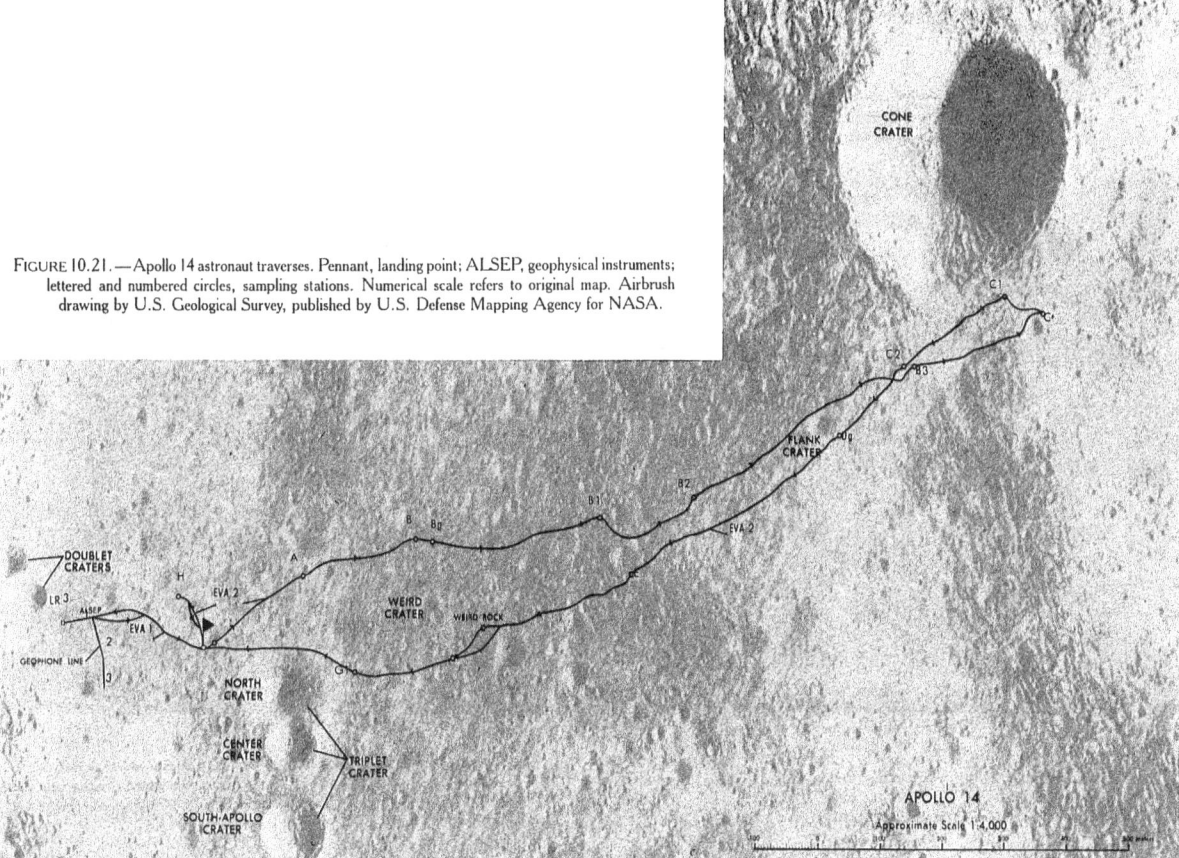

emplaced, hot ejecta blanket (fig. 10.22; Engelhardt and others, 1972; Quaide and Wrigley, 1972; Wilshire and Jackson, 1972). Heat-induced textures were first interpreted as the products of metamorphism or annealing (Quaide and Wrigley, 1972; Warner, 1972; Wilshire and Jackson, 1972) but are now widely considered to be the result of crystallization of impact melts (Engelhardt and others, 1972; Ryder and Bower, 1976; Lofgren, 1977). The impact-melt rock is crystalline-matrix breccia that grades from clast-free (fig. 2.8) to heavily laden with fragments (fig. 10.24). The most common types of this gradational series consist of more than 15 percent clasts in gray, coherent,

FIGURE 10.22.—Large boulders near rim of Cone Crater, containing light and dark materials. From Swann and others (1977, fig. 44).

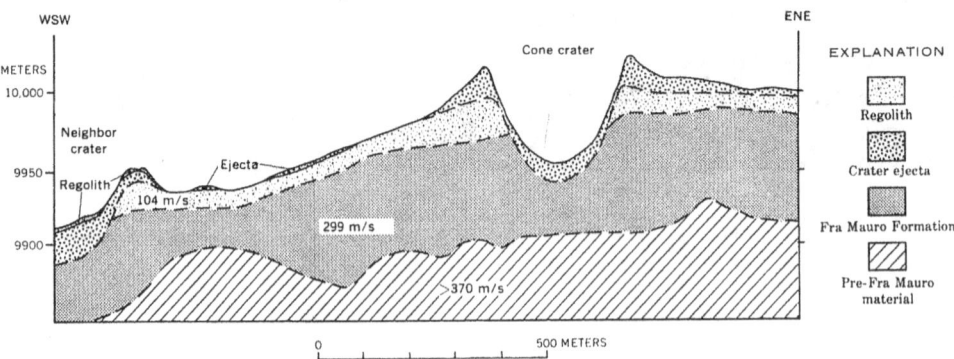

FIGURE 10.23.—Schematic geologic cross section of Apollo 14 landing site (from Chao, 1973). Section is drawn just north of landing site in west to Cone Crater in east. Thicknesses of geologic units are based mostly on active-seismic data (Kovach and Watkins, 1972). Compressional-wave velocities of some units are indicated. Elevations in vertical scale are based on reference elevation chosen to correspond to mean radius of Moon, 1,738 km.

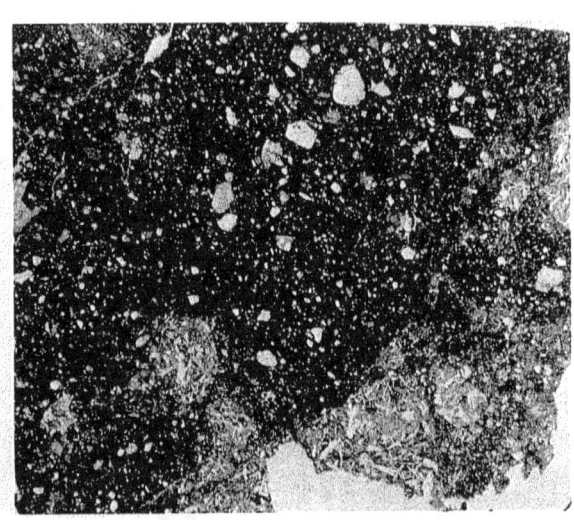

FIGURE 10.24.—Large (8.9 kg) Fra Mauro breccia sample 14321, "Big Bertha," from rim of Cone Crater, Apollo 14 landing site: a polymict breccia displaying multiple episodes of brecciation (Duncan and others, 1975; Grieve and others, 1975).
A. Entire hand specimen, showing complex relations of clasts and matrix.
B. Thin section of sample 14321,197, showing abundant mineral and lithic clasts set in dark, fine-grained matrix. Light-colored, coarsely crystalline materials along right and bottom edges are clasts of high-Al mare basalt (table 9.5). Nonpolarized light; field of view, about 1.5 cm.

melt-derived matrices (Chao and others, 1972; Engelhardt and others, 1972; Wilshire and Jackson, 1972; Chao, 1973; Simonds and others, 1977). These clast-laden breccia samples were collected from both Cone Crater and the smooth terrain; because they characterize the landing-site region, they have been termed *Fra Mauro breccias* (Chao and others, 1972). Light-colored friable breccia samples ("white rocks") are also common but are probably clasts previously included in the crystalline-matrix breccia (Chao and others, 1972). Samples of almost fragment-free impact melt (Green and others, 1972; Taylor and others, 1972; James, 1973; McKay and others, 1978, 1979) were collected from the regolith on the smooth terrain; the largest are samples 14073, 14276, and 14310 (fig. 2.8). At one time, these samples were widely considered to be volcanic (see chap. 8).

The samples are typically rich in KREEP. Most of the crystalline-matrix breccia seems to be relatively uniform in composition; 12 of the 16 samples analyzed by Simonds and others (1977) cluster together on a plot of MgO content versus Al_2O_3 content as does "vitric matrix" breccia, consisting of consolidated regolith fragments; not further discussed here). Clast-free rock samples 14276 and 14310 contain less MgO and more Al_2O_3 than the cluster. McKay and others (1979) found that the chemical and isotopic differences among the KREEP-rich impact-melt samples are consistent with an origin in a single melt sheet.

Emplacement process

A debate about emplacement mechanism persists even among those who agree that most samples are from the Fra Mauro Formation and that the Fra Mauro is somehow related to the Imbrium basin (fig. 10.25). One view is that the rocks that compose the Fra Mauro are parts of the primary ejecta of the basin (Eggleton and Offield, 1970; Chao and others, 1972; Dence and Plant, 1972; Engelhardt and others, 1972; Quaide and Wrigley, 1972; Wilshire and Jackson, 1972; Chao, 1973). That is, they were parts of the target material that traveled at least 550 km to their site of deposition (hypothesis b, fig. 10.25). This view is consistent with the generally uniform composition of the Fra Mauro breccia samples. Either thorough mixing of target materials or derivation from a compositionally uniform source, such as a KREEP-rich crustal layer or a KREEP-basalt section in the Procellarum basin, could have produced this uniformity. The primary-impact mechanism also allows for variations in composition and shock grades because of the general disorder of the basin-forming process. The differing compositions and shock grades of the friable, light-colored materials and the Fra Mauro breccia are readily comprehended in this model as the results of formation in different parts of the shock-engulfed target zone, followed by mixing during ejection or emplacement (Engelhardt and others, 1972). That is, some of the already heterogeneous bedrock-breccia, regolith, and volcanic deposits in the target zone were little altered by the Imbrium impact, whereas others were more highly shocked, melted, and homogenized.

The opposing view is that the crystalline-matrix breccia originated at or near the deposition site (Oberbeck and others, 1974; Morrison and Oberbeck, 1975; Head and Hawke, 1975; Oberbeck, 1975; Stöffler and others, 1976; Hawke and Head, 1977b, 1978a, b; Simonds and others, 1977). The proponents of this "local" hypothesis are troubled by the diversity of textures and compositions of the breccia samples and by the great distance of the Apollo 14 landing site from the Imbrium basin. They believe that the terrestrial analogs (Chao and others, 1972; Dence and Plant, 1972; Simonds and others, 1977) show that the abundant melt found in the breccia deposits could not have been transported so far. In the local-derivation models, the Fra Mauro Formation (and analogous deposits of Orientale and smaller craters) consists of mixtures of the primary ejecta with material derived from the local terrane by secondary impacts (hypothesis a, fig. 10.25). The radial ridges of the Fra Mauro that are interpreted as flows of primary ejecta in the first model are considered in the local-derivation model to be analogous to the herringbone texture visible among secondary craters of Copernicus-type craters (fig. 3.4), or to surges of debris that continued to flow outward after their generation by secondary impacts. Such coalesced debris flows may have erased the original secondary-ejecta pattern. On the basis of the upper equation in figure 10.26, Morrison and Oberbeck (1975) calculated that only 15 to 20 percent of the Fra Mauro Formation at the Apollo 14 landing site should consist of primary ejecta. Head and Hawke (1975) pointed out craters and basins whose melted and fragmental deposits are reasonable sources of the local materials. They considered the KREEP-rich compositions to have been derived from volcanic extrusions in the vicinity of the landing site (Hawke and Head, 1978b). They stated that mineralogically recorded shock pressures in the narrow range 20–30 GPa should be of Imbrium-basin origin and that less highly shocked and, possibly, more highly shocked material should be of local origin (Hawke and Head, 1978a).

Both the primary-ejecta model and the secondary-impact, local-material model probably contain some truth. Great diversity in shock grade and other lithologic attributes is surely possible at a given point in the ejecta deposit of a large impact. On the other hand, the pervasiveness of secondary impacts at great ranges from basins cannot be denied, and incorporation of the impacted local material is to be expected. The issue is thus one of proportion. I believe that primary ejecta is the dominant component and that the proponents of secondary emplacement of the entire Fra Mauro Formation have overstated their case, for several reasons (Wilhelms and others, 1980):

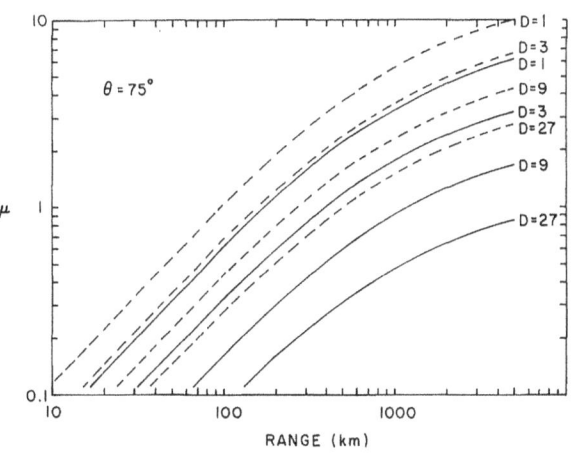

FIGURE 10.26.—Mass ratio of primary versus local material as a function of range from source of incoming ejecta. Dashed lines are based on upper equation, which assumes scaling of $D = KE^{-3.4}$ (Oberbeck and others, 1975); solid lines are based on lower equation, which assumes scaling of $D = KE^{-3.6}$ (H.J. Moore, written commun., 1980). μ, ratio of mass ejected from secondary crater to mass of ejecta projectile that created crater; θ, angle of impact (equal to angle of ejection), measured from local surface normal; D, diameter of secondary crater (in kilometers); R, range or distance along Moon's surface from point of ejection within primary crater or basin to secondary crater (in kilometers). Courtesy of H.J. Moore.

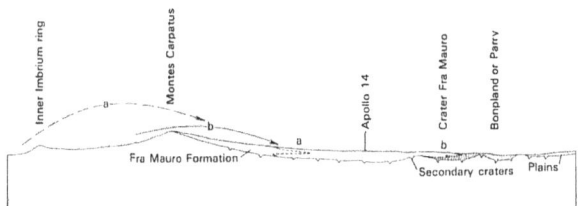

FIGURE 10.25.—Geologic cross section illustrating secondary-ejecta (a) and primary-ejecta (b) hypotheses for emplacement of the Fra Mauro Formation, drawn along north-south line in figure 10.18. In most secondary-ejecta models, Imbrium cavity is bounded by an inner ring, ejection trajectories are steep (dashed arc), and transition between primary and secondary deposits is close to rim (dashed sawteeth beneath right-hand letter a). In primary-ejecta model preferred here, basin rim is at Montes Carpatus, ejection trajectories are shallow (solid arc), and transition from primary to secondary deposits occurs south of Apollo 14 landing site (solid sawteeth beneath right-hand letter b). Secondary craters form in both models (diagrammatically indicated by notches in substrate): in hypothesis a, their ejecta forms sampled part of the Fra Mauro Formation; in hypothesis b, they are deeply buried by primary ejecta at Apollo 14 landing site.

1. The topography of the landing-site region is dominated by ridges formed by flow of a thick deposit that has obscured the secondary craters. This flow originated either at the basin rim or at secondary-impact sites much closer to the basin than its present resting place. Either circumstance would change the "range" input in the equation (fig. 10.26).
2. The equation in figure 10.26 is model-dependent in other ways. The assumed ejection angle, size of the Imbrium basin, point of derivation within the basin, size of secondary craters (which are difficult to see and measure near the landing site), and diameter-energy scaling laws are all questionable. An apparently small change in the exponent of the energy in the scaling relation from -3.4 to -3.6 changes the equation to the lower form in figure 10.26 and greatly affects the results of calculations. Moreover, the equation is based on single-body impacts (Morrison and Oberbeck, 1975, p. 2527), whereas most basin ejecta at great ranges is likely to be comminuted (Schultz and Mendell, 1978). Many small low-density fragments impacting over an extended time at low velocities will build up a deposit of their own material rather than excavate local materials greater than their mass (Chao and others, 1975; Morgan and others, 1977, p. 672).
3. Photogeologic relations indicate the presence of abundant melt in basin-ejecta deposits. (a) Material ponded in depressions in the inner Orientale and Imbrium ejecta displays "mud crack" patterns of fine fissures, probably resulting from shrinkage of a coherent material (figs. 4.4E, 10.14; Moore and others, 1974). These fissured ponds resemble the external impact-melt pools of large craters, which increase in abundance and distance from the rim with increasing crater size and occur as far as two crater radii from some rims (chap. 3; Howard and Wilshire, 1975; Hawke and Head, 1977a). (b) Crater ejecta is coated by additional melt in the form of veneers that are gradational with the ponded melt (fig. 3.36; Howard, 1975; Howard and Wilshire, 1975; Hawke and Head, 1977a). (c) Large tongues of nontextured material apparently have segregated from the radially lineate Hevelius Formation and have formed distant lobes (fig. 4.4D). The lobate, commonly leveed morphology of these segregates suggests emplacement as viscous fluids, for which melt is a reasonable, though not exclusive, explanation (Eggleton and Schaber, 1972; Moore and others, 1974). Finally, (d) pools in the floors of many secondary craters appear to have once been molten, an appearance suggesting considerable ranges for ejected melt (figs. 3.9A, 4.4C; Moore and others, 1974; Schultz and Mendenhall, 1979).
4. Scaling laws suggest that large, basin-forming impacts generate more melt than small, crater-size impacts on either the Moon or the Earth (chap. 3; Rehfuss, 1974; O'Keefe and Ahrens, 1977; Lange and Ahrens, 1979; Melosh, 1980, p. 77). An increase in the ratio of projectile mass to crater size is predicted with increasing magnitude of impact (Baldwin, 1963, chap. 8; O'Keefe and Ahrens, 1975; Gaffney, 1978). The impact energy of large bodies is coupled into the target longer, so that the shock energy is distributed over a larger volume of rock. In contrast, small fast bodies strike and rapidly release their kinetic energy into small, intensely heated volumes of rock. The result is that large impacts generate larger volumes of less highly heated melt than do small impacts. As a corollary, melt of the large impacts would presumably be less thoroughly homogenized than that of small impacts.
5. Finally, the target may have been hot already, so that any impact would create more melt than would an impact into a cold target (Wood, 1975d, p. 515). Evidence of heat at crustal depths is contained in the coarse grain size of Apollonian metamorphic rocks (Stewart, 1975) and the textures of granulitic impactites, ascribed to accumulated heat from hot ejecta blankets (Warner and others, 1977). Radiogenic heat may have melted KREEP-rich magmas before (see chap. 9) and after (Apennine Bench Formation) the Imbrium impact, and liquid magma may even have been ejected (Schultz and Mendenhall, 1979).

These points are offered not as proof of a primary-ejecta origin but as reasonable alternatives to the local-origin models currently widely accepted. The data of Simonds and others (1976b, 1977) also seem to be consistent with a primary-ejecta origin, although those authors favored the local-origin model. Simonds and others (1976a, b) suggested that the Fra Mauro crystalline-matrix breccia formed as superheated impact melt plus cold fragmental debris and that the differences in clast content are solely responsible for the textural differences. As discussed above, the compositional data (Simonds and others, 1977; McKay and others, 1979) also seem to be more consistent with a limited range of source compositions than with the diverse "local" target cited by many authors. The main support for the local-derivation model seems to be (1) terrestrial-analog craters whose ejecta has not been identified (Canadian-shield craters), and (2) the upper equation in figure 10.26 (Morrison and Oberbeck, 1975), which is highly model-dependent and is qualitatively contradicted by the absence of conspicuous Imbrium-secondary craters in the appropriate uprange positions to have been the sources of the Fra Mauro Formation.

Absolute age

Two age clusters have been identified in the returned samples by both the ^{40}Ar-^{39}Ar (Turner, 1977) and Rb-Sr (Papanastassiou and Wasserburg, 1971b) methods, which agree well for these rocks (table 10.1). The younger cluster was obtained mainly for some clast-poor impact-melt rock samples (14073, 14276, 14310), for a few samples of Fra Mauro breccia, and for small fragments from mixed regolith samples. The measured ages of the clast-free rocks average 3.82 aeons; of the clast-free fragments, 3.83 aeons; and of other melt-rock samples of uncertain origin, including some that may belong to the older cluster, 3.84 aeons (table 10.1). If the dated samples are of Imbrium-basin impact-melt rock, as I believe, the basin formed 3.82 to 3.84 aeons ago, with an uncertainty of only a few tens of millions of years. However, the uncertain field relations between samples from the smooth terrain and from Cone Crater do not preclude the possibility that some samples are derived from melts of a post-Imbrium crater.

The best dated older group of Apollo 14 samples (extremes and stated analytical errors larger than 0.1 aeon eliminated) yields ages of 3.87 to 3.96 aeons (tables 9.5, 10.1). All three large samples (14053, 14072, and clasts of 14321) came from the Cone Crater flank, and all are of high-Al mare basalt (fig. 10.24; table 9.5; Grieve and others, 1975; Ridley, 1975; Taylor, 1975; Simonds and others, 1977; Ryder and Spudis, 1980). They differ further from the younger group in their lower initial Sr-isotope ratios (Papanastassiou and Wasserburg, 1971b). Although error bars overlap, these rocks form a distinct age group from the younger rocks. Clasts of breccia sample 14066 and some fragments from regolith samples 14161 and 14167, all collected within 400 m of the LM, may also belong to the older cluster (lithologies uncertain; listed with the younger group in table 10.1). The older clasts and regolith fragments appear to be fragments of pre-Imbrian rock, including volcanic basalt, incorporated into the Imbrium-ejecta breccia before its emplacement near the future landing site (chap. 9; Mark and others, 1975; Wetherill, 1977b). This basalt-in-breccia relation is consistent with either the primary- or the secondary-ejecta model; beds of basalt could have existed before the Imbrium impact either in the Imbrium target area (Wilshire and Jackson, 1972) or in the vicinity of the Apollo 14 landing site (Hawke and Head, 1978b). In either model, the age of the Imbrium basin is constrained as equal to (Compston and others, 1972) or less than (Wetherill, 1977b) 3.87 aeons.

Conclusions

The following working hypotheses (fig. 10.25) are consistent with all data from the samples known to me and with Moon-wide relations. (1) The crystalline-matrix clast-laden Fra Mauro breccia excavated from Cone Crater is sufficiently uniform structurally and compositionally to indicate derivation from a single major rock unit. (2) That unit is part of the photogeologically observable ridgy unit, the Fra Mauro Formation, which is composed mostly of primary ejecta of the Imbrium basin. (3) Part of the target material of the Imbrium impact consisted of mare basalt. (4) Ages of basalt clasts in the Fra Mauro breccia deposits constrain the age of the formation to 3.87 aeons or younger. (5) Ages of clast-poor impact-melt rocks in other parts of the landing-site region constrain the age of the formation to 3.82 aeons or older. (6) These results are consistent with the Apollo 15 ages (table 10.1). (7) The likely time of the Imbrium-basin impact is narrowed to 3.845 ± 0.03 aeons ago by these combined results.

IMBRIUM-BASIN SECONDARY CRATERS

Like the Hevelius Formation at Orientale, the Fra Mauro Formation grades outward from a thick and continuous coarsely textured blanket, through plains deposits, to a wide zone of secondary craters. Southeast of the Imbrium basin, the transition from plains to secondaries occurs about at Sinus Medii, 450 to 750 km from the Apenninus crest or 1,150 to 1,450 km from the basin center (figs. 10.16, 10.27, 10.28). Due south, the Fra Mauro Formation and mare materials have mostly buried the secondary craters, some of whose rims protrude around the margins of the Fra Mauro peninsula (fig. 10.19). In the north, this transition is also poorly defined because so many rugged linear chains of the Vallis Bouvard type (fig. 4.4D) are visible through the Fra Mauro continuous ejecta (figs. 10.6, 10.15). The transition is estimated to occur about 300 to 400 km north of Mare Frigoris or 1,000 to 1,100 km from the basin center (pl. 3). In all observed sectors, the transition occurs farther from the basin than for Orientale, as expected for a topographically larger basin.

The most striking characteristic of the zone of secondary craters is the "Imbrium sculpture" (figs. 10.15, 10.27–10.29). Most of this sculpture, once thought to be fault-controlled (see chap. 6), is now thought to consist of chains of elliptical secondary-impact craters (compare figs. 10.27, 10.29). Although some of the finer outer sculpture seems to have formed by surface flow (fig. 10.17), as in the Fra Mauro closer to the basin and in the Hevelius Formation, the origins of individual grooves and ridges cannot be determined so readily as around Orientale (Wilhelms, 1980).

Abundant secondary-impact craters appear beyond the groovelike chains at distances of about 1,700 km from the basin center (figs. 10.30, 10.31). Like the sculpture, the more nearly circular secondaries were once considered to be volcanic or primary-impact craters. One basis for this interpretation was the morphologic diversity within a given cluster (figs. 10.28, 10.29). However, individual satellitic craters of Orientale and of large craters also differ from one another (figs. 3.9, 4.4). Good photographs show that a range in morphologies from very sharp to almost obliterated results from differential burial of a coeval cluster by deposits that flowed from nearer the basin (figs. 4.4G, H; Oberbeck, 1975; Wilhelms, 1976; Wilhelms and

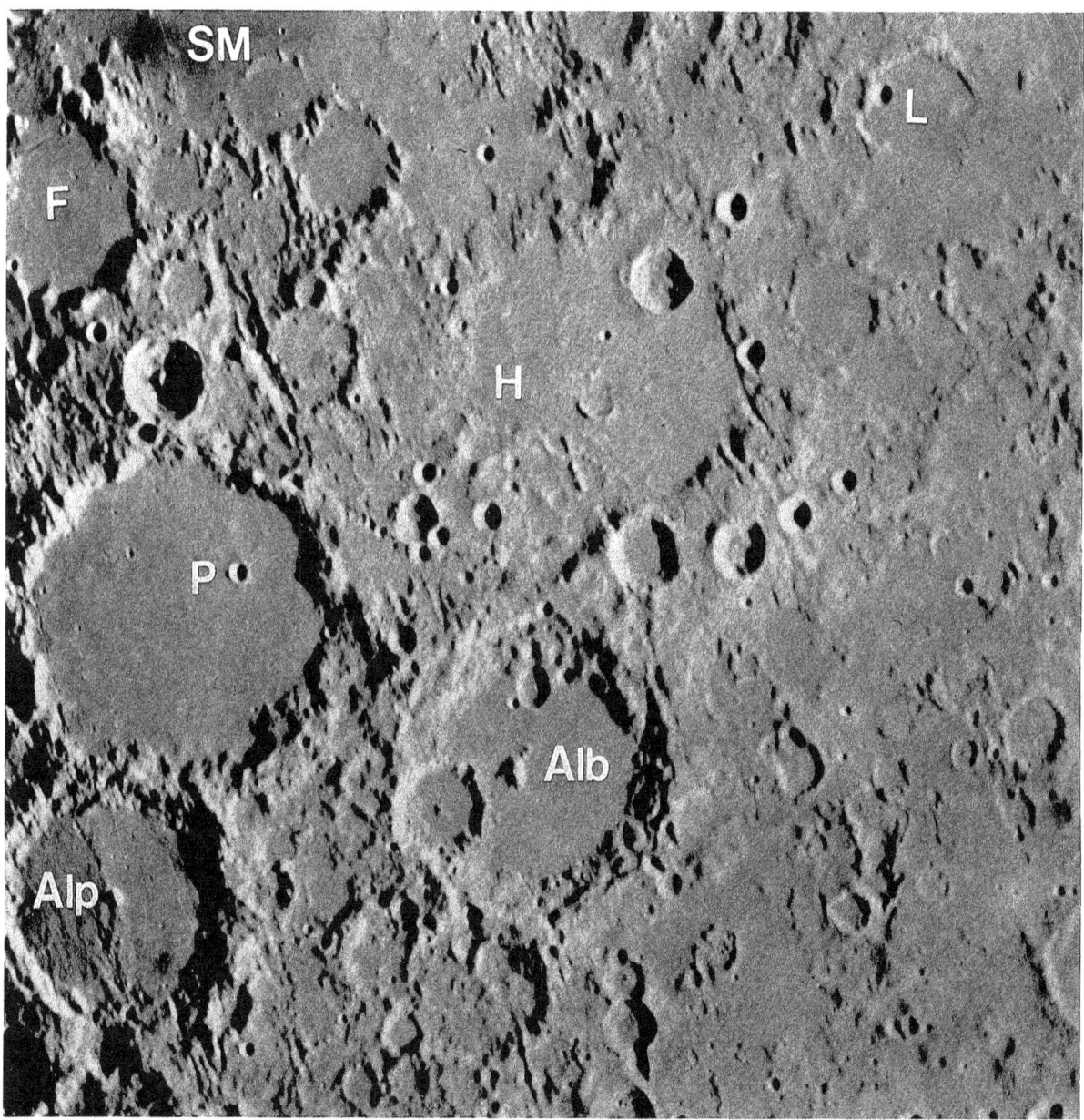

FIGURE 10.27.—Imbrium sculpture in south-central nearside highlands south of Sinus Medii (SM). Craters cut by sculpture include Lade (L; compare figs. 10.16, 10.17), Flammarion (F; 75 km), Hipparchus (H; 151 km), Ptolemaeus (P; 153 km), Albategnius (Alb; 136 km, compare figs. 10.29, 10.36), and Alphonsus (Alp; 119 km). Planar fill apparently transects sculpture. Subcircular secondary craters of Imbrium are visible below latitude of Ptolemaeus. Catalina Observatory photograph No. 1907.

others, 1978, p. 3749). The septa between craters and "domes" superposed on their intersections (figs. 3.9B, 4.4C, 10.29–10.31), which were once cited as evidence of volcanism, resemble forms created in similar positions by the interaction of ejecta in laboratory experiments (fig. 3.7). Only the gross form of the domes and intercrater ridges is observed in most of the Imbrium-secondary field. Study of Orientale analogs and the laboratory experiments were required before the secondary-impact origin of these domes and ridges could be supported.

An Imbrium-basin origin of furrows or pits outside the main concentration of secondaries (figs. 10.32, 10.33) is also likely. A large tract west of Mare Humorum (fig. 9.24A), which contains more rugged, hilly topographic elements than most terrain at the same distance from Imbrium (2,000 km from the basin center), was interpreted as Humorum-basin ejecta (Titley, 1967) (part of the Vitello Formation) or as volcanic (McCauley, 1973; Wilshire, 1973). Secondary-impact origin emerged as a reasonable alternative during mapping of the whole circum-Imbrium zone. This tract is well within the reach of Imbrium ejecta projectiles and contains numerous soft-appearing, north-northeast-trending Imbrium-radial gougelike depressions as long as 30 km. It also adjoins, and is faintly gradational with, pitted terrain related to Orientale (fig. 4.4H). The tract is likely to be a composite unit consisting of Humorum ejecta scoured by Imbrium-basin secondaries and then mantled by thin Orientale ejecta; it is included in the Imbrium-related zone on plates 3 and 8. A somewhat similar terrane, previously thought to be volcanic (Wilhelms, 1972a; Olson and Wilhelms, 1974), is superposed on the south Crisium-basin rim flank, including the vicinity of the Luna 20 landing site (figs. 10.33, 11.16). Imbrium-secondary origin is suggested by radiality of the bands of pits and furrows.

Imbrium-basin effects persist farther outward. The first clustered craters that were identified as secondary to the basin (Scott, 1972b) occur near the craters Riccius and Maurolycus, 1,900 km from Montes Apenninus or 2,600 km from the basin center (fig. 3.10D). Scattered secondary craters of Imbrium occur beyond 3,000 km from the basin center and, possibly, in all parts of the lunar terrae. Probable secondaries have been identified at least 3,300 km from the center

FIGURE 10.29.—Terrain east of crater Albategnius (truncated by left edge of photograph; compare figs. 10.27, 10.28). Imbrium sculpture is generally linear but consists of elliptical craters. Domelike intersections of craters and other small hills (arrows) were created by multiple secondary impacts (see chap. 3; compare fig. 3.9B), not volcanism. Apollo 16 frame M–392.

FIGURE 10.28.—Strip from Mare Vaporum (top) southward to lat 23.5 S., showing transition from smooth, continuous Fra Mauro deposits (compare fig. 10.16) to highly diverse secondary craters below latitude of Ptolemaeus and Albategnius (compare figs. 10.27, 10.36). Mosaic of Orbiter 4 frames H–101 (bottom) and H–102 (top); overlaps figures 7.7 and 10.16.

on the southeast nearside (fig. 9.2; Stuart-Alexander, 1971; Wilhelms and others, 1978, 1979). Linear chains on the farside are so closely radial to the basin as to indicate Imbrium origin (fig. 8.13; Wilhelms and El-Baz, 1977).

Finally, peculiar grooves at the antipode of Imbrium on the farside (figs. 10.32, 10.33; Stuart-Alexander, 1978) have been ascribed to convergence of Imbrium-basin ejecta (Moore and others, 1974) or to convergence of seismic waves from the Imbrium impact shock (Schultz and Gault, 1975a, b). Similar terrain north of Mare Marginis antipodal to Orientale was noted in chapter 4 (fig. 4.7). All grooved terrain may, in fact, be related to basins (fig. 10.33). The Serenitatis and Crisium antipodes are similarly grooved (fig. 9.21A; Hood and others, 1979). Strangely, all four of these antipodal grooved regions are also characterized by anomalous concentrations of bright swirls and by high remanent magnetism as observed from orbit (Hood and others, 1979). The cause-and-effect relations of these peculiar coincidences of old topography (the grooves), very young features superposed on maria and lacking intrinsic relief (the swirls), and magnetism have yet to be determined. Hood and others (1979) tentatively suggested that the basin impact magnetized the antipodal terrain, either by secondary impact or seismicity, so as to deflect the solar wind during the rest of geologic time and permit the maintenance of high albedo in the otherwise-fragile swirls.

LIGHT-PLAINS MATERIALS

Around both the Orientale and Imbrium basins, light-colored plains are concentrated in the transition zones between the continuous deposits and the secondary craters (figs. 4.4, 10.34, 10.35). The clearly visible Orientale relations and the fact that the samples of presumably typical plains material returned by Apollo 16 were emplaced by impact have combined to shift interpretations from a volcanic to an impact origin for terra plains in general (see chaps. 2, 4). The superposition of plains on the "Imbrium sculpture" (figs. 10.27, 10.35, 10.36), which was once considered prime evidence for a volcanic origin (Howard and Masursky, 1968; Milton, 1968a; Wilhelms, 1968, 1970b), has been shown to be consistent with impact emplacement on the basis of similar relations at Orientale. Plains occupying depressions thousands of kilometers from Imbrium may also be of impact and not volcanic origin because they generally occur downrange from secondary-crater clusters (fig. 4.6). Although the appearance and crater densities of many plains remain equally consistent with volcanic and impact origins (Neukum and others, 1975a; Neukum, 1977), and although volcanic substrates may underlie the surficial ejecta in some plains (chap. 9; Eggleton and Marshall, 1962; Schultz and Spudis, 1979), impacts evidently formed most of the visible plains.

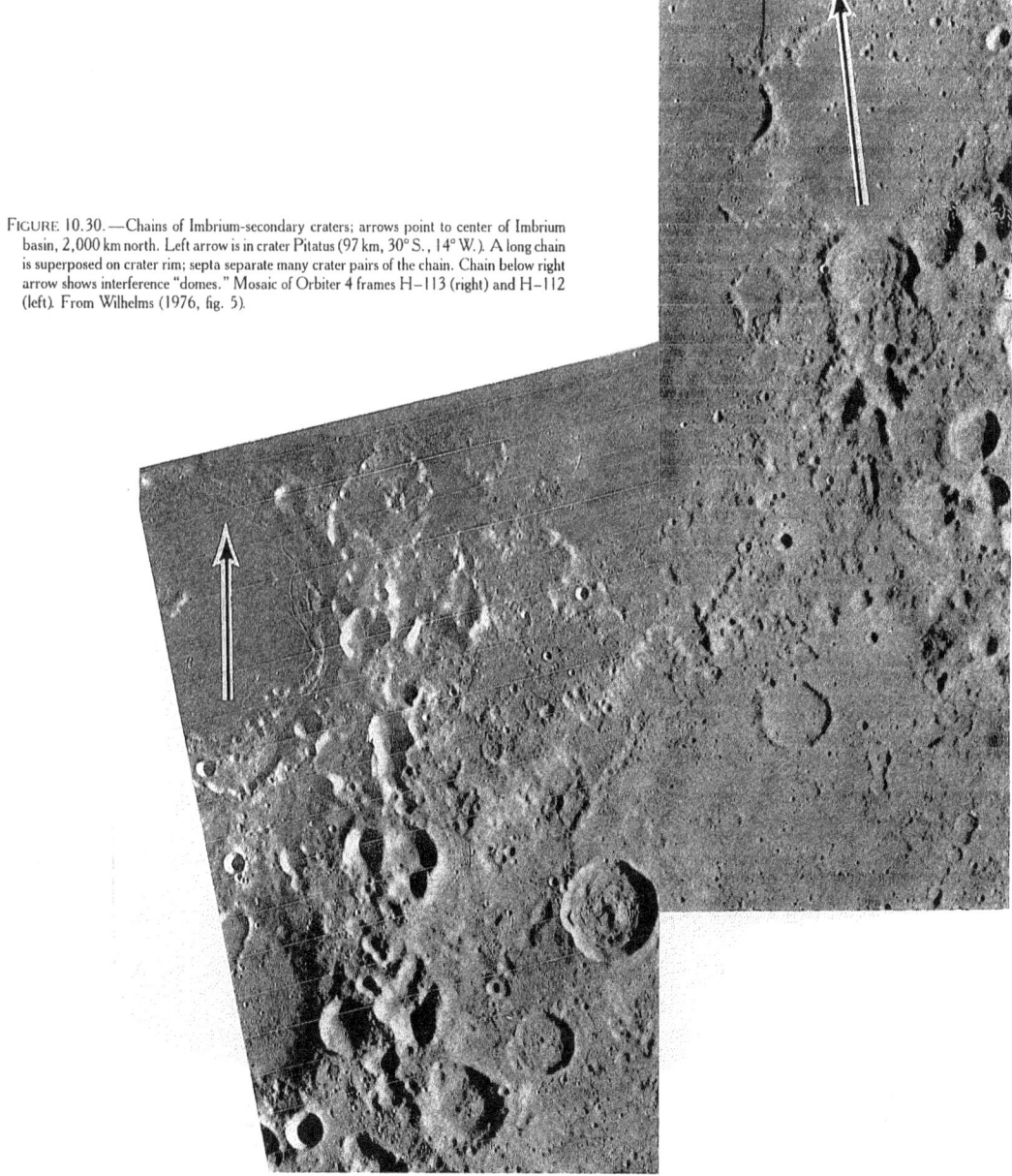

FIGURE 10.30.—Chains of Imbrium-secondary craters; arrows point to center of Imbrium basin, 2,000 km north. Left arrow is in crater Pitatus (97 km, 30° S., 14° W.). A long chain is superposed on crater rim; septa separate many crater pairs of the chain. Chain below right arrow shows interference "domes." Mosaic of Orbiter 4 frames H-113 (right) and H-112 (left). From Wilhelms (1976, fig. 5).

More patches of well-defined plains are Early Imbrian than of any other single age. Burial of the Apennine Bench Formation inside the Imbrium-basin rim by the deposits of the craters Archimedes and Cassini suggests an Early Imbrian age for this plains unit, and the radiometric dates point to an earliest Early Imbrian age. An earliest Early Imbrian age is suggested for the plains north of Mare Frigoris (fig. 10.15) by textural gradation with the Fra Mauro Formation, crater size-frequency values (figs. 7.15, 10.37), and visual inspection of crater densities. The partly planar Maunder Formation inside the Orientale basin almost certainly consists of impact melt contemporaneous with that basin (see chap. 4) and thus is latest Early Imbrian.

D_L values of 1,000 to 1,200 m have been identified with the plains north of Frigoris and, therefore, with the Imbrium basin (fig. 7.15; Boyce and others, 1974; Moore and others, 1980b). Similar values, however, have also been obtained for other plains (Boyce and others, 1974) that are clearly older than the basin (figs. 7.15, 8.11, 9.26, 9.28; Neukum, 1977; Wilhelms and El-Baz, 1977; Stuart-Alexander, 1978; Lucchitta, 1978). The inability of D_L values to discriminate among old deposits (fig. 7.15) may be due to several factors: (1) Such material properties as low cohesion of thick regoliths, and (2) the large sizes of superposed craters obviously present (fig. 9.26). For these large craters, (a) the production function may be less than the -2 cumulative assumed in D_L models (Soderblom, 1970; Soderblom and Lebofsky, 1972), and (b) initial crater shapes and wall slopes may differ from those of the small craters measured during D_L determinations. Stratigraphic relations and size-frequency counts are preferred here to D_L measurements as indicators of terra-plains age.

The relative age of the Cayley Formation is important to interpretations of the returned materials. On Lunar Orbiter 4 photographs, the Cayley generally seems to have flatter surfaces and fewer, sharper craters than the earliest Early Imbrian plains north of Mare Frigoris that are clearly contemporaneous with the Imbrium basin. (The Cayley seems to be less planar at higher resolution [fig. 10.38] and on the surface as seen by the astronauts.) D_L values of 475 to 640 m (fig. 7.15; Soderblom and Boyce, 1972; Boyce and others, 1974), some crater-density studies (fig. 10.37; Boyce and others, 1974; Neukum and others, 1975b), and visual inspection also suggest a clear temporal hiatus between the plains north of Mare Frigoris and the latest Early Imbrian plains at Orientale. The Cayley and circum-Orientale plains look alike on Lunar Orbiter 4 photographs. Some crater counts and D_L measurements also show them to be contemporaneous with each other and with the Hevelius and Maunder Formations (Soderblom and Boyce, 1972; Boyce and others, 1974). Therefore, an Orientale age has been suggested for the Cayley plains in the central nearside. Suggested mechanisms for creating Orientale-age surfaces on plains that are spatially related to Imbrium include mantling of older, Imbrium-basin-age (and Nectarian) plains by far-thrown Orientale ejecta (Hodges and others, 1973; Chao and others, 1975) and degradation induced by seismic waves from the Orientale impact (Schultz and Gault, 1975a). Either mechanism could theoretically have rejuvenated the cratered surface on the older material.

A dissenting view, based on different evaluations of the same data, is that the Cayley Formation is contemporaneous with the Imbrium basin. Other crater counts (fig. 10.37; Neukum and others, 1975a, b; Neukum, 1977), the inability of D_L determinations to date old plains, the plains' concentration around Imbrium, and the improbability that Orientale ejecta was sufficiently thick to cover all Cayley patches lead me to support this interpretation.

CAYLEY AND DESCARTES FORMATIONS

Introduction

The problem of determining whether the materials composing a geologic unit of impact origin were formed before or during the unit's emplacement is especially acute for the two units sampled by Apollo 16. Chapter 9 showed that the distribution and absolute ages of many Apollo 16 samples, particularly those from the Descartes Formation, are consistent with a Nectaris-basin origin. The overall mountainous topography of the Descartes could also be an expression of the Nectaris ejecta. As discussed in the previous section, however, the distribution and planar morphology of the Cayley Formation originated in the Imbrian Period. The paucity of overlying primary-impact craters and the topographic sharpness of the furrows (fig. 9.9) suggest that at least the surficial part of the Descartes Formation is also of Imbrian age. Because few lunar terra surfaces are shaped without addition of material, the presence of some Imbrian material in the Descartes is expected.

This section discusses the events in the units' histories that might have taken place during the Imbrian Period. Three possible ramifications of an Imbrium-basin origin are explored first: that the

FIGURE 10.31.—Short chains of Imbrium secondary craters near craters Stöffler (S; 126 km) and Licetus (L; 75 km, 47° S., 7° E.), 2,600 km from center of Imbrium basin. Orbiter 4 frame H-107.

FIGURE 10.32.—Approximate antipode of Imbrium basin (arrow; 38° S., 160° E.). Grooves score rims of old craters in vicinity of antipode (Stuart-Alexander, 1978). Bright swirls above arrow are in Mare Ingenii. Figure-8 crater at right edge is Van de Graaff (234 km long, probably Nectarian). Orbiter 2 frame M-75.

units consist (1) mostly of Imbrium-basin ejecta, (2) mostly of Nectaris-basin ejecta reworked during the Imbrian Period, or (3) of basin ejecta of different provenance (fig. 10.39). Then follows a consideration of "local" origins as (1) reworked pre-Imbrian crater materials or (2) post-Imbrian materials—origins that I consider subordinate in importance to a basin origin at the Apollo 16 landing site.

Imbrium-basin origin

Spatial distribution, relative age, and comparison with the Orientale-basin geology, as already discussed, suggest that the agent which created the planar surface of the Cayley Formation was the Imbrium impact. Similarly, the furrows of the Descartes Formation

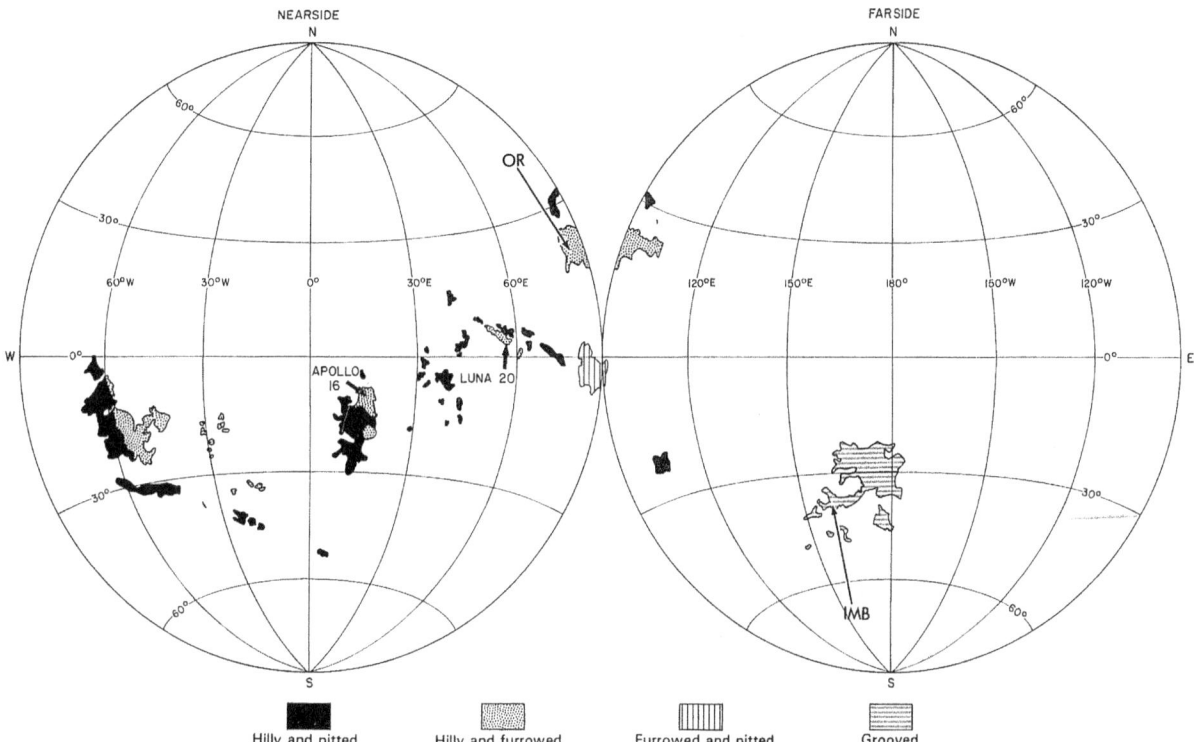

FIGURE 10.33.—Four types of lunar terrain with small negative features (after Howard and others, 1974, fig. 13). Original caption states that terrains are not clearly related to basins, but text offers basin-related, nonvolcanic interpretations. Antipodal effects may explain patches IMB (Imbrium) and OR (Orientale) (compare figures 4.7 and 10.32). Pitted and furrowed deposit at long 90° E. (in Smythii basin) may be related to Crisium basin.

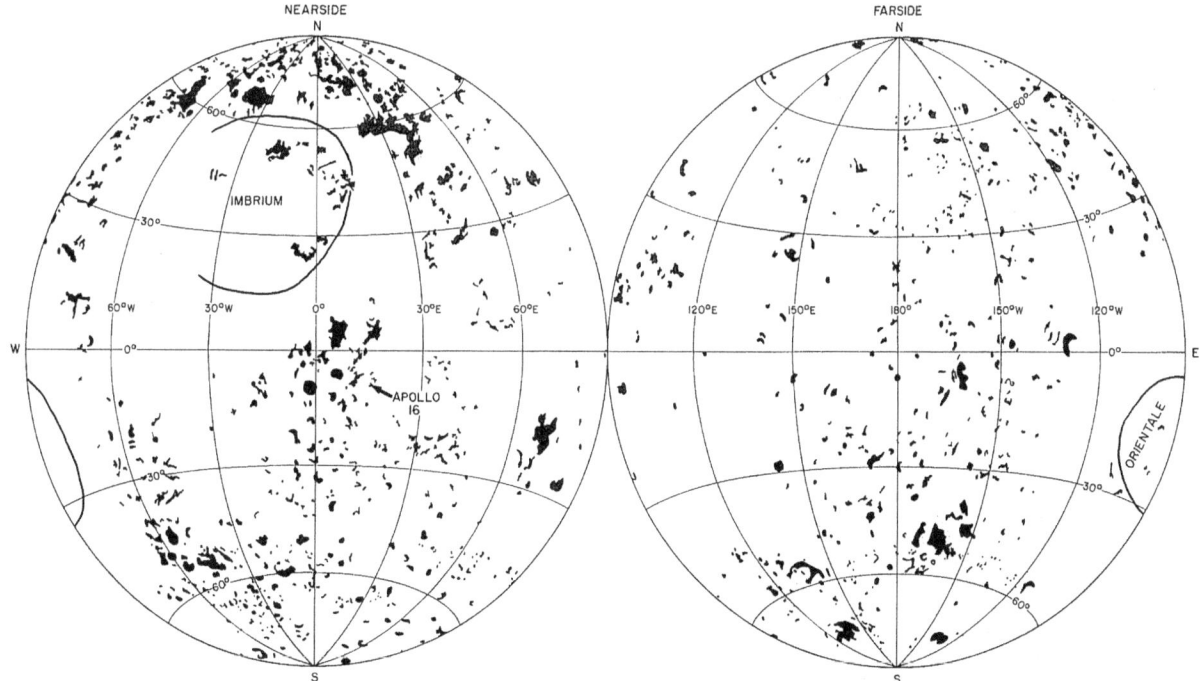

FIGURE 10.34.—Terra plains (Howard and others, 1974, fig. 12). Most nearside and eastern farside plains surround Orientale and Imbrium basins except in a one-basin-radius zone occupied by continuous basin ejecta; plains occur as far as one-quarter of Moon's circumference from source basin. Other large patches are associated with Nectaris and other Nectarian basins and have been reassigned a Nectarian or Nectarian-Imbrian age since this map was prepared (Wilhelms and El-Baz, 1977; Wilhelms and others, 1979).

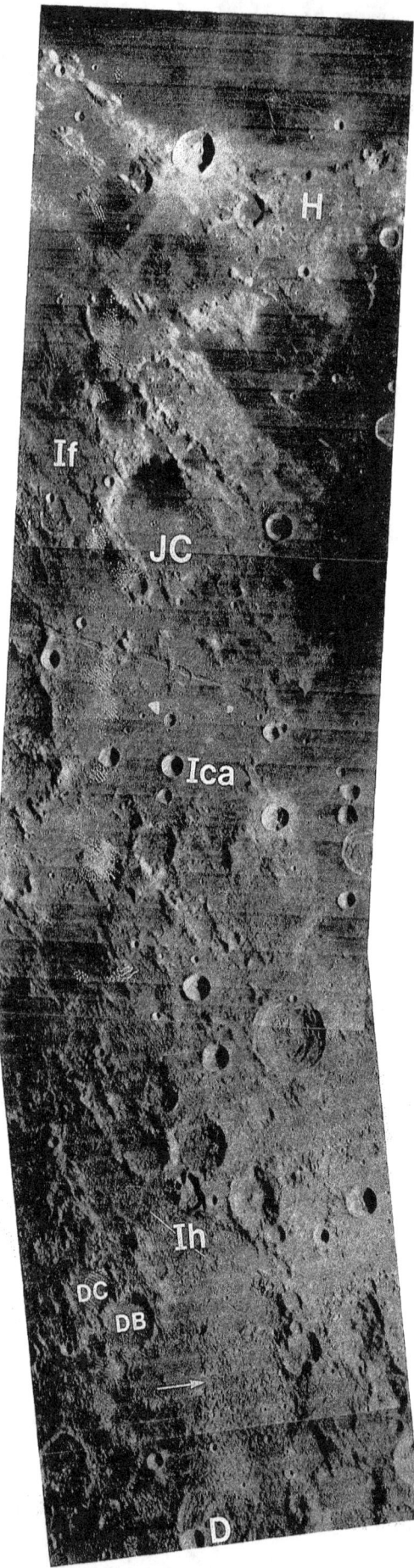

seem to be somehow related to the Imbrium basin (Eggleton and Marshall, 1962); the grooves of the northern (Smoky Mountain) facies of the Descartes are closely radial to Imbrium.

These relations do not specify whether primary- or secondary-impact mechanisms were more important in the units' emplacement. Whereas gradation with the Hevelius Formation indicates that some of the Cayley-like Orientale plains deposits consist of pooled primary basin ejecta that segregated from the Hevelius (fig. 4.4D; Eggleton and Schaber, 1972), similar plains are distal to Orientale-secondary craters (fig. 4.4G). The Descartes Formation, though exceptionally conspicuous, is similar in general morphology to other circum-Imbrium hilly-and-furrowed units already interpreted as part of the secondary-crater retinue of the Imbrium basin (fig. 10.33). Many grooves and scarps of the southern (Stone Mountain) facies, however, are transverse to the Imbrium-radial direction (fig. 9.9) and are less like secondary craters than like the dunelike deposits of Orientale that originated by surface flow of basin ejecta (figs. 4.4F–H). Therefore, the major issue being debated, as was the case with the Apollo 14 materials, is the proportion of Imbrium ejecta to "local" material excavated by the Imbrium secondaries.

Diversion by the Kant Plateau of material flowing from Imbrium has been suggested as the cause of the transverse trend of the Stone Mountain facies (Hodges, 1972; Hodges and others, 1973; Moore and others, 1974; Hodges and Muehlberger, 1981). Hodges and Muehlberger proposed that a long trough radial to Imbrium (fig. 10.35) facilitated the flow. Flow elsewhere in the region is indicated by a thick deposit, evidently from Imbrium, in the crater Andĕl M west of the Apollo 16 landing site (Moore and others, 1974). The Apollo 16 landing site and many plains-dunes associations around Orientale lie two-thirds of a basin diameter from the rims of their respective basins.

Most investigators, however, doubt that much primary ejecta could have reached the Apollo 16 landing site all the way from the Imbrium rim (Montes Apenninus) by surface flow. This viewpoint is supported by the abundant evidence for Imbrium secondaries in the region. Many pits and deeply etched grooves that seem clearly to be secondary to Imbrium are adjacent to the two sampled units (fig. 10.35). Clustering of a group of craters on the Kant Plateau immediately east of the landing site suggests an origin as Imbrium secondaries. Alignment of the nearly equant pair Dollond B and Dollond C northwest of the landing site is also typical of secondary fields, and these 32- to 37-km-diameter craters are similar in size to Imbrium secondaries in other regions. Any Imbrium material at the site may, therefore, have arrived in impacting Imbrium projectiles (Wilhelms, 1972c) or in a debris flow initiated by secondary impacts of the projectiles (Morrison and Oberbeck, 1975; Oberbeck, 1975; Oberbeck and others, 1975).

As for the Apollo 14 landing-site region, an intermediate view is more likely than an all-primary or all-secondary origin. A flow initiated by secondary impacts would contain mixed primary and secondary ejecta of Imbrium and could create the same morphologies and regional relations as a flow of purely primary material. Morrison and Oberbeck (1975) calculated that a deposit dislodged at the Apollo 16 landing site by Imbrium projectiles would contain only 13 to 18 percent of Imbrium-primary ejecta. At the landing site itself, however, whatever secondaries were formed are buried by the Cayley and Descartes deposits. Therefore, these deposits either are primary (like those at the Apollo 14 site) or originated from secondary craters northwest of the Apollo 16 site. Much more Imbrium material than 13 to 18 percent would be present if the deposit originated to the northwest (fig. 10.26).

FIGURE 10.35.—Strip of eastern nearside terra, 900 km long, from Montes Haemus (H) to crater Descartes (D; 48 km, 12° S., 16° E.). Montes Haemus is 1,000 km from center of Imbrium basin, and Descartes 1,800 km. North half of photographed area is distinctly inundated by Imbrium-basin deposits (If, Fra Mauro Formation); formation has accumulated against southeast wall of Julius Caesar (JC). The Cayley Formation in its type area (Ica; compare fig. 10.38) is apparently gradational with the Fra Mauro Formation; other patches, as at Apollo 16 landing site (arrow), seem more isolated, yet are situated near Imbrium-radial lineations, as at Dollond B (DB; 37 km) and Dollond C (DC; 32 km). Hilly and furrowed material of the Descartes Mountains (around and south of Apollo 16 landing site to bottom of photograph), though also seemingly isolated, forms part of Imbrium-radial system north of Apollo 16 site and is similar to hilly material farther north (Ih). Mosaic of Orbiter 4 frames H–89 (below) and H–90 (above).

Any of the debris-flow mechanisms could emplace the Cayley and the Descartes Formations either simultaneously or sequentially (fig. 10.39). Simultaneous emplacement (fig. 10.39A) seems to be consistent with the Orientale relations because the Orientale plains and dunelike deposits appear to be gradational (figs. 4.4F–H). If simultaneously emplaced, the differing morphologies might arise either because (1) a higher melt content (chap. 9) caused the Cayley deposits to pool in depressions (Hodges and Muehlberger, 1981; Ulrich and Reed, 1981) or because (2) the Descartes material, but not the Cayley material, reached and was crumpled into dunelike topography by the Kant Plateau. Alternatively, an embankment of ejecta may have been emplaced by the Nectaris basin, followed at the beginning of the Imbrian Period by secondary impacts forming the Descartes furrows and a debris flow forming the Cayley (fig. 10.39B). The measured absolute ages and the apparent lithologic differences between the Cayley and the Descartes (see chap. 9; table 9.1) are consistent with this sequence.

Among the absolute ages of the sampled impact-melt fragments are some that are distinctly younger than the 3.92 ± 0.03-aeon age tentatively identified here with the Nectaris impact (table 9.1). A few of these ages might indicate an Imbrium-basin origin of materials in the units or, at least, in the melt-rich Cayley. Although the methodologic problems are severe (see chap. 9), these ages are at least consistent with an Imbrian age of the younger impact-melt rocks.

"Local" origin

In the Imbrium debris-flow models, strata of the Cayley and Descartes Formations would possess considerable lateral continuity, though also considerable lithologic diversity. In "local origin" models, in contrast, the units formed piecemeal and were derived from various relatively small deposits quite diverse in lithology, absolute age, and composition. Sample diversity is commonly cited as requiring such diverse "local" sources.

Local origin of the Cayley and of plains in general is commonly envisioned as dislodgement of debris from crater walls and redeposition in adjacent depressions (Morrison and Oberbeck, 1975; Oberbeck and others, 1974, 1975, 1977; Oberbeck, 1975; Oberbeck and Morrison, 1976). An early suggestion of this type was that the plains consist of accumulated secondary ejecta of craters (Oberbeck and others, 1974). The concentration of plains around Imbrium in many additional localities indicates, however, that if secondary impacts emplaced the Cayley, they were from Imbrium.

One of the most widely favored hypotheses has been that the deposits originated as material of pre-Imbrian "local" craters (Head, 1974d; Oberbeck and others, 1974, 1975). Head (1974d) suggested that an "unnamed crater B" dominates the landing site's geology and petrology (fig. 10.40). The Cayley is supposed to consist of melt rock from the fallback on this crater's floor, and melt-poor rocks are supposed to have come from the less highly shocked rim materials of unnamed crater B and other old craters. Head (1974d) thought that unnamed crater B transects the Stone Mountain facies (of Nectaris-basin origin) and is transected by the Smoky Mountain facies. Unnamed crater B, however, is a vague circular feature or coalescence of subcircular features that lacks the continuity, terraces, and other hallmarks of a crater young enough to affect the surface geology; it interrupts no furrows. If it exists at all, it is too deeply buried by both Descartes facies and the Cayley to have furnished much material to the bedrock-breccia deposits of either unit. Furthermore, no other craters older than the Cayley and the Descartes Formations are sufficiently near to have furnished much material to either bedrock deposit (pls. 6, 7). If "local" material constitutes these formations, it is Nectaris-basin primary ejecta.

Although a "local" origin in the sense generally meant seems to be refuted by the photogeologic relations, several processes that might be defined as local probably did contribute to the observed diversity of the samples. First, secondary impacts of Imbrium ejecta dislodged material, mostly Nectaris-basin primary ejecta but possibly including some crater material, that lay northwest of the landing site. Second, the reworked Nectaris-basin ejecta itself consists of materials that were derived from diverse small-scale deposits of craters and, possibly, of volcanic flows that were in the Nectaris target area before the Nectaris impact. Any later impact would have incorporated, shocked, melted, and reworked such earlier impact and volcanic materials. Most of the chemical, chronologic, and lithologic diversity of the Apollo 16 samples probably results from this pre-Nectarian diversity, which is a feature of all impact deposits.

The preceding discussion pertains to the bedrock deposits. All hypotheses acknowledge that several local processes contributed to

FIGURE 10.36.—Plains materials superposed on Imbrium-secondary craters, probably consisting of debris dislodged and ejected by secondary impacts (Oberbeck and others, 1974) superposed on north (near-Imbrium) wall of crater Albategnius (136 km). Visibility of secondary chains buried by plains on floor is enhanced by Sun elevation, only 5° above horizontal (from right). Neukum (1977) suggested that plains postdate Imbrium and Orientale basins and are volcanic. Apollo 16 frame M–449.

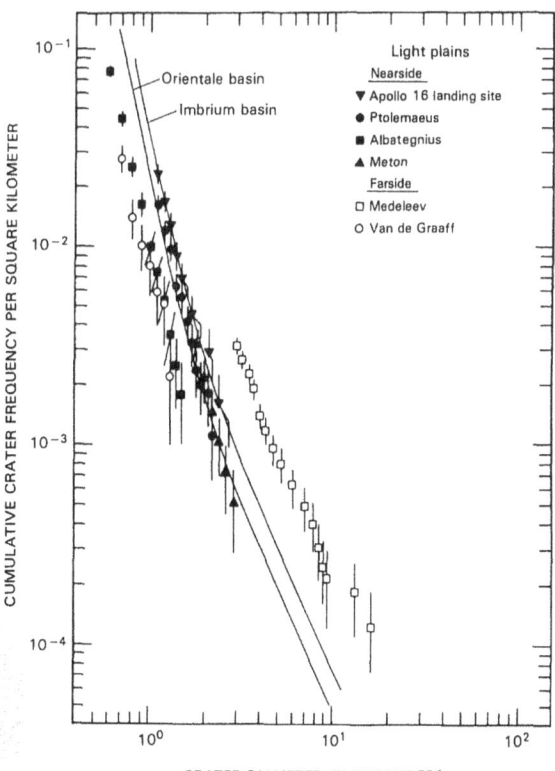

FIGURE 10.37.—Size-frequency distributions of craters superposed on terra plains of several ages (Neukum and others, 1975b; compare fig. 11.2). Curves represent, from oldest to youngest: plains in Mendeleev (compare fig. 9.29); Imbrium basin and plains at Apollo 16 landing site, indistinguishable; plains in Ptolemaeus; Orientale basin, similar to Ptolemaeus plains; plains in Albategnius, considerably younger than Orientale basin; and mare materials in Van de Graaff.

the diversity of the overlying relatively thin regolith at the landing site:

1. Large craters probably pierced the Cayley Formation and introduced some of the underlying material into the regolith. Seismic data indicate that the Cayley is about 200 m thick (Hodges and others, 1973; Cooper and others, 1974). Impacts on the order of North Ray Crater in size (1 km) could penetrate this thickness. Many subdued craters of this size are visible and probably "contaminated" the regolith during the Imbrian Period (fig. 9.9). The introduced material probably was mostly Nectaris ejecta but may also have included some crater materials.
2. The regolith developed on each formation contains some fragments derived from the other formation by post-Imbrium impacts. Origin as Orientale-basin ejecta seemed to be supported by three post-Imbrium Ar-Ar ages (table 9.1) and a precise Rb-Sr age of 3.76 ± 0.01 aeons (Papanastassiou and Wasserburg, 1972b) for Cayley sample 68415 (Chao and others, 1975). An alternative origin more in keeping with the regional relations is that this rock (samples 68415, 68416) was derived from impact melt of an Imbrian crater that formed nearby a few tens of millions of years after the Cayley and Descartes Formations. This origin is supported by the fact that its composition is similar to that of the station 11 regolith, which accumulated on the Descartes Formation (P.D. Spudis, oral commun., 1983).
3. Like all lunar regoliths, the Apollo 16 regolith undoubtedly contains some material derived by small post-Imbrium impacts from other nearby units. For example, fragments of mare basalt were probably introduced into Theophilus ejecta (see chap. 11; Delano, 1975).

Conclusions

1. The relation of surface morphologies to the Imbrium basin suggests that the Cayley Formation and at least part of the Descartes Formation are also genetically related to that basin (fig. 10.39).
2. I favor origin of the Cayley by some sort of Imbrium debris-flow emplacement, and origin of the Descartes by modification of Nectaris-basin deposits by the same debris flow or by Imbrium-secondary impacts (fig. 10.39).
3. Secondary impacts are more likely to have contributed to the deposits than at the Apollo 14 landing site because secondary craters are more evident and primary-ejecta features less evident than at that site. However, I consider neither the relative role of primary and secondary ejecta of the Imbrium basin nor the origin of the overall mountainous topography of the Descartes Formation to be established.
4. The absolute ages of certain samples support a Nectaris-basin origin for much of the Descartes material, and other absolute ages suggest addition of some Imbrium material (table 9.1). However, the relative amounts of Imbrium and Nectaris ejecta in the Descartes and Cayley Formations are unknown.
5. Pre-Imbrium "local" craters contributed little material to the site except as parts of the Nectaris ejecta, as minor parts of the Imbrium debris flow, and as regolith fragments.
6. Post-Imbrium craters melted some local materials.
7. Although this list of unsolved problems is unsatisfyingly long, at least the dominant role of basin deposits over "local" crater deposits appears to be well established.

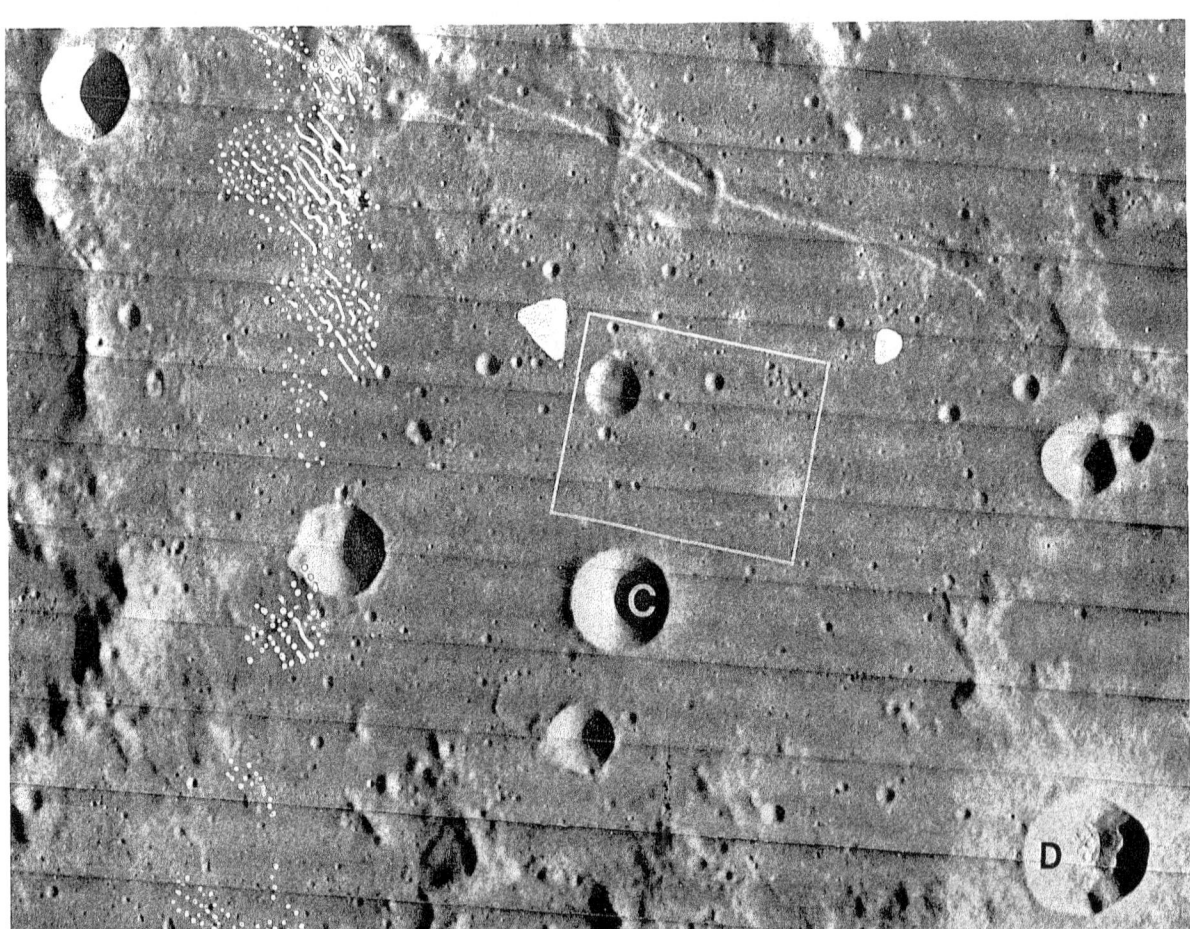

FIGURE 10.38.—Type area of the Cayley Formation (Morris and Wilhelms, 1967).
  A. Regional view. Type area (3.5–4.5° N., 15.5–16.25° E.) is partly within and partly below rectangle outlining B. Crater Cayley (C; 14 km), moderately fresh, with lightly cratered ejecta, is probably Eratosthenian. Dionysius (D; 18 km), very fresh, is Copernican. More highly degraded and shallower craters above, below, and left of crater Cayley are probably Imbrian. Intercrater septum of pair near right edge of photograph indicates simultaneous (primary or secondary) impact (see chap. 3). Part of Rima Ariadaeus is at top (see chap. 6). Lineated, pre-Imbrian craters truncated by left edge of photograph are also truncated by right edge of figure 10.16. Orbiter 4 frame H–90.

OTHER BASIN AND CRATER MATERIALS

Recognition criteria

The units discussed heretofore in this chapter are parts of extensive deposits of the Imbrium basin, and the Orientale deposits described in chapter 4 constitute another widespread group of contemporaneous units. Like all other lunar time-stratigraphic units, the Lower Imbrian Series also contains randomly scattered primary-crater materials (table 10.2).

The regional units establish Imbrian ages of the scattered units over a large area of the Moon. Imbrian craters are sandwiched between deposits of the Imbrium basin and Upper Imbrian mare materials. Examples inside the Imbrium basin include Archimedes, Plato, the Iridum crater, and Cassini. Outlying craters, such as Lansberg, whose ejecta is superposed on Imbrium deposits but embayed by mare materials, are also Imbrian (fig. 10.41).

Furthermore, stratigraphic relations divide some craters into Lower and Upper Imbrian. Craters superposed on Imbrium deposits and overlain by Orientale deposits are Lower Imbrian (fig. 10.2). Other fresh-appearing craters buried by Orientale ejecta or secondaries are also likely to be Lower Imbrian (fig. 10.42), as are craters overlain by relatively old Upper Imbrian crater materials (fig. 10.43).

Counts of craters of the large sizes required to date pre-Nectarian and Nectarian surfaces (see chap. 7) are statistically valid for deposits of two Imbrian basins and marginally definitive for a third (fig. 10.4). I determined cumulative frequencies of 28 and 22 craters larger than 20 km in diameter per million square kilometers on Imbrium-basin and Orientale-basin deposits, respectively. The large statistical samples suggest that these are good values for the lowermost Imbrian and

TABLE 10.2.—*Possible Lower Imbrian craters*

[Cross rules divide diameter ranges mapped differently in plate 8: smaller than 30 km, unmapped; 30 to 59 km, interiors mapped; 60 km and larger, exterior deposits mapped where exposed. UI?, possibly Upper Imbrian]

Crater	Diameter (km)	Center (lat)	Center (long)	Figure	Remarks
Protagoras	22	56° N.	7° E.	10.15	---
Ukert	23	8° N.	2° E.	10.1	---
Inghirami A	34	45° S.	65° W.	10.42	Covered by Orientale.
La Condamine	37	53° N.	28° W.	3.8, 10.43	Covered by Iridum.
Crüger	46	17° S.	67° W.	4.4H, 7.8	---
Maupertuis	46	50° N.	27° W.	3.8, 10.43	Covered by Iridum.
Campanus	48	28° S.	28° W.	6.9	---
Cassini	57	40° N.	5° E.	1.6, 10.12	UI?
Macrobius	64	21° N.	46° E.	9.11	---
Lomonosov	93	27° N.	98° E.	1.2, 4.7, 8.13, 9.28	Typical; mare fill.
Arzachel	97	18° S.	2° W.	1.8, 7.7	UI?
Letronne	119	11° S.	42° W.	6.3, 12.3C	UI?, half-buried.
Compton	162	56° N.	105° E.	4.2B, 9.4, 9.23, 10.6	UI?, interior ring.
Keeler	169	10° S.	162° E.	1.4, 3.1, 8.7	---
Petavius	177	25° S.	60° E.	4.2C, 9.5	Floor and rim faulted.

B. Area outlined in A. The Cayley Formation is more densely cratered than a mare surface and is saturated by small subdued craters, as expected on a Lower Imbrian surface. Many secondary clusters are also superposed—large in upper right, small and linear elsewhere (probable source, Dionysius). Relief is enhanced by low Sun illumination, 11° above horizon (from right). AB, crater Ariadaeus B (8 km), whose flat floor is visible in A but is obscured here; Imbrian age is confirmed by crater density similar to that on the Cayley Formation. Orbiter 2 frame M–61.

the boundary between the two Imbrian series, respectively. Sparse crater counts indicate that the Schrödinger basin is also Imbrian (fig. 10.4; density, 20 such craters). Schrödinger is the topographically freshest lunar basin of its size and thus appears to be relatively young (fig. 10.44). It had been considered Nectarian because of its moderately large density of small craters, although no definite Nectarian craters are known to be superposed (Wilhelms and others, 1979). If it is Imbrian, superposition of some probable Orientale secondaries (fig. 10.44) restricts its age to Early Imbrian.

Otherwise, counts of craters smaller than 20 km in diameter are required to date Imbrian surfaces not in contact with definite Imbrian deposits. Figure 10.4 plots the counts by Neukum and others (1975a) of craters as small as 1.4 km in diameter on Montes Apenninus, which fall on the projection of the curve determined here for craters larger than 20 km in diameter. Crater frequencies for the plains at the Apollo 16 landing site nearly coincide with those for the Apennines. The counts by Neukum and others (1975a) and myself indicate that the Orientale and Imbrium curves are approximately parallel for both small and large superposed craters.

Craters that are thought to be too fresh to be Nectarian and that are overlain by relatively old mare materials are likely to be Early Imbrian—for example, Lomonosov (fig. 9.28). Stratigraphically isolated Early Imbrian craters are difficult to distinguish from Nectarian and Late Imbrian craters, and so the compilation here (pl. 8; table 10.2) is based mostly on qualitative assessments of morphology and some sparse size-frequency counts.

The distinction between small craters of the Lower and Upper Imbrian Series is still less clear. Craters smaller than about 20 km in diameter in the terrae are hard to date by morphology (Wilhelms and others, 1978). Debris from the walls has filled Imbrian craters smaller than about 8 km in diameter more deeply than their younger counterparts, and commonly has created shallow bowl-shaped or flat-floored interiors from originally steeper more nearly conical profiles. However, detection of this distinction requires good photographic resolution and Sun illumination (figs. 10.38, 10.41). Fresh Imbrian-age primary craters may be indistinguishable from sharp Orientale- or Imbrium-basin secondary craters of similar sizes. Small flat-floored or otherwise subdued Imbrian primary craters may resemble some Nectarian primaries or partly buried secondaries of one of the Imbrian basins. Imbrian craters disappear at diameters of about 1 km on the Fra Mauro Formation (fig. 7.14; Trask, 1971).

Frequency

Despite the difficulties, Lower Imbrian craters are here distinguished from Nectarian and Upper Imbrian craters as well as possible and are mapped in plate 8. In a farside area of 14.2×10^6 km^2 that excludes the Orientale deposits which overlie all other materials of the series, 73 unburied craters at least 30 km in diameter are interpreted as Lower Imbrian. For this density (5.1 craters/10^6 km^2), 195 craters of this size were formed over the whole Moon.

OTHER TERRA MATERIALS

The discussions of the Orientale and Imbrium basins have offered basin-impact alternatives (Howard and others, 1974) to the volcanic origins previously considered or assumed for many terra landforms. However, volcanic origins for some terra plains (Neukum, 1977), as well as for a few isolated domelike structures, are still being entertained.

The "domes" have a volcanolike morphology in the form of summit or flank pits and unusually red spectra (fig. 10.45; Wilhelms and McCauley, 1971; McCauley, 1973; Scott and Eggleton, 1973; Malin, 1974; Head and McCord, 1978; Basaltic Volcanism Study Project, 1981, p. 762–763). The most densely concentrated are the Mairan and Gruithuisen domes near the Iridum crater, between Mare Imbrium and northern Oceanus Procellarum or Sinus Roris (Scott and Eggleton, 1973). They may represent rare differentiated volcanic materials akin to KREEP-rich material or certain still more highly

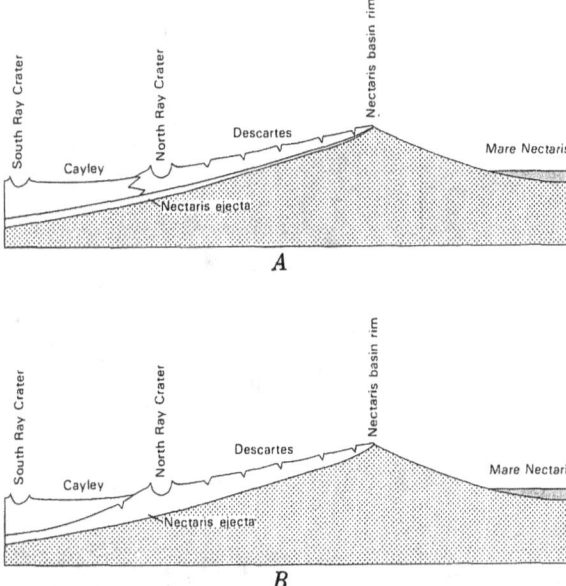

FIGURE 10.39.—Diagrammatic geologic cross sections drawn across region of Apollo 16 landing site, illustrating alternative interpretations.
A. The Descartes and Cayley Formations were both emplaced in an Imbrium debris surge.
B. The Descartes was emplaced as Nectaris ejecta and furrowed by impacts of Imbrium ejecta. The Cayley was emplaced in an Imbrium debris surge. The Cayley would consist mostly of Imbrium ejecta if the surge originated near Imbrium, and mostly of Nectaris ejecta if the surge originated nearer the Apollo 16 landing site.

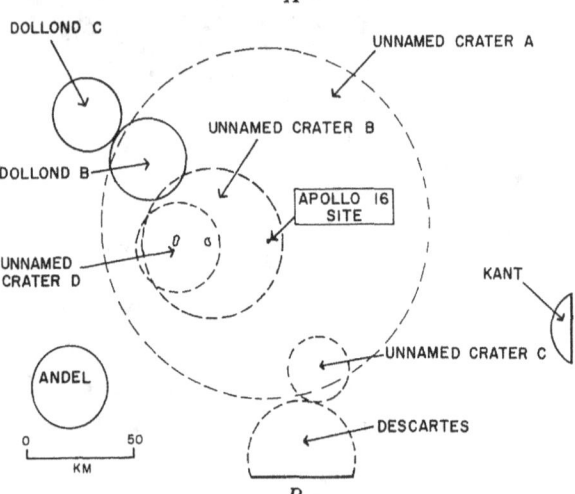

FIGURE 10.40.—Vicinity of Apollo 16 landing site.
A. Telescopic photograph.
B. "Unnamed" craters believed by Head (1974d) to constitute area of A and to have contributed materials to Apollo 16 samples.

differentiated "granitic" or "quartz-monzodioritic" compositions that appear in small amounts in lunar samples (Ryder and others, 1975a; Ryder, 1976). The craters atop some of these domes argue for volcanic origins (fig. 10.45A). Experience with other seemingly volcanic landforms, however, recommends a search for impact origins. Some "domes" may result from blanketing of Imbrium massifs and Iridum-secondary craters by Iridum ejecta (P.D. Spudis, oral commun., 1980).

MARE MATERIALS

Whether any Lower Imbrian mare materials are exposed is uncertain. None could remain exposed within reach of the Orientale ejecta. Judging from the mapped extent of the Orientale deposits and secondary craters on the west-limb and farside terrae, the nearside maria probably would have been heavily mantled or cratered if their surface units predated Orientale. Orientale and other large basins might even cover the whole Moon with debris (Moore and others, 1974; Chao and others, 1975). However, some outlying points may be spared a cover because outer impact ejecta is concentrated in raylike stringers (pl. 3; fig. 4.6).

Certain heavily cratered maria on the east limb may be of pre-Orientale, Early Imbrian age. One example is the patchily distributed planar material in the Australe basin (fig. 10.46). In albedo, these plains are intermediate between typical mare and typical circumbasin terra plains, and the superposed crater densities are similar to those of Orientale (J.M. Diaz and A.B. Watkins, written commun., 1976). These marelike materials are older than the Imbrian craters Humboldt and Jenner (Wilhelms and El-Baz, 1977). Dark-haloed craters are evident on the marelike plains (Schultz and Spudis, 1979). A thin dusting of mare basalt by Orientale ejecta may explain the plains' albedo. Other densely cratered materials with

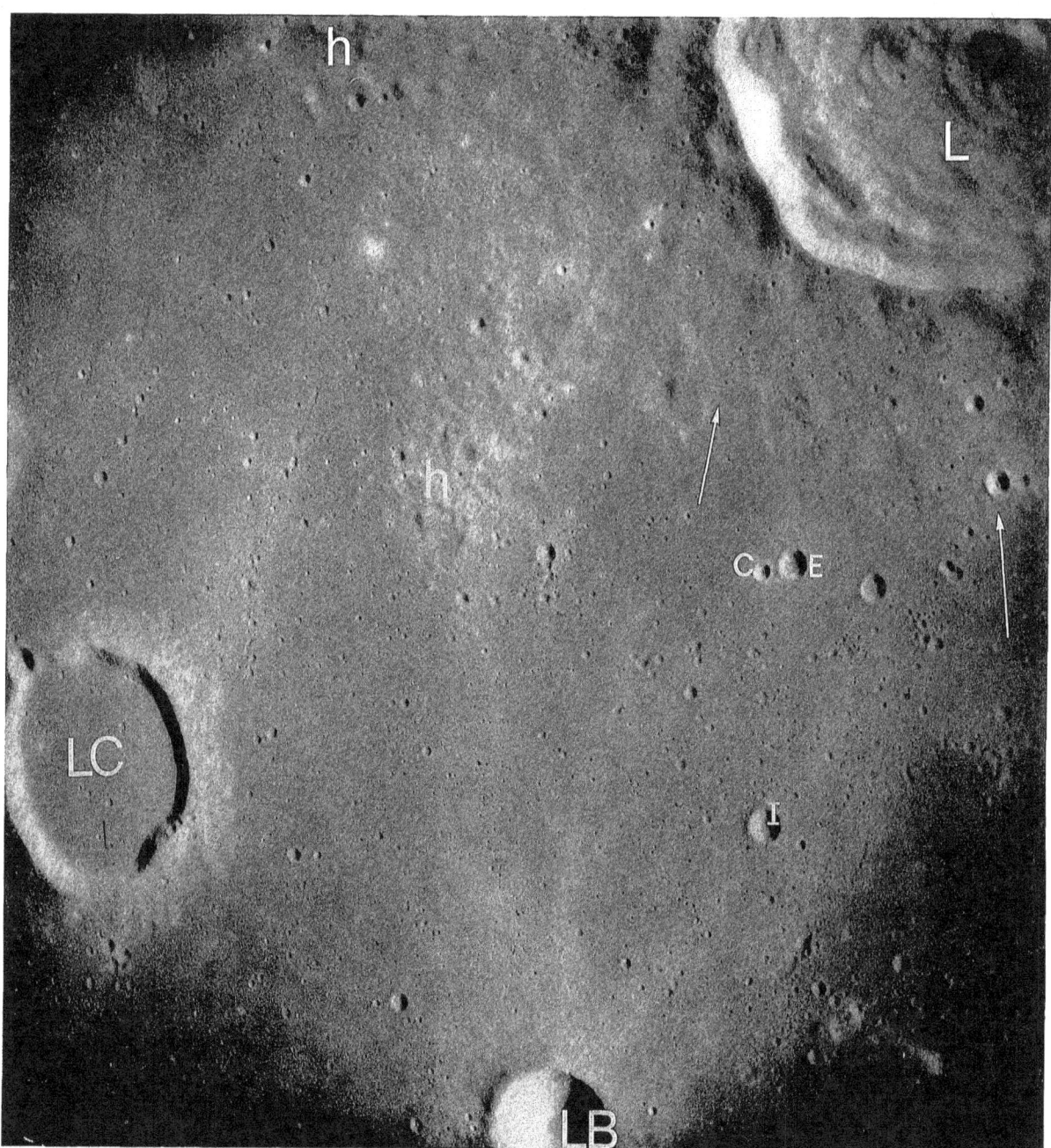

FIGURE 10.41.—Imbrian craters on periphery of Imbrium basin. Lansberg (L; 39 km, 0.3° S., 26.5° W.) is superposed on hummocky Imbrium ejecta (h) and embayed by mare materials (arrows). Interior and periphery of Lansberg C (LC; 17 km) are flooded more deeply. Rayed, probably Copernican, but somewhat degraded crater Lansberg B (LB; 9 km) is superposed on the mare. Craters in 1-km size range are probably Imbrian (I), Eratosthenian (E), and Copernican (C). Apollo 12 frame H-8095.

intermediate albedos lie in Mare Marginis (fig. 4.7) and at the distal margin of Mare Frigoris (Lucchitta, 1978).

Nevertheless, none of these criteria prove an Early Imbrian age. Even the obviously Late Imbrian materials near Orientale (fig. 10.3) are quite heavily cratered. All exposed mare materials may, therefore, be Upper Imbrian or younger and are so mapped here (pls. 9–11). They can be locally subdivided in considerable detail, as shown in the next two chapters.

CHRONOLOGY

Radiometric ages of Imbrium-basin materials are clustered within a short interval. The oldest are those of an older cluster of ages at the Apollo 14 landing site (3.87–3.96 aeons), but these ages pertain to pre-Imbrian rocks. The Apollo 15 KREEP basalt, 3.85 ± 0.04 aeons old, probably was derived from a deposit (Apennine Bench Formation) that is superposed on other basin materials. This basalt may be the best dated of the Apollo 14 and 15 terra materials (G. W. Lugmair, oral commun., 1982); it is contemporaneous with or younger than the Imbrium basin, depending on its origin. Black-and-white breccia believed to have been collected from the Imbrium massifs has been dated at 3.86 ± 0.04 aeons. A younger cluster of ages from the Fra Mauro Formation at the Apollo 14 landing site falls mostly between 3.82 and 3.84 aeons. The time of the Imbrium impact seems to be well constrained at from 3.82 to 3.87 aeons ago; the average and well-represented date of 3.85 aeons ago is tentatively adopted here.

One group of basalt samples from the Apollo 11 landing site yielded ages as old as 3.84 ± 0.03 aeons (table 11.1; Guggisberg and others, 1979). The source flows of these samples are not exposed (see chap. 11); they are probably Lower Imbrian.

No other Lower Imbrian units have been dated isotopically. The interpretation of ages reported for the Cayley Formation at the Apollo 16 landing site is highly uncertain, except that the Cayley materials are no more than 3.92 aeons old. Materials of the Descartes Formation, which may have been emplaced 3.92 ± 0.03 aeons ago, were modified and probably augmented by Imbrium-basin materials 3.85 aeons ago.

The upper age boundary of the series is limited by the ages of the oldest nearly exposed mare basalt. This material was recovered from the Apollo 17 landing site and is described in the next chapter; it is 3.72 aeons old. Therefore, the Orientale impact that ended the Early Imbrian Epoch occurred some time between 3.85 and 3.72 aeons ago. Relative crater frequencies (figs. 8.16, 10.4) suggest that it occurred about 3.8 aeons ago. Thus, the Early Imbrian Epoch spanned about 50 million years.

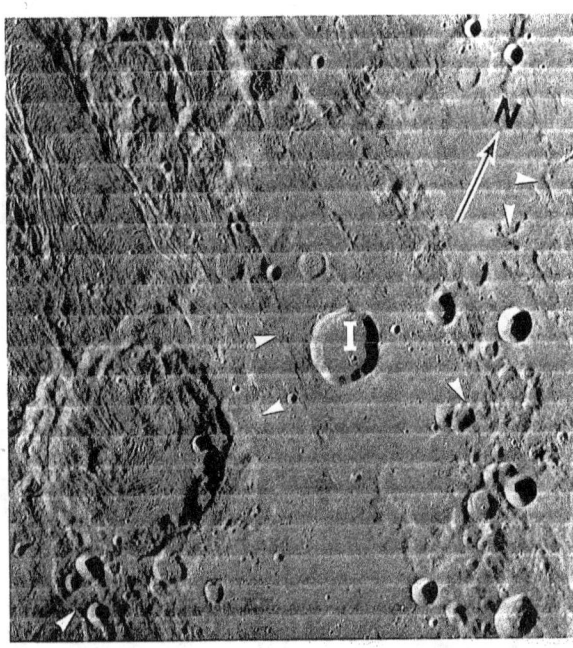

FIGURE 10.42.—Terrain centered 600 km southeast of Orientale-basin rim (Montes Cordillera). Largest crater is Inghirami (91 km, 48° S., 69° W.); other probable primary crater older than Orientale deposits is Inghirami A (I; 34 km). Relatively young ages for both craters are suggested by moderately sharp textural elements, especially smaller Inghirami A. If they are Imbrian and not Nectarian, these craters must be Early Imbrian because they are buried by Orientale deposits. Textured Hevelius Formation and plains intergrade. Arrowheads indicate embayment of Orientale-secondary craters by plains; most or all of clustered craters in lower right corner are Orientale secondaries. Orbiter 4 frame H–172.

FIGURE 10.43.—Terra between Mare Frigoris (F) and Mare Imbrium, underlain by deposits of Imbrium basin. Lower Imbrian craters La Condamine (L; 37 km, 53° N., 28° W.) and Maupertuis (M; 46 km) are swamped by deposits of Iridum crater (Ulrich, 1969), which grade outward to conspicuous secondary craters (sc) and to light plains deposits (white p). Secondary craters of crater Plato (black P), centered outside right edge of picture, are superposed on Iridum-secondary craters. Iridum crater (centered outside lower left corner of photograph) is slightly younger than Orientale (fig. 10.4) but older than Plato, and so is probably early Late Imbrian. The Straight Range (Montes Recti, R) is part of Imbrium ring overlain by Iridum deposits. Orbiter 4 frame H–139.

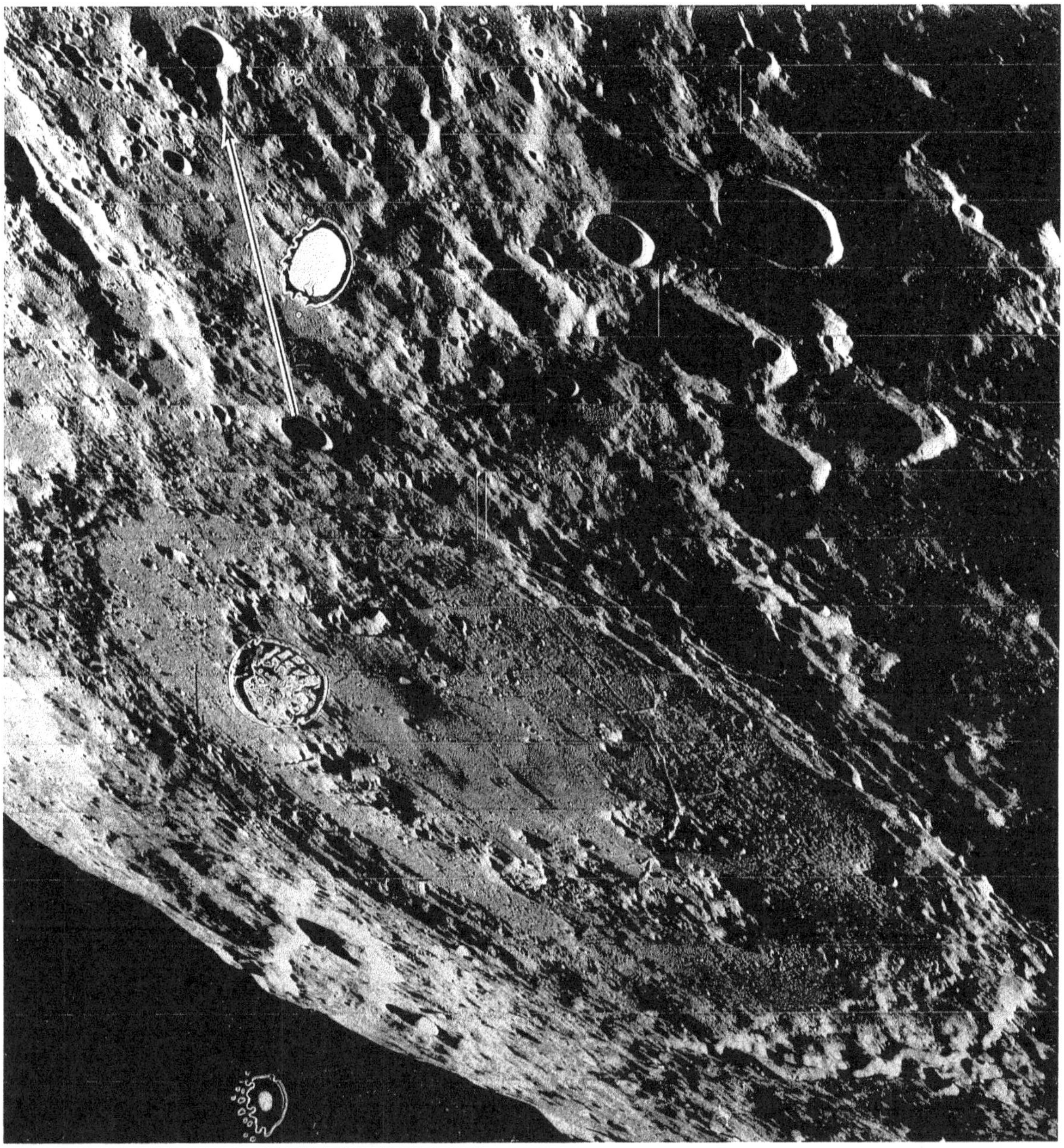

FIGURE 10.44.—Schrödinger basin (320 km). Arrow indicates probable secondary craters of Orientale, centered 2,150 km in direction of arrow; overlap and orientation of cluster are consistent with Orientale origin (Wilhelms and others, 1979). Large chain to left of arrow is secondary to Schrödinger. (White and black-and-white spots are blemishes.) Orbiter 5 frame H–21.

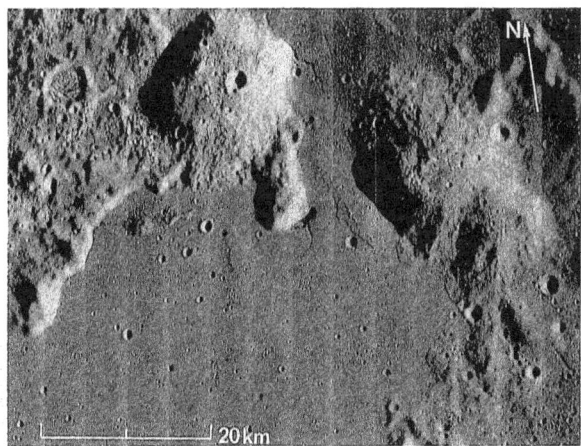

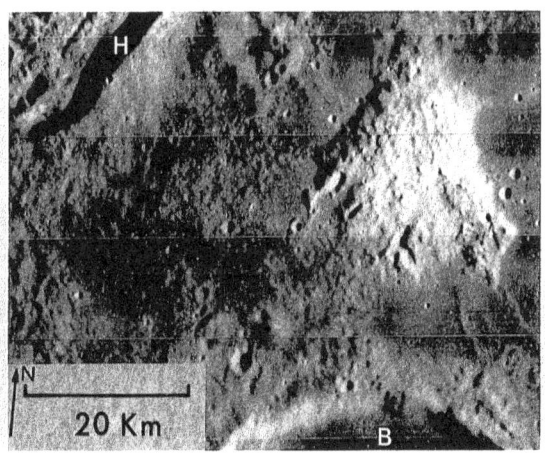

FIGURE 10.45.—Domical terra landforms of possible volcanic origin.
A. Gruithuisen domes (36° N., 40° W.). Orbiter 5 frame M–182.

B. Hansteen alpha (12° S., 50° W.). B, crater Billy; H, crater Hansteen. Orbiter 4 frame H–149.

FIGURE 10.46.—Stratigraphic relations of mare and crater materials in Mare Australe.
A. Deposits of Jenner (J; 72 km, 42° S., 96° W.) and Humboldt (H; 207 km, 27° S., 81° E.), both Upper Imbrian, are superposed on mare basalt (white-on-black arrowheads), whereas a younger mare unit encroaches on Humboldt secondaries (white arrowheads). C, pre-Nectarian crater Curie (139 km, 23° S., 92° E.); L, pre-Nectarian crater Lyot. Orbiter 4 frame M–9.

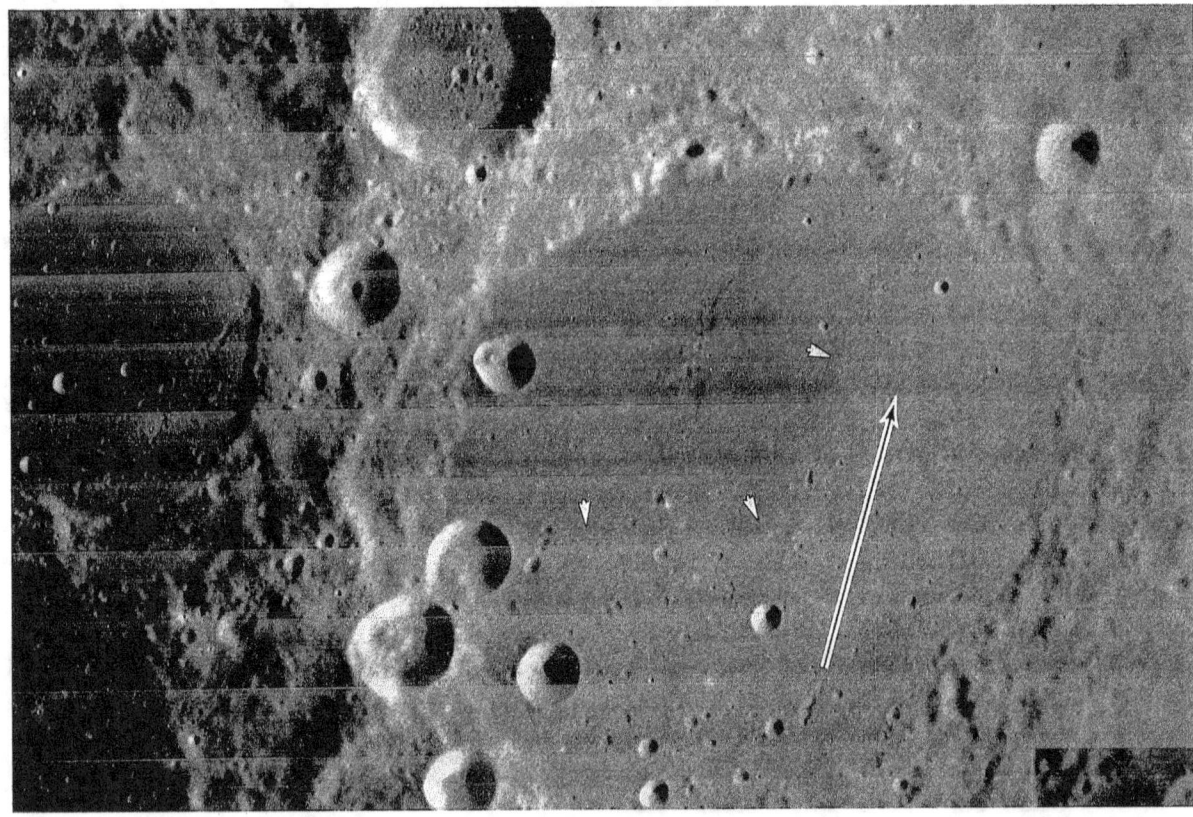

B. Two mare units in Lyot (141 km, 50° S., 84° W.). Younger unit encroaches on older (white arrowheads). Secondary craters of crater Humboldt are superposed on older unit, including at shaft end of black-and-white arrow, which points to Humboldt, 10° W. of N. Older unit may be Lower Imbrian; younger unit is Upper Imbrian. Orbiter 4 frame H–9.

11. UPPER IMBRIAN SERIES

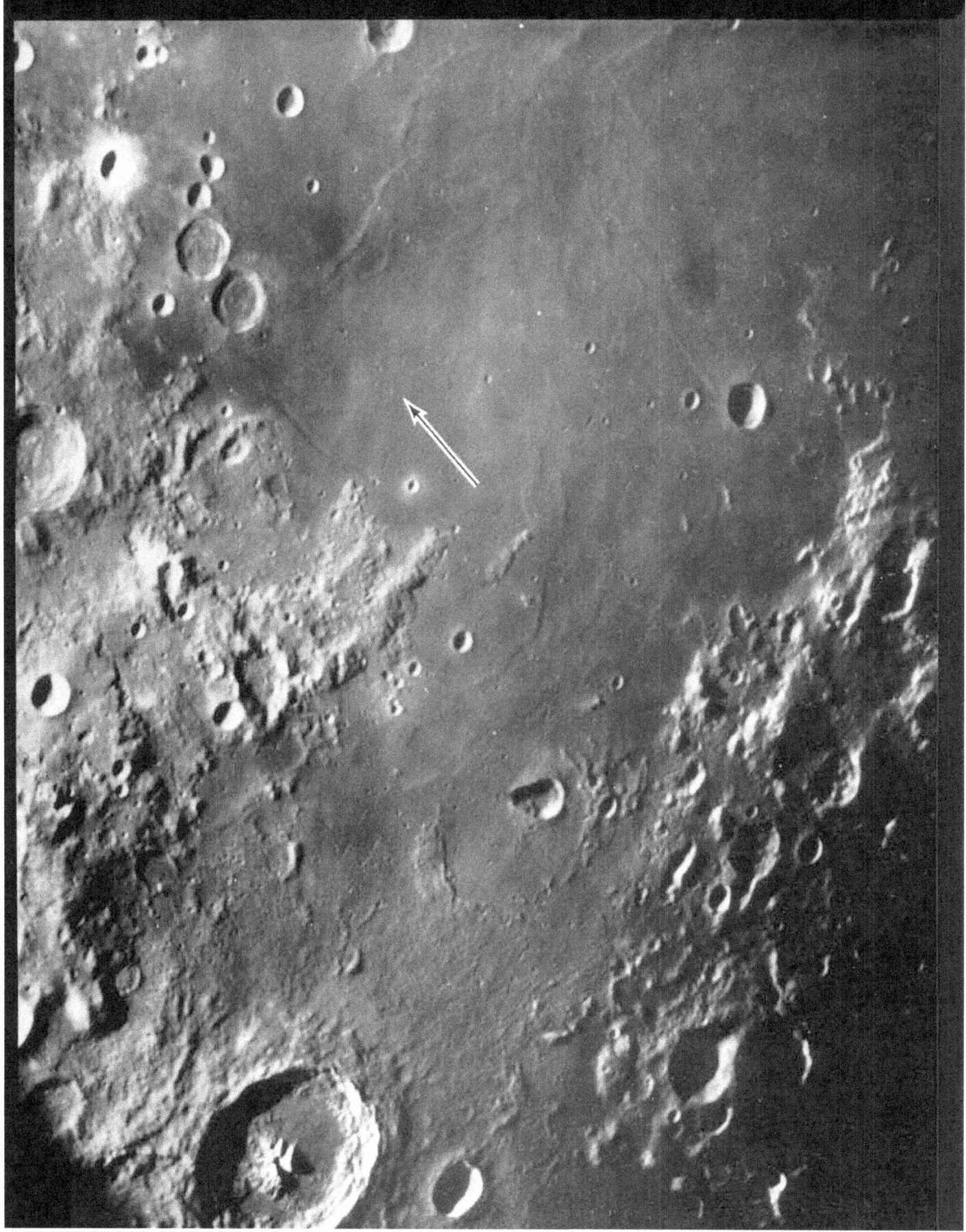

FIGURE 11.1 (OVERLEAF).—Region centered on Apollo 11 landing site in Mare Tranquillitatis (arrow). Large crater at bottom is Theophilus (100 km). Rings of Procellarum basin pass under mare-ridge feature Lamont (ring and radial pattern above arrow) and through rugged terra at right. Telescopic photograph by the Pic du Midi Observatory, France.

11. UPPER IMBRIAN SERIES

CONTENTS

	Page
Introduction	229
Definition	229
Crater materials	231
Recognition criteria	231
Frequency	232
Mare materials—general stratigraphy and distribution	232
Mare Tranquillitatis	235
General stratigraphy	235
Apollo 11 samples	235
Summary and interpretation	238
Mare Nectaris	238
General stratigraphy	238
Apollo 16 samples	238
Summary and interpretation	238
Mare Serenitatis	238
General stratigraphy	238
Apollo 17 samples	239
Summary and interpretation	241
Mare Fecunditatis	241
General stratigraphy	241
Luna 16 samples	241
Summary and interpretation	241
Mare Crisium	241
General stratigraphy	241
Luna 24 samples	241
Summary and interpretation	242
Mare Imbrium	243
General stratigraphy	243
Apollo 15 samples—Palus Putredinis	243
Summary and interpretation	244
Other maria inside the Procellarum basin	244
Near-limb maria outside the Procellarum basin	245
Farside maria	245
Chronology	245

INTRODUCTION

The Upper Imbrian Series contains extensive mare materials, subordinate crater materials, and no basins at all (pl. 9; fig. 11.1). This geologic style contrasts sharply with the exposed record of the earlier, impact-dominated era (see chaps. 8–10; pls. 6–8), which ended with the Orientale impact. Only on the farside and only if exterior deposits are considered (as shown for craters larger than 60 km in diameter on pl. 9) do Upper Imbrian crater deposits exceed Upper Imbrian maria in area.

Two-thirds of the nearside mare surface is Upper Imbrian. Because of this extent and because the Imbrium impact struck the nearside, beginning the Imbrian Period, the Imbrian System dominates the nearside (pls. 8, 9). Absence of an Imbrian cover leaves the pre-Nectarian and Nectarian impact materials, which doubtless formed randomly on the whole Moon, as the dominant deposits on the farside (pls. 6, 7).

Upper Imbrian mare units furnished more samples than any other class of lunar geologic unit, because of their accessibility to landings and their prominence on the nearside. Samples of mare basalt were returned from all landing sites (see chap. 5; table 1.2), and seven of these sites furnished data on absolute ages and composition that are useful in regional studies like the present one. Only one of these seven sampling missions, Apollo 12, returned basalt that is here regarded as younger than Imbrian (see chap. 12).

The emphasis in this chapter corresponds to the relative importance of the two main types of lunar materials in the Upper Imbrian Series. Craters are described briefly in the first section after the series definition. The rest of the chapter describes the general stratigraphy of each major mare or mare cluster and describes in detail the absolute ages and stratigraphy for those maria from which basaltic samples were collected. I show that the volume, sequence, and composition of basaltic eruptions were fundamentally predetermined in the foregoing era, when basin impacts thinned the crust, uplifted the mantle, and weakened the lithosphere.

DEFINITION

The bottom of the Upper Imbrian Series is the top of the Orientale-basin deposits (see chap. 10). Frequencies of 2.2×10^{-2} craters at least 1 km in diameter per square kilometer and D_L values of 550 m, therefore, are approximate maximums for the Upper Imbrian Series (figs. 10.37, 11.2; tables 7.3, 11.1). To my knowledge, the largest frequencies actually determined on Upper Imbrian materials are about 2×10^{-2} craters of this size per square kilometer, for old, probably unsampled units near the Apollo 11 landing site (Neukum and others, 1975b; compiled from data of Greeley and Gault, 1970). The large error bar for this determination (table 11.1) overlaps the crater frequency for the older Orientale basin and demonstrates the imprecision of such counts for small mare units. The largest D_L value measured on a mare unit is 390 m, for a spot in Mare Nectaris (Boyce and others, 1975).

No stratigraphic unit comparable to the Orientale deposits defines the top of the Upper Imbrian Series and the Imbrian System, because the individual mare and crater units that formed after Orientale do not cover a sufficient area to be globally useful as stratigraphic-datum horizons. Therefore, the boundary is defined on the basis of D_L values.

TABLE 11.1.—*Properties of the Upper Imbrian Series and adjacent time-stratigraphic units*

[References: B75, Boyce and others (1975); B76, Boyce (1976); M80b, Moore and others (1980b); N75a, Neukum and others (1975a); N75b, Neukum and others (1975b); SL72, Soderblom and Lebofsky (1972). Crater frequencies: Cumulative, determined by normalization, using a calibration curve and commonly based on craters smaller than 1 km in diameter (see fig. 7.10G; Neukum and others, 1975b). Ages: Radiometric ages obtained from samples collected from the listed rock-stratigraphic units (see tables 10.1, 11.3, 12.4)]

Time-stratigraphic unit	Rock-stratigraphic unit		Crater frequency (larger than 1 km in diameter per square kilometer)	Reference	D_L (m)	Reference	Age (aeons)
	Mare group	Formation					
Eratosthenian System	---	Apollo 12 mare	$2.4^{+0.5}_{-0.4} \times 10^{-3}$	N75b	210±20 210±30 215±45	B76 SL72 SL72	3.16
Upper Imbrian Series	3	Apollo 15 mare	$2.6^{+0.6}_{-0.4} \times 10^{-3}$	N75a, N75b	270±15	B76	3.25
	3	Eastern Mare Crisium	Approx 2.6×10^{-3}	N75a	---		
	3	Central Mare Serenitatis	3.9×10^{-3}	N75a	225–320	B75	---
	3	Luna 16 mare	---		270±30	B76	3.40
	2	Apollo 11 mare	$3.4^{+1.0}_{-1.1} \times 10^{-3}$	N75b	320±40 330±40 335±55	SL72 SL72 B76	3.57
	2	Apollo 17 mare	$9.0^{+2.2}_{-3.0} \times 10^{-3}$	N75b	360±30 365±30	B76 M80b	3.72
	2	Old flows near the Apollo 11 landing site	$2.0^{+1.2}_{-0.8} \times 10^{-2}$	N75b	---		
Lower Imbrian Series	---	Orientale basin	2.2×10^{-2}	N75a, N75b	550±50	Chapter 10	---
	---	Imbrium basin	$2.5–4.8 \times 10^{-2}$	N75a, N75b	---		3.85

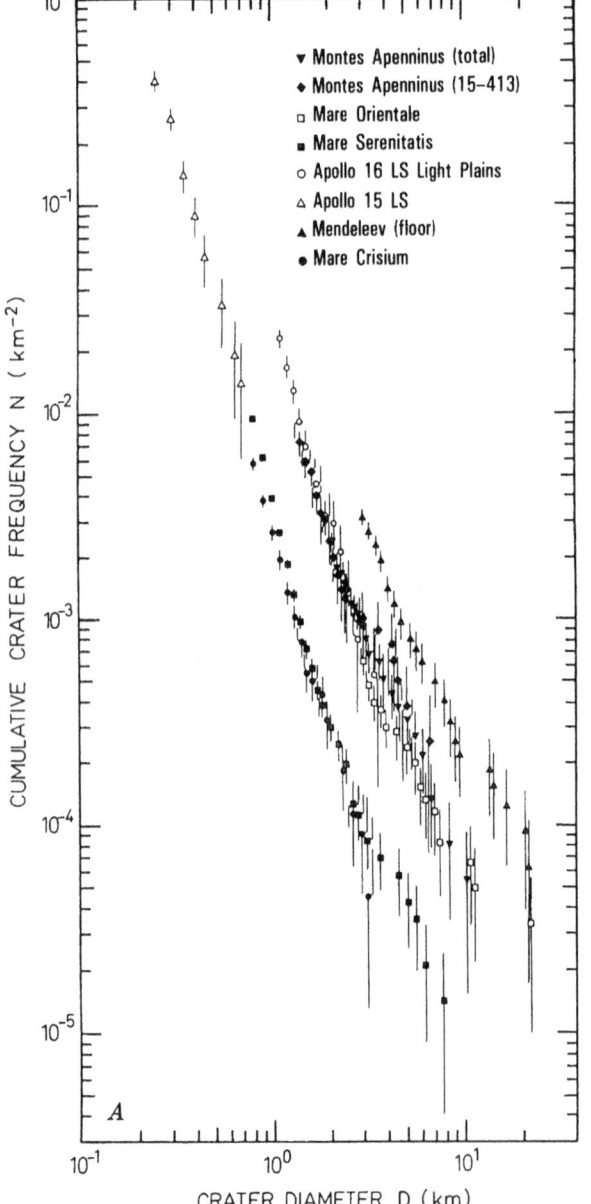

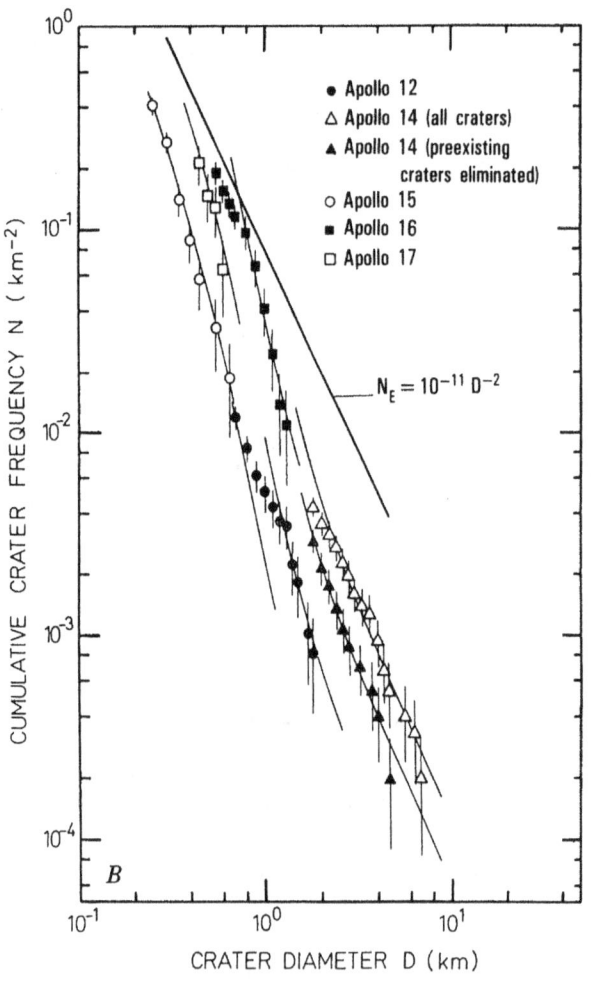

FIGURE 11.2.—Crater frequencies of units of Upper Imbrian and other ages.
A. Distinct curves for three time-stratigraphic units: (1) Nectarian System—plains on floor of Mendeleev; (2) Lower Imbrian Series—Montes Apenninus flank (two curves), light plains at Apollo 16 landing site (LS), and "Mare Orientale," actually Orientale basin (compare fig. 10.37); (3) Upper Imbrian Series—eastern Mare Crisium, Palus Putredinis near Apollo 15 landing site (both sides of Rima Hadley), and two-thirds of central Mare Serenitatis. Upper Imbrian units are near upper boundary of the series. From Neukum and others (1975a, fig. 6).
B. Upper Imbrian Apollo 15 mare in comparison with older and younger units. Apollo 17 mare (sparse data) is also Upper Imbrian. Offset in Apollo 12 curve shows that data for two units are combined—an older Upper Imbrian unit not sampled and an Eratosthenian unit that probably was sampled (see chap. 12); Eratosthenian unit is barely younger than Apollo 15 mare. Lower Imbrian units (see chap. 10) are Apollo 16 Cayley plains and Fra Mauro Formation in area of Apollo 14 landing site; one Apollo 14 count ("all craters") is hybrid of superposed and buried craters. Curve N_E is equilibrium (steady state) curve of Trask (1966); Apollo 16 plot shows rollover near steady state at about 800-m crater diameter (table 7.4); craters too small to count would show rollover for younger units. From Neukum and others (1975b, fig. 7).

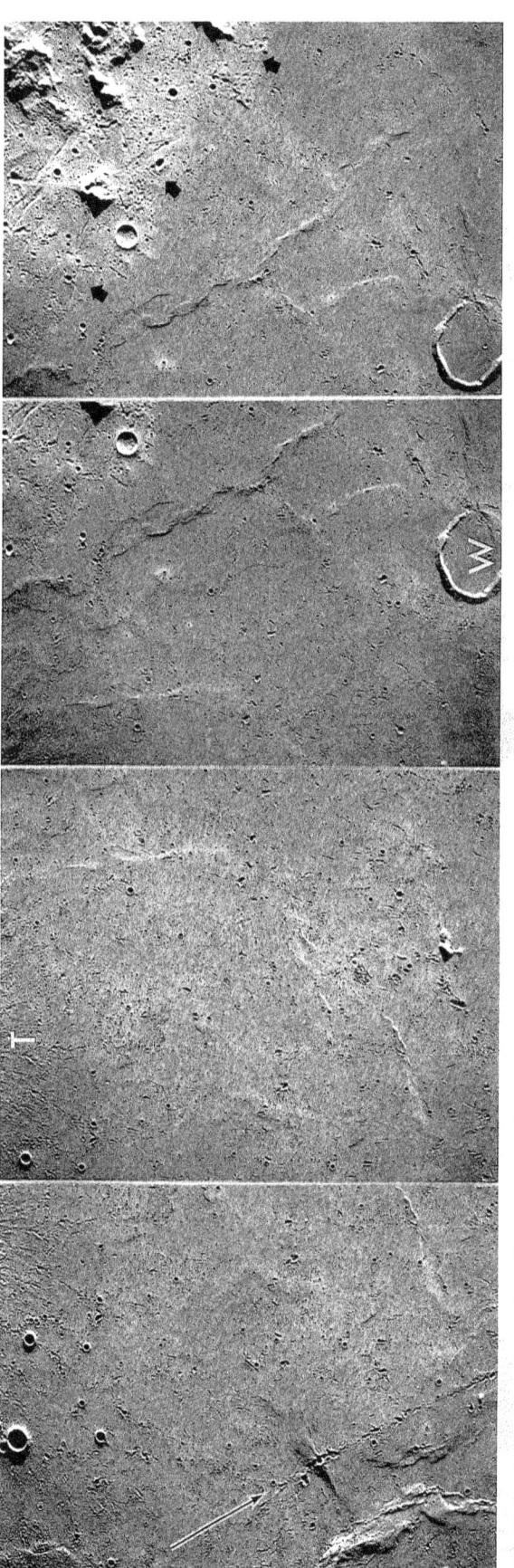

FIGURE 11.3.—Type area of top of Imbrian System. Most of area is covered by mare material older than superposed secondary craters of Eratosthenes (centered 250 km south in direction of arrow) and of Timocharis (T, fan-shaped pattern). D_L values average 270 m (Boyce, 1976); subunits are indistinct because of secondaries. Patches in, north of, and east of Wallace (W; 26 km) are probably Eratosthenian; patch north of (above) Wallace has $D_L = 220 \pm 30$ m (Boyce and others, 1975). Mare truncates Lower Imbrian Apennine Bench Formation and secondary craters of Archimedes superposed thereon (black arrows; compare fig. 10.1). Stereoscopic overlap of Apollo 17 frames M-2114, M-2115, M-2117, and M-2119 (from right to left).

Boyce and others (1975) determined that the mare units between Eratosthenes and Archimedes (figs. 1.6, 2.1, 10.1, 11.3) have D_L values from 190 to 340 m, including error ranges; midpoints range from 220 to 290 m. Wilhelms (1980) observed that units here and in the rest of southern Mare Imbrium which have D_L values less than 230 m match Eratosthenian units as traditionally mapped, and that units with D_L values greater than 250 m match Imbrian units. Therefore, he proposed these values as working criteria for the Imbrian-Eratosthenian boundary. Units with D_L values from 230 to 250 m were to be assigned on the weight of other evidence. This definition, though not formally rigorous, has served conveniently to divide a maximum number of units into Imbrian and Eratosthenian age groups that harmonize with existing mapping. This working criterion of $D_L = 240 \pm 10$ m for the upper boundary of the Imbrian System and of the Upper Imbrian Series is used here.

The characteristic size-frequency distribution for the upper boundary of the system and the series is less well defined than those for Imbrium- and Orientale-basin deposits because of the statistical unreliability of counts for the small units typical of maria. The similar values determined by Neukum and others (1975a, b) for central Mare Serenitatis, the region of the Apollo 15 landing site, and eastern Mare Crisium are probably representative of units near the upper boundary (fig. 11.2; table 11.1). Midpoints of these determinations range about from 2.6×10^{-3} to 3.9×10^{-3} craters at least 1 km in diameter per square kilometer.

Thus, the total range of observed Upper Imbrian cumulative frequencies of such craters per square kilometer is from about 2.2×10^{-2} for the Orientale basin to 2.6×10^{-3} for these young mare units (figs. 10.4, 11.2; tables 7.3, 11.1). Figures 7.15 and 12.8 and tables 7.3 and 11.1 give some equivalent D_L and crater-frequency values. As discussed in chapters 12 and 13, the values on terra-breccia and mare-basalt units are not directly comparable. Interpolations between values for breccia deposits and mare lavas are hazardous because the correction factor for the Imbrian System has not yet been determined. Thus, some of the units considered here to be Upper Imbrian may be Eratosthenian, and vice versa.

CRATER MATERIALS

Recognition criteria

Some large craters can readily be assigned to the Upper Imbrian Series by stratigraphic criteria (table 11.2). An example is Schlüter, which is superposed on Orientale deposits but slightly flooded by mare material (fig. 11.4).

TABLE 11.2.—*Representative Upper Imbrian craters*

[Cross rules divide diameter ranges mapped differently in plate 9: smaller than 30 km, unmapped; 30 to 59 km, interiors mapped; 60 km and larger, exterior deposits mapped where exposed. E?, possibly Eratosthenian; LI?, possibly Lower Imbrian]

Crater	Diameter (km)	Center (lat)	Center (long)	Figure	Remarks
St. George	2.3	25.9° N.	3.4° E.	5.11	Apollo 15.
Bonpland D	6	10° S.	18° W.	6.6	Faulted.
Ariadaeus B	8	5° N.	15° E.	10.38	Typical.
Lansberg C	17	2° S.	29° W.	10.41	Flooded.
Daedalus S	20	7° S.	173° E.	8.7	LI?
Krieger	22	29° N.	46° W.	11.6	Asymmetric (see table 7.2).
Vitruvius	30	18° N.	31° E.	9.13	LI?
Timaeus	33	63° N.	0.5° W.	10.15	---
Bianchini	38	49° N.	34° W.	3.8	---
Lansberg	39	0°	27° W.	10.41, 13.10	LI?
Mairan	40	42° N.	43° W.	3.8	---
Sharp	40	46° N.	40° W.	3.8	E?
Stiborius	44	34° S.	32° E.	11.5, 11.7	---
Cardanus	50	13° N.	72° W.	10.3, 11.8	Typical.
Krafft	51	17° N.	73° W.	10.3, 11.8	Twin of Cardanus.
Lindenau	53	32° S.	25° E.	11.5	---
Thebit	57	22° S.	4° W.	7.7	---
Jenner	72	42° S.	96° E.	1.3, 1.5, 10.46	Sandwiched by mare units.
Archimedes	83	30° N.	4° W.	1.6, 6.10, 10.1	Mare fill.
Atlas	87	47° N.	44° E.	1.7, 9.3	Floor uplifted.
Piccolomini	88	30° S.	32° E.	11.7	LI?
Schlüter	89	6° S.	83° W.	11.4	Typical.
Posidonius	95	32° N.	30° E.	1.7, 5.17, 2.50	LI?; floor uplifted.
Plato	101	52° N.	9° W.	1.6, 7.3	Mare fill.
Antoniadi	135	70° S.	172° W.	1.5, 4.2A	LI?; interior ring.
Tsiolkovskiy	180	20° S.	129° E.	1.3, 1.4, 3.24, 3.27	Mare fill.
Humboldt	207	27° S.	81° E.	1.5, 4.2E, 9.6, 10.46	Floor uplifted.
Iridum	260	44° N.	31° W.	1.6, 3.8, 10.43	Half-buried.

Other crater materials can be compared with the well-dated ones by means of crater densities where photographs are adequate and where exposed surfaces are large enough to underlie a statistically valid number of craters. These conditions are rarely met for craters of the Imbrian System. D.B. Snyder (written commun., 1980) determined that the Iridum crater (fig. 3.8) is Upper Imbrian and that Petavius (table 10.2) is Lower Imbrian. Most crater units, however, are too small or too poorly photographed to permit counts of a statistically significant number of superposed craters larger than the steady-state diameters (approx 300–800 m for the Imbrian System; see table 7.3).

Qualitative estimates of the age significance of erosional morphology are, therefore, used here more than stratigraphic or size-frequency criteria to determine the ages of craters larger than 2 or 3 km in diameter within the Imbrian System. Morphology reflects erosion by small impacts, many of whose individual craters are now obliterated because they are in a steady state. The radial ejecta and secondary-crater groups of many large craters considered to be Late Imbrian are conspicuous, though slightly blurred and visibly cratered; examples are Tsiolkovskiy (figs. 3.24, 3.27; Guest and Murray, 1969) and Lindenau (fig. 11.5). Krieger, embayed by Late Imbrian mare materials, is a very fresh Late Imbrian crater (fig. 11.6). In general, however, assignments to the Lower or Upper Imbrian Series are less secure than those to a system. For example, such outlying craters as Piccolomini that are more subdued than Eratosthenian craters and are superposed on Imbrium-basin secondaries must be Imbrian (fig. 11.7), but whether they are Lower or Upper Imbrian is less clear.

Frequency

Plate 9 shows a total of 162 unburied Upper Imbrian craters larger than 30 km in diameter within an area of 35.3×10^6 km² that excludes younger deposits from the total lunar area of 38×10^6 km². Extrapolation of this density (4.6 craters per million square kilometers) suggests that 174 such craters formed over the whole Moon during the Late Imbrian Epoch. These values are similar to those for the Lower Imbrian Series (5.1 such craters per million square kilometers or 195 such craters on the whole Moon; see chap. 10). As shown at the end of this chapter, however, the formation rates differed greatly because the two epochs were of unequal duration. The cumulative frequency of Lower and Upper Imbrian craters at least 5 km in diameter superposed on deposits of Imbrium, Orientale, and the Iridum crater—thus, of different maximum ages—is 10^{-4} per square kilometer (figs. 7.16, 10.4; Wilhelms and others, 1978).

MARE MATERIALS—GENERAL STRATIGRAPHY AND DISTRIBUTION

Imbrian mare materials are divided here (pl. 9; figs. 11.8, 11.9) into three groups: old (unit 1), intermediate-age (unit 2), and young (unit 3). Finer subdivision is possible in well-photographed areas but not in all maria.

Exposures of the old units are the least extensive; they cover only 0.24×10^6 km². They occupy elevated terra tracts in the outer Procellarum-basin shelf and outside this giant pre-Nectarian basin, places where their elevation or presence in a generally mare-poor area protected them from later inundation (figs. 10.3, 10.46, 11.8). They undoubtedly extend beneath the younger units but their extent is unknown. The western Procellarum remnants of the old group are spectrally blue, as are those on the east shelf of Mare Nubium (not distinguished on pl. 4). The more extensive exposures inside and near Oceanus Procellarum in the northeast quadrant of the nearside are spectrally red. Some units of the old group were discussed in chapter 10 because they may be Lower Imbrian (fig. 10.46); however, they are mapped here with other Upper Imbrian units because that is their

FIGURE 11.4.—Crater Schlüter (89 km, 6° S., 83° W.), superposed on Hevelius Formation at Montes Cordillera (scarp below and left of Schlüter; compare fig. 4.4). Late Imbrian age of Schlüter is indicated by its stratigraphic position between the Hevelius Formation and mare basalt; thus, the morphology and density of small superposed craters are useful for comparison with other areas (fig. 11.5). Orbiter 4 frame H–181.

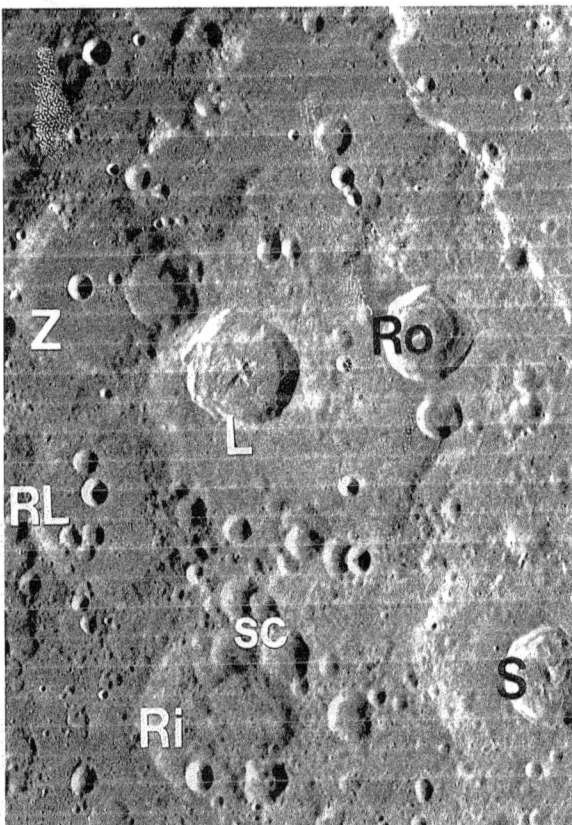

FIGURE 11.5.—Craters Lindenau (L; 53 km, 32° S., 25° E.) and Stiborius (S; 44 km, 34 S., 32 E.), probably Late Imbrian, judging by their similarity to Schlüter (fig. 11.4). Rothmann (Ro; 42 km, 31° S., 28° E.), generally similar though less cratered and slightly sharper despite its smaller size than Lindenau, is probably Eratosthenian. Pre-Nectarian craters buried by Nectaris ejecta include Zagut (Z; 84 km), Rabbi Levi (RL; 81 km), and Riccius (Ri; 71 km); sc, Imbrium-secondary craters superposed on Nectaris ejecta. Orbiter 4 frame H–83.

more likely age. D_L values of the old group are larger than 390 m, values that correspond to ages of 3.75 to 3.80 aeons (Boyce, 1976). Size-frequency counts are unavailable.

The intermediate-age and young groups occupy much more area, about 1.19×10^6 and 2.67×10^6 km², respectively (pl. 9). D_L values range from 290 to 390 m for the intermediate group and from 240 to 290 m for the young group. These values correpond to ages of 3.50 to 3.75 and 3.20 to 3.50 aeons, respectively (Boyce, 1976). Crater frequencies do not accurately discriminate these groups because the individual units are not sufficiently extensive for statistically valid counting of craters. Table 11.1 lists some typical frequencies for craters at least 1 km in diameter (Neukum and others, 1975a, b). An approximate dividing frequency is 3×10^{-3} such craters per square kilometer, but this value is not intended to be definitive for group assignment. Both groups occur in all nearside maria, although one group may not occupy enough area to be shown on plate 9. The intermediate-age group is abundant in Maria Fecunditatis, Nectaris, Australe, and other maria outside Procellarum, and on shelves and other elevated terrain of many maria. The young group dominates the surfaces of most maria. The intermediate-age and young groups are subequal in Mare Tranquillitatis. On the farside, photographic quality permits subdivision of mare units (Wilbur, 1978; Walker and El-Baz, 1982) only where they are covered by Lunar Orbiter H-frames (parts of Maria Australe and Orientale and in the crater Tsiolkovskiy) or by Apollo orbital photographs (most of the area north of lat 30° S. and west of long 180°; see pl. 2).

The original extent of the three groups is difficult to estimate. Assuming that Imbrian mare units underlie all exposed Eratosthenian and Copernican mare units and all the crater deposits that are superposed on maria (pls. 4, 10, 11), the total area of Imbrian mare units is 6.17×10^6 km². If each group covers buried terrain in proportion to its exposure, the total unburied and buried extent of each is: old group, 6.17×10^6 km²; intermediate-age group, 5.18×10^6 km²; and young group, 4.01×10^6 km². These calculations probably are reasonably accurate for the young group but less accurate for the intermediate-age group and quite uncertain for the old group, which may not have been extruded everywhere that maria are now exposed.

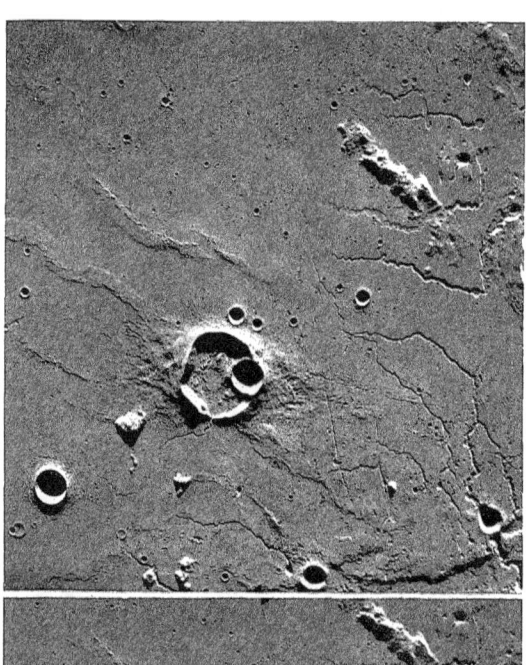
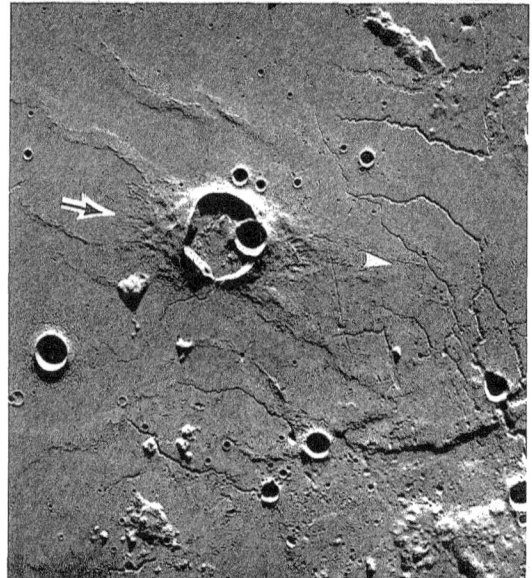
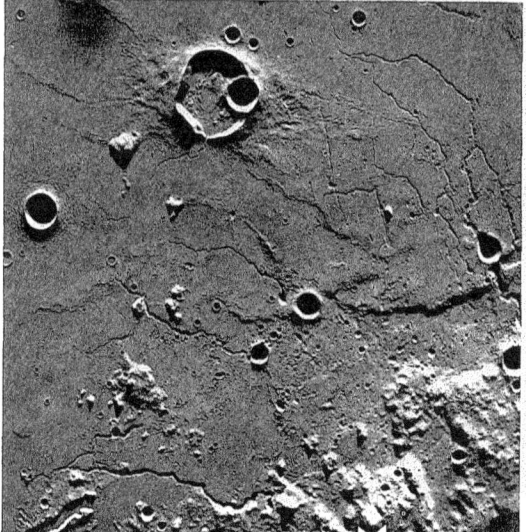

FIGURE 11.6.—Morphologically fresh crater Krieger (22 km, 29° N., 46° W.), both superposed on mare (white arrowhead) and flooded by other mare materials (black-on-white arrow). Crater's freshness suggests an Eratosthenian age, but superpositional relations and number of superposed smaller craters indicate an Imbrian age. Possible Orientale-secondary craters are embayed by dark-mantling materials (X, left) and Montes Harbinger (lower right) are visible (compare fig. 5.12). Stereoscopic pairs of Apollo 15 frames M–2082, M–2083, and M–2084 (from right to left).

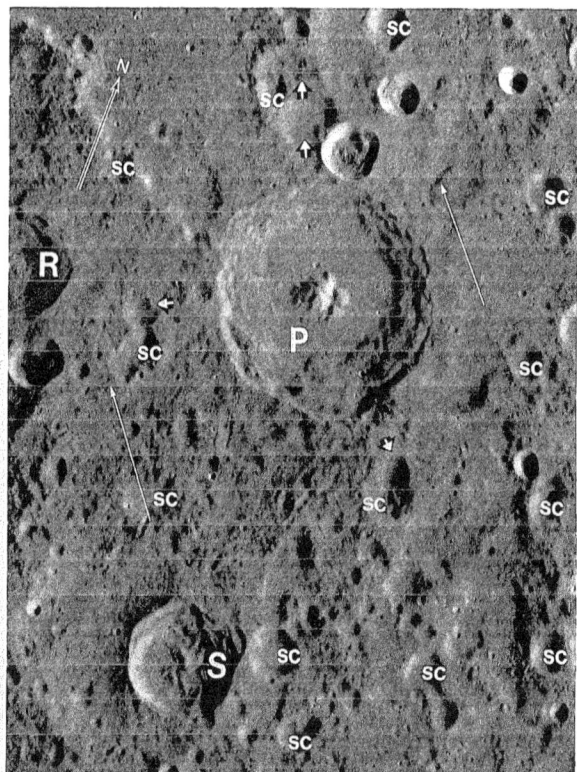

FIGURE 11.7—Crater Piccolomini (P; 88 km, 30° S., 32° E.), superposed on Rupes Altai (under north arrow). Piccolomini deposits are superposed on Imbrium-secondary craters (sc), as indicated by white-on-black arrows. Imbrium-basin rim (Montes Apenninus) is 1,800 km away in direction of white arrows. Craters Rothmann (R) and Stiborius (S) are also visible in figure 11.5. Orbiter 4 frame H–76.

Dark materials that mantle terra topography are also Late Imbrian, judging from the absence of Orientale effects on their surfaces (figs. 5.12, 5.15–5.17, 6.2, 11.6). Most known dark-mantling materials occur along the margins of maria, especially those near the center of the Procellarum basin (pl. 4). Carr (1965b; 1966a, b) observed superpositions of some dark-mantling deposits on mare surfaces (figs. 5.16–5.18) and, accordingly, surmised a volcanic origin (see chap. 5) and an Imbrian or Eratosthenian age. The smoothness and low crater densities of the deposit of dark mantle at the Apollo 17 landing site led later workers to conclude a Copernican age (Scott and Pohn, 1972; Scott and others, 1972; Greeley and Gault, 1973). Embayment relations in several areas showed impingement by mare material on dark-mantled hills (fig. 6.2B; Scott and others, 1972), but this evidence was not accepted as definitive of an old age. Subsequent

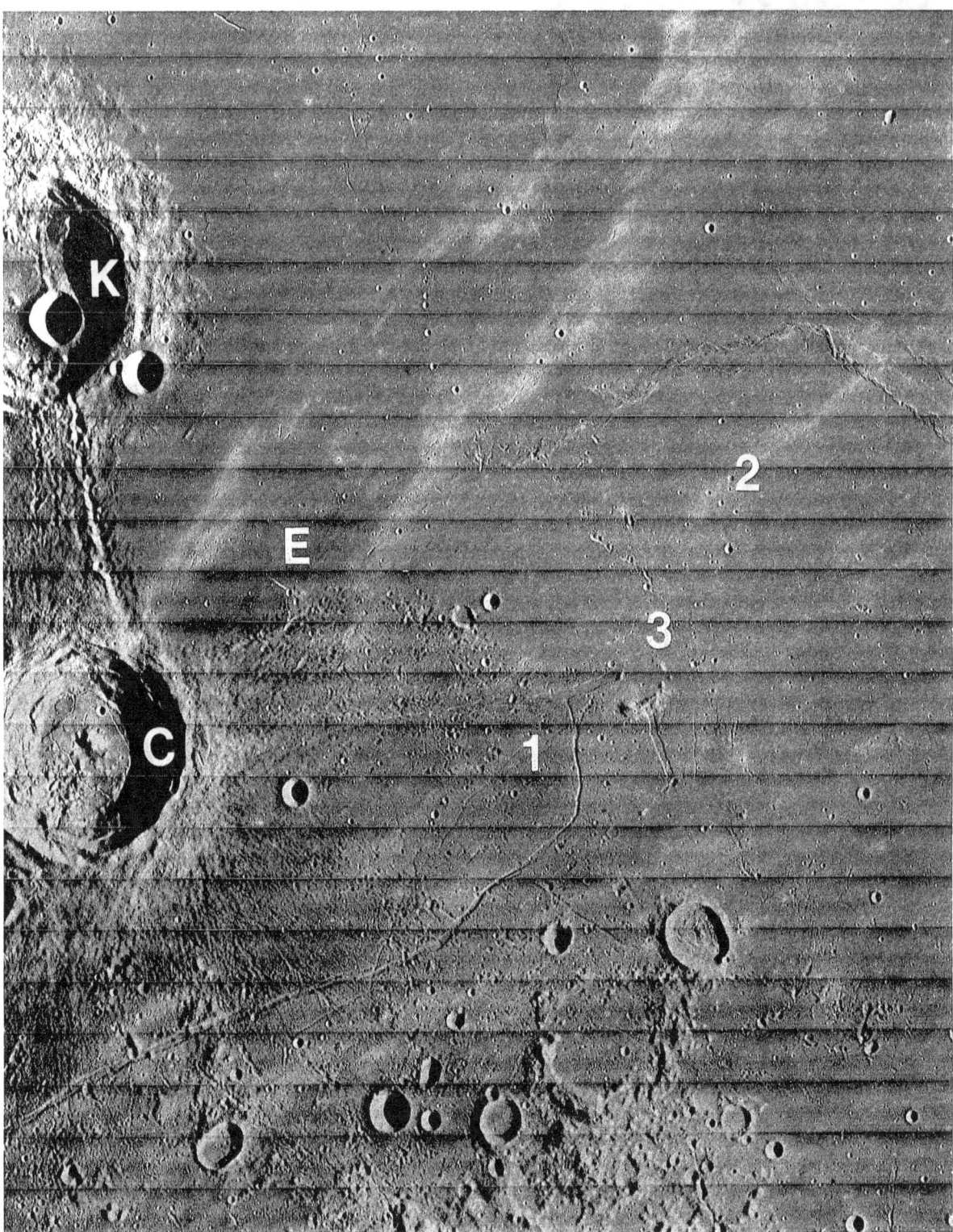

FIGURE 11.8.—Three mare units near craters Krafft (K; 51 km, 17° N., 73° W.) and Cardanus (C; 50 km, 13° N., 72° W.; compare fig. 10.3). Old unit (1) underlying Cardanus ejecta is sharply truncated by units of uppermost Upper Imbrian group (3) and Eratosthenian System (E; pl. 10). Patches too small to show on plate 9 may belong to middle Upper Imbrian group (2). Orbiter 4 frame H-169.

results, however, have confirmed the validity of the photogeologic observations of embayments; post-Imbrian dark-mantling deposits are rare.

The following six sections describe the sampled Upper Imbrian maria in order of decreasing absolute age of their oldest samples. The Apollo 11 suite contains basalt of both the old and intermediate-age Late Imbrian groups, as well as some Early Imbrian basalt. The next two suites, from the Apollo 16 (one fragment) and Apollo 17 landing sites, fall entirely within the intermediate-age group. The last three Imbrian suites are in the young group. The rest of this chapter discusses unsampled maria according to their position relative to the Procellarum basin, which, I believe, has controlled mare distribution (pls. 4, 9–11) by thinning the elastic lithosphere and terra crust as a prelude to the additional thinning inflicted by the immediate mare host basins.

MARE TRANQUILLITATIS

General stratigraphy

Mare Tranquillitatis occupies a pre-Nectarian basin, about 800 km in diameter, with irregular margins and without definite multiple-ring structure (fig. 11.1). The irregular topography in and near this basin results from the intersection of the Tranquillitatis, Nectaris, Crisium, Fecunditatis, and Serenitatis basins with two throughgoing rings of the Procellarum basin (pl. 3; fig. 11.1). The apparent double-basin structure of Tranquillitatis (Stuart-Alexander and Howard, 1970) results from impingement of the older and younger rings on the outline of Tranquillitatis.

Mare Tranquillitatis contains northern and southern belts of the intermediate-age group of Upper Imbrian basalt, separated by a belt of the young age group (pl. 9). Most of the central belt and the western parts of all three belts are of a blue spectral type that is more extensive here than in any other mare (pl. 4; table 5.1). The eastern parts of the northern and southern belts, however, are spectrally red. Sinus Amoris (fig. 9.11) contains thin highly cratered material that is bright and spectrally red. The southeast sector of Mare Tranquillitatis contains similar spectrally red or mottled material, whose wavy topography and degraded superposed craters lend a terralike appearance (Carr, 1970; Wilhelms, 1970a; Wilhelms and McCauley, 1971), but which result from a very thin cover of mare basalt on elevated basin structure (Wilhelms, 1972a). Depths of buried craters also indicate that the eastern Tranquillitatis basalt section is thinner than the western (fig. 5.21B; De Hon and Waskom, 1976). These differences correlate with a position in the Procellarum basin: thick and blue closest to the center, thin and red nearer the edge (pl. 4). Units of both age groups, as well as the west basin margin, are cut by arcuate rilles (figs. 6.11, 11.1).

The unique mare-ridge feature Lamont (fig. 11.1) has elicited considerable comment and has been diversely interpreted. It consists of both concentric and radial ridges (Morris and Wilhelms, 1967), is very conspicuous at low Sun elevations, has a small mascon (Scott, 1974; Dvorak and Phillips, 1979), and lies athwart a ring of the Procellarum basin (pl. 4; fig. 11.9; Whitaker, 1981). I suggest that a premare crater perched on this ring created the circular part of the ridge pattern, contains sufficient basalt to appear as a mascon, and remained elevated while subsidence crumpled the surrounding basalt flows into a radial ridge pattern.

Apollo 11 samples

The first rocks returned from the Moon are among the oldest mare-basalt samples. Apollo 11 landed in the southern intermediate-age belt on the oldest of three geologic units mapped near the landing site (figs. 9.8, 11.1, 11.10; Grolier, 1970a, b). Most of the samples obtained presumably came from this unit or from still older buried units. A crater frequency between those for the Fra Mauro Formation and the Apollo 15 mare (table 11.1) and a D_L value of 335 ± 55 m (Boyce, 1976) place the surface unit in the intermediate-age group mapped here (pl. 9). A nearby, older unit that was also dated by crater frequencies (table 11.1) has not been definitely correlated with any samples.

The landing point of the Lunar Module *Eagle* was about 400 m west of the 180-m-wide, 30-m-deep crater "West" between blocky rays of that Copernican crater (figs. 11.10B, 11.11; Shoemaker and others, 1969a; Schmitt and others, 1970). Thickness of the regolith is estimated at 3 to 6 m (fig. 11.12; Shoemaker and others, 1970). Beaty and Albee (1978, p. 431–432; 1980) suggested that almost all the samples were derived from the ejecta of West crater and thus came from no more than 30 m deep (fig. 11.12). Because most of the samples were exposed to the space environment for about 0.1 aeon (Eberhardt and others, 1970; Guggisberg and others, 1979), West is probably about 0.1 aeon old. At least one sample (10050) has an exposure age of about 0.5 aeon and thus was probably excavated by a crater older than West, possibly from a basalt unit distant from the landing site (Beaty and Albee, 1978, p. 431–432; 1980).

The returned basalt is characterized by its high Ti content. This property calibrated the telescopic spectral data (McCord, 1969) and showed that blue spectra indicate abundant Ti (Adams and McCord, 1970; McCord and Johnson, 1970). This Ti is contained in several modal minerals, mostly ilmenite (10–17 volume percent; James and Jackson, 1970).

Two types of basalt were recognized by megascopic differences during the first examination of the rock-size samples (larger than 1 cm) and were referred to as types A and B (U.S. National Aeronautics and Space Administration, 1969, p. 123–126; Schmitt and others, 1970). These two basalt types have proved to be distinct in composition, microscopic texture, and initial Sr-isotopic ratios as well. They also differ in crystallization age and display the largest spread in ages from any mare-sampling site.

Low-K, high-Ti.—Most basalt that was called type B in the original classification appears medium-grained and vuggy in hand specimen and ophitic or subophitic in thin section (compare fig. 2.6A). It is commonly referred to as Apollo 11 ophitic ilmenite basalt (James and Jackson, 1970) or Apollo 11 low-K basalt. Of the 10 rock samples of this compositional group, 7 have been well dated at from 3.66 to 3.85 aeons by the Ar-Ar method (table 11.3; Guggisberg and others, 1979). Although the Rb-Sr method is less reliable for low-K rocks, the ages

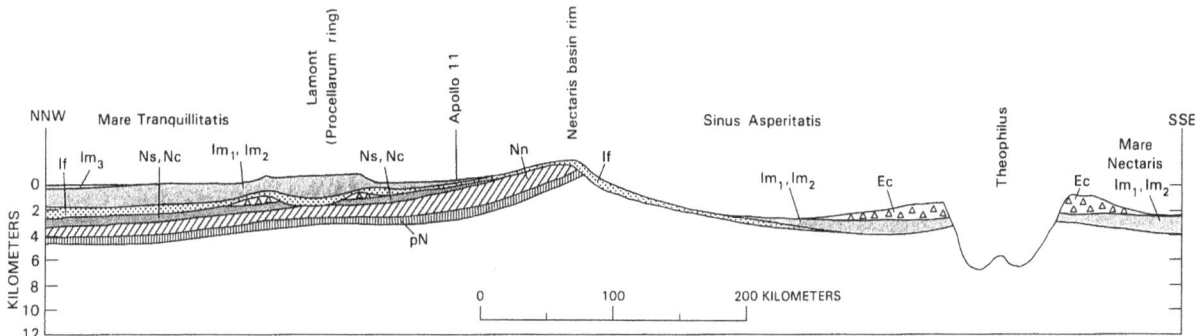

FIGURE 11.9.—Inferred geologic cross section in southern Tranquillitatis basin, Sinus Asperitatis, and Nectaris basin, drawn approximately from top to bottom of figure 11.1, with a southward extension. Units, from oldest to youngest: pN, pre-Nectarian basin, crater, and volcanic deposits, undivided; Nn, Nectaris-basin primary ejecta; Ns and Nc, primary and secondary ejecta of Serenitatis and Crisium basins; If, Fra Mauro Formation (Imbrium-basin ejecta); Im$_1$, Im$_2$, and Im$_3$, Upper Imbrian mare basalt; Ec, Theophilus ejecta. Lamont is drawn as a buried Nectarian crater superposed on a ring of Procellarum basin. An inner ring of Nectaris basin passes under Theophilus, thinning the basalt units. Vertical scale approximate; Theophilus measurements from Pike (1980b).

obtained by the Lunatic Asylum of the California Institute of Technology agree approximately with the Ar-Ar ages, as does one Sm-Nd age (table 11.3; Papanastassiou and others, 1977). Textures, modal mineralogy, and compositions differ sufficiently within the sample group to indicate at least three separate flows (fig. 11.12; Beaty and Albee, 1978). Ar-Ar ages suggest at least four flows (table 11.3; Geiss and others, 1977; Guggisberg and others, 1979), or three flows and an ejecta blanket containing sample 10050 (Beaty and Albee, 1980). Averages of the best Ar-Ar age determinations for each of these four units, performed on plagioclase separates where available, are 3.84, 3.79, 3.70, and 3.67 aeons.

An early conclusion that sample 10003 is older than the average for the compositional group (Turner, 1970) has been confirmed and extended to samples 10029 and 10062 (table 11.3). The 3.84 ± 0.03-aeon average Ar-Ar age obtained for plagioclase separates from samples 10003 and 10029 is close to the 3.85 ± 0.03-aeon age of the Imbrium basin, and the Rb-Sr and Sm-Nd ages of 10062 are older than 3.85 aeons. A pre-Imbrian age is unlikely, however, because Imbrium ejecta is evidently thicker on the adjacent terra than the total 30-m-thick basalt section from which the samples are believed to have been derived (Beaty and Albee, 1980). Also, the error ranges for all age determinations on the older basalt samples and the Imbrium

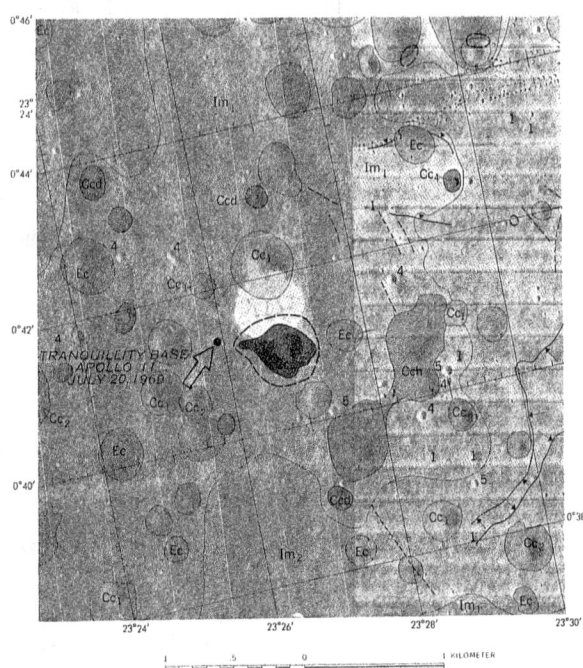

FIGURE 11.10.—Geologic maps of region of Apollo 11 landing site.
A. General region (Grolier, 1970b). Units, from oldest to youngest: Im_1, older Imbrian mare material, surrounded by younger mare unit (unsymbolized here); both units are parts of intermediate-age Upper Imbrian unit as defined in this chapter; Ec, Eratosthenian crater material; Ccd, Copernican "dimple crater" material; Crct, ray-cluster (secondary) craters of Theophilus. Small craters not separately mapped are indicated by letters (E) and numbers (1, 2, 5) corresponding to scheme of N.J. Trask (see chap. 7; figs. 7.12–7.14). Barbed arrows and dashed lines indicate minor mare topography.

B. Detailed map (Grolier, 1970a). Cc_5, young Copernican crater "West" (5 in A); heavy dashed line delimits greatest concentration of ejected blocks. Ec and low-numbered Cc, older craters; Cch, material of irregular crater chain (probably secondary); Im, same as in A.

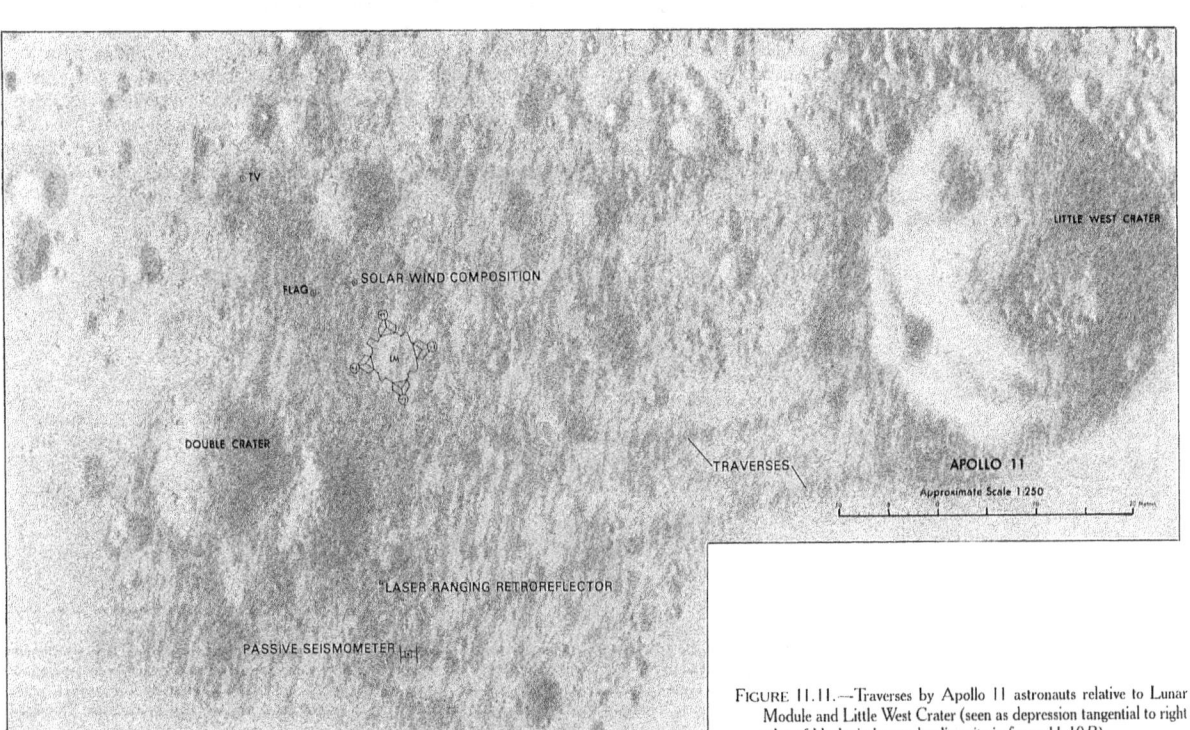

FIGURE 11.11.—Traverses by Apollo 11 astronauts relative to Lunar Module and Little West Crater (seen as depression tangential to right edge of black circle over landing site in figure 11.10B).

11. UPPER IMBRIAN SERIES 237

TABLE 11.3.—*Radiometric ages of large Upper Imbrian basalt samples and pyroclastic glasses*

[Types of samples: (r), small sample collected by rake. Apollo 11 stratigraphic classification after Beaty and Albee (1978, 1980; see fig. 11.12). Apollo 17 chemical classification after Rhodes and others (1977). Stations: LM, vicinity of landed Lunar Module. Apollo 15 stations (fig. 10.9): 1, Elbow Crater; 2 and 7, Apennine front; 3 and 8, on mare; 4, Dune Crater; 9A, rim of Rima Hadley. Apollo 16 station 6, base of Stone Mountain (fig. 9.9; Ulrich and others, 1981). Apollo 17 stations (fig. 9.12; Wolfe and others, 1981): 1, 1 km south of LM; 4, rim of Shorty Crater, 4.3 km west of LM; 5, 1.3 km west of LM; 9, 2 km northeast of LM. All Luna 16 and 24 samples were collected as regolith particles in a single core tube.
References: A80, Alexander and others (1980); AD74, Alexander and Davis (1974; incorporates revisions of earlier determinations by the same laboratory (University of California, Berkeley), which are not given here); BB1, Basaltic Volcanism Study Project (1981, p. 940-962, table 7.3.1); BA73, Birck and Allegre (1973); D73, de Laeter and others (1973; reports revisions of earlier determinations by the same laboratory (Australian National University), which are not given here); D78, DePaolo and others (1978); E73, Evensen and others (1973); E73a, Eberhardt and others (1973a); G79, Guggisberg and others (1979); HJ2, Huneke and others (1972); HJ3, Huneke and others (1973); HJ4, Huneke and others (1974); HJ5, Horn and others (1975); HJ6, Huneke (1978); HS73, Husain and Schaeffer (1973); HS74, Husain (1974); JJ5, Jessberger and others (1975); K73a, Kirsten and others (1973a); KH74, Kirsten and Horn (1974); L75, Lugmair and others (1975); L78, Lunatic Asylum (1978); M71, Murthy and others (1971); M72, Murthy and others (1972); MC76, Murthy and Coscio (1976); N74, Nyquist and others (1974); N75, Nyquist and others (1975); N76, Nyquist and others (1976; see also Nyquist, 1977); N79a, Nyquist and others (1979a); P70, Papanastassiou and others (1970); P72, Podosek and others (1972); P77, Papanastassiou and others (1977); PH73, Podosek and Huneke (1973); PW71a, Papanastassiou and Wasserburg (1971a); PW72a, Papanastassiou and Wasserburg (1972a); PW73, Papanastassiou and Wasserburg (1973); PW75, Papanastassiou and Wasserburg (1975); S73, Stettler and others (1973); S74, Stettler and others (1974); S78, Schaeffer and others (1978); SA78, Salto and Alexander (1978); SA79, Salto and Alexander (1979); SH73, Schaeffer and Husain (1973); SS77, Schaeffer and Schaeffer (1977); T73, Tatsumoto and others (1973); T74, Tera and others (1974); Tu70, Turner (1970); Tu71, Turner (1971); Tu73, Turner and others (1973); TC74, Turner and Cadogan (1974); TC75, Turner and Cadogan (1975; see also Turner, 1977); TW75, Tera and Wasserburg (1975); TW76, Tera and Wasserburg (1976); Y72, York and Others (1972). There is a question as to whether the values obtained by HS73, HS74, S78, SH73, and SS77 should be altered in accord with the decay constants recommended by the International Union of Geological Sciences (IUGS) (Steiger and Jäger, 1977; see table 4.1, headnote); the recalculated values are listed here and in BB1 for those and all other samples. p, analysis performed on plagioclase mineral separates and considered to be superior. Other methods: Sm, Sm-Nd; U, U-Pb-Th]

Group	Sample	Type	Station	Rb-Sr age (aeons)	Reference	Ar-Ar age (aeons)	Reference	Other age (aeons)	Method	Reference
Apollo 15 green glass.	15086	---	1			3.25±0.06	H74			
	15426	---	7			3.34±0.06	PH73			
Apollo 15 olivine basalt.	15016	---	3	3.22±0.05	E73	3.34±0.08	BB1			
	15385	(r)	7			3.35±0.05	Hs74			
	15555	---	9A	3.23±0.08	M72	3.22±0.03	AD74	3.3	U	TW75
		---	---	3.25±0.04	PW73	3.24±0.06	Hs74			
						3.27±0.03	P72 (p)			
						3.27±0.06	Y72			
						3.29±0.04	Hs74			
	15607	(r)	9A			3.23±0.12	Hs74			
	15633	(r)	9A			3.22±0.05	Hs74			
	15659	(r)	9A			3.30±0.04	Hs74			
	15668	(r)	9A			3.10±0.05	Hs74			
	15678	(r)	9A			3.34±0.05	Hs74			
	15683	(r)	9A			3.32±0.03	Hs74			
Apollo 15 pigeonite basalt.	15058	---	8			3.32±0.03	Hs74			
	15065	---	1	3.21±0.04	PW73					
	15075	---	1			3.41±0.20	SS77			
	15076	---	1	3.26±0.08	PW73	3.31±0.04	S73			
	15085	---	1	3.33±0.04	PW73					
	15117	(r)	2	3.28±0.04	PW73					
	15499	---	4			3.30±0.08	Hs74			
	15682	(r)	9A	3.37±0.07	PW73					
Luna 24 VLT basalt.	24077,13					3.33±0.21	S78			
	24077,63					3.26±0.06	S78			
	24170					3.30±0.04	L78	3.30±0.05	Sm	D78
						3.33±0.20	BB1			
	L24A					3.61±0.12	SA78			
Luna 16 feldspathic basalt.	L16-B1	---	---	3.35±0.18	PW72a	3.41±0.04	H72			
		---	---	3.4±0.2	BA73					
Apollo 17 orange-and-black glass.	74001	OB	4			3.65±0.02	SA79			
	74220	OB	4			3.50±0.05	H73	3.48±0.03	U	TW76
						3.60±0.04	H78	3.63	U	T73
						3.66±0.03	A80			
						3.66±0.06	HS73			
						3.55-3.66	E73a			
Apollo 17 low-K, high-Ti basalt.	70017	U	LM	3.60±0.18	N75					
	70035	U	LM	3.65±0.11	N74	3.69±0.07	S73			
		---	---	3.74±0.06	E73					
	70135	U	LM	3.67±0.09	N75			3.76±0.06	Sm	N79a
	70215	B	LM			3.79±0.04	KH74			
	70255	A	LM			3.79±0.02	SS77			
	71055	U	1	3.56±0.09	T74					
	74255	C	4	3.62±0.12	MC76					
		---	---	3.75±0.06	N76					
	75035	A	5	3.73±0.14	MC76	3.71±0.05	TC75			
	75055	A	5	3.69±0.06	T74	3.71±0.05	Tu73			
						3.73±0.04	H73			
						3.77±0.05	K73a			
	75075	U	5	3.74±0.06	MC76	3.67±0.06	J75	3.70±0.07	Sm	L75
		---	---	3.75±0.10	T73	3.69±0.02	H75			
		---	---	3.76±0.12	N75					
	79155	U	9			3.75±0.04	KH74			
Apollo 16 feldspathic basalt.	66043,2, 17	---	6			3.74±0.05	SH73			
Apollo 11 high-K, high-Ti basalt.	10017	A	LM	3.51±0.05	P70					
	10022	A	LM			3.54±0.06	G79			
						3.54±0.09	Tu70			
	10024	A	LM	3.53±0.07	PW71a					
	10031	A	LM			3.55±0.08	G79			
	10032	A	LM			3.53±0.06	G79			
	10057	A	LM	3.55±0.04	P70					
	10069	A	LM	3.60±0.06	P70					
	10071	A	LM	3.60±0.02	P70	3.47±0.06	S73 (p)			
	10072	A	LM	3.56±0.05	P77	3.48±0.05	Tu70	3.57±0.03	Sm	P77
		---	---	3.63±0.11	D73	3.57±0.05	G79 (p)			
Apollo 11 low-K, low-Ti basalt.	10044	B1	LM	3.63±0.11	P70	3.66±0.04	G79			
						3.68±0.05	Tu70			
	10047	B1	LM			3.69±0.03	S74, G79			
	10058	B1	LM	3.55±0.11	M71	3.66±0.04	G79			
		---	---	3.55±0.22	P70					
	10020	?	LM			3.72±0.04	G79			
	10050	?	LM			3.70±0.03	G79			
	10062	B2?	LM	3.92±0.11	P77	3.77±0.06	Tu70	3.88±0.06	Sm	P77
						3.79±0.04	G79 (p)			
	10003	B2	LM	3.76±0.08	PW75	3.85±0.03	S74 (p)			
						3.86±0.07	Tu70			
	10029	B2	LM			3.83±0.03	G79 (p)			

basin overlap and are large for the Rb-Sr and Sm-Nd ages. The 3.84 ± 0.03-aeon average Ar-Ar age of the plagioclase separates is probably close to the actual age of the flow (barring systematic errors in all Ar-Ar ages). Extrusion of the low-K, high-Ti basalt probably began soon after the Imbrium impact (fig. 11.9). A pre-Orientale (Early Imbrian) age is possible because Orientale ejecta may well be absent or very thin here.

High-K, high-Ti.—The other type of basalt (originally, type A) is richer in K_2O, as well as in TiO_2, Na_2O, CaO, and Al_2O_3, than the low-K group (Papike and others, 1976). Grain sizes are finer than those of the low-K group, and textures are commonly intersertal (unoriented crystal grains separated by small amounts of glass or crystallized glass); accordingly, the group is also called Apollo 11 intersertal ilmenite basalt (James and Jackson, 1970). For this sample group, Rb-Sr ages are more reliable than Ar-Ar ages (Stettler and others, 1974; Papanastassiou and others, 1977). Six rocks have been dated by this method (table 11.3). Rb-Sr ages from any single laboratory cluster tightly (for example, Papanastassiou and others, 1970, 1977). This consistency and the data on textures and bulk compositions are consistent with derivation of all the rocks of this type from a single flow (Papike and others, 1976; Beaty and Albee, 1978, 1980); only minor fractionation near the surface (Papike and others, 1976) or in shallow magma chambers (Beaty and Albee, 1978) is indicated. Most Ar-Ar ages are younger than the Rb-Sr ages, probably because argon was lost after crystallization (Turner, 1977). The average of the best Rb-Sr ages and the Sm-Nd age is 3.57 aeons.

Exposure ages (duration of exposure to cosmic rays) of the two compositional groups differ and have been interpreted as indicating that the high-K flow is the uppermost sampled flow and overlies the low-K flows (Eberhardt and others, 1970; Beaty and Albee, 1980), in accord with the absolute ages (fig. 11.12). Origin of the two basalt types from different magma reservoirs is indicated by the different initial Sr-isotopic ratios (Papanastassiou and others, 1977). I tentatively correlate the frequency of 3.4×10^{-3} craters at least 1 km in diameter per square kilometer and the D_L value of 335 m listed in table 11.1 (both with large error bars) with the high-K flow.

Summary and interpretation

If the source section of all the samples is, indeed, only 30 m thick, an incredibly low extrusion rate of one flow averaging 6 m in thickness every 54 million years is indicated. Other deposits presumably underlie the sampled flows because the mare surface here is relatively unbroken by islands belonging to buried craters or other terra (fig. 11.1). Nonetheless, a thick section of Upper Imbrian flows is unlikely, considering the very old age, within the Imbrian System, of some of the lavas that were sampled. Therefore, Lower Imbrian, Nectarian, and pre-Nectarian deposits must be present in the subsurface (fig. 11.9). These deposits include thick ejecta of the nearby Nectaris basin, thinner deposits from the Serenitatis and Imbrium basins, and possible deposits of the Crisium and Orientale basins. Pre-Nectarian, Nectarian, and Lower Imbrian basalt flows may be interbedded with this ejecta sequence. Regoliths formed by smaller craters (not shown in fig. 11.9) separate all the deposits in proportion to the timespan each was exposed to the surface.

Basalt extrusion was also relatively sluggish in the rest of Mare Tranquillitatis during the Late Imbrian Epoch. Although interbedded basalt flows and ejecta may reach a thickness of 1.25 km in one spot, most of the total section is thin (fig. 5.21B; De Hon and Waskom, 1976). Two somewhat indistinct color and age units are present.

In summary, the most recent history of Mare Tranquillitatis is characterized by thin and slowly extruded, yet areally extensive, lavas of intermediate and late Late Imbrian age. The western flows are thicker and richer in Ti than the eastern, except that some eastern flows of the young Late Imbrian group are also Ti-rich.

MARE NECTARIS

General stratigraphy

Mare Nectaris differs markedly from Mare Tranquillitatis. Although the Nectaris basin is much younger and slightly larger than the Tranquillitatis basin (860 versus 800 km diam), mare volcanism was even more sluggish. Nectaris is much the smaller mare (table 6.1). Little or no basalt of the young Late Imbrian group is present (pl. 9; fig. 11.9; Boyce, 1976; Whitford-Stark, 1981). The Nectaris basalt has subsided enough to open only a few arcuate grabens on the west margin of the mare (pl. 5). The spectral class (mBG-, table 5.1) indicates intermediate values of the spectrally measured properties. Orbital geochemical data are not available except in the narrow neck, Sinus Asperitatis, that connects Nectaris with Tranquillitatis (figs. 11.1, 11.9), and there the surface is contaminated by Theophilus ejecta. The Asperitatis basalt section is as thick as or thicker—possibly 1.5 km—than those in central Nectaris and Tranquillitatis (fig. 5.21B; De Hon and Waskom, 1976). It even coincides with a small mascon (Scott, 1974).

Apollo 16 samples

Although it did not land near a mare and was not designed to sample mare materials, Apollo 16 returned a few fragments of mare basalt. They were presumably thrown to the sampling site in the ejecta of an impact crater, probably Theophilus, in which case their source is Mare Nectaris (fig. 11.1; Delano, 1975). The fragments include both Ti- and Al-rich types. Ar-Ar analysis of one Ti-rich sample yielded an age of 3.74 aeons, consistent with the photogeologically old age of most of Mare Nectaris (table 11.3; Schaefer and Husain, 1973).

Summary and interpretation

The differences in age and position relative to the Procellarum basin between the Nectaris and Tranquillitatis basins may account for the differences between Maria Nectaris and Tranquillitatis. First, the fact that Mare Tranquillitatis is extensive, yet relatively thin (table 6.1; fig. 5.21B; De Hon and Waskom, 1976), probably results from premare leveling of the basin floor by thick Nectaris-basin and other ejecta deposits and, possibly, by pre-Imbrian basalt (fig. 11.9). Second, the thick lithosphere outside Procellarum may have prevented sufficient subsidence to erase the Nectaris mascon, to open large grabens, or to provide much space for basalt extrusion. Each condition was opposite in Tranquillitatis, where the weak lithosphere allowed extensive basalt extrusion and subsidence. This hypothetical significance of lithospheric differences is supported by the relatively extensive mare development in Sinus Asperitatis, which is outside the centers of Nectaris and Tranquillitatis but inside the rim of the Procellarum basin (pl. 4).

MARE SERENITATIS

General stratigraphy

Whereas Serenitatis is among the Moon's least well understood basins (see chap. 9), Mare Serenitatis displays clear stratigraphic and structural relations that have made it a model for other maria (Carr,

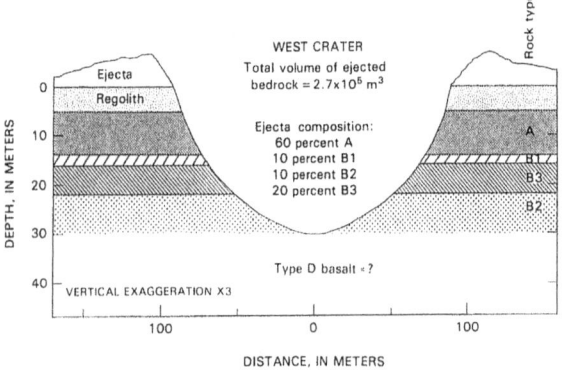

FIGURE 11.12.—Inferred basalt stratification penetrated by West Crater and yielding the Apollo 11 samples (after Beaty and Albee, 1980). A, high-K, high-Ti basalt; B1, B3, and B2, flows of low-K, high-Ti basalt (Beaty and Albee, 1978). B1 may be ejecta blanket. No definite B3 basalt is listed in table 11.3; samples 10050 and 10062 may be part of that flow (Beaty and Albee, 1978). Type D, possible additional basalt type not dated or discussed here.

1966a; Howard and others, 1973; Thompson and others, 1973; Muehlberger, 1974; Head, 1979c; Solomon and Head, 1979). The southeast margin contains a continuation of the intermediate-age, dark, spectrally blue unit that characterizes northwestern Mare Tranquillitatis. This border zone is faulted by conspicuous arcuate grabens (pls. 4, 5, 9; figs. 5.17, 6.2, 11.13). The southwest-marginal unit is younger (pl. 9; Howard and others, 1973), of a similar but less blue spectral class (hDWA, pl. 4), and unfaulted. The division between the two units falls along the middle ring of the Procellarum basin (pl. 4). The center of Mare Serenitatis is occupied by younger, relatively bright material of spectral class mISP (table 5.1), which, under the designation "MS-2," is the standard for normalizing color spectra (fig. 5.19). This red or "orange" material appears to be spectrally uniform even though it consists of numerous small flows (not mapped here). The central units are not cut by grabens but contain large mare ridges (see chap. 6; Muehlberger, 1974; Maxwell and others, 1975; Solomon and Head, 1979).

In early work, the dark units were thought to be the younger (Carr, 1966a, b), partly because the central mare seemed in telescopic photographs to have more craters than the border mare. Along the east and west margins, this relation has been confirmed; the dark border units there (spectral class hDWA, pl. 4) are younger than the light (pl. 10). Stereoscopic spacecraft photographs have shown, however, that the central mare is younger than the southeast- and southwest-border units; it sharply abuts most of the raised border zone and transects the rilles (figs. 5.16, 6.2, 11.13; Howard and others, 1973). The apparent abundance of craters in the central mare results from their greater brightness and sharpness, which properties are caused by excavation of fresh rock through the thinner regoliths on the younger lavas (Young, 1975); the older border mare has more large, subdued craters.

Serenitatis is a massive mare. The 1-km isopach encloses a large part of the mare (fig. 5.21B; table 6.1), and the radar-sounder and mascon data suggest depths of more than 2 km (see chap. 5). The large massive load of basalt depressed the basin and the earlier Ti-rich basalt and opened the many arcuate grabens (chap. 6; fig. 6.2; Muehlberger, 1974; Solomon and Head, 1979, 1980). Serenitatis, therefore, differs considerably from the maria already described. Its basalt units are strongly deformed, spectrally distinct, sharply divergent in age, thick, and extensive relative to basin size (figs. 11.13, 11.14).

Apollo 17 samples

The largest suite of lunar mare basalt was returned by Apollo 17, despite the emphasis of this last Apollo mission on the Serenitatis-basin massifs (see chap. 9). The landing site was on the dark floor of a marginal embayment of Mare Serenitatis called the Taurus-Littrow Valley (figs. 9.13, 11.13, 11.15; Wolfe and others, 1981). The "subfloor basalt" may be as thick as 1.4 km (Cooper and others, 1974); the returned samples were excavated by craters from as much as 100 m below the surface (fig. 11.15; Wolfe and others, 1981). They are as rich as or richer in TiO_2 than the Apollo 11 basalt samples and are generally similar petrologically. Two main types were at first recognized on the basis of major-element composition (Papike and others, 1976): (1) a dominant mafic Mg-rich olivine basalt especially rich in Ti (max 14 weight percent TiO_2), and (2) a less abundant, coarse-grained basalt very similar to the Apollo 11 low-K, high-Ti variety. The second type was obtained from the rim of the largest sampled crater, Camelot, 650 m in diameter (sta. 5), and thus may have lain at the greatest depth of all the sampled basalt—possibly deeper than 100 m (Wolfe and others, 1981, p. 109–114, 204–205).

Subsequent examination of a larger sample suite by Rhodes and others (1976) and Warner and others (1979) disclosed that the gaps in major-element composition are occupied and led to a fourfold classification into types A, B, C, and an unclassified type U that is based more on trace elements. These authors suggested that the coarse-grained, low-K, relatively low-Ti basalt from the Camelot rim was not derived from a separate magma but is a highly fractionated form of the type A basalt. Other type A materials are scattered over the collection area. Type B basalt was collected mostly in the east half of the collection area, especially in ejecta of the 600-m-diameter crater Steno (sta. 1). These rocks may have lain somewhat above the source of the Camelot samples (Wolfe and others, 1981, p. 35, 205). Type C basalt, the rarest and most distinctive, was collected only from the rim of the 110-m-diameter crater Shorty (sta. 4) and thus probably

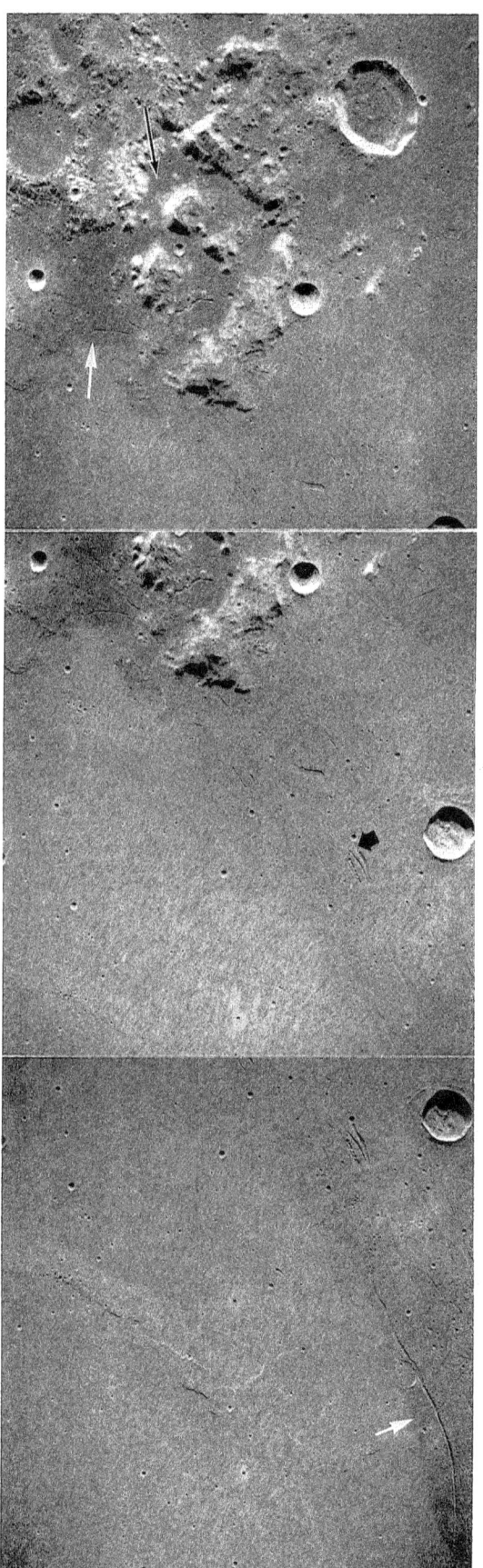

FIGURE 11.13.—Southeast border of Mare Serenitatis. Young light-colored basalt of mare center sharply abuts older dark mare and dark-mantling materials (white arrows). Older units are faulted (compare fig. 6.2). Black arrow, rilles and endogenic craters shown in figure 5.10A. Rayed Copernican crater below black arrow is Dawes (18 km, 17° N., 26° E.). Apollo 17 landing site is visible in right-hand frame (black-and-white arrow; compare fig. 9.13). Stereoscopic pairs of Apollo 17 frames M–5000, M–5002, and M–5004 (from right to left).

was derived from the shallowest depths (fig. 11.15). The mostly coarse-grained type U rocks came from all parts of the collection area and are the most numerous of the rock-size samples that were dated (table 11.3).

Rhodes and others (1976) suggested that each of the three basalt types A, B, and C differs sufficiently from the others to indicate extrusion as a separate lava flow. Variation within each type and, probably, among some or all of the type U rocks arose by fractional crystallization of one of the magmas on or near the surface.

The scatter in the absolute ages is so great (table 11.3) that no firm correlations with the A-B-C stratigraphic sequence or composition have been proposed. The Ar-Ar ages range from 3.67 to 3.79 aeons, and the Rb-Sr ages from 3.56 to 3.76 aeons; two Sm-Nd ages are 3.70 and 3.76 aeons (table 11.3). The average of the ages obtained by all methods having stated error ranges less than 0.1 aeon is 3.72 aeons. The absence of obvious correlations among age, composition, and sampling locality suggests that all the Apollo 17 basalt was extruded within a short time. The age of 3.72 aeons is, therefore, tentatively accepted here as the emplacement age of the entire sequence, more than 100 m thick. Partial melting of a heterogeneous source in a small zone of the mantle is indicated by the small ranges in age and composition (Rhodes and others, 1976).

A major Apollo 17 mission objective was to sample the dark-mantling materials, which were believed to be young (Scott and others, 1972; Hinners, 1973). They were recovered in the form of orange and black glass droplets ("beads") with a mean diameter of 0.04 mm. This material, which composes from a few percent to 20 percent of the regolith, was recovered in greater concentrations from the rim of the 19-million-year-old (Eugster and others, 1977) crater Shorty (Muehlberger and others, 1973; Heiken and others, 1974; Wolfe and others, 1975, 1981). Shorty's ejecta appears dark from orbit and black and orange on the surface. Orange droplets compose a bed, about 25 cm thick, overlying a bed of black droplets. The color difference of these droplets, whose compositions are the same, is due to more nearly complete crystallization of the black (Haggerty, 1974; Heiken and others, 1974). The devitrified black beads have a blue spectrum, and the glassy orange beads a red spectrum (Adams and others, 1974; Pieters and others, 1974). Presumably, the degree of devitrification accounts for similar spectra elsewhere, such as the red spectra (and orange-brown visual appearance) of the dark-mantling materials on the Aristachus Plateau (Zisk and others, 1977).

The telescopic spectrum of "pure" dark-mantling materials north of the Apollo 17 landing site is similar to the laboratory spectrum of the droplets from Shorty but is only partly like that of droplets from the regolith around the landing site (Adams and others, 1974). The regolith there probably contains a mixture of true dark-mantling materials, consisting of orange and black glass derived from a sheet originally at least 1.5 m thick, and of Ti-rich fragments derived from the subfloor basalt. The total thickness of the weakly bonded glass-bead deposit plus the regolith is 5 to 28 m, and the mean thickness 14 m (fig. 11.15; Wolfe and others, 1975, 1981).

A pyroclastic rather than an impact origin is indicated by the chemistry and absence of fragments in the glasses and by the distribution of dark-mantling deposits at the margins of several addi-

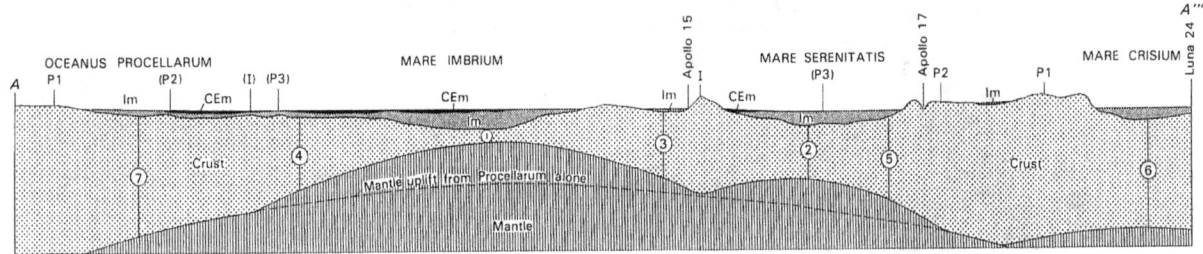

FIGURE 11.14.—Geologic cross section, drawn along line A–A''' in figure 8.4, showing relative depths of mare basalt and inferred shape of crust-mantle boundary; vertical scale arbitrary. P1, P2, and P3, rings of Procellarum basin (pl. 3); I, rim of Imbrian basin; symbols in parentheses, buried rings; Im, Imbrian mare basalt; CEm, Copernican and Eratosthenian mare basalt (pls. 10, 11). Superposed mantle uplifts resulting from superposed basin impacts are indicated diagrammatically; total uplift is thought to depend on crustal thickness remaining after formation of Procellarum basin and on size of superposed basins. Numbers 1 through 7 indicate pathways from mantle source of mare materials to surface at time of eruptions, from shortest to longest.

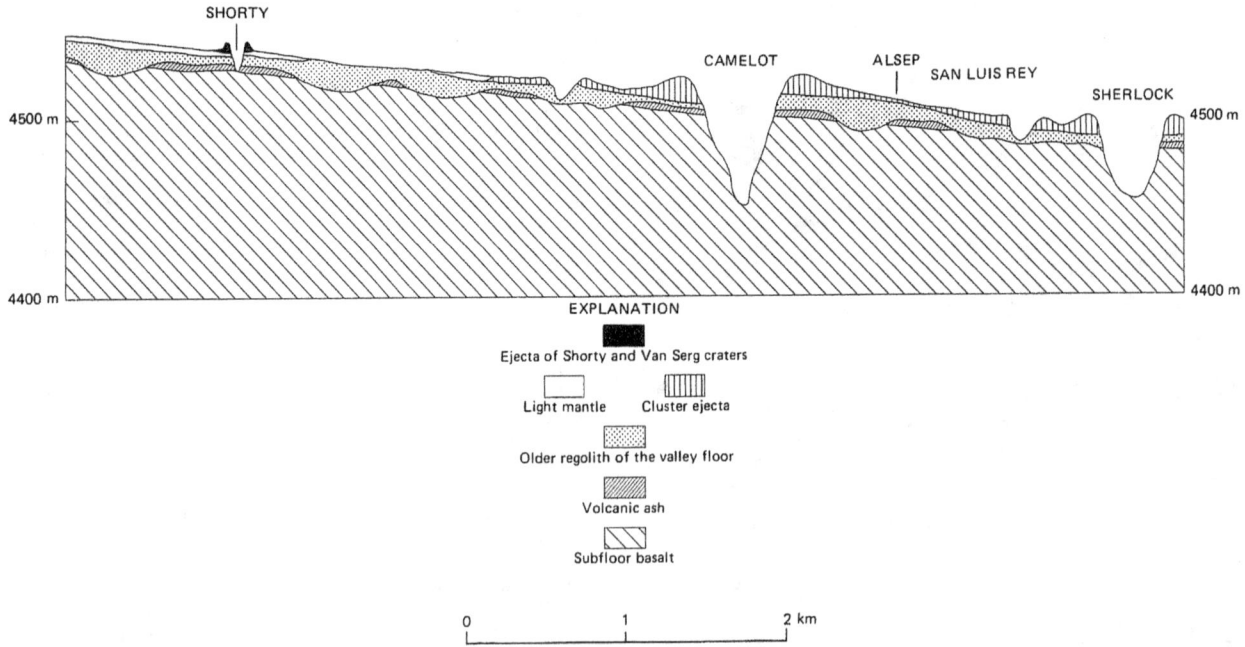

FIGURE 11.15.—Geologic cross section, drawn approximately east-west in area of figure 9.17, showing regolith, mare, and glass-droplet ("volcanic ash") deposits in Taurus-Littrow Valley. Light mantle, landslide from South Massif; cluster ejecta, ejecta of "Central Cluster" craters believed to be secondary to Tycho; ALSEP, Apollo Lunar Surface Experiments Package (mostly geophysical instruments). Vertical elevations above arbitrary datum. From Wolfe and others (1981, fig. 248).

tional maria (Heiken and others, 1974). These glasses apparently were formed during an early "lava fountain" or "fire fountain" stage of mare eruption. At Taurus-Littrow, they were deposited on the subfloor basalt before most of the regolith formed (Head, 1974a; Heiken and others, 1974). The pyroclastic material was probably derived from a different magma than was the sampled basalt (Hubbard and Rhodes, 1976).

Although sampling confirmed the pyroclastic origin of the darkmantling materials, it refuted estimates of a young age. Determinations of absolute age by different laboratories range from 3.50 to 3.66 aeons (table 11.3); the average of the single determination for sample 74001 and the three best determinations for sample 74220 is 3.64 aeons, and this age is adopted here. All these determinations are consistent with superposition of the orange glass deposit on the subfloor basalt. The low crater density that led to estimates of a young age is due to initially subdued shapes and rapid degradation of impact craters in the noncohesive deposits (Lucchitta and Sanchez, 1975).

Summary and interpretation

The extrusional history of the Apollo 17 basalt was very different from that of its compositional relative, the Apollo 11 basalt. In the Taurus-Littrow embayment, a massive flow or related flows, more than 100 m thick, formed in a time so short that age differences are geochronologically unresolvable. Fountains driven by volatile material spewed forth liquid droplets that accumulated as deposits of glass. More than 100 million years before and after the lavas and pyroclastic materials poured out here, a few meters of basalt were being added to the margin of Mare Tranquillitatis. Then, after both margins were covered by the high-Ti basalt flows, the center of Serenitatis was filled by voluminous low-Ti basalt flows that depressed and faulted the older Ti-rich basalt and the basin. The spectral and stratigraphic uniformity of the many small flows of low-Ti basalt suggests high eruption rates.

Although Maria Serenitatis and Tranquillitatis are similar in area, the volume of Serenitatis is by far the greater—second only to that of Mare Imbrium (fig. 5.21; table 6.1). The voluminous volcanism is due to the relatively young age of the Serenitatis basin (see chap. 9), to its premare filling by only one major deposit (Imbrium ejecta), and, probably, to its superposition on a relatively weak part of the lithosphere relatively near the center of the Procellarum basin (figs. 11.9, 11.14).

MARE FECUNDITATIS

General stratigraphy

The poorly photographed Mare Fecunditatis occupies a pre-Nectarian basin, about 990 km in diameter, outside the Procellarum basin. The mare is shallow, spectrally mottled and bland, relatively aluminous in X-ray readings (Adler and others, 1972; Hubbard, 1979), and apparently middle to late Late Imbrian in age (pl. 9). Like Mare Tranquillitatis, Mare Fecunditatis occupies a large fraction of the basin diameter but is relatively thin (table 6.1). Its west edge is faulted by arcuate grabens.

Selenographically, Mare Fecunditatis includes a subcircular southward extension of the main mare. This second mare, about 250 km in diameter, may occupy an (unmapped) large crater or small basin superposed on the 990-km-diameter Fecunditatis basin. Alternatively, the Fecunditatis impact may have been double.

Luna 16 samples

The Luna 16 unmanned sample-return mission (September 1970) yielded 101 g of small regolith fragments from Mare Fecunditatis (lat 0.7° S, long 56.3° E.; fig. 11.16; McCauley and Scott, 1972). The dominant basalt is a feldspathic, high-Al type distinguished by a high FeO/MgO ratio from other mare basalt, KREEP-rich material, and high-Al mare-basalt clasts from the Apollo 14 breccia (Kurat and others, 1976). The Rb-Sr and Ar-Ar ages agree fairly well at an average of 3.38 aeons, but the Ar-Ar age of 3.41 ± 0.02 aeons has much the smaller error range (table 11.3); a weighted average of 3.40 aeons is adopted here. The sampling site is on a dark unit whose spectral class (hDW-) indicates a higher Ti content than most of Mare Fecunditatis (pl. 4). The Al-rich sample composition is consistent with the orbital detection of relatively high Al/Si ratios for the overflown part of the mare (Hubbard, 1979).

Summary and interpretation

The aluminous composition of the basaltic magmas may have favored their hydrostatic ascent and extrusion through the thick crust in the Fecunditatis region, because they probably are the marebasalt magmas closest in density to the terra crust (2.8–2.9 g/cm³; Solomon, 1975). Possibly, Al-rich magmas could be extruded where denser magmas would remain beneath the surface.

Of all the maria considered here that lie outside Procellarum, Mare Fecunditatis occupies the largest percentage of its basin's area (table 6.1). A possible explanation is that the younger Nectaris and Crisium basins "assisted" the volcanism. The grabens in western Fecunditatis lie within the rim of the Nectaris basin (pl. 5). Northern Mare Fecunditatis, including the compositionally anomalous (relatively Ti-rich) Luna 16 material, lies within the proposed 1,060-kmdiameter rim of the Crisium basin (pl. 9; fig. 11.16). Therefore, the lithospheric thinning required for mare extrusion and structural deformation is achieved here by superposition of basins other than Procellarum.

MARE CRISIUM

General stratigraphy

Mare Crisium occupies an elliptical area, 420 km north-south by 560 km east-west (pl. 4). The average diameter of this mare is less than half that of the containing Crisium basin in my interpretation (1,060 km; see chap. 9; tables 4.1, 6.1). Crisium is probably between the Serenitatis and Nectaris basins in age, closer to Serenitatis (table 9.3). Mare Crisium has properties in common with both Maria Serenitatis and Nectaris. It has a mascon, as do both of these maria. The basalt section is about as thick as that of Serenitatis and thicker than that of Nectaris (fig. 5.21B; De Hon and Waskom, 1976). In respect to its areal extent relative to the basin diameter, however, it more closely resembles Nectaris (if the Crisium basin is indeed as large as I suggest). Tectonically, Crisium is more like Nectaris than Serenitatis because no arcuate grabens have opened since the visible mare formed (pl. 5).

Mare Spumans, Mare Undarum, and other "lakes" surround the main Mare Crisium at considerably higher elevation (fig. 4.9). In chapter 4, this relation was interpreted as analogous to the steplike structure of the Orientale basin and its small superposed maria. The "lakes" are as high as or higher than the adjacent Mare Fecunditatis (fig. 6.23).

Investigative groups led by J.M. Boyce (Boyce and Johnson, 1977; Boyce and others, 1977) and J.W. Head (Head and others, 1978a) mapped several age units, but these two groups did not agree on the units' distribution. Photographs suitable for age determinations are available only in the southern part of the mare under the Apollo 17 orbits. To me, Crisium seems to be among the stratigraphically most uniform maria, and I map it here entirely as part of the youngest of the three Late Imbrian age categories (pl. 9). The "lakes" are older.

Head and others (1978a) and Pieters (1978) mapped several color units in Mare Crisium (pl. 4). An Mg-rich unit similar to the Luna 16 basalt is exposed at the southeast margin (hDWA, pl. 4); this may also be the unit exposed in the ejecta of the crater Picard (fig. 5.23; Andre and others, 1978; Head and others, 1978a). The unit sampled by Luna 24 is the type area of spectral unit LISP. Unit hDWA is moderately rich in Ti, but the other spectral units are relatively poor in Ti and rich in Fe (table 5.1). Hubbard (1979) interpreted data obtained by the orbiting X-ray spectrometer to mean that Mare Crisium is more aluminous than the maria farther west, about equal to Mare Fecunditatis, and less aluminous than only Mare Smythii among the major analyzed maria.

Luna 24 samples

The most recently (August 1976) collected lunar sample is a 160-cm-long drill core obtained from the regolith of Mare Crisium by the unmanned Soviet spacecraft Luna 24 (lat 12.7° N., long 62.2° E.; fig. 9.3; Butler and Morrison, 1977; Florensky and others, 1977). Most of the basalt fragments are very low in Ti (less than 1.5 weight percent TiO_2), a basalt type previously found only in small amounts in the Apollo 17 drill core (Vaniman and Papike, 1977). The Luna 24 VLT "ferrobasalt" is richer in FeO than the Apollo 17 VLT basalt and is also highly feldspathic (12.5–14 weight percent Al_2O_3; Papike and Vaniman, 1978). In addition, the Luna 24 core contains an MgO-rich olivine basalt and, possibly, several additional types or subtypes (for example, see Ma and others, 1978). The spectra indicate the presence of additional units of Mare Crisium that were not sampled. Discovery of the VLT basalt caused a rethinking of previous classification schemes (Papike and Vaniman, 1978). In general, current thinking favors a continuum of lunar-basalt types rather than the pigeonhole classifications that emerged from earlier missions.

Despite the small sample size, accurate absolute ages of the underlying bedrock apparently were obtained from the regolith fragments (table 11.3; Lunatic Asylum, 1978). Identical Sm-Nd and ^{40}Ar-^{39}Ar ages fix the crystallization age of the Luna 24 VLT ferrobasalt at 3.30 aeons (Lunatic Asylum, 1978). The Lunatic Asylum (1978) did not regard Rb-Sr ages as reliable for these basalt samples, and so none are given here. An Ar-Ar age for a different sample that is 0.3 aeon older (table 11.1; Stettler and Albarede, 1978), and the presence of the buried Mg-rich basalt (Andre and others, 1978; Head and others, 1978a), suggest an extended history of basalt extrusion in Mare Crisium.

Summary and interpretation

The properties of Mare Crisium are consistent with excavation of a large, young, deep basin in a previously thick crust. I speculate that relatively thick crustal material favored aluminous mare volcanism and resistance to deformation, as in Mare Fecunditatis. Unlike Fecunditatis, however, the Crisium basin was filled by thick basalt because it had not been filled previously by thick basin ejecta (fig. 11.17). An older and smaller basin filled by thick earlier deposits would resemble Fecunditatis. A smaller basin of the same (Nectarian) age as Crisium would contain a small mare like Nectaris if it were in the same position as Crisium, or a large mare like Serenitatis if it were inside the Procellarum basin.

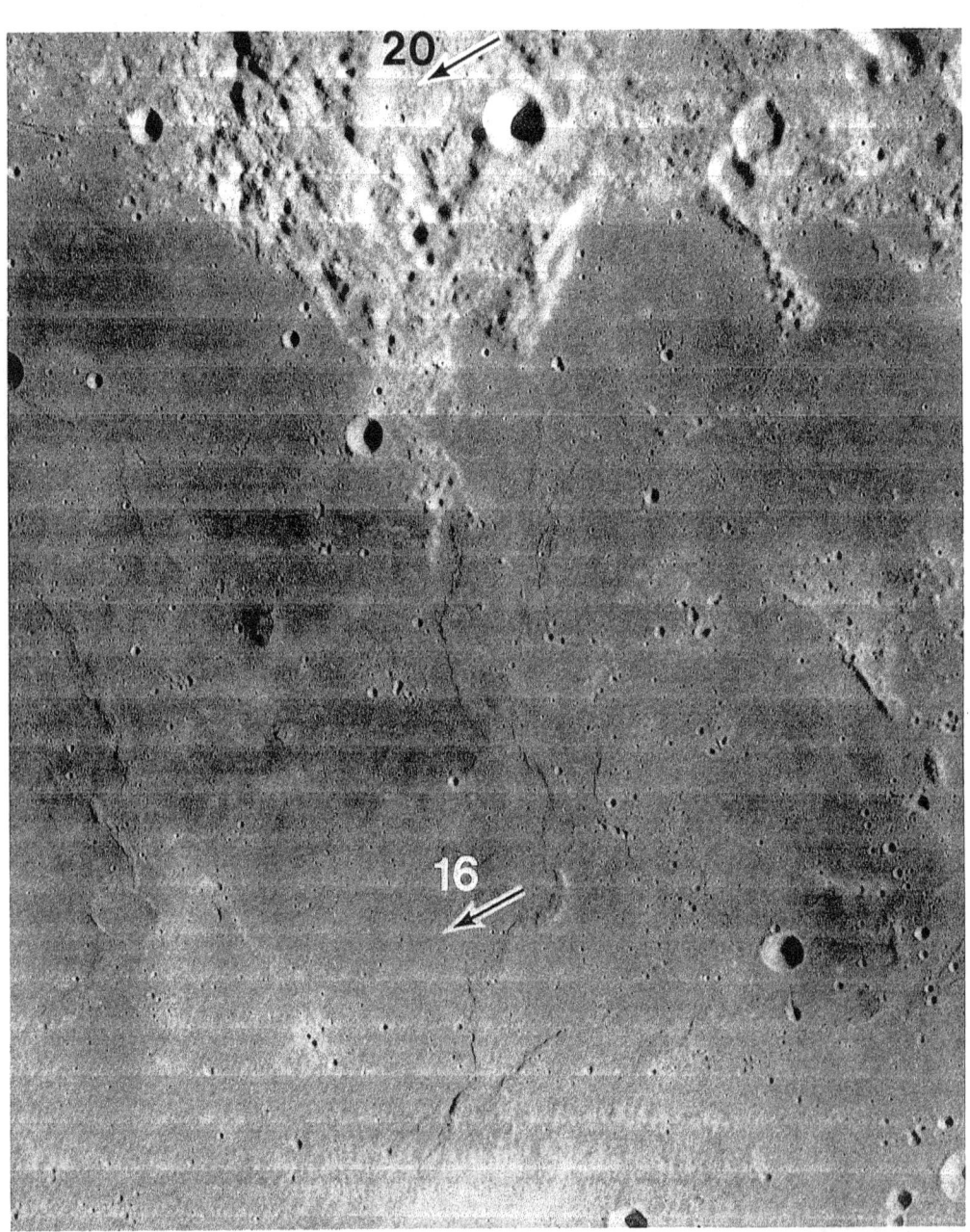

FIGURE 11.16.—Mare Fecunditatis and south Crisium-basin rim, showing Luna 16 and Luna 20 (see chap. 9) landing sites. Orbiter 1 frame M–33.

MARE IMBRIUM

General stratigraphy

The center of Mare Imbrium contains the Moon's thickest known mare-basalt section; the 1-km isopach extends to the largest diameter of all lunar maria and probably encloses several kilometers of basalt in the center (De Hon, 1979). Subsidence has opened large concentric grabens on the Apennine Bench and, in lesser degree, elsewhere (pl. 5; figs. 6.10, 10.7). Deformation was undoubtedly great in other, now-buried sectors, as is shown by Orientale, where most grabens would be buried if filled by a larger mare. Mare Imbrium has a large mascon, though not so large (table 6.1) as would be expected from the Moon's third largest basin (after Procellarum and South Pole-Aitken). The thin lithosphere may have allowed a more complete isostatic adjustment here than at Crisium.

The mare units of Imbrian age in Mare Imbrium are mostly spectrally red and bright (pl. 4; fig. 5.20; Head and others, 1978b). As in central Serenitatis, these units are more complex than can be shown in plate 9 and consist of a patchwork of units ranging in age from early to late Late Imbrian (D_L = 240–385 m; Boyce and others, 1975; Boyce, 1976; Wilhelms, 1980). Eratosthenian mare materials in Mare Imbrium are dark and relatively blue (see chap. 12). The spectrally red and blue types contrast sharply with each other but are each relatively uniform internally.

The Mare Frigoris-Sinus Roris trough is concentric with both the Imbrium and Procellarum basins (pls. 3, 4). It contains generally bright and thin mare materials of different spectral classes from most of Oceanus Procellarum but similar to some of the older units in the adjacent part of northern Mare Imbrium (pl. 4).

Apollo 15 samples—Palus Putredinis

Apollo 15, the first complex Apollo mission, was able to explore a wide area of both mare and terra systematically (see chap. 10). The peculiar sinuous rille Rima Hadley (figs. 5.11, 10.7, 10.9) helped attract attention to the landing site, which lies in the marginal mare of Mare Imbrium called Palus Putredinis. At one time, the rille or its source crater was thought to be a possible source of rare lunar volatile materials, although this hope faded with analyses of the volatile-free Apollo 11 basalt samples. The spectrally red mare of Palus Putredinis was a bonus to the more significant mission objective, materials of the Imbrium-basin rim (chap. 10; McCord and others, 1972b). This color indicates a low Ti content (Pieters, 1978), and the returned basalt samples contain much less of this element (1.5–2.0 weight percent TiO_2) than any other mare suite except that of Luna 24. The Palus Putredinis mare differs somewhat in remotely sensed properties from the spectrally red part of Mare Imbrium proper (Head and others, 1978b). Before the mission, the mare's age was correctly determined by the Trask method (see chap. 7; figs. 7.12–7.14) as near the Imbrian-Eratosthenian boundary (Carr and others, 1971). As was true for Apollo 17, more rock-size mare-basalt samples than terra materials were returned because the basalt was more accessible.

A twofold classification of the samples into an older, dominant type A and a younger, less abundant type B that was proposed upon preliminary examination of the field relations, modal mineralogy, and textures of the returned samples (Swann and others, 1972) has withstood further testing (for example, Rhodes and Hubbard, 1973; Lofgren and others, 1975). Type A is now generally referred to as quartz-normative or pigeonite basalt, and type B as olivine-normative or olivine basalt (table 5.2). Two conspicuous types of pyroclastic materials are also among the returned samples.

Pigeonite basalt.—The basalt type richer in silica is characterized by abundant modal pyroxene, mostly pigeonite with augite mantles. This basalt composes most of the rock-size samples collected by Apollo 15. It is exposed in boulders that are nearly in place at the rille lip (sta. 9A, fig. 10.9) and in ejecta of the major craters Elbow (400 m diam, max 80 m deep; sta. 1) and Dune (460 m diam, max 90 m deep; sta. 4). Textures are mostly porphyritic, and the groundmass of one subgroup is megascopically glassy (vitrophyric). The various textures are compatible with derivation from a single flow (Lofgren and others, 1975). Judging from the field relations, this flow constitutes most or all of the 60-m-thick section of layered lavas exposed in the walls of Hadley rille (figs. 1.10B, 11.18; Howard and others, 1972; Swann and others, 1972; Lofgren and others, 1975). The Apennine Bench Formation, which may be the source of small KREEP-rich fragments in the regolith (see chap. 10), may underlie this flow or a still older mare-basalt flow (fig. 11.18).

The best Ar-Ar ages of rocks and of smaller samples collected by a rake cluster between 3.30 and 3.32 aeons (table 11.3); Rb-Sr ages range from 3.21 to 3.37 aeons. Despite this spread in Rb-Sr ages, petrologic and chemical relations among the samples indicate moderate crystal fractionation of a single magma on or near the surface (Rhodes and Hubbard, 1973). The average crystallization age of the samples dated by both methods is 3.30 aeons.

Olivine basalt.—Rock-size olivine basalt was collected only from station 9A at the edge of Rima Hadley and from station 3 at Rhysling Crater (Swann and others, 1972). The near-surface position of the olivine basalt suggested by this distribution is corroborated by its greater abundance in smaller fragments, at all collection stations, than the pigeonite basalt (Swann and others, 1972; Rhodes and Hubbard, 1973; Lofgren and others, 1975). The petrologic and compositional data are consistent with derivation from a single flow different

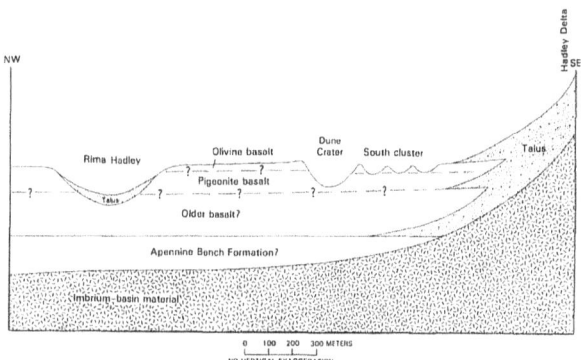

FIGURE 11.18.—Geologic cross section of mare materials at Apollo 15 landing site, drawn northwest-southeast on figure 10.7. Rima Hadley profile is from Howard and others (1972). Thicknesses of basalt units and presence of older basalt and Apennine Bench Formation are hypothetical.

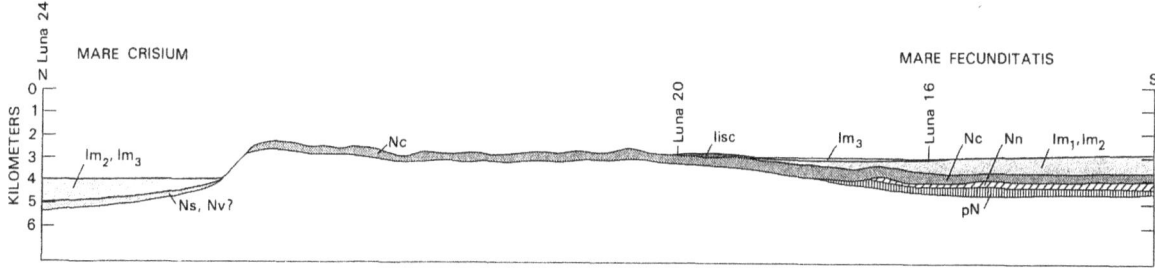

FIGURE 11.17.—Inferred geologic cross section in Crisium and Fecunditatis basins, drawn along line A'''–F in figure 8.4, with an extension to lat 5° S. Units, from oldest to youngest: pN, pre-Nectarian basin, crater, and volcanic deposits, undivided; Nn, Nectaris-basin ejecta; Nc, Crisium-basin ejecta (inner knobby and outer lineated types); Ns, Serenitatis-basin ejecta; Nv?, possible Nectarian volcanic deposits; Iisc, deposits of Imbrium-secondary craters; Im$_1$, Im$_2$, and Im$_3$, Imbrian mare units.

in parentage from the thicker pigeonite basalt flow (Rhodes and Hubbard, 1973).

The 9.6-kg porphyritic olivine-basalt boulder-size sample 15555 ("Great Scott") from station 9A has been dated by three geochronologic methods at 3.25 ± 0.05 aeons (table 11.3). One other rock (sample 15016) and numerous walnut-size rake samples from many collection stations average 3.27 aeons in age (Husain, 1974); the average by all methods is 3.26 aeons. This younger age is consistent with stratigraphic superposition on the pigeonite basalt but may not be significant in view of the spread in age determinations for each flow; the 40-million-year age difference between flows is just at the limit of geochronologic discriminability.

Pyroclastic green glass.—Apollo 15 also collected more pyroclastic material than did any other mission except Apollo 17. Green-glass droplets attracted the astronauts' attention and were collected from the regolith at two main stations (table 11.3). Emplacement by both impact and volcanism has been considered (see summary by Delano, 1979). Delano (1979) suggested that compositional data that were once seen as supporting evidence for an impact origin are more consistent with a volcanic origin. He proposed a derivation from heterogeneous sources rich in volatile material at 400 ± 50-km depth.

Although some older ages have been reported in the literature for the green glass, ages close to those of the lavas appear to be more likely (table 11.3). A dark-mantling deposit observed photogeologically near the landing site (Carr and others, 1971) could be the source of the green-glass samples (Carr and Meyer, 1974). However, the green glass, which is not dark in hand specimen, may be from a mantling deposit not detected either photogeologically or spectrally.

Pyroclastic red glass.—Droplets of red glass were also recovered by Apollo 15. Because this glass was found at only one station (6), it may be exotic to the landing site (Quaide, 1973). Delano (1980) suggested that its parent magma formed at the great depth of 480 km.

Summary and interpretation

Mare Imbrium is similar to western Mare Serenitatis in general stratigraphic sequence. Old, spectrally blue border basalt typical of eastern Serenitatis is not exposed. The main basin-filling sequence in both maria is thick, Ti-poor, relatively uniform spectrally, but complex in internal stratigraphy. Both sequences occasioned substantial basin subsidence, and both were followed by lesser amounts of spectrally blue Eratosthenian basalt erupted from the basin margins— more in Imbrium than in Serenitatis. These similarities probably result from superposition of the two relatively young basins on the thin lithosphere and crust in the central Procellarum basin (figs. 8.4, 11.14).

OTHER MARIA INSIDE THE PROCELLARUM BASIN

The outermost shelf of Oceanus Procellarum contains a large expanse of basalt of the youngest Late Imbrian age group (pl. 9). Most of this tract belongs to the same spectral class as central Serenitatis (mISP, pl. 4) but may be poorer in Fe (Basaltic Volcanism Study Project, 1981, p. 455). This young Late Imbrian basalt transects older Late Imbrian mare units and is overlain in places by Eratosthenian units (fig. 11.8). At the north end of the mare-filled part of the shelf and in the adjacent middle trough of the Procellarum basin, intermediate-age Upper Imbrian material of red spectral class LBG- overlaps the very old, spectrally blue units (class hDWA; Luna 16 type) perched on the mare margin. Class LBG- (unit 11, pl. 4) is also typical of Lacus Somniorum, in the eastern hemisphere, where intermediate-age and old Late Imbrian mare units overlie the same shelf and trough of the Procellarum basin (pls. 4, 9). Spectrally blue Eratosthenian units occupy the middle trough in much of the western nearside and locally extend onto the outer shelf (see chap. 12; pl. 10; fig. 12.11). The persistence of similar concentric compositional patterns in so much of the Procellarum basin, in otherwise different geologic settings, is remarkable.

The middle Procellarum trough is also the site of the Moon's three largest complexes of nonplanar volcanic deposits and unusual landforms (see chap. 5; pl. 4). The northernmost complex is the Rümker Hills (Mons Rümker), which consists of domes and dark-mantling deposits (fig. 5.9). Intermediate is the Aristarchus Plateau, which contains the Moon's largest and most numerous sinuous rilles and thick dark-mantling materials (fig. 5.12; Zisk and others, 1977). Farther south are the Marius Hills, characterized by diverse cones, domes, and sinuous rilles (fig. 5.7; McCauley, 1967a, 1968, 1969b). Although the morphologic freshness of the landforms in these three complexes suggests an Eratosthenian or even a Copernican age (Moore, 1965, 1967; McCauley, 1967a, b; Wilhelms and McCauley, 1971), they are partly flooded by late Late Imbrian lavas and so are mostly or entirely Imbrian in age (Zisk and others, 1977). The middle Procellarum trough apparently was conducive not only to late volcanism but also to extrusion of diverse magma types.

Oceanus Procellarum and other western-hemisphere maria east of long 40 W. have complex, mottled patterns of unit distribution that cannot be completely portrayed here (see Pieters and others, 1980; Whitford-Stark and Head, 1980; Wilhelms, 1980). Much of this region is covered by spectral class mIG- (Apollo 12 type), which represents a class of related compositional units (see chap. 5; Pieters, 1978). This complexity probably results from superposition on the larger Procellarum-basin structures of several old basins, such as Flamsteed-Billy, Insularum, Nubium (pl. 3), and the small basin or large crater that contains Mare Cognitum. The lithospheric and mantle structure resulting from these superpositions is complex and presumably influences a complex pattern of mare extrusion.

Oceanus Procellarum is generally stated to be a nonmascon mare, but, in fact, parts of its outer troughs contain small linear positive anomalies (Scott, 1974). Although Scott (1974) ascribed these anomalies to intrusion of dikes beneath linear mare ridges (fig. 6.2), I suggest that they are due to relatively thick sections of basalt in the troughs. They alternate with parallel negative anomalies aligned along the buried rings of the Procellarum basin (compare pl. 3 with fig. 2 of Scott, 1974).

The Procellarum pattern of mare distribution seems to persist into Mare Humorum, which has the third large mascon inside the Procellarum basin (besides Serenitatis and Imbrium; see table 6.1). Mare Humorum occupies the complex Nectarian Humorum basin (see chap. 9; figs. 4.3U, 9.24), which is superposed on the outer Procellarum shelf and periphery. The central mare is the most extensive, but maria are also abundant in parts of the north basin periphery that overlie the Procellarum basin (pl. 4). The presence of numerous conspicuous grabens (figs. 6.9, 9.24) indicates considerable subsidence. Mare Humorum contains mare units of diverse color (pl. 4) and age classes (Pieters and others, 1975; Boyce, 1976; Boyce and Johnson, 1978). The western part of the basin contains spectral class mISP, and the eastern part mIG-, as in Oceanus Procellarum. This relatively red unit is covered by a young (late Imbrian and (or) Eratosthenian), moderately blue unit originating outside central Mare Humorum. The southern contact between the mare and the basin is draped by a dark-mantling deposit called the Doppelmayer Formation (Titley, 1967).

The central highlands of the Moon apparently were a barrier to mare extrusion (pls. 4, 9; fig. 1.8). Between the Insularum-Nubium and Tranquillitatis-Asperitatis mare zones, hardly any maria occupy the zone between the Procellarum-basin rim and middle ring. Furthermore, the maria inside the middle ring are much smaller than elsewhere in the same radial position from the Procellarum center; they apparently required the additional lithospheric penetration afforded by such small basins or large craters as Vaporum, Aestuum (Wilhelms, 1968), and, possibly, a circular Sinus Medii structure (De Hon, 1979) not mapped here. The terra borders of these maria are rich in dark-mantling deposits (Wilhelms, 1968; Wilhelms and McCauley, 1971). Penetration by the Mutus-Vlacq and Werner-Airy basins was apparently insufficient to allow Late Imbrian volcanism, although pre-Nectarian basalt may be buried by ejecta plains in the Mutus-Vlacq basin (see chaps. 8, 9), and a small dark deposit in the crater Airy may be volcanic in origin (unless it is ejected Imbrium-basin impact melt). Thus, the south-central terra "backbone" may be a first-order crustal feature. A northward extension seems to block mare extrusion in the central part of the outer Procellarum trough (pl. 4). These major terra structures may have arisen during the "magma ocean" stage, or they may be remnants spared from excavation by giant undiscovered basins still older than Procellarum.

NEAR-LIMB MARIA OUTSIDE THE PROCELLARUM BASIN

Mare Orientale and Mare Smythii are similar in many respects. They are thin mascon maria (Solomon and Head, 1980), and their basins are of about the same size (table 4.1). They also are equally deep (Sjogren and Wollenhaupt, 1976; Strain and El-Baz, 1979)—a surprising fact at first glance, considering the difference in age of their basins (Imbrian and pre-Nectarian, respectively). Both basins contain floor-rebound craters (pl. 5); Smythii has more because it is the older basin and thus contains more craters of all types. The ring-and-trough structure is similar in the two basins (Strain and El-Baz, 1979). Smythii lacks arcuate grabens, and those of Orientale, though conspicuous and numerous, are confined to a smaller area of the basin than those in more deeply filled maria (pl. 5; table 6.1). The paucity of grabens is expected from their position on a thick lithosphere, which resisted vertical movements in both basins.

As in other maria superposed on thick terra crusts, the Smythii basalt appears to be highly aluminous (Conca and Hubbard, 1979; Hubbard, 1979), although this appearance may result from mixing of the thin mare with terra material (P.D. Spudis, oral commun., 1982). Like Crisium, Smythii is low in radioactivity (Metzger and others, 1977). Mare Orientale was not analyzed.

The two maria apparently had somewhat different patterns of basalt emplacement. Smythii contains late Imbrian units (pl. 9) thinly covered by Eratosthenian units (pl. 10; Boyce and Johnson, 1978). At Orientale, extrusion ended during the middle Late Imbrian in the central mare, which is comparable to all of Mare Smythii, but migrated to the inner and then to the outer crescentic maria in the concentric troughs (Lacus Veris and Autumni, respectively; Greeley, 1976).

Small, thin, old maria occupy some of the terrain between the Humboldtianum and Procellarum basins in the northeast quadrant of the nearside (pl. 9; fig. 9.3). A very old, still-undelineated basin may exist in this region.

Mare Australe probably contains a more nearly equal amount of all three Imbrian mare units than any other studied mare (pl. 9). Most of the maria occupy craters superposed on the Australe basin, and much less basalt probably would have erupted in this shallow weakly defined basin if these craters were absent. Mare ridges, but not grabens, are present (pl. 5; Wilhelms and others, 1979).

Mare Marginis is another near-limb mare that contains old (fig. 4.7) as well as young (Boyce and Johnson, 1978) mare units. Some of the basalt is KREEP-rich (Hawke and Spudis, 1980).

FARSIDE MARIA

Most maria that lie entirely on the farside are poorly characterized because of absence of telescopic spectra, paucity of orbital geochemical readings, and poor photographic coverage (pl. 2). No arcuate grabens are known, and they are probably absent (pl. 5). I know of only five farside mare patches that have been dated by crater counts (Wilbur, 1978; Walker and El-Baz, 1982), all of which indicate a middle Late Imbrian age (pl. 9).

The farside's most extensive maria, those in the South Pole-Aitken basin, are similar to those of Australe in that they occupy parts of small superposed basins (pls. 4, 9) and large craters. Thus, South Pole-Aitken played a lithosphere-thinning role on the farside comparable to that of the Procellarum basin on the nearside. The fact that maria are sparser in South Pole-Aitken than in Procellarum is probably due to: (1) the smaller size of South Pole-Aitken, 2,500 versus 3,200 km; (2) the absence of such large superposed basins as Imbrium and Serenitatis; and, possibly, (3) a first-order hemispheric dichotomy in crustal thickness, such as is commonly hypothesized (see chaps. 1, 6, 8).

Other farside maria are scattered in the few other large basins, in a few large craters, or in such seemingly isolated and unexplained spots as the crater Kohlschütter and Lacus Solitudinis (pl. 4; figs. 5.2, 5.3). The Moscoviense basin contains the largest mare patch outside the South Pole-Aitken basin; photographic coverage is inadequate for dating the mare, but it appears to be old. The mare's relatively large area may result from superposition of two basins (D.E. Davis, oral commun., 1982), only one of which is mapped here. The conspicuous mare in crater Tsiolkovskiy probably results from the considerable depth of this large Late Imbrian crater, aided by its superposition on the Tsiolkovskiy-Stark basin. The mare, which is covered by one of the best Lunar Orbiter photographs of the farside (an Orbiter 3 H-frame; pl. 2), is Imbrian in age. D_L values of 255 m (Boyce and others, 1975) suggest a late Late Imbrian age, whereas crater frequencies (Wilbur, 1978; Walker and El-Baz, 1982) suggest a middle Late Imbrian age. The basalt may be rich in Ti (Metzger and others, 1977; Wilbur, 1978). Finally, pre-Imbrian or Early Imbrian basaltic plains are likely to be hidden beneath the smooth, light-colored plains fillings of such basins as Hertzsprung and Korolev (see chap. 9).

CHRONOLOGY

Late Imbrian mare basalt is the most accurately dated lunar material. The Late Imbrian Epoch began immediately after the Orientale impact, which is estimated from superposed crater frequencies to have occurred 3.8 aeons ago, between the Imbrium impact 3.85 aeons ago and the extrusion of the Apollo 17 mare basalt 3.72 aeons ago (see chap. 10). The youngest sampled Imbrian mare basalt, from the Apollo 15 landing site, formed 3.26 aeons ago. Therefore, the Late Imbrian Epoch lasted at least 0.54 aeon and probably a little longer, allowing for extrusion of some unsampled units after 3.26 aeons (for example, in Maria Serenitatis and Crisium; table 11.1). An approximate date of 3.2 aeons ago, between the Apollo 15 and Apollo 12 dates, is adopted here for the end of the Late Imbrian Epoch and the Imbrian Period. Therefore, the Late Imbrian Epoch lasted about 0.60 aeons and the Imbrian Period about 0.65 aeons.

Exposed basalt units of the oldest of the three Late Imbrian age groups mapped here are all on the outer shelf of the Procellarum basin or outside that basin. Their age is estimated at from 3.75 to 3.80 aeons; the exposed flows were not sampled, but the 3.79-aeon Ar-Ar age of one sample from a buried unit at the Apollo 11 landing site (10062; table 11.1) lies within this age range.

Exposed flows of the intermediate-age group are represented by the Apollo 17 Ti-rich basalt (3.72 aeons) and the Apollo 11 high-K, Ti-rich basalt (3.57 aeons). The age range of this group suggested by these absolute ages, as calibrated with the relative stratigraphy, is 3.50 to 3.75 aeons. In addition, therefore, an Apollo 16 fragment probably from Mare Nectaris (3.74 aeons), two low-K, high-Ti Apollo 11 units (3.67 and 3.70 aeons), and the Apollo 17 pyroclastic deposit (3.64 aeons) also probably originated in middle Late Imbrian time. Orbital geochemistry indicates that compositions of this age are very diverse. Many units are deformed by grabens.

The youngest Late Imbrian age group, 3.20 to 3.50 aeons old, is represented by the Luna 16 feldspathic basalt (3.40 aeons), the Luna 24 VLT basalt (3.30 aeons), the Apollo 15 green glass (3.30 aeons), and the Apollo 15 low-Ti olivine basalt and pigeonite basalt (3.26 and 3.30 aeons, respectively). Few units of this group are cut by grabens, but most are deformed by mare ridges. Spectrally red or "orange" basalt units with medium to low Ti contents characterize the most extensive exposures of this age group, that is, those in Mare Serenitatis, Mare Imbrium, and western Oceanus Procellarum. Basalt of this composition may be the Moon's most voluminous basaltic type.

Dark-mantling materials lie at the margins of most maria but predominate around maria superposed on the center and intermediate trough of the Procellarum basin. Many of these dark-mantling deposits are older than the adjacent lavas. Spewing of magmas driven by volatile materials probably was important in the early filling of basins (Head, 1974a). At 3.64 aeons, the middle Late Imbrian Apollo 17 orange and black glass is somewhat younger than the subjacent Apollo 17 lavas (3.72 aeons). The 3.3-aeon age of the Apollo 15 green glass shows that pyroclastic materials were also erupted late in the Late Imbrian Epoch.

In the Imbrian Period, the cratering rate changed from the intensive early bombardment (fig. 8.16) to a much-reduced rate that characterizes the rest of lunar history. This rapid falloff in impact rate is recorded by successively lower crater densities for successively younger time-stratigraphic units (pls. 7–11; figs. 7.16, 8.6, 9.22, 10.4). During the total Imbrian Period, 9.7 craters larger than 30 km in diameter formed per million square kilometers (pls. 8, 9), in contrast to about 35 such craters in the Nectarian Period and more than 77 in pre-Nectarian time. If the Orientale basin formed 3.8 aeons ago and the Late Imbrian Epoch ended 3.2 aeons ago, formation of the 195 Lower Imbrian craters of this size on the whole Moon required 0.05 aeons, and that of the 174 Upper Imbrian craters required 0.6 aeons.

These values translate into rates of 3,900 and 290 impacts per aeon, respectively. These rates are transitional between the 19,000 impacts per aeon of the Nectarian Period (1,330 craters in 0.07 aeon) and the much lower Eratosthenian and Copernican rates of 42 impacts per aeon (133 craters in 3.2 aeons). Errors in the durations of the periods and epochs proposed here would change these values but would not alter the conclusion that the rate of impacts decreased sharply between 3.85 and 3.2 aeons ago.

12. ERATOSTHENIAN SYSTEM

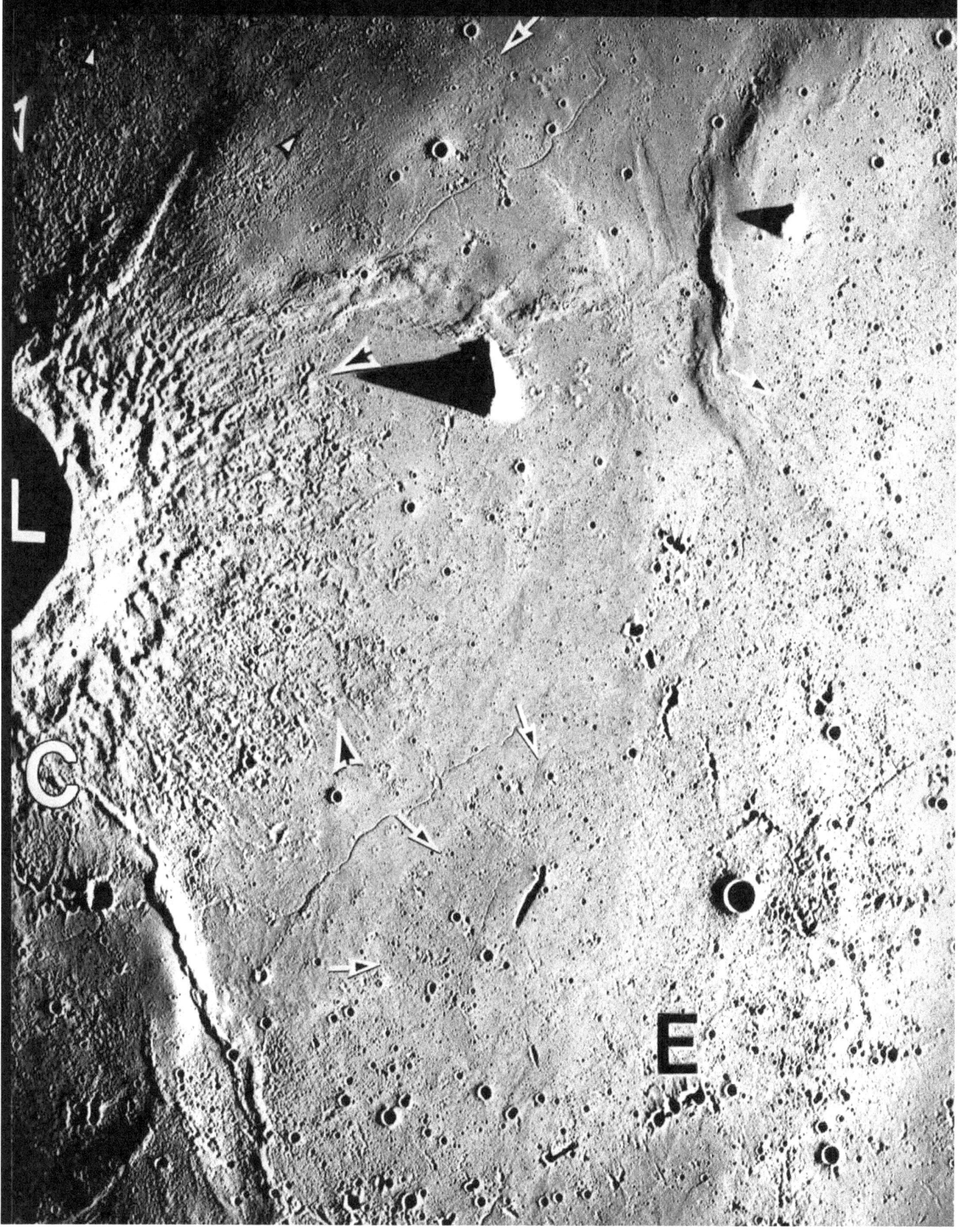

FIGURE 12.1 (OVERLEAF).—Eratosthenian crater and mare materials. L, Lambert (30 km), superposed on Imbrian mare units (small white-on-black arrowheads) and embayed by Eratosthenian mare units (large black-on-white arrowheads). Mare unit containing sinuous rille has a D_L value of 220 ± 45 m (Boyce, 1976), floods southeastern secondary craters of Lambert, and covers older mare unit (three black-on-white arrows pointing southeastward). Older unit underlies secondaries of Eratosthenes (E). Therefore, stratigraphic relations of Lambert and Eratosthenes are similar; crater densities (fig. 12.4) distinguish Lambert (older) from Eratosthenes (younger). Secondary craters of Copernicus (C) are superposed on Lambert. Apollo 15 frame M–1009.

12. ERATOSTHENIAN SYSTEM

CONTENTS

	Page
Introduction	249
Definition	249
Crater materials	250
Recognition criteria	250
Frequency	257
Mare materials—general stratigraphy and distribution	258
Mare Insularum	259
Setting of the Apollo 12 landing site	259
Apollo 12 samples	259
Summary and conclusions	262
Chronology	262

INTRODUCTION

The two time-stratigraphic units described here and in the next chapter are less extensive than the four already described. The contrast in rates of formation is even greater: The Eratosthenian and Copernican Systems combined required three times longer to form than did the four previous units combined. The Eratosthenian and, to a lesser extent, the Copernican Systems continue the geologic style of the Upper Imbrian Series. Eratosthenian mare and crater materials are both present; the maria are about twice as extensive as the crater deposits (pl. 10).

Because the Eratosthenian and Copernican rock units are relatively young, they are little modified. Each crater deposit is visible farther from the rim crest than are older crater deposits, because the topography has been less eroded and the materials less completely mixed with the substrate. Eratosthenian and Copernican units serve as models for interpretations of the older units because of their freshness and clear stratigraphic relations (fig. 12.1).

In a sense, all the sampling missions obtained samples of Eratosthenian and Copernican material because they returned regolith material (Heiken, 1975; Papike and others, 1982). Many minor impacts have been dated either individually or collectively by determining the duration of exposure of regolith particles (Arvidson and others, 1975; Burnett and Woolum, 1977; Crozaz, 1977). However, Eratosthenian and Copernican bedrock units (those visible on regional photographs) were only sparsely sampled. Mare-basalt flows believed to be Eratosthenian were sampled only at the Apollo 12 landing site. No large Eratosthenian craters were sampled, and large Copernican craters were sampled only indirectly (see chap. 13). Therefore, the reconstruction of post-Imbrian geologic history requires considerable photogeologic extrapolation—a process not yet complete.

DEFINITION

Shoemaker and Hackman (1962) established the craters Eratosthenes and Copernicus as the type areas of the Moon's two youngest time-stratigraphic systems because of stratigraphic relations evident on telescopic photographs. Ejecta of Eratosthenes is superposed on the nearby mare material, whereas rays of Copernicus are superposed on Eratosthenes (figs. 1.1, 1.6, 12.2). Eratosthenes is much the darker of the two craters on full-Moon photographs (fig. 12.2B) and was thought to be rayless. Although one faint ray, in fact, extends northwestward from the crater (formed by a long linear chain of secondary craters visible in fig. 11.3), Eratosthenes nevertheless does represent a class of postmare craters that essentially lack bright rays.

Accordingly, postmare craters have been operationally divided in most lunar geologic mapping into Eratosthenian and Copernican categories according to whether they are nonrayed or bright-rayed, respectively. Although ray visibility is neither exact nor correct as a criterion of age according to the stratigraphic code,[12.1] it has yielded many subsequently confirmed age estimates and is still useful where better data are unavailable.

Neither Shoemaker and Hackman (1962) nor later work have established the stratigraphic position of Eratosthenes and Copernicus within their respective systems. Wilhelms (1980) attempted to define or, at least, to better characterize the two systems on the basis of units in southern Mare Imbrium and northern Oceanus Procellarum. A stratigraphic sequence was established (fig. 7.1; table 7.2), and the existing age assignments of as many units as possible were retained. The definitions were based on mare units because, in this region, they are more extensive and lithologically uniform than impact deposits. Because time-stratigraphic units are contiguous, the working criterion for mare units at the top of the Imbrian System, $D_L = 240 \pm 10$ m, also characterizes the base of the Eratosthenian System (table 12.1). Mare units having these or smaller D_L values are post-Imbrian. Though not a formally acceptable criterion, the D_L values are at least a secondary (superposed) criterion that is a function of time.

Although fewer crater frequencies than D_L values are available for individual Eratosthenian units, the frequencies determined by Neukum and others (1975a, b) and Neukum and König (1976) probably represent the approximate range of Eratosthenian frequencies[12.2]. A plot of many of their determinations suggests that a crater frequency of 2.5×10^{-3} corresponds approximately to the borderline D_L value of 240 m (fig. 7.15) and thus should characterize the mare

---

[12.1] The stratigraphic code (American Commission on Stratigraphic Nomenclature, 1970) was partly adhered to as follows. Rays were considered to be a physical characteristic by which *rock-stratigraphic* units may be legitimately defined. Each ray deposit or ejecta deposit from which rays emanate was considered to be a formation and could be mapped by objective criteria, as should all formations. Then, the formations were assigned *time-stratigraphic* significance according to the physical property of the presence or absence of rays; all materials of postmare nonrayed craters were lumped on a map as Eratosthenian, and all materials of rayed craters as Copernican. Before Lunar Orbiter photographs were available in 1966-67, rayless but topographically fresh craters not in contact with mare units were commonly assigned an indefinite Imbrian or Eratosthenian age (map unit EIc). The fact that some crater ages were incorrectly estimated by the ray criterion illustrates the validity of the code's insistence on the separation of rock and time-rock categories of units.

[12.2] Except as noted, all crater frequencies given in this chapter refer to cumulative numbers of craters 1 km in diameter and larger per square kilometer. Neukum and his coworkers obtained these frequencies by normalizing their crater size-frequency curves at this 1-km intercept.

TABLE 12.1.—*Properties of craters superposed on units of the Eratosthenian System*

[Frequencies are those of craters at least 1 km in diameter per square kilometer, as determined by normalization to 1 km by means of a calibration curve (see fig. 7.10G; Neukum and others, 1975a); n.a., not available]

Boundary	Substrate	Frequency	Basis	D_L (m)	Basis	Age (aeons)	Basis
Upper	Crater material	10^{-3}	Figure 12.4	180±20	Table 12.3	1.1	Chapter 13.
	Mare basalt	n.a.	---	≤165±25	Youngest flows traditionally considered Eratosthenian		
Lower	Crater material	$1-2\times10^{-2}$	Figure 12.4	≥310	Table 12.3	3.2	Between Apollo 12 and 15 ages.
	Mare basalt	2.5×10^{-3}	Table 11.1	240±10	Consistent with traditional mapping of crater and mare units (Wilhelms, 1980)		

materials at the base of the Eratosthenian System (table 12.1). Some flows in Oceanus Procellarum and Mare Imbrium with D_L values smaller than 240 m were found to have frequencies of 1 to 2×10^{-3} (figs. 12.3, 12.4A, E; Neukum and König, 1976; Gerhard Neukum, written commun., 1979). A further indication that a crater frequency of 2.5×10^{-3} and a D_L value of 240 m are equivalent for the system's base is the finding (with a large uncertainty) of a crater frequency of 2.6×10^{-3} on the upper Upper Imbrian mare unit(s) at the Apollo 15 landing site, where $D_L=270\pm15$ m (table 11.1; figs. 11.2, 12.4B, C).

The top of the system is harder to characterize. Wilhelms (1980) employed a D_L value of 140 m as a working criterion for the Eratosthenian-Copernican boundary, based on values of 165 ± 25 m that were measured on the youngest mare units in Mare Imbrium traditionally called Eratosthenian (Boyce and others, 1975). However, smaller values may eventually be found. Similarly, the small frequencies found for some young flows (fig. 12.3) may not be the smallest that exist.

In summary, D_L values for Eratosthenian mare materials range from less than 140 m to 240 ± 10 m, and the frequencies of craters at least 1 km in diameter per square kilometer (normalized) range as high as to 2.5×10^{-3}. The values for crater materials of the same age differ (table 12.1), as shown in the next section.

CRATER MATERIALS
Recognition criteria

The best criteria for an Eratosthenian age of a crater deposit are stratigraphic relations with the mare materials. Eratosthenian mare and crater deposits interfinger on much of the central and western nearside. Excellent examples of these relations are in Mare Imbrium (figs. 7.4, 12.1, 12.5; table 7.2) and Oceanus Procellarum (fig. 12.6). Extensive flows with D_L values smaller than 230 m overlap Lambert and other craters that traditionally have been mapped as Eratosthenian (fig. 12.1; Wilhelms, 1980). However, the fresh bright craters Euler and Timocharis, previously considered to be Copernican, are also overlapped by such flows (fig. 7.4). These and other stratigraphically inconsistent assignments have been revised according to the philosophy of preserving a maximum number of existing assignments. The extensive mare materials remain Eratosthenian, and the few embayed bright craters become Eratosthenian (table 12.2).

Crater counts made marginally feasible by the best Lunar Orbiter and Apollo photographs (fig. 12.4; Neukum and König, 1976) can date favorably situated craters. Neukum and König (1976) found that the ages assigned to large rayed craters by the criteria of rays and

A

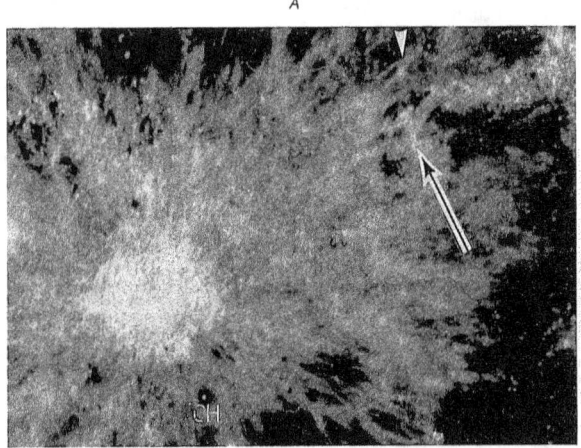

B

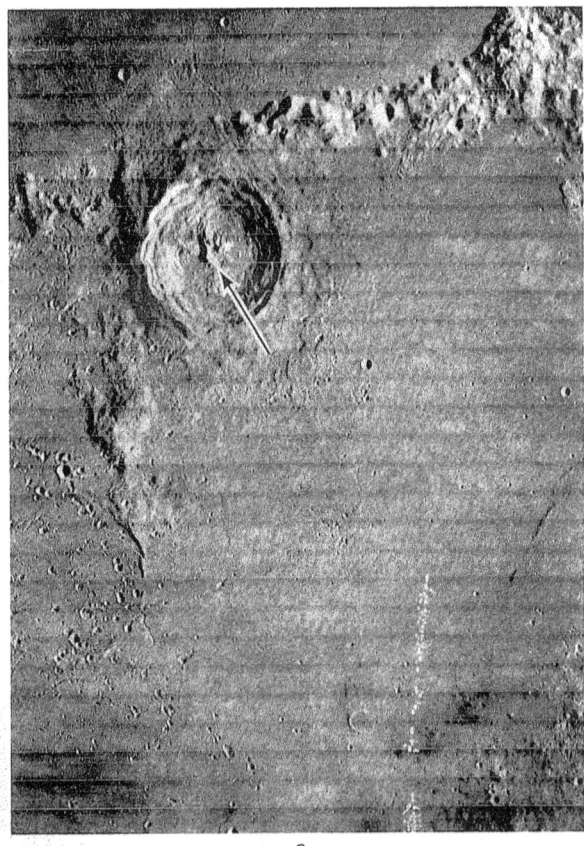

C

FIGURE 12.2.—Superposition of Copernicus rays and secondary craters on Eratosthenes; arrows mark same point in each photograph.
A. Low-Sun telescopic photograph.
B. High-Sun telescopic photograph. Eratosthenes is almost invisible. White arrowhead marks radial ray. CH, Copernicus H (compare fig. 13.6).
C. Part of Orbiter 4 frame H–114.

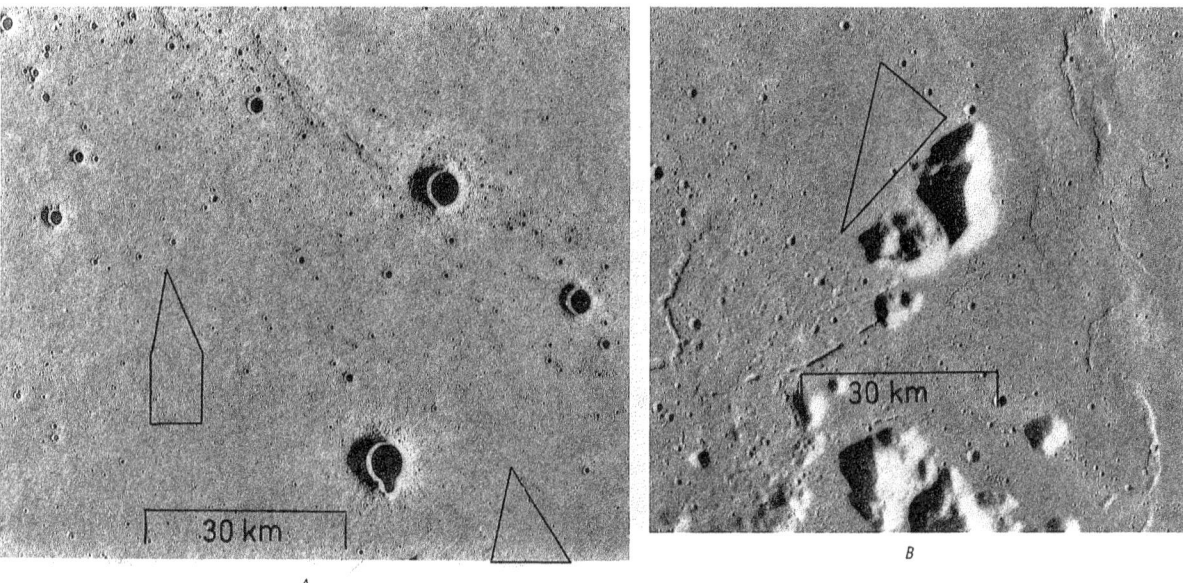
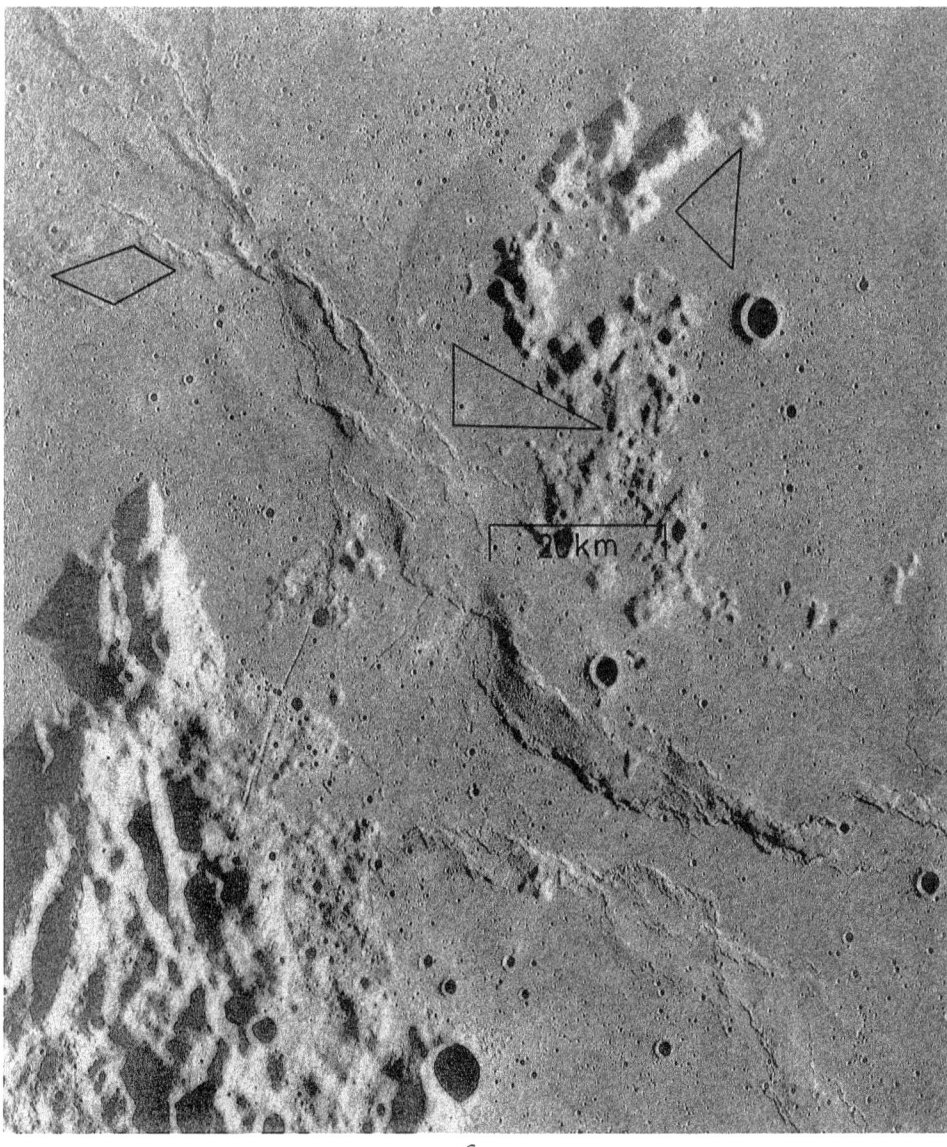

FIGURE 12.3.—Eratosthenian mare materials in Oceanus Procellarum, showing areas included in crater counts by Neukum and others (1975a, b) and Neukum and König (1976). Courtesy of Gerhard Neukum.
  A. Northern Procellarum. Scene centered at 26.5° N., 62.5° W.
  B. Montes Harbinger, at border between northern Procellarum and Mare Imbrium. Crater-count area centered at 27.5° N., 41.5° W.
  C. Southern Procellarum, near crater Letronne (large rugged area at lower left). Scale bar at 10° S., 38.5° W., over island probably belonging to Flamsteed-Billy basin.

morphology were consistently too young, probably because the rays and the sharpness and roughness of the coarse-textured crater subunits give the impression of youth. Langrenus (fig. 9.5), Theophilus (fig. 11.1), and several other large rayed craters are here accepted as Eratosthenian because their crater densities are similar to those on Lambert (fig. 12.1) and Bullialdus (fig. 12.7), which are sandwiched between Eratosthenian and young Imbrian mare materials.

Crater frequencies as high as 2×10^{-2} were found to be superposed on craters traditionally mapped as Eratosthenian (fig. 12.4). This and other frequencies for Eratosthenian craters are substantially larger than the frequencies of 2.5×10^{-3} superposed on mare materials at the base of the Eratosthenian System. The curve for the Apollo 15 units, just below the system's base, lies amidst the frequency plots for many typical Eratosthenian craters and even for some Copernican craters (fig. 12.4B, C). Different crater frequencies, therefore, seem to apply to mare and crater materials (Ahrens and Watt, 1980).

The 18-km-diameter crater Diophantus is a borderline case near the upper boundary of the Eratosthenian System. Diophantus has no visible rays (fig. 2.2), is younger than the youngest Eratosthenian flows shown in figure 2.2 by an unknown amount, and has not been flooded by younger mare units. It has been retained in the Eratosthenian System because that was its longstanding assignment (Wilhelms and McCauley, 1971) and because no reason for a change is apparent (Wilhelms, 1980). Diophantus is the youngest crater here considered to be Eratosthenian for which, to my knowledge, superposed crater frequencies have been accurately determined ($1.25 \pm 0.5 \times 10^{-3}$; fig. 12.4D, E; Neukum and König, 1976). Thus, this

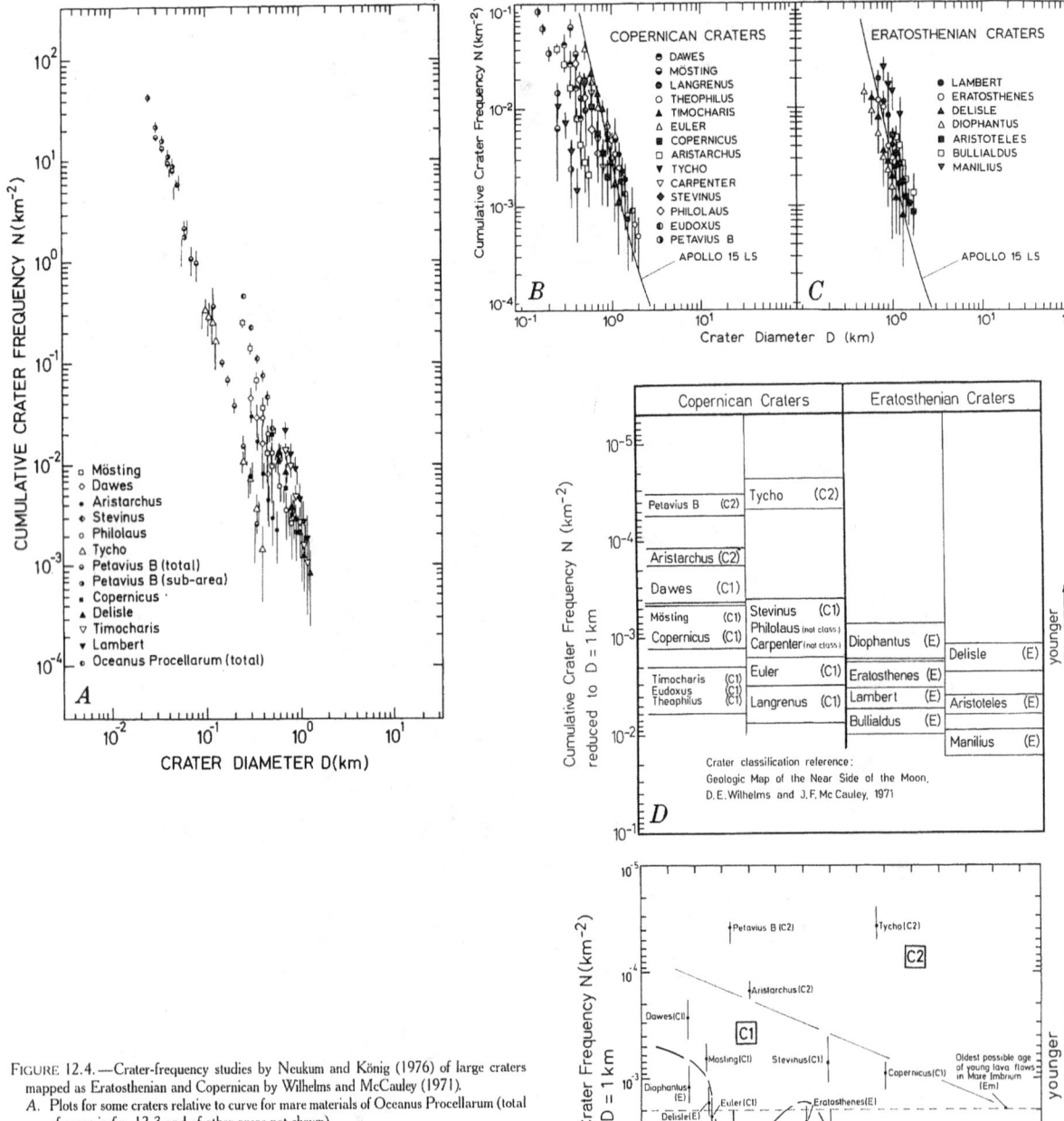

FIGURE 12.4.—Crater-frequency studies by Neukum and König (1976) of large craters mapped as Eratosthenian and Copernican by Wilhelms and McCauley (1971).
A. Plots for some craters relative to curve for mare materials of Oceanus Procellarum (total of areas in fig. 12.3 and of other areas not shown).
B. Plots for Copernican craters relative to curve for mare material at Apollo 15 landing site (LS; compare fig. 11.2). Mare curve is intermediate in group of apparent ages; mare and crater materials with same superposed-crater frequencies differ in age.
C. Similar plots for craters considered to be Eratosthenian by Wilhelms and McCauley (1971).
D. Age assignments (E, C1, C2) by Wilhelms and McCauley (1971); ranking by Neukum and König (1976), youngest at top.
E. Same craters plotted by age (youngest at top) and size (horizontal axis), showing probable bias to too-young age assignments of such large craters as Langrenus and Theophilus.

TABLE 12.2.—*Representative Eratosthenian craters*

[Cross rules divide diameter ranges mapped differently in plate 10: smaller than 30 km, unmapped; 30 to 59 km, interiors mapped; 60 km and larger, exterior deposits mapped. UI, possibly Upper Imbrian]

Crater	Diameter (km)	Center (lat)	Center (long)	Figure	Remarks
Maestlin G	3.5	2° N.	42° W.	12.6	---
Taruntius H	9	0.5° N.	50° E.	3.2B	---
Bessarion B	12	17° N.	42° W.	3.14A	Double impact.
Cayley	14	4° N.	15° E.	10.3B	---
Brayley	15	21° N.	37° W.	12.5	Rim flooded.
Diophantus	18	28° N.	34° W.	2.2	See figure 12.4; table 7.2.
Peirce	19	18° N.	54° E.	5.23, 9.11	---
Flamsteed	21	5° S.	44° W.	12.6A	Clear stratigraphy.
Picard	23	15° N.	55° E.	9.11	---
Delisle	25	30° N.	35° W.	2.2	See figure 12.4; table 7.2.
Arago	26	6° N.	21° E.	3.2C	---
Euler	28	23° N.	29° W.	1.6, 3.29, 7.4	See figure 12.4; table 7.2.
Lambert	30	26° N.	21° W.	1.6, 6.4, 12.1	See figure 12.4; tables 7.2, 12.3.
Reiner	30	7° N.	55° W.	12.10	Asymmetric.
Archytas	32	59° N.	5° E.	10.15	---
Timocharis	34	27° N.	13° W.	1.6, 11.3	See figure 12.4; tables 7.2, 12.3.
Stearns	37	35° N.	163° E.	12.9	Farside example.
Manilius	39	15° N.	9° E.	1.7, 10.16	See figure 12.4; table 12.3.
Herschel	41	6° S.	2° W.	7.7	See table 12.3.
Rothmann	42	31° S.	28° E.	11.5, 11.7	---
Plinius	43	15° N.	24° E.	5.17	See table 12.3.
Reinhold	43	3° N.	23° W.	13.10	Do.
Agrippa	44	4° N.	11° E.	10.16	Do.
Hainzel A	53	41° S.	34° W.	9.24	Do.
Maunder	55	15° S.	94° W.	3.15A, 4.4B	Typical.
Eratosthenes	58	15° N.	11° W.	1.6, 11.3, 12.2	Typical (see fig. 12.4; tables 7.2, 12.3).
Bullialdus	61	21° S.	22° W.	12.7	See figure 12.4; table 12.3.
Hercules	69	47° N.	39° E.	1.7, 9.3	Mare fill.
Werner	70	28° S.	3° E.	1.8, 3.26, 7.7, 9.27	Typical.
Fabricius	78	43° S.	42° E.	8.5	UI?
Aristoteles	87	50° N.	17° E.	1.7, 10.12	See figure 12.4.
Theophilus	100	11° S.	26° E.	11.1	Rayed (see fig. 12.4; table 12.3).
Pythagoras	130	64° N.	63° W.	1.6, 10.2	---
Langrenus	132	9° S.	61° E.	9.5	Rayed (see fig. 12.4).
Hausen	167	66° S.	88° W.	3.2E	Largest young crater.

TABLE 12.3.—*Properties of craters superposed on larger crater deposits*

[Orbiter frames listed are those used for D_L determinations and differ from those used for crater-frequency counts.
Ages: C, Copernican; E, Eratosthenian (pls. 10, 11).
Crater frequencies are from Neukum and König (1976) (see fig. 12.4); they refer to craters at least 1 km in diameter per square kilometer, as determined by normalization to 1 km by means of a calibration curve (see fig. 7.10G; Neukum and others, 1975a).
D_L values determined by J.M. Boyce]

Crater	Diameter (km)	Orbiter frame	Age	Crater frequency	D_L (m)
Tycho	85	5 H-128	C	3.7×10^{-5}	20±10
Aristarchus	40	5 H-201	C	1.5×10^{-4}	50±10
Kepler	32	5 H-138	C	---	75±10
Copernicus	93	5 H-146	C	9.0×10^{-4}	100±30
Gassendi A	33	5 H-180	C	---	100±15
Eudoxus	67	4 H-98	C	3.6×10^{-3}	120±30
Godin	35	4 H-97	C	---	130±20
Aristillus	55	4 H-103	C	---	140±30
Timocharis	34	4 H-114	E	3.0×10^{-3}	160±30
Bürg	40	4 H-86	C	---	175±35
Autolycus	39	4 H-103	C	---	180±20
Herschel	41	4 H-108	E	---	180±20
Manilius	39	4 H-97	E	1.3×10^{-2}	185±40
Horrocks	31	4 H-96	E	---	200±20
Bullialdus	61	4 H-125	E	7.0×10^{-3}	215±25
Plinius	43	4 H-85	E	---	215±35
Cichus	41	4 H-124	E	---	215±65
Reinhold	43	4 H-125	E	---	245±35
Lambert	30	4 H-126	E	4.8×10^{-3}	250±10
Theophilus	100	4 H-77	E	4.6×10^{-3}	260±40
Eratosthenes	58	5 H-134	E	2.7×10^{-3}	270±60
Lexell	63	4 H-107	E	---	260±20
Agrippa	44	4 H-97	E	---	265±55
Hainzel A	53	4 H-136	E	---	310±20

frequency approximates, but does not specify, the frequency of craters superposed on crater materials at the Eratosthenian-Copernican boundary. The total range for Eratosthenian crater deposits is, therefore, about 7.5×10^{-4} to 2×10^{-2}; most midpoint values lie between 10^{-4} and 10^{-3} craters.

D_L values for crater deposits scatter widely (fig. 12.8; table 12.3). Including error bars, the range for Eratosthenian units is 130 to 330 m; midpoints range from 160 to 310 m. Like the frequencies, therefore, the D_L values apparently are displaced toward larger values for crater substrates than for mare substrates (compare total range of 140 to 250 m, table 12.1).

Small craters are identified as Eratosthenian by the criteria of Trask (see chap. 7), which were also devised by observing the relations of craters to mare materials. Small Eratosthenian craters are buried in many places by mare flows on which other Eratosthenian craters are superposed (figs. 12.5, 12.6).

In most farside and some limb regions, where photographs are poor, crater assignments to the Eratosthenian are still based on qualitative estimates of a low crater density, coupled with fresh morphology and an absence of detectable rays. The fine texture of radial ejecta and secondary craters is probably the morphologic feature most sensitive to degradation by small impacts and thus is most

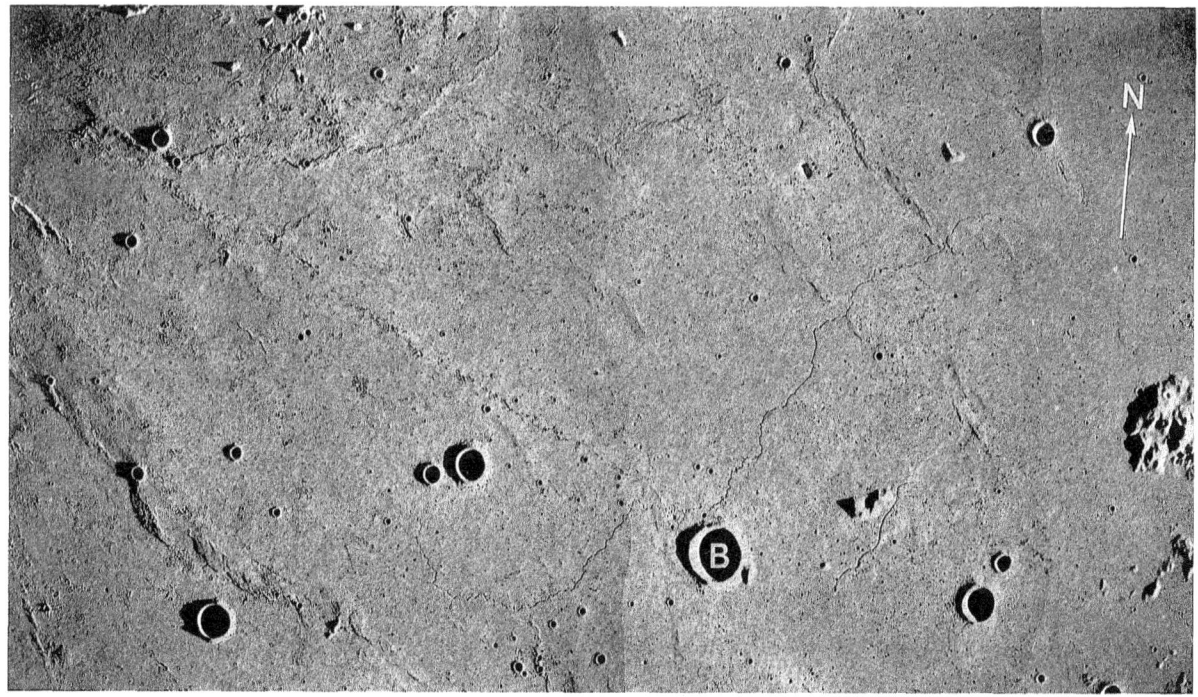

FIGURE 12.5.—Extensive Eratosthenian mare materials in southern Mare Imbrium, flooding rim materials of many small Eratosthenian craters, the largest of which is Brayley (B; 15 km, 21° N., 37° W.). Long chain of secondary craters extends northwestward of Brayley radial to Aristarchus, just outside left edge of photograph. Parts of Imbrium-basin rings are exposed as rugged terra islands. Mosaic of Apollo 17 frames M-2925, M-2928, and M-2931 (from right to left).

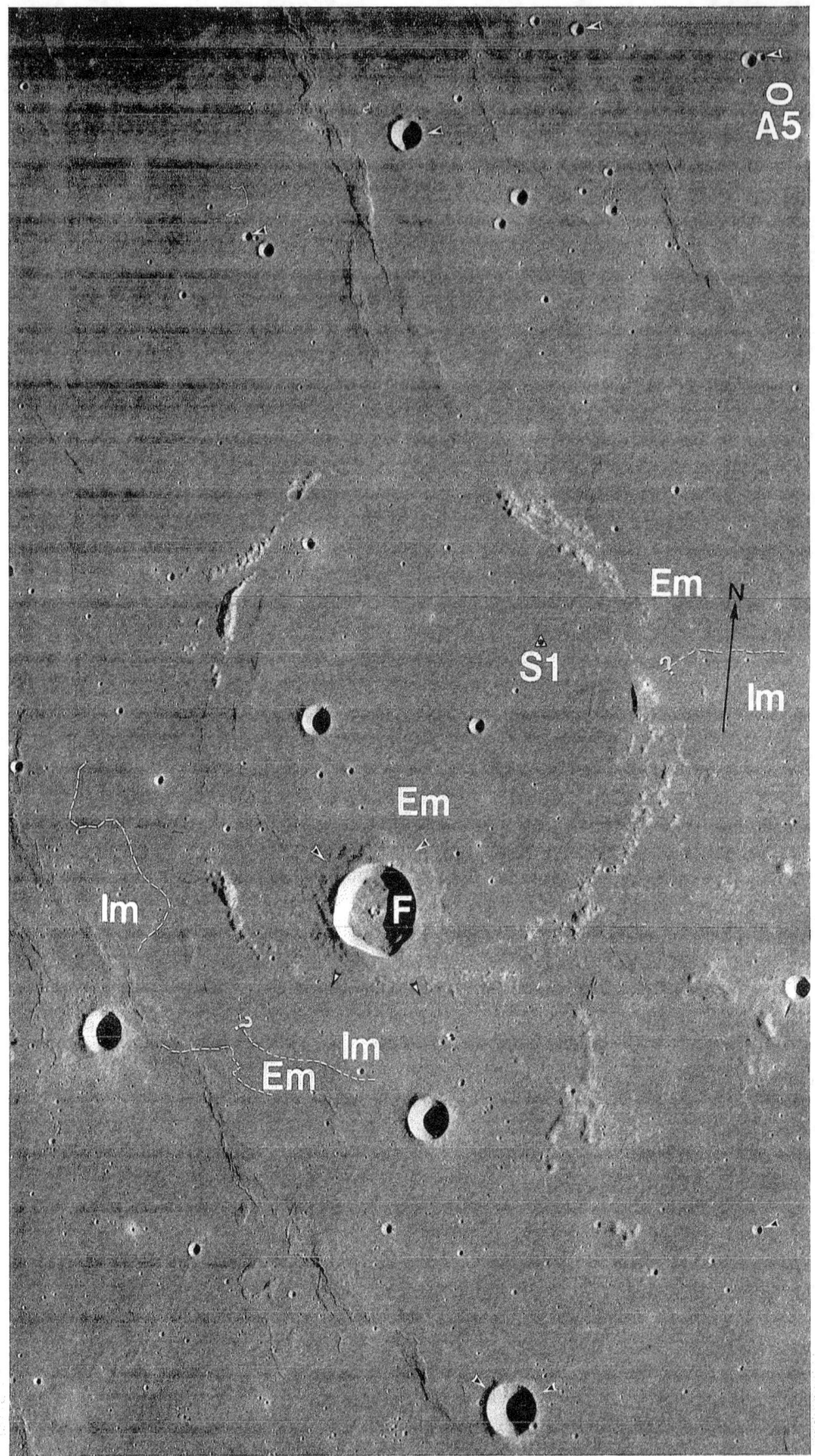

FIGURE 12.6.—Part of Oceanus Procellarum near crater Flamsteed.
A. Large ring of rounded hills is Flamsteed P (112 km), a nearly buried pre-Imbrian (Nectarian or pre-Nectarian) crater. Eratosthenian crater Flamsteed (F; 21 km) is superposed on Imbrian (Im) mare unit in south (white-on-black arrowheads) and embayed by younger Eratosthenian (Em) mare unit in north (black-on-white arrowheads). Other embayments of Eratosthenian or Imbrian craters are also shown by black-on-white arrowheads. Dashed lines indicate subtle contacts between Em and Im units; mare is closer to rims of small craters in Em than in Im, and crater density is lower in Em. S1, Surveyor 1 landing site; A5 (ellipse), potential Apollo landing site 5. Orbiter 4 frame H–143.

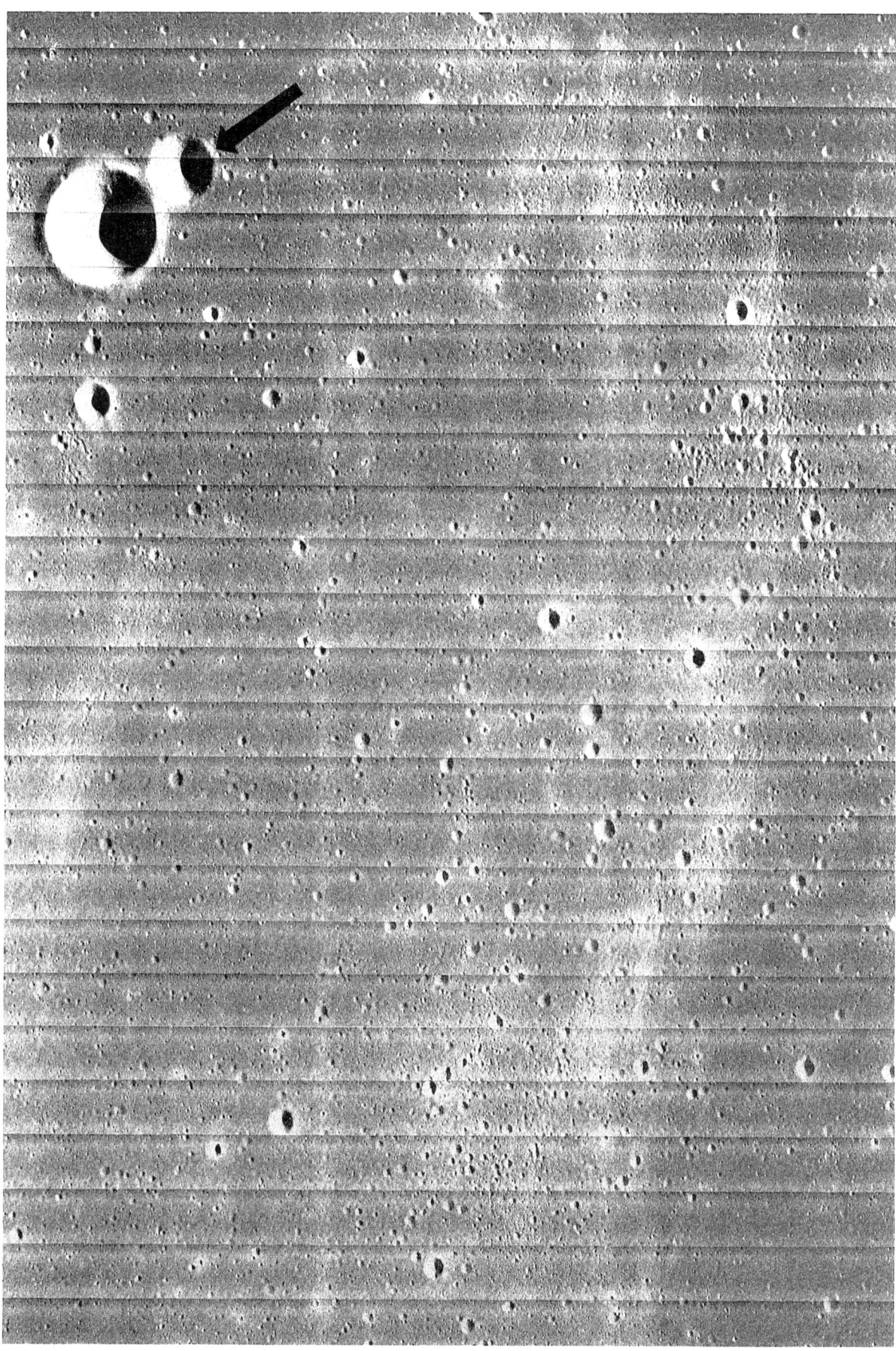

B. Arrow indicates embayment of pair of otherwise fresh-appearing small craters also shown by black-on-white arrowhead above ellipse A5 in A. Largest crater is Maestlin G (3.5 km). Extensive level terrain, despite numerous small craters, indicates that no craters larger than 200 m in diameter are degraded to saucer shapes, as would be true on an Imbrian mare surface. Heavily cratered streak extending over most of right half of photograph is ray of crater Kepler (Carr and Titley, 1969), 200 km away in direction indicated by herringbone pattern. Orbiter 2 frame M-207.

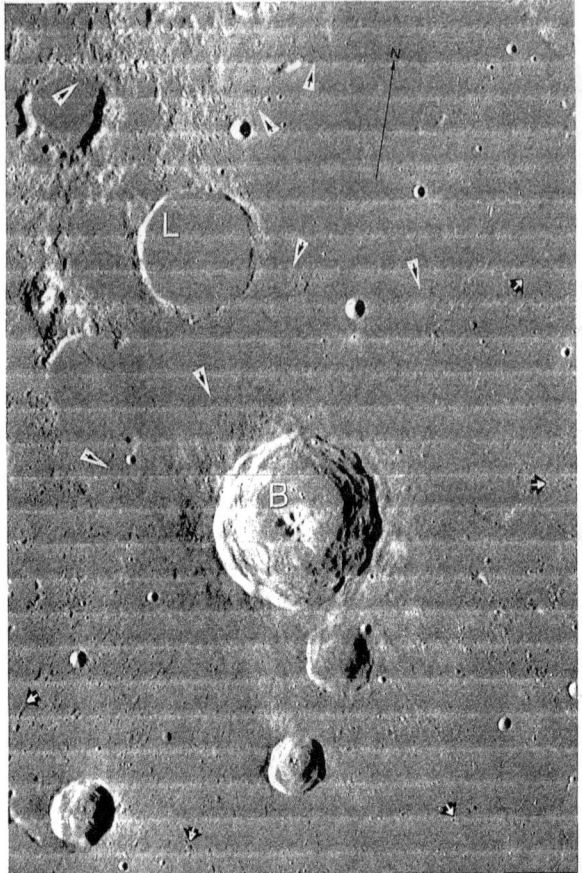

FIGURE 12.7.—Eratosthenian crater Bullialdus (B; 61 km), superposed on Imbrian mare materials (white-on-black arrows) and embayed by Eratosthenian mare materials (black-on-white arrows). Crater Lubiniezky (L; 44 km) is filled by thin mare unit younger than Bullialdus-secondary craters; embayment relations are traceable northwest and northeast of Lubiniezky. Orbiter 4 frame H–125.

diagnostic (fig. 12.9). Where neither this fine texture nor crater densities are visible, ray brightness is still used for dating. The inconsistencies in dating uncovered by Neukum and König (1976) illustrate that the morphologies of Late Imbrian, Eratosthenian, and early Copernican craters of subequal sizes differ only slightly when seen on any but the best photographs. Theophilus and Langrenus, for example, would still be considered Copernican if they were on parts of the farside not covered by Apollo photographs and if their rays were visible against the terra background.

Among the Eratosthenian craters dated by size-frequency studies is Cavalerius (fig. 12.10; D.B. Snyder, written commun., 1980). The peculiar bright feature in Oceanus Procellarum called Reiner gamma may be an exceptionally bright part of the otherwise-faint ray pattern of Cavalerius (fig. 12.10; Hood and others, 1979). Reiner gamma has been thought to be Copernican because of its brightness and superposition on other features (McCauley, 1967a). Interest in the origin of this enigmatic feature has been revived by the discovery that it has the highest magnetism yet observed from lunar orbit (Hood and others, 1979). The Eratosthenian age of Reiner gamma, if confirmed, would constrain speculations about the origin of lunar magnetism, one of the major unsolved problems raised by Apollo exploration. It also would illustrate a failure of the bright-ray criterion for distinguishing Eratosthenian and Copernican ages.

Some of the physical explanations first proposed for the implied fading of bright rays over time were solar radiation, cosmic-ray bombardment, and micrometeorite mixing with the darker substrate (Shoemaker, 1962b, p. 345; Shoemaker and Hackman, 1962, p. 299). Chapter 5 explained that the main darkening agent proved to be the accumulation of Fe- and Ti-rich agglutinates generated in the regolith by small impacts at the expense of brighter crystalline material. Rays that cross mare regoliths rich in these agglutinates are less conspicuous than those superposed on mare and terra materials that are poor in Ti and Fe.

The interiors of Eratosthenian craters smaller than about 8 km in diameter commonly are more nearly V-shaped than are both older and younger craters of the same size (figs. 10.41, 12.3, 12.5–12.7). Imbrian craters are shallower and commonly more nearly U-shaped (figs. 10.38, 10.41). Many Copernican craters still display their original floor topography. These differences result from time-dependent filling of crater bottoms by debris from the walls. The best Orbiter resolutions show that Eratosthenian craters smaller than about 3 km in diameter are smoother than their Copernican counterparts. The

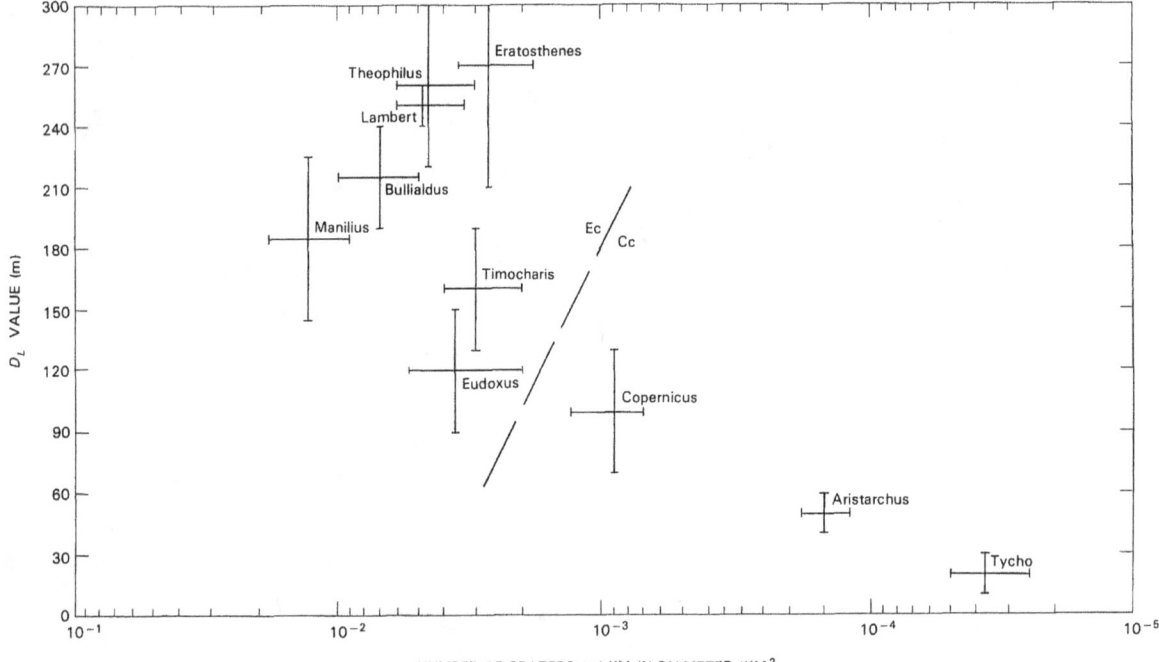

FIGURE 12.8.—Comparison of D_L values and frequencies of small (min 1 km diam) craters superposed on crater materials, based on figure 12.4 and table 12.3. Cc, Copernican craters; Ec, Eratosthenian craters.

oldest Eratosthenian craters disappear completely at sizes of about 300 to 400 m, and the youngest at about 75 m (fig. 7.14).

In summary, crater units of the Eratosthenian System are best recognizable where fine-scale features are visible: morphologies and frequencies of small superposed craters, fine-scale smoothing by invisible craters in the steady state, and superpositional relations with mare flows. The brightness of a ray is a function not only of its age but also of the material in which the ray-forming secondary craters were excavated. Better data might require revision of some of the age assignments given here (pl. 10), particularly on the farside and on the north, south, and east limbs of the nearside.

Frequency

Application of these qualitative criteria apparently has resulted in fairly consistent age assignments, judging by the even densities on the paleogeologic maps (pls. 10, 11) of Eratosthenian and Copernican craters larger than 30 km in diameter. A total of 88 Eratosthenian craters of this size are recognized.

Wilhelms and McCauley (1971) mapped 256 Eratosthenian craters larger than 10 km in diameter within an area of 12.07×10^6 km^2. In the same area, they mapped 145 Copernican craters of these sizes. Their criteria differed somewhat from those used here because brightness depends partly on substrate composition and because some sharp craters thought to be primary are actually secondary-impact craters (Wilhelms, 1976). Nevertheless, the criteria were applied consistently, and no more complete compilation of craters of this size exists. Smaller craters were mapped at 1:1,000,000 and larger scales, but total numbers or frequencies have not been determined; such an exercise would suffer from greatly differing data quality, the effects of differing substrates, and subjective differences among mappers.

Wilhelms and others (1978) determined the frequencies of Eratosthenian and Copernican craters, combined, larger than 4.5 km in diameter (fig. 7.16); craters of these two systems were not distinguished because of the uneven quality of data in the area studied. The slope of the cumulative-frequency curve that best fits postmare craters 10 to 80 km in diameter is -1.8 (Baldwin, 1964; Shoemaker, 1965; Neukum and others, 1975a; Basaltic Volcanism Study Project, 1981, p. 1051–1052). Slopes of the curves for smaller craters depend on diameter range, as shown in figures 7.10G and 7.16.

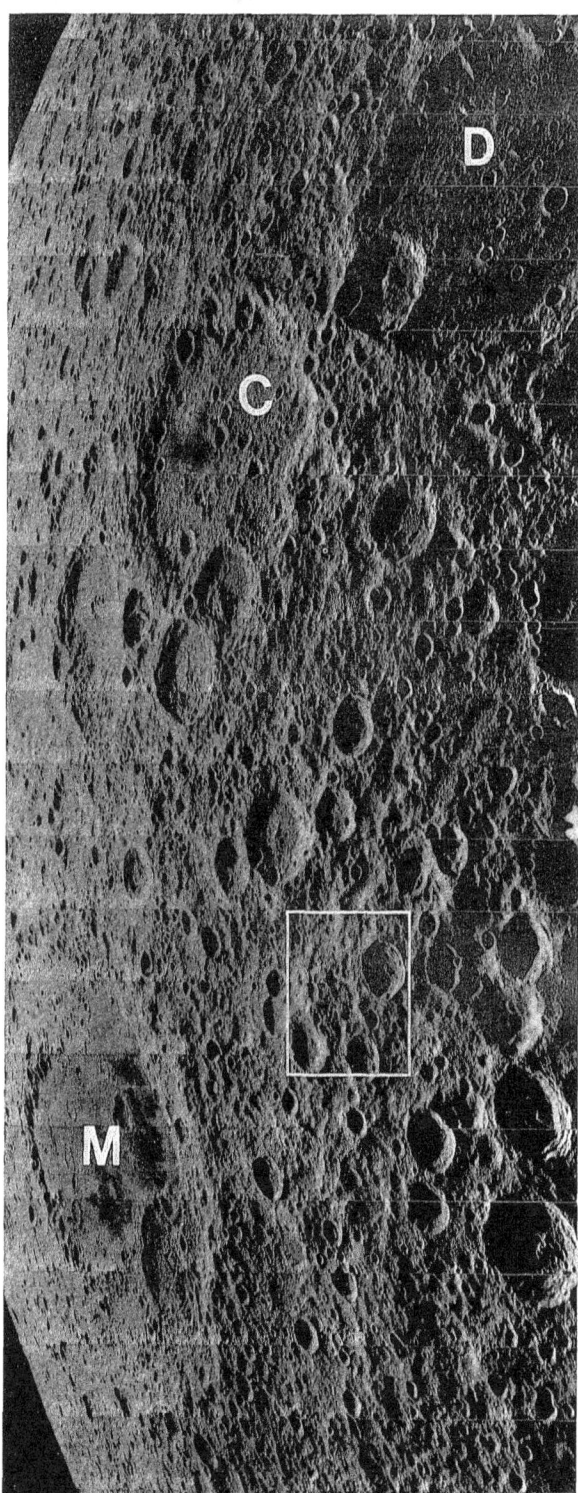

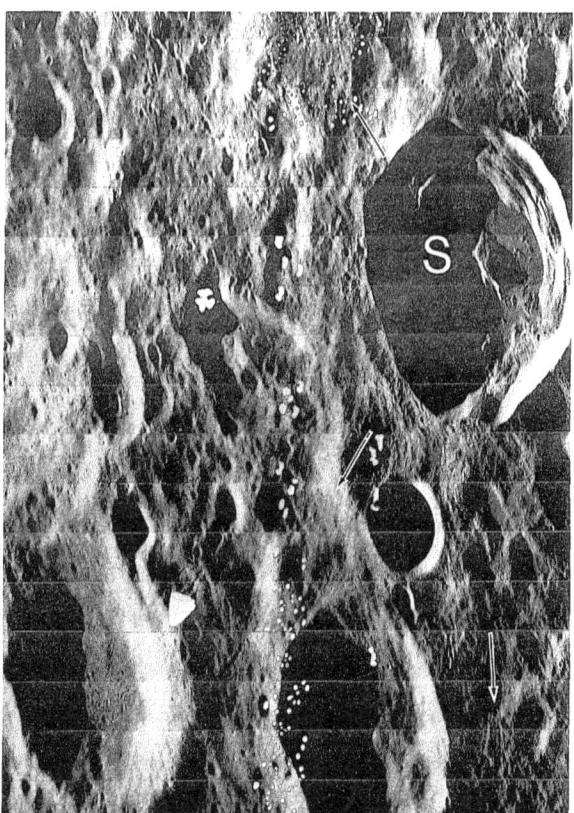

FIGURE 12.9.—Crater Stearns (S, 37 km) on lunar farside, identified as Eratosthenian by topographic sharpness.
 A. Stearns inside box (B). Basin partly filled by mare is Moscoviense (M; 26° N., 147° E.). Large craters are Campbell (C; 225 km, probably pre-Nectarian) and D'Alembert (D; 225 km, probably Nectarian). Orbiter 5 frame M–85.

B. Freshness of Stearns' (S) interior is evident, despite oblique viewing angle and obscuration by shadow. Smoothed radial ejecta and secondary craters are also preserved (arrows), despite rugged terra substrate; periphery of Stearns is pitted by small craters. No exterior textures would be visible around an Imbrian crater of same size and setting; fewer pits and sharper texture would characterize Copernican craters of same size and setting. Orbiter 5 frame H–85.

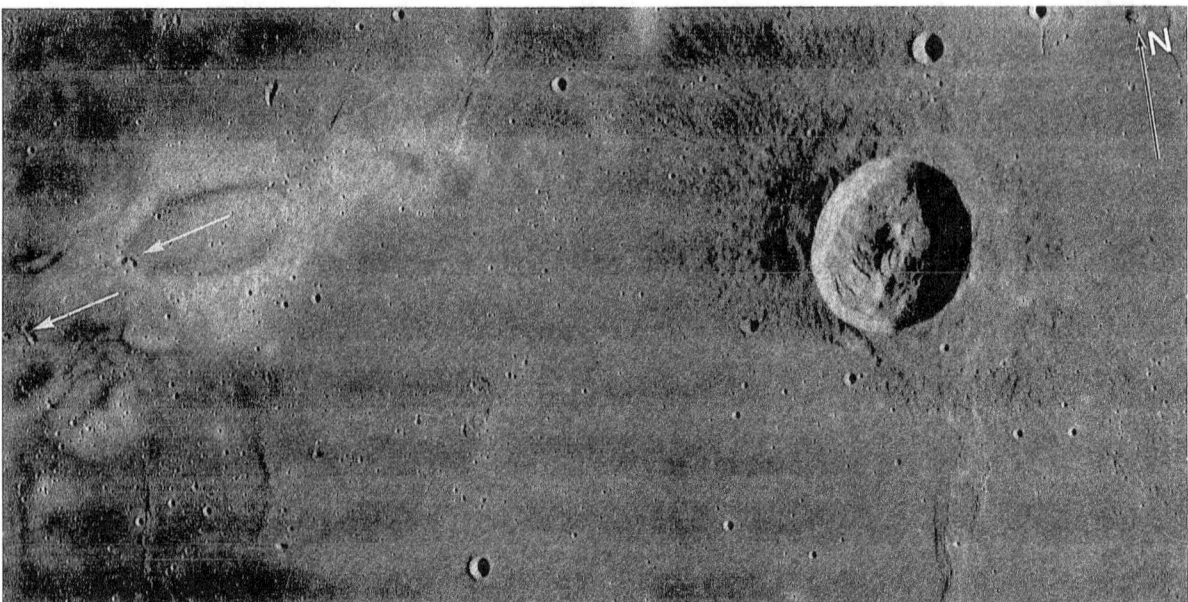

FIGURE 12.10.—Reiner gamma (tadpole-shaped bright patch). Arrows indicate secondary craters of crater Cavalerius, centered 220 km away in direction of arrows. Conspicuous crater is Reiner (30 km), whose superposition on mare material indicates Eratosthenian or Late Imbrian age; pinch in ejecta pattern northeast and southwest of crater is probably due to oblique impact (see chap. 3), but possibly is due to embayment by younger (Eratosthenian) mare. North arrow is adjacent to part of Marius Hills (see chap. 5). Orbiter 4 frame H-157.

MARE MATERIALS— GENERAL STRATIGRAPHY AND DISTRIBUTION

Eratosthenian mare units occupy two main settings—central Mare Imbrium and a crescentic zone extending from Mare Frigoris to southern Oceanus Procellarum (pl. 10; figs. 12.1, 12.3, 12.5–12.7, 12.11). The Imbrium and Procellarum tracts join in a zone south of Montes Harbinger. Smaller patches of Eratosthenian basalt are scattered in the rest of Oceanus Procellarum and in other maria.

The massive, dark, spectrally blue flows in Mare Imbrium described in chapter 5 extend northeastward from the connection with Oceanus Procellarum—near the edge of Mare Imbrium—to a point north of the Imbrium center (fig. 5.1; Schaber, 1969; Schaber and others, 1975). Their stratigraphic relations are among the Moon's clearest because they flood small and large Eratosthenian craters and cover Imbrian mare units (figs. 12.1, 12.5; Wilhelms, 1980). Before the Apollo missions, an Eratosthenian age was also established for mare units in a large area of equatorial Oceanus Procellarum by superpositional relations with small craters (fig. 12.6) and by evidence that the regolith there is thin. Thin regolith was indicated by the blocky surface revealed by Surveyor 1 at one point within this area (Shoemaker and Morris, 1970; Offield, 1972) 1970), and by an abundance of small blocky craters (Trask, 1969, 1971) and certain nested-crater configurations (Oberbeck and Quaide, 1968; Quaide and Oberbeck, 1968, 1969) that are visible on high-resolution Lunar Orbiter photographs over extensive additional parts of the area. Several sites within the area were considered for Apollo landings because of the mare's youth (Carr and Titley, 1969; Titley and Trask, 1969; Cummings, 1971; West and Cannon, 1971; Offield, 1972). Many flows in northern Oceanus Procellarum are also young; some west of the Aristarchus Plateau are only about twice as old as the crater Copernicus (Young, 1977).

The most extensive Eratosthenian mare units of Mare Imbrium and Oceanus Procellarum are spectrally blue and, thus, probably rich in Fe and Ti (see chap. 5); they also are more radioactive than most other mare units (Soderblom and others, 1977). The most common spectral type is hDSA, although type HDSA is present in several spots in Oceanus Procellarum, including the Flamsteed-Flamsteed P region (pl. 4; fig. 12.6; table 5.1; Pieters, 1978; Pieters and others, 1980). The spectral class (hDSA) and gamma-ray-spectrometer readings of the massive Imbrium flows indicate Ti contents in the range of about 2.0 to 4.5 weight percent TiO_2 (Pieters, 1978; Davis, 1980).

Small patches of Eratosthenian flows, most spectrally blue but some red, are scattered throughout southern Oceanus Procellarum and Maria Insularum and Nubium (pl. 10; Whitford-Stark and Head, 1980). Similar patches, including several with red spectra (Wilhelms, 1980), are scattered between long 25 W. and the central highlands "backbone" (fig. 1.8; see chap. 11).

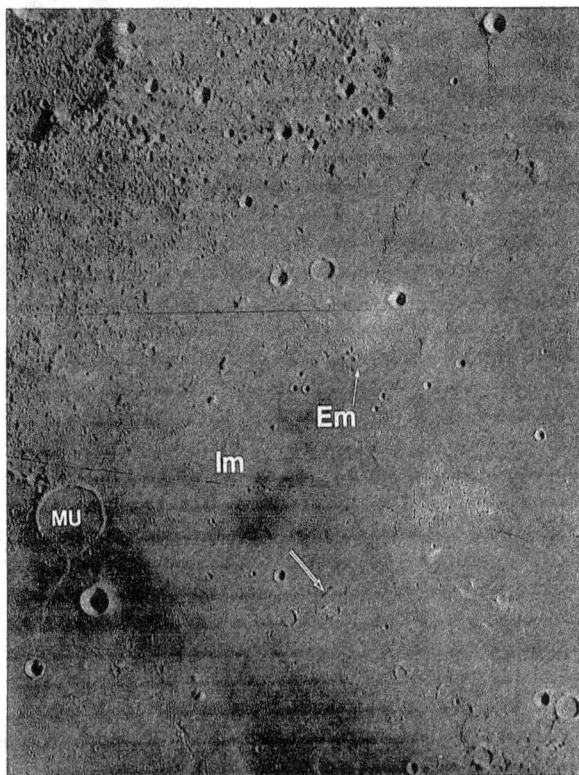

FIGURE 12.11.—Sinus Roris, showing flooding of Imbrian mare materials (Im) by dark, thin Eratosthenian mare materials (Em). Secondary craters of Eratosthenian crater Pythagoras (out of scene to upper left) overlie Imbrian units but are truncated by Eratosthenian unit (white arrow). Both mare units flood secondary craters of Iridum crater (black-and-white arrow). Large crater flooded by Imbrian units is Markov U (MU; 29 km, 52° N., 60° W.). Orbiter 4 frame H-170.

The most abundant spectrally red Eratosthenian flows are on opposite sides of the ring of the Imbrium basin that is overlain by the Iridum crater and Plato. The flows on the Mare Imbrium side are beyond the margins of the spectrally blue Imbrium flows (compare pls. 4, 10).

I suggest that the concentration of Eratosthenian as well as Imbrian eruptions in the Imbrium-Procellarum region is due to the thin lithosphere beneath the Procellarum basin (fig. 11.14). Conditions elsewhere also favored local late eruptions, however, judging by some patches of young basalt in Mare Smythii and other outlying maria (Boyce and Johnson, 1978). The factors controlling the distribution of spectral types are unclear. The main reason that high-Ti basalt dominates the Eratosthenian System may be its high radioactivity; the requisite heat for longlasting melting was greatest in this compositional type (Taylor, 1975, 1982).

Many of the Eratosthenian flows occupy, or flowed from, the margins of several maria in addition to Imbrium (pl. 10; Boyce and Johnson, 1978). Small patches are perched high on the peripheries of Imbrium, Serenitatis, Crisium, and, probably, other basins (figs. 5.10D, 5.16, 5.25). Solomon and Head (1980) suggested that the changing lunar stress pattern resulting from global cooling and consequent lithospheric thickening favored closing of conduits everywhere except at mare margins. However, the thick existing sections of Imbrian basalt in the basin centers may have deflected the rising Eratosthenian magmas to the mare margins.

MARE INSULARUM

Setting of the Apollo 12 landing site

Apollo 12 landed in a topographically complex region, rich in islands of terra material, which at that time (November 1969) was considered part of Oceanus Procellarum but is now known as Mare Insularum (lat 3.2° S., long 23.4° W.; pl. 1; fig. 12.12; U.S. National Aeronautics and Space Administration, 1970; Marvin and others, 1971). As mentioned above, the existence of the blue Eratosthenian mare material in the west-central equatorial belt of the nearside was known when the site was selected. These maria were called "western," as opposed to the "eastern" maria east of long 25° W., which are mostly Imbrian. After the succesful Apollo 11 mission to an "eastern" site, geologists wanted to explore a "western" mare. However, the Apollo 12 landing site ended up being selected entirely for operational reasons. The desire was to demonstrate the ability of the Apollo system to land a LM at a predesignated point, and the Surveyor 3 spacecraft, which had landed 2½ years earlier, provided an appropriate target. Surveyor 1 farther to the west offered a similar target, and its site was called Apollo landing site 6; but complex requirements of launch opportunity and other factors favored the Surveyor 3 site. As predictable from its position near the 25° W. dividing longitude, the geologic context of the site is less well known than that of the Eratosthenian Surveyor 1 mare (Offield, 1972).

The vicinity of the landing site and almost all the sampled terrain are occupied by the interiors or ejecta of several large craters (figs. 12.13–12.15). Most of this area is theoretically within the ejecta blanket of the largest of these, Middle Crescent Crater, 400 m across and originally about 80 m deep (Sutton and Schaber, 1971). Surveyor Crater (the 200-m crater in which Surveyor 3 landed) and Head Crater (110 m, the head of a crater configuration that resembles a snowman) also have strongly influenced the site. Samples were also taken from several spots on the ejecta of Bench Crater and near Halo Crater (fig. 12.14).

The overlapping ejecta blankets contribute to considerable uncertainty about the source strata of the basaltic samples and the significance of remotely sensed spectra. A spectrum centered over the landing site is of type mIG-, that is, intermediate in Ti, intermediate in albedo, and average or lacking in the two long-wavelength absorption bands (Pieters and McCord, 1976; Pieters, 1978). The TiO_2 content of 2 to 3 percent suggested by this spectrum approximately matches that of the average regolith sample and falls between those of the two main types of sampled basalt. This averaging is expectable wherever surface deposits are composed of the ejecta blankets of craters that penetrate more than one unit. The crater frequencies and D_L values that characterize the sampled material are also unclear because the few named craters dominate the local crater population. Soderblom and Lebofsky (1972) suggested that two units are present in the general vicinity—an older unit at the site proper and a younger one about 1 km both west and east of the site. The unit at the site proper has a D_L value of 210 to 215 m (table 11.1; Soderblom and Lebofsky, 1972; Boyce, 1976). The D_L value of the other unit is smaller by an undetermined amount (Soderblom and Lebofsky, 1972). These values fall in the Eratosthenian System as defined here. The midpoint of size-frequency counts in the unit at the site is 2.4×10^{-3} craters larger than 1 km in diameter per square kilometer (table 11.1, figs. 11.2, 12.12). The relative and absolute ages are the youngest from any sampled mare site, including the Apollo 15 units, which lie near the top of the Imbrian System. Therefore, I consider the Apollo 12 mare materials to be Eratosthenian.

Apollo 12 samples

A total of 36 rock-size basalt samples were returned by Apollo 12 (Warner, 1971; James and Wright, 1972). Four compositional types of basalt have been recognized, starting with the early studies. They are now thought to compose two major and one minor unit at the site (Rhodes and others, 1977).

Olivine and pigeonite basalt.—Two of the compositional groups—olivine basalt characterized by modal olivine, and pigeonite basalt characterized by especially abundant modal pigeonite—are probably parts of the same flow, which fractionated at the surface or in a shallow magma chamber (James and Wright, 1972; Rhodes and others, 1977). The pigeonite basalt has the finer textures, which indicate more rapid cooling, and so is presumably the upper fraction of the flow. The distribution of samples near the landing site is consistent with this stratigraphic relation (Rhodes and others, 1977). The olivine basalt apparently was excavated only by the largest crater, Middle Crescent (original depth, approx 80 m; fig. 12.15), whereas the smaller crater, Surveyor (original depth, approx 40 m), excavated the pigeonite basalt but not the olivine basalt.

The radiometric ages of samples of these two compositional groups have about the same broad range (table 12.4). Rb-Sr ages from a given laboratory favor an age 0.1 aeon older for the olivine than for the pigeonite basalt—3.25 versus 3.15 aeons (Papanastassiou and Wasserburg, 1970, 1971a), or 3.22 versus 3.12 aeons (Nyquist and

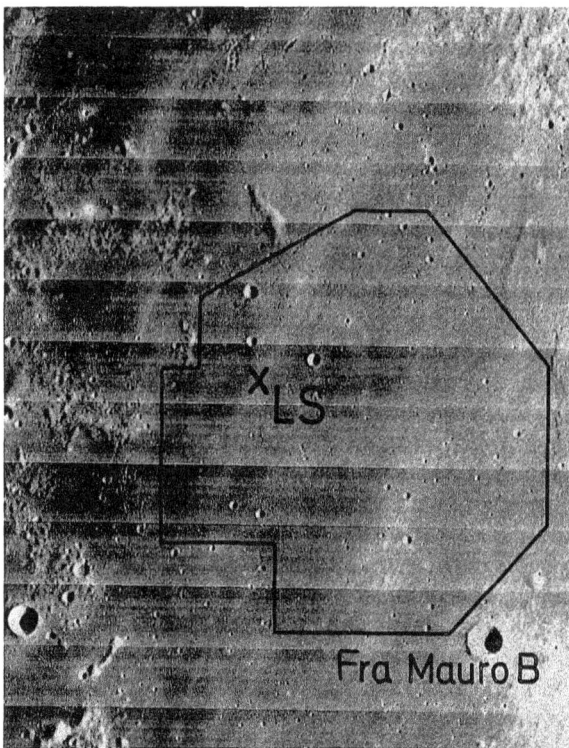

FIGURE 12.12.—Area around Apollo 12 landing site (LS) whose size-frequency distribution was determined by Neukum and others (1975b). Two units (at least) were included in counts (fig. 11.2B); less densely cratered unit around landing site gives lower frequency plotted in figure 11.2B. Fra Mauro B is 7 km across. Courtesy of Gerhard Neukum. Orbiter 4 frame H–125.

FIGURE 12.13.—Apollo 12 landing site and vicinity. Arrow marks location of Lunar Module in Surveyor Crater. Mare unit around landing site contains more large, subdued, old craters than unit outside line and thus is the older unit. Orbiter 3 frame H-154.

others, 1977, 1979a). The fewer Ar-Ar ages cluster more tightly, however, and those determined by Turner (1971, 1977) in each group are the same—3.21 aeons. In view of the overlapping analytical errors and the inconsistent discrepancies between the two methods (for example, sample 12065 is older by Ar-Ar, and sample 12002 older by Rb-Sr), the absolute ages may be considered consistent with the chemical, textural, and petrologic evidence for a common origin of the two compositional groups (James and Wright, 1972; Rhodes and others, 1977). Rhodes and others (1977) estimated a flow thickness of 30 to 50 m. The average age for the combined compositional groups, from determinations whose stated error ranges are less than 0.1 aeon, is 3.16 aeons, which is adopted here as the age of the flow.

Ilmenite basalt.—The ilmenite basalt contains 2.7 to 5.5 weight percent TiO_2, more than the other Apollo 12 groups, though less than many mare-basalt samples from other sites. Accordingly, it has been called either "low-Ti basalt" or "intermediate-Ti basalt" (table 5.2). Ilmenite basalt has been considered a minor component near the landing site, possibly introduced in crater ejecta from a nearby Ti-rich spectral unit approximately corresponding to the younger unit recognized by Soderblom and Lebofsky (1972). In this interpretation (Pieters and McCord, 1976), the olivine-pigeonite flow would be the local bedrock at the landing site. However, the restudy by Rhodes and others (1977) of the surficial distribution of recovered samples showed that ilmenite basalt is, in fact, quite common among the rock-size samples and is probably local to the site. The ilmenite basalt is apparently the only inplace basalt excavated by craters smaller than Surveyor, such as Head, whose original depth was about 20 m, whereas only the larger Surveyor and Middle Crescent Craters excavated inplace olivine and pigeonite basalt (fig. 12.15). Although the smaller craters also reejected some fragments of the olivine-pigeonite flow, these fragments resided in the ejecta of the larger craters.

Therefore, the ilmenite basalt flow overlies the olivine-pigeonite flow. On the basis of the likely original penetration depths of the craters (fig. 3.3), Rhodes and others (1977) estimated that the ilmenite basalt flow is about 40 m thick. Compositional data indicate derivation of this flow from a different magma from the parent magma of the olivine and pigeonite groups, which originated in a different mantle source (James and Wright, 1972; Rhodes and others, 1977). However, the ages for the ilmenite basalt listed in table 12.4, which average 3.15 aeons from the best determinations, indicate no large gap between times of eruption of the two magmas.

Feldspathic basalt.—Only one or, possibly, two rock-size samples of Al-rich, feldspathic basalt were recovered (12031?, 12038). The Rb-Sr ages overlap those of the other groups. The average of the best Rb-Sr age determinations (stated error ranges, less than 0.1 aeon) and the one young Ar-Ar age (3.08 aeons) is, again, 3.16 aeons. The petrologic relation of the feldspathic basalt to the other compositional groups is uncertain (Rhodes and others, 1977). The paucity of samples suggests a provenance in a distant unit or in a deeply buried unit at the landing site proper. The latter view (Nyquist and others, 1981b) is supported by collection of sample 12038 from the largest crater, Middle Crescent (by way of the superposed Bench Crater). Nyquist and others (1981b) favored derivation of this and other aluminous mare basalt from sources containing plagioclase in addition to the olivine and pyroxene generally thought to compose mare-source zones.

TABLE 12.4.—*Radiometric ages of large Eratosthenian basalt samples*

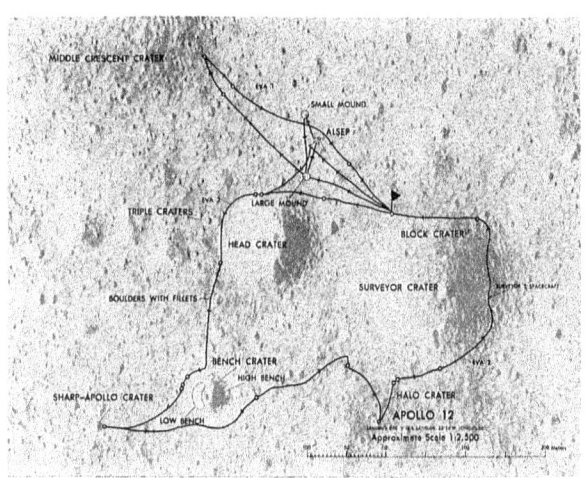

FIGURE 12.14.—Astronaut traverses at Apollo 12 landing site. Small circles, sampling localities; pennant, landing site; ALSEP, geophysical instruments. Airbrush map by U.S. Geological Survey.

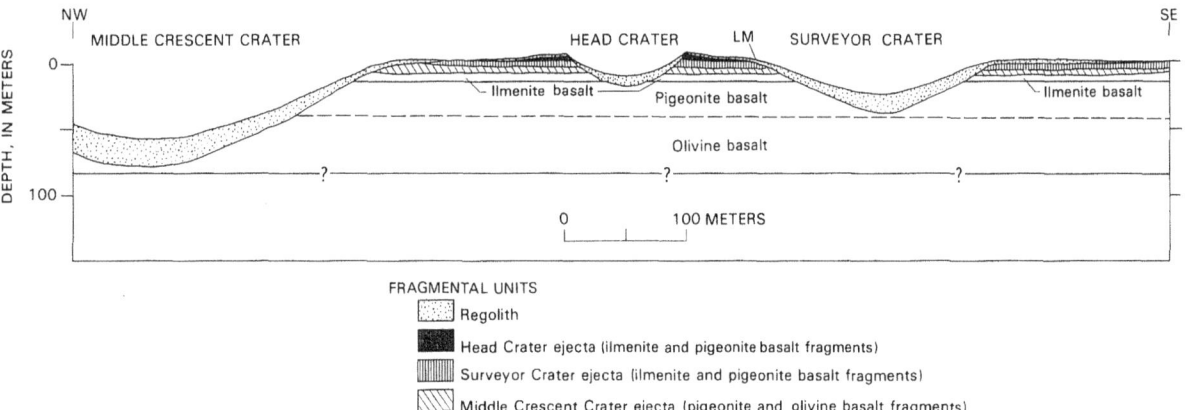

FIGURE 12.15.—Geologic cross section northwest-southeastward across area sampled by Apollo 12 (compare fig. 12.13). Interpretations after Sutton and Schaber (1971) and Rhodes and others (1977). Substrate of Fe-Mg-rich basalt may be another mare unit or Fra Mauro Formation.

Summary and conclusions

The best determined ages on Apollo 12 samples cluster between 3.08 and 3.26 aeons and average 3.16 aeons. The stratigraphic sequence, from lowest to highest, is olivine basalt, pigeonite basalt, and ilmenite basalt. The bottom two compositional groups are parts of the same flow, and the ilmenite basalt is from a different flow. Absolute ages of these compositional groups or of the rare feldspathic basalt, whose provenance is uncertain, are indistinguishable.

Each mare-sampling site has increased our knowledge of the petrogenetic processes and source materials of lunar mare basalt after intensive and diligent study lasting more than a decade. After Apollo 11, there was talk of a single episode of mare extrusion; after Apollo 12, the number of episodes rose to two; and after each subsequent mission, the history of lunar volcanism appeared increasingly complex. We should, therefore, be chastened by the fact that at least two-thirds of the major mare spectral types remain unsampled (Pieters, 1978).

In particular, Pieters (1978) cautioned against drawing conclusions about the rather poorly defined spectral class (mIG-), to which the Apollo 12 basalt samples apparently belong. This class of material is widespread on the Moon (pl. 4) and contains many age and spectral subunits. Nevertheless, its general distribution may be of interest in attempts to discern first-order features of the lunar crust. It lies in areas of known thick crust outside the Procellarum basin (Crisium, Fecunditatis), as well as inside Procellarum in zones nearest the central terra "backbone." Eratosthenian basalt is also less abundant there than elsewhere in Procellarum. Therefore, the lithosphere may have been thicker, and the crust or upper mantle different in composition, here than elsewhere in southern, western, and northern Procellarum, where Ti-rich basalt flows are abundant. The unusual north-south trend of the Eratosthenian patches and of such terra features as the "Fra Mauro peninsula" and the "central backbone" itself (fig. 1.8) may be additional evidence for a major structural trend along the central meridian of the nearside.

CHRONOLOGY

The Apollo 12 basalt flows, which lie near the base of the Eratosthenian System, average about 3.16 aeons in age. No younger or older Eratosthenian materials were sampled, and so absolute ages of other Eratosthenian units must be extrapolated from the dated samples by means of superposed-crater frequencies and D_L values. The Eratosthenian Period began some time between 3.16 aeons ago and the 3.26-aeon formational time of the youngest Apollo 15 basalt samples. An approximate intermediate date of 3.2 aeons ago is adopted here.

The end of the Eratosthenian Period is harder to estimate. As discussed in detail in the next chapter, a date of 1.1 aeons is adopted on the assumption that the cratering rate has been constant since 3.2 aeons ago. Thus, in this hypothesis, the Eratosthenian Period lasted 2.1 aeons—the longest of the six lunar time divisions adopted in this volume and nearly half the age of the Moon.

13. COPERNICAN SYSTEM

FIGURE 13.1 (OVERLEAF).—Fresh, sharp-textured Copernican crater Aristarchus (40 km, 24° N., 47° W.). Orbiter 5 frame M-198.

13. COPERNICAN SYSTEM

CONTENTS

	Page
Introduction	265
Crater materials	265
Mare materials	269
Structure	269
Chronology	269

INTRODUCTION

Most early astronomic and geologic observers who carefully studied the Moon realized that rays are the youngest lunar features because they are superposed on all other terrains. The stratigraphically minded geologists Shoemaker and Hackman (1962) furthermore knew that the rays' youth indicates that the topographically expressed ejecta blankets of the source craters are also among the Moon's youngest. They accordingly established the Copernican System as the Moon's youngest assemblage of rock units.

No formal definition of the base of the Copernican System exists because no extensive stratigraphic-datum horizons exist near the lower system boundary. The rays of Copernicus are good early Copernican markers, but Copernicus does not mark the base of the system; many craters traditionally mapped as Copernican are older. The young, Eratosthenian mare lavas in Mare Imbrium and Oceanus Procellarum are near, but evidently not exactly at, the base of the system. Chapter 12 and table 12.1 give tentative crater-frequency and D_L criteria for dividing Eratosthenian and Copernican units, but these criteria, also, are not definitive. This chapter further specifies the position of the system's base. The end of the Copernican Period is defined as the present; a crater formed today would be Copernican.

Copernican craters are sparsely scattered over the entire Moon; Copernican mare units are concentrated in northwestern Oceanus Procellarum (pl. 11). There are half as many craters larger than 30 km in diameter as in the Eratosthenian System (44 versus 88; pls. 10, 11), and still fewer than in older time-stratigraphic units. The rays, however, make the Copernican System more conspicuous than the number of its units would indicate. Some Copernican deposits are beautifully fresh (figs. 13.1, 13.2; table 13.1). They serve as analogs of more degraded features and hint at the hidden complexity of the older rock record. In other words, as in terrestrial geology, we can say that the present is the key to the past—if the lunar "present" means hundreds of millions of years.

CRATER MATERIALS

Rays and extreme topographic freshness remain the principal criteria for the assignment of a Copernican age to craters (table 13.1). In favorable circumstances, ages are assigned more rigorously. Superposition of a given unit on the rays of Copernicus establishes its Copernican age (fig. 7.4; table 7.2). Crater counts can correlate a few units with Copernicus or with other bright-rayed craters that also are certainly Copernican (figs. 12.4, 13.3; Neukum and König, 1976; Guinness and Arvidson, 1977; Young, 1977). Blocky ejecta and other features detectable at high resolutions that are used as age criteria by Trask (1969, 1971) distinguish small Copernican craters (see chap. 7; figs. 7.12–7.14, 13.2).

Problems in distinguishing Copernican from Eratosthenian craters arise when the craters are older than Copernicus, do not contact Copernicus, are faintly rayed, or are poorly photographed. D_L values and crater frequencies (figs. 12.4, 12.8; tables 12.1, 12.3) can be determined only on superior photographs (fig. 13.4); even the best nearside

TABLE 13.1.—*Representative Copernican craters*

[Cross rules divide diameter ranges mapped differently in plate 11: smaller than 30 km, unmapped; 30 to 59 km, interiors mapped; 60 km and larger, exterior deposits mapped]

Crater	Diameter (km)	Center (lat)	Center (long)	Figure	Remarks
South Ray	0.8	9.2° S.	15.3° E.	9.9	See table 13.2.
North Ray	1.0	8.8° S.	15.5° E.	9.9	See table 13.2.
Linné	2.5	28° N.	12° E.	3.2A	Young.
Copernicus H	4.3	7° N.	18° W.	12.2B, 13.6	Dark-haloed.
Lansberg B	9	3° S.	28° W.	10.41	---
Messier	11	2° S.	48° E.	3.11A	Elongate.
Gambart A	12	1° N.	19° W.	13.5	---
Goddard A	12	17° N.	90° E.	4.7	Very young.
Sulpicius Gallus	12	20° N.	12° E.	5.16	---
Dawes	18	17° N.	26° E.	11.13	See figure 12.4.
Dionysius	18	3° N.	17° E.	10.38	Young.
Lichtenberg	20	32° N.	68° W.	10.2, 13.7	Rim flooded.
Pytheas	20	21° N.	21° W.	3.4D	See table 7.2.
Thebit A	20	22° S.	5° W.	7.7	---
Conon	22	22° N.	2° E.	10.7, 10.14	---
Giordano Bruno	22	36° N.	103° E.	9.28	Very young.
Mösting	25	1° S.	6° W.	7.7	See figure 12.4.
Triesnecker	26	4° N.	4° E.	6.20	---
Proclus	28	16° N.	47° E.	3.33, 9.11, 9.15	Young; asymmetric rays.
Kepler	32	8° N.	38° W.	8.10	Young (see tables 7.2, 12.3).
Petavius B	33	20° S.	57° E.	13.4	Young (see fig. 12.4).
Godin	35	2° N.	10° E.	10.16	See table 12.3.
Autolycus	39	31° N.	2° E.	1.6, 1.7, 2.5B	Do.
Aristarchus	40	24° N.	47° W.	5.12, 7.4, 13.1	Young (see figs. 12.4, 13.3; tables 7.2, 12.3).
Olbers A	43	8° N.	78° W.	10.3	Young.
Crookes	49	10° S.	165° W.	9.21	---
Anaxagoras	51	73° N.	10° W.	3.34, 10.15	---
Aristillus	55	34° N.	1° E.	1.6, 1.7, 2.5B	See table 12.3.
Taruntius	56	6° N.	47° E.	6.13B, 6.1B, 9.3	Floor uplifted.
Eudoxus	67	44° N.	16° E.	1.7, 10.12	See figure 12.4; table 12.3.
King	77	5° N.	121° E.	1.2, 3.23, 3.32, 3.36	Farside example.
Tycho	85	43° S.	11° W.	1.1, 1.8, 2.3, 3.20	Young (see figs. 12.4, 13.3; tables 12.3, 13.2).
Copernicus	93	10° N.	20° W.	1.1, 1.6, 3.4, 3.30, 3.35, 7.4, 12.2, 13.10	Typical (see figs. 12.4, 13.3; tables 7.2, 12.3, 13.2).

Lunar Orbiter 4 H-frames are barely adequate to date Copernican crater units. Because the D_L method is valid only on level surfaces, the only crater materials it can date reliably are impact-melt pools, which are relatively small. The traditional criteria of ray brightness and topographic freshness are also hard to apply in poorly photographed areas, particularly on most of the farside poleward of lat 40° N. and lat 40° S. (pl. 2). Some observational bias may account for the excess of nearside over farside craters mapped in plate 11 (26 versus 18) and the excess of northern- over southern-hemisphere craters on the nearside (17 versus 9). Nevertheless, the identification, during two separate iterations, of half as many Copernican as Eratosthenian craters in two size ranges (min 10 km diam [see chap. 12] and min 30 km diam [pls. 10, 11]) suggests that the criteria have been consistently applied and correctly discriminate ages to a good approximation.

Copernican craters can commonly be identified by remote sensing. Temperatures measured in the infrared during a total lunar eclipse or from orbit are valuable for detecting young blocky craters and other fresh surfaces covered by little insulating fragmental material (see chap. 5, subsection entitled "Other Properties"). The eclipse infrared values were extensively used during the lunar geologic-mapping program before the advent of spacecraft photography to distinguish between Copernican ("hot") and Eratosthenian ("cool") craters (fig. 13.5). Some of the crater-age assignments in this volume still rest on this distinction (for example, Agrippa and Godin, fig. 10.16). Radar has also been used to detect blockiness in craters and thus to estimate cohesiveness and age-related state of degradation (Thompson and others, 1980, 1981). Color spectra may also reveal youthful surfaces in craters; enhanced reflectivity near 1 and 2 μm indicates high proportions of fresh crystalline material to agglutinitic glass, whose formation is a function of exposure age (see chap. 5). Color contrasts seen on color-difference images also are strongest around the youngest craters (fig. 5.20).

Age interpretations of low albedo have changed for crater materials as they have for mare materials. The dark color of the volcanic maria and the existence of the irregular, dark-haloed endogenic craters in Alphonsus (fig. 5.10F) led to volcanic interpretations for the dark halos of certain circular craters as well (Shoemaker and Hackman, 1962, p. 297; Shoemaker, 1964; Salisbury and others, 1968). Impact origin was also entertained for these dark-haloed circular craters (Carr, 1965b) and was later substantiated by the impactlike morphology (deep floor, rough ejecta) of a typical example, Copernicus H (figs. 12.2B, 13.6). Most circular craters with impactlike morphology and dark ejecta are superposed on bright rays or other thin bright materials that, in turn, overlie maria or dark terra plains (see chap. 9; figs. 13.5, 13.6). Thus, the halos are only dark by contrast with their surroundings and probably contain basaltic materials brought to the surface by impacts (Scott and others, 1971, p. 276–277; Lucchitta, 1972; Hodges, 1973a; Lucchitta and Schmitt, 1974; Wolfe and others, 1975; Schultz and Spudis, 1979). Whereas the interiors of most endogenic craters are dark, the interiors of dark-haloed impact craters are as bright as those of any other impact crater. The dark-haloed craters superposed on Copernican rays are, of course, Copernican.

Other strong contrasts in albedo may also indicate youth because albedos become neutral with advancing age (see chap. 5). The rim of the young crater Tycho is surrounded by both bright and dark zones (fig. 1.1), which are unobserved around most craters. Dark and bright rays were artificially generated by impacts of spacecraft on the Moon (Whitaker, 1972a) and presumably also surrounded many natural

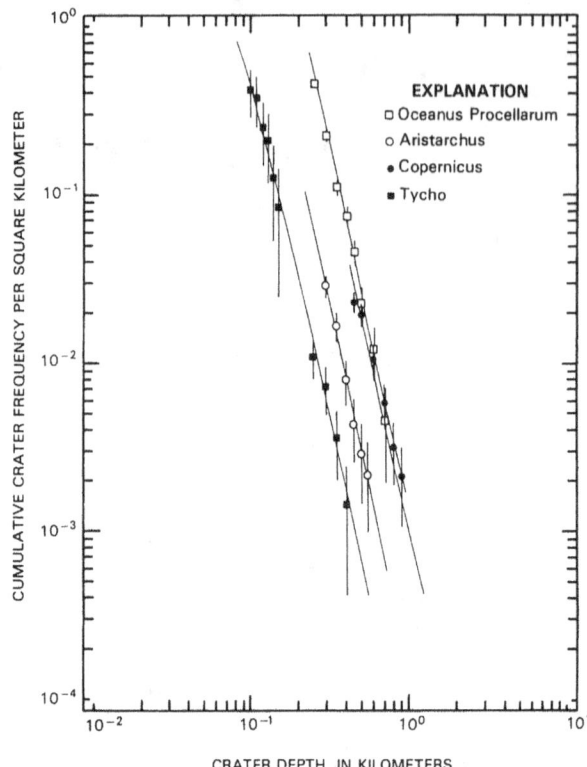

FIGURE 13.3.—Frequencies of craters superposed on three Copernican craters, in comparison with mare units in Oceanus Procellarum shown in figure 12.3 (compare fig. 12.4A). Courtesy of Gerhard Neukum.

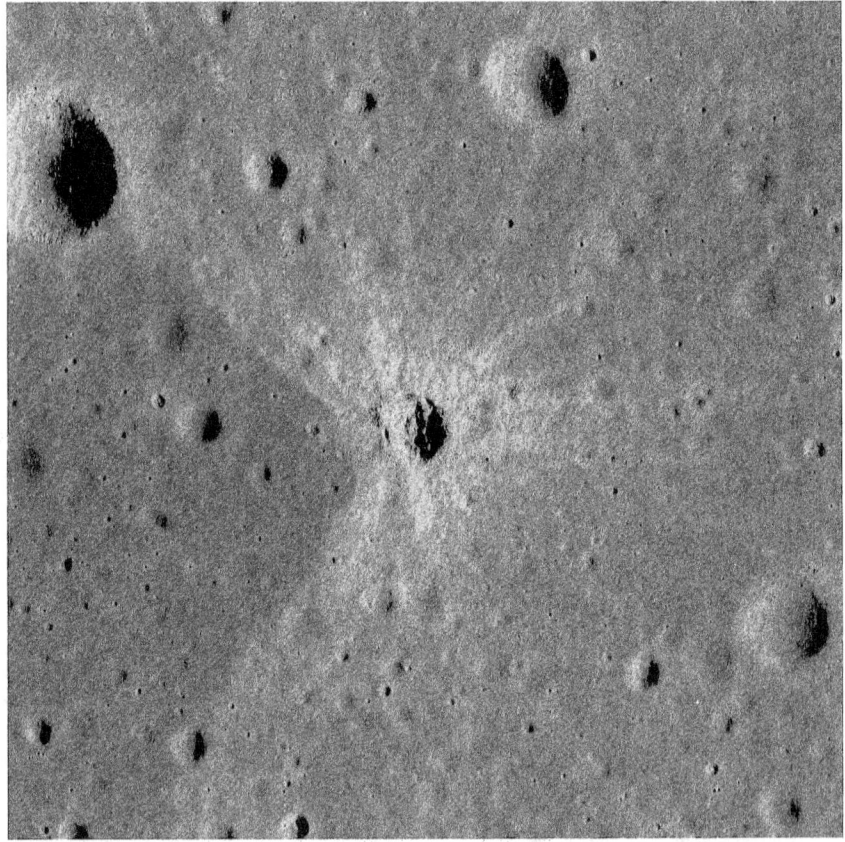

FIGURE 13.2.—Fresh Copernican crater, about 0.5 km in diameter, in central Mare Serenitatis. Blocks ejected from crater in direction of bright rays are preserved as far as three crater diameters from rim. In contrast, blocks are preserved only in interior and on rim crest of older, larger crater in upper left. Mare surface is brightened by Copernican rays except in sector of nondeposition left of crater (probably owing to oblique impact). Apollo 15 frame P–9337.

impacts, but most have been homogenized. Original albedos of target strata uplifted in the crater may be preserved in some young craters (fig. 3.29).

In summary, the brightest rays, most highly contrasting albedos of other crater materials, highest thermal anomalies, freshest morphologies, most coherent ejecta blocks, deepest floors, and fewest superposed craters indicate a Copernican crater age. The relative degree of development of these attributes also may be used to subdivide the Copernican System (Trask, 1969, 1971; Wilhelms and McCauley, 1971). Mapping and dating of older craters is aided by reference to the appearance of Copernican primary and secondary craters.

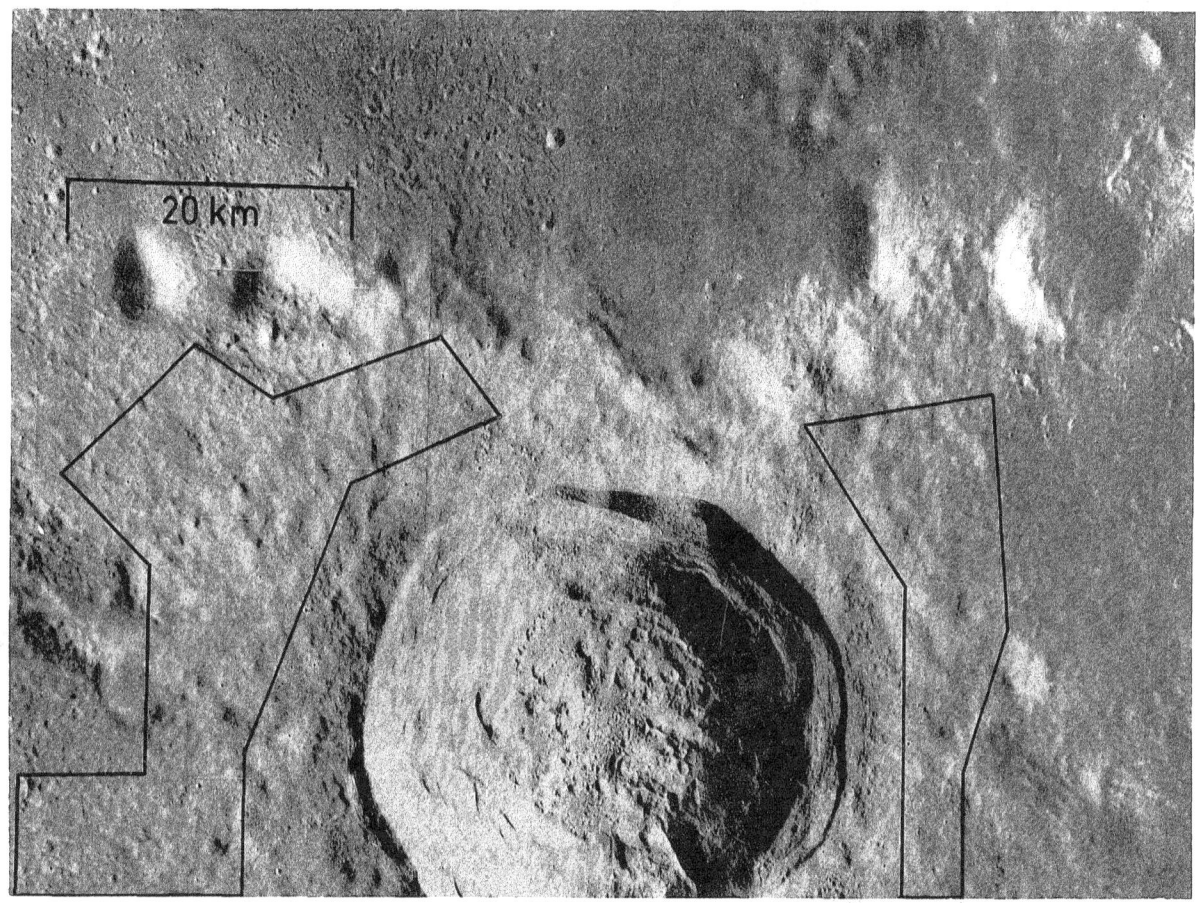

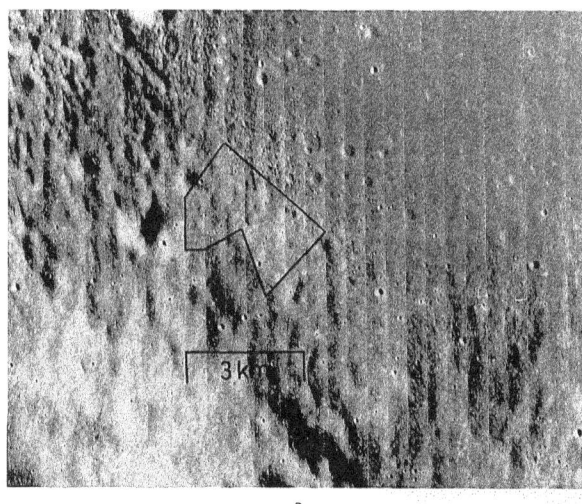

FIGURE 13.4.—Fresh Copernican crater Petavius B (33 km, 20° S., 57° E.), illustrating crater counts made on small areas from superior photographs. Counts made at each scale are combined and included in figure 12.4. Courtesy of Gerhard Neukum.
A. Regional view. Ejecta is indistinct in north sector because it is superposed on mare material and probably because of oblique primary impact. Orbiter 4 frame M-37.
B. Near northern "excluded zone." Orbiter 5 frame H-37.
C. Southeast sector. Orbiter 5 frame H-37.

FIGURE 13.5.—Small dark-haloed craters (arrows) superposed on rays of Copernican crater Gambart A (12 km), a thermal-infrared "hotspot." Mare material underlies Gambart A. Apollo 12 frame H-7737.

FIGURE 13.6.—Dark-haloed crater Copernicus H (4.3 km; compare fig. 12.2B), showing subconcentric dunes characteristic of small impact craters. Orbiter 5 frame H-147.

MARE MATERIALS

Northern Oceanus Procellarum contains some very young mare materials (pl. 11), the largest patch of which covers the southeastern ejecta of the bright-rayed crater Lichtenberg (fig. 13.7; Moore, 1967; Wilhelms, 1970b). Thus, the mare basalt is Copernican, and the Copernican Period had started by the time it was extruded. Three additional patches that I estimate to have the same or lower frequencies as this post-Lichtenberg unit are also mapped in plate 11. This part of northern Oceanus Procellarum was apparently the most active, or only, site of Copernican volcanism.

Dating of these patches would greatly help in determining the duration of lunar volcanism. The density of craters superposed on the patch near Lichtenberg has been determined to be about equal to that on Copernicus (P.H. Schultz and P.D. Spudis, written commun., 1982). If, as discussed previously, a given impact superposes a larger crater on breccia than on basalt, the cumulative size-frequency curve for a crater deposit will be displaced upward from that for a mare deposit of the same age (see chap. 12; fig. 12.4; tables 7.1, 12.1; Schultz and Spencer, 1979; Ahrens and Watt, 1980). The flow that embays Lichtenberg would thus be younger than Copernicus.

The smallest D_L value determined by Boyce and others (1975), 150 ± 20 m, probably pertains to one of the Copernican patches. This value is somewhat smaller than 165 ± 25 m, the D_L value for the extensive neighboring Eratosthenian flows.

The small Copernican mare units are among the spectrally bluest on the color-difference photograph reproduced in figure 5.20. They are slightly bluer than the adjacent Eratosthenian units, whose spectral class is hDSA (pl. 4); the Copernican units are too small to show in plate 4 or on the spectral map of Pieters (1978). These units overlie the extensive Eratosthenian flows in the middle trough of the Procellarum basin (pl. 10). Other Copernican mare units may eventually be found in maria superposed on the central Procellarum basin, for example, Mare Imbrium; or young magmas melted there may not have been able to rise through the thick section of Imbrian and Eratosthenian basalt already present.

Otherwise, no Copernican volcanic materials are known. Dark-mantling materials that were thought to be young on the basis of smoothness have been proved to be Imbrian in age (see chap. 11). Volcanism or emissions of gas were among the early-suggested explanations for Reiner gamma (fig. 12.10; McCauley, 1967b) and similar but more extensive bright "swirls" on the farside (fig. 4.7; El-Baz, 1972). Some connection with lunar magnetism and an origin by Copernican (Schultz and Srnka, 1980) or pre-Copernican (Hood and others, 1979) impacts have been suggested more recently (see chap. 10, section entitled "Imbrium-Secondary Craters," and chap. 12, section entitled "Crater Materials").

STRUCTURE

Copernican tectonism was minor. Narrow gashes in the mare west of the Apollo 17 landing site (fig. 13.8) are young but probably resulted from drainage of regolith into older voids (B.K. Lucchitta, in Masursky and others, 1978, p. 209). The existence of mascons indicates that isostatic compensation of mare-filled basins has ceased or is very sluggish. Continuing subsidence and the resulting mare-ridge compression (see chap. 6) are unlikely because most endogenic moonquakes do not correlate with maria (Nakamura and others, 1979, fig. 3; Solomon and Head, 1979). The only definite Copernican tectonic features are the uplifted floors of such craters as Taruntius (fig. 6.13B). Binder (1982), however, suggested that young-appearing lobate scarps on the farside terra (Masursky and others, 1978, p. 96) are thrust faults caused by currently increasing global contraction.

Lunar seismic energy of endogenic origin is minuscule, only 10^{-12} to 10^{-7} that of the Earth (French, 1977, p. 228; Lammlein, 1977, p. 266). The most energetic, but rarest, endogenic moonquakes probably originate in the upper mantle from small thermal stresses (Nakamura and others, 1979). The most common, but very small, moonquakes originate at depths of $1,000 \pm 100$ km from tidal stresses that do not deform the surface (Lammlein, 1977). The third type of moonquake, intermediate in frequency but unlimited in magnitude, is induced by impacts.

CHRONOLOGY

Establishing the chronology of the Copernican Period is hindered by the small number of well-dated stratigraphic units of regional extent. At 1.29 aeons old (Bernatowicz and others, 1978), sample 15405 from the Apollo 15 landing site is the youngest rock-size lunar sample and the oldest that could be Copernican (fig. 13.9; table 13.2). This sample contains KREEP-rich "granitic" and "monzodioritic" material (Ryder and others, 1975a; Ryder, 1976) that is exotic to the landing site. A source in the Copernican craters Aristillus or Autolycus, centered 250 and 150 km north of the landing site, respectively, is considered possible by most investigators. Autolycus is marginally favored for several reasons: (1) It formed at least partly on the required target, KREEP-rich plains (Metzger and others, 1979), whereas mare basalt probably constitutes the uppermost target material of Aristillus; (2) it is closer than Aristillus to the collection site, easing somewhat the objection that the 1-m boulder from which sample 15405 was taken should have disintegrated in a long flight; and (3) it is older than Aristillus, making it the more likely source of 1.29-aeon-old material if Aristillus is younger than Copernicus (as tentatively suggested by Guinness and Arvidson, 1977) and if Copernicus is about 0.8 aeon old.

Most estimates of the post-Imbrian cratering rate are based on an 0.8- to 0.85-aeon age for Copernicus that was determined on light-colored KREEP-rich material dug from a shallow trench in the Apollo 12 mare regolith (Hubbard and Gast, 1971; Hubbard and others, 1971; Marvin and others, 1971; Meyer and others, 1971). This material was identified with Copernicus by the situation of the landing site along a Copernicus ray (fig. 13.10; Pohn, 1971) and by the fact that part of the Copernicus target material was the KREEP-rich (see chap. 10) Fra Mauro Formation (Schmitt and others, 1967). U-Pb-Th systematics yielded a date of 0.85 ± 0.10 aeon for a thermal event affecting regolith fragments (Silver, 1971). Although U-Pb-Th techniques have proved to be ambiguous in many lunar stratigraphic applications, the age of this event seemed to be substantiated by Ar-Ar determinations of 0.81 ± 0.04 aeon on sample 12033 (Eberhardt and others, 1973b; Alexander and others, 1976). Alexander and others (1977) obtained similar results for another KREEP-rich sample, 12032.

The identification of this approximately 0.8-aeon age with Copernicus has been doubted from several standpoints. Wasson and Baedecker (1972) doubted that primary ejecta would be preserved so far (340 km) from Copernicus. Quaide and others (1971, p. 715)

TABLE 13.2.—*Absolute ages of Copernican rock units*

[Methods: Ar-Ar, the ^{40}Ar-^{39}Ar method, applicable to old rocks (Turner, 1977). Kr-Kr, dating of duration of exposure to the space environment (exposure age) by measuring the ratio of the radioactive isotope ^{81}Kr generated by cosmic-ray spallation to stable krypton isotopes, generally ^{83}Kr; applicable to relatively recent exposures (Marti, 1967; Arvidson and others, 1975). U-Th-Pb, applicable to major re-equilibrations in old rocks; the entire system of U, Th, and Pb isotopes was examined by Silver (1971).
References: A76/77, Alexander and others (1976, 1977); B73, Behrmann and others (1973); B78, Bernatowicz and others (1978); C72, Crozaz and others (1972); D74, Drozd and others (1974); D77, Drozd and others (1977); E73b, Eberhardt and others (1973b); E77, Eugster and others (1977); LM72, Lugmair and Marti (1972); M73, Marti and others (1973); S71, Silver (1971). Summarized by Arvidson and others (1975)]

Unit	Apollo sampling site	Sample	Age (m.y.)	Method	Reference
South Ray Crater	16	Several	2.04 ± 0.08	Kr-Kr	D74
Shorty Crater	17	do.	19	Kr-Kr	E77
Cone Crater	14	14306 (dark)	23.4 ± 1.4	Kr-Kr	C72
		14306 (light)	25.4 ± 2.9	Kr-Kr	C72
		14321,FM 1+2	23.8 ± 0.6	Kr-Kr	LM72
		14321,FMS	27.2 ± 0.5	Kr-Kr	LM72
North Ray Crater	16	Several	48.9 ± 1.7	Kr-Kr	B73, M73
			50.3 ± 0.8	Kr-Kr	D74
Tycho (bright landslide)	17	do.	109 ± 4	Kr-Kr	D77
Copernicus (ray)	12	12033	810 ± 40	Ar-Ar	E73b
			810^{+400}_{-50}	Ar-Ar	A76/77
		Others	850 ± 100	U-Th-Pb	S71
Autolycus or Aristillus (ray)	15	15405,90	$1,290 \pm 40$	Ar-Ar	B78

observed that the Copernicus ray begins at a cluster of secondary craters only 45 km north of the landing site; the samples would thus include material excavated at that point (fig. 13.10). Alexander and others (1977) pointed out that non-KREEP particles yield the same ages as the KREEP-rich samples and questioned the identity of the high-temperature event. Finally, Alexander and others (1976, 1977), though favoring the 0.8-aeon age as the more significant, showed that a plateau around 1.2 to 1.5 aeons is also quite well defined both in their data and in those of Eberhardt and others (1973b).

If the samples are not from Copernicus or if one of these older ages is that of the crater, most current estimates for the cratering rate would have to be revised. Earlier discussions in this volume suggest, however, that abandonment of the 0.8-aeon age would be premature. The argument that primary Copernicus ejecta is not present at the Apollo 12 landing site is based largely on the upper equation in figure 10.26, which is highly questionable (see chap. 10, section entitled "Fra Mauro Formation"). Spectral studies show that primary ejecta is present in rays (see chap. 3; Pieters and others, 1982). Mixtures of diverse shock grades and compositions (KREEP-rich and KREEP-poor) are the rule rather than the exception in impact deposits (see chaps. 3, 8–10). Furthermore, the secondary impacts 45 km north of the site could have reexcavated the Fra Mauro Formation (fig. 13.10) and reset its age (although resetting by secondary impacts is doubted by most investigators). In summary, the presence of 0.8-aeon-old primary Copernicus ejecta or some other influence of Copernicus on the age of the analyzed material is possible, despite the objections.

The best established date of a large Copernican crater is probably the one that appears to rest on the flimsiest evidence. The landslide at the Apollo 17 landing site was probably triggered by the impact of projectiles from crater Tycho, 2,250 km away (fig. 13.11; Wolfe and others, 1975; Arvidson and others, 1976; Lucchitta, 1977a). The duration of the landslide's exposure to cosmic rays and thus its time of formation has been accurately bracketed at 0.1 aeon (Arvidson and others, 1976; Drozd and others, 1977). Similar exposure ages of regolith materials in the central area of the landing site may date excavation by the "Central Cluster" craters (for example, Camelot and Sherlock, figs. 9.17, 11.15), which are thought to be additional Tycho secondaries (Wolfe and others, 1975, 1981; Lucchitta, 1977a). The age of Tycho is, therefore, widely accepted as about 100 million years and probably can be pinpointed to 109 million years (Drozd and others, 1977).

The three best established Copernican ages (Arvidson and others, 1975) were obtained from the exposure ages of small craters (table 13.1). Two young craters at the Apollo 16 landing site, North Ray and South Ray, are 50 million and 2 million years old, respectively. Cone Crater at the Apollo 14 landing site lies between the two, at about 25 million years (fig. 13.12). These absolute age differences are consistent with the scheme of Trask (1969, 1971; Moore and others, 1980a). Several other craters have been dated by exposure ages, including Shorty at the Apollo 17 landing site (19 million years; Eugster and others, 1977) and, less directly, West at the Apollo 11 landing site (100 million years; see chap. 11; Beaty and Albee, 1980).

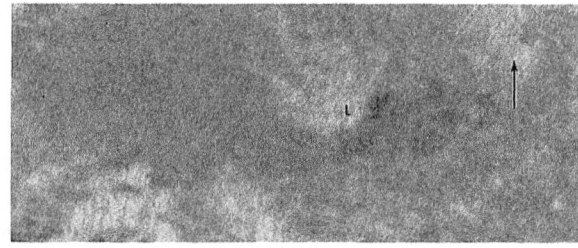

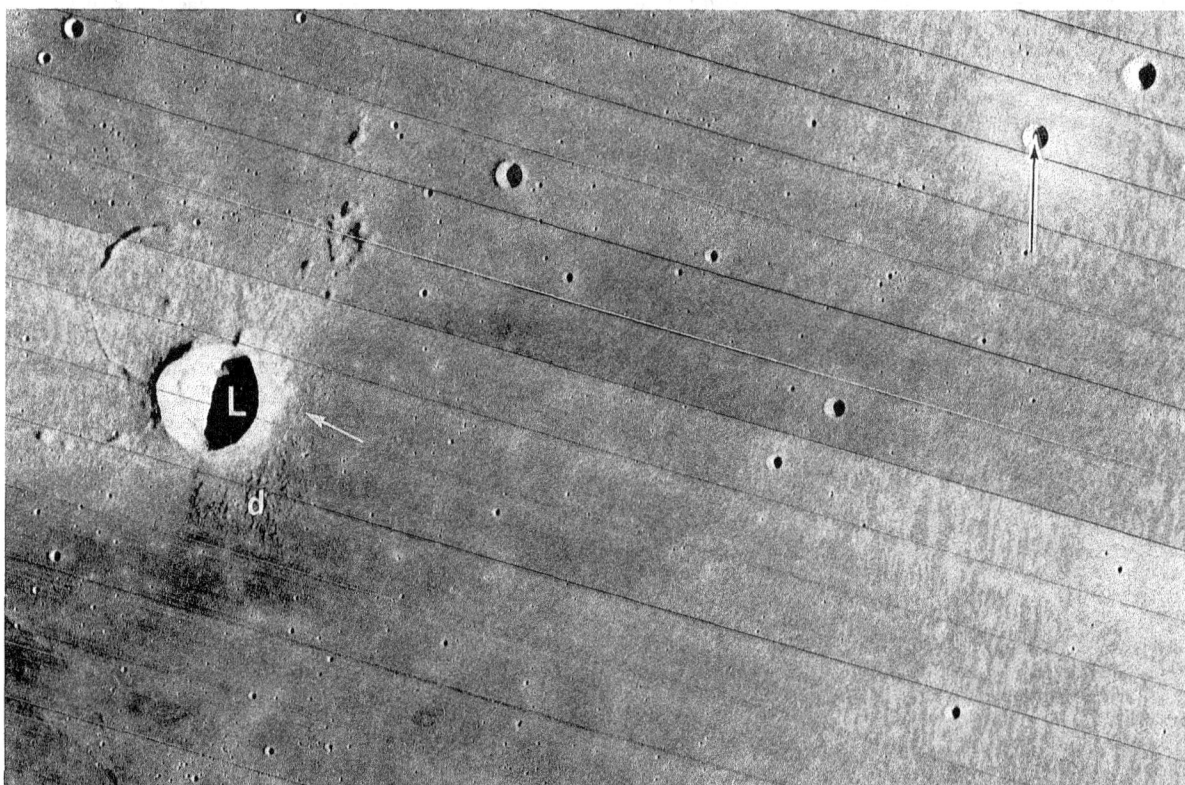

FIGURE 13.7.—Crater Lichtenberg (L; 20 km), northwestern Oceanus Procellarum. Black-and-white north arrow shows same point in A and B.
 A. Telescopic view suggesting sharp truncation of Lichtenberg materials by discrete dark unit.
 B. Lichtenberg ejecta flooded by mare basalt (white arrow) and thinly mantled by dark-mantling material (d). Orbiter 4 frame H-170.

Because no stratigraphic units older than 0.8 or 1.29 aeons and younger than the 3.16-aeon-old Apollo 12 mare units have been absolutely dated even tentatively, the ages of the many stratigraphic units that formed during this long gap in the record must be derived by interpolation based on the cratering-rate curve (fig. 13.13). However, the shape of this curve is poorly constrained because the known ages are so widely separated and the best Copernican ages are too young to affect it significantly.

Independent knowledge of the cratering rate would provide the necessary basis for interpolation. The terrestrial cratering rate is used to calibrate the impact rate in the Earth-Moon system, with corrections for the sampling problem caused by a geologically active Earth and for Earth-Moon differences in gravity and target properties (Baldwin, 1949, 1963, 1964, 1971; Öpik, 1960; Shoemaker and others, 1962a, 1979; Hartmann, 1965b; Grieve and Dence, 1979; Grieve and Robertson, 1979; Basaltic Volcanism Study Project, 1981, chap. 8; Shoemaker, 1981). Some authors believe that the cratering rate has been constant since the Late Imbrian (Hartmann, 1972c; Soderblom and Lebofsky, 1972; Neukum and König, 1976, p. 2881; Guinness and Arvidson, 1977; Young, 1977). The shapes of the lunar and terrestrial curves suggest to others that the rate has declined during the past aeon (Trask, 1972; Soderblom and Boyce, 1972; Neukum and König, 1976, p. 2880; Shoemaker and others, 1979).

A curve corresponding to a constant rate since 3.2 aeons ago and fitted to a point between the Apollo 12 and 15 crater-frequency midpoints (curve a, fig. 13.13A) passes near, but not through, the bar representing the 0.8-aeon age and the frequencies of small craters superposed on Copernicus. The fit would be better if (1) the cratering rate has varied (curve b, fig. 13.13A), (2) the cratering rate was constant but has declined more steeply than shown because the crater

FIGURE 13.8.—Narrow gashes in surface of eastern Mare Serenitatis just west of Apollo 17 landing site. Width on the order of tens of meters and apparent transection of rim of bright blocky crater (arrow) suggest a Copernican age. Gashes probably formed by drainage of regolith into larger bedrock grabens (B.K. Lucchitta, in Masursky and others, 1978, p. 209). Apollo 17 frame P-2313.

C. Detail. Apollo 15 frame P-0370.

FIGURE 13.9.—Sample 15405, the youngest dated lunar rock, from Apennine front (sta. 6A), Apollo 15 landing site.

frequencies determined for the Apollo 12 and 15 basalt units are too small (curve c, fig. 13.13A), (3) the crater-frequency values determined for Copernicus are too large, or (4) Copernicus is about 1.2 aeons old (point a′, fig. 13.13A), one of the possibilities mentioned by Alexander and others (1976, 1977). The Copernicus and Apollo 12 and 15 frequencies could be reconciled if substrate properties affect the size-frequency curves, as has been suggested. For example, if the frequency of small craters (normalized to min 1 km diam) superposed on Copernicus were about 6.5×10^{-4} per square kilometer, Copernicus would fall on the constant-rate curve, which applies to mare substrates, without any adjustment of the absolute age (point a″, fig. 13.13A). The D_L values could be reconciled in similar ways (fig. 13.13B). Because of the many uncertainties in both the absolute and relative ages of Copernicus, the currently available data appear to be equally consistent with either a constant or a varying cratering rate since 3.2 aeons ago. I tentatively accept the simpler model, the constant rate.

Granted a constant rate, wide latitude still remains in assessing the age of the Eratosthenian-Copernican boundary because of the gap in the record and the uncertain significance of the crater frequencies and D_L values. Because most assessments of the stratigraphic significance of the crater frequencies and D_L values given in this volume are based on previous calibrations with the large craters traditionally mapped as Eratosthenian or Copernican, I return to this basis for estimating the duration of the two periods. If the craters larger than 30 km in diameter mapped here (pls. 10, 11) formed at a constant rate since 3.2 aeons ago, formation of the 88 Eratosthenian and 44 Copernican craters would require 2.13 and 1.07 aeons, respectively. Similarly, the 256 Eratosthenian and 145 Copernican craters larger than 10 km in diameter mapped on the nearside by Wilhelms and McCauley (1971) (see chap. 12) would require 2.04 and 1.16 aeons, respectively. Averages of 2.1 and 1.1 aeons are tentatively adopted here for the Eratosthenian and Copernican Periods.

In summary, a 1.1-aeon duration for the Copernican Period is consistent with: (1) a constant impact rate for primary craters of all sizes since the beginning of the Eratosthenian Period 3.2 aeons ago, between the emplacement of the Apollo 12 and 15 basalt units; (2) existing age assignments of craters based on the criteria of stratigraphic superpositions, superposed craters, rays, morphology, and remote sensing; (3) a 0.8-aeon age for Copernicus; and (4) a position of Copernicus somewhat above the base of the Copernican System, in accord with stratigraphic estimates (fig. 7.1; table 7.2). Future data may demonstrate that the Copernican Period was longer or shorter than 1.1 aeons.

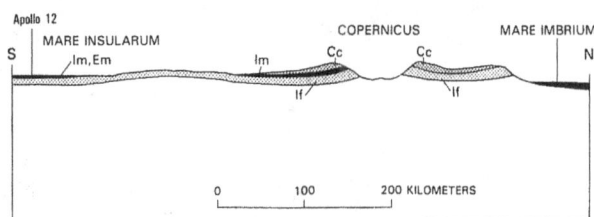

FIGURE 13.10.—North-south strip from crater Copernicus to Apollo 12 landing site.
A. Arrow indicates Apollo 12 landing site and points along ray toward Copernicus (C). Other craters include Reinhold (R; 43 km, 3° N., 23° W., probably Eratosthenian), and Lansberg (L; 39 km, 0°, 27° W., Upper Imbrian). Montes Carpatus (MC), part of Imbrium basin rim, divide Maria Imbrium (above) and Insularum (below). Orbiter 4 frame H-126.

B. Geologic cross section drawn along length of A from base of arrow to upper right corner. Units, from oldest to youngest: If, Fra Mauro Formation; Im, Imbrian mare basalt; Em, Eratosthenian mare basalt; Cc, Copernicus ejecta. Fra Mauro Formation composes submare section from Montes Carpatus to Apollo 12 landing site and thus would be excavated by postmare craters, including Copernicus and its secondaries, anywhere along this profile.

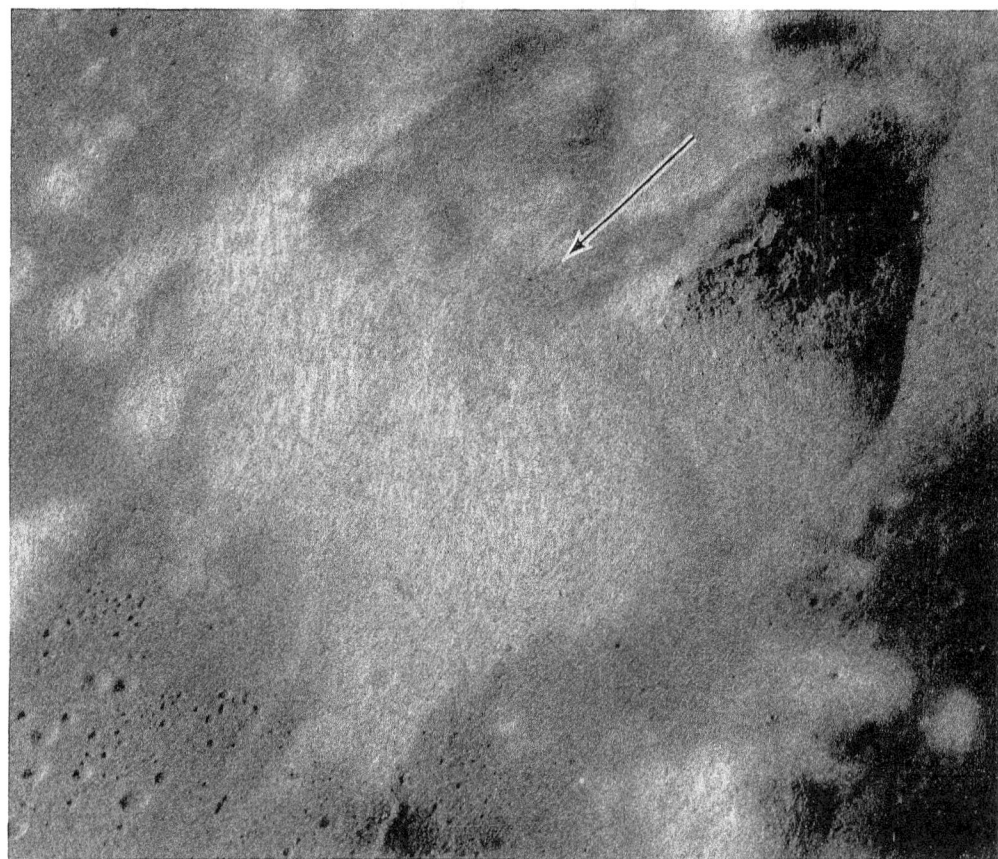

FIGURE 13.11.—Landslide (light-colored area at bottom) triggered from slope of South Massif, bordering the Taurus-Littrow Valley, by impact of projectiles from young Copernican crater Tycho, 2,200 km to southwest; arrow indicates probable Tycho secondaries (Lucchitta, 1977a). Massif is about 2,000 m high. South at top. Apollo 15 frame P-9297.

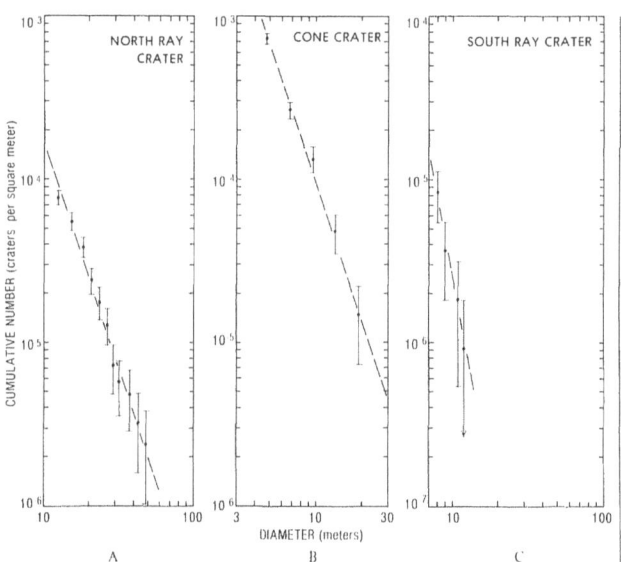

FIGURE 13.12.—Size-frequency distributions of very small craters superposed on the three sampled and absolutely dated Copernican craters, North and South Ray Craters at Apollo 16 landing site and Cone Crater at Apollo 14 landing site (Moore and others, 1980a).

FIGURE 13.13.—Alternative cratering rates since 3.2 aeons ago. Eratosthenian-Copernican boundary is estimated from numbers of large Eratosthenian and Copernican craters, assuming a constant cratering rate.

A. Based on frequencies of craters at least 1 km in diameter. Copernicus and Tycho frequencies from fig. 12.4, absolute ages from table 13.2. Mare-basalt frequencies from table 11.1, absolute ages from tables 11.3 and 12.4. Curve a, constant cratering rate since 3.2 aeons ago, based on crater frequencies on Apollo 12 and 15 mare units. Point a′, intersection with 1.2-aeon age of Copernicus; point a″, intersection with 0.81-aeon age of Copernicus (Eberhardt and others, 1973b; Alexander and others, 1976). Curve b, varying cratering rate since 3.2 aeons ago, passing through midpoint of Copernicus age. Curve c, constant cratering rate since 3.2 aeons ago, passing through midpoint of Copernicus age.

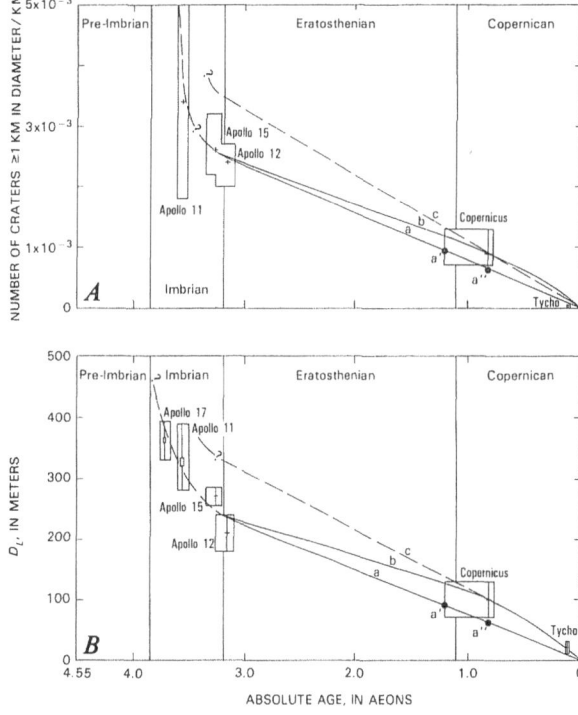

B. Based on D_L values. Copernicus and Tycho values from table 12.3 and Moore and others (1980b); mare-basalt values from table 11.1. Absolute ages as in A. Curve a, constant cratering rate since 3.2 aeons ago, based on average of Apollo 12 and 15 D_L values. Curve b, varying cratering rate since 3.2 aeons ago, drawn through midpoint of Copernicus age. Curve c, one of many possible curves representing D_L values on crater substrates (compare fig. 12.8).

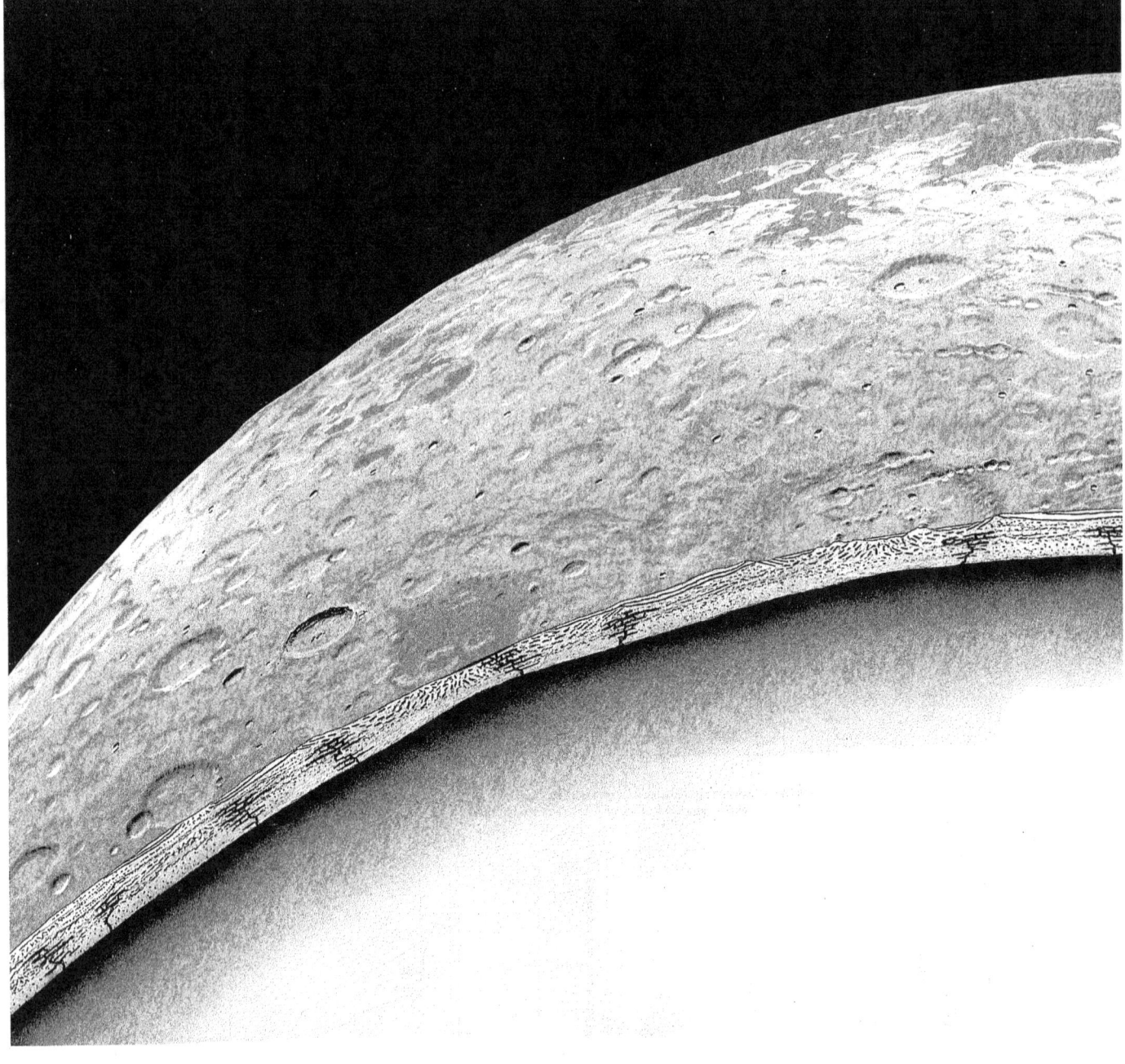

FIGURE 14.1.—Representative area of lunar surface (composite of several regions) and inferred subsurface structure. Mare basalt, gray in plan view and black in cross section, overlies impact melt in basins. Basin and crater ejecta blankets (white in cross section) overlap in several places, in accord with superposed-crater densities and degradational morphologies of surface exposures. Deformation of crustal material beneath basin rings is shown in accord with model in chapter 4; crust is thin, and mantle correspondingly uplifted, beneath basins. Stippling denotes possible unstratified lower crust that may have been reached only by the very largest impacts. True curvature; no vertical exaggeration. Painting by Donald E. Davis, courtesy of the artist.

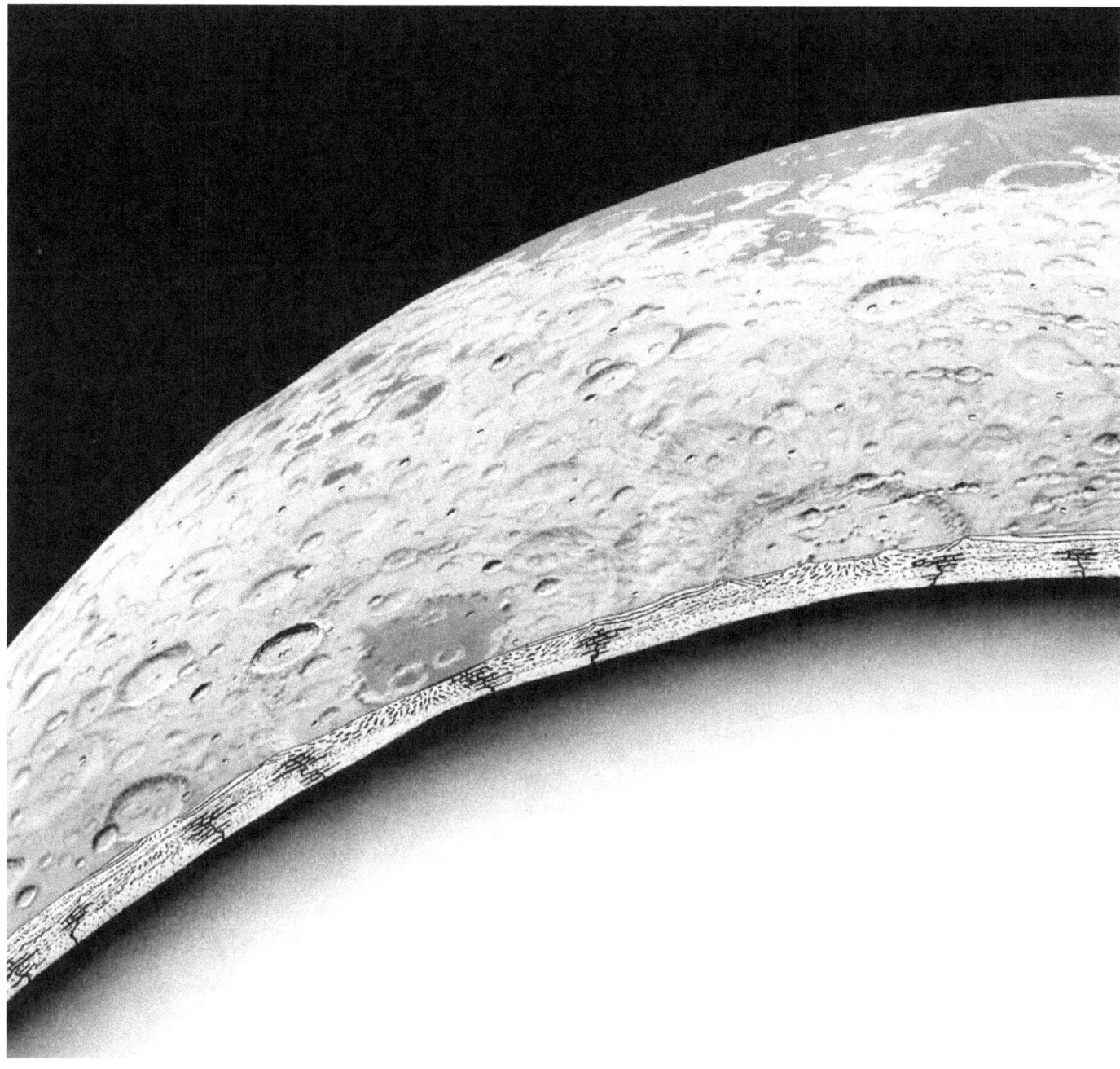

FIGURE 14.1. — Representative area of lunar surface (composite of several regions) and inferred subsurface structure. Mare basalt, gray in plan view and black in cross section, overlies impact melt in basins. Basin and crater ejecta blankets (white in cross section) overlap in several places, in accord with superposed-crater densities and degradational morphologies of surface exposures. Deformation of crustal material beneath basin rings is shown in accord with model in chapter 4; crust is thin, and mantle correspondingly uplifted, beneath basins. Stippling denotes possible unstratified lower crust that may have been reached only by the very largest impacts. True curvature; no vertical exaggeration. Painting by Donald E. Davis, courtesy of the artist.

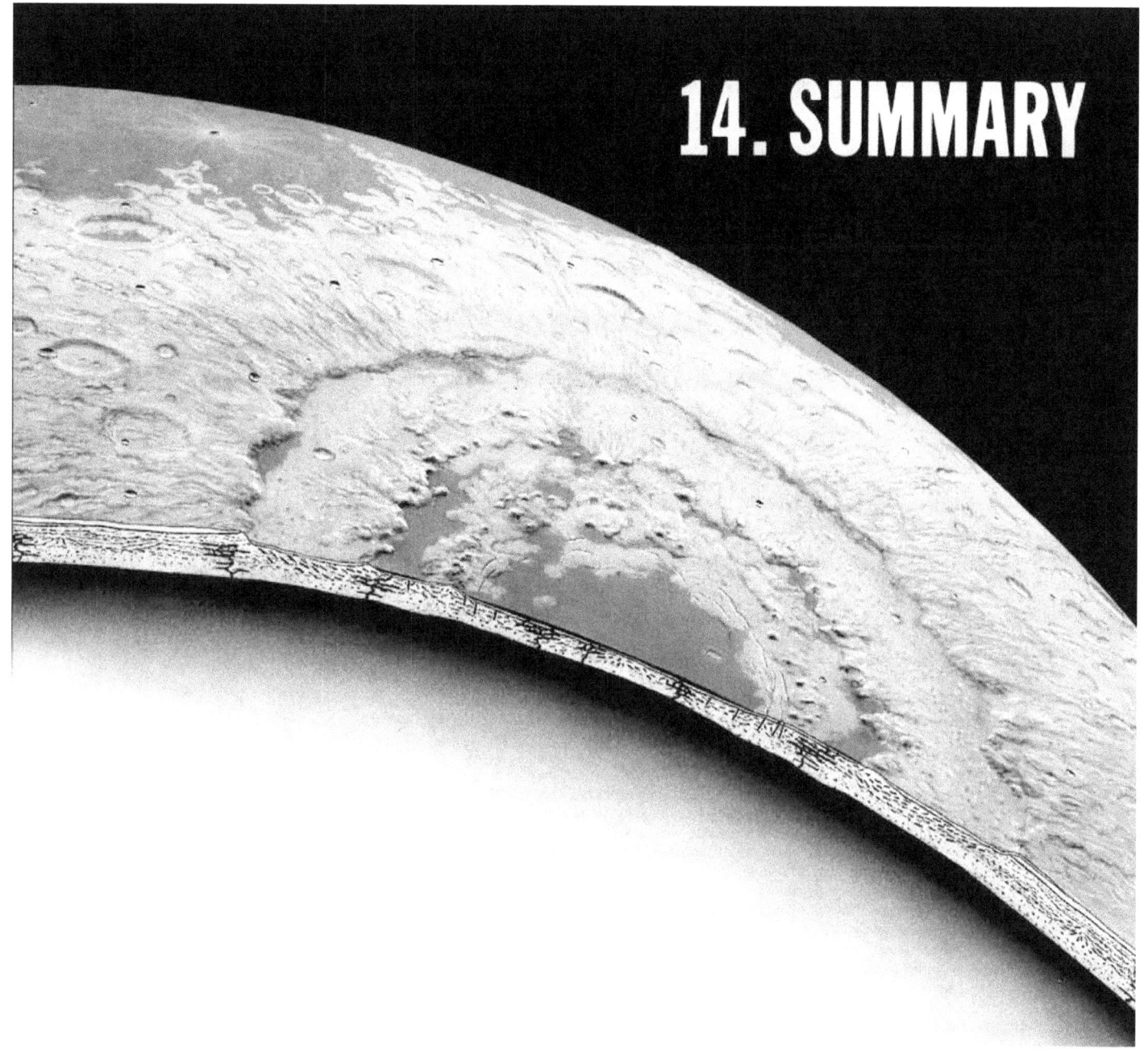

14. SUMMARY

CONTENTS

	Page
Geologic style of the Moon	276
Pre-Nectarian period	277
Nectarian Period	278
Early Imbrian Epoch	279
Late Imbrian Epoch	279
Eratosthenian Period	279
Copernican Period	280
Remaining problems	280

GEOLOGIC STYLE OF THE MOON

Two decades of study have shown that two major processes, impact and basaltic volcanism, have shaped the major physical features of the present lunar crust. Impact deposits and basalt presumably also constitute much of the subsurface. Tectonism has modified the original depositional geometry far less than on the Earth.

A typical section of the upper lunar crust consists of laterally continuous, interfingering beds of basin and crater materials that are pierced, deformed, and redistributed by later impacts (fig. 14.1). Deposits of ringed impact basins are the main constituents of this section. Crustal material originally created by extensive plutonism was redistributed far and wide over the surface, partly preserved and partly newly modified, by the great impacts. The upper crust was removed as far as each topographic basin rim. Ejected material, emplaced largely by turbulent surface flow, thickly mantled the surface outside the topographic basin rims for average distances of one basin radius. Beyond this distance, secondary-impact craters, as large as 2 to 4 percent of the basin size, were formed in abundance. Their ejecta incorporated some of the primary basin ejecta that formed them and greater amounts of the circumbasin terrain that they excavated. The primary-ejecta deposits contain much melted and incompletely melted target material, mixed with less highly shocked debris, whereas the secondary deposits contain little or no newly melted material except that from the primary ejecta which formed them. Plains deposits containing the same breccia types as other primary and secondary ejecta are concentrated in the primary/secondary transition zone and among the secondary craters.

Basin interiors contain rings uplifted by deformation of the basin floor, knobby deposits ejected late in the excavation sequence, and impact-melt rocks. The rings increase in spacing, number, and complexity with increasing basin size. The outermost zones of large basins' interiors are especially complex. The knobby deposits also extend locally beyond the rim.

Smaller primary impacts have disturbed smaller amounts of the crust and redistributed it as lesser strata and fragmental regoliths. These ejecta deposits are proportional to those of basins in average extent; ejecta both of basins and of these smaller craters commonly have asymmetric, lobate or bow-tie map patterns. The secondary craters of basins and craters have similar size-frequency distributions, and the secondary-to-primary size ratio is only slightly smaller for the secondaries of basins. Primary-impact craters larger than 16 to 21 km in diameter have central peaks consisting of severely deformed material derived during crater excavation from beneath the crater center. Unlike the chaotic ejecta of larger craters and basins, ejecta of small simple craters may preserve the bedding of the target, inverted from the original sequence.

Basaltic volcanism probably has altered this basic impact architecture throughout lunar history. Extrusions of Fe-rich, alkali-poor, dry, and reduced basaltic magmas generated in the mantle not only formed the visible maria but also were probably extruded earlier. Relatively silicic basalt, rich in trace elements characteristic of KREEP and originating in the crust, also may have contributed surface extrusions. Basaltic magmas that did not reach the surface must have solidified as extensive dikes and sills in the crust (fig. 14.1). Most lunar basalt has formed planar lava flows, some with lobate frontal scarps; minor eruptions have constructed positive landforms, such as domes and cones.

The position and volume of the extrusions were determined by the crustal and mantle structure established by the basins. The mantle rose beneath each basin and, probably, beneath each large crater to compensate for loss of the crustal overburden. Thus, the crust-mantle interface probably resembles a series of domelike swells, proportional in size to the basins and superposed where basins are superposed. Far more basalt was extruded into basin depressions through the thin crusts that cap the mantle uplifts than through the thick crusts outside the large basins. Except for the Apollo 11 suite from southern Mare Tranquillitatis (fig. 14.2), each suite of basalt samples returned by the Apollo and Luna missions apparently was formed by magmas extruded within a short time from a small, compositionally heterogeneous zone of the mantle.

Pyroclastic glass fountained as liquid droplets from fissures and irregular craters at the mare margins. These pyroclastic deposits are now observed as dark mantles on the nearby terra and mare lavas,

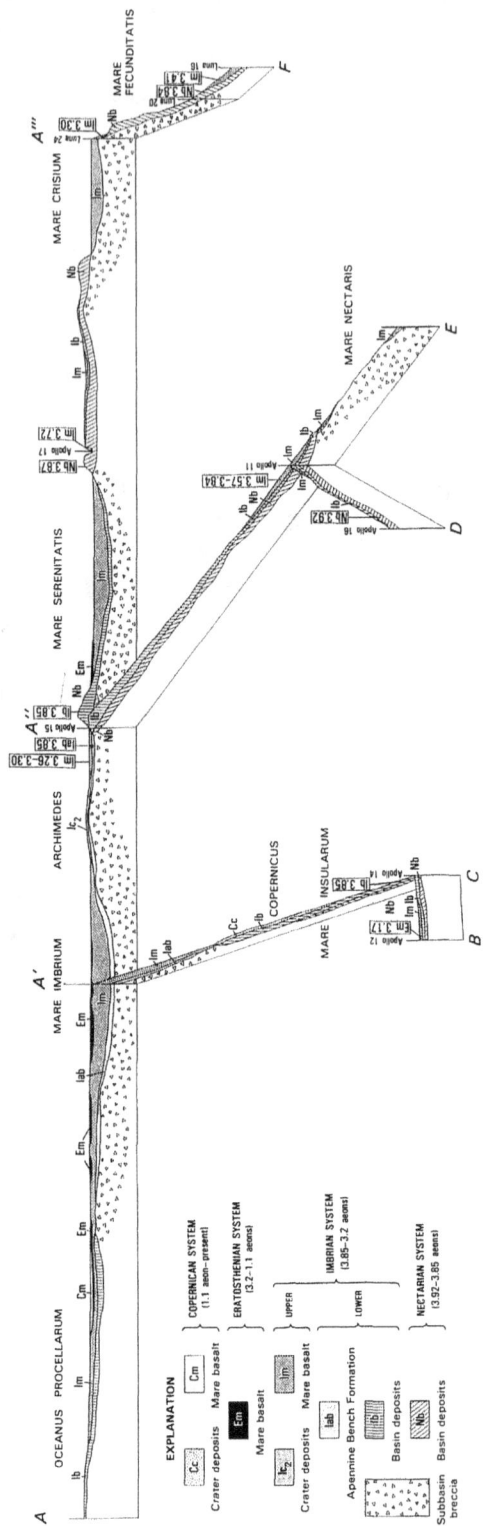

FIGURE 14.2.—Geologic cross section, drawn along lines indicated in figure 8.4 and based on plate 12. Boxes enclose symbols and ages (in aeons) of major sampled and dated geologic units. Brecciated crustal material is shown beneath basins whose deposits are mapped; rest of crust is probably also brecciated. Plains material (Ip, pl. 12) intersected here is Apennine Bench Formation, interpreted as extending over most of floor of Imbrium basin. Other basins probably have similar plains deposits on floors (impact melt and, possibly, volcanic basalt).

and probably also interfinger with the lavas. The total volume of extruded mare basalt and pyroclastic material amounts to only 0.05 to 0.2 percent of the upper 75 km of the Moon.

Most lunar faulting was caused by the subsidence of mare basalt, which was greatest where basin excavations had thinned and weakened the elastic lithosphere. The lithosphere was probably equivalent to the petrologic terra crust during most of the volcanic and tectonic activity. Arcuate grabens were opened by stretching of the margins of the subsiding basalt sections. Mare surfaces were folded into complex archlike and spinelike ridges where the surface area was shortened; the shortening apparently was greatest where basalt sections of unequal thickness settled differentially. The degree of subsidence is also recorded in the gravity structure of maria; maria and basins that have been fully compensated isostatically are gravitationally neutral, whereas others appear as mascons that represent enduring superisostatic loads of basalt on the lithosphere. Crater floors were uplifted and fractured most extensively where the lithosphere was weakest. Otherwise, internal activity contributed very little to crater development except by generating the mare fillings of many craters and creating small vents for the pyroclastic eruptions.

In summary, geologic history at any given point on the Moon has advanced in a series of catastrophic impacts separated by much longer periods of gradual degradation and intermittent volcanism. Regoliths accumulated on each of the larger deposits in proportion to the duration of exposure and the impact flux at the time. Innumerable small impacts blurred depositional textures unremittingly, but subsurface geometry persisted until disturbed by large impacts. The ejecta from large impacts incorporated and mixed materials from the crater and basin units, the regoliths, the volcanic lavas, and subsurface intrusions. Lunar geology is simple in general style, though commonly very complex in detail.

This account pictures a uniformitarian Moon in which old features once resembled the more recent. Since crustal solidification, the face of the Moon has always displayed craters and ringed basins, and probably has always been spotted by dark maria. Major changes have occurred, however, in the proportions of units. The details of mare distribution constantly shifted as new basins appeared, were flooded, and were covered by later ejecta. Because volcanism continued after the impact rate had decreased sharply, more basalt remained exposed late than early in lunar history (fig. 14.3). The rest of this chapter traces the changing scene from pre-Nectarian to Copernican time and concludes with a summary of unsolved problems.

PRE-NECTARIAN PERIOD

Lunar and pre-Nectarian history began with the formation of the Moon about 4.55 aeons ago. Similar oxygen-isotope ratios of the Earth and the Moon favor accretion in the same general part of the Solar System (Clayton and Mayeda, 1975). The process of accretion has not been agreed upon (Wood, 1979; Cadogan, 1981; Glass, 1982; Taylor, 1982; Hartmann, 1983), despite an intense search for chemical clues in the lunar samples (Dreibus and others, 1977; Kaula, 1977; Anders, 1978; Delano and Ringwood, 1978; Wänke and others, 1978). The accreted material is commonly assumed to have been chemically similar to that of chondritic meteorites, depleted in volatile elements.

The newly accreted material thoroughly differentiated. A feldspathic crust, now averaging about 75 km in thickness, and the upper several hundred kilometers of a denser, ultramafic, much more voluminous mantle are believed to have separated from an extensive early pre-Nectarian magma system. This magma is commonly considered to have formed a global ocean. However, recent modeling of crustal petrogenesis envisions a succession of bodies, which may or may not have been emplaced into the crystallization products of a primordial magma ocean. A chemically distinct core may or may not exist (Levin and Mayeva, 1977; Wiskerchen and Sonett, 1977; Taylor, 1982). A barrage of impacts must have deposited impact-melt pools, heated and mixed the environment of the crustal magmas to considerable depths, and generally influenced the igneous petrogenesis and the sites of magma migration. The differentiation ended at a time commonly estimated at from 4.4 to 4.2 aeons ago; it may have ended at different times in different places.

The cumulative effect of a heavy impact rate during pre-Nectarian time is evident in the highly brecciated, impact-melted, siderophile-rich material that constitutes the Nectarian and Imbrian deposits from which the returned samples were collected. The sample record of individual pre-Nectarian events that affected the solid terra crust, however, is less complete than the photogeologic record.

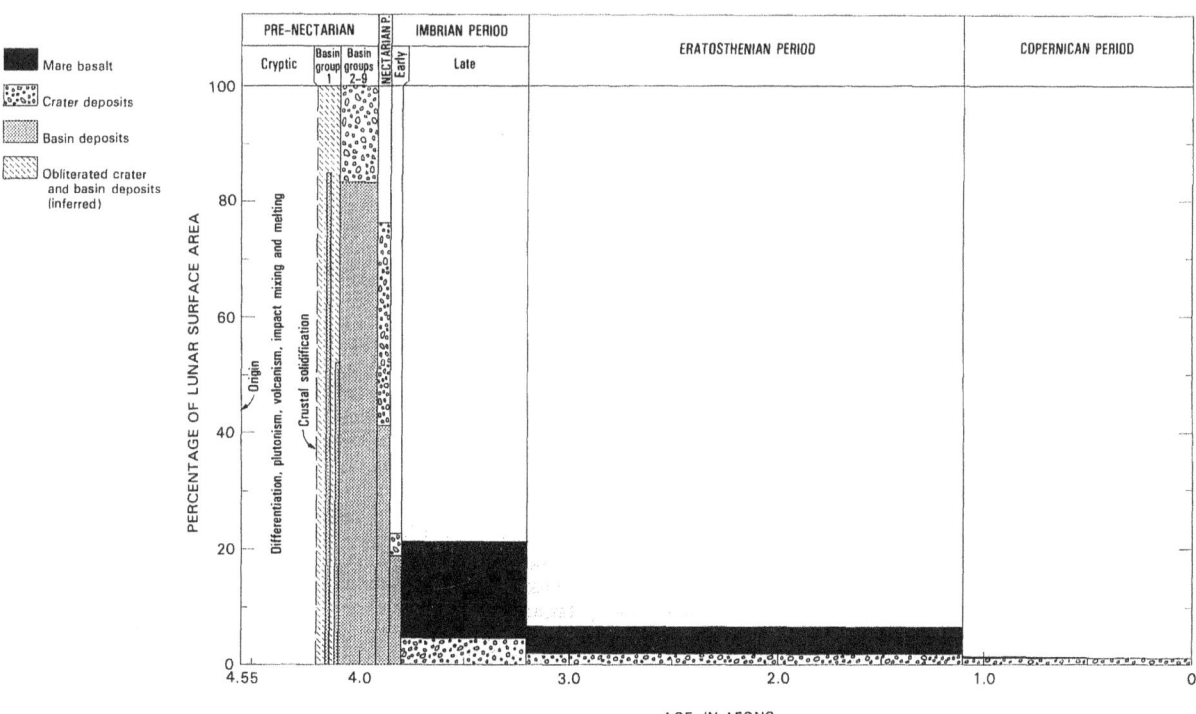

FIGURE 14.3.—Evolution of the lunar surface. Vertical axis, surface area inferred to have been originally covered by each group of deposits. Basin and crater deposits (approximately, the continuous ejecta) are assumed to extend one diameter from the crater or basin center; mare deposits of each age are assumed to extend beneath entire area of younger maria. Procellarum and South Pole-Aitken basins (pre-Nectarian basin group 1) are shown diagrammatically by narrow vertical bars; ages are estimated. Observed and inferred pre-Nectarian crater deposits cover more than 100 percent of the Moon. Early peak in lunar geologic activity is evident.

Mountainous rings of the giant Procellarum and South Pole-Aitken basins apparently constitute the earliest preserved record. Together, these two basins excavated crustal material from 40 percent of the Moon's area, and this material must have covered or impacted most of the Moon. This redistribution of material apparently resulted in terra surfaces more magnesian than average inside the basins and more aluminous outside; furthermore, KREEP-rich material of the deep crust may have been exposed in the deepest parts of Procellarum or of both basins. These newly exposed compositional provinces later became the targets of other basin impacts. The mantle does not seem to have been reached by impacts except, possibly, by the combined excavations of Procellarum and Imbrium. The crustal structure resulting from the giant impacts also set the stage for later mare volcanism and tectonism by determining local thicknesses of the lithosphere and the depths to the mare source regions in the underlying mantle.

Thirty pre-Nectarian basins, including Procellarum and South Pole-Aitken, have been identified or tentatively identified (table 8.2). They are ranked according to crater-density and superposition data into nine age groups. The oldest basins are obscure, whereas the youngest display textures and secondary craters similar to those of Nectarian and Imbrian basins. The spatial distribution of the pre-Nectarian basins seems to be random (pl. 6).

The ages of these basins and of the oldest preserved surface features are only approximately known because few ancient units have been dated absolutely and because extrapolations of frequencies of craters that are saturated to large crater diameters are uncertain. If the impact rate was constant during the pre-Nectarian and Nectarian Periods and if the Nectarian Period lasted 0.07 aeon, about 28 basins (groups 2–9) and 3,400 craters larger than 30 km in diameter formed from about 4.1 aeons to 3.92 aeons ago. This estimate is too young if the Nectarian Period was longer than 0.07 aeon, and too old if the early impact rate declined steeply as is, in fact, commonly assumed. The Procellarum and South Pole-Aitken basins are older than the group-2 basins by an unknown amount; to construct figure 14.3, I assume ages of 4.15 and 4.10 aeons, respectively, for these two giant basins. Many additional craters and basins that are now obscured also formed between the time of crustal solidification and the first group-2 basins.

Although volcanic materials are probably interbedded with the impact deposits, the pre-Nectarian volcanic record is much less evident than the rich impact record. The absence of observed tectonic deformation of pre-Nectarian units suggests that the crust had cooled before the oldest observed basins formed. The lithosphere presumably thickened throughout the pre-Nectarian.

NECTARIAN PERIOD

The Nectarian System continues the geologic style of the visually detectable part of the pre-Nectarian, but with fewer units (pl. 7). By definition, the Nectarian Period began with the impact that created the Nectaris basin and ended with the Imbrium-basin impact 3.85 aeons ago. Geochronology and photogeology have suggested, but not definitely established, that feldspathic, melt-poor Nectaris-basin material was sampled and dated at the Apollo 16 landing site (fig. 14.2). The age of Nectaris is tentatively estimated at 3.92 aeons on the basis of one interpretation of the ages determined on small samples of this material (table 14.1). If correct, this age implies that the Nectarian Period lasted 0.07 aeon and the pre-Nectarian 0.63 aeon.

From 10 to 12 basins formed during the Nectarian Period; the doubtful ones are Mendel-Rydberg, which may be pre-Nectarian, and Sikorsky-Rittenhouse, which may not actually be a basin. The Nectarian basins are ranked individually by age, but only two distinct age groups are recognized (table 9.4). If 11 basins are Nectarian and the Nectarian Period lasted 0.07 aeon, these basins formed at the rate of 157 per aeon. An extrapolation from the observed number of craters to the whole Moon indicates that 1,330 craters from 30 to 300 km in diameter formed during the Nectarian Period. Their formation rate was 19,000 per aeon, and 121 craters formed for each Nectarian basin. These Nectarian impact rates and crater-to-basin ratio are the basis for my estimates of pre-Nectarian ages and areal coverage (fig. 14.3).

At least one, possibly two, Nectarian basins in addition to Nectaris were sampled (fig. 14.2). Abundant fragment-laden impact melt was collected by Apollo 17 from massifs near the rim of the Serenitatis basin. Isotopically requilibrated clasts have been dated at 3.87 aeons. Although some doubts persist about the provenance of the best dated material, 3.87 aeons is probably the age of Serenitatis (table 14.1). Luna 20 returned small feldspathic fragments of probable Crisium-basin material. Their determined age is a little younger than that of Serenitatis but not so precise.

These young isotopic ages, a concentration of other isotopic ages of terra samples between 3.85 and 3.95 aeons, the relative paucity of older ages, and the apparently old morphologic age of Serenitatis led to the suggestion that a "terminal impact cataclysm" formed most lunar basins. However, the large number of identified and inferred pre-Nectarian and Nectarian basins would create the same effects and would intensely rework the lunar crustal material, as is observed. Although the record does not require a terminal cataclysm, basins may have formed in groups because clusters of projectiles derived by breakup of a single large object temporarily maintained similar orbits after breakup (Wetherill, 1981); these projectiles may have impacted approximately along lunar paleoequators (Runcorn, 1982).

An undetermined amount of volcanism occurred during the Nectarian and pre-Nectarian Periods. If the Nectarian surface was covered by mare basalt at the same average rate as was the Late Imbrian surface (approx 6.4×10^6 km² in 0.6 aeon), 0.75×10^6 km² of the lunar surface was covered from 3.92 to 3.85 aeons ago. This is enough basalt to cover a circular area about 295 km in diameter—slightly smaller than the smallest basin—in each of the 11 Nectarian basins. Large basins and all basins inside the giant basins would receive more than this average, other basins less. Extrapolation of this rate to the pre-Nectarian depends on the source of heat for the volcanism. If most of it was leftover primordial heat (Hubbard and Minear, 1975), volcanism declined steadily since the crust solidified, and the pre-Nectarian basins would have been filled somewhat more extensively than the Nectarian basins. Failure to detect abundant pre-Nectarian volcanic geologic units and samples would then be explainable by the intensity of the early impact bombardment. If, alternatively, most of the heat came from radioactivity, premare volcanism may have been relatively uncommon (Taylor, 1982, p. 317), and the pre-Nectarian basins would contain relatively little pre-Nectarian basalt. The data are insufficient to choose definitively between these possible early rates of volcanism. Sparse data from mare-basalt clasts found in breccia deposits suggest that more Nectarian than pre-Nectarian

TABLE 14.1.—*Adopted absolute ages of major sampled lunar geologic units*

[Sampling sites: L, Luna; numbers only, Apollo]

System or series	Unit or rock type	Sampling site	Age (aeons)	Basis	Reference
Copernican	Small craters	14, 16, 17	≤0.05	Exposure ages	Table 13.2.
	Tycho	17	.109	do	Do.
	Copernicus	12	.81	Ar-Ar ages	Do.
Eratosthenian	Basalt flows	12	3.16	Average of best Rb-Sr and Ar-Ar ages for all flows.	Table 12.4.
Upper Imbrian	Olivine basalt	15	3.26	do	Table 11.3.
	Pigeonite basalt	15	3.30	do	Do.
	Pyroclastic glass	15	3.30	Ar-Ar ages	Do.
	VLT basalt	L24	3.30	Ar-Ar age	Do.
	Feldspathic basalt	L16	3.40	do	Do.
	High-K, high-Ti basalt	11	3.57	Rb-Sr and Sm-Nd ages.	Do.
	Pyroclastic glass	17	3.64	Ar-Ar ages	Do.
	Low-K, high-Ti basalt	11	3.67 3.70	Averages of best Ar-Ar ages for two flows.	Do.
	Thick basalt section	17	3.72	Average of best Rb-Sr and Ar-Ar ages.	Do.
	Low-K, high-Ti basalt	11	3.79	Average of best Ar-Ar ages.	---
Lower Imbrian	Low-K, high-Ti basalt	11	3.84	Average of best Ar-Ar ages.	Table 11.3.
	Apennine Bench Formation (KREEP-rich basalt).	15	3.85	Average of best Rb-Sr, Sm-Nd, and U-Th-Pb ages.	Table 10.1.
	Imbrium-basin ejecta	14, 15	3.85	Young cluster (Apollo 14) and black-and-white breccia (Apollo 15).	Do.
Nectarian	Crisium basin	L20	3.84±0.04	Ar-Ar age	Chapter 9.
	Serenitatis basin	17	3.87	Youngest Ar-Ar ages.	Table 9.2
	Nectaris basin	16	3.92	Average age of feldspathic fragment-laden impact-melt rock from North Ray Crater.	Table 9.1.
Nectarian and pre-Nectarian.	Volcanic-rock clasts	14, 17	3.85–3.98	Individual ages	Table 9.5.
Pre-Nectarian	Plutonic-rock clasts	15, 16, 17	4.17–4.54	Individual ages	Table 8.4.

basalt was formed (table 9.5). Furthermore, more craters with dark, Mg-rich ejecta, indicating burial of basalt flows by basin debris, are superposed on light-colored planar deposits of Nectarian than of pre-Nectarian age. The buried basalt, however, may be pre-Nectarian.

EARLY IMBRIAN EPOCH

During the Early Imbrian, two large impacts left effects on the terra that still dominate its appearance (pl. 8). The creation of Imbrium, the Moon's third largest known basin (after Procellarum and South Pole-Aitken), uplifted rings that are the largest lunar mountains after those of South Pole-Aitken, and launched ejecta over most of the nearside and part of the farside. Thick primary ejecta was deposited within several hundred kilometers of the basin, and some probably reached more than 1,000 km from the rim. Extensive light-colored plains consisting of primary and secondary ejecta were deposited beyond the primary ejecta. Secondary craters formed over vast outlying areas, possibly even at the antipode of the impact point. The KREEP-rich breccias recovered from the Fra Mauro Formation at the Apollo 14 landing site and lesser amounts of breccia from Montes Apenninus at the Apollo 15 landing site yield an average age of about 3.85 aeons. The same age was obtained for fragments of KREEP-rich volcanic basalt or impact melt probably derived from a planar deposit near the Apollo 15 landing site (Apennine Bench Formation), which is contemporaneous with or slightly younger than the basin (tables 10.1, 14.1). The Imbrian Period, therefore, began 3.85 aeons ago. The Imbrium basin was the site for the most voluminous preserved extrusions of lunar mare basalt.

The other large Early Imbrian basin, Orientale, formed about 50 million years later on the west limb of the Moon and ended the era of enormous impacts. In accord with the younger relative age of the Orientale basin and its distance from later maria, its effects on the terra are even more striking than those of Imbrium. The Orientale ridged ejecta, impact melt, ejecta plains, secondary craters, and ring structure serve as models for the interpretation of older basin materials. A similar role at the small end of the size series of impact basins is played by the third Lower Imbrian basin, Schrödinger.

The Early Imbrian Epoch was too brief, and the effects of Orientale and of later mare extrusions too severe, for smaller impacts or volcanism to have left a conspicuous record. The oldest high-K, high-Ti flow excavated from the subsurface at the Apollo 11 landing site may be Early Imbrian (fig. 14.2; table 14.1). Orientale formed on a thick terra crust not previously thinned by a giant basin, and so, unlike Imbrium, it was not the site of abundant later basalt extrusions. Volcanism of terra composition may have formed some terralike plains and domical landforms during the Early Imbrian Epoch.

LATE IMBRIAN EPOCH

The record of the rest of lunar geologic history differs greatly from that of the 0.75 aeon already summarized. The cessation of giant impacts enabled the mare-basalt flows, which were continuously extruded, finally to remain preserved on the lunar surface. Previously, they had been catastrophically excavated and redistributed along with terra materials of the target or, if not disrupted, at least were blanketed by impact ejecta so as to loose their distinctive surface albedo. More visible mare units are Late Imbrian than any other age. Basalt of this age constitutes parts of all the larger maria (pl. 9).

Late Imbrian craters are also numerous and are surrounded by moderately well preserved ejecta blankets (pl. 9). Late Imbrian basalt is the Moon's most abundantly sampled and most reliably dated material (fig. 14.2; tables 11.3, 14.1). Three age groups are mapped here (pl. 9).

Units of the oldest group, which formed within about 0.05 aeon, occupy elevated terrain in the outer Procellarum-basin shelf and terrain outside Procellarum, localities where they were protected from later volcanism. Their subsurface extent is probably much greater. Samples 3.79 aeons old were obtained from a buried unit at the Apollo 11 landing site. Units of the intermediate-age group, which span the interval from about 3.75 to 3.50 aeons ago, lie at the surface in all maria and are especially abundant outside Procellarum. They probably formed in all basins but remained uncovered where a thick lithosphere blocked ascent of magmas and prevented sufficient basin subsidence to allow subsequent basalt extrusion. This and the older group of mare units commonly are faulted inside Procellarum but rarely outside, a further indication of relative subsidence. Apollos 11 and 17 collected Ti-rich basalt samples of this group 3.57 to 3.70 aeons old and 3.72 aeons old, respectively. The sampled flows were extruded intermittently and sparsely in the shallow pre-Nectarian Tranquillitatis basin but much more rapidly in the Taurus-Littrow Valley marginal to the Nectarian Serenitatis basin. Pyroclastic dark-mantling materials fountained from the marginal zones of Serenitatis and other basins while the intermediate-age and, probably, the older groups of basalt units were forming; samples of such pyroclastic material collected by Apollo 17 are 3.64 aeons old.

Addition of the third and youngest group of mare basalt units further depressed the older units and the underlying basin, further fracturing the older mare and the adjacent basin material. These young basalt units appear in all maria, most abundantly in those whose total basalt section is thick because of a thin lithosphere. Units of this young group cover as much area as those of the other two groups combined, but are probably not proportionally voluminous because the older units underlie many or all of their exposures. If the rate of volcanism declined continuously during the Late Imbrian Epoch, the young group of basalt units is only one-seventh as voluminous as the intermediate-age group. Al-rich mare-basalt extrusions outside Procellarum are represented by fragments collected by Luna 16 from Mare Fecunditatis (3.40 aeons old) and very-low-Ti fragments collected by Luna 24 from Mare Crisium (3.30 aeons old). Abundant Ti-poor basalt samples from an older, relatively silicic flow and a younger, thinner, relatively mafic flow were collected by Apollo 15 from a marginal mare of Imbrium (Palus Putredinis) and were dated at 3.26 to 3.30 aeons. Pyroclastic material also formed at that time. In the convention adopted in this volume, the Apollo 15 and Luna 24 units formed late in the Late Imbrian Epoch.

The Late Imbrian Epoch lasted about 0.6 aeon, from about 3.8 aeons (Orientale impact) to 3.2 aeons ago (about midway between the older Apollo 15 and younger Apollo 12 absolute ages). Imbrian mare-basalt units probably covered the entire area now covered by maria (including the mare area covered by younger craters), that is, about 17 percent of the lunar surface, or 6.5×10^6 km^2.

During the same epoch, the continuous deposits of craters larger than 30 km in diameter covered about 1.7×10^6 km^2 of the whole lunar surface, or about 25 percent of the mare area. Late Imbrian craters of this size formed at the rate of 280 per aeon, only about 1.5 percent of the Nectarian rate. If crater age assignments to the Lower Imbrian Series are correct (pl. 8), the Early Imbrian cratering rate (approx 3,000 per aeon) is transitional between the Nectarian and Late Imbrian rates, closer to the latter. Thus, the crater-size impacts declined simultaneously with the basin-size impacts.

ERATOSTHENIAN PERIOD

Volcanism continued during the Eratosthenian Period, generating extensive flows in Oceanus Procellarum and Mare Imbrium and less extensive flows in other maria. These lavas are predominantly Ti-rich except in northern Mare Imbrium and Mare Frigoris. At least the Ti-rich types were derived from sources rich in radioactive elements and were extruded through thin lithospheres. Graben formation had ceased by the Eratosthenian Period except in uplifted crater floors, but thick sections of basalt and the underlying basin floors continued to sink sufficiently to crumple the mare surfaces into ridges. The only absolutely dated Eratosthenian basalt samples are those of four compositional types collected from at least three flows at the Apollo 12 landing site. These rocks, which are 3.16 aeons old, are atypical of the mare units surrounding this site, most of which are Imbrian in age.

Interfingering of crater and mare materials is more obvious in the Eratosthenian System than in any other time-stratigraphic unit. Most Eratosthenian craters are rayless; some superposed on low-Fe, low-Ti mare materials and terrae are rayed. Exploration has confirmed an early hypothesis that a time-dependent exogenic process fades rays—probably the manufacture by small impacts of Fe- and Ti-rich agglutinitic glasses at the expense of crystalline materials.

Ages younger than those of the Apollo 12 basalt samples and older than Copernicus must be interpolated by data on small superposed craters calibrated with models of the cratering rate. Available terrestrial and lunar data are consistent with a constant impact rate since the start of the Eratosthenian Period until the present, if mare and crater substrates of the same age display different crater frequencies. This constant rate would create 41 of the 132 observed Eratosthenian and Copernican craters larger than 30 km in diameter per aeon, a rate that is substantially lower than even the Late Imbrian rate and that signals a final leveling out of the early bombardment which shaped so much of the Moon's face. The Eratosthenian Period thus lasted about 2.1 aeons. The only impacting bodies available to create craters since the beginning of the Eratosthenian Period were similar to those still present in Earth-Moon space today (Shoemaker and others, 1979; Shoemaker, 1981).

COPERNICAN PERIOD

By the time the Coperican Period began, volcanism had yielded the stage almost entirely back to impact. Copernican mare units are known in only a few spots in the intermediate trough of the Procellarum basin that were also the favored sites of Eratosthenian volcanism. Later, further global cooling, depletion of radioactive-heat sources by extrusion of magma, and consequent lithospheric thickening and strengthening halted surface volcanism. The only possible melted zones today are below the zone of deep moonquakes, about $1,000 \pm 100$ km deep, in the zone where seismic shear waves are attenuated (Goins and others, 1979). Whatever core dynamo may have generated an early magnetic field (Runcorn, 1980) has been extinct for aeons.

The beginning of the Copernican Period has been accurately dated on neither the relative nor the absolute time scale. Absolute ages thought to date Copernican units include a questionable age of 0.8 aeon for Copernicus, a probably accurate age of 0.1 aeon for Tycho, and accurate ages of 2 to 50 million years for small craters dated by Apollo 14 and 16 samples (tables 13.2, 14.1). An estimate for the duration of the Copernican Period, assuming a constant impact rate and a nonequivalence of the crater frequencies on basalt and crater substrates, is 1.1 aeon. If this calculation is correct, limited volcanism lasted until at least 1 aeon ago—a late date by lunar, though not by terrestrial, standards.

Today, the Moon's surface is changing very slowly. Regoliths have formed during the past aeon at the sluggish rate of less than 1 mm per million years (Shoemaker, 1970; Quaide and Oberbeck, 1975). Downslope movement of debris is continuing to degrade slopes, but at a rate too slow to obliterate even such fine-scale features as the secondary-impact craters of Tycho (fig. 3.6), which formed during the Earth's Mesozoic Era. Freshly exposed crystalline material is being very gradually converted to darker glasses. No Copernican units except a few small craters are being added to the Moon. The meteorite impacts recorded by the Apollo seismometers prove that impacts still do occur; one large enough to probe the deep interior of the Moon struck the farside in July 1972 (Nakamura and others, 1973). One to three craters at least 10 km in diameter may form every 10 million years (Shoemaker, 1981). Volcanism and almost all tectonism have ceased.

REMAINING PROBLEMS

This account has shown that much has been learned about the Moon during the past two decades. The origin of the maria and of most craters is settled. Key deposits exposed at the surface have been dated on both the relative and absolute time scales, and the antiquity of the Moon's face has been established. The composition of the Moon and many of its geologic units has been learned to a good first approximation.

It should also be clear that some key questions remain unanswered. Lunar remotely based studies and direct exploration, though in a lull, have not completed their job. The Moon's mode and place of origin are still unknown. Its subsurface structure is very poorly known even in relatively well explored areas. Because Imbrium is the only basin that is well dated on both the relative and absolute time scales, the premare impact rate is too poorly calibrated before 3.85 aeons ago to establish such important points as the time of crustal solidification and the origin and lifetime of large Solar-System projectiles. The dates of volcanism before 3.8 and after 3.2 aeons ago are uncertain. The ages of most of the rayed craters are uncertain within broad limits. The relation among composition, source depth, and extrusion site of the mare basalt flows is hypothetical. The compositions of the farside maria are unknown. Terra compositions are known only very crudely from a few spot samples extrapolated from orbital measurements made at low resolutions of a small percentage of the surface. The origin of central peaks and shallow floors of complex craters is uncertain. The origin of basin rings and even of the position of the boundary of basin excavation—central questions in studies of impact mechanics, lunar petrology, and stratigraphy—is frustratingly elusive.

Some of these questions can probably be answered by continued experimental and field study on the Earth. The origin of complex craters and basin rings might yield to further study of terrestrial craters, laboratory and large-scale explosive experiments, and physical theory. Other questions, such as the distribution of mare-basalt compositions, can be partly answered by continued geologic mapping, crater-frequency counts, telescopic spectral studies, and petrologic theory. Still other questions may be answerable by continued examination of the 80 percent of lunar samples that have not yet been thoroughly analyzed.

Answers to most of the remaining geologic questions, however, require resumption of lunar spaceflights. A global orbiter could gather important data concerning: (1) mare compositions; (2) terra compositions (an even more serious gap from the point of view of the petrologist, the geochemist, and the cosmogonist attempting to learn the origin of the Moon and the Solar System); (3) the Moon's gravitational field; (4) the topography of basins, especially Orientale; (5) the puzzling problem of lunar magnetism; and (6) the stratigraphy of poorly photographed regions (particularly the polar regions above lat 40°, a zone along long 100°–120° W., and the east limb on both the nearside and farside hemispheres). Knowledge of the Moon's third dimension could be greatly improved by this remote exploration. Mars is better photographed than the Moon.

Other problems will require additional samples from the Moon itself. Table 14.2 lists my recommendations for the landing sites of unmanned sample-returning spacecraft that could answer some of the key geologic questions. Objectives fall into five main categories: (1) absolute ages needed to calibrate the stratigraphic column, (2) compositions and textures to decide such genetic problems as the hypothetical terra volcanism, (3) crustal compositions at points of known stratigraphic context that can be extrapolated to larger areas, (4) mantle compositions inferrable from samples of currently unsampled color and age units of mare basalt, and (5) compositions and ages of premare volcanic basalt. Data from most of the selected targets can be extrapolated by means of currently available or future orbital sensing. Petrologists and geochemists might have a different list. The highest priority is given here to dating the Nectaris basin, whose relative age is well known and which, therefore, would provide the needed calibration for the premare cratering rate. Manned missions could achieve multiple goals at single landing sites but are not likely to be resumed within the foreseeable future.

Whether these programs should be undertaken is, of course, a matter of priorities. Very few spaceflights are undertaken by any nation or consortium of nations for purely scientific purposes. Moreover, some scientists give the Moon a low priority despite the easy accessibility of this primitive, cool, internally inactive, silicate planet unmodified by air or water that has preserved an unsurpassed record of the early Solar System. These recommendations are offered here in the belief that the Moon contains additional clues to our understanding of the Solar System, including our own home planet.

14. SUMMARY

TABLE 14.2.—*Potential landing sites for future unmanned lunar sampling missions*

[Each probe is considered capable of returning a single sample of regolith randomly selected from within the designated area. Contributions from Paul D. Spudis]

Priority	Stratigraphic unit	Landing area	Figure	Objective
1	Nectaris basin	(a) Ejecta, near lat 35° S., long 42° E (b) Plains (impact melt?), near lat 22° S., long 41° E	9.1 9.2A	(a) Absolute age. (b) Composition.
2	Copernican mare	Southeast of Lichtenberg, near lat 31° N., long 67° W	13.7	(a) Absolute age. (b) Composition of source.
3	Terra plains (see also priority 11)	(a) Albategnius (b) Ptolemaeus	1.8, 10.36 1.8, 7.7	Nonmare volcanism or buried mare basalt flows(?).
4	Terra domes	(a) Gruithuisen gamma or delta (b) Hansteen alpha	10.45A 10.45B	Nonmare volcanism(?).
5	Farside mare	(a) Floor of Tsiolkovskiy (b) Mare Ingenii	1.3, 3.27A 1.4	Composition of source.
6	Maunder Formation	South of Mare Orientale	4.4B	(a) Age of Orientale basin. (b) Composition of thick crust.
7	Copernicus	Impact melt on floor	3.35	(a) Absolute age. (b) Deep crustal composition(?).
8	King	Impact melt on floor	3.32	(a) Farside crustal composition. (b) Absolute age.
9	Ancient crust	Near lat 30° N., long 160° E	8.1	(a) Composition. (b) Absolute age.
10	South Pole-Aitken massifs	South of Korolev, lat 21.5° S., long 160° W	9.21A	(a) Farside crustal composition. (b) Absolute age.
11	Pre-Late Imbrian mare(?) basalt	(a) Center of Schickard (b) North of Balmer	7.6 1.5, 9.5	(a) Absolute age. (b) Composition.
12	Early Late Imbrian mare	Mare Marginis, in Ibn Yunus	4.7	(a) Absolute age. (b) Composition (KREEP-rich?).
13	Eratosthenian mare	(a) Southwestern Mare Imbrium (a) Surveyor 1 region, near lat 2.5° S., long 43.5° W	5.1 12.6	(a) Absolute age. (b) Calibration of color spectra.
14	Central Mare Serenitatis	Between Bessel and Dawes	11.13	(a) Calibration of color spectra (standard spectrum). (b) Absolute age (near Imbrian-Eratosthenian boundary).
15	Orientale lobate ejecta	Near lat 53° S., long 79° W	4.4D	Impact melt or other ejecta.
16	Alpes Formation	Southeast of Vallis Alpes, near lat 45° N., long 5° E	10.12	(a) Impact-melt/debris content. (b) Composition of deep ejecta.
17	Apennine Bench Formation	Near lat 27° N., long 8° W	6.10	(a) Impact melt or KREEP-rich volcanic materials(?). (b) Absolute age. (c) Calibration of gamma-ray data.
18	Fissured crater-floor deposits	Floor of Murchison, lat 1° W., long 5° N	10.28	Ejected impact melt (Imbrium?).

SUPPLEMENTARY TABLE.—*Craters superposed on basins*

[Diameters of superposed craters larger than 100 km are listed individually. A/K, Al-Khwarizmi/King; Ap, Apollo; B, Birkhoff; C, Crisium; F-S, Freundlich-Sharonov; Hb, Humboldtianum; Hu, Humorum; Hz, Hertzsprung; I, Imbrium; K, Korolev; K-H, Keeler-Heaviside; L, Lorentz; M-R, Mendel-Rydberg; Md, Mendeleev; Mo, Moscoviense; N, Nectaris; O, Orientale; Sc, Schrödinger; Sm, Smythii. Photographic coverage: G, good; GP, half of area good and half poor (Orientale, two-thirds good); P, poor; M, numbers of craters probably diminished by mutual obliteration; R, most reliable data because of large sample, good photographs, and absence of mutual obliteration or blanketing]

	Pre-Nectarian							Nectarian									Imbrian		
Crater	A-K	K-H	Sm	L	B	F-S	Ap	N	M-R	K	Mo	Hb	Md	C	Hu	Hz	I	Sc	O
Crater count (/10^6 km^2)	0.320	0.371	0.445	0.208	0.401	0.629	0.480	1.286	0.247	1.113	0.609	0.515	0.569	0.843	0.428	0.883	2.491	0.594	3.001
Coverage	G, M	G, M	P, M	P, M	G, M?	P	GP	G, R	P	G, R	P	GP	G	GP	G	G, R	G, R	G	GP, R
Diameter (km)																			
≥100	118, 115, 106, 13	177, 172, 125, 156, 7, 135, 126, 122, 112	152, 125, 110, 106	123, 120	164, 148, 135, 117	110	209	126, 117, 100	-	-	-	127	122, 103	138	110	105	260, 105, 102, 100	-	-
99	-	-	-	-	-	1	-	2	-	-	-	-	-	-	-	-	1	-	-
98	-	-	-	-	-	-	-	-	-	-	-	-	-	-	-	-	-	-	-
97	1	-	-	-	-	-	-	-	-	-	-	-	-	-	-	-	-	-	-
96	-	-	-	-	-	-	-	-	-	-	-	-	-	-	-	-	-	-	-
95	-	-	1	-	-	-	-	-	-	-	-	-	-	-	-	-	-	-	-
94	-	-	-	1	1	-	-	-	-	1	-	-	-	-	1	-	-	-	-
93	-	-	1	-	1	1	-	-	-	-	-	-	-	-	-	-	-	-	-
92	-	-	-	-	-	1	-	-	-	2	-	-	-	-	-	-	1	-	-
91	-	-	-	1	-	-	-	-	-	-	-	-	-	-	-	1	-	-	-
90	-	-	-	-	-	-	-	1	-	-	-	-	-	-	-	-	-	-	1
89	-	1	-	-	-	-	-	-	1	-	-	-	-	-	-	-	-	-	-
88	-	-	-	1	-	-	-	-	-	-	-	-	-	-	-	-	-	-	-
87	-	-	-	-	-	-	1	2	-	-	-	-	-	-	-	-	-	-	-
86	-	-	-	-	-	-	-	-	-	-	-	-	-	-	-	-	1	-	1
85	-	-	-	-	-	-	-	1	-	-	-	-	-	-	1	-	-	-	1
84	-	-	-	-	-	-	-	-	-	-	-	-	-	-	-	-	-	1	-
83	1	1	-	-	-	-	1	-	-	-	-	-	1	1	-	-	-	1	-
82	-	1	-	-	-	-	-	-	-	1	-	-	-	-	1	-	1	-	-
81	1	1	-	-	-	-	-	1	-	-	-	-	-	-	-	-	-	-	-
80	-	-	-	-	-	2	-	-	-	-	-	-	-	-	-	-	-	-	-
79	-	-	1	-	1	-	-	-	-	-	-	-	-	-	-	-	-	-	-
78	-	2	-	-	-	-	-	-	-	-	1	-	-	-	-	-	-	-	-
77	1	-	-	-	1	-	-	-	-	1	-	-	-	-	-	-	-	-	-
76	-	-	-	-	-	1	-	-	-	-	-	-	-	?	-	-	-	-	-
75	1	-	-	-	-	-	-	-	-	-	-	-	1	1	-	-	-	-	-
74	-	-	-	1	-	1	-	-	-	-	-	-	-	-	-	-	-	-	-
73	1	2	1	-	-	-	1	1	-	-	-	-	-	-	-	-	-	-	-
72	1	-	-	-	-	-	-	-	-	1	-	-	-	-	-	1	-	-	-
71	-	1	-	-	-	-	-	2	-	1	-	-	-	-	-	-	-	-	?
70	-	-	-	-	-	1	-	-	1	-	-	1	-	-	-	-	-	-	?
69	1	1	-	-	-	-	-	1	1	1	-	-	1	-	-	-	1	-	-
68	1	-	-	2	1	1	1	-	-	1	1	-	-	-	-	-	-	-	-
67	1	-	1	-	1	-	1	1	-	-	-	-	-	1	-	-	-	-	-
66	1	-	-	-	-	-	-	-	-	-	1	-	2	-	-	-	1	-	-
65	-	-	1	-	-	1	-	1	-	-	-	1	-	-	-	-	1	-	-
64	-	-	1	-	-	-	-	-	-	-	-	-	-	-	-	-	-	-	-
63	-	2	-	1	-	-	-	1	-	-	1	1	-	-	-	-	-	-	2
62	2	-	-	-	-	1	-	-	-	1	-	1	-	-	1	-	-	-	-
61	1	-	1	-	-	1	1	-	-	1	-	-	-	-	-	-	-	-	-
60	1	-	2	-	-	-	-	-	-	-	-	-	1	1	-	-	-	-	-
59	-	-	1	-	-	1	1	-	-	-	-	2	-	-	-	-	-	-	-
58	-	1	1	-	2	2	-	-	-	?	-	-	-	-	-	-	-	1	-
57	1	1	-	1	-	-	-	-	-	3	-	1	-	-	-	1	-	-	-
56	-	-	-	-	-	2	-	-	-	1	1	-	?	1	-	2	-	-	3
55	1	-	-	-	-	-	1	-	-	1	-	-	-	-	-	1	1	-	-
54	1	1	1	-	1	-	-	1	1	1	-	-	-	-	-	-	-	-	3
53	2	-	3	-	-	2	2	-	-	-	-	1	-	1	-	-	-	-	1
52	1	1	2	-	-	-	-	-	-	3	-	-	-	1	-	-	-	-	2
51	-	-	1	-	-	-	1	2	-	1	-	-	2	-	1	1	-	-	2
50	2	1	-	-	-	?	-	1	1	-	1	-	-	-	-	-	-	1	3
49	1	-	-	1	1	-	1	2	-	1	1	1	-	-	-	1	-	-	-
48	2	1	1	-	1	-	-	1	-	2	-	-	2	2	1	-	1	-	-
47	1	-	1	-	-	-	-	3	-	-	-	-	2	-	1	1	1	-	-
46	-	3	1	-	2	2	-	-	-	1	-	-	1	-	1	2	-	-	-
45	-	-	1	-	-	1	-	2	-	1	-	1	-	-	-	-	-	-	-
44	-	-	-	1	1	-	1	5	1	2	1	-	-	-	-	1	1	-	-
43	1	-	2	2	1	2	-	1	-	-	-	-	1	3	-	-	1	-	2
42	2	1	2	-	1	1	-	2	-	-	-	-	-	3	-	1	1	-	1
41	2	-	-	-	-	1	1	2	-	-	-	-	2	-	1	1	-	-	1
40	1	1	1	-	2	-	2	4	1	2	1	-	3	1	1	1	3	-	3
39	1	2	2	-	1	2	1	1	-	-	1	1	-	-	1	1	3	-	2
38	1	4	-	1	1	-	3	2	1	2	1	-	-	2	-	1	5	1	1
37	1	1	-	-	3	2	1	-	1	2	3	-	-	-	1	-	-	-	-
36	-	-	1	-	1	3	-	2	1	2	2	-	-	1	-	1	1	-	3
35	2	1	5	-	4	2	4	1	1	3	2	-	-	1	1	3	2	-	4
34	1	1	1	2	-	-	-	3	-	1	-	-	-	1	-	1	1	1	1
33	-	-	1	-	-	2	5	2	-	1	1	1	1	1	-	1	2	-	-
32	2	1	-	-	2	3	-	8	-	2	1	2	2	3	2	3	4	-	1
31	2	-	1	1	-	1	4	4	-	1	1	1	-	-	1	1	2	-	-
30	1	3	1	3	3	7	2	2	2	4	2	-	1	1	-	1	1	-	1
29	1	2	3	1	-	-	-	4	-	5	4	-	1	2	-	1	2	-	-
28	1	3	5	1	1	3	4	6	1	1	3	-	4	-	1	1	2	1	1
27	2	2	2	-	3	5	1	4	1	5	-	1	2	-	5	2	1	-	4
26	-	4	3	-	3	3	-	3	-	1	1	3	2	-	-	1	1	-	-
25	4	1	2	2	3	4	1	2	1	4	2	2	-	2	1	1	4	1	2
24	1	3	1	-	1	2	3	5	-	-	3	2	1	1	2	1	3	-	3
23	3	3	2	1	1	4	3	2	-	7	4	-	1	4	2	4	4	1	2
22	-	5	3	1	-	3	2	2	-	7	3	1	2	4	1	4	4	-	4
21	2	2	4	2	1	1	3	2	1	8	5	2	1	1	1	4	3	-	5
20	2	2	3	4	-	5	2	4	1	2	2	2	-	3	2	2	3	1	9

REFERENCES CITED

[Abbreviations: ASAPR, Astrogeologic Studies Annual Progress Report; ASSPR, Astrogeologic Studies Semiannual Progress Report; LPS, Lunar and Planetary Science (Abstracts of Papers Submitted to Lunar and Planetary Science Conferences); LPSCP, Lunar and Planetary Science Conference Proceedings; LS, Lunar Science (Abstracts of Papers Submitted to Lunar Science Conferences); LSCP, Lunar Science Conference Proceedings; NASA, U.S. National Aeronautics and Space Administration; USGS, U.S. Geological Survey]

Adams, J.B., 1975, Interpretation of visible and near-infrared diffuse reflectance spectra of pyroxenes and other rock-forming minerals, *in* Infrared and Raman spectroscopy of lunar and terrestrial minerals: New York, Academic Press, p. 91–116.

Adams, J.B., and McCord, T.B., 1970, Remote sensing of lunar surface mineralogy: Implications from visible and near-infrared reflectivity of Apollo 11 samples: LSCP 1, v. 3, p. 1937–1945.

———1971, Alteration of lunar optical properties: Age and composition effects: Science, v. 171, no. 3971, p. 567–571.

———1973, Vitrification darkening in the lunar highlands and identification of the Descartes material at the Apollo 16 site: LSCP 4, v. 1, p. 163–177.

Adams, J.B., Pieters, C.M., and McCord, T.B., 1974, Orange glass: Evidence for regional deposits of pyroclastic origin on the moon: LSCP 5, v. 1, p. 171–186.

Adler, Isidore, Gerard, J., Trombka, J.I., Schmadebeck, R., Lowman, P., Blodget, H., Yin, L., Eller, E., Lamothe, R., Gorenstein, P., Bjorkholm, P., Harris, B., and Gursky, H., 1972, The Apollo 15 x-ray fluorescence experiment: LSCP 3, v. 3, p. 2157–2178.

Adler, Isidore, and Trombka, J.I., 1977, Orbital chemistry—lunar surface analysis from the X-ray and gamma-ray remote sensing experiments: Physics and Chemistry of the Earth, v. 10, no. 1, p. 17–43.

Ahrens, T.J., and O'Keefe, J.D., 1972, Shock melting and vaporization of lunar rocks and minerals: The Moon, v. 4, no. 1-2, p. 214–249.

Ahrens, T.J., and Watt, J.P., 1980, Dynamic properties of mare basalts: Relation of equations of state to petrology: LPSCP 11, v. 3, p. 2059–2074.

Albritton, C.C., ed., 1963, The fabric of geology: Reading, Mass., Addison-Wesley, 372 p.

———1967, Uniformity and simplicity: A symposium on the principle of the uniformity of nature: Geological Society of America Special Paper 89, 99 p.

Alexander, E.C., Jr., Bates, Allan, Coscio, M.R., Jr., Dragon, J.C., Murthy, V.R., Pepin, R.O., and Venkatesan, T.R., 1976, K/Ar dating of lunar soils II: LSCP 7, v. 1, p. 625–648.

Alexander, E.C., Jr., Coscio, M.R., Jr., Dragon, J.C., Pepin, R.O., and Saito, Kazuo, 1977, K/Ar dating of lunar soils III: Comparison of ^{39}Ar-^{40}Ar and conventional techniques; 12032 and the age of Copernicus: LSCP 8, v. 3, p. 2725–2740.

Alexander, E.C., Jr., Coscio, M.R., Jr., Dragon, J.C., and Saito, Kazuo, 1980, K/Ar dating of lunar soils IV: Orange glass from 74220 and agglutinates from 14259 and 14163: LPSCP 11, v. 2, p. 1663–1677.

Alexander, E.C., Jr., and Davis, P.K., 1974, ^{40}Ar-^{39}Ar ages and trace element contents of Apollo 14 breccias: An interlaboratory cross-calibration of ^{40}Ar-^{39}Ar standards: Geochimica et Cosmochimica Acta, v. 38, no. 6, p. 911–928.

Alexander, E.C., Jr., and Kahl, S.B., 1974, ^{40}Ar-^{39}Ar studies of lunar breccias: LSCP 5, v. 2, 1353–1373.

Alter, Dinsmore, 1963, Pictorial guide to the Moon: New York, Crowell, 183 p.

American Commission on Stratigraphic Nomenclature, 1970, Code of stratigraphic nomenclature: Tulsa, Okla., American Association of Petroleum Geologists, 22 p.

Anders, Edward, 1978, Procrustean science: Indigenous siderophiles in the lunar highlands, according to Delano and Ringwood: LPSCP 9, v. 1, p. 161–184.

Andersson, L.E., and Whitaker, E.A., 1982, NASA catalog of lunar nomenclature: NASA Reference Publication 1097, 183 p.

Andre, C.G., Hallam, M.E., Weidner, J.R., Podwysocki, M.H., Philpotts, J.A., Clark, P.E., and Adler, Isidore, 1975, Correlation of Al/Si X-ray fluorescence data with other remote sensing data from the Taurus-Littrow area: LSCP 6, v. 3, p. 2739–2748.

Andre, C.G., Wolfe, R.W., and Adler, Isidore, 1978, Evidence for a high-magnesium subsurface basalt in Mare Crisium from orbital X-ray fluorescence data, *in* Lunar and Planetary Institute, compiler, Mare Crisium: The view from Luna 24: Conference on Luna 24, Houston, Tex., 1977, Proceedings: New York, Pergamon (Geochimica et Cosmochimica Acta, supp. 9), p. 1–12.

———1979a, Are early magnesium-rich basalts widespread on the moon?: LPSCP 10, v. 2, p. 1739–1751.

Andre, C.G., Wolfe, R.W., Adler, Isidore, and Clark, P.E., 1979b, Mare basalt depths from orbital X-ray data [abs.]: LPS X, pt. 1, p. 38–40.

Andre, C.G., Wolfe, R.W., Adler, Isidore, Clark, P.E., Weidner, J.R., and Philpotts, J.A., 1977, Chemical character of the partially flooded Smythii Basin based on Al/Si orbital X-ray data: LSCP 8, v. 1, p. 925–931.

Andrews, R.J., 1977, Characteristics of debris from small-scale cratering experiments, *in* Roddy, D.J., Pepin, R.O., and Merrill, R.B., eds., Impact and explosion cratering: Planetary and terrestrial implications: New York, Pergamon, p. 1089–1100.

Apollo Soil Survey, 1971, Apollo 14: Nature and origin of rock types in soil from the Fra Mauro Formation: Earth and Planetary Science Letters, v. 12, no. 1, p. 49–54.

———1974, Phase chemistry of Apollo 14 soil sample 14259: Modern Geology, v. 5, no. 1, p. 1–13.

Arthur, D.W.G., 1962, Some systematic visual lunar observations, *in* Kopal, Zdeněk, and Mikhailov, Z.K., eds., The Moon: London, Academic Press, p. 317–324.

Arthur, D.W.G., Agnieray, A.P., Horvath, R.A., Wood, C.A., and Chapman, C.R., 1963, The system of lunar craters, quadrant I: Tucson, University of Arizona, Lunar and Planetary Laboratory Communications, v. 2, no. 30.

Arvidson, Raymond, Crozaz, Ghislaine, Drozd, R.J., Hohenberg, C.M., and Morgan, C.J., 1975, Cosmic ray exposure ages of features and events at the Apollo landing sites: The Moon, v. 13, no. 1–3, p. 259-276.

Arvidson, Raymond, Drozd, R.J., Guinness, E., Hohenberg, C.M., Morgan, C.J., Morrison, R.H., and Oberbeck, V.R., 1976, Cosmic ray exposure ages of Apollo 17 samples and the age of Tycho: LSCP 7, v. 3, p. 2817–2832.

Baldwin, R.B., 1949, The face of the Moon: Chicago, University of Chicago Press, 239 p.

———1963, The measure of the Moon: Chicago, University of Chicago Press, 88 p.

———1964, Lunar crater counts: Astronomical Journal, v. 69, no. 5, p. 377–392.

———1965, A fundamental survey of the Moon: New York, McGraw-Hill, 149 p.

———1968, Lunar mascons: Another interpretation: Science, v. 162, no. 3860, p. 1407–1408.

———1969, Ancient giant craters and the age of the lunar surface: Astronomical Journal, v. 74, no. 4, p. 570–571.

———1970, A new method of determining the depth of the lava in lunar maria: Astronomical Society of the Pacific Publications, v. 82, no. 488, p. 857-864.

———1971, On the history of lunar impact cratering: The absolute time scale and the origin of planetesimals: Icarus, v. 14, no. 1, p. 36–52.

———1972, The tsunami model of the origin of ring structures concentric with large lunar craters: Physics of the Earth and Planetary Interiors, v. 5, no. 5, p. 327–339.

———1974a, On the origin of the mare basins: LSCP 5, v. 1, p. 1–10.

———1974b, Was there a "terminal lunar cataclysm" 3.9–4.0 × 10^9 years ago?: Icarus, v. 23, no. 2, p. 157–166.

———1978, An overview of impact cratering: Meteoritics, v. 13, no. 4, p. 364–379.

Barabashov, N.P., Mikhailov, A.A., and Lipsky, Y.N., eds., 1961, An atlas of the Moon's far side: The Lunik III reconnaissance: New York, Interscience, 143 p.

Basaltic Volcanism Study Project, 1981, Basaltic volcanism on the terrestrial planets: Houston, Tex., Lunar and Planetary Institute, 1,286 p.

Basu, Abhijit, and McKay, D.S., 1979, Petrography and provenance of Apollo 15 soils: LPSCP 10, v. 2, p. 1413–1424.

Beaty, D.W., and Albee, A.L., 1978, Comparative petrology and possible genetic relations among the Apollo 11 basalts: LPSCP 9, v. 1, p. 359-463.

———1980, The geology and petrology of the Apollo 11 landing site: LPSCP 11, v. 1, p. 23–35.

Behrmann, C.J., Crozaz, G., Drozd, R.J., Hohenberg, C.M., Ralston, C.E., Walker, R.M., and Yuhas, D.E., 1973, Cosmic-ray exposure history of North Ray and South Ray material: LSCP 4, v. 2, p. 1957–1974.

Bernatowicz, T.J., Hohenberg, C.M., Hudson, B., Kennedy, B.M., and Podosek, F.A., 1978, Argon ages for lunar breccias 14064 and 15405: LSCP 9, v. 1, p. 905–919.

Bickel, C.E., and Warner, J.L., 1978, Survey of lunar plutonic and granulitic lithic fragments: LPSCP 9, v. 1, p. 629–652.

Bielefeld, M.J., Reedy, R.C., Metzger, A.E., Trombka, J.I., and Arnold, J.R., 1976, Surface chemistry of selected lunar regions: LSCP 7, v. 3, p. 2661–2676.

Bills, B.G., and Ferrari, A.J., 1977, A lunar density model consistent with topographical, gravitational, librational, and seismic data: Journal of Geophysical Research, v. 82, no. 8, p. 1306–1314.

Binder, A.B., 1980, On the origins of lunar pristine crustal rocks, *in* Lunar and Planetary Institute, compiler, Conference on the Lunar Highlands Crust, Houston, Tex., 1979, Proceedings: New York, Pergamon (Geochimica et Cosmochimica Acta, supp. 12), p. 71–79.

———1982, Post-Imbrian global lunar tectonism: Evidence for an initially totally molten Moon: The Moon and the Planets, v. 26, no. 2, p. 117–133.

Binder, A.B., Lange, M.A., Brandt, H.-J., and Kähler, Susanne, 1980, Mare basalt units and the compositions of their magmas: The Moon and the Planets, v. 23, no. 4, p. 445–481.

Binzel, R.P., and Van Flandern, T.C., 1979, Minor planets: The discovery of minor satellites: Science, v. 203, no. 4383, p. 903–905.

Birck, J.L., and Allegre, C.J., 1973, ^{87}Rb-^{87}Sr age of fragments and soils from the lunar Sea of Fertility: Geochimica et Cosmochimica Acta, v. 37, no. 9, p. 2025–2031.

Bogard, D.D., Nyquist, L.E., Bansal, B.M., Wiesmann, H.W., and Shih, C.-Y., 1975, 76535: An old lunar rock: Earth and Planetary Science Letters, v. 26, no. 1, p. 69–80.

Bowen, N.L., 1928, The evolution of the igneous rocks: Princeton, N.J., Princeton University Press, 332 p.

Bowin, Carl, Simon, Bruce, and Wollenhaupt, W.R., 1975, Mascons: A two-body solution: Journal of Geophysical Research, v. 80, no. 35, p. 4947–4955.

Bowker, D.E., and Hughes, J.K., 1971, Lunar Orbiter photographic atlas of the Moon: NASA Report SP–206, 41 p., 675 pls.

Boyce, J.M., 1976, Ages of flow units in the lunar nearside maria based on Lunar Orbiter IV photographs: LSCP 7, v. 3, p. 2717–2728.

Boyce, J.M., and Dial, A.L., Jr., 1973, Relative ages of some near-side mare units based on Apollo 17 metric photographs, pt. C *of* Stratigraphic studies, [chap.] 29, *of* Apollo 17 preliminary science report: NASA Report SP-330, p. 29–26 to 29–28.

———1975, Relative ages of flow units in Mare Imbrium and Sinus Iridum: LSCP 6, v. 3, p. 2585–2595.

Boyce, J.M., and Johnson, D.A., 1977, Ages of flow units in Mare Crisium based on crater density: LSCP 8, v. 3, p. 3495–3502.

———1978, Ages of flow units in the far eastern maria and implications for basin-filling history: LPSCP 9, v. 3, p. 3275–3283.

Boyce, J.M., Dial, A.L., Jr., and Soderblom, L.A., 1974, Ages of the lunar nearside light plains and maria: LSCP 5, v. 1, p. 11–23.

———1975, A summary of relative ages of lunar nearside and farside plains: Flagstaff, Ariz., USGS Interagency Report: Astrogeology 66, 26 p.

Boyce, J.M., Schaber, G.G., and Dial, A.L., Jr., 1977, Age of Luna 24 mare basalts based on crater studies: Nature, v. 265, no. 5589, p. 38–39.

Brennan, W.J., 1975, Modification of premare impact craters by volcanism and tectonism: The Moon, v. 12, no. 4, p. 449-461.

———1976, Multiple ring structures and the problem of correlation between lunar basins: LSCP 7, v. 3, p. 2833–2843.

Brown, W.E., Jr., Adams, G.F., Eggleton, R.E., Jackson, P., Jordan, R., Kobrick, M., Peeples, W.J., Phillips, R.J., Porcello, L.J., Schaber, G.G., Sill, W.R., Thompson, R.W., Ward, S.H., and Zelenka, J.S., 1974, Elevation profiles of the moon: LSCP 5, v. 3, p. 3037–3048.

Bryan, W.B., 1973, Wrinkle-ridges as deformed surface crust on ponded mare lava: LSCP 4, v. 1, p. 93–106.

Burnett, D.S., and Woolum, D.S., 1977, Exposure ages and erosion rates for lunar rocks: Physics and Chemistry of the Earth, v. 10, no. 2, p. 63–101.

Butler, Patrick, and Morrison, D.A., 1977, Geology of the Luna 24 landing site: LSCP 8, v. 3, p. 3281–3301.

Cadogan, P.H., 1974, Oldest and largest lunar basin?: Nature, v. 250, no. 5464, p. 315–316.

———1981, The Moon—our sister planet: Cambridge, U.K., Cambridge University Press, 391 p.

Cadogan, P.H., and Turner, Grenville, 1976, The chronology of the Apollo 17 Station 6 boulder: LSCP 7, v. 2, p. 2267–2285.

Carlson, R.W., and Lugmair, G.W., 1979, Sm-Nd constraints on early lunar differentiation and the evolution of KREEP: Earth and Planetary Science Letters, v. 45, no. 1, p. 123–132.

———1981a, Sm-Nd age of lherzolite 67667: Implications for the processes involved in lunar crustal formation: Earth and Planetary Science Letters, v. 56, no. 1, p. 1–8.

———1981b, Time and duration of lunar highlands crust formation: Earth and Planetary Science Letters, v. 52, no. 2, p. 227–238.

Carr, M.H., 1965a, Dark volcanic materials and rille complexes in the north-central region of the Moon: USGS ASAPR, July 1, 1964 to July 1, 1965, pt. A, p. 35–43.

———1965b, Geologic map and section of the Timocharis region of the Moon: USGS Map I-462 (LAC–40), scale 1:1,000,000.

———1966a, Geologic map of the Mare Serenitatis region of the Moon: USGS Map I-489 (LAC–42), scale 1:1,000,000.

———1966b, The geology of the Mare Serenitatis region of the Moon: USGS ASAPR, July 1, 1965 to July 1, 1966, pt. A, p. 11–16.

———1969, Geologic map of the Alphonsus region of the Moon: USGS Map I–599 (RLC-14), scale 1:250,000.

———1970, Geologic map of the Maskelyne DA region of the Moon, Lunar Orbiter site II P-2, southwestern Mare Tranquillitatis, including Apollo landing site 1: USGS Map I-616 [ORB II-2(100)], scale 1:100,000.

———1974, The role of lava erosion in the formation of lunar rilles and martian channels: Icarus, v. 22, no 1, p. 1–23.

Carr, M.H., and Meyer, C.E., 1974, The regolith at the Apollo 15 site and its stratigraphic implications: Geochimica et Cosmochimica Acta, v. 38, no. 7, p. 1183–1197.

Carr, M.H., and Titley, S.R., 1969, Geologic map of the Maestlin G region of the Moon, Lunar Orbiter site II P-13, Oceanus Procellarum, including Apollo landing site 5: USGS Map I-622 [ORB II-13 (100)], scale 1:100,000.

Carr, M.H., Howard, K.A., and El-Baz, Farouk, 1971, Geologic maps of the Apennine-Hadley region of the Moon; Apollo 15 pre-mission maps: USGS Map I-723, scales 1:250,000, 1:50,000, 2 sheets.

Casella, C.J., 1976, Evolution of the lunar fracture network: Geological Society of America Bulletin, v. 87, no. 2, p. 226–234.

Chabai, A.J., 1977, Influence of gravitational fields and atmospheric pressures on scaling of explosion craters, *in* Roddy, D.J., Pepin, R.O., and Merrill, R.B., eds., Impact and explosion cratering: Planetary and terrestrial implications: New York, Pergamon, p. 1191–1214.

Chao, E.C.-T., 1967, Shock effects in certain rock-forming minerals: Science, v. 156, no. 3772, p. 192–202.

———1973, Geologic implications of the Apollo 14 Fra Mauro breccias and comparison with ejecta from the Ries crater, Germany: USGS Journal of Research, v. 1, no. 1, p. 1–18.

———1974, Impact cratering models and their application to lunar studies—a geologist's view: LSCP 5, v. 1, p. 35–52.

———1977, The Ries crater of southern Germany, a model for large basins on planetary surfaces: Geologisches Jahrbuch, ser. A, no. 43, 85 p.

Chao, E.C.-T., and Minkin, J.A., 1977, Impact cratering phenomenon for the Ries multiring structure based on constraints of geological, geophysical, and petrological studies and the nature of the impacting body, *in* Roddy, D.J., Pepin, R.O., and Merrill, R.B., eds., Impact and explosion cratering: Planetary and terrestrial implications: New York, Pergamon, p. 405–424.

Chao, E.C.-T., Hodges, C.A., Boyce, J.M., and Soderblom, L.A., 1975, Origin of lunar light plains: USGS Journal of Research, v. 3, no. 4, p. 379–392.

Chao, E.C.-T., Minkin, J.A., and Best, J.B., 1972, Apollo 14 breccias: General characteristics and classification: LSCP 3, v. 1, p. 645–659.

Chao, E.C.-T., Minkin, J.A., and Thompson, C.L., 1976, The petrology of 77215, a noritic impact ejecta breccia: LSCP 7, v. 2, p. 2287–2308.

Chao, E.C.-T., Shoemaker, E.M., and Madsen, B.M., 1960, First natural occurrence of coesite: Science, v. 132, no. 3421, p. 220–222.

Chapman, C.R., and Haeffner, R.R., 1967, A critique of methods for analysis of the diameter-frequency relation for craters with special application to the Moon: Journal of Geophysical Research, v. 72, no. 2, p. 549–577.

Chapman, C.R., Aubele, J.C., Roberts, W.J., and Cutts, J.A., 1979, Sub-kilometer lunar craters: Origins, ages, processes of degradation, and implications for mare basalt petrogenesis [abs.]: LPS X, pt. 1, p. 190–191.

Charette, M.P., McCord, T.B., Pieters, C.M., and Adams, J.B., 1974, Application of remote spectral reflectance measurements to lunar geology classification and determination of titanium content of lunar soils: Journal of Geophysical Research, v. 79, no. 11, p. 1605–1613.

Cintala, M.J., Head, J.W., III, and Veverka, Joseph, 1978, Characteristics of the cratering process on small satellites and asteroids: LPSCP 9, v. 3, p. 3803–3830.

Cintala, M.J., Wood, C.A., and Head, J.W., 1977, The effects of target characteristics on fresh crater morphology: Preliminary results for the moon and Mercury: LSCP 8, v. 3, p. 3409–3425.

Clark, P.E., and Hawke, B.R., 1981, Compositional variation in the Hadley Apennine region: LPSCP 12, pt. B, sec. 1, p. 727–749.

Clayton, R.N., and Mayeda, T.K., 1975, Genetic relations between the moon and meteorites: LSCP 6, v. 2, p. 1761–1769.

Cliff, R.A., Lee-Hu, C., and Wetherill, G.W., 1971, Rb-Sr and U, Th-Pb measurements on Apollo 12 materials: LSCP 2, v. 2, p. 1493–1502.

Colton, G.W., Howard, K.A., and Moore, H.J., 1972, Mare ridges and arches in southern Oceanus Procellarum, pt. S *of* Photogeology, [chap.] 29 *of* Apollo 16 preliminary science report: NASA Report SP-315, p. 29–90 to 29–93.

Compston, W., Berry, H., Vernon, M.J., Chappell, B.W., and Kaye, M., 1971, Rubidium-strontium chronology and chemistry of lunar material from the Ocean of Storms: LSCP 2, v. 2, p. 1471–1485.

Compston, W., Foster, J.J., and Gray, C.M., 1975, Rb-Sr ages of clasts within boulder 1, station 2, Apollo 17: The Moon, v. 14, no. 3–4, p. 445–462.

Compston, W., Vernon, M.J., Berry, H., Rudowski, R., Gray, C.M., Ware, N.G., Chappell, B.W., and Kaye, M.J., 1972, Apollo 14 mineral ages and the thermal history of the Fra Mauro Formation: LSCP 3, v. 2, p. 1487–1501.

Conca, James, and Hubbard, N.J., 1979, Evidence for early volcanism in Mare Smythii: LPSCP 10, v. 2, p. 1727–1737.

Conel, J.E., and Nash, D.B., 1970, Spectral reflectance and albedo of Apollo 11 samples: Effects of irradiation and vitrification and comparison with telescopic observations: LSCP 1, v. 3, p. 2013–2023.

Cooper, H.F., Jr., 1977, A summary of explosion cratering phenomena relevant to meteor impact events, *in* Roddy, D.J., Pepin, R.O., and Merrill, R.B., eds., Impact and explosion cratering: Planetary and terrestrial implications: New York, Pergamon, p. 11–44.

Cooper, H.F., and Sauer, F.M., 1977, Crater-related ground motions and implications for crater scaling, *in* Roddy, D.J., Pepin, R.O., and Merrill, R.B., eds., Impact and explosion cratering: Planetary and terrestrial implications: New York, Pergamon, p. 1133–1163.

Cooper, M.R., Kovach, R.L., and Watkins, J.S., 1974, Lunar near-surface structure: Reviews of Geophysics and Space Physics, v. 12, no. 3, p. 291–308.

Crater Analysis Techniques Working Group, 1979, Standard techniques for presentation and analysis of crater size-frequency data: Icarus, v. 37, no. 2, p. 467–474.

Croft, S.K., 1978, Lunar crater volumes: Interpretation by models of impact cratering and upper crustal structure: LPSCP 9, v. 3, p. 3711–3733.

———1980, Cratering flow fields: Implications for the excavation and transient expansion stages of crater formation: LPSCP 11, v. 3, p. 2347–2378.

———1981, The modification stage of basin formation: Conditions of ring formation, *in* Multi-ring basins: LPSCP 12, pt. A, p. 227–257.

Crozaz, Ghislaine, 1977, The irradiation history of the lunar soil: Physics and Chemistry of the Earth, v. 10, no. 3, p. 197–214.

Crozaz, Ghislaine, Drozd, R.J., Hohenberg, C.M., Hoyt, H.P., Ragan, D., Walker, R.M., and Yuhas, D., 1972, Solar flare and galactic cosmic ray studies of Apollo 14 and 15 samples: LSCP 3, v. 3, p. 2917–2931.

Cruikshank, D.P., and Wood, C.A., 1972, Lunar rilles and Hawaiian volcanic features: Possible analogues: The Moon, v. 3, no. 4, p. 412–446.

Cummings, David, 1971, Geologic map of the Wichmann CA region of the Moon, Lunar Orbiter site III P-11, Oceanus Procellarum, including Apollo landing sites 4 and 4R: USGS Map I-624 [ORB III-P11 (110)], scale 1:100,000.

———1972, Geologic map of the Clavius quadrangle of the Moon: USGS Map I-706 (LAC-126), scale 1:1,000,000.

Danes, Z.F., 1965, Rebound processes in large craters: USGS ASAPR, July 1, 1964 to July 1, 1965, pt. A, p. 81–100.

Davis, P.A., Jr., 1980, Iron and titanium distribution on the moon from orbital gamma ray spectrometry with implications for crustal evolutionary models: Journal of Geophysical Research, v. 85, no. B6, p. 3209–3224.

De Hon, R.A., 1971, Cauldron subsidence in lunar craters Ritter and Sabine: Journal of Geophysical Research, v. 76, no. 23, p. 5712–5718.

———1974, Thickness of mare material in the Tranquillitatis and Nectaris basins: LSCP 5, v. 1, p. 53–59.

———1979, Thickness of the western mare basalts: LPSCP 10, v. 3, p. 2935–2955.

De Hon, R.A., and Waskom, J.D., 1976, Geologic structure of the eastern mare basins: LSCP 7, v. 3, p. 2729–2746.

De Laeter, J.R., Vernon, M.J., and Compston, W., 1973, Revision of lunar Rb-Sr ages: Geochimica et Cosmochimica Acta, v. 37, no. 3, p. 700–702.

Delano, J.W., 1975, Petrology of the Apollo 16 mare component: Mare Nectaris: LSCP 6, v. 1, p. 15–47.

———1979, Apollo 15 green glass: Chemistry and possible origin: LPSCP 10, v. 1, p. 275–300.

———1980, Chemistry and liquidus phase relations of Apollo 15 red glass: Implications for deep lunar interior: LPSCP 11, v. 1, p. 251–288.

Delano, J.W., and Ringwood, A.E., 1978, Siderophile elements in the lunar highlands: Nature of the indigenous component and implication for the origin of the Moon: LPSCP 9, v. 1, p. 111–159.

Dence, M.R., 1964, A comparative structural and petrographic study of probable Canadian meteorite craters: Meteoritics, v. 2, p. 249–270.

———1965, The extraterrestrial origin of Canadian craters, *in* Whipple, H.E., ed., Geological problems in lunar research: New York Academy of Science Annals, v. 123, art. 2, p. 941–969.

———1968, Shock zoning at Canadian craters: Petrography and structural implications, *in* French, B.M., and Short, N.M., eds., Shock metamorphism of natural materials: Baltimore, Mono, p. 169–184.

———1971, Impact melts: Journal of Geophysical Research, v. 76, no. 23, p. 5552–5565.

Dence, M.R., and Grieve, R.A.F., 1979, The formation of complex impact structures [abs.]: LPS X, p. 292–294.

Dence, M.R., and Plant, A.G., 1972, Analysis of Fra Mauro samples and the origin of the Imbrium Basin: LSCP 3, v. 1, p. 379–399.

Dence, M.R., Grieve, R.A.F., and Robertson, P.B., 1977, Terrestrial impact structures: Principal characteristics and energy considerations, *in* Roddy, D.J., Pepin, R.O., and Merrill, R.B., eds., Impact and explosion cratering: Planetary and terrestrial implications: New York, Pergamon, p. 247–275.

Dennis, J.G., 1971, Ries structure, southern Germany, a review: Journal of Geophysical Research, v. 76, no. 23, p. 5394–5406.

De Paolo, D.J., McCulloch, M.T., Papanastassiou, D.A., Huneke, J.C., and Wasserburg, G.J., 1978, Ages, evolution and neutron effects of Luna 24 samples [abs.]: LPS IX, p. 244–246.

Dodd, R.T., Jr., Salisbury, J.W., and Smalley, V.G., 1963, Cratering frequency and interpretation of lunar history: Icarus, v. 2, no. 5–6, p. 466–480.

Dowty, Eric, Keil, Klaus, Prinz, Martin, Gros, J., and Takahashi, H., 1976, Meteorite-free Apollo 15 crystalline KREEP: LSCP 7, v. 2, p. 1833–1844.

Dowty, Eric, Prinz, Martin, and Keil, Klaus, 1974a, Ferroan anorthosite: A widespread and distinctive lunar rock type: Earth and Planetary Science Letters, v. 24, no. 1, p. 15–25.

———1974b, "Very high alumina basalt": A mixture and not a magma type: Science, v. 183, no. 4130, p. 1214–1215.

Dreibus, Gerlind, Kruse, H., Spettel, B., and Wänke, H., 1977, The bulk composition of the moon and the eucrite parent body: LSCP 8, v. 1, p. 211–227.

Drozd, R.J., Hohenberg, C.M., Morgan, C.J., Podosek, F.A., and Wroge, M.L., 1977, Cosmic-ray exposure history at Taurus-Littrow: LSCP 8, v. 3, p. 3027–3043.

Drozd, R.J., Hohenberg, C.M., Morgan, C.J., and Ralston, C.E., 1974, Cosmic-ray exposure history at the Apollo 16 and other lunar sites: Lunar surface dynamics: Geochimica et Cosmochimica Acta, v. 38, no. 10, p. 1625–1642.

Duncan, A.R., Grieve, R.A.F., and Weill, D.F., 1975, The life and times of Big Bertha: Lunar breccia 14321: Geochimica et Cosmochimica Acta, v. 39, no. 3, p. 265–273.

Dvorak, John, and Phillips, R.J., 1978, Lunar Bouguer gravity anomalies: Imbrian age craters: LPSCP 9, v. 3, p. 3651–3668.

———1979, Gravity anomaly and structure associated with the Lamont region of the moon: LPSCP 10, v. 3, p. 2265–2275.

Dymek, R.F., Albee, A.L., and Chodos, A.A., 1975, Comparative petrology of lunar cumulate rocks of possible primary origin: Dunite 72415, troctolite 76535, norite 78235, and anorthosite 62237: LSCP 6, v. 1, p. 301–341.

———1976, Petrology and origin of Boulders #2 and #3, Apollo 17 Station 2: LSCP 7, v. 2, p. 2335–2378.

Eberhardt, P., Geiss, J., Graf, H., Grögler, N., Krähenbühl, U., Schwaller, H., Schwarzmüller, J., and Stettler, A., 1970, Correlation between rock type and irradiation history of Apollo 11 igneous rocks: Earth and Planetary Science Letters, v. 10, no. 1, p. 67–72.

Eberhardt, P., Geiss, J., Grögler, N., Maurer, P., and Stettler, A., 1973a, ^{39}Ar-^{40}Ar *ages of lunar material* [abs.]: *Meteoritics*, v. 8, no. 4, p. 360–361.

Eberhardt, P., Geiss, J., Grögler, N., and Stettler, A., 1973b, How old is the crater Copernicus?: The Moon, v. 8, no. 1–2, p. 104–114.

Eggleton, R. E., 1964, Preliminary geology of the Riphaeus quadrangle of the Moon and definition of the Fra Mauro Formation: USGS ASAPR, August 25, 1962 to July 1, 1963, pt. A, p. 46–63.

———1965, Geologic map of the Riphaeus Mountains region of the Moon: USGS Map I-458 (LAC-76), scale 1:1,000,000.

Eggleton, R.E., and Marshall, C.H., 1962, Notes on the Apenninian Series and pre-Imbrian stratigraphy in the vicinity of Mare Humorum and Mare Nubium: USGS ASSPR, February 26, 1961 to August 24, 1961, p. 132–137.

Eggleton, R.E., and Offield, T.W., 1970, Geologic maps of the Fra Mauro region of the Moon: Apollo 14 premission maps: USGS Map I-708, scales 1:250,000, 1:25,000, 2 sheets.

Eggleton, R.E., and Schaber, G.G., 1972, Cayley Formation interpreted as basin ejecta, pt. B *of* Photogeology, [chap.] 29 *of* Apollo 16 preliminary science report: NASA Report SP-315, p. 29–7 to 29–16.

Eggleton, R.E., Schaber, G.G., and Pike, R.J., 1974, Photogeologic detection of surfaces buried by mare basalts [abs.]: LS V, pt. 1, p. 200–202.

Eichhorn, G., McGee, J.J., James, O. B., and Schaeffer, O.A., 1979, Consortium breccia 73255: Laser ^{39}Ar-^{40}Ar dating of aphanite samples: LPSCP 10, v. 1, p. 763–788.

El-Baz, Farouk, 1972, The Alhazen to Abul Wafa swirl belt: An extensive field of light-colored, sinuous markings, pt. T *of* Photogeology, [chap.] 29 *of* Apollo 16 preliminary science report: NASA Report SP-315, p. 29–93 to 29–97.

El-Baz, Farouk, and Roosa, S.A., 1972, Significant results from Apollo 14 lunar orbital photography: LSCP 3, v. 1, p. 63–83.

El-Baz, Farouk, Worden, A.M., and Brand, V.D., 1972, Astronaut observations from lunar orbit and their geologic significance: LSCP 3, v. 1, p. 85–104.

Eliason, E.M., and Soderblom, L.A., 1977, An array processing system for lunar geochemical and geophysical data: LSCP 8, v. 1, p. 1163–1170.

Elston, D.P., 1972, Geologic map of the Colombo quadrangle of the Moon: USGS Map I-714 (LAC-79), scale 1:1,000,000.

Engelhardt, Wolf von, 1967, Neue Beobachtungen im Nördlinger Ries [New observations in the Nördlingen Ries]: Geologische Rundschau, v. 57, no. 1, p. 165–188.

———1971, Detrital impact formations: Journal of Geophysical Research, v. 76, no. 23, p. 5566–5574.

Engelhardt, Wolf von, and Stengelin, Rudolf, 1979, Normative composition and classification of lunar igneous rocks and glasses, I. Lunar igneous rocks: Earth and Planetary Science Letters, v. 42, no. 2, p. 213–222.

Engelhardt, Wolf von, Arndt, J., Stöffler, D., and Schneider, H., 1972, Apollo 14 regolith and fragmental rocks, their compositions and origin by impacts: LSCP 3, v. 1, p. 753–770.

Eugster, O., Eberhardt, P., Geiss, J., Grögler, N., Jungck, M., and Mörgeli, M., 1977, The cosmic ray exposure history of Shorty Crater samples: The age of Shorty Crater: LSCP 8, v. 3, p. 3059–3082.

Evans, R.E., and El-Baz, Farouk, 1973, Geological observations from lunar orbit, [chap.] 28 *of* Apollo 17 preliminary science report: NASA Report SP-330, p. 28–1 to 28–32.

Evensen, N.M., Murthy, V.R., and Coscio, M.R., 1973, Rb-Sr ages of some mare basalts and their isotopic and trace element

systematics in lunar fines: LSCP 4, v. 2, p. 1707–1724.
Ferrari, A.J., 1977, Lunar gravity: A harmonic analysis: Journal of Geophysical Research, v. 82, no. 20, p. 3065–3084.
Ferrari, A.J., Nelson, D.L., Sjogren, W.L., and Phillips, R.J., 1978, The isostatic state of the lunar Apennines and regional surroundings: Journal of Geophysical Research, v. 83, no. B6, p. 2863–2871.
Fielder, Gilbert, 1961, Structure of the moon's surface: New York, Pergamon, 266 p.
———1965, Lunar geology: London, Lutterworth, 184 p.
Firsoff, V.A., 1961, Surface of the Moon: Its structure and origin: London, Hutchinson, 128 p.
Floran, R.J., and Dence, M.R., 1976, Morphology of the Manicouagan Ring-Structure, Quebec [Canada], and some comparisons with lunar basins and craters: LSCP 7, v. 3, p. 2845–2865.
Floran, R.J., Grieve, R.A.F., Phinney, W.C., Warner, J.L., Simonds, C.H., Blanchard, D.P., and Dence, M.R., 1978, Manicouagan impact melt, Quebec [Canada], 1, stratigraphy, petrology, and chemistry: Journal of Geophysical Research, v. 83, no. B6, p. 2737–2759.
Florensky, C.P., Basilevsky, A.T., Ivanov, A.V., Pronin, A.A., and Rode, O.D., 1977, Luna 24: Geologic setting of landing site and characteristics of sample core (preliminary data): LSCP 8, v. 3, p. 3257–3279.
Freeman, V.F., 1981, Regolith of the Apollo 16 site, [chap.] F *of* Ulrich, G.E., Hodges, C.A., and Muehlberger, W.R., eds., Geology of the Apollo 16 area, central lunar highlands: USGS Professional Paper 1048, p. 147–159.
French, B.M., 1977, The moon book: New York, Penguin, 287 p.
French, B.M., and Short, N.M., eds., 1968, Shock metamorphism of natural materials: Baltimore, Mono, 644 p.
Frondel, J.W., 1975, Lunar mineralogy: New York, Wiley, 323 p.
Fudali, R.F., Milton, D.J., Fredriksson, K., and Dube, A., 1980, Morphology of Lonar Crater, India: Comparisons and implications: The Moon and the Planets, v. 23, no. 4, p. 493–515.
Gaffney, E.S., 1978, Effects of gravity on explosion craters: LPSCP 9, v. 3, p. 3831–3842.
Gall, Horst, Müller, Dieter, and Stöffler, Dieter, 1975, Verteilung, Eigenschaften und Entstehung der Auswurfsmassen des Impaktkraters Nördlinger Ries [Distribution, properties, and origin of the ejecta of the Nördlingen Ries impact crater]: Geologische Rundschau, v. 64, p. 915–947.
Gault, D.E., 1970, Saturation and equilibrium conditions for impact cratering on the lunar surface: Criteria and implications: Radio Science, v. 5, no. 2, p. 273–291.
———1974, Impact cratering, *in* Impact craters, chap. 5 *of* Greeley, Ronald, and Schultz, P.H., eds., A primer in lunar geology (comment ed.): NASA Technical Memorandum TM-X-62359, p. 137–175.
Gault, D.E., and Heitowit, E.D., 1963, The partition of energy for hypervelocity impact craters formed in rock: Symposium on Hypervelocity Impact, 6th, Cleveland, Ohio, 1963, Proceedings, v. 2, pt. 2, p. 420–456.
Gault, D.E., and Moore, H.J., 1965, Scaling relationships for microscale to megascale impact craters: Symposium on Hypervelocity Impact, 7th, Tampa, Fla., 1964, Proceedings, v. 6, p. 341–351.
Gault, D.E., and Wedekind, J.A., 1977, Experimental hypervelocity impact into quartz sand—II, Effects of gravitational acceleration, *in* Roddy, D.J., Pepin, R.O., and Merrill, R.B., eds., Impact and explosion cratering: Planetary and terrestrial implications: New York, Pergamon, p. 1231–1244.
———1978, Experimental studies of oblique impact: LPSCP 9, v. 3, p. 3843–3875.
Gault, D.E., Adams, J.B., Collins, R.J., Kuiper, G.P., Masursky, Harold, O'Keefe, J.A., Phinney, R.A., and Shoemaker, E.M., 1968a, Lunar theory and processes, [chap.] 9 *of* Surveyor VII mission report. Part II. Science results: Pasadena, California Institute of Technology, Jet Propulsion Laboratory Technical Report 32-1264, p. 267–313.
Gault, D.E., Guest, J.E., Murray, J.B., Dzurisin, Daniel, and Malin, M.C., 1975, Some comparisons of impact craters on Mercury and the moon: Journal of Geophysical Research, v. 80, no. 17, p. 2444–2460.
Gault, D.E., Quaide, W.L., and Oberbeck, V.R., 1968b, Impact cratering mechanics and structures, *in* Cratering mechanics and meteorite impact craters, pt. 3 *of* French, B.M., and Short, N.M., eds., Shock metamorphism of natural materials: Baltimore, Mono, p. 87–99.
Gault, D.E., Shoemaker, E.M., and Moore, H.J., 1963, Spray ejected from the lunar surface by meteoroid impact: NASA Technical Note TN D-1767, 39 p.
Geiss, J., Eberhardt, P., Grögler, N., Guggisberg, S., Maurer, P., and Stettler, A., 1977, Absolute time scale of lunar mare formation and filling, *in* The Moon—a new appraisal from space missions and laboratory analyses: Royal Society of London Philosophical Transactions, ser. A, v. 285, p. 151–158.
Gilbert, G.K., 1893, The Moon's face, a study of the origin of its features: Philosophical Society of Washington Bulletin, v. 12, p. 241–292.
Gilvarry, J.J., 1970, Internal temperature of the Moon: Nature, v. 224, no. 5223, p. 968–970.
Glass, B.P., 1982, Introduction to planetary geology: Cambridge, U.K., Cambridge University Press, 469 p.
Goins, N.R., Toksöz, M.N., and Dainty, A.M., 1979, The lunar interior: A summary report: LPSCP 10, v. 3, p. 2421–2439.
Gold, Thomas, 1955, The lunar surface: Royal Astronomical Society Monthly Notices, v. 115, p. 585–604.
———1971, Evolution of mare surface: LSCP 2, v. 3, p. 2675–2680.
Golombek, M.P., 1979, Structural analysis of lunar grabens and the shallow crustal structure of the Moon: Journal of Geophysical Research, v. 84, p. 4657–4666.
Gooley, R.C., Brett, Robin, Warner, Jeff, and Smyth, J.R., 1974, A lunar rock of deep crustal origin: Sample 76535: Geochimica et Cosmochimica Acta, v. 38, no. 9, p. 1329–1339.
Greeley, Ronald, 1971, Observations of actively forming lava tubes and associated structures, Hawaii: Modern Geology, v. 2, no. 3, p. 207–233.
———1973, Comparative geology of crater Aratus CA (Mare Serenitatis) and Bear Crater (Idaho), pt. A *of* Volcanic studies, [chap.] 30 *of* Apollo 17 preliminary science report: NASA Report SP 330, p. 30-1 to 30-6.
———1976, Modes of emplacement of basalt terrains and an analysis of mare volcanism in the Orientale basin: LSCP 7, v. 3, p. 2747–2759.
Greeley, Ronald, and Carr, M.H., eds., 1976, A geological basis for the exploration of the planets: NASA Report SP-417, 109 p.
Greeley, Ronald, and Gault, D.E., 1970, Precision size-frequency distributions for craters for 12 selected areas of the lunar surface: The Moon, v. 2, no. 1, p. 10–77.
———1973, Crater frequency age determinations for the proposed Apollo 17 site at Taurus-Littrow: Earth and Planetary Science Letters, v. 18, no. 1, p. 102–108.
———1979, Endogenic craters on basaltic lava flows: Size frequency distributions: LPSCP 10, v. 3, p. 2919–2933.
Greeley, Ronald, and Spudis, P.D., 1978, Mare volcanism in the Herigonius region of the Moon: LPSCP 9, v. 3, p. 3333–3349.
Green, D.H., Ringwood, A.E., Hibberson, W.O., and Ware, N.G., 1975, Experimental petrology of Apollo 17 basalts: LSCP 6, v. 1, p. 871–893.
Green, D.H., Ringwood, A.E., Ware, N.G., and Hibberson, W.O., 1972, Experimental petrology and petrogenesis of Apollo 14 basalts: LSCP 3, v. 1, p. 197–206.
Green, Jack, 1971, Copernicus as a lunar caldera: Journal of Geophysical Research, v. 76, no. 23, p. 5719–5731.
———1976, *Review of* "Planetary Geology," by N.M. Short: Sky and Telescope, v. 51, no. 6, p. 417–420.
Grieve, R.A.F., 1975, Petrology and chemistry of the impact melt at Mistastin Lake crater, Labrador [Canada]: Geological Society of America Bulletin, v. 86, no. 12, p. 1617–1629.
———1978, The melt rocks at Brent Crater, Ontario, Canada: LPSCP 9, v. 2, p. 2579–2608.
———1980, Cratering in the lunar highlands: Some problems with the process, record and effects, *in* Lunar and Planetary Institute, compiler, Conference on the Lunar Highlands Crust, Houston, Tex., 1979, Proceedings: New York, Pergamon (Geochimica et Cosmochimica Acta, supp. 12), p. 173–196.
Grieve, R.A.F., and Dence, M.R., 1979, The terrestrial cratering record. II. The crater production rate: Icarus, v. 38, no. 2, p. 230–242.
Grieve, R.A.F., and Floran, R.J., 1978, Manicouagan impact melt, Quebec, 2. Chemical interrelations with basement and formational processes: Journal of Geophysical Research, v. 83, no. B6, p. 2761–2771.
Grieve, R.A.F., and Robertson, P.B., 1979, The terrestrial cratering record. I. Current status of observations: Icarus, v. 38, no. 2, p. 212–229.
Grieve, R.A.F., Dence, M.R., and Robertson, P.B., 1977, Cratering processes: As interpreted from the occurrence of impact melts, *in* Roddy, D.J., Pepin, R.O., and Merrill, R.B., eds., Impact and explosion cratering: Planetary and terrestrial implications: New York, Pergamon, p. 791–814.
Grieve, R.A.F., McKay, G.A., Smith, H.D., and Weill, D.F., 1975, Lunar polymict breccia 14321: A petrographic study: Geochimica et Cosmochimica Acta, v. 39, no. 3, p. 229–245.
Grieve, R.A.F., Plant, A.G., and Dence, M.R., 1974, Lunar impact melts and terrestrial analogs: Their characteristics, formation, and implications for lunar crustal evolution: LSCP 5, v. 1, p. 261–273.
Grolier, M.J., 1970a, Geologic map of Apollo site 2 (Apollo 11), part of Sabine D region, southwestern Mare Tranquillitatis: USGS Map I–619 [ORB II–6 (25)], scale 1:25,000.
———1970b, Geologic map of the Sabine D region of the Moon, Lunar Orbiter site II P–6, southwestern Mare Tranquillitatis, including Apollo landing site 2: USGS Map I–618 [ORB II–6 (100)], scale 1:100,000.
Guest, J.E., 1971, Centers of igneous activity in the maria, *in* Fielder, Gilbert, ed., Geology and physics of the moon: A study of some fundamental problems: Amsterdam, Elsevier, p. 41–53.
———1973, Stratigraphy of ejecta from the lunar crater Aristarchus: Geological Society of America Bulletin, v. 84, no. 9, p. 2873–2893.
Guest, J.E., and Greeley, Ronald, 1977, Geology on the Moon: London, Wykeham, 235 p.
Guest, J.E., and Murray, J.B., 1969, Nature and origin of Tsiolkovsky Crater, lunar farside: Planetary and Space Science, v. 17, p. 121–141.
———1976, Volcanic features of the nearside equatorial lunar maria: Geological Society of London Journal, v. 132, no. 3, p. 251–258.
Guggisberg, S., Eberhardt, P., Geiss, J., Grögler, N., Stettler, A., Brown, G.M., and Peckett, A., 1979, Classification of the Apollo-11 mare basalts according to Ar^{39}-Ar^{40} ages and petrologic properties: LPSCP 10, v. 1, p. 1–39.
Guinness, E.A., and Arvidson, R.E., 1977, On the constancy of the lunar cratering flux over the past 3.3×10^9 yr: LSCP 8, v. 3, p. 3475–3494.
Hackman, R.J., 1962, Geologic map and sections of the Kepler region of the Moon: USGS Map I–355 (LAC–57), scale 1:1,000,000.
———1964, Stratigraphy and structure of the Montes Apenninus region of the Moon: USGS ASAPR, August 25, 1962 to July 1, 1963, pt. A, p. 1–8.
———1966, Geologic map of the Montes Apenninus quadrangle of the Moon: USGS Map I–463 (LAC–41), scale 1:1,000,000.
Hackman, R.J., and Mason, A.C., 1961, Engineer special study of the surface of the moon: USGS Map I–351, scale 1:3,800,000, 4 sheets.
Haggerty, S.E., 1974, Apollo 17 orange glass: Textural and morphological characteristics of devitrification: LSCP 5, v. 1, 193–205.
Haines, E.L., and Metzger, A.E., 1980, Lunar highland crustal models based on iron concentrations: Isostasy and center-of-mass displacement: LPSCP 11, v. 1, p. 689–718.
Haines, E.L., Etchegaray-Ramirez, M.I., and Metzger, A.E., 1978, Thorium concentrations in the lunar surface. II: Deconvolution modeling and its application to the regions of Aristarchus and Mare Smythii: LPSCP 9, v. 3, p. 2985–3013.
Hale, Wendy, and Head, J.W., 1979, Central peaks in lunar craters: Morphology and morphometry: LPSCP 10, v. 3, p. 2623–2633.
Hall, J.L., Solomon, S.C., and Head, J.W., 1981, Lunar floor-fractured craters: Evidence for viscous relaxation of crater topography: Journal of Geophysical Research, v. 86, no. B10, p. 9537–9552.
Hall, R.C., 1977, Lunar impact: A history of Project Ranger: NASA Report SP–4210, 450 p.
Hansen, T.P., 1970, Guide to Lunar Orbiter photographs: NASA Report SP–242, 125 p.
Hapke, Bruce, Cassidy, William, and Wells, Edward, 1975, Effects of vapor-phase deposition processes on the optical, chemical, and magnetic properties of the lunar regolith: The Moon, v. 13, no. 1–3, p. 339–353.
Hartmann, W.K., 1963, Radial structures surrounding lunar basins, I: The Imbrium system: Tucson, University of Arizona, Lunar and Planetary Laboratory Communications, v. 2, no. 24, p. 1–15.
———1964a, On the distribution of lunar crater diameters: Tucson, University of Arizona, Lunar and Planetary Laboratory Communications, v. 2, no. 38, p. 197–203.
———1964b, Radial structures surrounding lunar basins, II: Orientale and other systems; conclusions: Tucson, University of Arizona, Lunar and Planetary Laboratory Communications, v. 2, no. 36, p. 175–191.
———1965a, Secular changes in meteoritic flux through the history of the solar system: Icarus, v. 4, no. 2, p. 207–213.
———1965b, Terrestrial and lunar flux of large meteorites in the last two billion years: Icarus, v. 4, no. 2, p. 157–165.
———1966, Early lunar cratering: Icarus, v. 5, no. 4, p. 406–418.
———1967, Lunar crater counts. II: Three lunar surface type areas: Tucson, University of Arizona, Lunar and Planetary Laboratory Communications, v. 6, no. 81, p. 39–41.
———1968, Lunar crater counts. VI: The young craters Tycho, Aristarchus, and Copernicus: Tucson, University of Arizona, Lunar and Planetary Laboratory Communications, v. 7, no. 119, p. 145–156.
———1972a, Interplanet variations in scale of crater morphology—Earth, Mars, Moon: Icarus, v. 17, no. 3, p. 707–713.
——— 1972b, Moons and planets: Belmont, Calif., Wadsworth, 404 p.
——— 1972c, Paleocratering of the Moon: Review of post-Apollo data: Astrophysics and Space Science, v. 17, p. 48–64.
——— 1973, Ancient lunar mega-regolith and subsurface structure: Icarus, v. 18, no. 4, p. 634–636.
——— 1975, Lunar "cataclysm": A misconception?: Icarus, v. 24, no. 2, p. 181–187.
——— 1980, Dropping stones in magma oceans: Effects of early lunar cratering, *in* Lunar and Planetary Institute, compiler, Conference on the Lunar Highlands Crust, Houston, Tex., 1979, Proceedings: New York, Pergamon (Geochimica et Cosmochimica Acta, supp. 12), p. 155–171.
——— 1981, Discovery of multi-ring basins: Gestalt perception in planetary science, *in* Multi-ring basins: LPSCP 12, pt. A, p. 79–90.
——— 1983, Moons and planets (2d ed): Belmont, Calif., Wadsworth, 509 p.
Hartmann, W.K., and Kuiper, G.P., 1962, Concentric structures surrounding lunar basins: Tucson, University of Arizona, Lunar and Planetary Laboratory Communications, v. 1, no. 12, p. 51–66.
Hartmann, W.K., and Wood, C.A., 1971, Moon: Origin and evolution of multi-ring basins: The Moon, v. 3, no. 1, p. 3–78.
Hartmann, W.K., and Yale, F.G., 1968, Lunar crater counts. IV: Mare Orientale and its basin system: Tucson, University of Arizona, Lunar and Planetary Laboratory Communications, v. 7, no. 117, p. 131–137.
Hawke, B.R., and Head, J.W., 1977a, Impact melt on lunar crater rims, *in* Roddy, D.J., Pepin, R.O., and Merrill, R.B., eds., Impact and explosion cratering: Planetary and terrestrial implications: New York, Pergamon, p. 815–841.
——— 1977b, Pre-Imbrian history of the Fra Mauro region and Apollo 14 sample provenance: LSCP 8, v. 3, p. 2741–2761.
——— 1978a, Criteria for the identification of the Imbrium ejecta and local components in the Apollo 14 samples [abs.]: LPS IX, pt. 1, p. 477–479.
——— 1978b, Lunar KREEP volcanism: Geologic evidence for history and mode of emplacement: LPSCP 9, v. 3, p. 3285–3309.

Hawke, B.R., and Spudis, P.D., 1980, Geochemical anomalies on the eastern limb and farside of the moon, *in* Lunar and Planetary Institute, compiler, Conference on the Lunar Highlands Crust, Houston, Tex., 1979, Proceedings: New York, Pergamon (Geochimica et Cosmochimica Acta, supp. 12), p. 467–481.

Head, J.W., 1972, Small-scale analogs of the Cayley Formation and Descartes Mountains in impact-associated deposits, pt. C *of* Photogeology, [chap.] 29 *of* Apollo 16 preliminary science report: NASA Report SP–315, p. 29–16 to 29–20.

——— 1974a, Lunar dark-mantle deposits: Possible clues to the distribution of early mare deposits: LSCP 5, v. 1, p. 207–222.

——— 1974b, Morphology and structure of the Taurus-Littrow highlands (Apollo 17): Evidence for their origin and evolution: The Moon, v. 9, no. 3–4, p. 355–395.

——— 1974c, Orientale multi-ringed basin interior and implications for the petrogenesis of lunar highland samples: The Moon, v. 11, no. 3–4, p. 327–356.

——— 1974d, Stratigraphy of the Descartes region (Apollo 16): Implications for the origin of samples: The Moon, v. 11, no. 1–2, p. 77–99.

——— 1976a, Lunar volcanism in space and time: Reviews of Geophysics and Space Physics, v. 14, no. 2, p. 265–300.

——— 1976b, The significance of substrate characteristics in determining morphology and morphometry of lunar craters: LSCP 7, p. 2913–2929.

——— 1977, Origin of outer rings in lunar multi-ringed basins: Evidence from morphology and ring spacing, *in* Roddy, D.J., Pepin, R.O., and Merrill, R.B., eds., Impact and explosion cratering: Planetary and terrestrial implications: New York, Pergamon, p. 563–573.

——— 1979a, Lava flooding of early planetary crusts: Geometry, thickness, and volumes of flooded impact basins [abs.]: LPS X, pt. 2, p. 516–518.

——— 1979b, Lava flooding of early planetary crusts: Geometry, thickness, and volumes of flooded lunar highland terrain [abs.]: LPS X, pt. 2, p. 519–521.

——— 1979c, Serenitatis multi-ringed basin: Regional geology and basin ring interpretation: The Moon and the Planets, v. 21, no. 4, p. 439–462.

Head, J.W., and Gifford, Ann, 1980, Lunar mare domes: Classification and modes of origin: The Moon and the Planets, v. 22, no. 2, p. 235–258.

Head, J.W., and Goetz, A.F.H., 1972, Descartes region: Evidence for Copernican-age volcanism: Journal of Geophysical Research, v. 77, no. 8, p. 1368–1374.

Head, J.W., and Hawke, B.R., 1975, Geology of the Apollo 14 region (Fra Mauro): Stratigraphic history and sample provenance: LSCP 6, v. 3, p. 2483–2501.

Head, J.W., and McCord, T.B., 1978, Imbrian-age highland volcanism on the Moon: The Gruithuisen and Mairan domes: Science, v. 199, no. 4336, p. 1433–1436.

Head, J.W., and Wilson, Lionel, 1979, Alphonsus-type dark-halo craters: Morphology, morphometry and eruption conditions: LPSCP 10, v. 3, p. 2861–2897.

Head, J.W., Adams, J.B., McCord, T.B., Pieters, C.M., and Zisk, S.H., 1978a, Regional stratigraphy and geologic history of Mare Crisium, *in* Lunar and Planetary Institute, compiler, Mare Crisium: The view from Luna 24: Conference on Luna 24, Houston, Tex., 1977, Proceedings: New York, Pergamon (Geochimica et Cosmochimica Acta, supp. 9), p. 43–74.

Head, J.W., Pieters, C.M., McCord, T.B., Adams, J.B., and Zisk, S.H., 1978b, Definition and detailed characterization of lunar surface units using remote observations: Icarus, v. 33, no. 1, p. 145–172.

Head, J.W., Robinson, Edmund, and Phillips, Roger, 1981, Topography of the Orientale basin [abs.]: LPS XII, pt. 2, p. 421–423.

Head, J.W., Settle, Mark, and Stein, R.S., 1975, Volume of material ejected from major lunar basins and implications for the depth of excavation of lunar samples: LSCP 6, v. 3, p. 2805–2829.

Heiken, G.H., 1975, Petrology of lunar soils: Reviews of Geophysics and Space Physics, v. 13, no. 4, p. 567–587.

Heiken, G.H., McKay, D.S., and Brown, R.W., 1974, Lunar deposits of possible pyroclastic origin: Geochimica et Cosmochimica Acta, v. 38, no. 11, p. 1703–1718.

Herbert, Floyd, Drake, M.J., Sonett, C.P., and Wiskerchen, M.J., 1977, Some constraints on the thermal history of the lunar magma ocean: LSCP 8, v. 1, p. 573–582.

Hertogen, Jan, Janssens, M.-J., Takahashi, H., Palme, Herbert, and Anders, Edward, 1977, Lunar basins and craters: Evidence for systematic compositional changes of bombarding population: LSCP 8, v. 1, p. 17–45.

Herzberg, C.T., 1978, The bearing of spinel cataclasites on the crust-mantle structure of the moon: LPSCP 9, v. 1, p. 319–336.

Herzberg, C.T., and Baker, M.B., 1980, The cordierite- to spinel-cataclasite transition: Structure of the lunar crust, *in* Lunar and Planetary Institute, compiler, Conference on the Lunar Highlands Crust, Houston, Tex., 1979, Proceedings: New York, Pergamon (Geochimica et Cosmochimica Acta, supp. 2), p. 113–132.

Hess, P.C., Rutherford, M.J., and Campbell, H.W., 1977, Origin and evolution of LKFM basalt: LSCP 8, v. 2, p. 2357–2373.

Hess, W.N., Menzel, D.H., and O'Keefe, J.A., eds., 1966, The nature of the lunar surface: Proceedings of the 1965 IAU-NASA Symposium: Baltimore, Johns Hopkins Press, 320 p.

Hinners, N.W., 1972, Apollo 16 site selection, [chap.] 1 *of* Apollo 16 preliminary science report: NASA Report SP–315, p. 1–1 to 1–3.

———1973, Apollo 17 site selection, [chap.] 1 *of* Apollo 17 preliminary science report: NASA Report SP–330, p. 1–1 to 1–5.

Hodges, C.A., 1972, Descartes highlands: Possible analogs around the Orientale Basin, pt. D *of* Photogeology, [chap.] 29 *of* Apollo 16 preliminary science report: NASA Report SP–315, p. 29–20 to 29–23.

———1973a, Geologic map of the Langrenus quadrangle of the Moon: USGS Map I–739 (LAC–80), scale 1:1,000,000.

———1973b, Geologic map of the Petavius quadrangle of the Moon: USGS Map I–794 (LAC–98), scale 1:1,000,000.

Hodges, C.A., and Muehlberger, W.R., 1981, A summary and critique of geological hypotheses, [chap.] K *of* Ulrich, G.E., Hodges, C.A., and Muehlberger, W.R., eds., Geology of the Apollo 16 area, central lunar highlands: USGS Professional Paper 1048, p. 215–230.

Hodges, C.A., and Wilhelms, D.E., 1978, Formation of lunar basin rings: Icarus, v. 34, no. 2, p. 294–323.

Hodges, C.A., Muehlberger, W.R., and Ulrich, G.E., 1973, Geologic setting of Apollo 16: LSCP 4, v. 1, p. 1–25.

Holcomb, Robin, 1971, Terraced depressions in lunar maria: Journal of Geophysical Research, v. 76, no. 23, p. 5703–5711.

Holt, H.E., 1974, Geologic map of the Purbach quadrangle of the Moon: USGS Map I–822 (LAC–95), scale 1:1,000,000.

Hood, L.L., Coleman, P.J., Jr., and Wilhelms, D.E., 1979, Lunar nearside magnetic anomalies: LPSCP 10, v. 3, p. 2235–2257.

Horn, P., Kirsten, T., and Jessberger, E.K., 1975, Are there A 12 mare basalts younger than 3.1 b.y. Unsuccessful search for A12 mare basalts with crystallization ages below 3.1 b.y. [abs.]: Meteoritics, v. 10, no. 4, p. 417–418.

Hörz, Friedrich, 1978, How thick are lunar mare basalts?: LPSCP 9, v. 3, p. 3311–3331.

———1981, The "Bunte Breccia" of the Ries: Implications for the Apollo 16 site, *in* James, O.B., and Hörz, Friedrich, eds., Workshop on Apollo 16: Houston, Tex., Lunar and Planetary Institute Technical Report 81–01, p. 53–57.

Hörz, Friedrich, and Banholzer, G.S., Jr., 1980, Deep seated target material in the continuous deposits of the Ries Crater, Germany, *in* Lunar and Planetary Institute, compiler, Conference on the Lunar Highlands Crust, Houston, Tex., 1979, Proceedings: New York, Pergamon (Geochimica et Cosmochimica Acta, supp. 12), p. 211–231.

Hörz, Friedrich, and Ostertag, Rolf, 1979, The transient crater of the Ries crater, Germany [abs.]: LPS X, pt. 2, p. 570–572.

Howard, K.A., 1970, Mascons, mare rock and isostasy: Nature, v. 226, no. 5249, p. 924–925.

———1974, Fresh lunar impact craters: Review of variations with size: LSCP 5, v. 1, p. 61–69.

———1975, Geologic map of the crater Copernicus: USGS Map I–840, scale 1:250,000.

Howard, K.A., and Masursky, Harold, 1968, Geologic map of the Ptolemaeus quadrangle of the Moon: USGS Map I–566 (LAC–77; RLC–13), scale 1:1,000,000.

Howard, K.A., and Wilshire, H.G., 1975, Flows of impact melt at lunar craters: USGS Journal of Research, v. 3, no. 2, p. 237–257.

Howard, K.A., Carr, M.H., and Muehlberger, W.R., 1973, Basalt stratigraphy of southern Mare Serenitatis, pt. A *of* Stratigraphic studies, [chap.] 29 *of* Apollo 17 preliminary science report: NASA Report SP–330, p. 29–1 to 29–12.

Howard, K.A., Head, J.W., and Swann, G.A., 1972, Geology of Hadley rille: LSCP 3, v. 1, p. 1–14.

Howard, K.A., Wilhelms, D.E., and Scott, D.H., 1974, Lunar basin formation and highland stratigraphy: Reviews of Geophysics and Space Physics, v. 12, no. 3, p. 309–327.

Hubbard, N.J., 1979, Regional chemical variations in lunar basaltic lavas: LPSCP 10, v. 2, p. 1753–1774.

Hubbard, N.J., and Gast, P.W., 1971, Chemical composition and origin of nonmare lunar basalts: LSCP 2, v. 2, p. 999–1020.

Hubbard, N.J., and Minear, J.W., 1975, A physical and chemical model of early lunar history: LSCP 6, v. 1, p. 1057–1085.

Hubbard, N.J., Meyer, Charles, Jr., Gast, P.W., and Wiesmann, Henry, 1971, The composition and derivation of Apollo 12 soils: Earth and Planetary Science Letters, v. 10, no. 3, p. 341–350.

Hubbard, N.J., Rhodes, J.M., and Gast, P.W., 1973, Chemistry of lunar basalts with very high alumina contents: Science, v. 181, no. 4097, p. 339–342.

Hubbard, N.J., Rhodes, J.M., Wiesmann, H., Shih, C.-Y., and Bansal, B.M., 1974, The chemical definition and interpretation of rock types returned from the non-mare regions of the Moon: LSCP 5, v. 2, p. 1227–1246.

Hubbard, N.J., Vilas, Faith, and Keith, J.E., 1978, From Serenity to Langemak: A regional chemical setting for Mare Crisium, *in* Lunar and Planetary Institute, compiler, Mare Crisium: The view from Luna 24: Conference on Luna 24, Houston, Tex., 1977, Proceedings: New York, Pergamon (Geochimica et Cosmochimica Acta, supp. 9), p. 13–32.

Huneke, J.C., 1978, ^{40}Ar-^{39}Ar microanalysis of single glass balls and 72435 breccia clasts: LPSCP 9, v. 2, p. 2345–2362.

Huneke, J.C., and Wasserburg, G.J., 1975, Trapped ^{40}Ar in troctolite 76535 and evidence for enhanced ^{40}Ar-^{39}Ar age plateaus [abs.]: LS VI, pt. 1, p. 417–419.

Huneke, J.C., Jessberger, E.K., Podosek, F.A., and Wasserburg, G.J., 1973, ^{40}Ar/^{39}Ar measurements in Apollo 16 and 17 samples and the chronology of metamorphic and volcanic activity in the Taurus-Littrow region: LSCP 4, v. 2, p. 1725–1756.

Huneke, J.C., Jessberger, E.K., and Wasserburg, G.J., 1974, The age of metamorphism of a highland breccia (65015) and a glimpse at the age of its protolith [abs.]: LS V, pt. 1, p. 375–377.

Huneke, J.C., Podosek, F.A., and Wasserburg, G.J., 1972, Gas retention and cosmic-ray exposure ages of a basalt fragment from Mare Fecunditatis: Earth and Planetary Science Letters, v. 13, no. 2, p. 375–383.

Husain, Liaquat, 1974, ^{40}Ar-^{39}Ar chronology and cosmic ray exposure ages of the Apollo 15 samples: Journal of Geophysical Research, v. 79, no. 17, p. 2588–2606.

Husain, Liaquat, and Schaefer, O.A., 1973, Lunar volcanism: Age of the glass in the Apollo 17 orange soil: Science, v. 180, no. 4093, p. 1358–1360.

———1975, Lunar evolution: The first 600 million years: Geophysical Research Letters, v. 2, no. 1, p. 29–32.

Husain, Liaquat, Schaeffer, O.A., Funkhouser, J., and Sutter, J.F., 1972, The ages of lunar material from Fra Mauro, Hadley Rille, and Spur Crater: LSCP 3, v. 2, p. 1557–1567.

Irving, A.J., 1975, Chemical, mineralogical and textural systematics of non-mare melt rocks: Implications for lunar impact and volcanic processes: LSCP 6, v. 1, p. 363–394.

———1977, Chemical variation and fractionation of KREEP basalt magma: LSCP 8, v. 2, p. 2433–2448.

Jackson, E.D., 1971, The origin of ultramafic rocks by cumulus processes: Fortschritte der Mineralogie, v. 48, no. 1, p. 128–174.

Jackson, E.D., Sutton, R.L., and Wilshire, H.G., 1975, Structure and petrology of a cumulus norite boulder sampled by Apollo 17 in Taurus-Littrow valley, the Moon: Geological Society of America Bulletin, v. 86, no. 4, p. 433–442.

James, O.B., 1972, Lunar anorthosite 15415: Texture, minerology, and metamorphic history: Science, v. 175, no. 4020, p. 432–436.

———1973, Crystallization history of lunar feldspathic basalt 14310: USGS Professional Paper 841, 29 p.

———1976, Petrology of aphanitic lithologies in consortium breccia 73215: LSCP 7, v. 2, p. 2145–2178.

———1977, Lunar highlands breccias generated by major impacts, *in* Pomeroy, J.H., and Hubbard, N.J., eds., The Soviet-American Conference on Cosmochemistry of the Moon and Planets: NASA Report SP–370, p. 637–658.

———1980, Rocks of the early lunar crust: LPSCP 11, v. 1, p. 365–393.

———1981, Petrologic and age relations of the Apollo 16 rocks: Implications for the subsurface geology and the age of the Nectaris basin: LPSCP 12, pt. B, sec. 1, p. 209–233.

James, O.B., and Hörz, Friedrich, eds., 1981, Workshop on Apollo 16: Houston, Tex., Lunar and Planetary Institute Technical Report 81–01, 157 p.

James, O.B., and Jackson, E.D., 1970, Petrology of the Apollo 11 ilmenite basalts: Journal of Geophysical Research, v. 75, no. 29, p. 5793–5824.

James, O.B., and McGee, J.J., 1980, Petrology of mare-type basalt clasts from consortium breccia 73255: LPSCP 11, v. 1, p. 67–86.

James, O.B., and Wright, T.L., 1972, Apollo 11 and 12 mare basalts and gabbros: Classification, compositional variations, and possible petrogenetic relations: Geological Society of America Bulletin, v. 83, no. 8, p. 2357–2382.

James, O.B., Hedenquist, J.W., Blanchard, D.P., Budahn, J.R., and Compston, W., 1978, Consortium breccia 73255: Petrology, major- and trace-element chemistry, and Rb-Sr systematics of aphanitic lithologies: LPSCP 9, v. 1, p. 789–819.

Jessberger, E.K., Dominik, B., Kirsten, T., and Staudacher, T., 1977a, New ^{40}Ar-^{39}Ar ages of Apollo 16 breccias and 4.42 AE old anorthosites [abs.]: LS VIII, pt. 1, p. 511–513.

Jessberger, E.K., Horn, P., and Kirsten, T., 1975, ^{39}Ar-^{40}Ar dating of lunar rocks: A methodical investigation of mare basalt 75075 [abs.]: LS VI, pt. 1, p. 441–443.

Jessberger, E.K., Huneke, J.C., Podosek, F.A., and Wasserburg, G.J., 1974, High resolution argon analysis of neutron-irradiated Apollo 16 rocks and separated minerals: LSCP 5, v. 2, p. 1419–1449.

Jessberger, E.K., Kirsten, T., Staudacher, T., 1977b, One rock and many ages—further K-Ar data on consortium breccia 73215: LSCP 8, v. 2, p. 2567–2580.

Jessberger, E.K., Staudacher, T., Dominik, B., and Kirsten, T., 1978, Argon-argon ages of aphanite samples from consortium breccia 73255: LPSCP 9, v. 1, p. 841–854.

Johnson, T.V., Matson, D.L., Phillips, R.J., and Saunders, R.S., 1975, Vidicon spectral imaging: Color enhancement and digital maps: LSCP 6, v. 3, p. 2677–2688.

Johnson, T.V., Saunders, R.S., Matson, D.L., and Mosher, J.L., 1977, A TiO_2 abundance map for the northern maria: LSCP 8, v. 1, p. 1029–1036.

Jones, E.M., and Sandford, M.T., II, 1977, Numerical simulation of a very large explosion at the earth's surface with possible application to textites, *in* Roddy, D.J., Pepin, R.O., and Merrill, R.B., eds., Impact and explosion cratering: Planetary and terrestrial implications: New York, Pergamon, p. 1009–1024.

Jones, G.H.S., 1977, Complex craters in alluvium, *in* Roddy, D.J., Pepin, R.O., and Merrill, R.B., eds., Impact and explosion cratering: Planetary and terrestrial implications: New York, Pergamon, p. 163–183.

Karlstrom, T.N.V., 1974, Geologic map of the Schickard quadrangle of the Moon: USGS Map I–823 (LAC–110), scale 1:1,000,000.

Kaula, W.M., 1977, On the origin of the moon, with emphasis on bulk composition: LSCP 8, v. 1, p. 321–331.

Kaula, W.M., Schubert, G., Lingenfelter, R.E., Sjogren, W.L., and Wollenhaupt, W.R., 1973, Lunar topography from Apollo 15 and 16 laser altimetry: LSCP 4, v. 3, p. 2811–2819.

———1974, Apollo laser altimetry and inferences as to lunar structure: LSCP 5, v. 3, p. 3049–3058.

Keil, Klaus, Kurat, Gero, Prinz, Martin, and Green, J.A., 1972,

Lithic fragments, glasses and chondrules from Luna 16 fines: Earth and Planetary Science Letters, v. 13, no. 2, p. 243–256.
Kesson, S.E., and Lindsley, D.H., 1976, Mare basalt petrogenesis—a review of experimental studies: Reviews of Geophysics and Space Physics, v. 14, p. 361–373.
Kieffer, S.W., and Simonds, C.H., 1980, The role of volatiles and lithology in the impact cratering process: Reviews of Geophysics and Space Physics, v. 18, no. 1, p. 143–181.
King, E.A., 1976, Space geology: An introduction: New York, Wiley, 349 p.
Kirsten, T., and Horn, P., 1974, Chronology of the Taurus-Littrow region III: Ages of mare basalts and highland breccias and some remarks about the interpretation of lunar highland rock ages: LSCP 5, v. 2, 1451–1475.
Kirsten, T., Horn, P., and Heymann, D., 1973a, Chronology of the Taurus-Littrow region I: Ages of two major rock types from the Apollo 17-site: Earth and Planetary Science Letters, v. 20, no. 1, p. 125–130.
Kirsten, T., Horn, P., and Kiko, J., 1973b, ^{39}Ar-^{40}Ar dating and rare gas analysis of Apollo 16 rocks and soils: LSCP 4, v. 2, p. 1757–1784.
Knowles, C.P., and Brode, H.L., 1977, The theory of cratering phenomena, an overview, *in* Roddy, D.J., Pepin, R.O., and Merrill, R.B., eds., Impact and explosion cratering: Planetary and terrestrial implications: New York, Pergamon, p. 869–895.
Kopal, Zdeněk, ed., 1962, Physics and astronomy of the moon: New York, Academic Press, 538 p.
Kopal, Zdeněk, and Mikhailov, Z.K., eds., 1962, The Moon: London, Academic Press, 571 p.
Kosofksy, L.J., and El-Baz, Farouk, 1970, The Moon as viewed by Lunar Orbiter: NASA Report SP-200, 152 p.
Kovach, R.L., and Watkins, J.S., 1972, The near-surface velocity structure of the Moon [abs.]: LS III, p. 461–462.
Koyama, Junji, and Nakamura, Yosio, 1979, Re-examination of the lunar seismic velocity structure based on the complete data set [abs.]: LPS X, pt. 2, p. 685–687.
Kreyenhagen, K.N., and Schuster, S.H, 1977, Review and comparison of hypervelocity impact and explosion cratering calculations, *in* Roddy, D.J., Pepin, R.O., and Merrill, R.B., eds., Impact and explosion cratering: Planetary and terrestrial implications: New York, Pergamon, p. 983–1002.
Kuiper, G.P., 1959, The exploration of the Moon, *in* Alperin, Morton, and Hollingsworth, F.G., eds., Vistas in astronautics: London, Pergamon, v. 2, p. 273–313.
———1965, Interpretation of *Ranger VII* records, [chap.] 3 *of Ranger VII*. Part II. Experimenters' analyses and interpretations: Pasadena, California Institute of Technology, Jet Propulsion Laboratory Technical Report 32–700, p. 9–73.
Kuiper, G.P., Strom, R.G., and Le Poole, R.S., 1966, Interpretation of the *Ranger* records, [chap.] 3 *of Ranger VIII* and *IX*. Part II. Experimenters' analyses and interpretations: Pasadena, California Institute of Technology, Jet Propulsion Laboratory Technical Report 32–800, p. 35–248.
Kurat, Gero, Kracher, A., Keil, K., Warner, R., and Prinz, M., 1976, Composition and origin of Luna 16 aluminous mare basalts: LSCP 7, v. 2, p. 1301–1321.
Lammlein, D.R., 1977, Lunar seismicity and tectonics: Physics of the Earth and Planetary Interiors, v. 14, no. 3, p. 224–273.
Lange, M.A., and Ahrens, T.J., 1979, Impact melting early in lunar history: LPSCP 10, v. 3, p. 2707–2725.
Langseth, M.G., Keihm, S.J., and Peters, Kenneth, 1976, Revised lunar heat-flow values: LSCP 7, v. 3, p. 3143–3171.
Latham, G.V., Dorman, H.J., Horvath, P., Ibrahim, A.K., Koyama, J., and Nakamura, Y., 1978, Passive seismic experiment: A summary of current status: LPSCP 9, v. 3, p. 3609–3613.
Leich, D.A., Kahl, S.B., Kirschbaum, A.R., Niemeyer, S., and Phinney, D., 1975, Rare gas constraints on the history of Boulder 1, Station 2, Apollo 17: The Moon, v. 14, p. 407–444.
Levin, B.J., and Mayeva, S.V., 1977, Riddles about the origin and thermal history of the Moon, *in* Pomeroy, J.H., and Hubbard, N.J., eds., The Soviet-American Conference on Cosmochemistry of the Moon and Planets: NASA Report SP-370, p. 367–385.
Lindsay, J.F., 1976, Lunar stratigraphy and sedimentology: Amsterdam, Elsevier, 302 p.
Lipsky, Y.N., 1965, Zond-3 photographs of the Moon's far side: Sky and Telescope, v. 30, no. 6, p. 338–341.
Lofgren, G.E., 1977, Dynamic crystallization experiments bearing on the origin of textures in impact-generated liquids: LSCP 8, v. 2, p. 2079–2095.
Lofgren, G.E., Donaldson, C.H., and Usselman, T.M., 1975, Geology, petrology, and crystallization of Apollo 15 quartz-normative basalts: LSCP 6, v. 1, p. 79–99.
Longhi, John, 1980, A model of early lunar differentiation: LPSCP 11, v. 1, p. 289–315.
Longhi, John, and Boudreau, A.E., 1979, Complex igneous processes and the formation of the primitive lunar crustal rocks: LPSCP 10, v. 2, p. 2085–2105.
Lowman, P.D., Jr., 1969, Lunar panorama—a photographic guide to the geology of the Moon: Zürich, Reinhold Müller, 101 p.
Lucchitta, B.K., 1972, Geologic sketch map of the candidate Proclus Apollo landing site, pt. K of Orbital-science investigations, [chap.] 25 *of* Apollo 15 preliminary science report: NASA Report SP-289, p. 25–76 to 25–80.
———1976, Mare ridges and related highland scarps—result of vertical tectonism?: LSCP 7, v. 3, p. 2761–2782.
———1977a, Crater clusters and light mantle at the Apollo 17 site: A result of secondary impact from Tycho: Icarus, v. 30, no. 1, p. 80–96.
———1977b, Topography, structure, and mare ridges in southern Mare Imbrium and northern Oceanus Procellarum: LSCP 8, v. 3, p. 2691–2703.
———1978, Geologic map of the north side of the Moon: USGS Map I-1062, scale 1:5,000,000.
Lucchitta, B.K., and Sanchez, A.G., 1975, Crater studies in the Apollo 17 region: LSCP 6, v. 3, p. 2427–2441.
Lucchitta, B.K., and Schmitt, H.H., 1974, Orange material in the Sulpicius Gallus Formation at the southwestern edge of Mare Serenitatis: LSCP 5, v. 1, p. 223–234.
Lucchitta, B.K., and Watkins, J.A., 1978, Age of graben systems on the moon: LPSCP 9, v. 3, p. 3459–3472.
Lugmair, G.W., and Carlson, R.W., 1978, The Sm-Nd history of KREEP: LPSCP 9, v. 1, p. 689–704.
Lugmair, G.W., and Marti, Kurt, 1972, Exposure ages and neutron capture record in lunar samples from Fra Mauro: LSCP 3, v. 2, p. 1891–1897.
Lugmair, G.W., Marti, Kurt, Kurtz, J.P., and Scheinin, N.B., 1976, History and genesis of lunar troctolite 76535 or: How old is old?: LSCP 7, v. 2, p. 2009–2033.
Lugmair, G.W., Scheinin, N.B., and Marti, Kurt, 1975, Sm-Nd age and history of Apollo 17 basalt 75075: Evidence for early differentiation of the lunar interior: LSCP 6, v. 2, p. 1419–1429.
Lunatic Asylum, 1978, Petrology, chemistry, age and irradiation history of Luna 24 samples *in* Lunar and Planetary Institute, compiler, Mare Crisium: The view from Luna 24: Conference on Luna 24, Houston, Tex., 1977, Proceedings: New York, Pergamon (Geochimica et Cosmochimica Acta, supp. 9), p. 657–678.
Ma, M.-S., Schmitt, R.A., Taylor, G.J., Warner, R.D., Lange, D.E., and Keil, Klaus, 1978, Chemistry and petrology of Luna 24 lithic fragments and <250 micrometer soils: Constraints on the origin of VLT mare basalts, *in* Lunar and Planetary Institute, compiler, Mare Crisium: The view from Luna 24: Conference on Luna 24, Houston, Tex., 1977, Proceedings: New York, Pergamon (Geochimica et Cosmochimica Acta, supp. 9), p. 569–592.
Mackin, J.H., 1969, Origin of lunar maria: Geological Society of America Bulletin, v. 80, no. 5, p. 735–747.
Malin, M.C., 1974, Lunar red spots: Possible pre-mare materials: Earth and Planetary Science Letters, v. 21, no. 4, p. 331–341.
Marcus, A.H., 1970, Comparison of equilibrium size distributions for lunar craters: Journal of Geophysical Research, v. 75, no. 26, p. 4977–4984.
Mark, R.K., Cliff, R.A., Lee-Hu, C.-N., and Wetherill, G.W., 1973, Rb-Sr studies of lunar breccias and soils: LSCP 4, v. 2, p. 1785–1795.
Mark, R.K., Lee-Hu, C.-N., and Wetherill, G.W., 1974, Rb-Sr ages of lunar igneous rocks 62295 and 14310: Geochimica et Cosmochimica Acta, v. 38, p. 1643–1648.
———1975, More on Rb-Sr in lunar breccia 14321: LSCP 6, v. 2, p. 1501–1507.
Markov, A.V., ed., 1962, The Moon—a Russian view: Chicago, University of Chicago Press, 391 p.
Marshall, C.H., 1961, Thickness of the Procellarian system, Letronne region of the Moon, art. 361 *of* Short papers in the geologic and hydrologic sciences: USGS Professional Paper 424–D, p. D208–D211.
———1963, Geologic map and sections of the Letronne region of the Moon: USGS Map I–385 (LAC–75), scale 1:1,000,000.
Marti, Kurt, 1967, Mass-spectrometric detection of cosmic-ray-produced Kr^{81} in meteorites and the possibility of Kr-Kr dating: Physical Review Letters, v. 18, no. 7, p. 264–266.
Marti, Kurt, Lightner, B.D., and Osborn, T.W., 1973, Krypton and xenon in some lunar samples and the age of North Ray Crater: LSCP 4, v. 2, p. 2037–2048.
Marvin, U.B., Wood, J.A., Taylor, G.J., Reid, J.B., Powell, B.N., Dickey, J.S., and Bower, J.F., 1971, Relative proportions and possible sources of rock fragments in the Apollo 12 soil samples: LSCP 2, v. 1, p. 679–699.
Mason, A.C., and Hackman, R.J., 1962, Photogeologic study of the Moon, *in* Kopal, Zdeněk, and Mikhailov, Z.K., eds., 1962, The Moon: London, Academic Press, p. 301–315.
Mason, Roger, Guest, J.E., and Cooke, G.N., 1976, An Imbrium pattern of graben on the Moon: Geologists' Association (London) Proceedings, v. 87, pt. 2, p. 161–168.
Masursky, Harold, Colton, G.W., and El-Baz, Farouk, eds., 1978, Apollo over the Moon: A view from orbit: NASA Report SP–362, 255 p.
Maurer, P., Eberhardt, P., Geiss, J., Grögler, N., Stettler, A., Brown, G.M., Peckett, A., and Krähenbühl, U., 1978, Pre-Imbrian craters and basins: Ages, compositions and excavation depths of Apollo 16 breccias: Geochimica et Cosmochimica Acta, v. 42, no. 11, p. 1687–1720.
Maxwell, T.A., and Andre, C.G., 1981, The Balmer Basin: Regional geology and geochemistry of an ancient lunar impact basin: LPSCP 12, pt. B, sec. 1, p. 715–725.
Maxwell, T.A., and Phillips, R.J., 1978, Stratigraphic correlation of the radar-detected subsurface interface in Mare Crisium: Geophysical Research Letters, v. 5, no. 9, p. 811–814.
Maxwell, T.A., El-Baz, Farouk, and Ward, S.H., 1975, Distribution, morphology, and origin of ridges and arches in Mare Serenitatis: Geological Society of America Bulletin, v. 86, no. 9, p. 1273–1278.
McCall, G.J.H., 1965, The caldera analogy in selenology, *in* Whipple, H.E., ed., Geological problems in lunar research: New York Academy of Sciences Annals, v. 123, art. 2, p. 843–875.
———1980, Impact and volcanism in planetology: The state of the lunar controversy in 1979: British Astronomical Association Journal, v. 90, p. 346–368.
McCallum, I.S., Okamura, F.P., and Ghose, Subrata, 1975, Mineralogy and petrology of sample 67075 and the origin of lunar anorthosites: Earth and Planetary Science Letters, v. 26, no. 1, p. 36–53.
McCauley, J.F., 1964a, A preliminary report on the geology of the Hevelius Quadrangle: USGS ASAPR, August 25, 1962 to July 1, 1963, pt. A, p. 74–85.
———1964b, The stratigraphy of the Mare Orientale region of the Moon: USGS ASAPR, August 25, 1962 to July 1, 1963, pt. A, p. 86–98.
———1967a, Geologic map of the Hevelius region of the Moon: USGS Map I–491 (LAC–56), scale 1:1,000,000.
———1967b, The nature of the lunar surface as determined by systematic geologic mapping, *in* Runcorn, S.K., ed., Mantles of the earth and terrestrial planets: New York, Interscience, p. 431–460.
———1968, Geologic results from the lunar precursor probes: American Institute of Aeronautics and Astronautics Journal, v. 6, p. 1991–1996.
———1969a, Geologic map of the Alphonsus region GA region of the Moon: USGS Map I–586 (RLC–15), scale 1:50,000.
———1969b, The domes and cones in the Marius Hills region—evidence for lunar differentiation? [abs.]: Eos (American Geophysical Union Transactions), v. 50, no. 4, p. 229.
———1973, Geologic map of the Grimaldi quadrangle of the Moon: USGS Map I–740 (LAC–74), scale 1:1,000,000.
———1977, Orientale and Caloris: Physics of the Earth and Planetary Interiors, v. 15, no. 2–3, p. 220–250.
McCauley, J.F., and Scott, D.H., 1972, The geologic setting of the Luna 16 landing site: Earth and Planetary Science Letters, v. 13, no. 2, p. 225–232.
McCord, T.B., 1969, Color differences on the lunar surface: Journal of Geophysical Research, v. 74, no. 12, p. 3131–3142.
McCord, T.B., and Johnson, T.V., 1970, Lunar spectral reflectivity (0.30 to 2.50 microns) and implications for remote mineralogical analysis: Science, v. 169, no. 3948, p. 855–858.
McCord, T.B., Charette, M.P., Johnson, T.V., Lebofsky, L.A., and Pieters, C.M., 1972a, Lunar spectral types: Journal of Geophysical Research, v. 77, no. 8, p. 1349–1359.
———1972b, Spectrophotometry (0.3 to 1.1 μ) of visited and proposed Apollo lunar landing sites: The Moon, v. 5, no. 1–2, p. 52–89.
McCord, T.B., Clark, R.N., Hawke, B.R., McFadden, L.A., Owensby, P.D., Pieters, C.M., and Adams, J.B., 1981, Moon: Near-infrared spectral reflectance, a first good look: Journal of Geophysical Research, v. 86, no. B11, p. 10883–10892.
McCord, T.B., Pieters, C.M., and Feierberg, M.A., 1976, Multispectral mapping of the lunar surface using ground-based telescopes: Icarus, v. 29, no. 1, p. 1–34.
McGetchin, T.R., Settle, Mark, and Head, J.W., 1973, Radial thickness variations in impact crater ejecta: Implications for lunar basin deposits: Earth and Planetary Science Letters, v. 20, no. 2, p. 226–236.
McGill, G.E., 1971, Attitude of fractures bounding straight and arcuate lunar rilles: Icarus, v. 14, no. 1, p. 53–58.
———1977, Craters as "fossils": the remote dating of planetary surface materials: Geological Society of America Bulletin, v. 88, no. 8, p. 1102–1110.
McKay, G.A., and Weill, D.F., 1977, KREEP petrogenesis revisited: LSCP 8, v. 2, p. 2339–2355.
McKay, G.A., Wiesmann, H., Bansal, J.L., and Shih, C.-Y., 1979, Petrology, chemistry, and chronology of Apollo 14 KREEP basalts: LPSCP 10, v. 1, p. 181–205.
McKay, G.A., Wiesmann, H., Nyquist, L.E., Wooden, J.L., and Bansal, J.L., 1978, Petrology, chemistry, and chronology of 14078: Chemical constraints on the origin of KREEP: LPSCP 9, v. 1, p. 661–687.
McKinnon, W.B., 1981, Application of ring tectonic theory to Mercury and other solar system bodies, *in* Multi-ring basins: LPSCP 12, pt. A, p. 259–273.
Melosh, H.J., 1977, Crater modification by gravity: A mechanical analysis of slumping, *in* Roddy, D.J., Pepin, R.O., and Merrill, R.B., eds., Impact and explosion cratering: Planetary and terrestrial implications: New York, Pergamon, p. 1245–1260.
———1978, The tectonics of mascon loading: LPSCP 9, v. 3, p. 3513–3525.
———1980, Cratering mechanics—observational, experimental, and theoretical: Annual Review of Earth and Planetary Science, v. 8, p. 65–93.
Mendell, W.W., and Low, F.J., 1975, Infrared orbital mapping of lunar features: LSCP 6, v. 3, p. 2711–2719.
Metzger, A.E., and Parker, R.E., 1979, The distribution of titanium on the lunar surface: Earth and Planetary Science Letters, v. 45, no. 1, p. 155–171.
Metzger, A.E., Haines, E.L., Parker, R.E., and Radocinski, R.G., 1977, Thorium concentrations in the lunar surface. I: Regional values and crustal content: LSCP 8, v. 1, p. 949–999.
Metzger, A.E., Haines, E.L., Etchegaray-Ramirez, M.I., and Hawke, B.R., 1979, Thorium concentrations in the lunar surface: III. Deconvolutions of the Apenninus region: LPSCP 10, v. 2, p. 1701–1718.
Metzger, A.E., Trombka, J.I., Peterson, L.E., Reedy, J.C., and Arnold, J.R., 1973, Lunar surface radioactivity: Preliminary results of the Apollo 15 and Apollo 16 gamma-ray spectrometer experiments: Science, v. 179, no. 4075, p. 800–803.
Meyer, Charles, Jr., 1977, Petrology, mineralogy and chemistry of KREEP basalt: Physics and Chemistry of the Earth, v. 10, no. 4, p. 239–260.
Meyer, Charles, Jr., Brett, Robin, Hubbard, N.J., Morrison, D.A., McKay, D.S., Aitken, F.K., Takeda, H., and Schonfeld, Ernest, 1971, Mineralogy, chemistry and origin of the KREEP compo-

nent in soil samples from the Ocean of Storms: LSCP 2, v. 1, p. 393–411.

M'Gonigle, J.W., and Schleicher, David, 1972, Geologic map of the Plato quadrangle of the Moon: USGS Map I–701 (LAC–12), scale 1:1,000,000.

Middlehurst, B.M., and Kuiper, G.P., eds., 1963, The moon, meteorites, and comets, v. 4 of The solar system: Chicago, University of Chicago Press, 810 p.

Milton, D.J., 1967, Slopes on the Moon: Science, v. 156, no. 3778, p. 1135.

———1968a, Geologic map of the Theophilus quadrangle of the Moon: USGS Map I–546 (LAC–78), scale 1:1,000,000.

———1968b, Structural geology of the Henbury meteorite craters, Northern Territory, Australia: USGS Professional Paper 599–C, p. C1–C17.

———1969, Astrogeology in the 19th century: Geotimes, v. 14, no. 6, p. 22.

Milton, D.J., and Hodges, C.A., 1972, Geologic maps of the Descartes region of the Moon, Apollo 16 pre-mission maps: USGS Map I–748, scales 1:250,000, 1:50,000, 2 sheets.

Milton, D.J., and Michel, F.C., 1965, Structure of a ray crater at Henbury, Northern Territory, Australia, in Geological Survey research, 1965: USGS Professional Paper 525–C, p. C5–C11.

Milton, D.J., and Roddy, D.J., 1972, Displacements within impact craters: International Geological Congress, 24th, Montreal, Quebec, Canada, 1972, sec. 15, p. 119–124.

Milton, D.J., Barlow, B.C., Brett, Robin, Brown, A.R., Glikson, A.Y., Manwaring, E.A., Moss, F.J., Sedmik, E.C.E., Van Son, J., and Young, G.A., 1972, Gosses Bluff impact structure, Australia: Science, v. 175, no. 4027, p. 1199–1207.

Minear, J.W., 1980, The lunar magma ocean: A transient lunar phenomenon?: LPSCP 11, v. 3, p. 1941–1955.

Minkin, J.A., Thompson, C.L., and Chao, E.C.-T., 1977, Apollo 16 white boulder consortium samples 67455 and 67475: Petrologic investigation: LSCP 8, v. 2, p. 1967–1986.

———1978, The Apollo 17 Station 7 boulder: Summary of study by the International Consortium: LPSCP 9, v. 1, p. 877–903.

Moore, H.J., 1964a, Density of small craters on the lunar surface: USGS ASAPR, August 25, 1962 to July 1, 1963, pt. D, p. 34–51.

———1964b, The geology of the Aristarchus quadrangle of the Moon: USGS ASAPR, August 25, 1962 to July 1, 1963, pt. A, p. 33–45.

———1965, Geologic map of the Aristarchus region of the Moon: USGS Map I–465 (LAC–39), scale 1:1,000,000.

———1967, Geologic map of the Seleucus quadrangle of the Moon: USGS Map I–527 (LAC–38), scale 1:1,000,000.

———1971, Craters produced by missile impacts: Journal of Geophysical Research, v. 76, no. 23, p. 5750–5755.

———1976, Missile impact craters (White Sands Missile Range, New Mexico) and applications to lunar research: USGS Professional Paper 812–B, p. B1–B47.

Moore, H.J., Boyce, J.M., and Hahn, D.A., 1980a, Small impact craters in the lunar regolith—their morphologies, relative ages, and rates of formation: The Moon and the Planets, v. 23, no. 2, p. 231–252.

Moore, H.J., Boyce, J.M., Schaber, G.G., and Scott, D.H., 1980b, Lunar remote sensing and measurements: USGS Professional Paper 1046–B, p. B1–B78.

Moore, H.J., Gault, D.E., and MacCormack, R.W., 1963, Fluid impact craters and hypervelocity—high velocity impact experiments in metals and rocks: USGS ASAPR, August 25, 1961 to August 24, 1962, pt. B, p. 80–101.

Moore, H.J., Hodges, C.A., and Scott, D.H., 1974, Multiringed basins—illustrated by Orientale and associated features: LSCP 5, v. 1, p. 71–100.

Moore, H.J., Lugn, R.V., and Gault, D.E., 1961, Experimental hypervelocity impact craters in rock: Symposium on Hypervelocity Impact, 5th, Denver, Colo., 1961, Proceedings, v. 1, pt. 2, p. 625–643.

Morgan, J.W., Ganapathy, R., Higuchi, Hideo, and Anders, Edward, 1977, Meteoritic material on the Moon, in Pomeroy, J.H., and Hubbard, N.J., eds., The Soviet-American Conference on Cosmochemistry of the Moon and Planets: NASA Report SP–370, p. 659–689.

Morris, E.C., and Shoemaker, E.M., 1970, Geology: Craters: Icarus, v. 12, no. 2, p. 167–172.

Morris, E.C., and Wilhelms, D.E., 1967, Geologic map of the Julius Caesar quadrangle of the Moon: USGS Map I–510 (LAC–60), scale 1:1,000,000.

Morrison, R.H., and Oberbeck, V.R., 1975, Geomorphology of crater and basin deposits—emplacement of the Fra Mauro Formation: LSCP 6, v. 3, p. 2503–2530.

———1978, A composition and thickness model for lunar impact crater and basin deposits: LPSCP 9, v. 3, p. 3763–3785.

Muehlberger, W.R., 1974, Structural history of southeastern Mare Serenitatis and adjacent highlands: LSCP 5, v. 1, p. 101–110.

Muehlberger, W.R., Hörz, Friedrich, Sevier, J.R., and Ulrich, G.E., 1980, Mission objectives for geological exploration of the Apollo 16 landing site, in Lunar and Planetary Institute, compiler, Conference on the Lunar Highlands Crust, Houston, Tex., 1979, Proceedings: New York, Pergamon (Geochimica et Cosmochimica Acta, supp. 12), p. 1–49.

Müller, H.W., Plieninger, T., James, O.B., and Schaeffer, O.A., 1977, Laser probe ^{39}Ar–^{40}Ar dating of materials from consortium breccia 73215: LSCP 8, v. 3, p. 2551–2565.

Muller, P.M., and Sjogren, W.L., 1968, Mascons: Lunar mass concentrations: Science, v. 161, no. 3842, p. 680–684.

Murray, J.B., 1971, Sinuous rilles, in Fielder, Gilbert, ed., Geology and physics of the moon: A study of some fundamental problems: Amsterdam, Elsevier, p. 27–39.

———1980, Oscillating peak model of basin and crater formation: The Moon and the Planets, v. 22, no. 3, p. 269–291.

Murthy, V.R., and Coscio, M.R., 1976, Rb-Sr ages and isotopic systematics of some Serenitatis mare basalts: LSCP 7, v. 2, p. 1529–1544.

Murthy, V.R., Evensen, N.M., Jahn, B.-M., and Coscio, M.R., 1971, Rb-Sr ages and elemental abundances of K, Rb, Sr, and Ba in samples from the Ocean of Storms: Geochimica et Cosmochimica Acta, v. 35, no. 11, p. 1139–1153.

———1972, Apollo 14 and Apollo 15 samples: Rb-Sr ages, trace elements, and lunar evolution: LSCP 3, v. 2, p. 1503–1514.

Mutch, T.A., 1970, Geology of the Moon—a stratigraphic view: Princeton, N.J., Princeton University Press, 324 p.

Mutch, T.A., and Saunders, R.S., 1972, Geologic map of the Hommel quadrangle of the Moon: USGS Map I–702 (LAC–127), scale 1:1,000,000.

Nabelek, P.I., Taylor, L.A., and Lofgren, G.E., 1978, Nucleation and growth of plagioclase and the development of textures in a high-alumina basaltic melt: LPSCP 9, v. 1, p. 725–741.

Nakamura, Noboru, and Tatsumoto, Mitsunobu, 1977, The history of the Apollo 17 Station 7 boulder: LSCP 8, v. 2, p. 2301–2314.

Nakamura, Noboru, Tatsumoto, Mitsunobu, Nunes, P.D., Unruh, D.M., Schwab, A.P., and Wildeman, T.R., 1976, 4.4 b.y.-old clast in Boulder 7, Apollo 17: A comprehensive chronological study by U-Pb, Rb-Sr, and Sm-Nd methods: LSCP 7, v. 2, p. 2309–2333.

Nakamura, Yosio, 1981, Geophysical data on structure and tectonics of the Apollo 16 landing site, in James, O.B., and Hörz, Friedrich, eds., Workshop on Apollo 16: Houston, Tex., Lunar and Planetary Institute Technical Report 81–01, p. 87–94.

Nakamura, Yosio, Lammlein, D.R., Latham, G.V., Ewing, Maurice, Dorman, H.J., Press, Frank, and Toksöz, Nafi, 1973, New seismic data on the state of the deep lunar interior: Science, v. 181, no. 4094, p. 49–51.

Nakamura, Yosio, Latham, G.V., Dorman, H.J., Ibrahim, A.-B.K., Koyama, Junji, and Horvath, Peter, 1979, Shallow moonquakes: Depth, distribution and implications as to the present state of the lunar interior: LPSCP 10, v. 3, p. 2299–2309.

Nash, D.B., and Conel, J.E., 1973, Vitrification darkening of rock powders: Implications for optical properties of the lunar surface: The Moon, v. 8, no. 3, p. 346–364.

Neukum, Gerhard, 1977, Different ages of lunar light plains: The Moon, v. 17, no. 4, p. 383–393.

Neukum, Gerhard, and König, Beate, 1976, Dating of individual lunar craters: LSCP 7, v. 3, p. 2867–2881.

Neukum, Gerhard, König, Beate, and Arkani-Hamed, Jafar, 1975a, A study of lunar impact crater size-distributions: The Moon, v. 12, no. 2, p. 201–229.

Neukum, Gerhard, König, Beate, Fechtig, H., and Storzer, D., 1975b, Cratering in the earth-moon system: Consequences for age determination by crater counting: LSCP 6, v. 3, p. 2597–2620.

Norman, M.D., and Ryder, Graham, 1979, A summary of the petrology and geochemistry of pristine highlands rocks: LPSCP 10, v. 1, p. 531–559.

Nunes, P.D., Tatsumoto, Mitsunobu, and Unruh, D.M., 1975, U-Th-Pb systematics of anorthositic gabbros 78155 and 77017—implications for early lunar evolution: LSCP 6, v. 2, p. 1431–1444.

Nyquist, L.E., 1977, Lunar Rb-Sr chronology: Physics and Chemistry of the Earth, v. 10, no. 2, p. 103–142.

Nyquist, L.E., Bansal, B.M., and Wiesmann, H., 1975, Rb-Sr ages and initial $^{87}Sr/^{86}Sr$ for Apollo 17 basalts and KREEP basalt 15386: LSCP 6, v. 2, p. 1445–1465.

———1976, Sr isotopic constraints on the petrogenesis of Apollo 17 mare basalts: LSCP 7, v. 2, p. 1507–1528.

Nyquist, L.E., Bansal, B.M., Wiesmann, H., and Jahn, B.-M., 1974, Taurus- Littrow chronology: Some constraints on early lunar crustal development: LSCP 5, v. 2, p. 1515–1539.

Nyquist, L.E., Bansal, B.M., Wooden, J.L., and Wiesmann, H., 1977, Sr-isotopic constraints on the petrogenesis of Apollo 12 mare basalts: LSCP 8, v. 2, p. 1383–1415.

Nyquist, L.E., Reimold, W.U., Bogard, D.D., Wooden, J.L., Bansal, B.M., Wiesmann, H., and Shih, C.Y., 1981a, A comparative Rb-Sr, Sm-Nd, and K-Ar study of shocked norite 78236: Evidence of slow cooling in the lunar crust?: LPSCP 12, pt. B, sec. 1, p. 67–97.

Nyquist, L.E., Shih, C.-Y., Wooden, J.L., Bansal, B.M., and Wiesmann, H., 1979a, The Sr and Nd isotopic record of Apollo 12 basalts: Implications for lunar geochemical evolution: LPSCP 10, v. 1, p. 77–114.

Nyquist, L.E., Wiesmann, H., Wooden, J.L., Bansal, B.M., and Shih, C.-Y., 1979b, Age and REE abundances of anorthositic norite from 15455 [abs.], in Papers presented to the Conference on the Lunar Highlands Crust: Lunar and Planetary Institute Contribution 394, p. 122–124.

Nyquist, L.E., Wooden, J.L., Shih, C.-Y., Wiesmann, H., and Bansal, B.M., 1981b, Isotopic and REE studies of lunar basalt 12038: Implications for petrogenesis of aluminous mare basalts: Earth and Planetary Science Letters, v. 55, no. 3, p. 335–355.

Oberbeck, V.R., 1971a, Laboratory simulation of impact cratering with high explosives: Journal of Geophysical Research, v. 76, no. 23, p. 5732–5749.

———1971b, Simultaneous impact and lunar craters: The Moon, v. 6, no. 1–2, p. 83–92.

———1975, The role of ballistic erosion and sedimentation in lunar stratigraphy: Reviews of Geophysics and Space Physics, v. 13, p. 337–362.

———1977, Application of high explosion cratering data to planetary problems, in Roddy, D.J., Pepin, R.O., and Merrill, R.B., eds., Impact and explosion cratering: Planetary and terrestrial implications: New York, Pergamon, p. 45–65.

Oberbeck, V.R., and Morrison, R.H., 1973a, On the formation of the lunar herringbone pattern: LSCP 4, v. 1, p. 107–123.

———1973b, The lunar herringbone pattern, pt. D of Crater studies, [chap.] 32 of Apollo 17 preliminary science report: NASA Report SP–330, p. 32–15 to 32–29.

———1974, Laboratory simulation of the herringbone pattern associated with lunar secondary crater chains: The Moon, v. 9, no. 3–4, p. 415–455.

———1976, Candidate areas for in situ ancient lunar materials: LSCP 7, v. 3, p. 2983–3005.

Oberbeck, V.R., and Quaide, W.L., 1967, Estimated thickness of a fragmental surface layer in Oceanus Procellarum: Journal of Geophysical Research, v. 72, no. 18, p. 4697–4704.

———1968, Genetic implications of lunar regolith thickness variations: Icarus, v. 9, no. 3, p. 446–465.

Oberbeck, V.R., Hörz, Friedrich, Morrison, R.H., Quaide, W.L., and Gault, D.E., 1975, On the origin of the lunar smoothplains: The Moon, v. 12, no. 1, p. 19–54.

Oberbeck, V. R., Morrison, R. H., Hörz, Friedrich, Quaide, W. L., and Gault, D. E., 1974, Smooth plains and continuous deposits of craters and basins: LSCP 5, v. 1, p. 111–136.

Oberbeck, V.R., Quaide, W.L., Arvidson, R.E., and Aggarwal, H.R., 1977, Comparative studies of lunar, martian, and mercurian craters and plains: Journal of Geophysical Research, v. 82, no. 11, p. 1681–1698.

Oberbeck, V.R., Quaide, W.L., and Greeley, Ronald, 1969, On the origin of lunar sinuous rilles: Modern Geology, v. 1, no. 1, p. 75–80.

Offield, T.W., 1971, Geologic map of the Schiller quadrangle of the Moon: USGS Map I–691 (LAC–125), scale 1:1,000,000.

———1972, Geologic map of the Flamsteed K region of the Moon, Lunar Orbiter site III P–12, Oceanus Procellarum: USGS Map I–626 [ORB III–12(100)], scale 1:100,000.

Offield, T.W., and Pohn, H.A., 1970, Lunar crater morphology and relative-age determination of geologic units—part 2. Applications, in Geological Survey research, 1970: USGS Professional Paper 700–C, p. C163–C169.

———1979, Geology of the Decaturville impact structure, Missouri: USGS Professional Paper 1042, 48 p.

O'Keefe, J.A., ed., 1963, Tektites: Chicago, University of Chicago Press, 288 p.

O'Keefe, J.A., and Cameron, W.S., 1962, Evidence from the moon's surface features for the production of lunar granites: Icarus, v. 1, no. 3, p. 271–285.

O'Keefe, J.A., Lowman, P.D., and Cameron, W.S., 1967, Lunar ring dikes from Lunar Orbiter I: Science, v. 155, no. 3758, p. 77–79.

O'Keefe, J.D., and Ahrens, T.J., 1975, Shock effects from a large impact on the moon: LSCP 6, v. 3, p. 2831–2844.

———1977, Impact-induced energy partitioning, melting, and vaporization on terrestrial planets: LSCP 8, v. 3, p. 3357–3374.

———1978, Impact flows and cratering scaling on the moon: Physics of the Earth and Planetary Interiors, v. 16, no. 4, p. 341–351.

Olson, A.B., and Wilhelms, D.E., 1974, Geologic map of the Mare Undarum quadrangle of the Moon: USGS Map I–837 (LAC–62), scale 1:1,000,000.

Onorato, P.I.K., Uhlmann, D.R., and Simonds, C.H., 1976, Heat flow in impact melts: Apollo 17 Station 6 Boulder and some applications to other breccias and xenolith laden melts: LSCP 7, v. 2, p. 2449–2467.

———1978, The thermal history of the Manicouagan impact melt sheet, Quebec [Canada]: Journal of Geophysical Research, v. 83, no. B6, p. 2789–2798.

Öpik, E.J., 1960, The lunar surface as an impact crater: Royal Astronomical Society Notices, v. 120, p. 404–411.

Orphal, D.L., 1977, Calculations of explosion cratering—II. Cratering mechanics and phenomenology, in Roddy, D.J., Pepin, R.O., and Merrill, R.B., eds., Impact and explosion cratering: Planetary and terrestrial implications: New York, Pergamon, p. 907–917.

Page, N.J, 1970, Geologic map of the Cassini quadrangle of the Moon: USGS Map I–666 (LAC–25), scale 1:1,000,000.

Papanastassiou, D.A., and Wasserburg, G.J., 1970, Rb-Sr ages from the Ocean of Storms: Earth and Planetary Science Letters, v. 8, no. 4, p. 269–278.

———1971a, Lunar chronology and evolution from Rb-Sr studies of Apollo 11 and 12 samples: Earth and Planetary Science Letters, v. 11, no. 1, p. 37–62.

———1971b, Rb-Sr ages of igneous rocks from the Apollo 14 mission and the age of the Fra Mauro Formation: Earth and Planetary Science Letters, v. 12, no. 1, p. 36–48.

———1972a, Rb-Sr age of a Luna 16 basalt and the model age of lunar soils: Earth and Planetary Science Letters, v. 13, no. 2, p. 368–374.

———1972b, The Rb-Sr age of a crystalline rock from Apollo 16: Earth and Planetary Science Letters, v. 16, no. 2, p. 289–298.

———1973, Rb-Sr ages and initial strontium in basalts from Apollo 15: Earth and Planetary Science Letters, v. 17, no. 2, p. 324–337.

———1975, Rb-Sr study of a lunar dunite and evidence for early lunar differentiates: LSCP 6, v. 2, p. 1467–1489.

———1976, Rb-Sr age of troctolite 76535: LSCP 7, v. 2, p. 2035–2054.

Papanastassiou, D.A., DePaolo, D.J., and Wasserburg, G.J., 1977, Rb-Sr and Sm-Nd chronology and genealogy of mare basalts

from the Sea of Tranquility: LSCP 8, v. 2, p. 1639–1672.
Papanastassiou, D.A., Wasserburg, G.J., and Burnett, D.S., 1970, Rb-Sr ages of lunar rocks from the Sea of Tranquillity: Earth and Planetary Science Letters, v. 8, no. 1, p. 1–19.
Papike, J.J., and Vaniman, D.T., 1978, Luna 24 ferrobasalts and the mare basalt suite: Comparative chemistry, mineralogy, and petrology, *in* Lunar and Planetary Institute, compiler, Mare Crisium: The view from Luna 24: Conference on Luna 24, Houston, Tex., 1977, Proceedings: New York, Pergamon (Geochimica et Cosmochimica Acta, supp. 9), p. 371–401.
Papike, J.J., Hodges, F.N., Bence, A.E., Cameron, Maryellen, and Rhodes, J.M., 1976, Mare basalts: Crystal chemistry, mineralogy, and petrology: Reviews of Geophysics and Space Physics, v. 14, no. 4, p. 475–540.
Papike, J.J., Simon, S.B., and Laul, J.C., 1982, The lunar regolith: Chemistry, mineralogy, and petrology: Reviews of Geophysics and Space Physics, v. 20, p. 761–826.
Patterson, Claire, 1956, Age of meteorites and the earth: Geochimica et Cosmochimica Acta, v. 10, no. 4, p. 230–237.
Peeples, W.J., Sill, W.R., May, T.W., Ward, S.H., Phillips, R.J., Jordan, R.L., Abbott, E.A., and Killpack, T.J., 1978, Orbital evidence for lunar subsurface layering in Maria Serenitatis and Crisium: Journal of Geophysical Research, v. 83, no. B7, p. 3459–3468.
Phillips, R.J., and Dvorak, John, 1981, The origin of lunar mascons: Analysis of the Bouguer gravity associated with Grimaldi, *in* Multi-ring basins: LPSCP 12, pt. A, p. 91–104.
Phillips, R.J., Conel, J.E., Abbot, E.A., Sjogren, W.L., and Morton, J.B., 1972, Mascons: Progress toward a unique solution for mass distribution: Journal of Geophysical Research, v. 77, no. 35, p. 7106–7114.
Phinney, D., Kahl, S.B., and Reynolds, J.H., 1975, ^{40}Ar-^{39}Ar dating of Apollo 16 and 17 rocks: LSCP 6, v. 2, p. 1593–1608.
Phinney, R.A., Gault, D.E., O'Keefe, J.A., Adams, J.B., Kuiper, G.P., Masursky, Harold, Shoemaker, E.M., and Collins, R.J., 1970, Lunar theory and processes: Discussion of chemical analysis: Icarus, v. 12, no. 2, p. 213–223.
Phinney, W.C., and Simonds, C.H., 1977, Dynamical implications of the petrology and distribution of impact melt rocks, *in* Roddy, D.J., Pepin, R.O., and Merrill, R.B., eds., Impact and explosion cratering: Planetary and terrestrial implications: New York, Pergamon, p. 771–790.
Phinney, W.C., Dence, M.R., and Grieve, R.A.F., 1978, Investigation of the Manicouagan impact crater, Quebec [Canada]: An introduction: Journal of Geophysical Research, v. 83, B6, p. 2729–2735.
Phinney, W.C., Warner, J.L., and Simonds, C.H., 1977, Lunar highland rock types: Their implications for impact-induced fractionation of vitric and clastic matrix breccias, *in* Pomeroy, J.H., and Hubbard, N.J., eds., The Soviet-American Conference on Cosmochemistry of the Moon and Planets: NASA Report SP-370, pt. 1, p. 91–126.
Piekutowski, A.J., 1977, Cratering mechanisms observed in laboratory-scale high-explosive experiments, *in* Roddy, D.J., Pepin, R.O., and Merrill, R.B., eds., Impact and explosion cratering: Planetary and terrestrial implications: New York, Pergamon, p. 67–102.
Pieters, C.M., 1977, Characterization of lunar mare basalt types—II: Spectral classification of fresh mare craters: LSCP 8, v. 1, p. 1037–1048.
———1978, Mare basalt types on the front side of the moon: A summary of spectral reflectance data: LPSCP 9, v. 3, p. 2825–2849.
Pieters, C.M., and McCord, T.B., 1976, Characterization of lunar mare basalt types: I. A remote sensing study using reflection spectroscopy of surface soils: LSCP 7, p. 2677–2690.
Pieters, C.M., Adams, J.B., Head, J.W., McCord, T.B., and Zisk, S.H., 1982, Primary ejecta in crater rays [abs.]: LPS XIII, pt. 2, p. 623–624.
Pieters, C.M., Head, J.W., Adams, J.B., McCord, T.B., Zisk, S.H., and Whitford-Stark, J.L., 1980, Late high-titanium basalts of the western maria: Geology of the Flamsteed region of Oceanus Procellarum: Journal of Geophysical Research, v. 85, no. B7, p. 3913–3938.
Pieters, C.M., Head, J.W., McCord, T.B., Adams, J.B., and Zisk, S.H., 1975, Geochemical and geological units of Mare Humorum: Definition using remote sensing and lunar sample information: LSCP 6, v. 3, p. 2689–2710.
Pieters, C.M., McCord, T.B., Charette, M.P., and Adams, J.B., 1974, Lunar surface: Identification of the dark mantling material in the Apollo 17 soil samples: Science, v. 183, no. 4130, p. 1191–1194.
Pieters, C.M., McCord, T.B., Zisk, S.H., and Adams, J.B., 1973, Lunar black spots and the nature of the Apollo 17 landing area: Journal of Geophysical Research, v. 78, no. 26, p. 5867–5875.
Pike, R.J., 1971, Genetic implications of the shapes of martian and lunar craters: Icarus, v. 15, no. 3, p. 384–395.
———1974, Depth/diameter relations of fresh lunar craters: Revision from spacecraft data: Geophysical Research Letters, v. 1, no. 7, p. 291–294.
———1976, Geologic map of the Rima Hyginus region of the Moon: USGS Map I–945, scale 1:250,000.
———1980a, Control of crater morphology by gravity and target type: Mars, Earth, Moon: LPSCP 11, v. 3, p. 2159–2189.
———1980b, Formation of complex impact craters: Evidence from Mars and other planets: Icarus, v. 43, p. 1–19.
———1980c Geometric interpretation of lunar craters: USGS Professional Paper 1046–C, p. C1–C77.
Pike, R.J., and Wilhelms, D.E., 1978, Secondary-impact craters on the Moon: Topographic form and geologic processes [abs.]: LPS IX, pt. 2, p. 907–909.
Podosek, F.A., and Huneke, J.C., 1973, Argon in Apollo 15 green glass spherules (15426): ^{40}Ar-^{39}Ar age and trapped argon: Earth and Planetary Science Letters, v. 19, no. 4, p. 413–421.
Podosek, F.A., Huneke, J.C., Gancarz, A.J., and Wasserburg, G.J., 1973, The age and petrography of two Luna 20 fragments and inferences for widespread lunar metamorphism: Geochimica et Cosmochimica Acta, v. 37, no. 4, p. 887–904.
Podosek, F.A., Huneke, J.C., and Wasserburg, G.J., 1972, Gas-retention and cosmic-ray exposure ages of lunar rock 15555: Science, v. 175, no. 4020, p. 423–425.
Pohl, Jean, Stöffler, Dieter, Gall, Horst, and Ernstson, Kord, 1977, The Ries impact crater, *in* Roddy, D.J., Pepin, R.O., and Merrill, R.B., eds., Impact and explosion cratering: Planetary and terrestrial implications: New York, Pergamon, p. 343–404.
Pohn, H.A., 1971, Geologic map of the Lansberg P region of the Moon, Lunar Orbiter site III P–9, Oceanus Procellarum, including Apollo landing site 7 (Apollo 12): USGS Map I–627 [ORB III–9 (100)], scale 1:100,000.
———1972, Geologic map of the Tycho quadrangle of the Moon: USGS Map I–713 (LAC–112), scale 1:1,000,000.
Pohn, H.A., and Offield, T.W., 1970, Lunar crater morphology and relative age determination of lunar geologic units—part 1. Classification, *in* Geological Survey research, 1970: USGS Professional Paper 700–C, p. C153–C162.
Pohn, H.A., and Wildey, R.L., 1970, A photoelectric-photographic study of the normal albedo of the Moon, *accompanied by an* Albedo map of the Moon, by H.A. Pohn, R.L. Wildey, and G.E. Sutton: USGS Professional Paper 599–E, p. E1–E20.
Prinz, Martin, and Keil, Klaus, 1977, Mineralogy, petrology and chemistry of ANT-suite rocks from the lunar highlands: Physics and Chemistry of the Earth, v. 10, no. 4, p. 215–237.
Prinz, Martin, Dowty, Eric, Keil, Klaus, and Bunch, T.E., 1973, Mineralogy, petrology and chemistry of lithic fragments from Luna 20 fines: Origin of the cumulate ANT suite and its relationship to high-alumina and mare basalts: Geochimica et Cosmochimica Acta, v. 37, no. 4, p. 979–1006.
Quaide, W.L., 1965, Rilles, ridges, and domes—clues to maria history: Icarus, v. 4, no. 4, p. 374–389.
———1973, Provenance of Apennine Front regolith materials [abs.]: LS IV, p. 606–608.
Quaide, W.L., and Oberbeck, V.R., 1968, Thickness determinations of the lunar surface layer from lunar impact craters: Journal of Geophysical Research, v. 73, no. 16, p. 5247–5270.
———1969, Geology of the Apollo landing sites: Earth-Science Reviews, v. 5, no. 5, p. 255–278.
———1975, Development of the mare regolith: Some model considerations: The Moon, v. 13, no. 1–3, p. 27–55.
Quaide, W.L., and Wrigley, Robert, 1972, Mineralogy and origin of Fra Mauro fines and breccias: LSCP 3, v. 1, p. 771–784.
Quaide, W.L., Gault, D.E., and Schmidt, R.A., 1965, Gravitative effects on lunar impact structures, *in* Whipple, H.E., ed., Geological problems in lunar research: New York Academy of Sciences Annals, v. 123, art. 2, p. 563–572.
Quaide, W.L., Oberbeck, V.R., Bunch, Theodore, and Polkowski, George, 1971, Investigations of the natural history of the regolith at the Apollo 12 site: LSCP 2, v. 1, p. 701–718.
Raedeke, L.D., and McCallum, I.S., 1980, A comparison of fractionation trends in the lunar crust and the Stillwater Complex, *in* Lunar and Planetary Institute, compiler, Conference on the Lunar Highlands Crust, Houston, Tex., 1979, Proceedings: New York, Pergamon (Geochimica et Cosmochimica Acta, supp. 12), p. 133–153.
Reed, V.S., 1981, Geology of the areas near South Ray and Baby Ray craters, [chap.] D3 *of* Ulrich, G.E., Hodges, C.A., and Muehlberger, W.R., eds., Geology of the Apollo 16 area, central lunar highlands: USGS Professional Paper 1048, p. 82–105.
Reed, V.S., and Wolfe, E.W., 1975, Origins of the Taurus-Littrow massifs: LSCP 6, v. 3, p. 2443–2461.
Reedy, R.C., 1978, Planetary gamma-ray spectroscopy: LPSCP 9, v. 3, p. 2961–2984.
Rehfuss, D.E., 1974, Glass production differences for equal-diameter impact craters: The Moon, v. 11, no. 1–2, p. 19–28.
Reid, A.M., Duncan, A.R., and Richardson, S.H., 1977, In search of LKFM: LSCP 8, v. 2, p. 2321–2338.
Reid, A.M., Ridley, W.I, Harmon, R.S., Warner, J.L., Brett, Robin, Jakes, P., and Brown, R.W., 1972a, Highly aluminous glasses in lunar soils and the nature of the lunar highlands: Geochimica et Cosmochimica Acta, v. 36, no. 8, p. 903–912.
Reid, A.M., Warner, J.L., Ridley, W.I., and Brown, R.W., 1972b, Major element composition of glasses in three Apollo 15 soils: Meteoritics, v. 7, no. 3, p. 395–415.
Rhodes, J.M., and Hubbard, N.J., 1973, Chemistry, classification, and petrogenesis of Apollo 15 mare basalts: LSCP 4, v. 2, p. 1127–1148.
Rhodes, J.M., Blanchard, D.P., Dungan, M.A., Brannon, J.C., and Rodgers, K.V., 1977, Chemistry of Apollo 12 mare basalts: Magma types and fractionation processes: LSCP 8, v. 2, p. 1305–1338.
Rhodes, J.M., Hubbard, N.J., Wiesmann, H., Rodgers, K.V., Brannon, J.C., and Bansal, B.M., 1976, Chemistry, classification, and petrogenesis of Apollo 17 mare basalts: LSCP 7, v. 2, p. 1467–1489.
Ridley, W.I., 1975, On high-alumina mare basalts: LSCP 6, v. 1, p. 131–145.
Rinehart, J.S., 1975, Stress transients in solids: Santa Fe, N. Mex., HyperDynamics, 230 p.
Ringwood, A.E., and Essene, E., 1970, Petrogenesis of Apollo 11 mare basalts, internal constitution and origin of the moon: LSCP 1, v. 1, p. 769–799.
Roberts, W.A., 1966, Shock—a process in extraterrestrial sedimentology: Icarus, v. 5, no. 5, p. 459–477.
———1968, Shock crater characteristics, *in* French, B.M., and Short, N.M., eds., Shock metamorphism of natural materials: Baltimore, Mono, p. 101–114.
Robertson, P.B., and Grieve, R.A.F., 1977, Shock attenuation at terrestrial impact structures, *in* Roddy, D.J., Pepin, R.O., and Merrill, R.B., eds., Impact and explosion cratering: Planetary and terrestrial implications: New York, Pergamon, p. 687–702.
Roddy, D.J., 1968, The Flynn Creek crater, Tennessee, *in* French, B.M., and Short, N.M., eds., Shock metamorphism of natural materials: Baltimore, Mono, p. 291–322.
———1976, High-explosion cratering analogs for bowl-shaped, central-uplift, and multiring impact craters: LSCP 7, v. 3, p. 3027–3056.
———1977, Pre-impact conditions and cratering processes at the Flynn Creek Crater, Tennessee, *in* Roddy, D.J., Pepin, R.O., and Merrill, R.B., eds., Impact and explosion cratering: Planetary and terrestrial implications: New York, Pergamon, p. 277–308.
———1979, Structural deformation at the Flynn Creek impact crater, Tennessee: A preliminary report on deep drilling: LPSCP 10, v. 3, p. 2519–2534.
Ronca, L.B., 1965, Selenology vs geology of the Moon etc: Geotimes, v. 9, no. 9, p. 13.
Rowan, L.C., 1971a, Geologic map of the Oppolzer A region of the Moon, Lunar Orbiter site II P–8, Sinus Medii, including Apollo landing sites 3 and 3R: USGS Map I–620 [ORB II–8 (100)], scale 1:100,000.
———1971b, Geologic map of the Rupes Altai quadrangle of the Moon: USGS Map I–690 (LAC–96), scale 1:1,000,000.
Runcorn, S.K., 1980, An iron core in the moon generating an early magnetic field?: LPSCP 10, v. 3, p. 2325–2333.
———1982, Primeval displacements of the lunar pole: Physics of the Earth and Planetary Interiors, v. 29, p. 135–147.
Ryder, Graham, 1976, Lunar sample 15405: Remnant of a KREEP basalt-granite differentiated pluton: Earth and Planetary Science Letters, v. 29, no. 2, p. 255–268.
———1979, The chemical components of highlands breccias: LPSCP 10, v. 1, p. 561–581.
———1981, Apollo 16 basaltic impact melts: Chemistry and relationships, *in* James, O.B., and Hörz, Friedrich, eds., Workshop on Apollo 16: Houston, Tex., Lunar and Planetary Institute Technical Report 81–01, p. 109–111.
Ryder, Graham, and Bower, J.F., 1976, Poikilitic KREEP impact melts in the Apollo 14 white rocks: LSCP 7, v. 2, p. 1925–1948.
———1977, Petrology of Apollo 15 black-and-white rocks 15445 and 15455— fragments of the Imbrium impact melt sheet?: LSCP 8, v. 2, p. 1895–1923.
Ryder, Graham, and Norman, M.D., 1980, Catalog of Apollo 16 rocks: NASA, Johnson Space Center, Curatorial Branch Publication 52, 3 pts.
Ryder, Graham, and Spudis, P.D., 1980, Volcanic rocks in the lunar highlands, *in* Lunar and Planetary Institute, compiler, Conference on the Lunar Highlands Crust, Houston, Tex., 1979, Proceedings: New York, Pergamon (Geochimica et Cosmochimica Acta, supp. 12), p. 353–375.
Ryder, Graham, and Taylor, G.J., 1976, Did mare-type volcanism commence early in lunar history?: LSCP 7, v. 2, p. 1741–1755.
Ryder, Graham, and Wood, J.A., 1977, Serenitatis and Imbrium impact melts: Implications for large-scale layering in the lunar crust: LSCP 8, v. 1, p. 655–668.
Ryder, Graham, Stoeser, D.B., Marvin, U.B., and Bower, J.F., 1975a, Lunar granites with unique ternary feldspars: LSCP 6, v. 1, p. 435–449.
Ryder, Graham, Stoeser, D.B., Marvin, U.B., Bower, J.F., and Wood, J.A., 1975b, Boulder 1, station 2, Apollo 17: Petrology and petrogenesis: The Moon, v. 14, no. 3–4, p. 327–357.
Ryder, Graham, Stoeser, D.B., and Wood, J.A., 1977, Apollo 17 KREEPy basalt: A rock type intermediate between mare and KREEP basalts: Earth and Planetary Science Letters, v. 35, no. 1, p. 1–13.
Saari, J.M., Shorthill, R.W., and Deaton, T.K., 1966, Infrared and visible images of the eclipsed Moon of December 19, 1964: Icarus, v. 5, no. 6, p. 635–659.
Sabaneyev, P.F., 1962, Some results deduced from simulation of lunar craters, *in* Kopal, Zdeněk, and Mikhailov, Z.K., eds., The Moon: London, Academic Press, p. 419–431.
Saito, Kazuo, and Alexander, E.C., 1979, ^{40}Ar-^{39}Ar studies of lunar soil 74001 [abs.]: LPS X, pt. 3, p. 1049–1051.
Salisbury, J.W., and Glaser, P.E., eds., 1964, The lunar surface layer: Materials and characteristics: New York, Academic Press, 532 p.
Salisbury, J.W., Adler, J.E.M., and Smalley, V.G., 1968, Dark-haloed craters on the Moon: Royal Astronomical Society Monthly Notices, v. 138, p. 245–249.
Sanchez, A.G., 1981, Geology of Stone mountain, [chap.] D4 *of* Ulrich, G.E., Hodges, C.A., and Muehlberger, W.R., eds., Geology of the Apollo 16 area, central lunar highlands: USGS Professional Paper 1048, p. 106–126.
Sato, Motokai, 1979, The driving mechanism of lunar pyroclastic eruptions inferred from the oxygen fugacity behavior of Apollo 17 orange glass: LPSCP 10, v. 1, p. 311–325.
Saunders, R.S., and Wilhelms, D.E., 1974, Geologic map of the Wilhelm quadrangle of the Moon: USGS Map I–824 (LAC–111), scale 1:1,000,000.
Schaber, G.G., 1969, Geologic map of the Sinus Iridum quadrangle of the Moon: USGS Map I–602 (LAC–24), scale 1:1,100,000.

——1973, Lava flows in Mare Imbrium: Geologic evaluation from Apollo orbital photography: LSCP 4, v. 1, p. 73–92.
——1981, Field geology of Apollo 16 central region, [chap.] D1 *of* Ulrich, G.E., Hodges, C.A., and Muehlberger, W.R., eds., Geology of the Apollo 16 area, central lunar highlands: USGS Professional Paper 1048, p. 21-44.
Schaber, G.G., Boyce, J.M., and Moore, H.J., 1976, The scarcity of mappable flow lobes on the lunar maria: Unique morphology of the Imbrium flows: LSCP 7, v. 3, p. 2783–2800.
Schaber, G.G., Thompson, T.W., and Zisk, S.H., 1975, Lava flows in Mare Imbrium: An evaluation of anomalously low Earth-based radar reflectivity: The Moon, v. 13, p. 395–423.
Schaeffer, G.A., and Schaeffer, O.A., 1977, ^{39}Ar-^{40}Ar ages of lunar rocks: LSCP 8, v. 2, p. 2253–2300.
Schaeffer, O.A., and Husain, Liaquat, 1973, Early lunar history: Ages of 2 to 4 mm soil fragments from the lunar highlands: LSCP 4, v. 2, p. 1847–1863.
——1974, Chronology of lunar basin formation: LSCP 5, v. 2, p. 1541–1555.
Schaeffer, O.A., Bence, A.E., Eichhorn, G., Papike, J.J., and Vaniman, D.T., 1978, ^{39}Ar-^{40}Ar and petrologic study of Luna 24 samples 24077,13 and 24077,63: LPSCP 9, v. 2, p. 2363–2373.
Schaeffer, O.A., Husain, Liaquat, and Schaeffer, G.A., 1976, Ages of highland rocks: The chronology of lunar basin formation revisited: LSCP 7, v. 2, p. 2067–2092.
Schmidt, R.M., 1977, A centrifuge cratering experiment: Development of a gravity-scaled yield parameter, *in* Roddy, D.J., Pepin, R.O., and Merrill, R.B., eds., Impact and explosion cratering: Planetary and terrestrial implications: New York, Pergamon, p. 1261–1278.
——1981, Scaling crater time-of-formation [abs.]: Eos (American Geophysical Union Transactions), v. 62, no. 45, p. 944.
Schmitt, H.H., Lofgren, G., Swann, G.A., and Simmons, G., 1970, The Apollo 11 samples: Introduction: LSCP 1, v. 1, p. 1–54.
Schmitt, H.H., Trask, N.J., and Shoemaker, E.M., 1967, Geologic map of the Copernicus quadrangle of the Moon: USGS Map I–515 (LAC–58), scale 1:1,000,000.
Schonfeld, Ernest, 1976, Chronology of the early lunar crust: LSCP 7, v. 2, p. 2093–2105.
——1977, Comparison of orbital chemistry with crustal thickness and lunar sample chemistry: LSCP 8, v. 1, p. 1149–1162.
Schonfeld, Ernest, and Bielefeld, M.J., 1978, Correlation of dark mantle deposits with high Mg/Al ratios: LPSCP 9, v. 3, p. 3037–3048.
Schonfeld, Ernest, and Meyer, Charles, Jr., 1972, The abundance of components of the lunar soils by a least-squares mixing model and the formation age of KREEP: LSCP 3, v. 2, p. 1397–1420.
Schultz, P.H., 1976a, Floor-fractured lunar craters: The Moon, v. 15, no. 3–4, p. 241–273.
——1976b, Moon morphology: Austin, University of Texas Press, 626 p.
Schultz, P.H., and Gault, D.E., 1975a, Seismic effects from major basin formations on the Moon and Mercury: The Moon, v. 12, no. 2, p. 159–177.
——1975b, Seismically induced modification of the lunar surface features: LSCP 6, v. 3, p. 2845–2862.
Schultz, P.H., and Mendell, Wendell, 1978, Orbital infrared observations of lunar craters and possible implications for impact ejecta emplacement: LPSCP 9, v. 3, p. 2857–2883.
Schultz, P.H., and Mendenhall, M.H., 1979, On the formation of basin secondary craters by ejecta complexes [abs.]: LPS X, pt. 3, p. 1078–1080.
Schultz, P.H., and Spencer, J., 1979, Effects of substrate strength on crater statistics [abs.]: LPS X, pt. 3, p. 1081–1083.
Schultz, P.H., and Spudis, P.H., 1979, Evidence for ancient mare volcanism: LPSCP 10, v. 3, p. 2899–2918.
Schultz, P.H., and Srnka, L.J., 1980, Cometary collisions on the Moon and Mercury: Nature, v. 284, no. 5751, p. 22–26.
Schultz, P.H., Greeley, Ronald, and Gault, D.E., 1977, Interpreting statistics of small lunar craters: LSCP 8, v. 3, p. 3539–3564.
Scott, D.H., 1972a, Geologic map of the Eudoxus quadrangle of the Moon: USGS Map I–705 (LAC–26), scale 1:1,000,000.
——1972b, Geologic map of the Maurolycus quadrangle of the Moon: USGS Map I–695 (LAC–113), scale 1:1,000,000.
——1973, Mare Serenitatis cinder cones and terrestrial analogs, pt. B *of* Volcanic studies, [chap.] 30 *of* Apollo 17 preliminary science report: NASA Report SP–330, p. 30–7 to 30–8.
—— 1974, The geologic significance of some lunar gravity anomalies: LSCP 5, v. 3, p. 3025–3036.
Scott, D.H., and Eggleton, R.E., 1973, Geologic map of the Rümker quadrangle of the Moon: USGS Map I–805 (LAC–23), scale 1:1,000,000.
Scott, D.H., and Pohn, H.A., 1972, Geologic map of the Macrobius quadrangle of the Moon: USGS Map I–799 (LAC–43), scale 1:1,000,000.
Scott, D.H., Diaz, J.M., and Watkins, J.A., 1977, Lunar farside tectonics and volcanism: LSCP 8, v. 1, p. 1119–1130.
Scott, D.H., Lucchitta, B.K., and Carr, M.H., 1972, Geologic maps of the Taurus-Littrow region of the Moon; Apollo 17 premission maps: USGS Map I–800, scales 1:50,000, 1:250,000, 2 sheets.
Scott, D.H., McCauley, J.F., and West, M.N., 1977, Geologic map of the west side of the Moon: USGS Map I–1034, scale 1:5,000,000.
Scott, D.H., West, M.N., Lucchitta, B.K., and McCauley, J.F., 1971, Preliminary geologic results from orbital photography, pt. B *of* Orbital-science photography, [chap.] 18 *of* Apollo 14 preliminary science report: NASA Report SP–272, p. 274–283.
Sekiguchi, Naosuke, 1970, On the fissions of a solid body under influence of tidal force: The Moon, v. 1, no. 4, p. 429–439.
Settle, Mark, and Head, J.W., 1979, The role of rim slumping in the modification of lunar impact craters: Journal of Geophysical Research, v. 84, no. B6, p. 3081–3096.
Shih, C.-Y., 1977, Origins of KREEP basalts: LSCP 8, v. 2, p. 2375–2401.
Shoemaker, E.M., 1960, Penetration mechanics of high velocity meteorites, illustrated by Meteor Crater, Arizona, *in* Kvale, Anders, and Metzger, A.A., eds., Structure of the earth's crust and deformation of rocks: International Geological Congress, 21st, Copenhagen, 1960, Report, pt. 18, p. 418–434.
—— 1962a, Exploration of the Moon's surface: American Scientist, v. 50, no. 1, p. 99–130.
—— 1962b, Interpretation of lunar craters, *in* Kopal, Zdeněk, ed., Physics and astronomy of the moon: New York, Academic Press, p. 283–359.
—— 1963, Impact mechanics at Meteor Crater, Arizona, *in* Middlehurst, B.M., and Kuiper, G.P., eds., The moon, meteorites, and comets, v. 4 *of* The solar system: Chicago, University of Chicago Press, p. 301–336.
—— 1964, The geology of the Moon: Scientific American, v. 211, no. 6, p. 38–47.
—— 1965, Preliminary analysis of the fine structure of the lunar surface in Mare Cognitum, [chap.] 4 *of Ranger VII*. Part II. Experimenters' analyses and interpretations: Pasadena, California Institute of Technology, Jet Propulsion Laboratory Technical Report 32–700, p. 75–134.
—— 1966, Progress in the analysis of fine structure and geology of the lunar surface from the Ranger VIII and IX photographs, [chap.] 4 *of Ranger VIII* and *IX*. Part II. Experimenters' analyses and interpretations: Pasadena, California Institute of Technology, Jet Propulsion Laboratory Technical Report 32–800, p. 249–337.
—— 1971, Origin of fragmental debris on the lunar surface and the history of bombardment of the Moon: Barcelona, Spain, Universidad de Barcelona, Instituto de Investigaciones Geológicas de la Diputación Provincial, v. 25, p. 27–56.
—— 1972, Cratering history and early evolution of the Moon [abs.]: LS III, p. 696–698.
—— 1981, The collision of solid bodies, [chap.] 4 *of* Beatty, J.K., O'Leary, Brian, and Chaikin, Andrew, eds., The new solar system: Cambridge, Mass., Sky, p. 33–44.
Shoemaker, E.M., and Chao, E.C.-T., 1961, New evidence for the impact origin of the Ries basin, Bavaria, Germany: Journal of Geophysical Research, v. 66, no. 10, p. 3371–3378.
Shoemaker, E.M., and Hackman, R.J., 1962, Stratigraphic basis for a lunar time scale, *in* Kopal, Zdeněk, and Mikhailov, Z.K., eds., The Moon: London, Academic Press, p. 289–300.
Shoemaker, E.M., and Morris, E.C., 1970, Geology: Physics of fragmental debris: Icarus, v. 12, no. 2, p. 188–212.
Shoemaker, E.M., Bailey, N.G., Batson, R.M., Dahlem, D.H., Foss, T.H., Grolier, M.J., Goddard, E.N., Hait, M.H., Holt, H.E., Larson, K.B., Rennilson, J.J., Schaber, G.G., Schleicher, D.L., Schmitt, H.H., Sutton, R.L., Swann, G.A., Waters, A.C., and West, M.N., 1969a, Geologic setting of the lunar samples returned by the Apollo 11 mission, [chap.] 3 *of* Apollo 11 preliminary science report: NASA Report SP–214, p. 41–83.
Shoemaker, E.M., Batson, R.M., Holt, H.E., Morris, E.C., Rennilson, J.J., and Whitaker, E.A., 1967a, Television observations from Surveyor III, [chap.] 3 *of* Surveyor III: A preliminary report: NASA Report SP–146, p. 9–59.
—— 1967b, Television observations from *Surveyor V*, [chap.] 3 *of Surveyor V* mission report. Part II: Science results: Pasadena, California Institute of Technology, Jet Propulsion Laboratory Technical Report 32–1246, p. 7–42.
—— 1968, Television observations from *Surveyor VII*, [chap.] 3 *of Surveyor VII* mission report. Part II: Science results: Pasadena, California Institute of Technology, Jet Propulsion Laboratory Technical Report 32–1264, p. 9–76.
—— 1969b, Observations of the lunar regolith and the earth from the television camera on Surveyor 7: Journal of Geophysical Research, v. 74, no. 25, p. 6081–6119.
Shoemaker, E.M., Hackman, R.J., and Eggleton, R.E., 1962a, Interplanetary correlation of geologic time: Advances in the Astronautical Sciences, v. 8, p. 70–89.
Shoemaker, E.M., Hackman, R.J., Eggleton, R.E., and Marshall, C.H., 1962b, Lunar stratigraphic nomenclature: USGS ASSPR, February 26, 1961 to August 24, 1962, p. 114–116.
Shoemaker, E.M., Hait, M.H., Swann, G.A., Schleicher, D.L., Schaber, G.G., Sutton, R.L., Dahlem, D.H., Goddard, E.N., and Waters, A.C., 1970, Origin of the regolith at Tranquillity base: LSCP 1, v. 3, p. 2399–2412.
Shoemaker, E.M., Williams, J.G., Helin, E.F., and Wolfe, R.F., 1979, Earth-crossing asteroids: Orbital classes, collision rates with Earth, and origin, *in* Gehrels, Tom, ed., 1979, Asteroids: Tucson, University of Arizona Press, p. 253–282.
Short, N.M., 1975, Planetary Geology: Englewood Cliffs, N.J., Prentice-Hall, 361 p.
Short, N.M., and Forman, M.L., 1972, Thickness of impact crater ejecta on the lunar surface: Modern Geology, v. 3, no. 2, p. 69–91.
Shorthill, R.W., 1973, Infrared atlas of the eclipsed Moon: The Moon, v. 7, no. 1, p. 22–45.
Shorthill, R.W., and Saari, J.M., 1969, Infrared observation on the eclipsed Moon: Seattle, Boeing Scientific Research Laboratories Document D1–82–0778, 73 p.
Silver, L.T., 1971, U-Th-Pb isotope systems in Apollo 11 and 12 regolithic materials and a possible age for the Copernican impact [abs.]: Eos (American Geophysical Union Transactions), v. 52, no. 7, p. 534.
Simonds, C.H., 1975, Thermal regimes in impact melts and the petrology of the Apollo 17 Station 6 boulder: LSCP 6, v. 1, p. 641–672.
——1979, Low speed (8 km/sec) impact-accretional heating and early fractionation of the Moon [abs.], *in* Papers presented to Conference on the Early Highlands Crust: Lunar and Planetary Institute Contribution 394, p. 148–150.
Simonds, C.H., Floran, R.J., McGee, P.E., Phinney, W.C., and Warner, J.L., 1978a, Petrogenesis of melt rocks, Manicouagan impact structure, Quebec [Canada]: Journal of Geophysical Research, v. 83, no. B6, p. 2773–2788.
Simonds, C.H., Phinney, W.C., McGee, P.E., and Cochran, Ann, 1978b, West Clearwater, Quebec impact structure, part I: Field geology, structure and bulk chemistry: LPSCP 9, v. 2, p. 2633–2658.
Simonds, C.H., Phinney, W.C., and Warner, J.L., 1974, Petrography and classification of Apollo 17 non-mare rocks with emphasis on samples from the Station 6 boulder: LSCP 5, v. 1, p. 337–353.
Simonds, C.H., Phinney, W.C., Warner, J.L., McGee, P.E., Geeslin, Jill, Brown, R.W., and Rhodes, J.M., 1977, Apollo 14 revisited, or breccias aren't so bad after all: LSCP 8, v. 2, p. 1869–1893.
Simonds, C.H., Warner, J.L., and Phinney, W.C., 1973, Petrology of Apollo 16 poikilitic rocks: LSCP 4, v. 1, p. 613–632.
——1976a, Thermal regimes in cratered terrain with emphasis on the role of impact melt: American Mineralogist, v. 61, no. 7–8, p. 569–577.
Simonds, C.H., Warner, J.L., Phinney, W.C., and McGee, P.E., 1976b, Thermal model for impact breccia lithification: Manicouagan and the Moon: LSCP 7, v. 2, p. 2509–2528.
Sjogren, W.L., and Smith, J.C., 1976, Quantitative mass distribution models for Mare Orientale: LSCP 7, v. 3, p. 2639–2648.
Sjogren, W.L., and Wollenhaupt, W.R., 1976, Lunar global figure from mare surface elevations: The Moon, v. 15, no. 1–2, p. 143–154.
Sjogren, W.L., Wimberly, R.N., and Wollenhaupt, W.R., 1974, Lunar gravity via the Apollo 15 and 16 subsatellites: The Moon, v. 9, no. 1–2, p. 115–128.
Smith, E.I., and Hartnell, J.A., 1978, Crater size-shape profiles for the Moon and Mercury: Terrain effects and interplanetary comparisons: The Moon and the Planets, v. 19, no. 4, p. 479–511.
Smith, E.I., and Sanchez, A.G., 1973, Fresh lunar craters: Morphology as a function of diameter, a possible criterion for crater origin: Modern Geology, v. 4, no. 1, p. 51–59.
Smith, J.V., 1974, Lunar mineralogy: A heavenly detective story. Presidential address, part I: American Mineralogist, v. 59, no. 3–4, p. 231–243.
——1979, Mineralogy of the planets: A voyage in space and time: Mineralogical Magazine, v. 43, p. 1–89.
——1982, Heterogeneous growth of meteorites and planets, especially the Earth and Moon: Journal of Geology, v. 90, p. 1–48.
Smith, J.V., and Steele, I.M., 1975, Lunar mineralogy: A heavenly detective story. Part II: American Mineralogist, v. 61, no. 11–12, p. 1059–1116.
Smith, J.V., Anderson, A.T., Newton, R.C., Olsen, E.J., Wyllie, P.J., Crewe, A.V., Isaacson, M.S., and Johnson, D., 1970, Petrologic history of the moon inferred from petrography, mineralogy, and petrogenesis of Apollo 11 rocks: LSCP 1, v. 1, p. 897–925.
Soderblom, L.A., 1970, A model for small-impact erosion applied to the lunar surface: Journal of Geophysical Research, v. 75, no. 14, p. 2655–2661.
Soderblom, L.A., and Boyce, J.M., 1972, Relative ages of some near-side and far-side terra plains based on Apollo 16 metric photography, pt. A *of* Photogeology, [chap.] 29 *of* Apollo 16 preliminary science report: NASA Report SP–315, p. 29–2 to 29–6.
——1976, Distribution and evolution of global color provinces on the moon [abs.]: LS VII, pt. 2, p. 822–824.
Soderblom, L.A., and Lebofsky, L.A., 1972, Technique for rapid determination of relative ages of lunar areas from orbital photography: Journal of Geophysical Research, v. 77, no. 2, p. 279–296.
Soderblom, L.A., Arnold, J.A., Boyce, J.M., and Lin, R.P., 1977, Regional variations in the lunar maria: Age, remanent magnetism, and chemistry: LSCP 8, v. 1, p. 1191–1199.
Solomon, S.C., 1975, Mare volcanism and lunar crustal structure: LSCP 6, v. 1, p. 1021–1042.
——1978, The nature of isostasy on the moon: How big a Pratt-fall for Airy models?: LPSCP 9, v. 3, p. 3499–3511.
Solomon, S.C., and Head, J.W., 1979, Vertical movement in mare basins: Relation to mare emplacement, basin tectonics, and lunar thermal history: Journal of Geophysical Research, v. 84, no. B4, p. 1667–1682.
——1980, Lunar mascon basins: Lava filling, tectonics, and evolution of the lithosphere: Reviews of Geophysics and Space Physics, v. 18, p. 107–141.
Spudis, P.D., 1978a, Composition and origin of the Apennine Bench Formation: LPSCP 9, v. 3, p. 3379–3394.
——1978b, Origin and distribution of KREEP in Apollo 15 soils [abs.]: LPS IX, pt. 2, p. 1089–1091.
Spudis, P.D., and Hawke, B.R., 1981, Chemical mixing model studies of lunar orbital geochemical data: Apollo 16 and 17 highlands compositions: LPSCP 12, pt. B, sec. 1, p. 781–789.
Spudis, P.D., and Head, J.W., 1977, Geology of the Imbrium Basin Apennine Mountains and relation to the Apollo 15 landing site: LSCP 8, v. 3, p. 2785–2797.
Spudis, P.D., and Ryder, Graham, 1981, Apollo 17 impact melts and their relation to the Serenitatis basin, *in* Multi-ring

basins: LPSCP 12, pt. A, p. 133–148.
Staudacher, T., Jessberger, E.K., Flohs, I., and Kirsten, T., 1979, ^{40}Ar-^{39}Ar age systematics of consortium breccia 73255: LPSCP 10, v. 1, p. 745–762.
Steele, I.M., and Smith, J.V., 1973, Mineralogy and petrology of some Apollo 16 rocks and fines: General petrologic model of the moon: LSCP 4, v. 1, p. 519–536.
———1976, Mineralogy and petrology of complex breccia 14063,14: LSCP 7, v. 2, p. 1949–1964.
Steiger, R.H., and Jäger, Emilie, compilers, 1977, Subcommission on Geochronology: Convention on the use of decay constants in geo- and cosmochronology: Earth and Planetary Science Letters, v. 36, no. 3, p. 359–362.
Stettler, A., and Albarede, F., 1978, ^{39}Ar-^{40}Ar systematics of two millimeter-sized rock fragments from Mare Crisium: Earth and Planetary Science Letters, v. 38, no. 2, p. 401–406.
Stettler, A., Eberhardt, P., Geiss, J., and Grögler, N., 1978, Chronology of the Apollo 17 Station 7 boulder and the South Serenitatis impact [abs.]: LPS IX, pt. 2, p. 1113–1115.
Stettler, A., Eberhardt, P., Geiss, J., Grögler, N., and Maurer, P., 1973, Ar^{39}-Ar^{40} ages and Ar^{37}-Ar^{38} exposure ages of lunar rocks: LSCP 4, pt. 2, p. 1865–1888.
———1974, On the duration of lava flow activity in Mare Tranquillitatis: LSCP 5, v. 2, p. 1557–1570.
Stewart, D.B., 1975, Apollonian metamorphic rocks—the products of prolonged subsolidus equilibration [abs.]: LS VI, pt. 2, p. 774–776.
Stoeser, D.B., Marvin, U.B., Wood, J.A., Wolfe, R.W., and Bower, J.F., 1974, Petrology of a stratified boulder from South Massif, Taurus-Littrow: LSCP 5, v. 1, p. 355–377.
Stöffler, Dieter, 1981, Cratering mechanics: Data from terrestrial and experimental craters and implications for the Apollo 16 site, *in* James, O.B., and Hörz, Friedrich, eds., Workshop on Apollo 16: Houston, Tex., Lunar and Planetary Institute Technical Report 81–01, p. 132–141.
Stöffler, Dieter, Dence, M.R., Graup, G., and Abadian, M., 1974, Interpretation of ejecta formations at the Apollo 14 and 16 sites by a comparative analysis of experimental, terrestrial, and lunar craters: LSCP 5, v. 1, p. 137–150.
Stöffler, Dieter, Gault, D.E., Wedekind, J.A., and Polkowski, G., 1975, Experimental hypervelocity impact into quartz sand: Distribution and shock metamorphism of ejecta: Journal of Geophysical Research, v. 80, no. 29, p. 4062–4077.
Stöffler, Dieter, Knöll, H.-D., and Maerz, U., 1979, Terrestrial and lunar impact breccias and the classification of lunar highland rocks: LPSCP 10, v. 1, p. 639–675.
Stöffler, Dieter, Knöll, H.-D., Marvin, U.B., Simonds, C.H., and Warren, P.H., 1980, Recommended classification and nomenclature of lunar highland rocks—a committee report, *in* Lunar and Planetary Institute, compiler, Conference on the Lunar Highlands Crust, Houston, Tex., 1979, Proceedings: New York, Pergamon (Geochimica et Cosmochimica Acta, supp. 12), p. 51–70.
Stöffler, Dieter, Knöll, H.-D., Reimold W.-U., and Schulien, S., 1976, Grain-size statistics, composition, and provenance of fragmental particles in some Apollo 14 breccias: LSCP 7, v. 2, p. 1965–1985.
Stöffler, Dieter, Ostertag, R., Reimold, W.U., Borchardt, R., Malley, J., and Rehfeldt, A., 1981, Distribution and provenance of lunar highland rock types at North Ray Crater, Apollo 16: LPSCP 12, pt. B, sec. 1, p. 185–207.
Strain, P.L., and El-Baz, Farouk, 1979, Smythii basin topography and comparisons with Orientale: LPSCP 10, v. 3, p. 2609–2621.
———1980, The geology and morphology of Ina: LPSCP 11, v. 3, p. 2437–2446.
Strangway, D.W., Gose, W.A., Pearce, G.W., and McConnell, R.K., 1973, Lunar magnetic anomalies and the Cayley Formation: Nature, Physical Science, v. 246, no. 155, p. 112–115.
Strom, R.G., 1964, Analysis of lunar lineaments, I: Tectonic maps of the Moon: Tucson, University of Arizona, Lunar and Planetary Laboratory Communications, v. 2, no. 39, p. 205–216.
———1971, Lunar mare ridges, rings and volcanic ring complexes: Modern Geology, v. 2, no. 2, p. 133–157.
Strom, R.G., and Fielder, Gilbert, 1971, Multiphase eruptions associated with the craters Tycho and Aristarchus, *in* Fielder, Gilbert, ed., 1971, Geology and physics of the Moon: Amsterdam, Elsevier, p. 55–92.
Stuart-Alexander, D.E., 1971, Geologic map of the Rheita quadrangle of the Moon: USGS Map I–694 (LAC–114), scale 1:1,000,000.
———1978, Geologic map of the central far side of the Moon: USGS Map I–1047, scale 1:5,000,000.
Stuart-Alexander, D.E., and Howard, K.A., 1970, Lunar maria and circular basins—a review: Icarus, v. 12, no. 3, p. 440–456.
Stuart-Alexander, D.E., and Tabor, R.W., 1972, Geologic map of the Fracastorius quadrangle of the Moon: USGS Map I–720 (LAC–97), scale 1:1,000,000.
Stuart-Alexander, D.E., and Wilhelms, D.E., 1975, The Nectarian System, a new lunar time-stratigraphic unit: USGS Journal of Research, v. 3, no. 1, p. 53–58.
Sutton, R.L., and Schaber, G.G., 1971, Lunar locations and orientations of rock samples from Apollo missions 11 and 12: LSCP 2, v. 1, p. 17–26.
Sutton, R.L., Hait, M.H., and Swann, G.A., 1972, Geology of the Apollo 14 landing site: LSCP 3, v. 1, p. 27–38.
Swann, G.A., Bailey, N.G., Batson, R.M., Eggleton, R.E., Hait, M.H., Holt, H.E., Larson, K.B., Reed, V.S., Schaber, G.G., Sutton, R.L., Trask, N.J., Ulrich, G.E., and Wilshire, H.G., 1977, Geology of the Apollo 14 landing site in the Fra Mauro highlands: USGS Professional Paper 880, 103 p.
Swann, G.A., Bailey, N.G., Batson, R.M., Freeman, V.L., Hait, M.H., Head, J.W., Holt, H.E., Howard, K.A., Irwin, J.B., Larson, K.B., Muehlberger, W.R., Reed, V.S., Rennilson, J.J., Schaber, G.G., Scott, D.R., Silver, L.T., Sutton, R.L., Ulrich, G.E., Wilshire, H.G., and Wolfe, E.W., 1972, Preliminary geologic investigation of the Apollo 15 landing site, [chap.] 5 *of* Apollo 15 preliminary science report: NASA Report SP-289, p. 5–1 to 5–112.
Swann, G.A., Trask, N.J., Hait, M.H., and Sutton, R.L., 1971, Geologic setting of the Apollo 14 samples: Science, v. 173, no. 3998, p. 716–719.
Swift, R.P., 1977, Material strength degradation effect on cratering dynamics, *in* Roddy, D.J., Pepin, R.O., and Merrill, R.B., eds., Impact and explosion cratering: Planetary and terrestrial implications: New York, Pergamon, p. 1025–1042.
Tatsumoto, Mitsunobu, Nunes, P.D., Knight, R.J., Hedge, C.E., and Unruh, D.M., 1973, U-Th-Pb, Rb-Sr, and K measurements of two Apollo 17 samples [abs.]: Eos (American Geophysical Union Transactions), v. 54, no. 6, p. 614–615.
Tatsumoto, Mitsunobu, Nunes, P.D., and Unruh, D.M., 1977, Early history of the Moon: Implications of U-Th-Pb and Rb-Sr systematics, *in* Pomeroy, J.H., and Hubbard, N.J., eds., The Soviet-American Conference on Cosmochemistry of the Moon and Planets: NASA Report SP-370, p. 507–523.
Taylor, G.J., Drake, M.J., Wood, J.A., and Marvin, U.B., 1973, The Luna 20 lithic fragments, and the composition and origin of the lunar highlands: Geochimica et Cosmochimica Acta, v. 37, no. 4, p. 1087–1106.
Taylor, S.R., 1975, Lunar science: A post-Apollo view: New York, Pergamon, 372 p.
———1978, Geochemical constraints on melting and differentiation of the Moon: PLPSC 9, v. 1, p. 15–23.
———1982, Planetary science: A lunar perspective: Houston, Tex., Lunar and Planetary Institute, 481 p.
Taylor, S.R., and Jakeš, Petr, 1974, The geochemical evolution of the Moon: PLSC 5, v. 2, p. 1287–1305.
———1977, Geochemical evolution of the moon revisited: PLSC 8, v. 1, p. 433–446.
Taylor, S.R., Gorton, M.P., Muir, Patricia, Nance, W., Rudowski, R., and Ware, N., 1973, Lunar highlands composition: Apennine Front: PLSC 4, v. 2, p. 1445–1459.
Taylor, S.R., Kaye, Maureen, Muir, Patricia, Nance, W., Rudowski, R., and Ware, N., 1972, Composition of the lunar uplands: Chemistry of Apollo 14 samples from Fra Mauro: PLSC 3, v. 2, p. 1231–1249.
Tera, Fouad, and Wasserburg, G.J., 1974, U-Th-Pb systematics on lunar rocks and inferences about lunar evolution and the age of the Moon: PLSC 5, v. 2, p. 1571–1599.
———1975, The evolution and history of mare basalts inferred from U-Th-Pb systematics [abs.]: LS VI, pt. 2, p. 807–809.
———1976, Lunar ball games and other sports [abs.]: LS VII, pt. 2, p. 858–860.
Tera, Fouad, Papanastassiou, D.A., and Wasserburg, G.J., 1974, Isotopic evidence for a terminal lunar cataclysm: Earth and Planetary Science Letters, v. 22, no. 1, p. 1–21.
Thompson, T.W., 1974, Atlas of lunar radar maps at 70-cm wavelength: The Moon, v. 10, no. 1, p. 51–85.
———1979, A review of Earth-based radar mapping of the Moon: The Moon and the Planets, v. 20, no. 2, p. 179–198.
Thompson, T.W., Cutts, J.A., Shorthill, R.W., and Zisk, S.H., 1980, Infrared and radar signatures of lunar craters: Implications about crater evolution, *in* Lunar and Planetary Institute, compiler, Conference on the Lunar Highlands Crust, Houston, Tex., 1979, Proceedings: New York, Pergamon (Geochimica et Cosmochimica Acta, supp. 12), p. 483–499.
Thompson, T.W., Howard, K.A., Shorthill, R.W., Tyler, G.L., Zisk, S.H., Whitaker, E.A., Schaber, G.G., and Moore, H.J., 1973, Remote sensing of Mare Serenitatis, pt. A *of* Remote sensing and photogrammetric studies, [chap.] 33 *of* Apollo 17 preliminary science report: NASA Report SP-330, p. 33–3 to 33–10.
Thompson, T.W., Masursky, Harold, Shorthill, R.W., Tyler, G.L., and Zisk, S.H., 1974, A comparison of infrared, radar, and geologic mapping of lunar craters: The Moon, v. 10, no. 1, p. 87–117.
Thompson, T.W., Zisk, S.H., Shorthill, R.W., Schultz, P.H., and Cutts, J.A., 1981, Lunar craters with radar bright ejecta: Icarus, v. 46, no. 2, p. 201–225.
Thurber, C.H., and Solomon, S.C., 1978, An assessment of crustal thickness variations on the lunar near side: Models, uncertainties, and implications for crustal differentiation: PLPSC 9, v. 3, p. 3481–3497.
Tilton, G.R., and Chen, J.H., 1979, Lead isotope systematics of three Apollo 17 mare basalts: PLPSC 10, v. 1, p. 259–274.
Titley, S.R., 1967, Geologic map of the Mare Humorum region of the Moon: USGS Map I–495 (LAC–93), scale 1:1,000,000.
Titley, S.R., and Eggleton, R.E., 1964, Description of an extensive hummocky deposit around the Humorum basin: USGS ASAPR, July 1, 1963 to July 1, 1964, pt. A, p. 85–89.
Titley, S.R., and Trask, N.J., 1969, Geologic map of Apollo landing site 5, part of Maestlin G region, Oceanus Procellarum: USGS Map I–623 [ORB II–13(25)], scale 1:25,000.
Todd, Terry, Richter, D.A., Simmons, Gene, and Wang, Herbert, 1973, Unique characterization of lunar samples by physical properties: PLSC 4, v. 3, p. 2639–2662.
Toksöz, M.N., 1974, Geophysical data and the interior of the Moon: Annual Review of Earth and Planetary Sciences, v. 2, p. 151–177.
Toksöz, M.N., and Johnston, D.H., 1974, The evolution of the Moon: Icarus, v. 21, no. 4, p. 389–414.
Toksöz, M.N., Dainty, A.M., Solomon, S.C., and Anderson, K.R., 1974, Structure of the Moon: Reviews of Geophysics and Space Physics, v. 12, no. 4, p. 539–567.
Trask, N.J., 1966, Size and spatial distribution of craters estimated from *Ranger* photographs, *in* Progress in the analysis of the fine structure and geology of the lunar surface from the *Ranger VIII* and *IX* photographs, [chap.] 4 *of Ranger VIII* and *IX*. Part II. Experimenters' analyses and interpretations: Pasadena, California Institute of Technology, Jet Propulsion Laboratory Technical Report 32-800, p. 252–263.
———1969, Geologic maps of early Apollo landing sites, *explanatory pamphlet accompanying* USGS Maps I–616 through I–627, 4 p.
———1971, Geologic comparison of mare materials in the lunar equatorial belt, including Apollo 11 and Apollo 12 landing sites, *in* Geological Survey research, 1971: USGS Professional Paper 750–D, p. D138–D144.
———1972, The contributions of Ranger photographs to understanding the geology of the Moon: USGS Professional Paper 599–J, p. J1–J16.
Trask, N.J., and McCauley, J.F., 1972, Differentiation and volcanism in the lunar highlands: Photogeologic evidence and Apollo 16 implications: Earth and Planetary Science Letters, v. 14, no. 2, p. 201–206.
Trask, N.J., and Titley, S.R., 1966, Geologic map of the Pitatus region of the Moon: USGS Map I–485 (LAC–94), scale 1:1,000,000.
Trulio, J.G., 1977, Ejecta formation: Calculated motion from a shallow-buried nuclear burst, and its significance for high velocity impact cratering, *in* Roddy, D.J., Pepin, R.O., and Merrill, R.B., eds., Impact and explosion cratering: Planetary and terrestrial implications: New York, Pergamon, p. 919–957.
Turkevich, A.L., 1971, Comparison of the analytical results from the Surveyor, Apollo, and Luna missions: PLSC 2, v. 2, p. 1209–1215.
Turkevich, A.L., Anderson, W.A., Economou, T.E., Franzgrote, E.J., Griffin, H.E., and others, 1969, The alpha-scattering chemical analysis experiment on the Surveyor lunar missions, [chap.] 8 *of* Surveyor program results: Final report: NASA Report SP-184, p. 271–350.
Turner, Grenville, 1970, Argon-40/argon-39 dating of lunar rock samples: PLSC 1, v. 2, p. 1665–1684.
———1971, ^{40}Ar-^{39}Ar ages from the lunar maria: Earth and Planetary Science Letters, v. 11, no. 3, p. 169–191.
———1977, Potassium-argon chronology of the Moon: Physics and Chemistry of the Earth, v. 10, no. 3, p. 145–195.
Turner, Grenville, and Cadogan, P.H., 1975, The history of lunar bombardment inferred from ^{40}Ar-^{39}Ar dating of highland rocks: PLSC 6, v. 2, p. 1509–1538.
Turner, Grenville, Cadogan, P.H., and Yonge, C.J., 1973, Argon selenochronology: PLSC 4, v. 2, p. 1889–1914.
Turner, Grenville, Huneke, J.C., Podosek, F.A., and Wasserburg, G.J., 1971, ^{40}Ar-^{39}Ar ages and cosmic ray exposure ages of Apollo 14 samples: Earth and Planetary Science Letters, v. 12, no. 1, p. 19–35.
———1972, Ar^{40}-Ar^{39} systematics in rocks and separated minerals from Apollo 14: PLSC 3, v. 2, p. 1589–1612.
Ulrich, G.E., 1969, Geologic map of the J. Herschel quadrangle of the Moon: USGS Map I–604 (LAC–11), scale 1:1,000,000.
———1973, A geologic model for North Ray Crater and stratigraphic implications for the Descartes region: PLSC 4, v. 1, p. 27–39.
———1981, Geology of North Ray crater, [chap.] D2 *of* Ulrich, G.E., Hodges, C.A., and Muehlberger, W.R., eds., Geology of the Apollo 16 area, central lunar highlands: USGS Professional Paper 1048, p. 45–81.
Ulrich, G.E., Hodges, C.A., and Muehlberger, W.R., eds., 1981a, Geology of the Apollo 16 area, central lunar highlands: USGS Professional Paper 1048, 539 p.
Ulrich, G.E., Moore, H.J., Reed, V.S., Wolfe, E.W., and Larson, K.B., 1981b, Ejecta distribution model, South Ray crater, [chap.] G *of* Ulrich, G.E., Hodges, C.A., and Muehlberger, W.R., eds., Geology of the Apollo 16 area, central lunar highlands: USGS Professional Paper 1048, p. 160–173.
Urey, H.C., 1951, The origin and development of the earth and other terrestrial planets: Geochimica et Cosmochimica Acta, v. 1, no. 4–6, p. 209–277.
———1952, The planets: New Haven, Conn., Yale University Press, 245 p.
U.S. Geological Survey, 1971, Geologic map of the Bonpland H region of the Moon: USGS Map I–693 (RLC–3), scale 1:100,000.
———1978, Shaded relief of the Mare Orientale area of the Moon: USGS Map I–1089, scale 1:5,000,000.
U.S. National Aeronautics and Space Administration, 1969, Preliminary examination of lunar samples, [chap.] 5 *of* Apollo 11 preliminary science report: NASA Report SP–214, p. 123–142.
———1970, Apollo 12 preliminary science report: NASA Report SP–235, 227 p.
Van Dorn, W.G., 1968, Tsunamis on the Moon?: Nature, v. 220, no. 5172, p. 1102–1107.
———1969, Lunar maria: Structure and evolution: Science, v. 165, no. 3894, p. 693–695.
Vaniman, D.T., and Papike, J.J., 1977, Very low Ti (VLT) basalts: A new mare rock type from the Apollo 17 drill core: PLSC 8, v. 2, p. 1443–1471.
———1980, Lunar highland melt rocks: Chemistry, petrology and silicate mineralogy, *in* Lunar and Planetary Institute, compiler, Conference on the Lunar Highlands Crust,

Houston, Tex., 1979, Proceedings: New York, Pergamon (Geochimica et Cosmochimica Acta, supp. 12), p. 271–337.
Walker, David, Longhi, John, Grove, T.L., Stolper, Edward, and Hays, J.F., 1973, Experimental petrology and origin of rocks from the Descartes Highlands: PLSC 4, v. 1, p. 1013–1032.
Walker, David, Longhi, John, and Hays, J.F., 1972, Experimental petrology and origin of Fra Mauro rocks and soil: PLSC 3, v. 1, p. 797–817.
———1975, Differentiation of a very thick magma body and implications for the source regions of mare basalts: PLSC 6, v. 1, p. 1103–1120.
Walker, A.S., and El-Baz, Farouk, 1982, Analysis of crater distributions in mare units on the lunar far side: The Moon and the Planets, v. 27, p. 91–106.
Wänke, H., Dreibus, Gerlind, and Palme, H., 1978, Primary matter in the lunar highlands: The case of the siderophile elements: PLPSC 9, v. 1, p. 83–110.
Warner, J.L., 1971, Lunar crystalline rocks: Petrology and geology: PLSC 2, v. 1, p. 469–480.
———1972, Metamorphism of Apollo 14 breccias: PLSC 3, v. 1, p. 623–643.
Warner, J.L., Phinney, W.C., Bickel, C.E., and Simonds, C.H., 1977, Feldspathic granulitic impactites and pre-final bombardment lunar evolution: PLSC 8, v. 2, p. 2051–2066.
Warner, J.L., Simonds, C.H., and Phinney, W.C., 1974, Impact-induced fractionation in the lunar highlands: PLSC 5, v. 1, p. 379–397.
———1976, Genetic distinction between anorthosite and Mg-rich plutonic rocks [abs.]: LS VIII, pt. 2, p. 915–917.
Warner, R.D., Taylor, G.J., Conrad, G.H., Northrup, H.R., Barker, S., Keil, K., Ma, M.-S., and Schmitt, R., 1979, Apollo 17 high-Ti mare basalts: New bulk compositional data, magma types, and petrogenesis: PLPSC 10, v. 1, p. 225–247.
Warren, P.H., and Wasson, J.T., 1977, Pristine nonmare rocks and the nature of the lunar crust: PLSC 8, v. 2, p. 2215–2235.
———1979a, Effects of pressure on the crystallization of a "chondritic" magma ocean and implications for the bulk composition of the Moon: PLPSC 10, v. 2, p. 2051–2083.
———1979b, The compositional-petrographic search for pristine nonmare rocks: Third foray: PLPSC 10, v. 1, p. 583–610.
———1979c, The origin of KREEP: Reviews of Geophysics and Space Physics, v. 17, no. 1, p. 73–88.
———1980a, Early lunar petrogenesis, oceanic and extraoceanic, *in* Lunar and Planetary Institute, compiler, Conference on the Lunar Highlands Crust, Houston, Tex., 1979, Proceedings: New York, Pergamon (Geochimica et Cosmochimica Acta, supp. 12), p. 81–99.
———1980b, Further foraging for pristine nonmare rocks: Correlations between geochemistry and longitude: PLPSC 11, v. 1, p. 431–470.
Wasserburg, G.J., and Papanastassiou, D.A., 1971, Age of an Apollo 15 mare basalt; lunar crust and mantle evolution: Earth and Planetary Science Letters, v. 13, no. 1, p. 97–104.
Wasson, J.T., and Baedecker, P.A., 1972, Provenance of Apollo 12 KREEP: PLSC 3, v. 2, p. 1315–1326.
West, Mareta, and Cannon, P.J., 1971, Geologic map of Apollo landing sites 4 and 4R, part of Wichmann CA region, Oceanus Procellarum: USGS Map I–625 [ORB III–11 (25)], scale 1:25,000.
Wetherill, G.W., 1971, Of time and the Moon: Science, v. 173, no. 3995, p. 383–392.
———1975a, Late heavy bombardment of the Moon and terrestrial planets: PLSC 6, v. 2, p. 1539–1561.
———1975b, Possible slow accretion of the moon and its thermal and petrological consequences, *in* Lunar Science Institute, compiler, Papers presented to the Conference on Origins of Mare Basalts and Their Implications for Lunar Evolution: Houston, Tex., Lunar Science Institute Contribution 234, p. 184–188.
———1976, The role of large bodies in the formation of the earth and moon: PLSC 7, v. 3, p. 3245–3257.
———1977a, Evolution of the earth's planetesimal swarm subsequent to the formation of the earth and moon: PLSC 8, v. 1, p. 1–16.
———1977b, Pre-mare cratering and early solar system history, *in* Pomeroy, J.H., and Hubbard, N.J., eds., The Soviet-American Conference on Cosmochemistry of the Moon and Planets: NASA Report SP–370, p. 553–567.
———1981, Nature and origin of basin-forming projectiles, *in* Multi-ring basins: LPSCP 12, pt. A, p. 1–18.
Whitaker, E.A., 1963, Evaluation of the Soviet photographs of he moon's far side, chap. 4 *of* Middlehurst, B.M., and Kuiper, G.P., eds., The moon, meteorites, and comets, v. 4 *of* The solar system: Chicago, University of Chicago Press, p. 123–128.
———1966, The surface of the moon, [chap.] 3 *of* Hess, W.N., Menzel, D.H., and O'Keefe, J.A., eds., The nature of the lunar surface: Proceedings of the 1965 IAU Syposium: Baltimore, Johns Hopkins Press, p. 79–98.
———1972a, Artificial lunar impact craters: Four new identifications, pt. 1 *of* Photogeology, [chap.] 29 *of* Apollo 16 preliminary science report: NASA Report SP–315, p. 29–39 to 29–45.
———1972b, Lunar color boundaries and their relationship to topographic features: A preliminary survey: The Moon, v. 4, p. 348–355.
———1978, Galileo's lunar observations and the dating of the composition of "Sidereus Nuncius": Journal for the History of Astronomy, v. 9, p. 155–169.
———1981, The lunar Procellarum basin, *in* Multi-ring basins: LPSCP 12, pt. A, p. 105–111.
Whitford-Stark, J.L., 1981, The evolution of the lunar Nectaris multiring basin: Icarus, v. 48, p. 393–427.
Whitford-Stark, J.L., and Head, J.W., 1980, Stratigraphy of Oceanus Procellarum basalts: Sources and styles of emplacement: Journal of Geophysical Research, v. 85, no. B11, p. 6579–6609.
Wilbur, C.L., 1978, Volcano-tectonic history of Tsiolkovskij [abs.]: LPS IX, pt. 2, p. 1253–1255.
Wilhelms, D.E., 1964, Major structural features of the Mare Vaporum quadrangle: USGS ASAPR, July 1, 1963 to July 1, 1964, pt. A, p. 1–16.
———1968, Geologic map of the Mare Vaporum quadrangle of the Moon: USGS Map I–548 (LAC–59), scale 1:1,000,000.
———1970a, Geologic map of Apollo landing site 1: USGS Map I–617 [ORB II–2 (25)], scale 1:25,000.
———1970b, Summary of lunar stratigraphy—telescopic observations: USGS Professional Paper 599–F, p. F1–F47.
———1972a, Geologic map of the Taruntius quadrangle of the Moon: USGS Map I–722 (LAC–61), scale 1:100,000.
———1972b, Geologic mapping of the second planet: Flagstaff, Ariz., USGS Interagency Report: Astrogeology 55, 36 p.
———1972c, Reinterpretations of the northern Nectaris basin, pt. F *of* Photogeology, [chap.] 29 *of* Apollo 16 preliminary science report: NASA Report SP–315, p. 29–27 to 29–30.
———1976, Secondary impact craters of lunar basins: PLSC 7, v. 3, p. 2883–2901.
———1980, Stratigraphy of part of the lunar near side: USGS Professional Paper 1046–A, p. A1–A71.
Wilhelms, D.E., and El-Baz, Farouk, 1977, Geologic map of the east side of the Moon: USGS Map I–948, scale 1:5,000,000.
Wilhelms, D.E., and McCauley, J.F., 1971, Geologic map of the near side of the Moon: USGS Map I–703, scale 1:5,000,000.
Wilhelms, D.E., and Trask, N.J., 1965, Compilation of geology in the lunar equatorial belt: USGS ASAPR, July 1, 1964 to July 1, 1965, pt. A, p. 29–34.
Wilhelms, D.E., Hodges, C.A., and Pike, R.J., 1977, Nested-crater model of lunar ringed basins, *in* Roddy, D.J., Pepin, R.O., and Merrill, R.B., eds., Impact and explosion cratering: Planetary and terrestrial implications: New York, Pergamon, p. 539–562.
Wilhelms, D.E., Howard, K.A., and Wilshire, H.G., 1979, Geologic map of the south side of the Moon: USGS Map I–1162, scale 1:5,000,000.
Wilhelms, D.E., Oberbeck, V.R., and Aggarwal, H.R., 1978, Size-frequency distributions of primary and secondary lunar impact craters: LPSCP 9, v. 3, p. 3735–3762.
Wilhelms, D.E., Stuart-Alexander, D.E., and Howard, K.A., 1969, Preliminary interpretations of lunar geology, *in* Initial photographic analysis, [chap.] 2 *of* Analysis of Apollo 8 photography and visual observations: NASA Report SP–201, p. 16–21.
Wilhelms, D.E., Ulrich, G.E., Moore, H.J., and Hodges, C.A., 1980, Emplacement of Apollo 14 and 16 breccias as primary basin ejecta [abs.]: LPS XI, pt. 3, p. 1251–1253.
Wilshire, H.G., 1973, Geologic map of the Byrgius quadrangle of the Moon: USGS Map I–755 (LAC–92), scale 1:1,000,000.
Wilshire, H.G., and Jackson, E.D., 1972, Petrology and stratigraphy of the Fra Mauro Formation at the Apollo 14 site: USGS Professional Paper 785, 26 p.
Wilshire, H.G., and Moore, H.J., 1974, Glass-coated lunar rock fragments: Journal of Geology, v. 82, no. 4, p. 403–417.
Wilshire, H.G., Offield, T.W., Howard, K.A., and Cummings, David, 1972a, Geology of the Sierra Madera cryptoexplosion structure, Pecos County, Texas: USGS Professional Paper 599–H, p. H1–H42.
Wilshire, H.G., Schaber, G.G., Silver, L.T., Phinney, W.C., and Jackson, E.D., 1972b, Geologic setting and petrology of Apollo 15 anorthosite (15415): Geological Society of America Bulletin, v. 83, no. 4, p. 1083–1092.
Wilson, Lionel, and Head, J.W., 1981, Ascent and eruption of basaltic magma on the earth and moon: Journal of Geophysical Research, v. 86, no. B4, p. 2971–3001.
Winter, D.F., 1970, The infrared moon: Data, interpretations, and implications: Radio Science, v. 5, no. 2, p. 229–240.
Winzer, S.R., Nava, D.F., Schuhmann, P.J., Lum, R.K.L., Schuhmann, S., Lindstrom, M.M., Lindstrom, D.J., and Philpotts, J.A., 1977, The Apollo 17 "melt sheet": Chemistry, age and Rb/Sr systematics: Earth and Planetary Science Letters, v. 33, no. 3, p. 389–400.
Wise, D.U., and Yates, M.T., 1970, Mascons as structural relief on a lunar "Moho": Journal of Geophysical Research, v. 75, no. 2, p. 261–268.
Wiskerchen, M.J., and Sonett, C.P., 1977, A lunar metal core?: PLSC 8, v. 1, p. 515–535.
Wisotski, John, 1977, Dynamic ejecta parameters from high-explosive detonations, *in* Roddy, D.J., Pepin, R.O., and Merrill, R.B., eds., Impact and explosion cratering: Planetary and terrestrial implications: New York, Pergamon, p. 1101–1121.
Wolfe, E.W., and Reed, V.S., 1976, Geology of the massifs at the Apollo 17 landing site: USGS Journal of Research, v. 4, no. 2, p. 171–180.
Wolfe, E.W., Lucchitta, B.K., Reed, V.S., Ulrich, G.E., and Sanchez, A.G., 1975, Geology of the Taurus-Littrow valley floor: PLSC 6, v. 3, p. 2463–2482.
Wolfe, E.W., Bailey, N.G., Lucchitta, B.K., Muehlberger, W.R., Scott, D.H., Sutton, R.L., and Wilshire, H.G., 1981, The geologic investigation of the Taurus-Littrow valley: Apollo 17 landing site: USGS Professional Paper 1080, 280 p.
Wood, C.A., and Andersson, Leif, 1978, New morphometric data for fresh lunar craters: PLPSC 9, v. 3, p. 3669–3689.
Wood, C.A., and Head, J.W., 1976, Comparison of impact basins on Mercury, Mars and the Moon: PLSC 7, v. 3, p. 3629–3651.
Wood, J.A., 1970, Petrology of the lunar soil and geophysical implications: Journal of Geophysical Research, v. 75, no. 32, p. 6497–6513.
———1972a, Fragments of terra rock in the Apollo 12 soil samples and a structural model of the Moon: Icarus, v. 16, no. 3, p. 462–501.
———1972b, Thermal history and early magmatism in the Moon: Icarus, v. 16, no. 2, p. 229–240.
———1973, Bombardment as a cause of lunar asymmetry: The Moon, v. 8, no. 1–2, p. 73–103.
———1975a, Glass compositions as a clue to unsampled mare basalt lithologies, *in* Lunar Science Institute, compiler, Papers presented to the Conference on Origins of Mare Basalts and Their Implications for Lunar Evolution: Houston, Tex., Lunar Science Institute Contribution 234, p. 194–198.
———1975b, Lunar petrogenesis in a well-stirred magma ocean: PLSC 6, v. 1, p. 1087–1102.
———1975c, The moon: Scientific American, v. 233, no. 3, p. 92–102.
———1975d, The nature and origin of boulder 1, station 2, Apollo 17: The Moon, v. 14, no. 3–4, p. 505–517.
———1977, A survey of lunar rock types and comparison of the crusts of Earth and Moon, *in* Pomeroy, J.H., and Hubbard, N.J., eds., The Soviet-American Conference on Cosmochemistry of the Moon and Planets: NASA Report SP–370, p. 35–53.
———1979, The solar system: Englewood Cliffs, N.J., Prentice-Hall, 196 p.
Wood, J.A., Dickey, J.S., Jr., Marvin, U.B., and Powell, B.N., 1970, Lunar anorthosites and a geophysical model of the moon: PLSC 1, v. 1, p. 965–988.
Woodford, A.O., 1965, Historical geology: San Francisco, Freeman, 512 p.
Woronow, A.E., 1977, Crater saturation and equilibrium: A Monte Carlo simulation: Journal of Geophysical Research, v. 82, no. 17, p. 2447–2456.
———1978, A general cratering-history model and its implications for the lunar highlands: Icarus, v. 34, no. 1, p. 76–88.
Wright, F.E., Wright, F.H., and Wright, Helen, 1963, The lunar surface: Introduction, chap. 1 *of* Middlehurst, B.M., and Kuiper, G.P., eds., The moon, meteorites, and comets, v. 4 *of* The solar system: Chicago, University of Chicago Press, p. 1–56.
York, Derek, Kenyon, W.J., and Doyle, R.J., 1972, ^{40}Ar-^{39}Ar ages of Apollo 14 and 15 samples: PLSC 3, v. 2, p. 1613–1622.
Young, R.A., 1975, Mare crater size-frequency distributions: Implications for relative surface ages and regolith development: PLSC 6, v. 3, p. 2645–2662.
———1977, The lunar impact flux, radiometric age correlation, and dating of specific lunar features: PLSC 8, v. 3, p. 3457–3473.
Young, R.A., Brennan, W.J., and Nichols, D.J., 1974, Problems in the interpretation of lunar mare stratigraphy and relative ages indicated by ejecta from small impact craters: PLSC 5, v. 1, p. 159–170.
Zisk, S.H., Hodges, C.A., Moore, H.J., Shorthill, R.W., Thompson, T.W., and Wilhelms, D.E., 1977, The Aristarchus-Harbinger region of the Moon: Surface geology and history from recent remote-sensing observations: The Moon, v. 17, no. 1, p. 59–99.
Zisk, S.H., Pettengill, G.H., and Catuna, G.W., 1974, High-resolution radar maps of the lunar surface at 3.8 cm wavelength: The Moon, v. 10, no. 1, p. 17–50.

INDEX

[Italic page numbers refer to principal discussions]

A

Abbe, pls. 4, 9
Abulfeda, 41, 180, pl. 7
Abulfeda chain, 41, 115
Adams, 167, 180, pl. 7
Aeon, defined, viii, 19
Aeronautical Chart and Information Center, 124
Age, meaning, 19-22, 49, 129, 139, 163, 277
Ages, relative, 121-136. *See also* Crater dating, Crater frequencies, D_L method, D_L values, Superpositions, *and individual series and systems*
Agglutinates, 95, 96, 256, 266, 279. *See also* Regolith
Agrippa, 205, 253, 266, pl. 10
Airy, 155, 244, pl. 5
Airy model of isostasy, 143
Aitken, 120, pls. 4, 9
Al Biruni, pls. 4, 9
Al-Khwarizmi, 154, 180, pl. 7
Al-Khwarizmi/King basin, 65, 148, 154, pls. 3, 6
  crater frequency, 136, 146, 148, 157, 160
Albategnius, 10, 213, 214, 219, pls. 7, 8
Albedo, 2, 3, 49, 85, *94*, 256, 265-267
  color, 96, 97
  magnetism, 215, 256
  rays, 94-96, 249, 250, 256, 265-267, 279
  slopes, 48, 49, 96
Aliacensis, 48, 127, 180, 189, pl. 7
Alkalis, mare basalt, 101, 276
Alpes Formation, 15, 73, 81, 82, 119, 168, 171, 174, 200, *203*, pl. 8
Alphonsus, 10, 89, 90, 94, 96, 128, 213, pls. 5, 7
Alpine Valley, 113-116, 119, 204
ALSEP, defined, 23, 165, 240
Altai scarp, 164, 233
Altimetry, Apollo and Zond, 77, 120, 145
Alumina (aluminum)
  Apollo 16 units, 165, 168
  mare basalt, 101-103
  orbital sensing, 98-103, 139-144, 165, 168
  terra crust, 139-144
Amundsen-Ganswindt basin, 7, 65, 148, pls. 3, 6
Anaxagoras, 51, 204, pl. 11
Andĕl M, 218
Anorthite, 140
Anorthosite and anorthositic norite, 140-143, 156, 165, 168
ANT suite, 140-143, 171, 201
Antipodes, basins, 67, 76, 82, 181, 215-217, 279
Antoniadi, 7, 58, 65, 81, 231, pls. 4, 9
Apennine Bench Formation, 8, 73, 113, 193, *197*, 231
  age, *198*, 212, 224, 243, 278
  geologic maps and sections, 15, 19, 200, 243, 276, pl. 8
  samples, *197*, 224, 278, 279
Apenninian Series, 123-125
Aphanitic texture, 23, 174, 177, 178
Apianus, 189, pl. 7
Apollo basin, 55, *60*, *64*, 79-82, 103, 145, *147*, *160*, 178-181, pls. 3, 6
  mare filling, 103, pls. 4, 9
Apollo missions, 12, 19-23, 82, 101, 140, *163*, *195*, *229*, 235, 249, 277-280, pls. 2-4, 7-11. *See also individual missions*
  unvisited sites, 125, 131, 254, 255, 258, 259
Apollo photographs, types, 4, pl. 2
Apollo 8, 12, 145, 152, 160
Apollo 10, 12
Apollo 11
  crater frequency and D_L values, 230, 235, 238
  crater morphology, 135
  landing site, 12, 13, 99, 101, *169*, 227, *235*, pls. 2, 4, 9
  mare samples, 19, 85, 101, 224, *235*, 245, 276-279
  regolith, 235
  spectral reflectance, 96, 99
  terra samples, 140, 142
Apollo 12
  crater frequency and D_L values, 230, 259, 271-273
  crater morphology, 135
  landing site, 12, 99, 101, 132, 207, *259*, 272, pls. 2, 4, 10
  mare samples, 19, 23, 101, 102, 249, *259*, 276-279
  objectives, 259
  spectral reflectance, 96, 99, 261
  terra and Copernicus-ray samples, 140, 269, 270, 278
Apollo 13, 12, 195
Apollo 14
  crater frequency, 136, 230
  Fra Mauro Formation samples, 82, 195, 200, *204*, 276-279
  landing site, 10, 12, *22*, 82, 144, *204*, pls. 2, 3, 8
  mare-basalt clasts, 101, 190, 200, 278
  objectives, 195, 204
  sampling stations, 190, 200, 204, 205, 209
  terra samples, general, 19, 23, 140-144
Apollo 15
  Apennine Bench Formation samples, *197*, 224, 278, 279
  crater frequency and D_L values, 136, 197, 230, 231, 252, 262, 271-273
  landing site, ii, 8, 9, 12-15, *22*, 88, 91, 99, 101, 193, *200*, pls. 2-4, 8, 9
  mare samples, 19-23, 101, 102, 237, *243*, 276-279
  massif samples, 19, 23, 49, 82, 140-144, 195, *198*, 224, 276-279
  objectives, 88, 198
  sampling stations, 201, 237, 243, 244
  spectral reflectance, 96, 99
Apollo 16
  crater frequency, 197, 219, 222, 230
  crustal thickness, 12, 143
  landing site, 12, 21, *22*, 101, *164*, *218*, pls. 2, 3, 7, 8
  mare samples, 101, 237, 238, 245, 276, 278
  objectives, 21, 195
  sampling stations, 165, 168-170, 237
  significance, 21, 57, 72, 127, 146, 164, 216
  spectral reflectance, 97
  terra samples, 21, 57, 82, 140-144, *163*, 216-220, 276
Apollo 17
  crater frequency and D_L values, 230, 273
  landing site, 9, 12, *22*, 93, 99, 101, 105, 109, *171*, 239, pls. 2-4, 7-9
  mare and dark-mantle samples, 101, 102, 237, *239*, 245, 276-279
  massif samples, 21, 46, 49, 82, 140-144, 156, 157, 163, *173*, *190*, 195, 201, 270, 276, 278
  objectives, 173
  sampling stations, 174, 176, 178, 190, 237, 239, 240
  spectral reflectance, 96-99, 240
Apollonian metamorphism, 212
Apparent crater, 43
Arago, 28, 114
Aratus CA, 90
Archimedes, 8, 15, *19*, 21, 113, *124*, 125, 156, *193*, 216, 221, 231, pl. 9
Archimedian Series, 123, 124
Archytas, 204, pl. 10
Ariadaeus B, 220, 221
Aristarchus, 50, *91*, 121, 125, 126, 156, 252, 253, *263*, pl. 11
Aristarchus Plateau, 86, 89, *91*, 125, 233, 240, 244, pl. 4
Aristillus, 8, 9, *22*, 48, 253, *269*, pl. 11
Aristoteles, 9, 202, 252, pl. 10
Artamonov, pls. 4, 9
Arzachel, 10, 128, 221, pls. 5, 8
Asclepi, 41, pl. 6
Ash-flow tuff, 21, 85
Asthenosphere, 115
Astroblemes, 77
Astrogeology, vii
Atlas, 9, 32, 166, 231, pls. 5, 9
Augite, 140, 143, 243
Australe basin, 7, *62*, *64*, 103, 145, *148*, 179, 245, pls. 3, 6
Autolycus, 8, 9, *22*, 200, 253, *269*, pl. 11

B

Baco, 75, pl. 6
Bailly basin, *11*, *59*, *64*, 80, 179, 180, pls. 3, 7
Balmer-Kapteyn basin, 7, 65, 148, 165, *167*, 190, pls. 3, 6
Barbier, 149, 151
Barocius G, 37
Barringer, 181, pl. 7
Barrow, 149, 199, 204, pl. 6
Basalt. *See* Fra Mauro basalt, KREEP basalt, Mare-basalt samples
  terrestrial, 101, 102
Base surge, 43
Basin-mare distinction, 3, 19, 85, 86
Basin materials, 55, 276-279, pls. 3, 6-8, 12. *See also basins listed on p. 64, 65, 148, 179*
  ages, relative, 64, 65, *148*, 157, *179*, 190, 224, 276-279, pls. 3, 6-9, 12
  ages, absolute, 157, 160, 168-171, 177, 178, 186, 190, 191, 200, 201, 212, 224, *276*
  asymmetry, 66, 81, *82*, 127, 171, 173, 179, 276, pls. 3, 6-8
  crater frequencies, 148, 149, 160, 179, 186, 293
  degradation, 65, 81, 82, 145, 146, 163-165
  dunelike deposits ("deceleration dunes"), 67, 71, 130, 204, 218, 219
  ejection angles, 67, 73, 80, 81
  emplacement, 66-82, 174-177, 195, 202-204, 210-212, 218-220, 276
  excluded zone, 66, 164, 165
  extent, 64-82, 123, 127, 129, 145-147, 171, 179, 190, 195, 213-215, 218, 276, pls. 3, 6-8, 12

Basin materials—Continued
  Imbrian, 178, 179, 195, 279, pls. 3, 8, 12. *See also* Imbrium basin, Orientale basin, Schrödinger basin
  impact melt, 66-70, 73, 76, 77, 82, 203, 211, 212, 276
  knobby, 73, 81, 82, 168, 171, 172, 276
  lateral transitions, 66, 67, 72, 73, 81, 82, 163, 164, 190, 204, 205, 214, 215, 218, 219, 224
  lineate, 66, 67, 80-82, 127, 147, 150, 163, 164, 179, 180, 188, 189, 204, 206, 209
  mapping conventions, 127-129, pl. 6
  massifs, 6-11, 49, 66-73, 76-82, 105, 127-129, 145, 146, 152, 155, 171-178, 181, 193, 198-203
  Nectarian, *161*, 276-278, pls. 3, 7, 12
  pre-Nectarian, 139, *143*, 178, 180, 277, 278, pls. 3, 6, 12
  sampling points, 82, 144, 163, 171, 195, 276, pls. 3, 7, 8. *See also* Apollo 14, Apollo 15, Apollo 16, Apollo 17, Luna 20
  secondary-crater relations. *See* Basin-secondary craters
  stratigraphic mapping conventions, 121-129, pl. 6
  superpositions. *See* Superpositions
  thickness, 66, 82
Basin rings, 57-73, 77, 110, 276, 280, pl. 3. *See also* Basin materials, massifs
  external, 78-81
Basin-secondary craters, *32*, *67*, *74*, 81, 137, 151, 159, 160, 181-186, 196, 213-215, 276. *See also* Imbrium basin, secondary craters; Nectaris basin, secondary craters; Orientale basin, secondary craters
  distribution and extent, *32*, *64*, 127, 163, 171, 213-215, pls. 3, 6-8
  ejecta relation, 42, 66, 67, 72, 81, 82, 211-213, 217-220, 276
  linearity and radiality, *3*, *32*, 67, 80, 81, 127, 147, 161-170, 179, 186-188, 196, 213-216, 225
  mapping conventions, 127-129, pl. 6
  plains relation, 67, 71, 72, 128, 146, 164, 190, 215-220, 224
  provenance of material, 42, 43, 81, 82, *129*, 211, 212, 217-220, 276
  "rays," 74, 75, 82, 127
  stratigraphic relations, *127*, 145-151, *158*, 161-167, 175, *179*, *196*, pls. 6-8
Basins (ringed basins, multi-ringed basins), 3, 19-23, *55*, 276, pls. 3, 6-8. *See also other basin headings and basins listed on p. 64, 65, 148, 179*
  antipodes, 67, 76, 82, 181, 215-217, 279
  craters, comparisons, 57, 58, 65, 66
  crater-uplift control, 145, 149
  crustal composition, 143-145
  defined, 3, 57, 65
  deposits. *See* Basin materials
  depth, 77, 80, 143-145
  ejecta. *See* Basin materials
  excavation cavity, 66, 77-81
  gravity, 77, 79, 117
  inventory, 57, 64, 65, 81, 148, 179, pl. 3
  mare-extrusion control, 102, 103, 145, 229, 238-245
  Moho control, 77, 80, 102, 103, 115, 120, 143-145, 229, 240-245, 276
  rings. *See* Basin rings
  secondary-impact craters. *See* Basin-secondary craters
  size series, 59-65
  topographic rim, defined, 66
  topography, 77, 100, 280
Bedrock, 12, 13, 21, 45
Beer, 8, 113
Bench Crater, 259, 261
Berosus, 166, pl. 7
Berzelius, 175
Bessarion B, 39
Bianchini, 34, pl. 9
Big Backside Basin. *See* South Pole-Aitken basin
Birkhoff basin, *59*, *64*, 147, *148*, *158*, 179, 188, 293, pls. 3, 6
Birkhoff X, 159
"Black-and-white rocks," *23*, 165, *203*, 224, 278
Blancanus, pl. 7
Blanchinus, 149, 189
Blocks, ejected, 29, 46, 47, 99, 168, 258, *265*
Bonpland, *22*, 189, 208
Bonpland D, 112, 221
Boscovich Formation, 205
Boulder tracks, 174, 178
Bowen's reaction series, 140, 142
Boyce-Dial crater-dating technique, 133. *See also* D_L method, D_L values
Brayley, 253
Breccia, 12, 45-47
  complexities, 21-23, 45-47, 177
  dating, 22, 47, 156, 169, 177, 178
  dikes, 44, 46

295

Breccia—Continued
  dilithologic (dimict), 46, 165, 168
  friable fragmental feldspathic, 165–171
  granulitic, 165, 168–170, 174
  lens, crater-floor, 43
  melt-poor, 46, 47, 165, 168
  melt-rich, 47, 82, 157, 165, 168
  mixing and recycling, 21–23, 139–143, 157, 163
  provenance, 22, 23, 47, 205, 211, 212, 217–220, 276
  regolith, 45
  samples, 45–47
  shock grades, 44–47
  terrestrial, 45
  textures, 23, 45–47, 171, 174
"Bright swirls," 76, 215, 216, 256, 258, 269
Bullialdus, 252, 253, 256, pl. 10
Bunte Breccia, 45, 46, 73
Bürg, 253, pl. 11
Buys-Ballot, pl. 4

C

Calcium, 101, 102, 139, 140
Calderas, 17, 32, 40, 88, 113
Camelot Crater, 239, 270
Campanus, 112, pl. 8
Campbell, 58, 257, pls. 4, 6, 9
Cardanus, 197, 231, 234, pl. 9
Carnot, 159, 180, pl. 7
Caroline Herschel, 8, 32
Carpenter, 252, pl. 11
Cassini, 8, 202, 216, 221, pl. 8
Cataclastic anorthosite, 141, 165, 168
Cataclysm, 190, 191, 278
Catena (pl., catenae), defined, 3
Catena Artamonov, 156
Catena Dziewulski, 156
Catena Mendeleev, 188
Catharina D, 37
Cauchy, 119
Cauchy structures, 115, 119, pl. 5
Cavalerius, 256, 258, pl. 10
Cayley, 220
Cayley Formation (Cayley plains), 21, *165*, 195, *216*, 224, pl. 8
  crater frequency, 197, 219, 221, 230
  samples, 165–170
  thickness, 220
  type area, 218–221
Central Cluster, 240, 270
Central peaks. See Crater materials, peak; Crater processes, peak formation
Chao, E.C.-T., vii, 42, 43
Chaplygin, 6, 120, 180, pl. 7
Chemistry
  mare basalt, 101, 102, 276
  orbital, 97, 139, 142, 143, 156, 190, 198, 241, 245, 258, pl. 2
  terra crust, 139, 140
Chretien, pls. 4, 9
Cichus, 253, pl. 10
Clavius, 58, 74, 180, pls. 7, 8
Clearwater craters, Quebec, 32, 45
Cleomedes, 166, 180, pls. 4, 5, 7
Clinopyroxene, 101, 102, 140
Collapse craters, 87–90
Colluvium, 49, 174, 201, 202
Color, 86, 96–99. See also Spectral reflectance
Compass directions, lunar conventions, 2
Compton, 58, 65, 81, 166, 186, 199, pls. 4, 5, 8
Cone Crater, 190, 205, 209–212, 269, 270, 273
Cones, 86–89
Congreve and Congreve U, 180, 184, pl. 7
Conon, 200, 203
Continuous deposits, defined, 43
Copernican Period
  defined, 221, 265
  duration, 249, 269–273, 280
  impact rate, 246, 271–273, 280
Copernican System, *263*, 280, pl. 11
  chronology, *269*, 280
  crater frequency and D_L values, 130, 136, 249–253, 256, *265*
  crater materials, *130*, 236, 249, 252, 253, *263*, 279, 280, pl. 11
  defined, 121–125, 249, 265
  distribution, 265, 280, pl. 11
  mare materials, 196, 265, *269*, pls. 11, 12
  remote-sensing properties, 265–267
  structures, *269*, 271
Copernicus, 1, 8, *29*, 49, 52, 92, 250, 265, *269*, pl. 11
  age, 269–273, 278, 280
  composition, 156, 269, 270
  crater frequency and D_L value, 136, 252, 253, 266, 271–273
  depth-diameter ratio, 80
  impact energy, 29
  impact melt, 49, 50, 52, 76
  rays, *1*, 29, 30, 48, *250*, 269–272
  secondary craters, 8, *29*, 125, 126, 247, *250*, *270*
  stratigraphic relations, 121, 125, 126, 247, *250*, 265, 272
Copernicus H, 31, 250, 265–268
Cordillera ring. See Montes Cordillera
Core, 12, 277, 280
Coriolis L, 149, 151, pl. 6

Coulomb, 158, 159, pl. 7
Coulomb-Sarton basin, *60*, *64*, 72, 79, 145, 148, *158*, 179, pls. 3, 6
Crater dating, 129–136
  "counts." See Crater frequencies
  D_L method, *133*, 136, 216, 250, 253, 265, 272, 273
  morphologic basis, *129*, 145, 149, 180, 222, 232, 249, 253–257
  Pohn-Offield method, 129–131, 143, 145
  superpositional basis, 17, 27, 47, *125*. See also Superpositions
  Trask's method, 131–135, 236, 243, 253
Crater frequencies, 18, 19, 129–136, 145–147, 180, 221, 230–232. See also individual Apollo and Luna missions, basins, formations, maria, series, and systems
  basins, 136, *148*, 160, *179*, *186*, *293*
  diameter dependence, 129, 132, 136, 257
  D_L correlations, 130, 136, 216, 230, 256, 273
  endogenic craters, 29
  secondary-impact craters, 29, 32, 132, 276
  slopes of size-frequency curves, 29, 32, 129–134, 145, 146, 216, 257
  small-on-large superpositions, 129, 133–135, 145, 250–253, 266, 267, 273
  substrate influence, 231, 234, 239, 241, 250, 252, 269, 272
  time-stratigraphic correlations, 130, 136
Crater materials, 3, 17–19, 25–53. See also *individual craters, series, and systems, and craters listed on p.* 149, 180, 221, 231, 253
  asymmetric, 32, 38, 39, 258, 266, 267
  dunelike, 27, 28, 268
  endogenic interpretations, 17–19
  erosional degradation, 19, 20, 27, 47, 48, *129*, 145, 149, 180, 185, 189, 222, 232, 249, 253, 256, 265–267
  extent, 27, 127
  floor, 27, 28, 43, 45, 50–52, 113, 116, 117
  geologic mapping, 15, 27, 47–50, 67, 125, 127, pls. 6–12
  herringbone, 28–35, 47, 48, 67, 125, 211
  mixing, 44–48
  morphology, 3, *27*, 149, 180, 222, 223, 232, 253–256
  origin, 17–19
  peak, 27, 43–51, 58, 65, 77–80, 280
  remote-sensing properties, 96–99, 265–267
  rim, 27–29, 42, 45–48
  shock zoning, 41, 42, 45–48
  subunits, 27, 28, 45–53
  wall, 27–29, 45–51
Crater processes. See also Ejection process
  floor uplift, 19, 58, 65, 93, *113*, 129, 145, 149, 196, 245, pl. 5
  formation mechanics, 33, 40–45
  formation times, 43, 45
  impact melting, 44. See also Impact melt
  interference, 29–32, 35, 37, 48, 220
  origin, general, 17–19, 27, 32, 33, 41, 47
  peak formation, 43, 44, 49, 50, 77
  "push" and "pull" mechanisms, 43
  terrace formation, 43, 44, 48, 77
  thrusting, 42, 47
Crater properties, 3, 17–19, 27–33
  circularity, 29
  depth-diameter ratios, 27, *29*, 42, 43, 66, 80
  excavation cavity, 41–43
  morphology, 3, 27–29
  simple-to-complex transition, 3, 27–29, 43, 65
  size classes, 27–29, 57, 58
  size-frequency distributions. See Crater frequencies
  spatial distributions, 17–19, 29, 32, 139, 163, 265, 276, pls. 6–12
  target influence, 43, 44, 47, 133, 239, 241
Crater types
  atypical, 19, *32*, *41*
  collapse, 87–92
  complex, 25–29, 43, 44, 47–53, 77, 78, 280
  dark-haloed. See Dark-haloed craters
  delta-rim, 32, 33, 92
  dimple, 236
  double-ring, 113, 116
  endogenic. See Endogenic craters
  explosion, 32, 77–81
  gravity, 43, 81
  hybrid, 33
  irregular, 32
  laboratory, 31–33, 38, 39
  missile, 29, 39
  nested, 78, 113, 116, 258
  noncircular, 32
  paired, 8, 32, 39
  primary, defined, 17
  ringed, 116
  satellitic (secondary). See Secondary-impact craters
  simple, 25–29, 33
  smooth-rimmed, 32, 33, 40
  strength, 43, 81
Cratering flow, 41, 80
Cratering rate. See Impact rate
Craters, 3. See also *crater and cratering headings, individual craters, and craters listed on p.* 149, 180, 221, 231, 253, 265
Crisium basin, 7, *63*, *65*, 100, 103, *165*, *170*, 242, pls. 3, 5, 7
  age, 171, 179, 186, 278
  antipode, 215–217
  composition, 144
  crater frequency, 179, 186, 293
  cross sections, 82, 192, 235, 243, 276

Crisium basin—Continued
  deposits, 65, 164–167, 171–175, 190, 238, 243, pls. 3, 7
  impact, 82, 171
  mare fill, 103. See also Mare Crisium
  mascon, 117
  rim and rings, *65*, 78–81, 100, 103, 110, 117, *171*, 214, 235, 241, 242, pl. 3
  samples, 82, 163, *171*, 276, 278
  secondary craters, 167, *171*, 175
  stratigraphic relations, 148, *165*, *171*, *179*, 192, 235, 243, pls. 7, 12
  structures, tectonic, 112, 115
  topography, 77, 78, 100
Crocco, pls. 4, 9
Crookes and Crookes D, 185, pl. 11
Cross sections
  diagrammatic (schematic), 44, 45, 82, 100, 102, 112, 115, 118, 120, 144, 200, 210, 211, 222, 240, 274, 275
  geophysical, 13, 79
  local, 19, 200, 210, 238, 240, 243, 261
  regional, 21, 144, 192, 235, 240, 243, 272, 274–276
Crozier, 40
Crüger, 67, 71, 127, 130, pls. 4, 8
Crust (terra)
  age (time of solidification), 156, 157, 277, 280
  basin effects, 77, 80, 102, 103, 115, 120, 143–145, 229, 240–245, 276
  composition, *139*, 165, 201, 280
  density, 12, 140, 143
  differentiation, 142, 143, 156, 157
  igneous and impact processes, importance, 140–143
  layering, 12, 13, 99, *143*, 274–276
  mare-extrusion influence, 102, 103, 115, 145, 229, 238–245
  mineralogy, 140
  mixing, 139–143
  petrology and petrogenesis, 140–143
  thickness, *12*, 77, 102, 103, 115, 143, 145, 278, 279
  zonation, 143
Cumulates, 141–143, 174
Curie, 226, pl. 6
Cyrano, 149, 151, pl. 6

D

"D caldera," 90, 203
Daedalus R, S, U, 149, 151, 231, pl. 6
D'Alembert, 257, pl. 7
Damoiseau, pls. 5, 9
Daniell, 40, pl. 5
Dark-haloed craters
  endogenic, 89, 90, 266
  exogenic (impact), 19, 176, 190, 191, 223, 266, 268, 279
Dark-mantling materials, 3, 8, 9, 40, *89*, 109, 113, 116, 117, 174, 203, 205, 233, *239*, 270, pl. 4. See also Glass
  age, 174, 234–237, *241*, 245, 269, 270, 278, 279
  Apollo 17, 234, 237–241, 278
  color (spectral reflectance), 96, 97, 240
  composition, 101
  distribution, 15, 89, 94, 95, 113, 174, 234, 240–245, 276, 279, pl. 4
  mantle source, 102
  origin and eruptive mode, 89, 102, 240, 241, 245, 276
  radar properties, 99
  source vents, 89–93, 102, 113, 234, 240, 276
  stratigraphic relations, 93–95, 109, 233–235, 239, 240, 245, 276–279
Darwin, 130, pl. 6
Davy, pls. 5, 9
Dawes, 239, 252
Debris surge (flow). See Ground surge
Deceleration dunes, 67, 130, 204, 218, 219
Delaunay, 37
Delisle, 18, 19, 121, 125, 252, pl. 10
Densities
  magma, 102, 103, 241
  mantle, 12
  mare basalt, 12, 102, 103
  Moon, 12
  terra crust, 12, 143
Depth-diameter ratio
  basins, 77, 80
  craters, 27, *29*, 42, 43, 66, 80
Descartes and Descartes A, 22, 218, pl. 7
Descartes Formation (Mountains), 21, 165–171, 195, *216*
Deslandres, 10, 149, pl. 6
Diameters, data source, 4
Differentiation, magmatic, 102, 142, 143, 156, 157
Dikes and veins
  breccia, 44, 46
  igneous, 101, 102, 157, 276
  impact melt, 44–47, 165
Dionysius, 220, 221, 265
Diophantus, 18, 19, 121, 125, *252*
D_L method, *133*, 136, 216, 250, 253, 265, 272, 273
D_L values
  Copernican-Eratosthenian boundary, 250, 265
  Copernican maria, 130, 136, 250, 269
  craters, 250, 253, 256, 273
  crater-frequency correlations, 130, 136, 216, 230, 256, 273
  Eratosthenian-Imbrian boundary, 229–231, 243, 250

INDEX

D_L values—Continued
    Eratosthenian maria, 125, 130, 136, 230, 231, 247–250, 253, 259, 269, 273
    farside maria, 245
    Imbrian maria, 125, 130, 136, 229–233, 243, 245, 250, 273
    Orientale basin, 216, 230
    plains, 136, 216
Dobrovol'skiy, 47
Dollond B and C, 218
Domes
    intercrater, 29, 35, 127, 214, 215
    mare, 86–89
    terra, 110, 111, 127, 146, 214, 215, 222–225
Doppelmayer, pls. 5, 7
Doppelmayer Formation, 244
Doppler, 184, pl. 8
Dorsum (pl., dorsa), defined, 3. *See also* Mare ridges
Downslope movement, gradual, 49, 89, 96, 110, 111, 133, 280. *See also* Colluvium, Slumping
Dune Crater, 201, 237, 243
Dunite, 143, 157, 174

E

Early Imbrian Epoch. *See also* Lower Imbrian Series
    defined, 121, 196
    duration, 224, 279
    igneous activity, 190, 198, 224, 232, 238, 279
    impact rate, 160, 246, 279
"Eclipse temperatures," 99, 113, 265–268
Einstein, 35, pl. 6
Ejecta, defined, 3. *See also* Basin materials, Crater materials
Ejection process
    angle, 41–43, 72, 80, 211
    ballistic, 41–43, 47, 48, 211, 212
    basins, 66, 67, 72, 73, 77–81
    complex craters, 47
    deposition, 42
    nonballistic, *43*, 46, 47, 202, 211, 212. *See also* Ground surge
    overturned flap, 42, 73
    range, 42, 48, 211, 212
    sequence, 41, 42
    target effects, 41, 47
    velocity, 42, 43, 72, 203
Elastic waves, 81
Elbow Crater, 201, 237, 243
Elements. *See also* individual elements and oxides
    large-ion-lithophile (LIL), 140
    major, 96–103, 139–144, 165, 280
    radioactive, 98, 99, 102, 140, 190, 198
    rare-earth. *See* Europium anomaly, KREEP, Rare-earth elements
    siderophile, 101, 141, 143, 157, 158, 202
    trace, 102, 140–142, 156, 276
    volatile, 101
Elevation
    average, 12
    Crisium (basin and mare), 77, 78, 99, 100
    farside, 12, 120
    maria, 93, 115, 120
    nearside, 12, 120
    Orientale (basin and mare), 77
Endogenic craters, 87–93, 239
    impact-crater distinction, 19, 27, 29, 32, 33, 47, 88, 89, 113, 266
    size-frequency distribution, 29
Endymion, 199, 166, 180, pls. 4, 7, 9
Energy coupling, 41, 43, 81, 212
Energy partitioning, 43, 44, 157, 212
Energy scaling. *See* Scaling laws
Engelhardt, 184, pl. 8
Epigenes, 204, pl. 7
Epoch, defined, 123
Equilibrium. *See* Steady state
Eratosthenes, 8, 121, 125, 193, 231, *247*, pl. 10
Eratosthenian Period
    defined, 121, 125
    duration, 249, 262, 280
    impact rate, 246, 271–273, 280
Eratosthenian System, *247*, pl. 10
    chronology, *262*, 279, 280
    crater frequency, 130, 136, 230, 249–253, 256, *257*, 265, 272, 280
    crater materials, *130*, 236, *247*, 265, 280, pl. 10
    defined, 121–125, 249, 250
    D_L values, 125, 130, 136, 229–231, 247–250, 253, 256, 259, 269, 273
    mare materials, 15, 121–125, 131–135, 207, 230, 231, 234, 243, 244, 247–256, *258*, 265, 269, 272, 278, 279, pls. 10, 12
    samples, 249, 259–262, 276–279
    structures, 279
Escape velocity, 29
Esnault-Pelterie, 149, 159
Eudoxus, 9, 202, 252, 253, pl. 11
Euler, 8, 49, 121, 125, *126*, 250–253, pl. 10
Europium anomaly, 101, 142
Excavation cavity
    basins, 23, 66, 77, 80, 280
    defined, 43
    depth and volume, 42, 43, 66, 77, 80
    excavation and growth, 41–43

Explorer 35, 12
Explosions, 41
Exposure ages, 235, 238, 269, 270

F

Fabricius, 147, 253, pl. 10
Fabry, 4, pl. 6
Farside
    crustal and lithospheric thickness, 12–14, 115, 145
    defined, 2, 3
    maria, 3, 103, 115, 233, 245, 280
    southern, 3
    structures, 107, 112, 115, 245, pl. 5
Faults, 107. *See also* Fractures, Rilles
    craters, 42, 47, 113–118
    mare ridges, 110–112
    solitary, 113, 115
    thrust, 115, 269
Fecunditatis basin, 40, 65, 117, 145, *146*, 148, 179, 235, *241*, 243, pls. 3, 6
Feldspar, crustal varieties, 140
Felsite, 177, 178
Fermi, 5, pl. 6
Ferroan anorthosite, 142, 143, 156
Fersman, 158, pl. 7
Feuillée, 8, 113
Fizeau, 179, pl. 9
Flammarion, 128, 213, pl. 6
Flamsteed and Flamsteed P, 111, 253, 254, 258, pl. 6
Flamsteed-Billy basin, 65, 146, 148, 244, 251, 254, 258, pls. 3, 6
Fleming, 4, 156, 180, 191, pl. 7
Flynn Creek Crater, Tenn., 43
Fontenelle, pls. 5, 8
Formation, defined, 123
Fowler, 158, 159, pl. 6
Fra Mauro, 22, 89, 149, 208, pl. 8
"Fra Mauro basalt," 140
Fra Mauro breccia, 200, 211, 212
Fra Mauro Formation, *204*, 218, pl. 8
    age, 200, *212*, 224, 279
    composition, 143, 144, 211, 269, 279
    crater frequency, 136, 230
    crater morphology, 131, 135
    definition and type area, 73, 124, 125, 196, 208
    distribution, 15, 204, 207, 213, pl. 8
    emplacement process, 195, 204, 211, 212
    Hevelius Formation analogy, 73, 82, *204*, 212, 213, 216, 218
    lateral gradations, 190, *204*, 208, 213
    samples, 195, 200, *204*, 278, 279
    stratigraphic relations, 127, 196, 207, 235, 272
    stratigraphic significance, 19, 21, 121, 125, 127, 195, 196
    thickness, 205, 210
Fra Mauro peninsula, 10, 22, 115, 262
Fracastorius, pls. 5, 7
Fractures (fissures, gashes). *See also* Faults, Rilles
    crater-floor, 113–118. *See also* Crater processes, floor uplift
    impact-melt, 113, 117
    regolith, 269, 271
Fraunhofer E and J, 147, 149
Freundlich, pl. 7
Freundlich-Sharonov basin, *60*, *64*, 82, *147*, 179, pls. 3, 6
Furnerius, 149, 167, pls. 4–6

G

Gabbro, 140
Gabbronorite, 142
Gagarin, 6, 31, 58, 120, 149, 191, pl. 6
Galois and Galois Q, 183, pl. 6
Gambart, 40, 92, 208
Gambart A, 268
Gamma-ray spectrometer, 97–99, 156
Gargantuan basin. *See* Procellarum basin
Gassendi, 113, 116, pls. 5, 7
Gassendi A, 253, pl. 11
Gaudibert, 113, 117, pl. 5
Gauss, 58, 166, pls. 5, 7
Geminus, 166, pl. 10
Geo-, vii
Geochronology
    breccia dating, 22, 47, 156, 169, 177, 178
    decay constants, 19, 168
    exposure ages, 235, 238, 269, 270
    mare-basalt dating, 101, 235–244, 261, 269, 270
    methods, 156, 177, 235–238, 269
    model ages, 156
    pyroclastic dating, 241, 244
Geologic cross sections. *See* Cross sections
Geologic mapping, vii, 17–21, 123–129
    basins, 15, 66, 67, 73, 76, 127, 129, 169, 174, 207, pls. 3, 6–8, 12
    craters, 27, 47, 50, 176, pls. 6–12
    maria, 86, 94, 95, 174, 236, pls. 4, 9–12
    systematic program, vii, 123–125
Geologic maps
    local, 176, 236
    regional, 15, 76, 94, 95, 169, 174, 207, 217, pls. 3–12
Geologic units, vii, 17–23, 121–129. *See also* Rock-stratigraphic units, Time-stratigraphic units, Time units

Geologic units—Continued
    correlation of types, 121–125
    correlation with samples, 21–23, 101, 139, 143, 276, 280
    interfingering, 27, 121, 124
    lateral continuity, 17, 21, 27
    nomenclature, 121–125
    superpositions, 17, 21, 125–127. *See also* Superpositions
    three-dimensionality, 17
Giordano Bruno, 191, 265
Glass. *See also* Dark-mantling material
    Apollo 15, green, 101, 237, 244, 245, 278
    Apollo 15, red, 101, 244
    Apollo 17, orange and black, 101, 237, 240, 241, 245, 278
    coatings, 46
    darkening, 95, 96, 280
    devitrification, 240
    regolith, 95, 96, 101
Goclenius, 119, pl. 7
Goddard A, 76, 265
Godin, 205, 253, 266, pl. 11
"Gold dust," 85
Goldschmidt, 49, 204, pl. 6
Gosses Bluff, Australia, 43
Grabens. *See* Rilles
Granite, 140, 223, 269
Granodiorite, 140
Granulitic and granoblastic texture, 46, 157
Granulitic breccia, 165, 168–170, 174
Granulitic impactite, 141, 212
Gravity, cratering effects, 43, 44, 81
Gravity anomalies, 77–80, 117. *See also* Mascons *and individual basins and maria*
    craters, 77, 113
Grimaldi (basin and mare), 11, *59*, *64*, 71, 77, 117, 120, 147, 148, pls. 3–6, 9, 10
Grissom-White basin, 65, 148, 160, 179, pls. 3, 6
Ground surge (debris flow, debris surge, ground flow, surface flow)
    basins, *66*, 164, 204, 211–213, 217–220, 222, 276
    craters, 42, 43, 47, 48
    secondary-crater relation, *42*, 67, 211–213, 217–220, 222, 276
Group, defined, 123
Gruithuisen domes, 222, 225
Gruithuisen K, 92

H

Hackman, R.J. *See* Shoemaker-Hackman stratigraphic scheme
Hadley rille, 13, 88, 91, 200, 201, 237, 243
Hahn, 166, pl. 8
Hainzel A, 187, 253, pl. 10
Halo Crater, 259, 261
Hansteen and Hansteen alpha, 225, pls. 5, 9
Hausen, 28, 253, pl. 10
Hayn, 199, pl. 11
Head Crater, 259, 261
Heat flow, 198
Heat sources, 102, 157, 190, 198, 212, 277–280
Heaviside, 6, 120, 150, pl. 6
Heis, 8, 32
Helicon, 8, 32
Henbury craters, Australia, 29, 31, 32
Hercules, 9, 32, 166, pl. 10
Herodotus, 91, pl. 9
Herschel, 128, 253, pl. 10
Hertzsprung basin, *60*, *64*, 79–82, 148, 149, 158, 159, 178–183, *188*, pls. 3, 7, 8
    crater frequency, 179, 180, 186, 293
    elevation, 120
    plains, 190, 245, pl. 8
Hevelius, 67, 71, 120, 180, pls. 5, 7
Hevelius Formation, 66–73, 79–82, pls. 3, 8. *See also* Orientale basin
    crater analogy, 57, 66, 79
    crater frequency, 216
    definition and type area, 66, 67, 71, 196, 197
    extent, 66, 67, 76, 79–82, pl. 8
    Fra Mauro Formation analogy, 73, 82, *204*, 212, 213, 216, 218
    inner facies, *66*, 76, 81, 197
    Montes Cordillera relation, 67–71, 79, 232
    nonlineated facies (member), 70, 72
    outer facies, *67*, 76, 82
    plains relation, 67, 164, 190, 215, 218, 219, 224
    stratigraphic significance, 66, 121–125, 196
    transverse facies, 67, 71, 73, 76, 204
"Highland basalt," 140, 141, 156
Hilbert, 4, 5, 154, 191, pl. 7
Hipparchus, 19, 37, 213, pls. 6, 8
Horrocks, 253
Hortensius-Milichius domes, 86
Hubble, pls. 4, 7, 9
Humboldt, 7, 58, 102, 113, 167, 223, 226, 231, pls. 4, 5, 9
Humboldtianum basin, *61*, *64*, 79–81, 117, 148, *166*, *179*, 186, 199, pls. 3–5
    crater frequency, 179, 186, 293
    impact, 79, 82, 171
    stratigraphic relations, 148, 171, 175, 179, 186, 199, 204, pl. 7
Humorum basin, 57, *61*, *64*, 82, 148, 153, *179*, 187, 192, 214, pls. 3–5, 7
    crater frequency, 179, 186, 293
    mascon, 117, 244

Humorum basin—Continued
  structures, tectonic, *112*, 117, 244, pl. 5
Hypervelocity, 17, 40

I

Ilmenite, 95, 101, 140, 235
Ilmenite basalt, 101, 235, 238, 261
Imbrian-Imbrium distinction, 124
Imbrian Period. *See also* Early Imbrian Epoch, Late Imbrian Epoch
  defined, 124, 125
  duration, 245, 279
  impact rate, 160, 191, 245, 246, 279
Imbrian System. *See also* Lower Imbrian Series, Upper Imbrian Series
  crater frequency, 136, 186, 197
  crater materials, 15, 131-136, 186, 221, 222, 232
  defined, 124, 125, 193, 196
  extent, 195, 229, 279, pls. 8, 9
  plains materials. *See* Plains
  type area, 196, 208, 231
Imbrium basin, 8, 19, 21, 36, 57, *63*, *65*, 79-82, *193*, pls. 3, 8, 12
  age, 186, 191, 201, 212, 224, 277, 279
  antipode, 215-217
  center, 103, 198, 202, 204
  composition, 143, 144, 168, 201, 211
  crater frequency, 135, 136, 160, 191, *197*, 219-222, 230, 293
  deposits, 19, 57, 65, 127, 129, 164, 165, 168, 174, 175, 190, *193*, 279, pls. 3, 8. *See also* Alpes Formation, Apennine Bench Formation, Fra Mauro Formation, "Material of Montes Apenninus"
  depth, 143, 240
  excavation cavity, 81, 197, 198, 204, 211
  fissured plains, 203
  impact angle, 82
  impact melt, 198, 203, 211, 212
  mapping conventions, 127, 129
  mare fill, 103, 240, 243-245, 258, 259, 269, 279. *See also* Mare Imbrium
  mascon, 117
  Orientale basin, comparisons, 196, 197, 202, 203, 213, 218
  plains, 3, 21, 164, 213, *215*, 279, pl. 8. *See also* Apennine Bench Formation, Cayley Formation
  plains/secondary-crater relation, 215-219
  radials. *See* Imbrium sculpture
  rim and rings, 57, *65*, 79, *81*, 110, 117, 193, *196*, 207, 211, 218, 240, 253, pl. 3. *See also* Montes Alpes, Montes Apenninus, Montes Archimedes, Montes Carpatus, Montes Caucasus
  samples, 19, 21, 82, 192, 200, 276, 278. *See also* Apennine Bench Formation, samples; Fra Mauro Formation, samples; Montes Apenninus, samples
  secondary craters, 32, *36*, 40, 41, 65, 74, 116, 127-131, 135, 156, 164-168, 172, 175, 187, 189, 196, 205, 206, *213*, 232, 233, 243, pls. 3, 8, 12
  secondary craters, sampling, 164, 165, 173, 174, 204, 205, *211*, *218*
  stratigraphic relations, 15, 19, 21, 121, 127, 129, 148, 173, 174, 179, 196, 199, 204-209, 213-219, 233, 243, 276-279, pls. 3, 8, 12
  stratigraphic significance, 19, 21, 121-125, 195, 196, 204, 221, 279
  structures, external, 81
  structures, tectonic, 112, *113*, 117, pl. 5
Imbrium sculpture, 10, 19, 37, 41, 57, 89, *113*, 115, 118, *127*, 187, 196, 199, 204, 208, *213*
Impact breccia. *See* Breccia
Impact melt
  basins, 66-70, 73, *76*, 82, 164, 198, 203, *211*, 274-276
  cooling time, 45
  ejected, 41, 47, 48, 113, 117, 174, 177, 212
  fissured (fractured), 50-52, 113, 117, 203, 212
  flows, 44, 47-50, 53
  homogenization, 50, 177, 178
  injected, 50, 44, 46
  origin, 40, *44*, 50, 212
  pools, 28, 44, 46-48, 50, 53, 212, 265
  projectile effects, 44, 47
  stratigraphic relations, 44, 45, 50
  target effects, 47, 212
  terrestrial, 45
  volcanic interpretation, 50, 76, 140
Impact-melt-rock samples, *22*, 23, *45*, 82, *140*, 157, 210
  aphanitic, 201
  Apollo 14, 211, 212, 278
  Apollo 15, 198, 201, 203, 278, 279
  Apollo 16, 165, 168-170, 219, 220, 278
  Apollo 17, 174, 177, 278
  bomb, 177, 178
  composition, 101, 140-142, 165, 174, 177, 201, 211, 212
  dating, 22, 168-170, 177, 178, 212, 278
  fragment-laden, 141, 165-170, 174, 177, 278
  mare-basalt samples, comparison, 101
  poikilitic, 168, 174, 177
  texture, 23, 46, 140, 165, 174, 177, 178, 201, 203, 210
  uncertain sources, 22, 23, 168-170, 174, 177, 211, 219, 220
Impact rates
  Copernican and Eratosthenian, 246, 271-273, 280
  Imbrian, 160, 191, 245, 246, 279

Impact rates—Continued
  Nectarian and pre-Nectarian, 151, 157, 160, 180, 190, 191, 246, 278-280
  present, 280
  terrestrial, 271
Impacts, 3, 17-19, 27, 33, 38-42. *See also* Projectiles *and basin and crater headings*
  clustered, 180
  crustal disruption, 142, 143, 157, 190, 279
  energy, 17, 29, 43
  experimental, 31, 33, 38, 39, 280
  layering effects, 40, 43, 44, 77, 78, 81
  missile, 29, 39
  oblique, 19, 32, 38, 39, 81, 82, 171, 173, 258, 266, 267
  petrogenetic effects, 143, 157, 277
  primary, defined, 17
  recent, 280
  scaling laws, 43, 81, 211, 212
  secondary, defined, 18. *See also* Secondary-impact craters
  simultaneous, 8, 19, 30-33, 39, 81, 171, 197
  spacecraft, 266
  velocities, 17, 18, 29
Infrared, 96, 97. *See also* "Eclipse temperatures"
Ingenii basin, 6, *60*, *64*, 146, 148, pls. 3, 6
Inghirami and Inghirami A, 71, 180, 221, 224, pls. 7, 8
Insularum basin, 65, 146, 148, 153, 190, 207, 244, pls. 3, 6
Intrusions, igneous, 101, 102, 113, 140-143, 157, 276, 277
Iridum crater, 8, 32, *34*, 125, 197, 221-224, 231, 232, 258, 259, pl. 9
Iron, 95-101, 140, 141, 276
Isaev, 4, 9
Isostasy, 77, 112-115, 143, 145, 245, 277. *See also* Mascons

J, K

J. Herschel, pls. 5, 6
Janssen, 37, 149, 162, 164, pls. 5, 6
Janssen Formation, 73, 121, 143, 163-167, pl. 7
Jenner, 5, 7, 223, 226, 231, pl. 9
Jetting, 41-46
Joliot, 76, 166, 191, pls. 4, 6, 9
Jules Verne, 6, pls. 4-6, 9
Julius Caesar, 9, 95, 218, pl. 6
Kant Plateau, 164, 165, 218
Karpinskiy, pls. 5, 8
Keeler, 6, 25, 120, 150, pl. 8
Keeler-Heaviside basin, 6, 25, *61*, *64*, 146, 148, *150*, 179, pls. 3, 6
Kepler, 121, 125, 253, 255, 265, pl. 11
Kibal'chich, 180, 183, pl. 7
King, 4, 46, 50, 53, 154, 265, pl. 11
Kipukas, 86, 110
Kirchhoff, 173, 175
Kohlschütter, 85, 245, pls. 4, 7
Kopff, 33, 40, pls. 5, 9
Korolev basin, *59*, *64*, 79-82, 148, 178, 179, *181*, 190, 245, pls. 3, 7, 8
Korolev M, 182
Kovalevskaya, 158, pl. 9
Krafft, 197, 231, 234, pl. 9
KREEP
  basalt, 140, 156, 190, 198, 202, 211, 224, 276-279
  composition, 140
  crustal component, 140, 143, 144, 156, 211
  defined, 140
  high-K (HKFM), 140
  low-K (LKFM), 140, 141, 168, 171, 174, 201
  medium-K (MKFM), 140, 198
  orbital chemistry, 143, 156, 190, 198, 245
  origin, 142, 198
  samples, 165, 168, 171, 174, 190, 198-203, 211, 269, 270
  volcanic, 144, 156, 198, 211
Krieger, 121, 125, 231-233
Krusenstern, 149, 189

L

La Caille, 128
La Condamine, 34, 221, 224, pls. 5, 8
Lacus, defined, 3
Lacus Autumni, 245, pls. 4, 9
Lacus Felicitatis, 90
Lacus Mortis, 96, pls. 4, 9
Lacus Odii, 93
Lacus Solitudinis, 86, 154, 245, pls. 4, 9
Lacus Somniorum, 40, 96, 99, 244, pls. 4, 9
Lacus Veris, 245, pls. 4, 9
Lade, 205, 206, 213, pl. 6
Lalande A, 49
Lambert, 8, 110, 121, 125, *247*, 250-253, pl. 10
Lambert R, 110, pl. 8
Lamont, 115, 227, 235
Landau, 58, 158, pl. 6
Landslide, Apollo 17 landing site, *105*, 174, 176, 240, 269, *270*, *273*
Langemak, 5, pls. 4, 9
Langrenus, 167, 252-256, pl. 10
Lansberg, 221, 223, 272, pl. 9
Lansberg B and C, 223, 231, 265
Lassell, 40, pl. 5
Late Imbrian Epoch. *See also* Upper Imbrian Series
  defined, 121, 196

Late Imbrian Epoch—Continued
  duration, 245, 246, 279
  impact rate, 246, 279
Latitude, equivalence in kilometers, 4, 98, pl. 1
Lava. *See also* Magma, Mare-basalt samples, Mare units
  channels and tubes, 89, 92
  flow lobes, 83, 86
Layered intrusions, 141, 143
Layering, effect on impacts, 40, 43, 44, 77, 78, 81
Lebedinskiy, 180, 184, pl. 8
Lee-Lincoln scarp, 105-107, 176
Leeuwenhoek, pls. 4, 7, 9
Leibnitz, 103, 160, pls. 4, 6, 9
Le Monnier, 109, 173, pl. 7
Letronne, 110, 221, 251, pl. 8
Le Verrier, 8, 32
Lexell, 253, pl. 10
Licetus, 216, pl. 6
Lichtenberg, 196, 265, 269-271, pl. 11
Light plains. *See* Plains
Limb, defined, 2, 4
Lindenau, 232, pl. 9
Lineated terrain, pre-Imbrian, 127, 205
Linné, 28, 265
Liouville DA, 117
Lithosphere, *12*, 77, 102, *115*, 145, 243-245, 259, 277-279
Littrow, 171-175
Lomonosov, 4, 76, 156, 191, 221, 222, pls. 4, 8, 9
Lomonosov-Fleming basin, *65*, 146, *148*, *156*, 179, 190, *191*, pls. 3, 6
Lonar Crater, India, 43
Longitude, lunar conventions, 2
Longomontanus, 74, pls. 7, 8
Lorentz basin, *59*, *64*, 72, 118, 147, 148, *158*, 179, 196, pls. 3, 6
Lower Imbrian Series, *193*, 279, pl. 8. *See also* Early Imbrian Epoch, Schrödinger, *and* Imbrium- *and* Orientale-*basin headings*
  chronology, 224, 279
  crater frequency, 130, 135, 160, 186, 197, 221, 222, 230, 246
  crater materials, 130-136, *221*, 246, 279, pl. 8
  definition and type area, 71, 121-123, *196*, 208
  domes, 222-225
  extent, 195, 279, pl. 8
  mare materials, 223, 224, 232, 238, 279
  plains materials, 156, 197, 198, 204, 208, *215*, 222, pl. 8
Lubiniezky, 256, pl. 7
Luna missions (U.S.S.R.), 12
Luna 3, 3, 12, 94
Luna 16
  crater frequency and D_L values, 230
  landing site, 12, 99, 101, *242*, pls. 2, 4, 9
  samples, 101, 140, 237, *241*, 245, 276-279
  spectral reflectance, 99
Luna 20
  landing site, 12, 101, *167*, 214, *242*, pls. 2, 3, 7
  samples, 82, 101, 140, 143, 144, *171*, 276-278
Luna 24
  landing site, 12, 99, 101, pls. 2, 4, 9
  samples, 101, 237, *241*, 245, 276-279
  spectral reflectance, 99
Lunar Astronautical Charts (LAC), 124
Lunar grid, 107
Lunar Module (LM), 235, 259
Lunar orbiter, future, 280
Lunar Orbiter missions (U.S.A.), 5, 12, 57, 196
Lunar Orbiter photographs, types, 5, pl. 2
Lunar Receiving Laboratory (LRL), 23
Lütke, 47
Lyot, 7, 103, 226, pls. 4, 6, 9

M

Macrobius, 172, 175, pl. 8
Maestlin G, 255
Maginus, 74, 149, pl. 6
Magma
  conduits and extrusion sites, 102, 103, 115, 240, 259
  density, 102, 103, 241
  differentiation, 102, 142, 143, 156, 157
  fluidity, 86
  fractionation, 102, 238, 243
  high-Al, 102, 103, 241
  high-Ti and low-Ti, 102, 103
  mixing, 102
  primary, mare, 102
  primary, terra, 141
Magma ocean, 142, 143, 244, 277
Magma pods (ponds), 143
Magnesium, 98-101, 140-143. *See also* Mg suite
Magnetism, 168, 215, 256, 269, 280
Main sequence, craters, 27, 32
Mairan 34, pl. 9
Mairan domes, 222
Maksutov, pls. 4, 9
Manicouagan Crater, Quebec, 45
Manilius, 8, 9, 205, 252, 253, pl. 10
Manned Spacecraft Center (MSC), Johnson Space Center, Houston, Tex., 23
Mantle
  composition, 102, 277, 280
  density, 12

INDEX

Mantle—Continued
    mare-source zones, *101*, 238, 240, 259, 261, 276, 280
    origin, 102
    samples, 73, 80, *143*, 278
    structure, 12, 13, 102, 103
    uplifts, 77, 102, 115, *120*, 145, *240*, 274–276
Maraldi, 111, 172–175, pl. 7
Maraldi B, 87
Mare (pl., maria), 3, *19*, 21, *83*, *117*, 229, 232–245, 249, 258–262, 269, pls. 4, 9–12. See also Mare-basalt samples, Mare units, *and lacus, mare, and sinus headings*
    area, 3, 85
    basin relations, 3, 85, 100–103, 115, 117, 145
    defined, 3, 85, 86
    elevations, 77, 78, 100, 115, 117, 120, 241
    morphology and landforms, 85–93
    origin and emplacement, 19, 21, 85, 86, *102*, 276
    thickness and volume, 12, 77, *99*, 117, 190, 277
Mare Australe, 7, *62*, 103, 167, *223*, *226*, 233, 245, pls. 4, 9
Mare-basalt samples, 19, 23, *101*. See also Mare Crisium, Mare Fecunditatis, Mare Imbrium, Mare Insularum, Mare Nectaris, Mare Serenitatis, Mare Tranquillitatis, Palus Putredinis
    ages, absolute, 235–245, 259, 261, 273, 278
    Al-rich (high-Al), 101–103, 190, 200, 210, 238, *241*, 242, 245, 279
    Apollo 11, 19, 85, 96, 101, 224, *235*, 245, 276–279
    Apollo 12, 19, 23, 96, 101, 102, 249, *259*, 276–279
    Apollo 15, 19–23, 96, 101, 102, 237, *243*, 276–279
    Apollo 16, 96, 101, 237, *238*, *245*, 278
    Apollo 17, 96, 101, 102, 239–241, 245, 276–279
    classification, *101*, 235, 239–242, 262
    clasts in breccia, 101, 156, *190*, 200, 210, 212, 278
    compositions, 96–103
    defined, 85, 101
    density, 12
    Eratosthenian. See Apollo 12 (above)
    europium anomaly, 142
    feldspathic, 101, 237, 241, 242, 245, *261*, 278
    high-K, high-Ti, 101, 237, 238, 245, 278
    ilmenite, 235, 238, 261, 262
    KREEP-rich, 156, 190, 198, 245
    low-K, high-Ti, 101, 235–241, 245, 278
    low-K, low-Ti, 101, 235–238, 278
    Lower (Early) Imbrian, 224, 279
    Luna 16, 96, 101, 237, *241*, 245, 276–279
    Luna 24, 96, 101, *241*, 245, 276–279
    mantle sources, 102, 103, 238, 240, 259, 261, 276, 280
    Mg-rich, 242
    mineralogy, 101
    olivine (olivine-normative), 237, 239, 243, 245, 259–262, 278
    pigeonite, 237, 243–245, 259–262
    pre-Imbrian (Nectarian and pre-Nectarian), 190, 200, 212, 238, 278, 280
    quartz-normative, 243
    reduction, 101
    terra-melt-rock samples, comparison, 101
    textures, *23*, 101, 102, 235–244, 259, 261
    Ti-poor (low-Ti), 96, 101–103, 235–238, 245, 278
    Ti-rich (high-Ti), 96, 101–103, 235–241, 245, 259, 262, 278
    Upper (Late) Imbrian. See Apollos 11, 15–17 *and* Lunas 16, 24 (*above*)
    very low titanium (VLT), 101, 102, 237, 242, 278
Mare Cognitum, 86, 112, 120, 244, pls. 4, 9
Mare Crisium, 4, *63*, 78, 100, 103, 165–167, *172*, *241*, pls. 4, 5, 9
    age, 230, 241, 245, 279
    color (spectral reflectance) and composition, 96, 99, 241, 242, pl. 4
    crater frequency and D_L values, 230, 231
    cross section, 100
    dark-mantling material, 89
    diameter, 103, 117
    elevation, 77, 78, 100, 117, 120
    Mare Imbrium, comparison, 243
    mascon, 99, 117, 241
    peripheral "lakes" and eruption sites, 78, 103, 241, 259
    samples, *241*, 245, 276–279
    stratigraphy, 100, *241*, 243, pl. 9
    structures, 117, 241, pl. 5
    thickness, 99, 100, 103, 117, 241, 242
Mare Fecunditatis, 119, 167, *241*, pls. 4, 9
    age, 241, 245, 279
    color (spectral reflectance), 96, 99, 241, pl. 4
    composition, 241
    crater frequency, 230
    diameter, 117, 241
    elevation, 78, 117, 120
    mascon, absence of, 99, 117
    samples, *241*, 245, 276–279
    stratigraphy, 233, *241*, pl. 9
    structures, 115–119, 241, pl. 5
    thickness, 99, 100, 103, 117, 242
Mare Frigoris, 8, 96, 99, 125, 199, 204, 224, 243, 258, 279, pls. 4, 9, 10
    basin-related trough, 81, 198, 243
Mare Humboldtianum, 9, 61, 103, 117, 166, pls. 4, 9
Mare Humorum, 5, 62, 117, 187, *244*, pls. 4, 5, 9, 10
    color (spectral reflectance), 96, 99, 244, pl. 4
    dark-mantling material, 89, 116
    mascon, 99, 117, 244
    structures, 107, 112, 116, 117, 244, pl. 5

Mare Humorum—Continued
    thickness, 99, 100, 117
Mare Imbrium, *8*, *63*, 83, 92, 99, 125–127, 193, 224, 231, *243*, 247, 253, 258, 259, 272, pls. 4, 5, 9, 10. See also Palus Putredinis
    age, 243–245
    color and composition, 96, 99, 243, 244, 258, 259, 279, pl. 4
    composition, 99, 244, 258, 259, 279
    diameter and elevation, 117, 120
    D_L values, 231, 243
    flow lobes, 83, 86, 258
    mascon, 99, 117, 243
    stratigraphy, 94, 121, 125–127, *243*, 247–253, 258, 259, 265, 269, pls. 9, 10
    structures, 113, 117, 243, pl. 5
    thickness, 100, 117, 243, 244
Mare Ingenii, 6, 60, 216, pls. 4, 9
Mare Insularum, 153, *259*, 272, pls. 4, 9, 10
    crater frequency and D_L values, 230, 259, 271–273
    domes, 86
    samples, *259*, 276
Mare Marginis, 4, 67, 76, 166, 191, 224, 245, pls. 4, 9, 10
Mare Moscoviense, 59, 245, 257, pls. 4, 9
Mare Nectaris, 1, 62, 164, *238*, pls. 4, 5, 9
    age, 245
    color (spectral reflectance), 96, 99, 238, pl. 4
    dark-mantling material, 89
    diameter and elevation, 115, 117, 238
    D_L value, 230
    mascon, 99, 115, 117, 238
    samples, *238*, 245, 276, 278
    stratigraphy, 233, *238*, pl. 9
    structures, 112–117, 238, pl. 5
    thickness, 100, 103, 117
Mare Nubium, *10*, 96, 103, 112, 116, 117, 120, *153*, 232, 244, 256, 258, pls. 4, 5, 9, 10
Mare Orientale, 11, 62, 69, 77, 99, 103, 115, 117, 233, *245*, pls. 4, 5, 9
    name, 2
Mare ridges (dorsa), 3, 8, 103, *107*, 153, 277, pl. 5
    age, 115
    basin-ring markers, 110, 196
Mare Serenitatis, *9*, 61, *92*, *108*, 171, 174, *238*, pls. 4, 5, 9, 10
    age, 230, 241, 245
    color (spectral reflectance), 96, 99, 239, 244, 245, pl. 4
    crater frequency and D_L values, 230, 231
    dark-mantling material, 89, 92–95, 239–241, 245
    diameter, 117
    elevation, 93, 117, 120
    geologic maps, 94, 95, 174, 176, pls. 9, 10, 12
    gravity structure (mascons), 79, 99, 117, 171, 239
    landforms, 86, 90, 239
    samples, 237, *239*, 245, 276–279
    sources, 259
    stratigraphy, 19, 94, 95, 99, *238*, 244, 245, pls. 9, 10
    structures, *107*, 117, 239, pl. 5
    thickness, 99, 100, 117, 173, 239, 241
Mare Smythii, 4, 7, *62*, 116, 154, *245*, pls. 4, 5, 9, 10
    composition, 99, 103, 245
    dark-mantling material, 89
    diameter, 117
    elevation, 115, 117, 120
    Mare Orientale, comparison, 245
    mascon, 99, 115, 117, 245
    stratigraphy, 245, 259, pls. 9, 10
    structures, 115–117, 245, pl. 5
    thickness, 99, 103, 117, 245
Mare Spumans, 78, 100, 241
Mare Tranquillitatis, 9, 39, 114, 119, 172, *227*, *235*, pls. 4, 5, 9
    color (spectral reflectance), 96, 99, 235, pl. 4
    crater frequency and D_L values, 230, 235, 238
    diameter, 117, 235
    elevation, 93, 115, 117, 120
    landforms, 86, 87
    mascon, absence of, 99, 117
    samples, 224, *235*, 245, 276
    stratigraphy, 233, *235*, 238, pl. 9
    structures, 39, 114–119, 117, pl. 5
    thickness, 99, 100, 117, 235, 238, 241
Mare Undarum, 78, 100, 241
Mare units. See also *individual lacus, mare, sinus, series, and system headings*
    ages, 96, 102, 124, 125, 190, 223, 224, 229–262, 265, 269, 278–280, pls. 9–12
    albedo, 94–96
    aluminous, 98, 99, 241, 242, 245
    basin relations, 223, 232–235, 238–245, 258, 259, 269, 276–280, pl. 4
    buried, 156, 190, 243, 245, 278, 279
    chemistry, orbital, 97, 139, 143, 156, 190, 198, 241, 245, 258
    color (spectral reflectance), 86, *96*, 235, 238–245, 258, 259, 279–281, pl. 4
    crater frequencies, 130, 136, 160, 229–235, 238, 252, 259, 271–273
    D_L values, 130, 136, 230–233, 243, 259, 271–273
    "eastern" and "western," 259
    eruption rates, 86, 238–241
    extent, 86, 127, 232, 233, 277
    farside, 3, 85, 103, 233, *245*, 280, pls. 4, 9–12
    flows, 83, 86, 127
    geologic mapping, 86, 94, 95, 125, 174, 232, 236, pls. 4, 9–12

Mare units—Continued
    KREEP-rich, 156, 190, 245
    mapping properties, 86–99
    Mg-rich, 100, 241
    oldest, 76, 197, 223
    radioactivity, 96–99, 245, 259
    remote-sensing properties, 94–99
    sources, *102*, 103, 240, 259
    stratigraphic relations. See Superpositions *and individual maria*
    subdivision, 85, 86, 94–96, 232–235
    thickness, 86, *99*, 153, 238–241, 261
    Ti-poor (low-Ti) and Ti-rich (high-Ti), 96, 99, 101–103, 235, 238–245, 258–262, pl. 4
    youngest, 197
Mare Vaporum, 86, 92, 99, *127*, 214, 244, pls. 9, 10
Marginis basin, *65*, 148, 179, pls. 3, 6
Marius Hills, 86, 88, 89, 244, 258, pl. 4
Mascons, 77, 99, 100, 112, *115*, *117*, 171, 235, 238–245, 269, 277
Mass wasting. See Downslope movement, Slumping, Talus, Terraces
Massifs. See Basin materials
"Material of Montes Apenninus," 15, 73, 200–203, pl. 8
Maunder, 30, 33, 40, 69, 76, 253, pl. 7
Maunder Formation, *69*, 76, 82, 164, 174, 198, 216, pl. 8
Maupertuis, 34, 221, 224, pls. 5, 8
Maurolycus, 155, 214, pl. 7
Maxwell, pls. 4, 7
McCauley, J.F., vii, 66–77
McClure, 40
Mechnikov, 180, 183, pl. 7
Megaregolith, 45
Megaterracing, 78, 79, 81
Member, defined, 123
Mendel-Rydberg basin, 11, *61*, *64*, 148, 178, 179, 186, 278, 293, pls. 3, 7
Mendeleev basin, 6, *59*, *64*, 85, 117, 120, 148, *179*, 186, 188, 219, 230, 293, 25 pls. 3, 7
Mercator, 112, pl. 7
Mersenius, 180, 187, pls. 5, 7
Messala, 166, pl. 6
Messier and Messier A, 32, 38, 171, 173, 265
Metamorphism
    Apollonian, 212
    shock, 45–48, 211
    textures, 46, 141
    thermal, 46, 157, 165
Meteor Crater, Ariz., vii
Meteorites, 101, 141, 142, 277. See also Projectiles
Metius, 147, 180, pl. 7
Mg suite, 141–144, 156, 157, 174, 201
Middle Crescent Crater, 259, 261
Milichius domes, 86, 87
Milne, 5, 58, 65, 145–149, 154, pl. 6
Minerals. See also Ilmenite, Olivine, Plagioclase, Pyroxene, Spinel
    crustal, 140
    mare-basalt, 101
    minor, 101, 140
    silica, 140
Mixing models, 142
Modal mineral (mode), defined, 102, 140
Modification stage, craters, 43
Moho (crust-mantle boundary), 12, 77–81, *102*, 115, *120*, *143*, 229, *240*, *245*, 262, 274–278
Mohorovičić A, 152
"Moldings," debris, 110, 111
Mons (pl., montes), defined, 3
Mons Rümker, 86, 244
Montes Alpes, 8, 9, 81, 125, 198, 199
Montes Apenninus, *8*, *9*, 13, 15, 22, 36, 81, 90, 108, *193*, *197*, pl. 8. See also "Material of Montes Apenninus"
    crater frequency, 136, 197, 222, 230
    samples, 19, 23, 49, 82, 140, 143, 144, 195, *198*, 224, 278, 279
    slumps, 81, 193, 198, 200
Montes Archimedes, 113, 193, 198
Montes Carpatus, 197, 198, 204, 207, 211, 272
Montes Caucasus, 8, 9, 81, 198, 199, *202*
Montes Cordillera, 66–73, 76–81, 179, 197
Montes Haemus, 9, 171, 174, 205, 218
Montes Harbinger, 89, 91, 233, 251, 258
Montes Recti, 224
Montes Riphaeus, 90, 112, pl. 6
Montes Rook, 66, 68–70, 73, 76, 77, 80, 171
Montes Rook Formation, *69*, 164, 168, 171–174, 203, pl. 8
Montes Taurus, 171–176
Monzodiorite, 140, 223, 269
Moon
    age, 156, 157, 277
    astronomic properties, vii
    comparative planetology, vii, 280
    cooling and contraction (global), 115, 269, 280
    geologic style, viii, 276, 277
    global properties, 12, 13
    origin, vii, 156, 157, 280
    subsurface, 12–14, 274, 275
    surface, 1–13
    surface area, 180
Moonquakes, 13, 115, 269, 280
Moscoviense basin, *59*, *64*, 81, 82, 148, 171, 178, *179*, *186*, 245, 257, 293, pls. 3, 7
Mösting, 128, 252

"Mug shot," defined, 23
Müller chain, 41, 115
Multi-ringed basin, defined, 3. See also basin headings and basins listed on p. 64, 65, 148, 179
Multispectral images, 99
Mutus, 155, pl. 6
Mutus-Vlacq basin, 65, 145-148, 155, 164, 179, 190, 244, pls. 3, 6

N

Nansen, 199, pl. 7
Nearch, 147, 149, pl. 6
Nearside
  defined, 2, 3
  mass offset, 103
Nectarian Period
  defined, 121, 163
  duration, 157, 190, 278
  geologic history, 190, 191, 277-279
  igneous activity, 190, 277-279
  impact rate, 151, 157, 160, 180, 190, 191, 246, 277-280
Nectarian System, 161, 276-278, pls. 3, 7
  basin materials, 161, 276-278, pls. 3, 7
  chronology, 190, 277-279
  crater frequency, 130, 136, 160, 178-180, 186, 191, 197, 230
  crater materials, 130, 136, 149, 180, 246, 278, pl. 7
  defined, 121-125, 163
  distribution, 163, pl. 7
  plains materials, 154-156, 163-167, 184, 188-191, 217, 219, 230, pl. 7
  structures, 278
  type area, 146, 147
  volcanic rocks, 190, 243, 278
Nectaris basin, 1, 3, 62, 64, 146, 161, 179, 238, pls. 3, 5, 7, 12
  age, 168, 178, 179, 186, 190, 278
  composition, 143, 144
  crater frequency, 136, 160, 179, 197, 293
  deposits, 64, 161, 189, 219, 220, 232, pls. 3, 7, 12
  impact angle, 82
  impact melt, 164
  knobby terrain, 168
  mascon, 117, 238
  Orientale basin, comparison, 163, 164, 168
  radials, 7, 131, 143, 189
  rim and rings, 22, 36, 64, 81, 155, 163, 235, pl. 3
  samples, 82, 163, 216, 220, 276
  secondary craters, 36, 37, 64, 74, 131, 143, 161
  stratigraphic relations, 121-125, 131, 143, 161, 179, 189, 192, 235, 243, 276-280, pls. 3, 7, 12
  stratigraphic significance, 3, 121-125, 143, 145, 163, 280
  structures, tectonic, 112, 115, 117, 238, pl. 5
Neper, 166, 180, pls. 4, 7, 9
Nernst, pls. 5, 7
Nicolai, 37, 164
Nishina, pls. 4, 9
Nomenclature
  petrologic, 12, 101, 102, 139-142, 156
  selenographic, 3
  stratigraphic, 121-125
Nonproportional growth, of craters, 43, 44
Norite, 140-142, 157, 174, 201, 203
Normal albedo, defined, 94
Normative mineral (norm), defined, 102
North Complex, 91
North Massif, 174, 175
North Ray Crater, 165, 168-171, 269, 270, 273
Nubium basin, 65, 103, 146, 148, 153, 179, 244, pls. 3, 6

O

Oceanus Procellarum, 11, 103, 115, 145, 153, 244, pls. 4, 5, 9-12
  color (spectral reflectance), 96, 99, 232, 244, 258, 259, 269, pl. 4
  crater frequency, 250-252, 266
  dark-mantling material, 89, 234
  elevation, 120
  landforms, 86, 89, 110
  localization factors, 103, 115, 145
  mascons, 99, 115, 244
  stratigraphy, 121-127, 135, 196-198, 232-234, 244, 250, 265, 269-271, pls. 9-12
  structures, 107, 110, 115, 118, 196, pl. 5
  thickness, 99, 100, 153
Olbers and Olbers A, 197, 265, pls. 5, 7, 11
Olivine, 101, 102, 140-142, 259
Olivine (olivine-normative) basalt, 101, 237, 239, 242-244, 261, 278
Ophitic texture, 23, 46, 174, 235
Oppenheimer, 160, pls. 5, 7
Orbital geochemistry. See Chemistry, orbital
Orientale basin, 3, 11, 55, 196, 279, pls. 3, 8, 12. See also Hevelius Formation, Maunder Formation, Montes Rook Formation
  age, 178, 186, 196, 224, 245, 277, 279
  antipode, 67, 76
  central uplift, 69, 81
  concentric facies, 66, 73
  corrugated facies, 73, 76
  crater frequency, 135, 136, 197, 216, 219-222, 230, 293
  D_L value, 216, 230
  domical facies, 73

Orientale basin—Continued
  dunelike deposits, 67, 71, 130, 218
  exterior, 66, 81, 82
  fissured facies, 66, 73
  formation, 78-82
  geologic maps, 76, pls. 3, 8, 12
  global effects, 67, 72, 190, 216, 220, 223
  gravity structure, 77, 80, 117
  grooved facies, 67, 73
  impact angle, 82
  impact melt, 66-72, 76, 77, 82, 212
  interior, 68-73, 76-82
  mare fill, 190, 245, 279, pls. 4, 9. See also Mare Orientale
  massifs, 66, 69, 73, 79, 80
  model for basins, 65, 66, 73, 77-82, 279
  pitted terrain, 67, 71
  plains, 3, 71, 128, 164, 179, 182, 184, 190, 197, 204, 213, 224, 279, pl. 8
  radial facies, 73
  "rays," 66, 74, 75, 127
  rings, 64-73, 77, pl. 3. See also Montes Cordillera, Montes Rook
  secondary craters, 32, 35, 66, 86, 127, 128, 151, 158, 159, 181-183, 188, 196, 205, 224, 225, 233, pls. 3, 8
  stratigraphic relations, 127, 128, 148, 158, 159, 179-183, 188, 196, 224, 225, 233, 277, pls. 3, 8, 12
  stratigraphic significance, 121-123, 195, 196, 221, 229, 279
  structures, tectonic, 117, 243, 245, pl. 5
  topography, 77, 280
Orientale Group, defined, 66, 72
Origin, meanings, 19, 21, 139, 163
Orontius, 10, 20, 149, pl. 6
Orthopyroxene, 140-142
Oscillatory uplift, 78-81
"Overturned flap," ejecta, 42, 73

P

Palus, defined, 3
Palus Putredinis, 8, 22, 113, 200, 243, pls. 4, 9
  age, 243, 244
  color and composition, 243
  crater frequency and D_L values, 136, 197, 230, 243
  elevation, 120
  samples, 243, 276-279
Parry, 22, 208, pl. 7
Partial melting, 101, 102, 140
Paschen, 183, pl. 6
Pasteur, 4, 120, 149, 154, 191, pl. 6
Pauli, 5, pls. 4, 8, 9
Pavlov, 5, pl. 7
Peak rings, 28, 50, 58, 65, 77-81
Peaks. See Basin materials, massifs; Crater materials, peak; Crater processes, peak formation
Peirce, 100, 172
Peirescius, 147, 149, pl. 6
Period, defined, 123
Petavius, 32, 58, 102, 143, 167, 232, pls. 4, 5, 8
Petavius B, 252, 265, 267, pl.11
Petrogenesis, early impacts, 157, 277
Petrologic nomenclature, 139, 140, 156
Philolaus, 252, pl. 11
Phocylides, pl. 7, 8
Photogeology, vii, 17, 72, 139, 143, 280
Photographs
  identification scheme, 4, 5, pls. 1, 2
  lighting effects, 2, 86, 94, 110, 133-135, 173, 219-221, 250
  quality, 145, 265, 280, pl. 2
  resolution effects, 12, 13
Picard, 100, 172, 241
Piccolomini, 231-233, pl. 9
Pictet, 20
Pigeonite, 101, 102, 140, 243, 259
Pigeonite basalt, 101, 102, 237, 243, 259, 278
Pingré-Hausen basin, 65, 148, 179, pls. 3, 6
Pitatus, 10, 116, 180, 215, pls. 5, 7
Pitiscus, 164, 180, pl. 7
Plagioclase. See also Anorthosite
  composition, 140, 142
  mare basalt, 101
  terra crust, 12, 140, 142
  textures, 23, 46, 141, 171
Plains (light plains, terra plains), 3, 4, 10, 11, 21, 71, 113, 128, 131, 154-156, 164, 172, 184, 188, 197, 200, 204, 213, 215, 276, 279, pls. 7, 8, 12
  crater frequencies, 191, 197, 216, 219, 222, 230
  distribution, 3, 15, 21, 169, 190, 215, 217, 276, pls. 7, 8
  D_L values, 136, 216
  Imbrian, 15, 156, 191, 198, 208, 213-224, 279, pl. 8. See also Apennine Bench Formation; Cayley Formation; Imbrium basin, plains; Lower Imbrian Series; Orientale basin, plains
  Nectarian, 154-156, 163, 164, 167, 184, 188-191, 217, 219, 230, pl. 7
  pitted, 146, 164
  pre-Nectarian, 146, 154, 156
  secondary-crater relations, 67, 71, 72, 128, 146, 164, 190, 213-220, 224
  volcanic, 139, 146, 190, 191, 243, 245, 279
Planck basin, 5-7, 59, 64, 103, 147, 148, pls. 3, 4, 6, 9
Planetology, defined, vii

Planté, 8, pl. 10
Plato, 8, 125, 221, 224, 231, 259, pls. 4, 9, 10
Playfair and Playfair G, 10, 189, pls. 6, 7
Plinius, 93, 253, pl. 10
Plutonic rocks, 140-143, 156, 157, 278
Pohn-Offield crater classification, 129-131, 143
Poikilitic texture, 23, 46, 165, 168, 174, 177
Poincaré basin, 6, 7, 59, 64, 103, 146, 148, pls. 3, 6
Pontecoulant, 167, 180, pl. 7
Posidonius, 9, 22, 93, 113, 231, pls. 5, 9
Potassium, 98, 99, 102, 140, 190. See also KREEP; Mare-basalt samples, high-K, low-K
Pratt model of isostasy, 143
Pre-Imbrian (materials and time), 143-145, 196
Pre-Nectarian period
  defined, 121-125, 139, 145
  duration, 157, 278
  geologic history, 156, 157, 277
  igneous activity, 139-143, 156, 157, 190, 277, 278
  impact rate, 151, 157, 160, 180, 191, 246, 277-280
Pre-Nectarian system, 137, 277, 278, pl. 6
  basin materials, 139, 143, 178, 180, 277, 278, pls. 3, 6
  chronology, 156, 277, 278
  crater frequency, superposed, 130, 136, 145-149, 186, 197
  crater materials, 130, 136, 139, 145, 180, 186-189, 278, pl. 6
  defined, 121-123, 125, 139, 143, 145
  distribution, 139, pl. 6
  plutonic rocks, 141-143, 156, 157, 278
  recognition criteria, 145-147
  samples, 139-143, 156, 157, 163, 195
  structures, 139, 278
  type area, 146, 147
  typical terrain, 137
  volcanic rocks, 156, 190, 238, 278, 279
Primary-impact crater, defined, 17
Prinz, 91, 125, pl. 8
Pristine rocks, 141-144, 156, 157
Procellarian System, 124, 125
Procellarum basin, 65, 103, 115, 120, 145, 153, pls. 3-6
  age, 157, 277, 278
  crust and lithosphere, 77, 103, 115, 120, 143, 235, 238-245, 259, 262, 278, 279
  depth, 80, 120, 143-145, 192
  hemispheric asymmetry, 103, 143, 145
  maria, 99, 103, 115, 145, 232-245, 259, 262, 278, 279
  rings and troughs, 65, 81, 103, 112, 115, 120, 145, 153, 171-173, 207, 227, 240, 243, 244, 278, pls. 3, 5
  structures, tectonic, 107, 112, 115, 145, 238, pl. 5
  volcanic filling, early, 156, 190, 211
Procellarum Group, 125
Proclus, 51, 172, 175, 265
Projectiles
  basin association, 143
  density, 43, 47
  ejection, 41
  mass-frequency distribution, 18, 129
  melting, 40
  multiple, 19, 30-33
  penetration, 40-44, 81
  present-day, 280
  secondary, 29, 211
  size effects, 43, 77, 81, 177
  velocities, 17, 18, 29, 177
Promontorium, defined, 3
Proportional growth, craters, 43
Protagoras, 204
Ptolemaeus, 10, 32, 128, 155, 213, 219, pls. 6, 8
Purbach, 10, 128, 149, pl. 6
Pyroclastic material. See Dark-mantling material, Glass
Pyroxene
  augite, 140, 143
  clinopyroxene, 101, 102, 140
  high-Ca, 97, 140
  low-Ca, 97, 102, 140
  orthopyroxene, 140-142
  pigeonite, 101, 102, 140, 243, 259
  textures, 23, 46, 141, 171, 177
Pythagoras, 8, 196, 258, pl. 10
Pytheas, 30, 121, 125

Q, R

Quartz, 140
Quartz monzodiorite, 140, 223, 269
Quartz-normative basalt, 102
Rabbi Levi, 232, pl. 6
Radar, 99, 100
Radioactivity, 98, 99, 143, 190, 198, 212, 259, 280
Ranger missions (U.S.A.), 12, 90, 125, pl. 2
Rare-earth elements, 101, 140, 142. See also Europium anomaly, KREEP
Raspletin, 31
Rays, 249-257, 265-267
  albedo, 94-96, 249, 250, 256, 265-267, 279
  asymmetric, 32, 38, 39, 266
  basins, 74, 75, 82, 127
  fading, 95, 96, 249, 256, 279
  length, 3, 29
  offcenter relation to primary crater, 29, 30
  origin, 29-32, 42, 48

Rays—Continued
 primary-ejecta content, 48, 269, 270
 samples, 269, 270
 stratigraphic relations, 125–127, 249–252, 265–267
Reduction, chemical, 101
Regiomontanus, 10, 128, pl. 6
Regolith (soil, surficial material), 12, 13, 95, 96
 age, 249
 agglutinates, 95, 96, 256, 266, 279
 albedo, 95, 96
 Apollo 11 landing site, 13, 235, 238
 Apollo 12 landing site, 259, 261, 269
 Apollo 14 landing site, 205
 Apollo 15 landing site, 13, 198, 243
 Apollo 16 landing site, 165, 168, 219, 220
 Apollo 17 landing site, 240, 241
 bedrock substrate, 12, 13, 21, 45
 blockiness, 99, 258
 breccia, 45
 buried, 238
 craters, 78, 132, 133, 258
 drainage, 269, 271
 formational rate, 280
 glasses, 95, 96, 101, 280
 Luna 16 landing site, 101, 241
 Luna 20 landing site, 171
 Luna 24 landing site, 101, 241, 242
 maturity, 95, 96
 origin, 17, 277
 sampling, 21, 249
 Surveyor I landing site, 258
 thickness, 12, 165, 205, 235, 239, 240, 258
Reimarus R, 147, 149
Reiner, 258, pl. 10
Reiner gamma, 256, 258, 269
Reinhold, 253, 272, pl. 10
Remote sensing, 94–99. *See also* Albedo; Chemistry, orbital; Color; Radar; Spectral reflectance
 defined, 86
 integration of types, 99
Repsold and Repsold C, pls. 5, 6
Residual liquids, 140
Rheita, 147, 161, 164, 180, pl. 7
Rhysling Crater, 243
Riccioli, 67, 71, 197, pls. 4, 6
Riccius, 37, 149, 214, 232, pl. 6
Richardson, 156, pls. 4, 6, 9
Ries Crater, Germany, vii, 43, 45, 73, 78, 165
Rilles (rimae), 3, 86–93, 107, 167, 168, 239. *See also individual rimae*
 ages, 115, 127
 arcuate, 3, 107, 145
 distribution, 3, 107, 112, pl. 5
 sinuous, 3, 18, 77, 86, 200, 233, 247
 straight, 3, 107, 113, 127, 145, 167, 168
Rima (pl., rimae), defined, 3
Rima Bode II, 90, 115, pl. 5
Rima Hadley, 13, 88, 91, 200, 201, 237, 243
Rima Hyginus, 9, 50, 88, 90, 115, 118, 120, 127, 205, pl. 5
Rima Sharp, 1, 117
Rimae Ariadaeus, 115, 118, 220, pl. 5
Rimae Cauchy, 115, 119, pl. 5
Rimae Hypatia, 39, pl. 5
Rimae Stadius, 30, 31
Rimae Sulpicius Gallus, 93, 108, pl. 5
Rimae Triesnecker, 115, 120, pl. 5
Ringed basins. *See basin headings and basins listed on p.* 64, 65, 148, 179
Ritter, 9, 32, 39, 113, pl. 8
Rocca, 71, 180, pl. 7
Roche, 5, pls. 4, 7
Rock-stratigraphic units, 121–129. *See also* Basin materials, Crater materials, Mare units, Orientale Group, *and individual formations*
 defined and named, 123–125
 interfingering, 27, 121, 124
 lateral continuity, 17, 21, 27
 sample correlations, 21–23, 101, 139
 three-dimensionality, 17
 time-stratigraphic relations, 121–125
Römer, 172, pl. 11
Römer A, 173, 175
Röntgen, 118, pl. 6
Rosenberger, Rosenberger B, Rosenberger C, 147, 149, pl. 6
Rothmann, 232, 233, pl. 10
Rumford, pls. 4, 9
Rümker hills, 86, 89, 244, pl. 4
Rupes, defined, 3
Rupes Altai, 131, 155, 164, 233
Rupes Recta, 10, 115, pl. 5
Rydberg, pl. 10

S

Sabine, 32, 39, 113, pl. 8
Sacrobosco, 131, 149, pl. 6
St. George Crater, 91
Samples (material from Moon), vii, 17–23, 280. *See also samples listed on p.* 168, 177, 200, 237, 253, 261
 correlation with stratigraphic units, 21, 101, 139, 143, 276, 280

Samples (material from Moon)—Continued
 mare basalt, 101, 102, 276, 278
 numbering system, 23, 168, 177, 190, 200
 terra breccia, 21–23, 45–47, 139–143, 276–279
 10003, 236, 237
 10029, 236, 237
 10050, 235–238
 10062, 236–238, 245
 12002, 261
 12031, 261
 12032, 269
 12033, 269
 12038, 261
 12051, 23
 12065, 261
 14053, 190
 14063, 101, 190
 14066, 212
 14072, 190
 14073, 211, 212
 14161, 212
 14167, 212
 14276, 211, 212
 14306, 269
 14310, 23, 211, 212
 14321, 190, 210, 269
 15016, 244
 15382, 198, 200
 15386, 198, 200, 202
 15405, 269, 271
 15445, 23, 200, 201
 15455, 156, 200–203
 15538, 23
 15555, 244
 60025, 141
 60315, 170
 62195, 168
 67015, 171
 67016, 169
 67435, 141, 156
 67603, 169
 67667, 156
 68415, 220
 68416, 220
 72255, 156
 72275, 190
 72415–72418, 156, 157, 174
 72435, 178
 73215, 177
 73255, 156, 177, 178
 74001, 241
 74220, 241
 76215, 177
 76535, 141, 143, 156, 157, 174
 77215, 156, 174
 78236, 156
Sarton, 158, 159, pl. 7
Satellitic craters, 30. *See also* Secondary-impact craters
Saussure, 20
"Scablands," 86
Scaling laws, impacts, 43, 81, 211, 212
Schickard, 11, 128, pls. 4, 6, 9
Schiller, 11, 32, 38, pl. 7
Schiller C, 128
Schiller-Zucchius basin, 11, 32, 38, 59, 64, 147, 148, 190, pls. 3, 6
Schlesinger, 170, pl. 6
Schlüter, 231, 232, pls. 4, 5, 9
Schrödinger basin, 7, 59, 64, 67, 79–82, 148, 155, 179, 186, 222, 225, 279, 293, pls. 3, 5, 8
Schwarzschild, 58, 65, 180, 199, pl. 7
Sculptured hills, 174, 176
Sechenov, 180, 183, pl. 7
Secondary-impact craters, of basins. *See* Basin-secondary craters; Imbrium basin, secondary-craters; Nectaris basin, secondary craters; Orientale basin, secondary craters
Secondary-impact craters, 3, 18, 19, 28, 125, 224, 226, 231, 247, 256–258, 276
 circularity, 29, 42, 130
 cratering process, 29, 31, 32
 endogenic hypotheses, 17, 29, 32
 formational sequence, 42, 48
 ground-surge relation, 42, 47
 impact velocities, 18, 29
 laboratory simulation, 31, 33
 primary-crater size, 29, 32, 211, 276
 primary-ejecta content, 42, 48, 276
 projectiles, 29, 42
 shock grades, 48
 size-frequency distribution, 29, 32, 129, 132, 276
 source direction, 125, 255
 spatial distribution, 18, 19, 29–35, 125, pls. 3, 6–11
 superpositions, 125–127. *See also* Superpositions
 target-structure influence, 29
Seismic velocities, 12, 13, 205, 210
Seismicity, 12, 13, 269
Seleno-, vii
Selenographic coordinates, 2
Septa, 29, 32, 33, 37, 39, 214, 220
Serenitatis basin, 3, 8, 61, 64, 93, 108, 109, 171, pls. 3, 5, 7
 age, absolute, 177, 190, 191, 278
 age, relative, 19, 173, 178–180, 190

Serenitatis basin—Continued
 antipode, 181, 215
 cross section, 82, 192, 240, 243, 276
 deposits, 174, pl. 7
 Imbrium-basin interaction, 81, 192, 198, 201
 mare extrusion, 103, 279
 mascons, 117, 171
 modification, 173
 Orientale basin, comparison, 171, 173
 rim and rings, 64, 81, 110, 117, 171, 218, pl. 3. *See also* North Massif, South Massif
 samples, 82, 143, 144, 157, 163, 173, 201, 278
 secondary craters, 172, 175
 stratigraphic relations, 148, 164, 173, 175, 179, 192, 235, 238, 243, 276, pls. 7, 12
 structures, tectonic, 108, 109, 112, 117, pl. 5
Series, defined, 123
Sharonov, pl. 11
Sharp, 34, pl. 9
Sherlock Crater, 270
Shield volcanos, 86
Shirakatsi, 47, pl. 10
Shock compression and decompression, 40–42
Shock grade, in ejecta, 41, 42, 45–48, 211, 212, 219
Shock melting. *See* Impact melt
Shock pressures, 40, 211
Shock waves, 40–42, 45, 80, 81
Shoemaker, E.M., vii, viii
Shoemaker-Hackman stratigraphic scheme, vii, 19, 123–125, 143, 249, 265
Shorty Crater, 237–240, 269, 270, 273
Siderophile elements, 101, 141, 143, 157, 198, 202
Sikorsky-Rittenhouse basin, 7, 65, 148, 179, 180, 278, pls. 3, 7
Silberschlag, 118
Silica
 mare basalt, 101, 102
 orbital detection, 98
 terra rocks, 140
Sills, 101, 102, 276
Sinuous rilles, 3, 18, 77, 86, 200, 233, 247
Sinus, defined, 3
Sinus Aestuum, 92, 244, pls. 4, 9
Sinus Amoris, 172, 235, pls. 4, 9
Sinus Asperitatis, 120, 235, 238, 244, pls. 4, 9
Sinus Iridum, 8, 34, pls. 4, 9. *See also* Iridum crater
Sinus Medii, 81, 86, 95, 115, 213, 244, pls. 4, 5, 9
Sinus Roris, 99, 117, 222, 243, 258, pls. 4, 9, 10
Size-frequency distributions. *See* Crater frequencies
Slope material, 49, 96, 110, 111
Slumping, 48, 51, 81, 193, 198, 200. *See also* Downslope movement, Megaterracing, Terraces
Smoky Mountain, 165, 170, 219
Smythii basin, 7, 62, 64, 77, 112–117, 146, 148, 154, 179, 245, pls. 3–6
Soda (sodium), mare basalt, 101
Soderblom-Lebofsky crater-dating technique, 133. *See also* D_L method
Soil, defined, 12. *See also* Regolith
South Massif, 105, 174, 175, 240, 273
South Pole-Aitken basin, 11, 65, 115, 143–148, 152, 181, 245, 278, pls. 3, 6
 age, 157, 277, 278
 crust and lithosphere, 115, 143, 245, 278
 deposits, 146, 278
 elevation, 120, 145
 mare extrusion, 103, 145
 rim and rings, 65, 145, 181, pl. 3
 stratigraphic relations, 145, 148
South Ray Crater, 165, 168, 170, 269, 270, 273
Spacecraft impacts, 266
Spaceflights, 12
Spalling, 81
Spectral classes, 96, pl. 4
 hDG-, 96, pl. 4
 HDSA, 96, 99, 258, pl. 4
 hDSA, 96, 99, 258, pl. 4
 hDSP, 96, 99, pl. 4
 HDWA, 96, 99, pl. 4
 hDWA, 96, 99, 239, 244, pl. 4
 LBG-, 96, 99, 244, pl. 4
 LBSP, 96, pl. 4
 LIG-, 96, 99, pl. 4
 LISP, 96, 99, pl. 4
 mBG-, 96, 99, 238, pl. 4
 mIG-, 96, 99, 244, 259, 262, pl. 4
 mISP, 96, 99, 239, 244, pl. 4
 sampling, 101, 262
 symbols, defined, 96
Spectral reflectance, 96–99, pl. 4
 craters, 266
 dark-mantling material, 97–99, 240
 maria, 96–99, 232, 235, 238–245, 258, 259, 269. *See also individual maria*
 terra soil, 97
Spinel, 140–143
Spur Crater, 203
Stadius chains, 30, 31
Steady state, cratering, 129–135, 157, 230
Stearns, 257, pl. 10
Stebbins, 149, 159, pl. 6
Steinheil, 147, 180, pl. 7

Steno Crater, 239
Steptoes, 86, 110
Stevinus, 252, pl. 11
Stiborius, 232, 233, pl. 9
Stillwater Complex, Mont., 143
Stöfler, 10, 216, pl. 6
Stone Mountain, 165, 170, 218, 219, 237
Straight Range, 224
Straight Wall, 10, 115
Stratigraphic code, 123-125
Stratigraphic column, 121-125, 280
Stratigraphic units, 17. *See also* Rock-stratigraphic units, Time-stratigraphic units, Time units, *and individual formations, series, and systems*
Stratigraphy, *vii, 15, 121. See also* Superpositions
    basin-mare distinction, 19
    crater interpretation, 17-19
    defined, vii
    nomenclature, 121-125
    regional context, 21
    Shoemaker-Hackman scheme, vii, 19, 123-125, 143, 249, 265
    terra-material interpretation, 19-21
Stratovolcanoes, 86
Stratton, 25, pl. 7
Strength crater, 43, 81
Structures, tectonic, *105, 277*, pl. 5. *See also* Faults, Fractures, Mare ridges, Rilles
    Copernican, 115, 269, 271
    crater-floor, 113-118, pl. 5
    distribution, 107, 112-115, pl. 5
    Eratosthenian, 279
    farside, 245
    Imbrian, 238, 241-245
    lithospheric thickness, 115, 277
    maria and basins, 107-115, pl. 5
    pre-Imbrian, 139, 278
Struve L, 35
Subophitic texture, 23, 46, 165, 168, 174, 177, 202, 235
Suevite, 45, 46, 73, 165
Sulpicius Gallus, 93
Superpositions (superpositional relations), *17, 125*
    basin-basin, *127, 145*, 175, *178*, 192, 196, 199, 222, 225, 235, 240, 243, 276, pls. 3, 6-8, 12
    crater-basin, *127*, 135-138, 145-151, 154-168, 172-175, 179-191, 193-199, 202-208, 213-234, 274, pls. 6-8, 12
    crater-crater, *18*, 25-27, 31, 34, 47, 48, *125*, *149*, 180, 185, 189, 191, 196, 224, 226, 248-253, 256, 266-268, 273
    mare-basin, *19, 21*, 77, 85, 94, 95, 108, 109, 115, *121*, 153, 165-169, 174, 193-200, 209, 223-227, 231-245, 259, 274-279, pls. 4, 12
    mare-crater, *18*, 21, 76, 85, 99, 121, *125*, 131-135, 191, 197, 200, 204, 205, 223, 226, 231-234, 238, 239, 247-261, 265-272, pls. 9-12
    mare-mare, 94, 95, 109, 110, 226, 231-245, 247, 254-262, 269
Surficial layer. *See* Regolith
Surveyor Crater, 259-261
Surveyor missions (U.S.A.), 12
    chemical analyses, 12, 85, 139
    landing sites, 12, 254, 258, 259, pl. 2
System, defined, 123
Szilard, 156, pl. 6

T

Talus, 89, 93
Tamm, 117
Taruntius, 113, *116*, 119, 166, 190, 265, *269*, pls. 5, 11
Taruntius H, 28
Taurus-Littrow Valley, 22, 171, 175, 239, 240, 273, 279
Tectonism. *See* Structures
Tektites, 45
Terminator, defined, 2
Terra (pl., terrae) (highlands, uplands), 3, 12, 19-23. *See also* Crust
    "backbone," 10, 19, 244, 258, 262

Terra (pl., terrae) (highlands, uplands)—Continued
    defined, 3
    regional differences, 3
    volcanism, 21, 146, 190, 215, 222, 223, 280
Terra plains. *See* Plains
Terraces
    basins, 64, 78-81
    craters, 27-29, 43, 44, 48
Textures
    aphanitic, 23, 174, 177, 178
    breccia, 23, 45-47, 171, 174
    cooling-rate relation, 46, 102, 174, 259
    cumulus, 141-143, 174
    fragment-content relation, 46, 165, 174, 177, 212
    granoblastic and granulitic, 46, 157
    igneous, terra rocks, 23, 46, 47, 140, 141, 174, 198
    impact melt, 23, 46, 165, 174, 177
    intergranular, 165, 168
    intersertal, 23, 202, 238
    mare basalt, 23, 101, 102, 235-244, 259, 261
    metamorphic, 46, 141
    ophitic, 23, 46, 174, 235
    poikilitic, 23, 46, 165, 168, 174, 177
    seriate, 141, 171
    shock intensity, 45-47, 165, 211
    subophitic, 23, 46, 165, 168, 174, 177, 202, 235
    vitrophyric, 243
Thalassoid, defined, 3
Thebit, 128, pls. 5, 9
Theophilus, 45, 227, 235-238, 252-256, pl. 10
Thorium, 98, 99, 102, 140, 190, 198
Tidal bulge and deformation, 115
Timaeus, 204, pl. 9
Time-stratigraphic units, 121-125, 130, 133, 136. *See also individual series and systems*
    areal extent, 163, 195, 229, 249, 265, 276-280, pls. 3, 4, 6-12
    defined, 121-123
Time units, defined, 121-123. *See also individual epochs and periods*
Timiryazev, 180, 183, pl. 7
Timocharis, 8, 121, 125, 231, 250-253, pl. 10
Tisserand, 172
Titanium, 95-103, 243. *See also* Mare-basalt samples, Mare units
Tranquillitatis basin, 227, 238, pls. 3, 6
    mare fill, 103, 235, 238, 279, pl. 9
    mascon, 115, 117
    rings, 174, 235, pl. 3
    samples, 157
    stratigraphic relations, 103, 145-148, 179, pl. 6
    structures, tectonic, 112-117, 235, pl. 5
Transient crater, 43
Trask, N.J., 131-135. *See also* Crater dating, Trask's method
Triesnecker, crater and rille system, 115, 120, pl. 5
Troctolite, 140-143, 156, 157, 174, 201
True crater, 43
Tsander, 183, pl. 6
Tsiolkovskiy, 4-6, 47, 48, 120, 188, 231-233, 245, pls. 4, 9
Tsiolkovskiy-Stark basin, 65, 148, 245, pls. 3, 6
Tsunami mechanism, 78
Tycho, *1*, 10, *20*, 28, 265, *270*, pl. 11
    age, 269, *270*, 273, 278, 280
    albedo, 266
    crater frequency and D_L values, 252, 253, 266, 273
    impact melt and knobs, 28, 73, 76
    rays and secondary craters, *1*, 20, 29, 32, 72, 125, 176, 240, *270*, 273, 280
    stratigraphic relations, 20, 121, 125, 270, 273

U

Ukert, 193
Uniformitarianism, 19, 20, 265
Unnamed basin A. *See* Sikorsky-Rittenhouse basin
Unnamed basin B. *See* Coulomb-Sarton basin

Unnamed craters A and B, 219, 222
Upper Imbrian Series, *227*, 279, pl. 9. *See also* Late Imbrian Epoch
    chronology, *245*, 279
    crater frequency, 130, 135, 136, 160, 221, 222, *229*
    crater materials, 130-136, 221, 222, *231*, 246, 254-256, 279, pl. 9
    dark-mantling materials, 109, 174, 233-241, 244, 245, 269, 278, 279
    definition and type area, 121-123, 229-231
    D_L values, 125, 130, 136, 229-233, 243, 245, 249, 250, 273
    extent, 229, 279
    mare materials, 121, 125, 135, 160, *227*, 252, 254, 256, 278, 279, pls. 9, 12. *See also lacus, mare, and sinus headings*
    structures, 245
Uranium, 98, 99, 140, 190

V

Vallis (pl., valles), defined, 3
Vallis Alpes, 113, 115, 116, 119, 204, pl. 5
Vallis Bohr, 35
Vallis Bouvard, 70, 73, 163, 187, 213
Vallis Palitzsch, 167, pl. 7
Vallis Rheita, 147, 161, 163, 164, 167
Vallis Schröteri, 91
Van de Graaff, 6, 39, 120, 180, 216, pls. 4, 7, 9
Van den Bos, 117
Vega, 147, 149, pl. 6
Vendelinus, pls. 4, 6
"Very high alumina (VHA) basalt," 141, 165, 168
Very low titanium (VLT) basalt, 101, 237, 242, 278
Viscous relaxation, 113, 115
Vitello, 113, 116, 187, pls. 5, 8
Vitello Formation, 214
Vitruvius, 86, 175, 221, pl. 9
Vitruvius front, 171, 175
Vlacq, 147, 155, pl. 6
Volatiles, 101, 102, 243, 245, 277
Volcanic craters. *See* Endogenic craters
Volcanic plains, buried, 139, 156, 164, *190*, 215, 232, 233, 238, 243, 245, 266, 276-281
Volcanic rocks. *See also* KREEP basalt, Mare-basalt samples
    Nectarian, 190, 243, 278
    pre-Nectarian, 156, 190, 243, 278
Volcanism, 19. *See also* Magma *and mare headings*
    eruptive styles, 86-89
    rates, 190, 238, 240, 241, 245, 278-280
    terra, 21, 146, 190, 215, 219, 222, 223, 280
Volcanotectonic craters, 32
Volta, pls. 5, 6
Von Kármán and Von Kármán M, 103, 160, pls. 4, 6, 9

W, X, Y, Z

W. Bond, 9, 199, 204, pl. 6
Wallace, 231
Walter, 10, pl. 7
Wargentin, 11, 72, pls. 7, 8
Water, 101
Waterdrop experiments, 80, 81
Werner, 10, *48*, 127, 128, 155, 189, 253, pl. 10
Werner-Airy basin, 65, 145, 148, *155*, 179, 244, pls. 3, 6
West Crater, 235-238, 270
Wrinkle ridges. *See* Mare ridges
X-ray-fluorescence spectrometer, 97-100
Young D, 147, 180, pl. 7
Zagut, 232, pl. 6
Zeno, 166, 171, pl. 7
Zero phase, 94
Zhukovskiy, 180, 184, pl. 7
Zond missions (U.S.S.R.), 12, 145
Zucchius, 38, pl. 11

PLATE 1. INDEX TO PHOTOGRAPHIC ILLUSTRATIONS

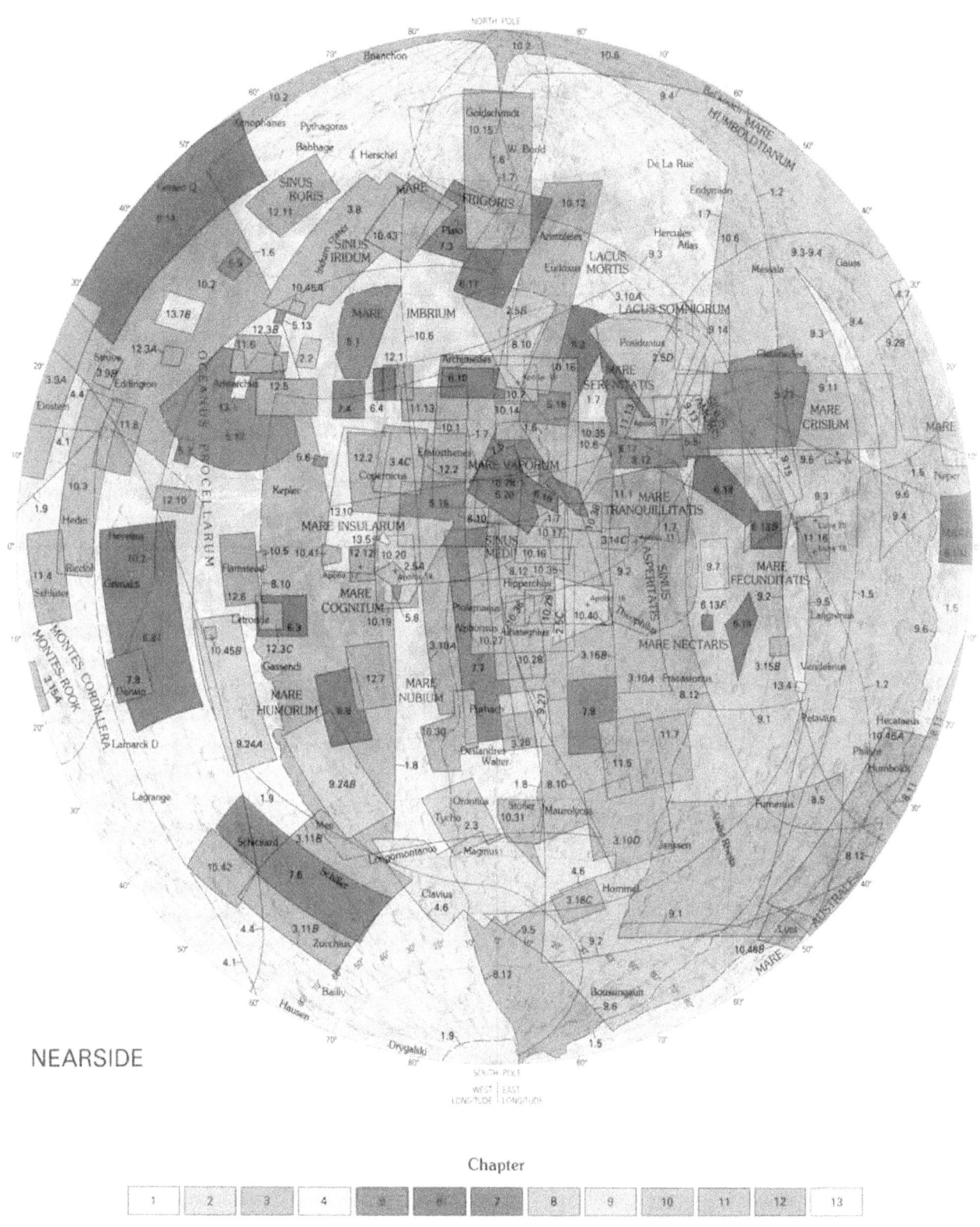

NEARSIDE

Chapter
| 1 | 2 | 3 | 4 | 5 | 6 | 7 | 8 | 9 | 10 | 11 | 12 | 13 |

Area of photograph and figure number

Chapter number precedes decimal point, figure number within chapter follows. Photographs of small areas, individual craters, and basin centers (fig. 4.3) are not shown. Boundaries dashed where approximate.

+ Landing site

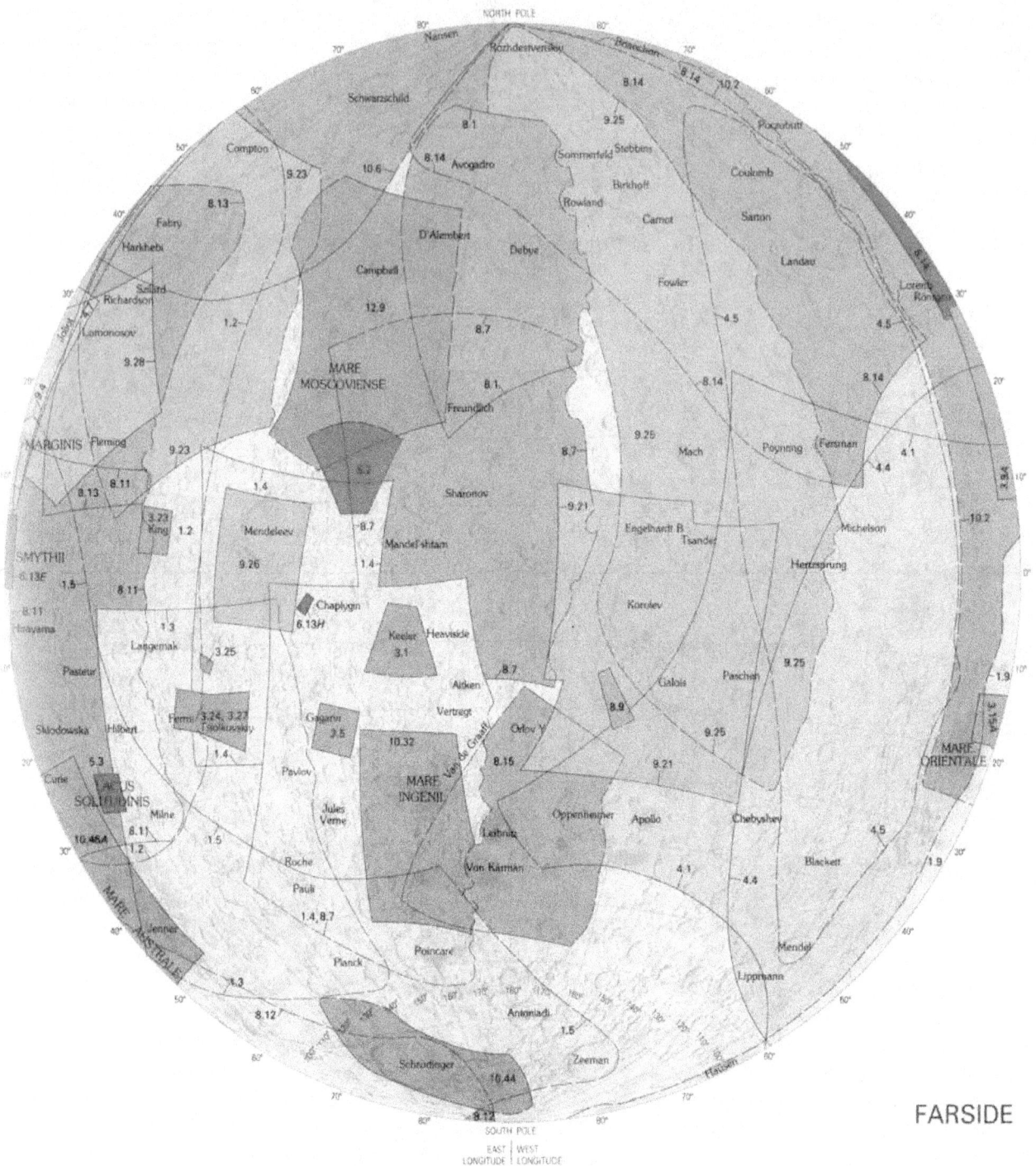

FARSIDE

Each 10° of latitude equals about 300 km

PLATE 2. PHOTOGRAPHIC COVERAGE OF THE MOON, SHOWING

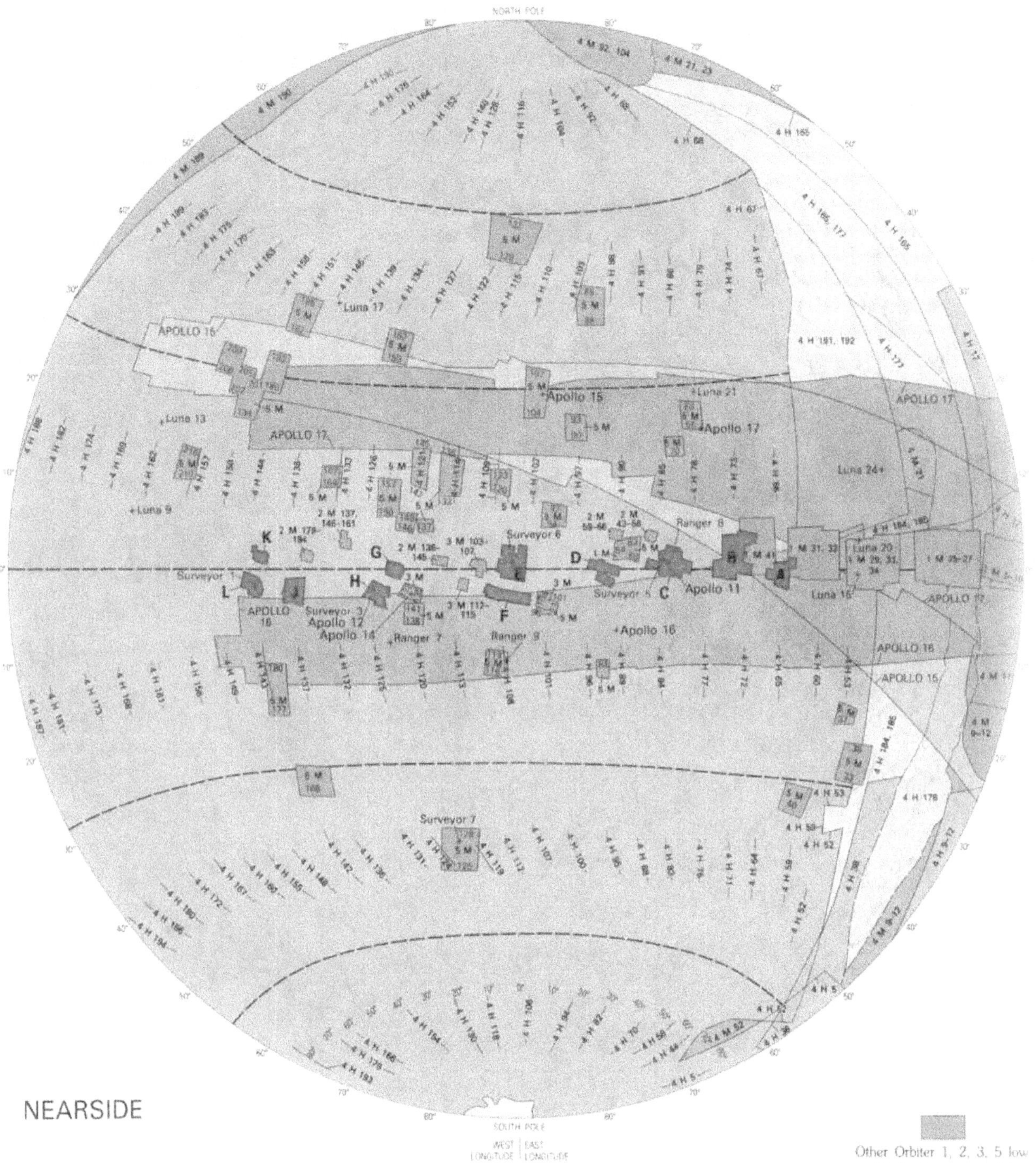

NEARSIDE

Areas multiply photographed by Lunar Orbiter, resolution, 10-20 m

- A: 1 M 44, 50-67; 3 M 25, 31, 33; 5 M 38, 41, 42, 44-51
- B: 1 M 48, 49, 68-83; 2 M 5, 7, 9, 11-20, 25-32, 35-42; 3 M 5-19 (odd nos.); 5 M 52, 55-63
- C: 1 M 85-100; 2 M 67-74, 76-91; 3 M 58, 60, 63, 66, 68, 70; 5 M 64, 71-78
- D: 1 M 105-112; 2 M 94; 3 M 80-83
- E: 1 M 118-133; 2 M 93, 95-111, 113-136; 3 M 84, 86-101; 5 M 108-115
- F: 1 M 141-148; 3 M 116-119
- G: 1 M 137, 139, 140; 2 M 163-178; 3 M 120, 124-131
- H: 1 M 157-172; 3 M 136-160
- I: 1 M 176-183; 3 M 171, 173-180; 5 M 169-176
- J: 2 M 195, 197-212; 3 M 161, 163-170
- K: 1 M 185-215; 3 M 172, 181-212

——— Actual boundary of photographs
——— Approximate boundary of best coverage

An H-frame is centered within each M-frame; each H-frame has an eightfold better resolution than the corresponding M-frame. X-ray spectrometer data acquired along Apollo 15 track between long 155° E and 40° W, and along Apollo 16 track between long 140° E and 30° W; gamma-ray spectrometer data acquired along entire Apollo 15 and 16 tracks including unphotographed regions (see chap. 5). Orbiter resolutions after Hansen (1970); complete plots of Orbiter coverage given by Hansen (1970), Kosofsky and El-Baz (1970), and Bowker and Hughes (1971). Apollo data after Masursky and others (1978).

Other Orbiter 1, 2, 3, 5 low-altitude frames; resolution 10-40 m. Numbers of Orbiter 5 M frames are inclusive

Orbiter 4 low-altitude H-frames, resolution, 60-100 m. Denoted by midpoint; adjacent rows overlap in zone centered on heavy dashed line

Other Orbiter, resolution, 150-300 m

Other Orbiter, resolution, >300 m

BEST PHOTOGRAPH OR SET OF PHOTOGRAPHS OF EACH AREA

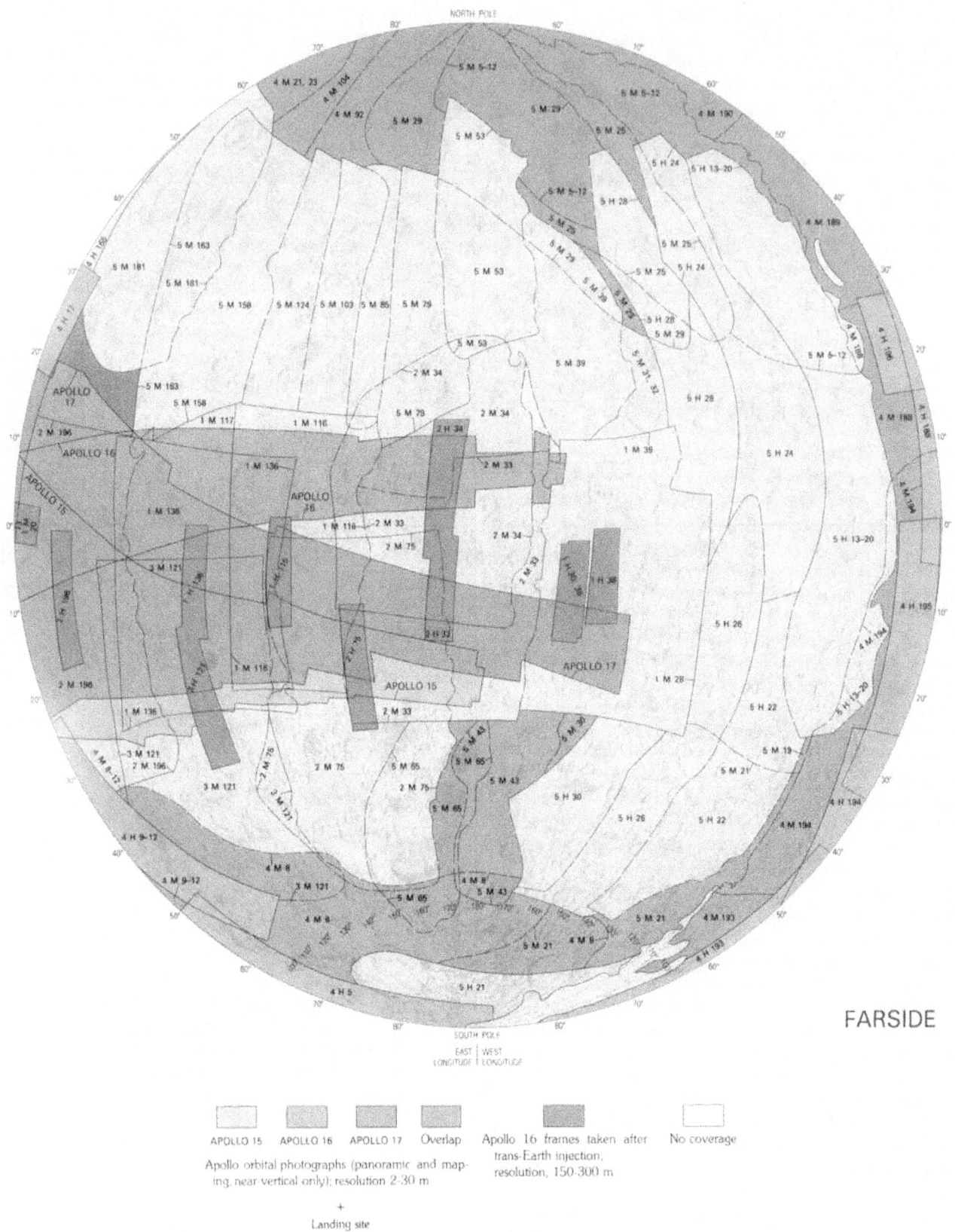

FARSIDE

PLATE 3. GEOLOGIC MAP OF RINGED BASINS

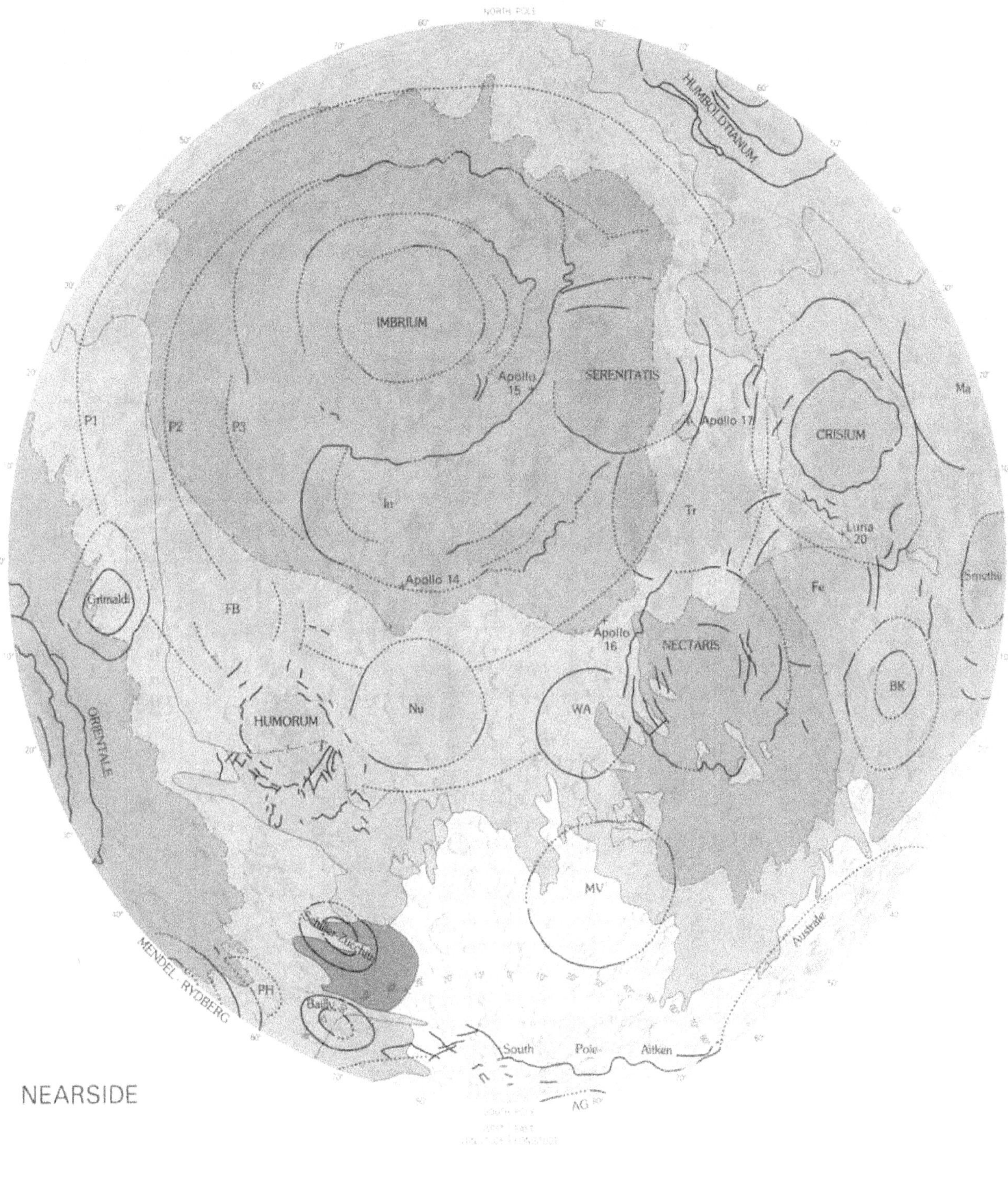

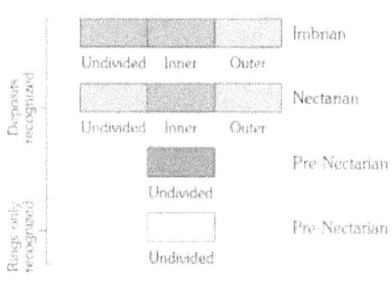

NEARSIDE

Basin ring or ring arc

Dotted where inferred. Names of definite basins with mappable deposits are in capital letters, and their deposits are shown in color; mapped units include all but minor deposits of each named basin and include inferred buried extensions. Names of definite basins without mappable deposits are in lowercase letters. Indefinite basins are identified by initials: AG, Amundsen-Ganswindt; AK, Al-Khwarizmi/King; BK, Balmer-Kapteyn; FB, Flamsteed-Billy; Fe, Fecunditatis; GW, Grissom-White; In, Insularum; LF, Lomonosov-Fleming; Ma, Marginis; MV, Mutus-Vlacq; Nu, Nubium; P1, P2, P3, Procellarum; PH, Pingre-Hausen; SR, Sikorsky-Rittenhouse; Tr, Tranquillitatis; Ts, Tsiolkovsky-Stark; WA, Werner-Airy.

3B FARSIDE

———?———
Contact
Queried where indefinite or buried

— — —?— — —
Approximate contact between inner and
outer deposits. Queried where doubtful

+
Landing site

Shaded-relief base by J. L. Inge, U.S. Geological Survey, reproduced
by courtesy of the National Geographic Society

PLATE 4. MARIA

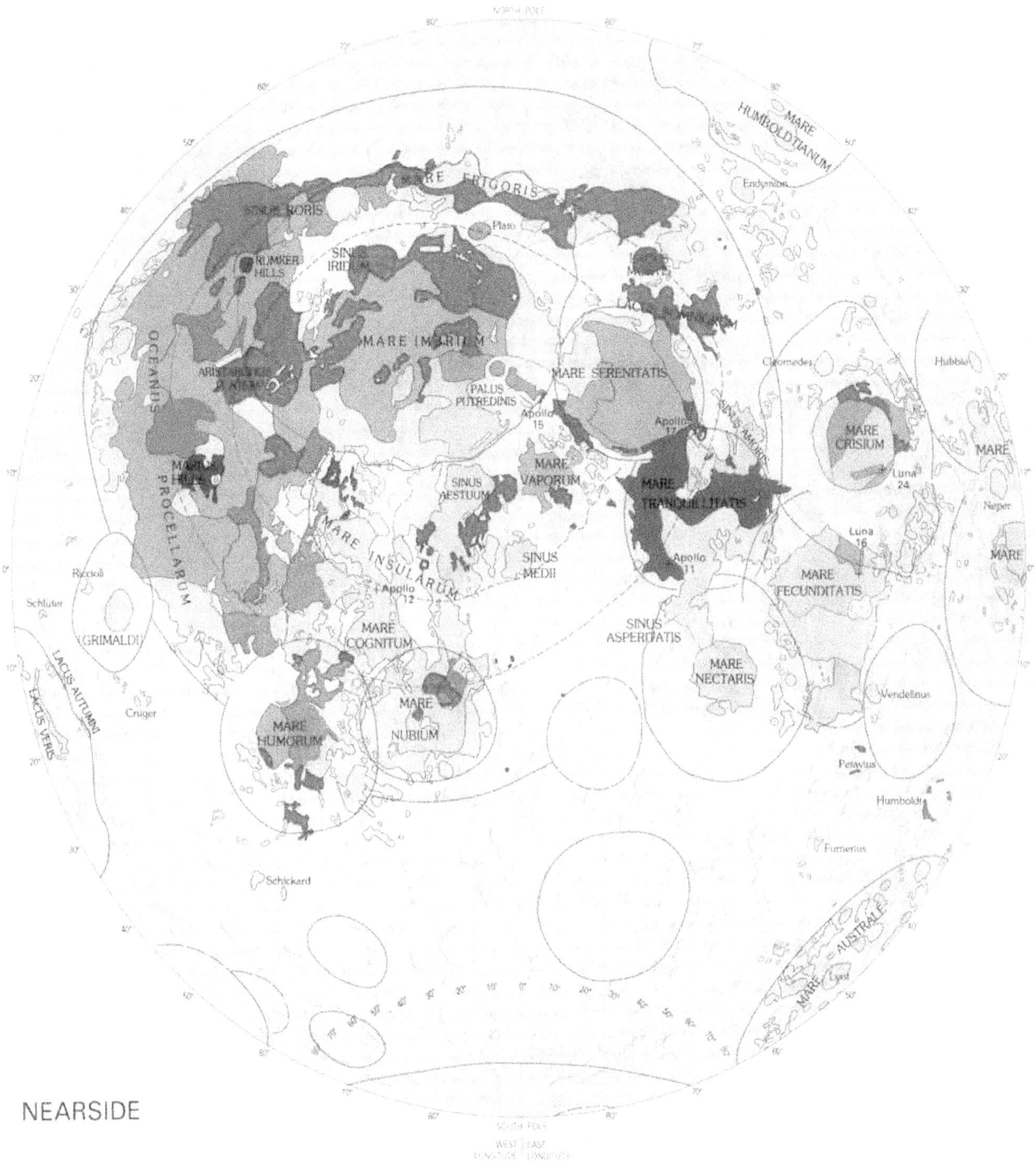

4A

NEARSIDE

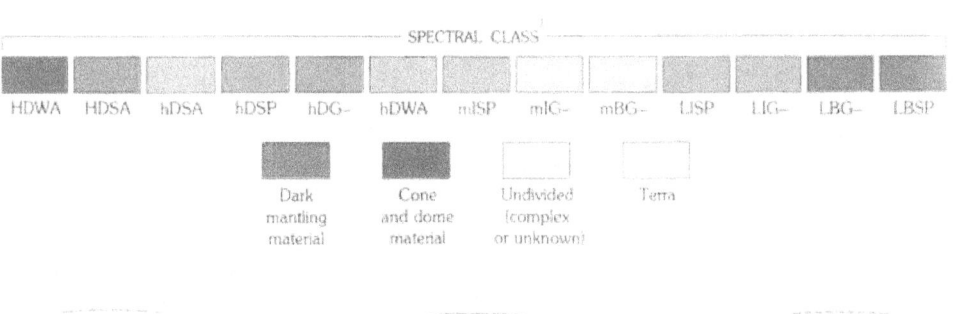

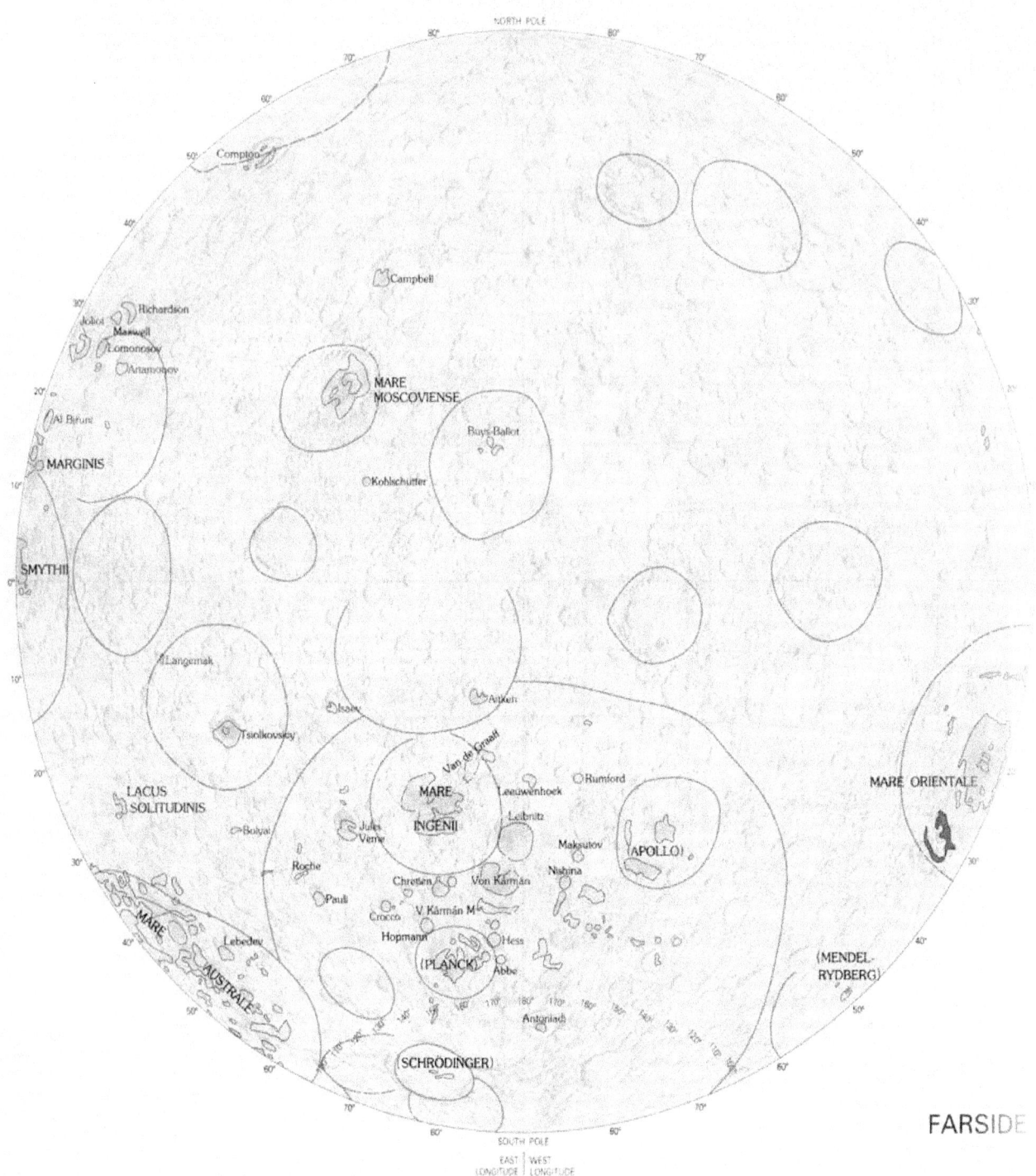

1/ NOTE
Spectral classes (see table 5.1) from Pieters (1978, see Basaltic Volcanism Study Project, 1981, pl. 2.8), redrawn here on the basis of color-difference image (fig. 5.20). Mare Humorum units from Pieters and others (1975). Mare Crisium units from Head and others (1978a). Lowercase names and names in parentheses refer to basins or craters containing otherwise unnamed maria.

+ Landing site

Shaded-relief base by J. L. Inge, U.S. Geological Survey; reproduced by courtesy of the National Geographic Society.

PLATE 5. STRUCTURAL FEATURES

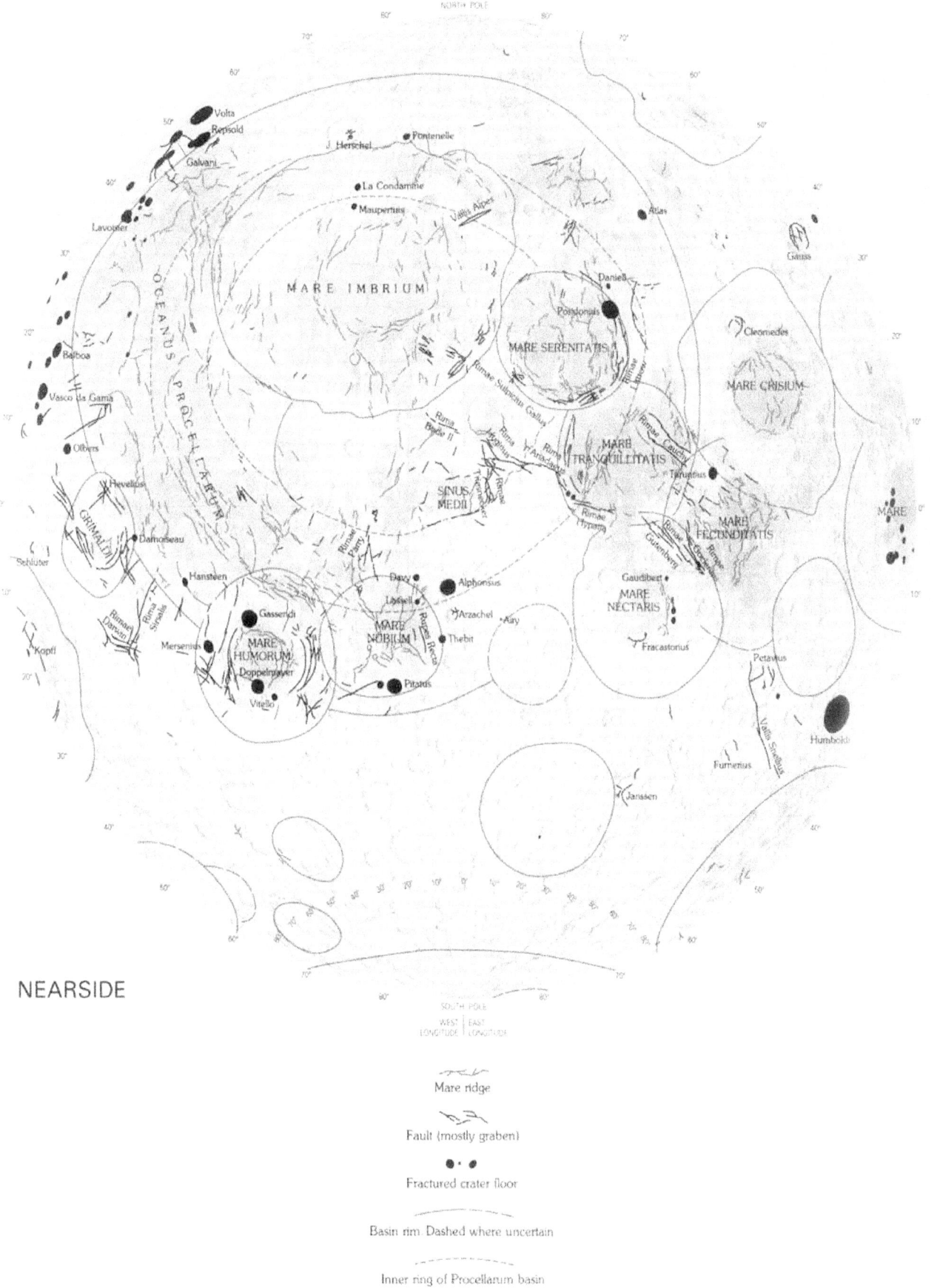

NEARSIDE

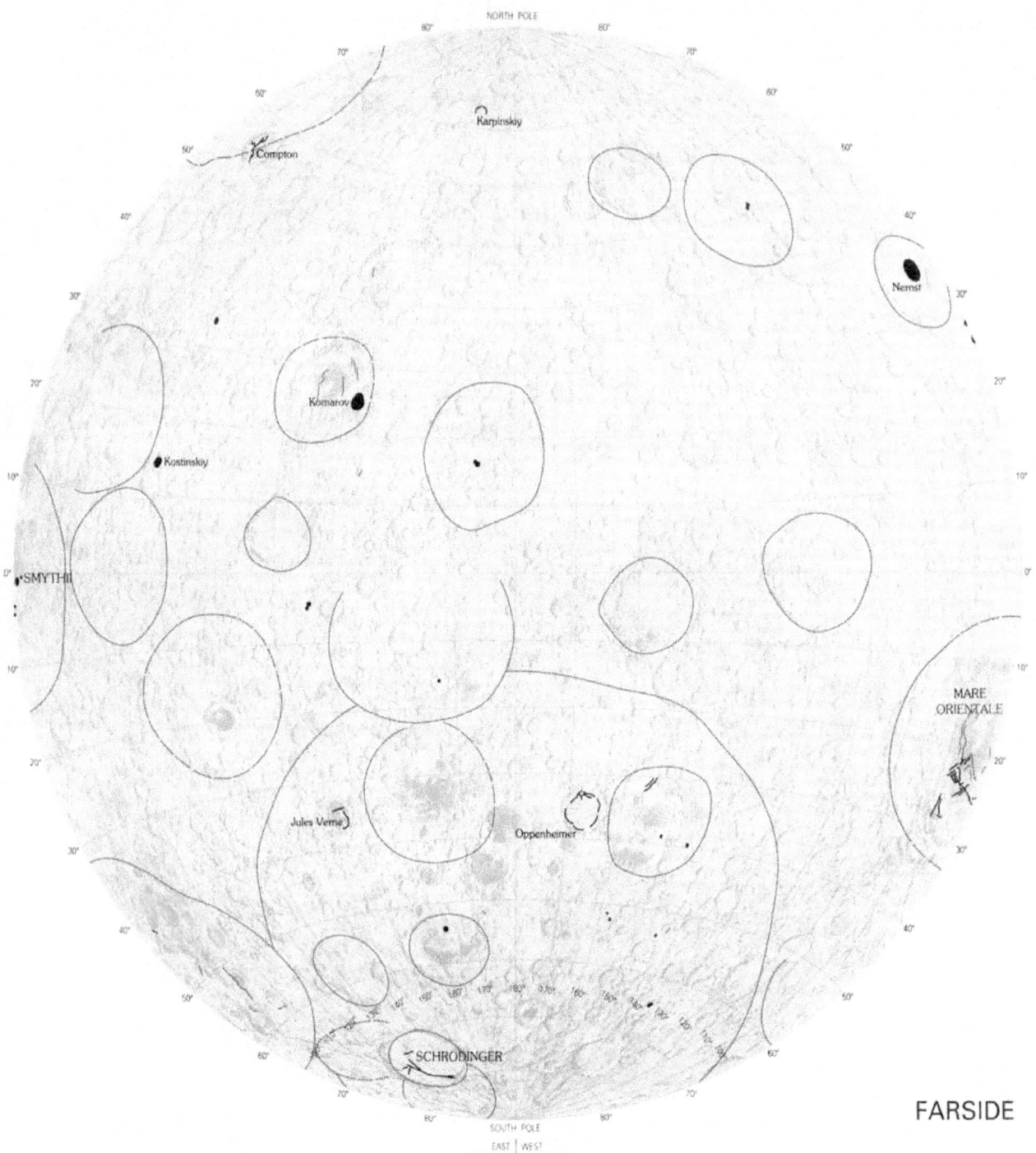

PLATE 6. PRE-NECTARIAN SYSTEM

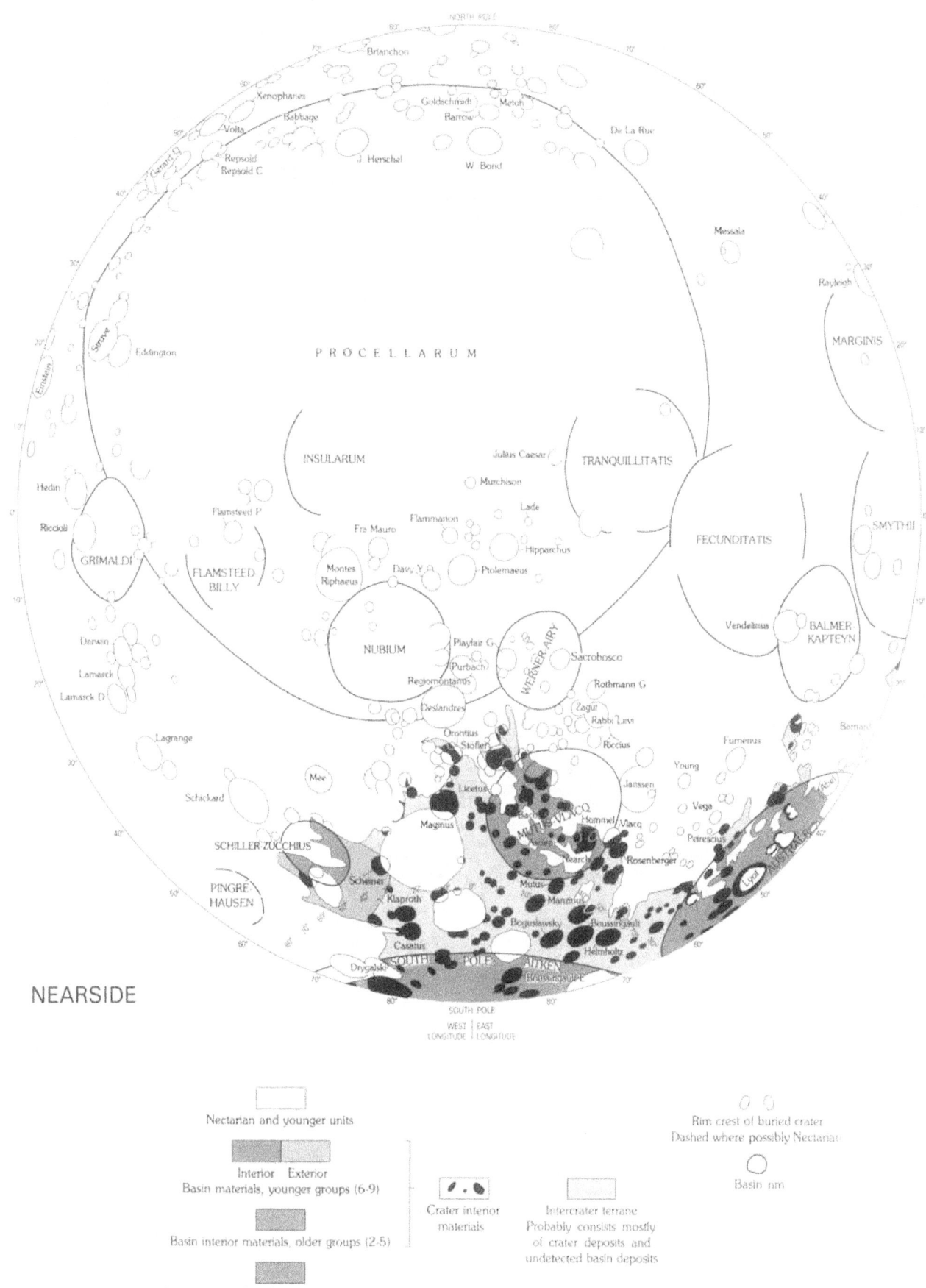

6A NEARSIDE

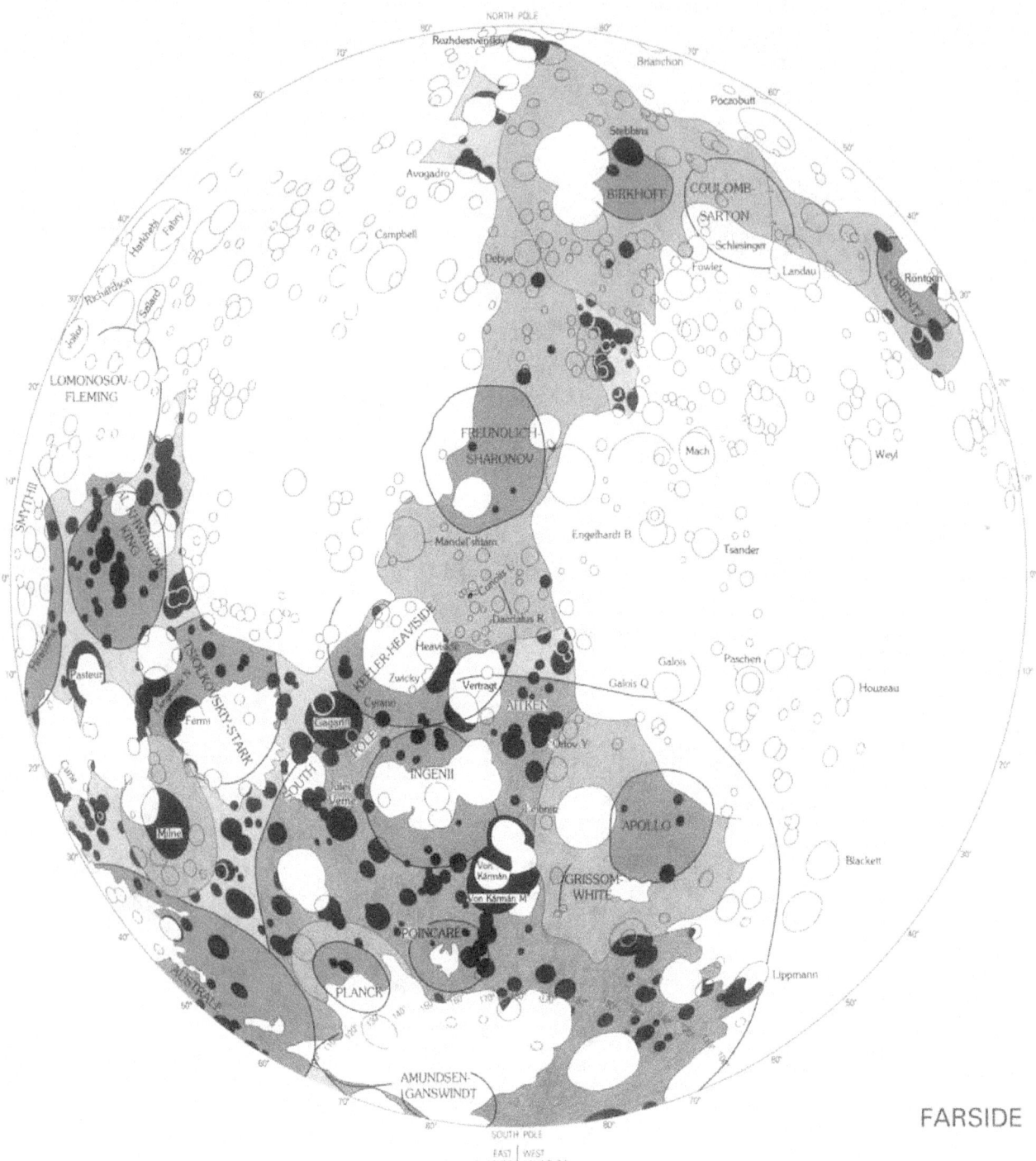

FARSIDE

Plates 6-11 are palaeogeologic maps that show units of the indicated age in color, older units only by the shaded-relief base, and younger units by blank areas. Exterior basin and crater deposits that are sufficiently well expressed to display stratigraphic relations with other units are mapped, divided where possible into near-rim deposits consisting of thick primary ejecta and outer discontinuous deposits including secondary craters and their ejecta, small patches of plains, terra-mantling material, and lineate ejecta. Craters >30 km in diameter are mapped; exterior deposits mapped around craters >120 km (pl. 7) or >60 km (pls. 8-11) in rim-crest diameter. Basin-interior materials (pls. 6-8) include massifs, knobby ejecta, and impact-melt plains. Number of superposed craters varies with age of underlying basin; number of visible buried craters varies with thickness of overlying deposit and with age of the substrate.

PLATE 7. NECTARIAN SYSTEM

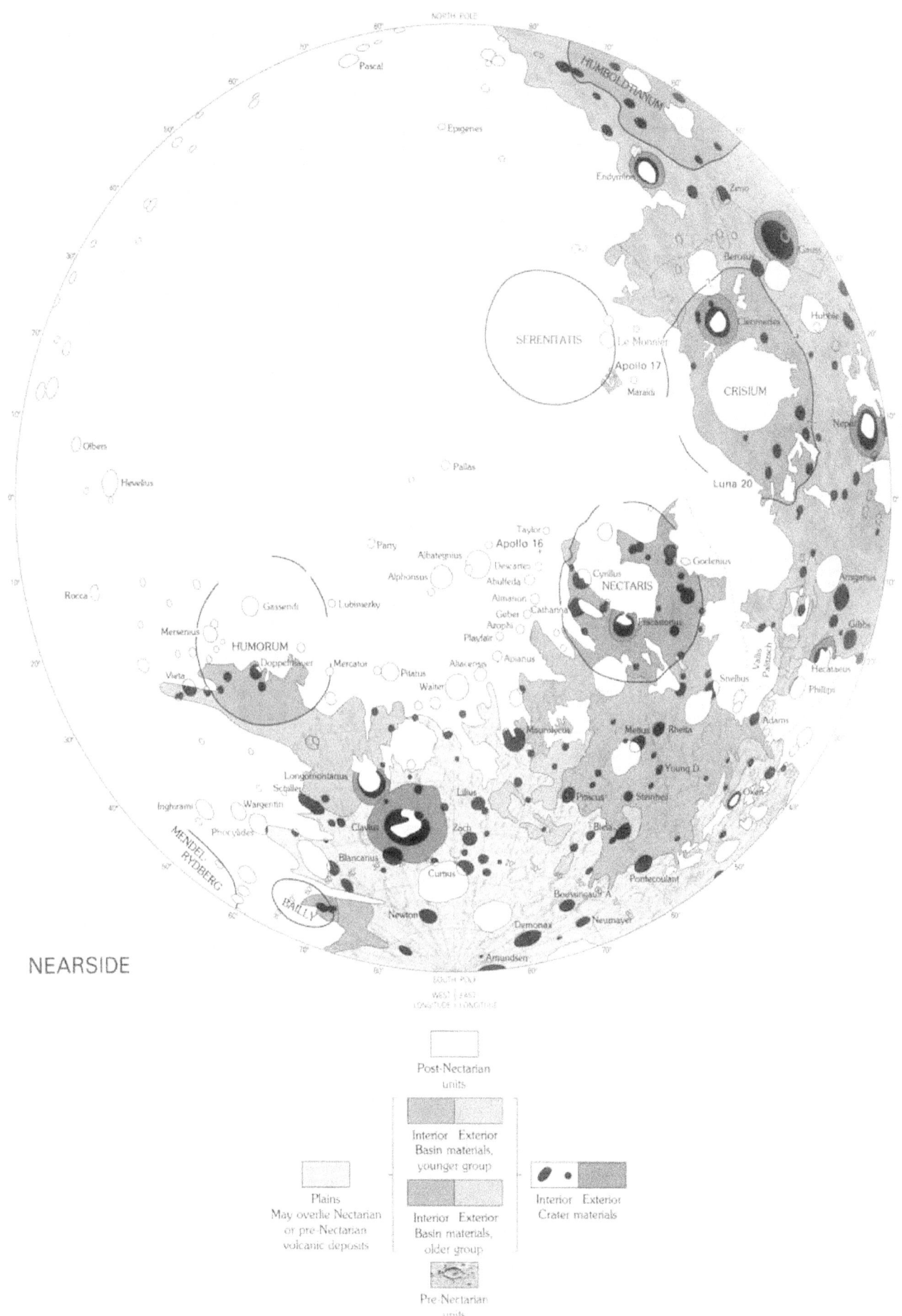

NEARSIDE

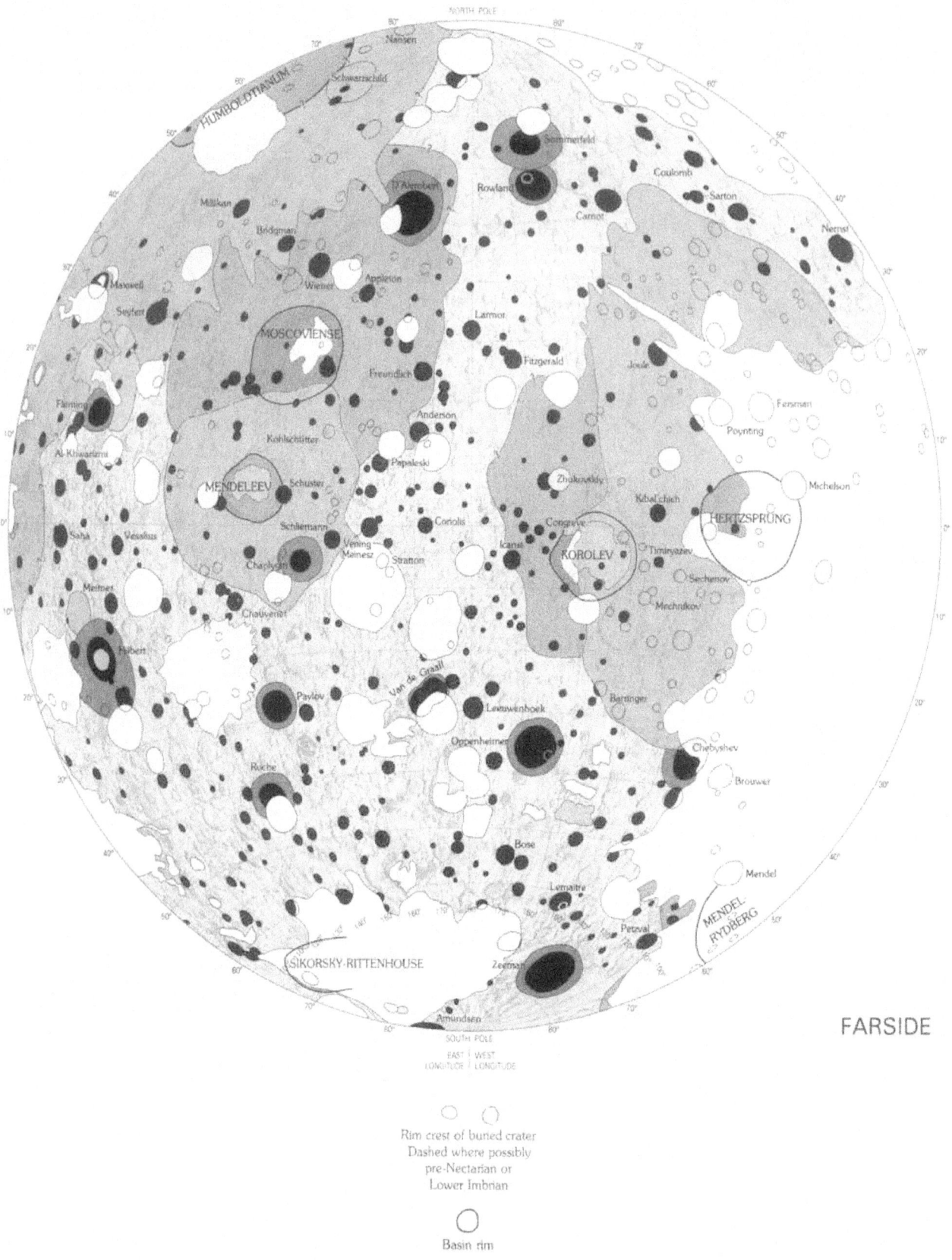

PLATE 8. LOWER IMBRIAN SERIES

8A

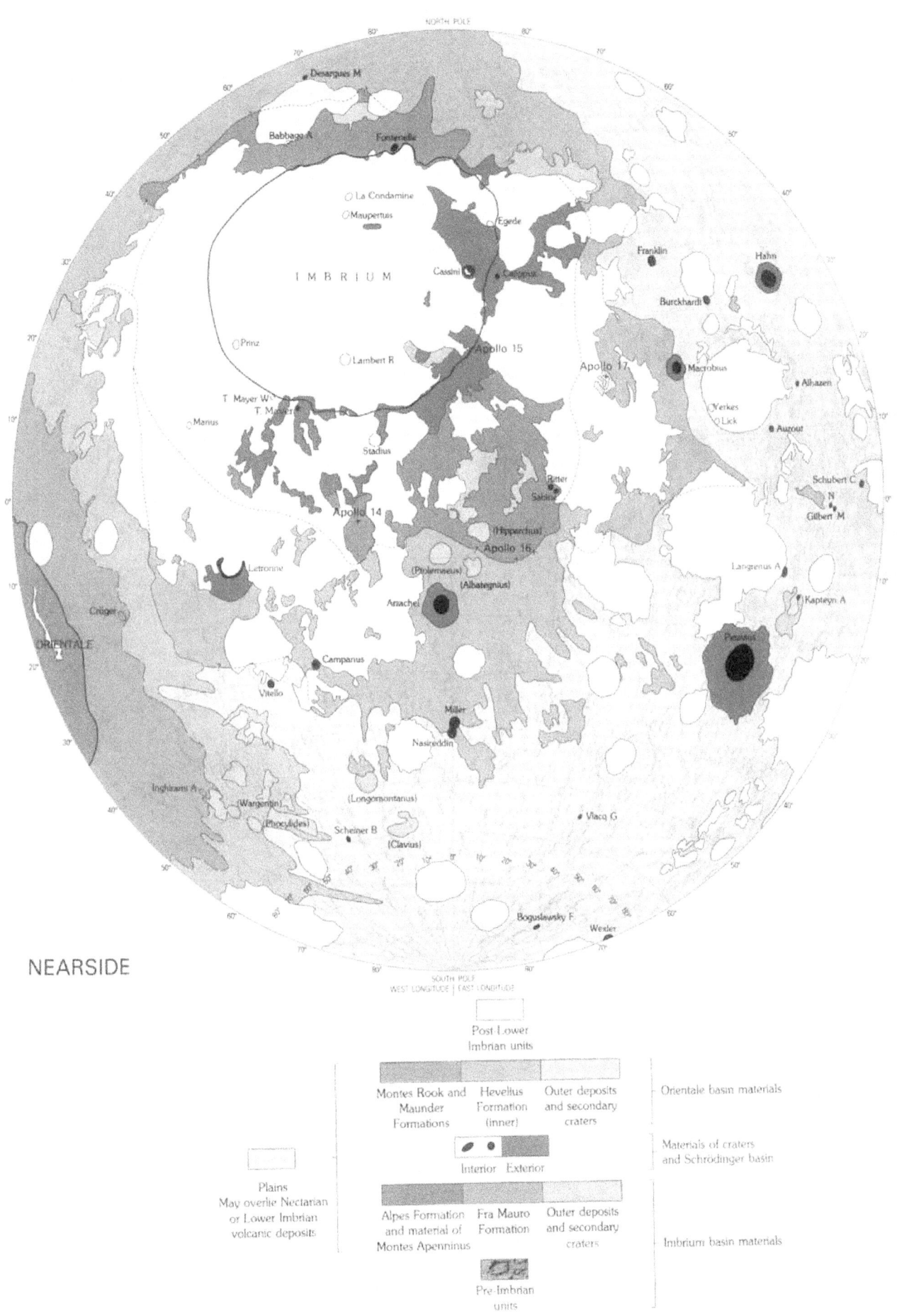

NEARSIDE

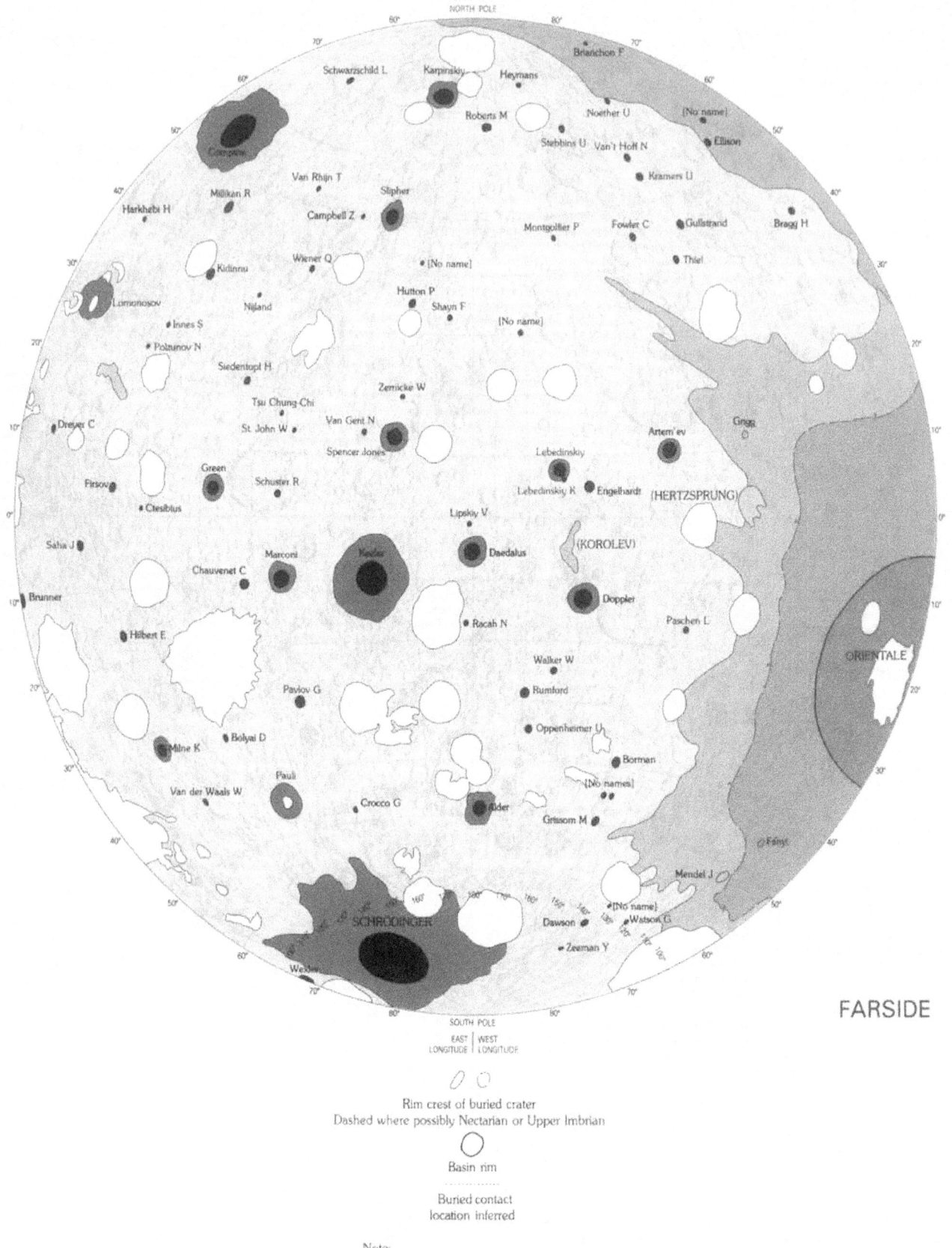

9A

PLATE 9. UPPER IMBRIAN SERIES

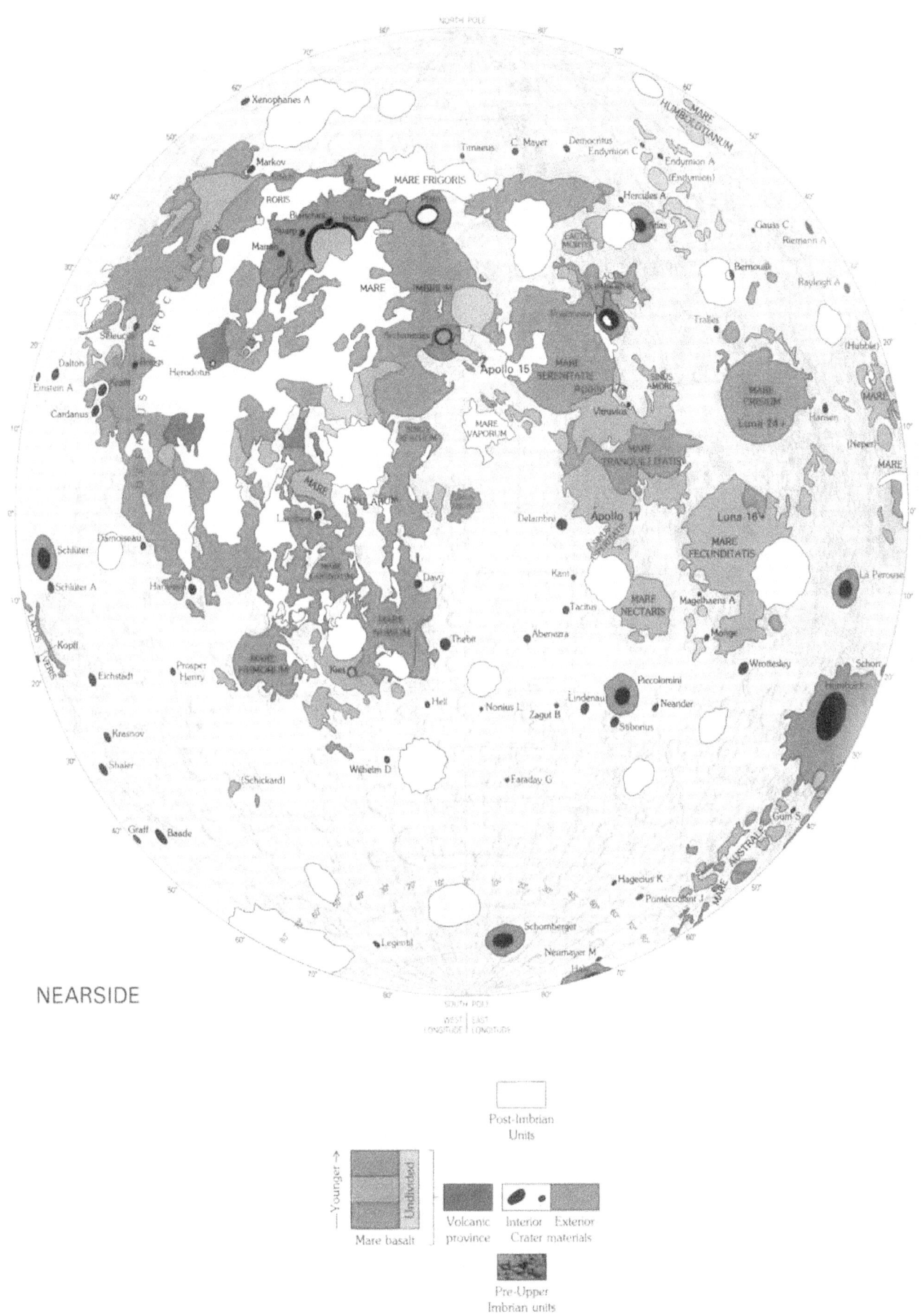

NEARSIDE

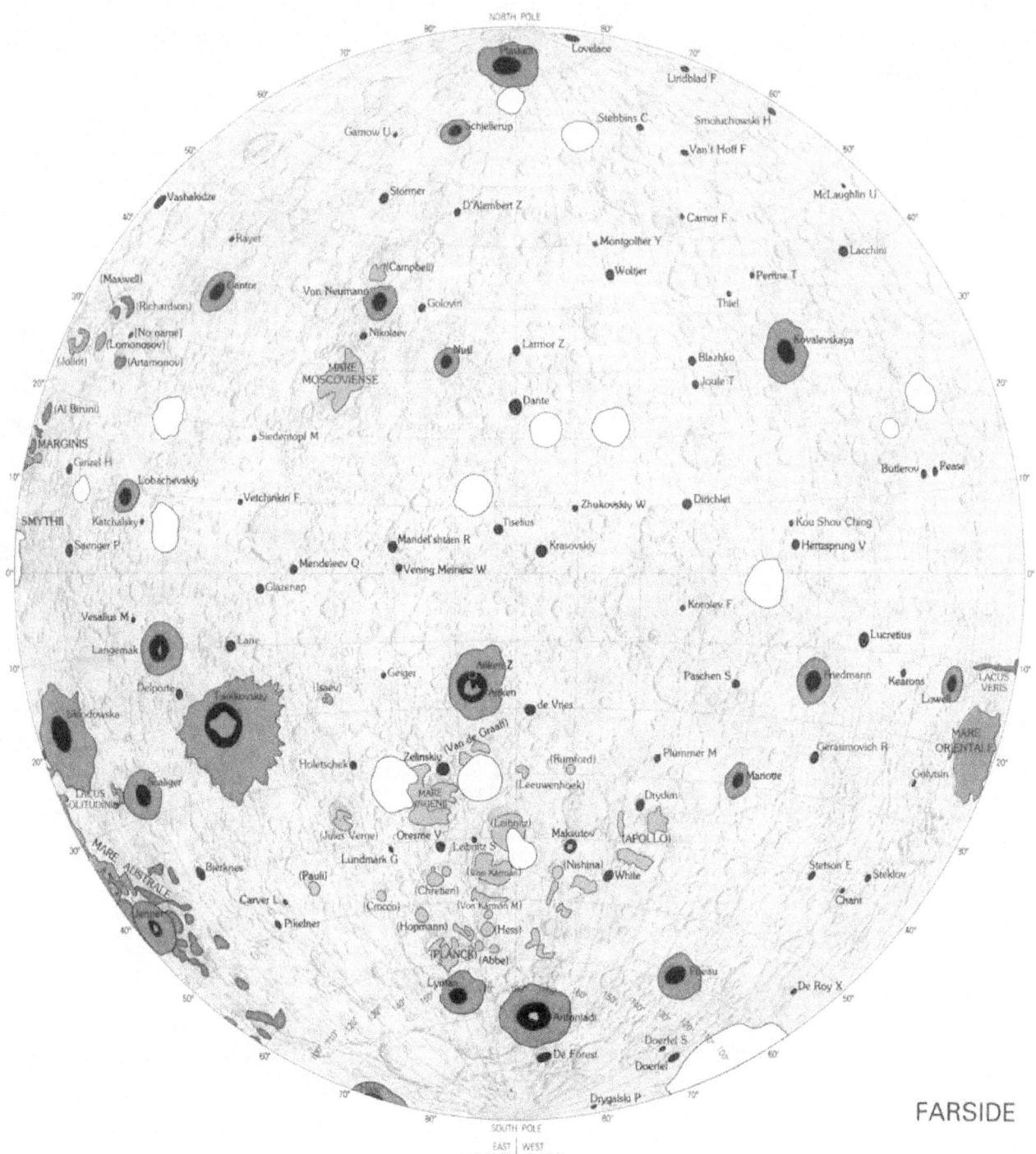

FARSIDE

NOTE
Mare mapping preliminary (pls. 9-12); contacts generalized from plate 4; small patches omitted; ages on central farside from Walker and El-Baz (1982). Volcanic provinces include numerous domes, cones, and dark-mantling deposits; other dark-mantling deposits are not shown (see pl. 4). Pre-Late Imbrian basins (capitals) and craters (lowercase) containing otherwise unnamed mare patches are named in parentheses.

Shaded-relief base by J. L. Inge, U.S. Geological Survey, reproduced by courtesy of the National Geographic Society

PLATE 10. ERATOSTHENIAN SYSTEM

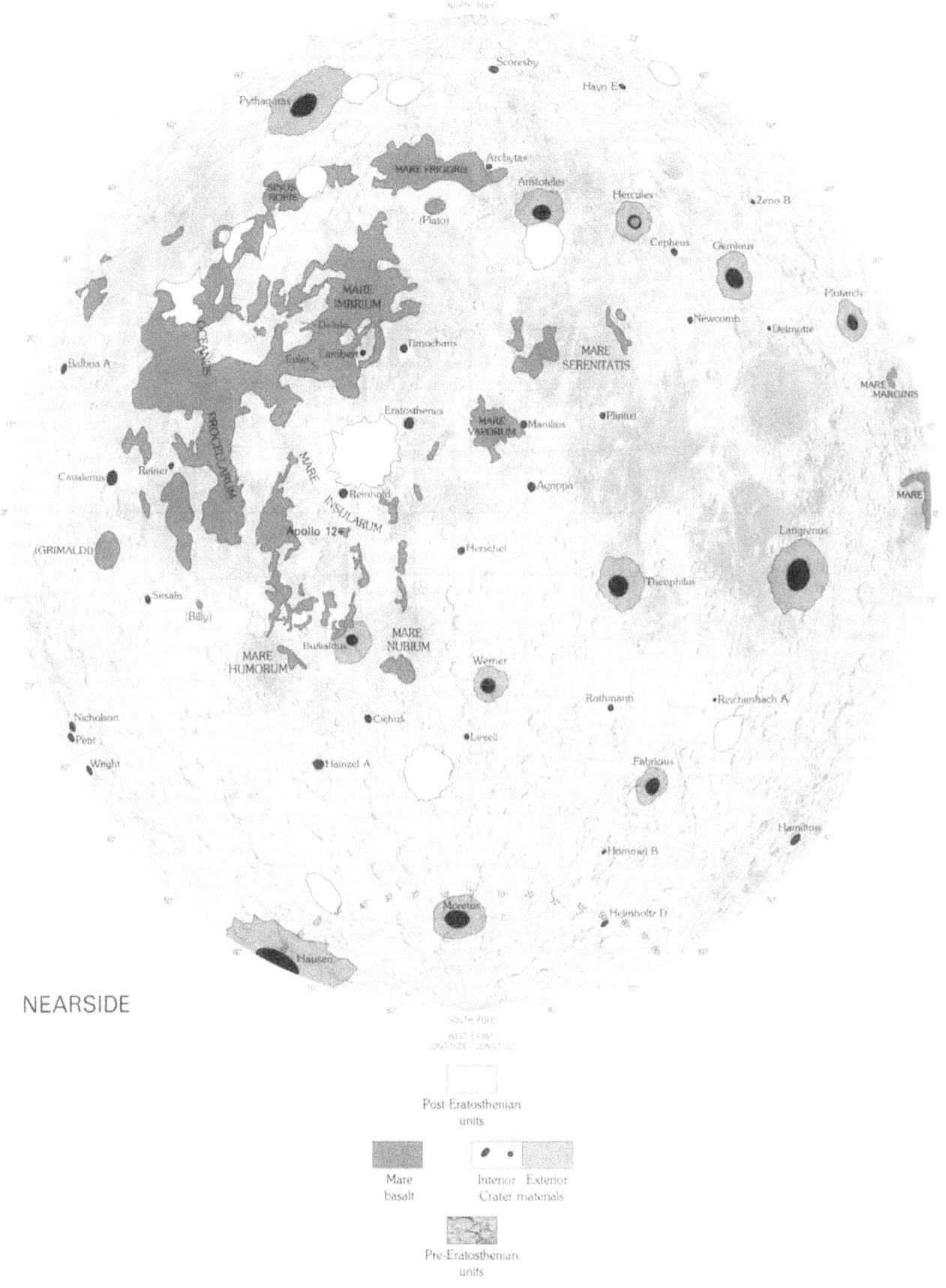

NEARSIDE

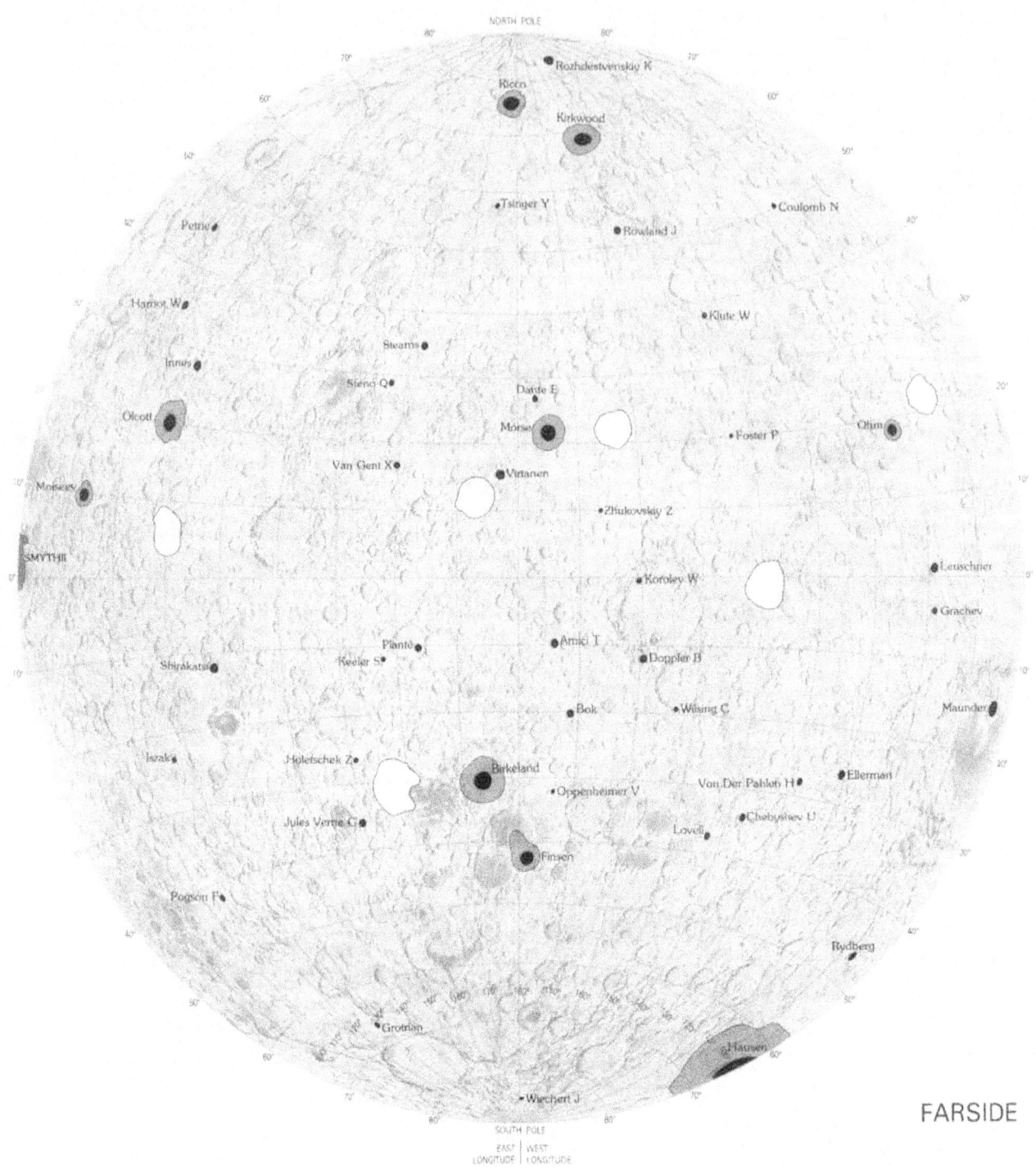

Note:
Pre-Eratosthenian basins (capitals) and craters (lowercase) containing otherwise unnamed mare patches are named in parentheses.

FARSIDE

Shaded-relief base by J. L. Inge, U.S. Geological Survey, reproduced by courtesy of the National Geographic Society

PLATE 11. COPERNICAN SYSTEM

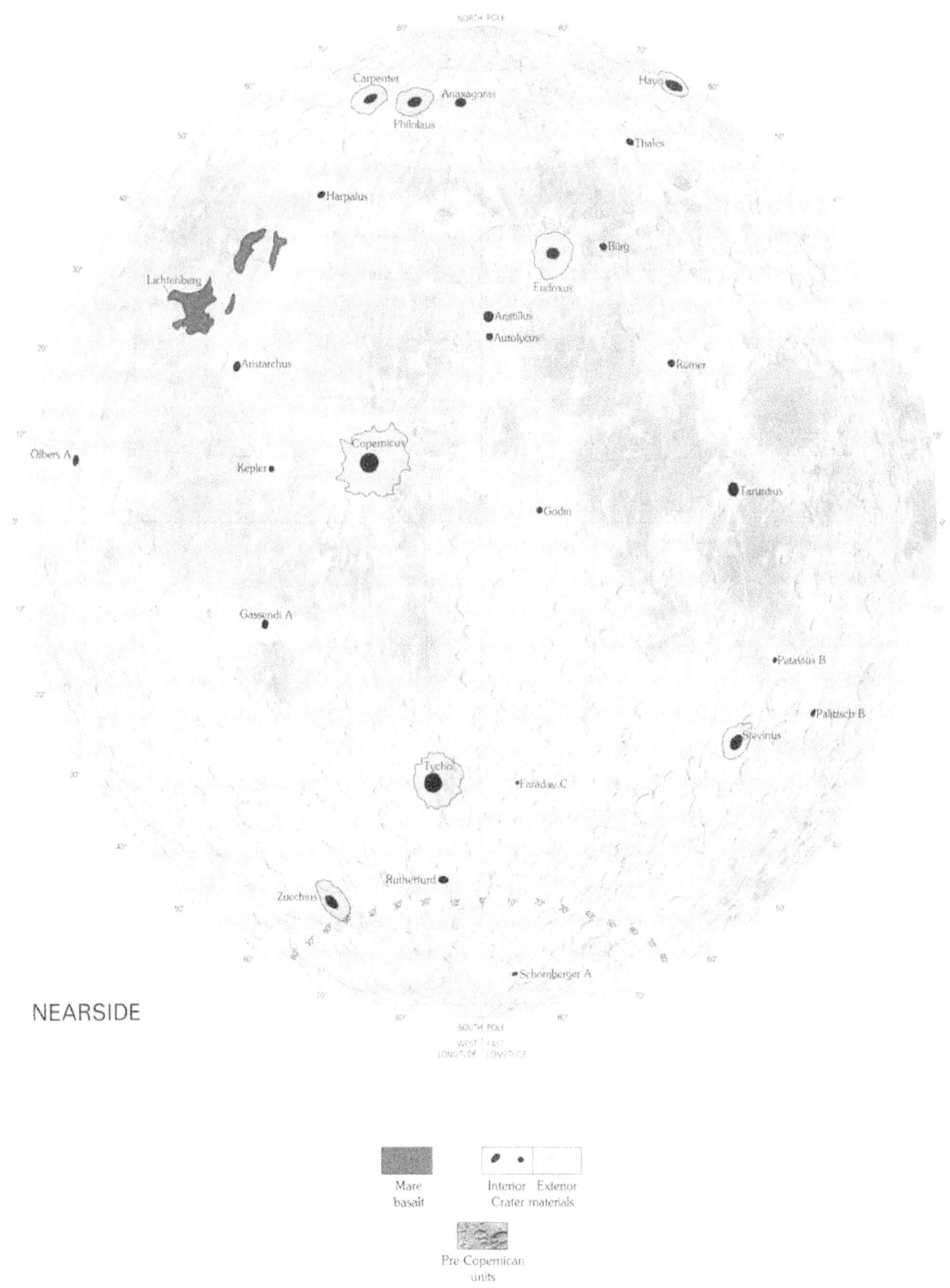

NEARSIDE

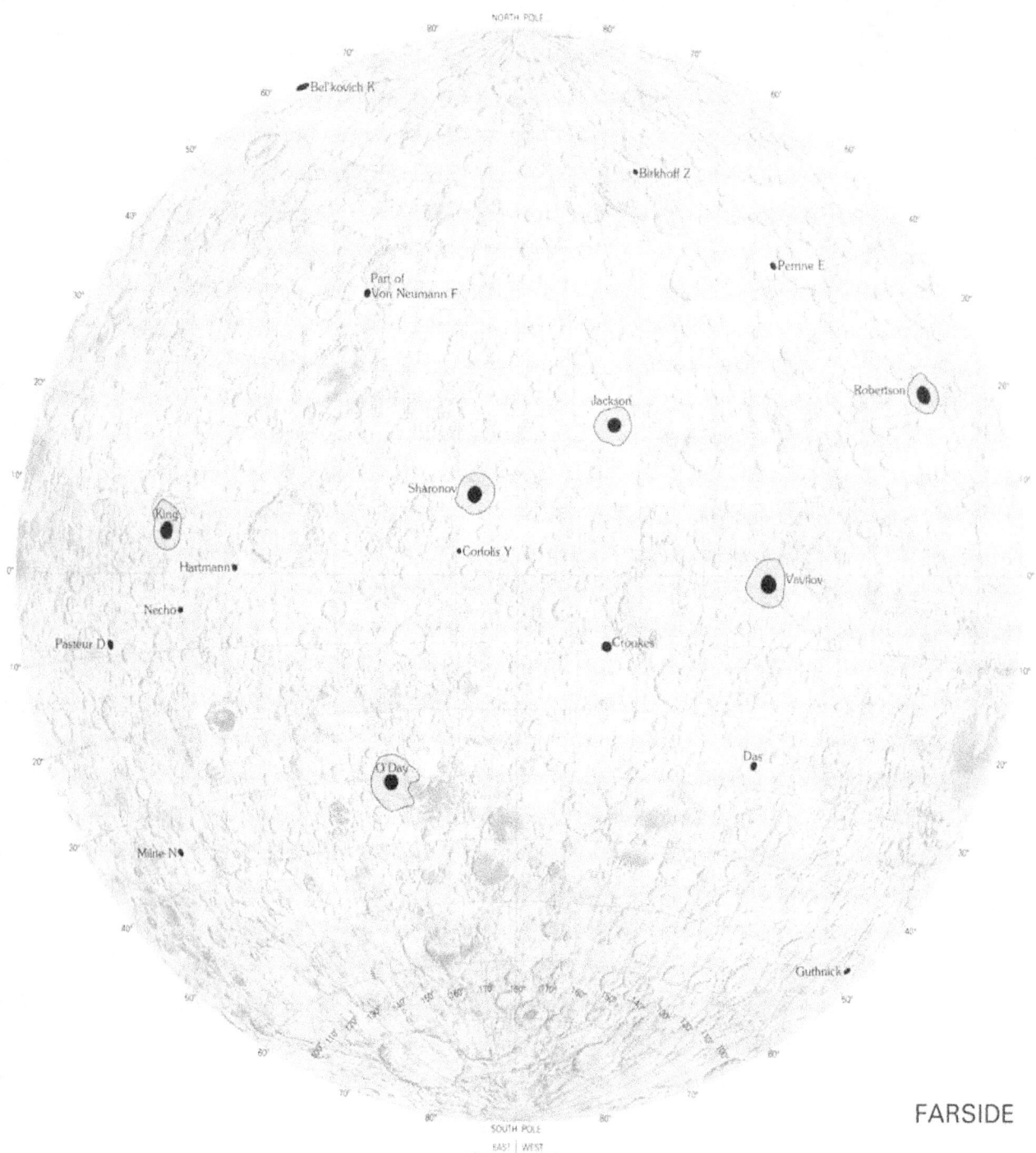

11B

FARSIDE

12A PLATE 12. GEOLOGIC MAP OF THE PRESENT MOON

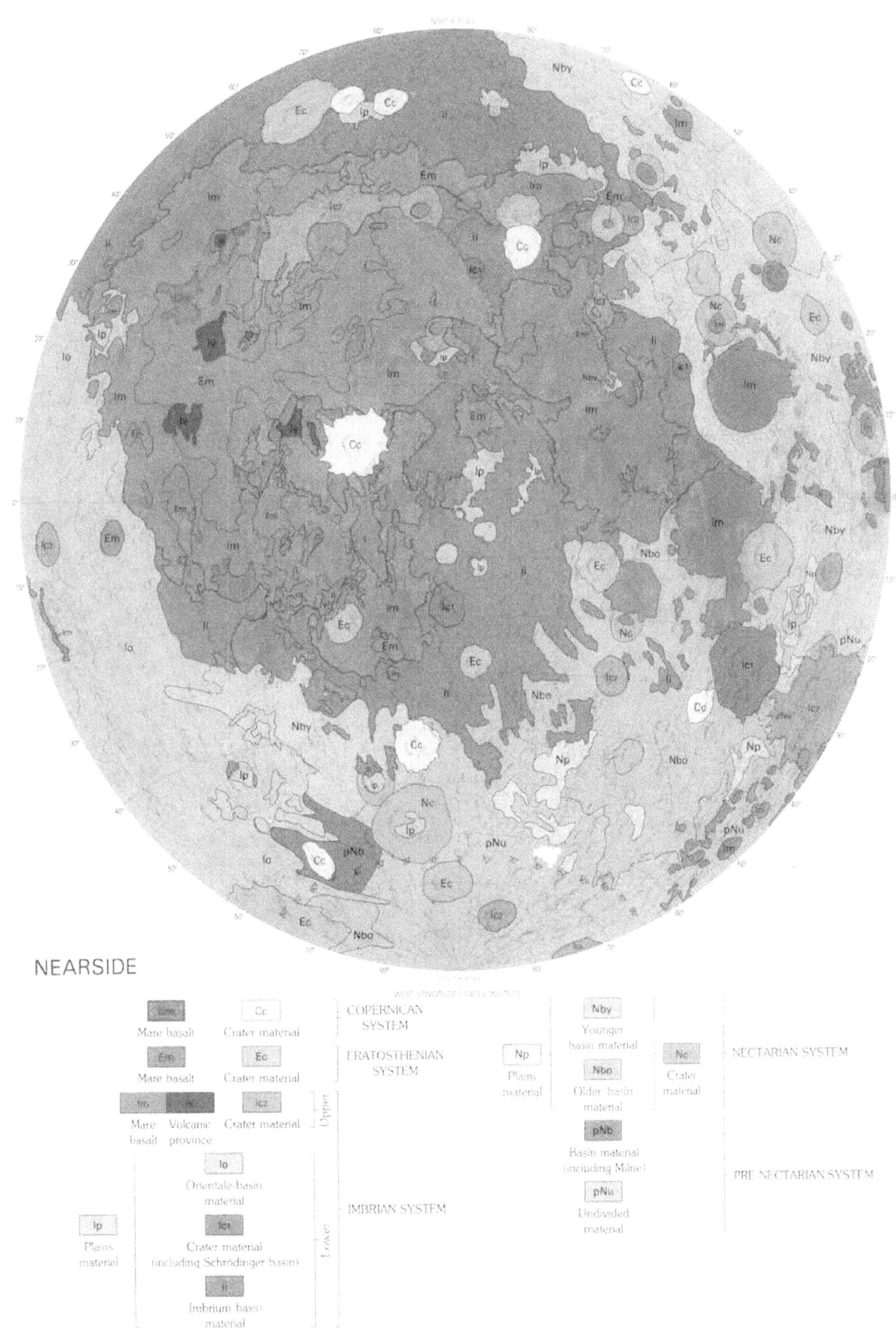

NEARSIDE

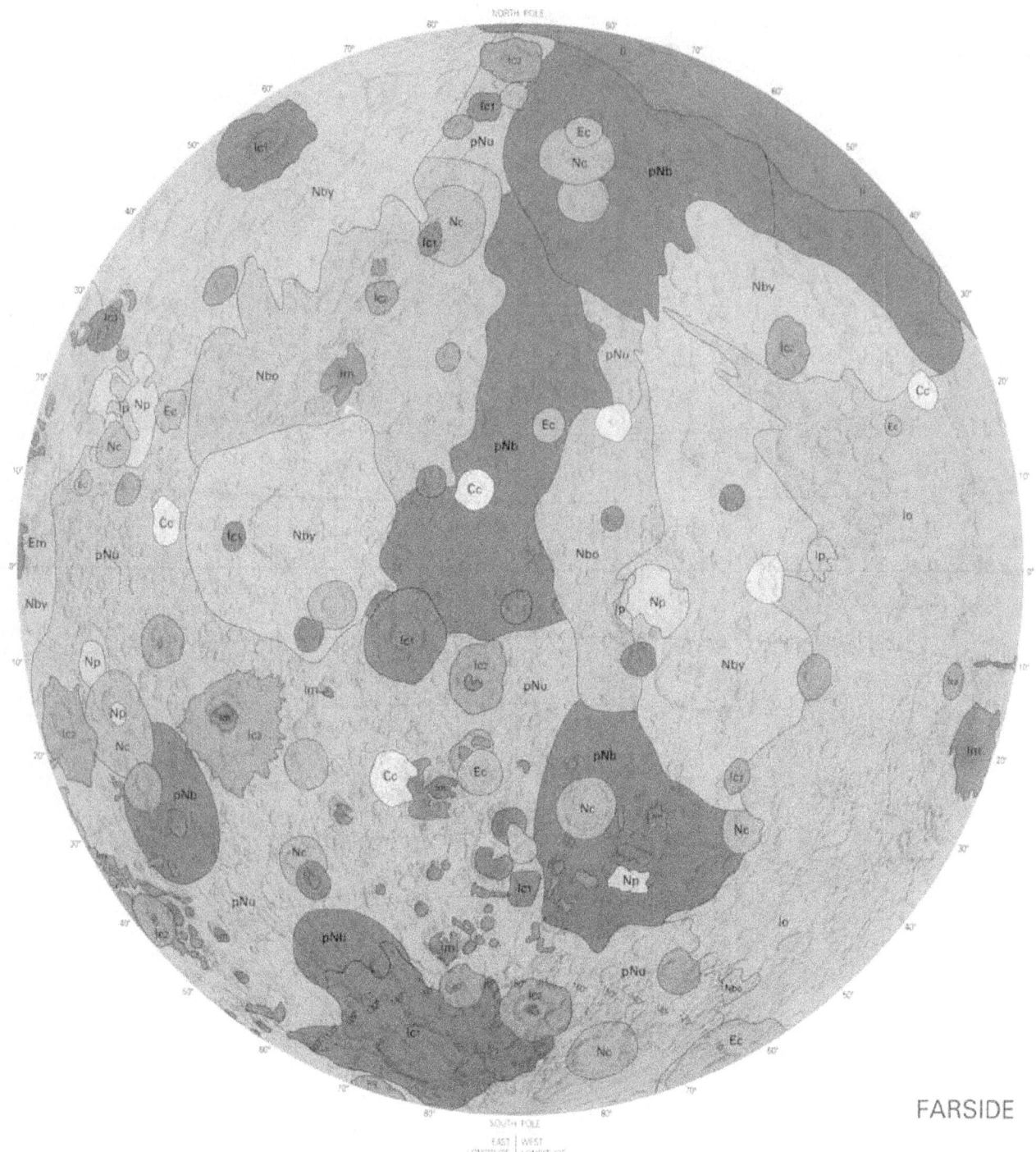

www.ingramcontent.com/pod-product-compliance
Lightning Source LLC
Chambersburg PA
CBHW081718170526
45167CB00009B/3618